Macromolecular Symposia

Symposium Editors: H.-J. P. Adler, K. Potje-Kamloth

Editor: I. Meisel
Associate Editors: K. Grieve, C.S. Kniep, S. Spiegel

Executive Advisory Board: M. Antonietti, M. Ballauff, H. Höcker,
S. Kobayashi, K. Kremer, T.P. Lodge,
H.E.H. Meijer, R. Mülhaupt,
A.D. Schlüter, H.W. Spiess, G. Wegner

187

pp. 1–958

September 2002

Macromolecular Symposia publishes lectures given at international symposia and is issued irregularly, with normally 14 volumes published per year. For each symposium volume, an Editor is appointed. The articles are peer-reviewed. The journal is produced by photo-offset lithography directly from the authors' typescripts.
Further information for authors can be obtained from:
Editorial office "Macromolecular Symposia"
Wiley-VCH
Boschstrasse 12, 69469 Weinheim,
Germany
Tel. +49 (0) 62 01/6 06-2 38 or -5 81; Fax +49 (0) 62 01/6 06-3 09 or 5 10;
Email: macromol@wiley-vch.de
http://www.ms-journal.de
Suggestions or proposals for conferences or symposia to be covered in this series should also be sent to the Editorial office at the address above.

Macromolecular Symposia:
Annual subscription rates 2002 (print only or online only)*
Germany, Austria € 1198; Switzerland SFr 1968; other Europe € 1198; outside Europe US $ 1448.
Macromolecular Package, including Macromolecular Chemistry & Physics (18 issues), Macromolecular Bioscience (9 issues), Macromolecular Rapid Communications (18 issues), Macromolecular Theory & Simulations (9 issues) is also available. Details on request.
* For a 5 % premium in addition to **Print Only** or **Online Only**, Institutions can also choose both print and online access.
Packages including Macromolecular Symposia and Macromolecular Materials & Engineering are also available. Details on request.
Single issues and back copies are available. Please, inquire for prices.

Orders may be placed through your bookseller or directly at the publishers:
WILEY-VCH Verlag GmbH & Co. KGaA, P. O. Box 10 11 61, 69451 Weinheim, Germany,
Tel. (0 62 01) 6 06–0, Telefax (0 62 01) 60 61 17, Telex 46 55 16 vchwh d.
E-mail: subservice@wiley-vch.de

Macromolecular Symposia (ISSN 1022-1360) is published with 14 volumes per year by WILEY-VCH Verlag GmbH & Co. KGaA, P. O. Box 10 11 61, 69451 Weinheim, Germany. Air freight and mailing in the USA by Publications Expediting Inc., 200 Meacham Ave., Elmont, NY 11003. Periodicals postage pending at Jamaica, NY 11431. POSTMASTER: send address changes to Macromolecular Symposia, Publications Expediting Inc., 200 Meacham Ave., Elmont, NY 11003.

© WILEY-VCH Verlag GmbH & Co. KGaA, Weinheim, Germany, 2002
Printing: Strauss Offsetdruck, Mörlenbach. Binding: J. Schäffer, Grünstadt

Quo Vadis – Coatings?

Lectures presented at the

XXVI FATIPEC Congress
European Organization of Paint Scientists and Engineers

*held at Dresden University of Technology,
Dresden, Germany,
September 9-11, 2002*

Editors

H.-J. P. Adler, K. Potje-Kamloth
Dresden University of Technology, Germany

Scientific Committee:

Prof. Dr. H.-J. P. Adler, Dresden University of Technology, Germany
(chairman of the congress, FATIPEC President)

Prof. Dr. E. L. J. Bancken, Technical University Eindhoven, The Netherlands
(FATIPEC board member)

Prof. Dr. P. Bonora, University of Trento, Italy (FATIPEC board member)

Dr. G. Botta, Torino, Italy

Prof. Dr. T. Brock, Fachhochschule Niederrhein, Krefeld, Germany

Dipl.-Ing.(FH) H.-D. Christian, Degussa AG, Hanau, Germany (chairman of VILF)

Dr. P. Klostermann, DuPont Performance Coatings, Wuppertal, Germany

Dr. A. Kovacs-Stahl, Research Institute for Paint Industry, Budapest, Hungary
(FATIPEC board member)

Dr. J. Koziel, S. I. T. P. C. H. E.M., Cieszyn, Poland (FATIPEC board member)

Dr. B. Lestarquit, Rohm and Haas, Valbonne, France

Dr. W. Ph. Oechsner, Forschungsinstitut für Pigmente und Lacke (FPL), Stuttgart, Germany

Dr. M. Osterhold, DuPont Performance Coatings, Wuppertal, Germany

Prof. Dr. H. Perrey, Bayer AG, Leverkusen, Germany

Prof. Dr. A. Revillon, NRS-LMOPS Lyon, France (FATIPEC board member)

Prof. Dr. A. Toussaint, St. Pieters Leeuw, Belgium (FATIPEC General Secretary)

Prof. Dr. J. Snuparek, University of Pardubice, Czech Republic (FATIPEC board member)

Dr. H. Streitberger, BASF Coatings AG, Münster, Germany

Dr. J. Vogelsang, Corporate Research, Sika Schweiz AG, Schwitzerland

Contents of Macromol. Symp. 187

XXVI FATIPEC Congress
Quo Vadis – Coatings
Dresden, Germany, 2002

Preface
H.-J. P. Adler

Advanced Technologies

Higher Speed

More Color & Appearance

* The asterisk indicates the name of the author to whom inquiries should be addressed

Author Index

Preface

The XXVI FATIPEC congress was held in Dresden from September 9–11, 2002. FATIPEC (Fédération d'Associations des Techniciens des Industries des Peintures, Vernis, Emaux et Encres d'Imprimerie de l'Europe Continentale), the European Organization of Paint Scientists and Engineers, has delegated the organization of the XXVI FATIPEC congress to the special branch "Coatings and Pigments" (Anstrichstoffe und Pigmente – APi) of the German Chemical Society (Gesellschaft Deutscher Chemiker – GDCh) together with the Association of the Paint Engineers and Technicians (Verband der Ingenieure des Lack-und Farbenfaches e. V. – VILF).

The main topic of the congress

"Quo Vadis – Coatings?"

reflects the actual challenges and opportunities in the broad field of the international coating technology.

Coatings are under permanent stress of cost, quality and environmental compliance. In spite of strong efforts to fulfill the regulatory requirements as well as own commitments the coating industry still has to fight for social acceptance. This results in even further investments in innovative technologies, where the aspects of product development, application techniques, film properties and environmental aspects have to be considered right from the start of customer demands. Consequently, an increased responsibility of the paint suppliers for the final result of the coating process is emerging.

Therefore this congress will give the participants a comprehensive overview, not only of new products and raw materials, but also of the driving forces of the market as well as of improved processes to meet the social requirements. The two basic functions of coatings "protection" and "color" in the triangle of cost, quality, and social responsibility are covered by eight sections representing the basic areas of activities in all different paint segments:

- Market & Trends
- New Substrates & Pretreatments
- Advanced Technologies
- Special Functions
- Higher Speed
- More Color & Appearance
- Better Eco-Efficiency
- Modern Characterization

We thank all who contributed to the success of the meeting and to its organization. In this place our special thanks to those who handed in their manuscripts for publication in this volume.

The XXVII FATIPEC congress will be held in France in 2004.

H.-J. P. Adler

Acknowledgements

The organizing committee of the XXVI FATIPEC congress would like to thank the FATIPEC board members, the VILF board, the GDCH and APi board, and the sponsoring companies for financial support.
On behalf of the scientific committee,

H.-J. P. Adler

K. Potje-Kamloth

Meeting the Challenge of Radical Change: Coatings R&D as We Enter the 21st Century

G. R. Pilcher

Akzo Nobel Coatings Inc., Columbus, Ohio, USA

In early 1980's, the late Dr. Marco Wismer, one of the coatings industry's more visionary members, enunciated "Six Strategic Goals for the Coatings Industry" which have never been far from our collective thoughts in the years since:[1]

WISMER'S SIX STRATEGIC GOALS
1.   *Corrosion Protection*
2.   *Elimination of Solvents*
3.   *Conservation of Energy*
4.   *Reduction of Toxic Wastes*
5.   *Cost Reduction*
6.   *Improved Durability*

Although stated over fifteen years ago, these six goals are more valid than ever in 2001, and it is clear that the various global coatings marketplaces, both industrial and consumer, are seeking (in some cases, *struggling*) to provide positive, proactive, and economically viable realizations for each of these goals.   The extent to which any given market segment is succeeding varies, of course, but these goals seem to be clear to almost everyone, almost everywhere.   The challenges are many, and the solutions demand, in many cases, new—even radical—responses from the coatings industry.   Of course the coatings industry cannot "go it alone," and we must engage the cooperative support of both our suppliers and our customers if we are to benefit from the full spectrum of creative insight regarding potential new approaches to these goals.   In this paper, I would like to offer my views regarding "where we are," as a global industry, "where we might be able to go," and "how we might be able to get there."

ENVIRONMENTAL CHALLENGES
Today, in the first year of the first decade of the first century of the Third Millennium, I am reminded of a statement made by my colleague, Dr. J. D. Remijnse, on the occasion of his presentation to the *Third International Paint Congress* in Sao Paulo, Brazil, on 6 September, 1993: "Faced with a great groundswell of public opinion and the first tangible effects of years of environmental neglect, governments around the world are tightening up environmental legislation at a rate that sometimes takes our breath away."   Environmental issues are no longer "topical"—they are here to stay, and they affect, to one degree or another, every aspect of the global coatings community.   Traditionally, "environmental" concerns have meant "lower VOC's," but today the phrase connotes a wide variety of issues, from elimination of the use of hexavalent chromium to identification and elimination of endocrine disrupters. Some of these issues are of greater concern in certain parts of the globe, and others in different parts—but all will need to be addressed eventually.

One of the major challenges facing the North American coatings industry, for instance, arises from a variety of local and national environmental concerns which are active in Canada, the United States and Mexico, but particularly in the U.S., on which these comments are principally based. Issues of major concern at the Federal level include Hazardous Air Pollutants (HAPS) and its subset of heavy metals, which are also the subject of separate State and Local legislation, as well. The HAPS list is comprised of ~188 materials (subject to change) targeted by the USEPA for substantial reduction, per the Clean Air Act Amendments (CAAA). This list includes many common coatings solvents (Volatile HAPS ["VHAPS"]), but also includes heavy metals, toluene diisocyanate (TDI), coke oven emissions, etc. VHAPS are volatile organic compounds (VOC's)—but not all VOC's are HAP's. Examples of the latter would include ethyl-3-ethoxyproprionate, propylene glycol ether acetate, dibasic ester, and several others. Under current U.S. laws, the EPA is required to develop a variety of MACT ("maximum available control technology") Standards and Proposed Rules by various deadlines for various industries, such as wood, coil coatings, general metals, et al. Although some of the Final Rules are pending, the program could be objectively characterized as being "on-track," but not without incident.

. . . The heavy metal issue, however—whether as addressed by HAPS legislation or other federal, state or local legislation—may be even more difficult to resolve. While it is not possible to be comprehensive in this paper, the metals which are subject to the most intense scrutiny in both North America and Europe are lead, cadmium, mercury and hexavalent chromium. At the moment, the most active legislation in the United States is taking place in the State of Minnesota, where paints and coatings containing these metals were banned as of 1 July, 1998. Of particular interest, not just in North America, but globally, has been the use of hexavalent chromium ion, particularly in the form of zinc and strontium chromate corrosion-inhibitive pigments in coil and spray industrial primers. Currently, hexavalent chromium is ubiquitous in two areas of industrial coating: pretreatments and primers. Because "chrome VI" ion possesses unique corrosion-inhibiting properties on a variety of metallic substrates, it is not at all clear where—or even if—its usage can be eliminated or substantially reduced in the North American marketplace. Both the pretreatment suppliers and the coatings community have been hard at work on "chrome-free" systems for the better part of twenty years, but only with mixed results. One of the serendipitous properties of chromate inhibitive pigments is that they tend to work over multiple substrates, but this has not proven true with either the so-called "white pigments" which are being proposed to replace hexavalent chromium pigments in chrome-free primers, or with the non-chromate chemistries which are being explored in "chrome-free" pretreatments. There has certainly been a certain amount of success in the aluminum arena—both chrome-free pretreatments and primers are commercially available from multiple sources, and are being used on a daily basis. Their quality seems to be acceptable, although they tend to be less user friendly than the chromate-containing products which they replaced.

Once we enter the steel arena, however, the scenery changes. In the coil coatings industry, a substantial user of both chrome-containing primers and pretreatments, there does not appear to be any commercial use of either chrome-free pretreatments or chrome-free primers over cold-rolled steel (CRS) in North America at the present time. This statement is basically true, with a few exceptions, with regard to hot dipped galvanized steel (HDG), although there is considerable research activity being conducted in this area by the pretreatment suppliers, and—perhaps to a lesser degree—by the primer suppliers. Since a certain level of success has been reported in Europe, particularly in Italy and Austria, it is to be hoped that the North American activity will prove successful, as well, but this is by no means assured. The North American, South American, Australian and European marketplaces frequently have different expectations of similar products, and it would not be surprising, therefore, if a "solution" for one marketplace were not automatically considered to be an acceptable solution for another. The current outlook for chrome-free pretreatments and primers for zinc/aluminum coated steel is not particularly positive, however. Certain combinations of chrome-containing pretreatments with chrome-free primers, and chrome-free pretreatments with chrome-containing primers give rise to cautious optimism, but the current state of performance for chrome-free primers over chrome-free pretreatments leaves much to be desired. (A pithy American phrase, "the wheels come off," is fairly descriptive of the current state-of-the-art.) Where will this lead? At the moment, no one can know. The state of Minnesota is granting both permanent and temporary exemptions, on a case-by-case basis, for the use of hexavalent chromium in both pretreatments and primers intended for steel surfaces. For at least the next few years, other rule-making governmental bodies are likely to do the same. At some point, however, U.S. industrial applications may be required to run "best available" technology, whatever that means. If such a requirement were applied uniformly throughout North America, this may be an acceptable solution, but if it were not—if "chrome-free" production were forced to compete with chrome-containing production—the suppliers of the chrome-free products may face insurmountable obstacles to successful competition in the North America marketplace. It would take true mastery of a crystal ball to predict where the "chrome-free" movement will end-up, and I suspect that, ten years from now, this will still not be a completely settled issue. Different parts of the globe will respond in different ways and at different rates, and will consider different scenarios to represent "the solution." Based upon everything generally known at present, it is likely that the EU countries will arrive at a "chrome-free" solution first, followed by North American, Australia, Japan and, finally, the "rest of the world." Part of the difficulty in making a confident transition to chrome-free status revolves around the fact that the anticipated performance of new systems is being assessed through the use of traditional accelerated "performance prediction" methodologies. Well-founded controversy, however, rages around the meaning and value of many such testing regimes.

ACCELERATED PERFORMANCE PREDICTION

George Orwell, in his famed 1949 novel, **Nineteen Eighty-four,** maintained that if one repeats a statement often enough, it will come to be believed *even if it is obviously and patently untrue.* The world saw many living examples of Orwell's premise when the iron-curtain countries of the world were at their zenith, and our own industry has succumbed, to a degree, to the Orwellian curse. Consider, for example, the following examples of conventional industry wisdom:

- Salt spray testing predicts "real world" corrosion performance
- Accelerated U-V devices predict "real world" durability
- "Falling Sand" tests predict "real world" coating erosion rates
- Etc., etc., etc.,—*ad nauseam*
  George Orwell was no dummy....

Even though the preponderance of evidence would suggest that none of the above are true statements, they are nonetheless believed by many coatings manufacturers and users, and no greater evidence of this exists than the presence of these tests in a wide variety of product specifications. This is partly understandable, insofar there are no other tests universally regarded as being more predictive of corrosion activity, exterior durability, abrasion resistance, etc.—but it does place the entire research community at the mercy of testing methodologies which (at best) are unreliable and (at worst) may be downright misleading. Attempts to predict the "real world," using accelerated methodologies are fundamentally flawed, although such tests—when used as part of a much larger research protocol, in conjunction with other accelerated tests, actual exterior exposure panels, and interpreted in the light of sound chemical and performance models—may produce helpful "pieces of the puzzle." Reliance on only one or two accelerated tests, such as salt spray, will serve no purpose other than to inhibit progress toward viable long-term solutions to the economic, performance and environmental goals which are being pursued by the global coatings community.

The area in which accelerated methods have traditionally been most used and abused has been in weathering prediction. In 1977, the Cleveland Society for Coatings Technology organized a symposium entitled "Accelerated Weathering: Myth vs. Reality." (In fact, one of my colleagues at Akzo Nobel, J. A. Chess, was a co-author of a paper presented at that meeting.[2]) The title of that symposium certainly spoke volumes about the state-of-the-art with regard to accelerated weathering two decades ago, but I doubt that some of the papers presented at that meeting so long ago would be seriously out of place at this meeting today. We've certainly generated a lot of data in the twenty-four years since that meeting, but do we know any more about accelerated weathering now than we did then? Perhaps. . . . . but we have only moved forward incrementally; no "silver bullets" have been discovered during the past quarter century, nor has anyone postulated a "unified weathering theory" pulling together all known information on the chemical and physical mechanisms by which diverse materials degrade, and how such degradation processes may be accelerated and correlated with "real world" results. Still, progress has been made in certain areas, and we must be prepared to use the products of this progress to optimum advantage. Based upon the results of correlation studies with outdoor weathering prepared by all of the companies that eventually comprised the North American Coil Coatings Business Unit of Akzo Nobel, my colleagues and I feel that—although no accelerated test is an infallible predictor of "real world" conditions—there are six statements that we believe to be not only true, but deeply important to the future health of our industry:

What We Believe About Durability Prediction[3]

- *The best predictor of durability is real-time exposure in the location where the coating will be installed.*
- *The second-best predictor is real-time exposure in Florida at 45° facing south (for roofs) and 90° facing south (for sidewalls).*
- *The (distant) third-best predictor is one of several available accelerated weathering devices; mounting evidence suggests that UV-A, Fresnel-type, and Xenon Arc are the most reliable.*
- *Accelerated weathering data is only valuable when it is placed into a proper context by skilled coatings scientists, who use it as one of several tools at their disposal to predict long-term weathering effects.*
- *No accelerated weathering device, when used in a "stand alone" fashion, can predict real world weathering with any level of accuracy.*
- *Even the "real world" is a kind of myth—ever changing climatic, atmospheric, environmental, chemical, thermal and other considerations render the "real world" a dynamic exposure site.*

This is our creed, based upon over thirty years' worth of serious attempts to correlate accelerated weathering tests with the "real world"—whatever that is. It applies specifically to coatings for metal building product components, which are applied by the coil coating process and thermally set or cured, although my guess is that our creed can probably be applied to all exterior coatings, regardless of chemical type or application method. Our work has benefited from the scientific inquiries and different practical approaches of several companies, since Akzo Nobel's current worldwide Coil Coatings organization is the product of a series of mergers, beginning in the late 1970's, which eventually involved the Wyandotte Paint Co., Pontiac Paint, Hanna Chemical Coatings Corp., Celanese, Reliance Universal,

Akzo Coatings, Midland Dexter, Svensk Färgindustri, Nobel Industrial Coatings, and Courtaulds. Each organization brought its own theoretical and experimental approach to the table, and each brought the results of its empirical testing, as well. Our current work is able to draw upon information generated by tens of thousands of exterior exposure panels exposed in over two dozen "UV," "aggressive chemical," and "corrosion" sites worldwide. It is upon these panels that we have based our current beliefs.

Not only is durability difficult to predict—it is difficult to define. "Durability" is often in the eyes of the beholder. There are parts of the world where rapid gloss loss of exterior coatings is valued because the coating has reached its "final" color very early in its life cycle and its appearance can be expected to remain relatively unchanged with time. In other areas of the world, however, rapid, precipitous gloss loss might cause the local denizens to clutch their chests because they prize the appearance of a glossy building and their concept of durability transcends mere color stability. Clearly, *appearance* depends upon far more than just color, and here's where the complications begin—"durability" is a concept which involves freedom from some combination of color change, gloss change, chalking, cracking, crazing, blistering, peeling, etc., etc., etc., but that "special combination" may differ from observer to observer, or market to market. Since no two observers can be reasonably counted upon to agree on the exact weighting of these various factors, they must all be tested for, and it is desirable that they all be "predictable."

Over the years, in pursuit of this "predictability," the coatings industry worked with just about every predictive methodology—and every piece of predictive equipment—which has come along. Witness the evaluation of accelerated testing methods from 1906 to the present:

- *1906—North Dakota Agricultural Experiment Station*
  This was the first test site for comparison of coatings in this country. Over time, this exterior test method moved to Florida and became ASTM G-7. Today, 5, 45, and 90-degree testing are common, along with black box, direct, and heated methods of testing.

- *1918—Fade-Ometer*
  First used for textiles, this was a dry test method and is no longer used for coatings. The carbon arc used did not achieve sufficiently short wavelengths for correlation with exterior exposure.

- *1927—Weather-Ometer*
  Water spray was added to the Fade-Ometer for improved results. However, the light distribution was still very poor.

- *1933—Sunshine Carbon ARC Lamp*
  This is the "Weather-Ometer" or "Atlas Weather-Ometer," as it has commonly been known for at least the past quarter century. The spectral light distribution from this equipment delivers considerably more energy than was obtainable before, but provides a poor simulation of natural sunlight in the UV region. Filters are, consequently, used to help improve the correlation. Two different types of filters are currently in use—the 2.5mm Corex 7058 and the newer 3mm Pyrex 7740 type. All types of carbon arc accelerated weathering are now covered by ASTM G-23.

- *1960—Xenon Arc Lamp*
  With filters, the "Xenon Arc" spectral light distribution places it in the position of being one of the best light sources for accelerated weathering. This method, covered by ASTM G-26, can reasonably approximate sunlight in both the UV and visible light range. The equipment, however, is both expensive to purchase and expensive to operate.

- *1960—Equatorial Mounts with Mirrors*
  Fresnel-type testing (commonly referred to as "EMMA" or "EMMAQUA") is a method of employing natural sunlight as the source of light but concentrating it with mirrors. With the addition of airflow for cooling and water spray for moisture, quite reasonable

correlation to natural weathering has been reported. The most important factor for correlation has been to record only UV light exposure as MJ/M^2, rather than as total light exposure, reported in Langleys. ASTM G-90 covers this test method; both "night time" and "day time" wetting methods are used, with "night time" giving results which are significantly closer to Florida exposure, in our opinion.

1965—Dew Cycle Weather-Ometer

The development of coatings with greatly improved durability made normal carbon arc exposure methods too time-consuming. The "Dew Cycle" method, covered by ASTM 3361, is based on the assumption that, when both moisture and radiant energy are present at the same time, the most rapid film degradation takes place. To maintain humidity during the dark period, cold water is sprayed on the back of the test panels. Also, to speed-up results, the filters are removed to increase the intensity of the light.

- ### 1970—Fluorescent UV Condensation
  This relatively inexpensive method of accelerated weathering uses fluorescent UV lights in a humidity condensation apparatus. It had, however, the same "excess UV light" drawback as an unfiltered Atlas Weather-Ometer. In 1987, a new bulb, QUV-A 340, was introduced which gives a good match to the short wave bands of UV in sunlight.

- ### 1985—Predicting Coating Durability with Analytical Methods
  Several papers have appeared, since 1985, which examine short-term exterior exposure panels with a variety of analytical methods used to follow chemical changes in the coating. The goal has been to predict the service life of the coating without using harsh or misleading acceleration factors. To date, the methods have worked best with basecoat/clearcoat systems. It is not yet certain how effective this approach will be with low gloss pigmented systems, but this work looks very promising.

The truth is that there is no "silver bullet"—no single, infallible predictor of outdoor durability, but there are some weathering devices which, when placed into the proper context and interpreted by expert coatings specialists, greatly increase the confidence level with which outdoor durability can be predicted. One such device is the QUV-A Weathering Tester, originally engineered and marketed by Q-Panel Lab Products, a leading company which has traditionally been on the cutting edge of accelerated testing methodology, whether radiation-related (UV-A and UV-B), cyclic corrosion, or condensing humidity. Q-Panel engineered both of the fluorescent-UV/condensation testers which the coatings industry commonly refers to as "QUV-A" and "QUV-B." Here is what Q-Panel has to say about these two devices:

> UV-A's are especially useful for tests comparing generically different types of polymers. Because UV-A lamps have no UV output below the normal solar cutoff of 295 nm, they usually do not degrade materials as fast as UV-B lamps. However, they usually give better correlation with actual outdoor weathering.[4]

At the time that QUV-A was introduced, "QUV" testing was being done by nearly everyone, but wasn't designated as "UV-B" testing. A lot of people felt that it *was* the silver bullet, and—for certain systems, under certain conditions—perhaps it was. Nonetheless, the opinion that the ubiquitous QUV-B tester was not perfect and might be improved upon was not an opinion unique to its creators; in the late 1980's, it was being echoed throughout the industry. After comparing the weathering of automotive clearcoats in three different accelerated weathering instruments (carbon arc, xenon arc, and UV-B), Dr. David R. Bauer, Ford Motor Company's internationally-reknowed coatings scientist, and his colleagues found that ". . . .since the degradation chemistry that occurs in these tests is unnatural," none of these devices were acceptable for the coatings that Ford was testing. Dr. Bauer concluded, "although acceleration factors can be calculated (based on amide II signal loss rates) they cannot be used reliably to predict service life."[5] This was reinforced, nearly a decade later, at the Spring, 1996, European Coil Coating Meeting, by Dr. G. C. Simmons of Becker Industrial Coatings: "It is now established fact that they [ASTM B117 salt spray and QUV-B] do not correlate well to natural exposures, and in some specific cases can lead to totally wrong conclusions being made."[6]

Nothing has occurred, in the fourteen years since Dr. Bauer's comments, to change my own thinking with regard to dissatisfaction with broad use of the weathering devices utilizing UV-B bulbs. In fact, it was reinforced by Ford's Dr. John L. Gerlock at a major scientific gathering in 1997 when he indicated that an FS40 bulb (UV-B) might be "good for studying the aging of the Taurus in low earth orbit," but not on the earth's surface.[7] Nor have such observations been limited to the automotive and coatings industries. In an important study by 3M Company, in which an attempt was made to develop an accelerated weathering test for films using two types each of carbon arc and UV-B bulbs, the researchers concluded that the results indicated "poor predictive ability using any of the laboratory devices."[8] They further noted that Spearman ranking of the results from the UV-B samples ranged from "perfect correlation (1.0) to an almost complete reversal (-0.8) with essentially random scatter in between."[9] 3M's conclusion was that "commonly used cycles in carbon arc and fluorescent UV-condensation [UV-B] test equipment exhibited generally unacceptable correlation levels for these materials."[10] A major study by Dr. Carl J. Sullivan of ARCO Chemical Company (UV-A, UV-B, carbon arc and xenon arc) seconded this opinion: "These four accelerated test procedures yield contradicting conclusions on the weatherability of these four resin systems."[11] Dr. Sullivan notes that "the Florida exposure data clearly corroborate conclusions drawn from A-340, Xenon, and EMMAQUA studies and contradict B-313 results."[12]

While it is true that UV-B testing certainly accelerates the aging and degradation processes, it may—depending upon the coatings system—be accelerating the wrong chemistry, thereby vitiating any value that the information might provide, and, consequently, discrediting the test. Our own work has repeatedly shown that UV-B testing is so riddled with anomalies that—even when its predictions are in the right church—they are only rarely (and possibly coincidentally) in the right pew. It was this general lack of correlation with real chemical reactions and authentic exterior weathering results that led to the development of the UV-A lamp testing devices, which correlate more closely with sunlight. This was a very important advancement because many of the most durable coatings in the "real world" are unnaturally damaged by the more destructive short wavelength of UV radiation below 295nm that is emitted by UV-B—radiation which **does not occur** in natural sunlight.

Even the American Society for Testing Materials (ASTM), which usually maintains a discreet silence on the appropriateness of its testing methods, allowed the inclusion of the following comments under the "non-mandatory information" section of Standard Method G53, "Standard Practice for Operating Light and Water Exposure Apparatus (Fluorescent UV-Condensation Type) for Exposure of Nonmetallic Materials," which is followed by laboratories around the world for running UV-A and UV-B accelerated testing:

> *All UV-B lamps emit UV below the normal sunlight cut-on.* [sic] *This short wavelength UV can produce rapid polymer degradation and **often causes degradation mechanisms that do not occur when materials are exposed to sunlight. This may lead to anomalous results.** . . . For certain applications, the longer wavelength spectrum emitted by UV-A lamps is useful. Because UV-A lamps typically have little or no UV output below 300 nm, they usually do not degrade materials as rapidly as UV-B lamps, but **they may allow enhanced correlation with actual outdoor weathering.***[13] (Emphasis the author's.)

In the course of the work which my colleagues and I have done at Akzo Nobel, we certainly haven't found the mythical "silver bullet" but, in UV-A testing, we have definitely found an accelerated device which has enabled us to make surprisingly accurate predictions of the "real world" behavior of experimental polymer systems, and—especially—of modifications to existing polymers. We have not, however, found UV-A accelerated exposure to be especially valuable for the prediction of the service performance of pigments, but we are not disappointed or disillusioned by this fact either. Although little generalization can be made about the huge group of different chemistries, both inorganic and organic, that make up the group of coatings components which are collectively referred to as "pigments," it seems clear that their performance can be so dramatically affected by chemical environments which are present at various places on the surface of the earth that it cannot be predicted with a high level of confidence by testing devices which lack these specific chemical atmospheres. (Xenon arc testing, which adds infrared and visible light components absent from UV-A testing, may prove helpful in evaluating the service life of pigments, albeit at a significantly higher operating cost—and still without benefit of the localized chemical environments at work in the "real world".) It would be nice to compare "brown coatings X, Y, and Z" from different sources based on different polymeric systems in a simple weathering device, but obtaining meaningful, valid data from such attempts is more wishful thinking than sound science.

Perhaps the most attractive aspect of UV-A testing, at least in our own program, has been the consistent absence of *truly damaging, seriously misleading* data. This was the most treacherous aspect of our past work with such devices as the dew cycle carbon arc tester and the UV-B tester. While not ideal, we can all live with systems that look bad in an accelerated test but good in "real life." The worst that can happen is that we fail to sell a good product. If we were to believe anomalous testing results, however, that predicted *good* weatherability and went to the market with what turned out to be a *bad* product, this could spell disaster. *This does happen, however.* We found many examples of coatings that looked good in an accelerated test, but poor under actual exterior service life conditions. This, of course, is the worst possible scenario. We have seen cases where dark brown plastisol films, based on poly(vinyl)chloride (PVC), have compared favorably in UV-B exposure to similarly pigmented dark brown fluoropolymer films, based on poly(vinylidene)fluoride (usually abbreviated "PVDF" or "PVF$_2$"). Plastisol coatings have a very definite specialized niche in the building product marketplace, which they serve admirably, but no one expects them to perform over time at the level of a fluoropolymer—nor do they. In another dramatic case involving new, experimental polymers, we ran multiple accelerated tests that we were able to correlate with Florida. As can be seen from the data, if we had taken samples No. 8 and 9 to the marketplace based upon UV-B results, we would have been making a grave mistake. *(See Figure)*

These disappointing results occurred because the new polymers were partially based on a relatively new, low-use cycloaliphatic monomer which does not absorb UV radiation in the 280-295nm range, where UV-B does its greatest damage. For this reason, these polymers looked wonderful in the UV-B test, even though they did not look acceptable under actual service conditions, where other factors—most likely the twin gremlins, "heat" and "humidity"—were apparently at work. The manufacturer of this monomer later included this *caveat* in its product literature: "The use of QUV-B 313 is not suggested as a screening tool because the low-wavelength portion of the exposure spectrum can lead to anomalous results."[14]

9 DIFFERENT BROWN POLYESTERS

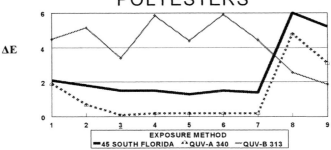

ΔE

EXPOSURE METHOD
45 SOUTH FLORIDA QUV-A 340 QUV-B 313

SAMPLES

The fact that even the UV-A failed to predict the full extent of the poor field performance of these polymers, based on new, unusual chemistry, leads us to another important aspect of "what we believe":

Accelerated Weathering Devices Only Have Proven Value—

- *When testing materials which are very similar to other materials for which correlation factors with real-time outdoor exposure at a variety of test sites have already been established;*
- *When they are used in conjunction with other "real-time, real conditions" test data;*
- *When the data which they yield are analyzed by an experts and compared against real-time, long-term exterior exposure data.*

We cannot stress this point too much or too often because a distressing new trend is emerging in the market place: Tools and data intended strictly for use in the scientific community, in the hands of skilled and experienced coatings scientists, are being moved into the marketplace, where they are being misused as marketing tools. This is clearly exemplified by cases where sales or marketing representatives are attempting to sell coatings, based on new chemistry, solely on the basis of their performance in an accelerated weathering device. The device of choice is often UV-B, probably because of its ubiquitous nature and "known destructiveness." Since almost any coating looks good in *some* accelerated test, there are those who arm themselves with such testing data, then imprudently and improperly enter the marketplace crying "Eureka" from the housetops. Potential customers then attempt to require competitive products—based upon completely different chemistry—to match the same set of test results, and the "accelerated testing wars" have begun. This is a dangerous development in the marketplace because it places the emphasis not on actual field performance but on accelerated testing data, which—in this context—is only so much *hocus-pocus.* Which brings us to the final major aspect of "what we believe":

Running unknowns, such as competitive products, in an accelerated testing device for the purpose of predicting actual field performance is extremely risky business and becomes "pseudo-science" in the hands of anyone other than expert coatings scientists.

ccelerated testing data has its place—but that place is in the laboratory, in the hands of skilled coatings scientists, and in the company of related data from other complementary testing regimes, which place the data in a proper context.

THE "BOTTOM LINE" ON DURABILITY PREDICTION

Science: There are no "silver bullets"—it takes time, expense, experience, expertise and a lot of hard work to develop a new coating.

        BOTTOM LINE: You get what you pay for.

Pseudo-Science: Anyone can compare coating "A" to coating "B" in some type of an accelerated testing regime, with meaningful results.

        BOTTOM LINE: Flip a coin; it's faster, cheaper, and generally just as accurate.

Wishing Thinking: Somewhere "out there" is a weathering device that will predict the durability of all systems under all conditions.

        BOTTOM LINE: There is no free lunch.

Now it is time to focus our attention on other archaic testing regimes, such as the "B 117 Salt Spray" test and the so-called "falling sand" test, among others. The results of such tests are questionable, at best—and dangerously misleading, at worst. Almost anything would be an improvement, and, as an industry, we owe ourselves such improvements. I am aware of no industry colleagues who do not feel that these testing protocols should be replaced by ones that are more reliable and—if, ten years from now, we are not looking back upon such tests through bemused eyes, wondering how we ever could have relied upon them for meaningful results—then shame on us.

WHAT COATINGS PRODUCTS WILL BE NEEDED IN THE FUTURE?
Growth in the coatings industry has been pretty well synonymous with "change" in the coatings industry, and—because we are so used to such a dynamic environment—it never occurs to most of us that this is not true in all industries. In 1948, however, when the author first drank cow's milk, it looked and tasted essentially the same as it does in 2000. The handling and processing of cow's milk may have changed over this 53-year time span, but such changes are transparent to the end-user, who only knows that the milk basically looks and tastes the same. Many of the products of the coatings industry today, however, are most decidedly *not* the same as they were twenty years ago, much less forty or fifty years ago. Today's coatings bear little resemblance in toughness, flexibility, durability and user-friendliness to their predecessors—the final products which result from the advanced technology being incorporated into our coatings and application equipment would have been unimaginable to all but the most visionary of our industry forefathers.

Even five years ago, who could have possibly guessed that the theme of the National Coil Coating Association's (NCCA) Annual Meeting in 1999 would be, **"Challenges, Changes and Choices on the Road Less Traveled: Coil Coating in the Automotive Market,"** or that five of the past six European Coil Coating Association (ECCA) Congresses would feature a paper on "coil automotive"? In the mid-1960's, the thought that we would one day be routinely coil coating refrigerator wrappers and microwave ovens ("critical surfaces"), louvered HVAC panels ("cut edges"), metal roofs in the shape of Spanish tiles ("too much 'work' on the metal"), and "self-lubricating" steel sheet which can undergo deep drawing without the need for press oil would have been met with skepticism, at best—and more likely with either disbelief or derision. The obstacles to realizing such products, using the coil coating process, were simply too great. Yet the obstacles were overcome, and limits were transcended, and—viewed from the year 2000, a mere thirty years later, when all of these wild-eyed concepts have become proven, commercial realities—our doubts of the past inspire disbelief that we could have had so little faith in ourselves and in our industry. These are just examples taken from one specialized segment of the coatings industry.[15] Many other examples of "limits transcended" and "obstacles overcome" have been documented in virtually all industrial and consumer segments of the global coatings community.

Replacement of Traditional Roofing Materials

Residential roofing is "established," insofar as the coatings technology is available. It is commonly used in Scandinavia, we see it on the occasional house in North America, and there are certified installers scattered around who are not afraid of it. This is a far cry from its potential penetration, however. The "best guess" information available suggests that only ~3% of the 17-18 billion square feet of residential roofing installed annually in the United States is made from painted metal.[16] This is big business, and it requires a carefully orchestrated plan of attack—a plan of attack which, happily, seems to be underway in North America, in the form of the *Metal Roofing Alliance,* and in Europe, where a similar concept is being organized by the ECCA. But will this be enough? Probably not. We are certainly going to need coatings which serve the needs of energy conserving programs, such as the joint USEPA/DOE "Energy Star" initiative, and the Cool Roof Rating Council. We're likely to need coatings which look even more like (and perhaps *feel* more like) cedar shakes and asphalt roofing than is currently the case. If we are smart, we will create a new roofing option that cannot be mimicked through the use of alternative materials, although creating a "new look" can be a scary business. Imagine, if you will, the look on the faces of consumers when they were first offered tar shingles to replace their slate roofs. Nonetheless, try we must, because this is one of the few ways in which we can positively affect the value chain of the coatings industry.

Solvent-Free Coatings

The opportunity for utilizing new application techniques to add value and to create both new looks for existing products, as well as to create actual new products, abound at the moment. Methods of applying powder coatings at undreamt-of speeds (>1000ft./min. have been reported) have emerged in the past few years, courtesy of the "Powder Cloud" concept from MSC/SMS, and the "Electromagnetic Brush" (EMB) concept from DSM. If the curing chemistry of the powder coatings themselves can catch up, we may be looking at powder coatings in a whole new way—as coatings which are not only environmentally friendly, but which can be applied at high speed, while producing thick, textured, functional and novelty finishes, as well. A glance at the current powder literature, which discusses new potential applications[17] and new approaches to curing mechanisms (such as U-V[18]) on an on-going basis, suggests that the future of powder is very bright, indeed. Likewise, "solid block" and "hot melt" coatings concepts hold great potential as "leap frog" technologies which may go nowhere, or which may catalyze a paradigm shift which will forever change the face of the coatings industry. Such technology holds the promise of new and novel patterns and designs, at both thin and thick films, in very small or very large runs with minimal clean-up and fast turnaround, on coating lines that produce very low emissions, require no coater room, have a very low energy demand and represent only a modest capital investment. Such lines were a dream in 1985, but became a reality in 1994, with construction of the first "solid block" line at BHP's Port Kembla facility in Australia. It's up to us to determine how (or even *if*) to utilize these new coatings and application technologies. If we want them to work for us, however, we can't just "wait and see"—there are many obstacles to the successful use of these technologies, and if we want them to work for us, we must *make* them work for us.

Radiation-curable Coatings

We hear more and more about "radiation curable coatings." Yet, when they are mentioned, most of the people in any given room look to the others, to see if they are interested. A survey of R&D managers of one-hundred thirty-two (132) U.S. manufacturers of paint and powder coatings indicated that 16.3% percent of the respondents are funding "radiation cure" projects of one stripe or another, although they are only investing an average of 3.9% of their total R&D budget in doing so.[19] U.V. curable coatings have found homes in the beverage can business, on high-end paper products and on certain types of wood products, such as household and office furniture, but they tend to have more novelty than reality for producers and users of coated metal products. (It is significant, however, that—in his 1999 presentation to the ECCA's June, 1999, General Meeting—DSM's Paul H. G. Binda described the application of a U.V.-curable powder coating, using electromagnetic brush technology.[20]) Even more removed from our thinking process tends to be electron beam (E/B) curing, but why? It's true that the coatings produced, using either U.V. or E/B curing methods, tend to be brittle, and are probably hampered by high internal stresses that work against adhesion. On the other hand, such coatings are generally very hard, extremely abrasion-resistant, and can withstand significant chemical attack. With the right work, we might see a variety of value-added products emerge from this technology, perhaps as a result of a two-stage cure, akin to what is often done in the plastics area—"step one" would cause the coating to harden sufficiently (yet remain sufficiently ductile) to be easily handled and formed (if necessary or desirable) then, in step two, high energy radiation (perhaps even visible light[21]) would be applied to give the fabricated object its final, "diamond-like" cure.

Automotive Challenges and Opportunities

"End of the line coating." "Mill-applied coating." I'm not sure that these are phrases which anyone, other than steel mills, find immediately comforting—and many of them probably have concerns of one type or another, as well. This is not a new idea, at least conceptually— Stein Heurtey, in France, claims development of this concept as early as 1984, and Kawasaki Steel's Chiba facility coupled electrogalvanizing and coil coating for certain automotive applications in 1988.[22] While this has not become widespread practice in the years since, it is probably significant that Danieli Wean announced last summer, on the Web, that they had received an order from A.G. Ban-Color S.A. for a new hot dip galvanizing line with an in-line painting section, to be commissioned in the second half of 2001 in Jerez de los Caballeros, Spain. Regardless of the relative lack of current, specific activity in this area, it is my impression that virtually all steel mills, as a result of actual or implied encouragement from segments of the automotive industry, are incorporating the concept of "mill-applied coatings" into their thinking process, as they contemplate the future—and many are including space and other key elements in their schematics for new metal coating lines. It is doubtful that anyone knows for sure what the future holds, with respect to this new approach, but it is equally doubtful that anyone can afford to risk being left "in the dust." Of course, there are a great many suppositions in this scenario. . . .

. . .Including the emergence of a significant need for precoated metal in the automotive industry, a supposition which is by no means a "given." There is, I am convinced, no element of a "done deal" about the future of what is increasingly referred to as "coil automotive." To bring the promises of the year 2000 to fruition in the year 2005 and beyond will require strategy, commitment, financial investment and good, old-fashioned hard work by the coil coaters, steel and aluminum suppliers, and coatings suppliers. We already have concrete examples of weldable, coil-coated primer being used as the base for electrodeposition coating, and rumors abound regarding the "automotive coil" plans of the various OEM's. In Europe, plans for the future go as far as to include a "fully coil coated" concept employing a combination of coatings and films. These are very encouraging signs—signs that were not in general evidence even five years ago, and still tend to be surrounded with an aura of secrecy and mystery. A few experimental cars were quietly built in the 1980's from coil-primed metal, and—thousands of miles and over a decade later—I took off my suit coat to climb over and under one of them, checking for corrosion and signs of other problems. It looked great—at least as good as a similar vehicle built using traditional materials, and possibly better. It *can* be done. Imagine the face of our industry in the year 2015, if automobiles are routinely coil coated, rather than electrocoated and spray-coated. Imagine the cost savings to the automotive industry if hem flanges are formed from coil coated steel, and no longer require wax and sealers. Imagine the labor savings if sanding and repair issues were to be reduced or eliminated. Times are changing, and opportunities of great magnitude are being created—but they bring with them equally great challenges, and the global coatings community must be prepared to meet these challenges with creative, economical, and environmentally, sound solutions.

HOW CAN THE R&D COMMUNITY RESPOND TO THE CHALLENGES AND OPPORTUNITIES OF THE FUTURE?

The past few years have seen the market introduction, in various parts of the globe, of an impressive variety of new high-performance coatings products, including self-cleaning coatings, I.R.–reflective roofing coatings, wrinkle finishes, 500μm textured plastisols, high-heat resistant coatings, and a host of others, in addition to those emerging technologies (powder, hot melts, solid block, chrome-free primers and weldable automotive primers) which I have already discussed. These have been the result of successful R&D programs conceived, appropriately funded and executed many years (usually 3-10) in advance of commercialization. But what about the products which will be needed to keep our industry both dynamic and viable five, ten and fifteen years from now? We need to be working on them today, but what is the state of those R&D programs? R&D programs cost money, and money is in short supply. It is no secret that less R&D (and, as a result of industry consolidation, less *diverse* R&D) is being funded today, compared to a decade ago, and that what *is* being funded has a much shorter-term focus. At the same time, the potential to investigate new chemistry, and place it in the service of the global coatings industry, has never been greater than it is today. The mind fairly reels at the availability of new chemistries and chemical concepts which are on the horizon: What, for instance, could our industry do with ceramic-like coatings ("ceramers") made using emerging sol-gel concepts? In the May 1, 2000, issue of *Chemical Engineering News*, published by the American Chemical Society, occurs the following, startling statement: ". . .according to government R&D planners, nanotechnology is nothing short of the next Industrial Revolution."[23] I agree. In my opinion, nanotechnology, which is the control of matter at the atomic and/or molecular level, has the potential to exert a profound influence on the physical and performance properties of products, such as coatings, which are engineered using these principals—properties such as hardness, flexibility, permeability, adhesion, corrosion-resistance and other parameters of critical interest to our industry. What if we were able to produce improved "self cleaning" coatings that worked equally well under all conditions, or better yet—coatings that don't actually get dirty to begin with? The ramifications of such a concept transcend mere aesthetics, and work being performed in a variety of laboratories, including the U.S. Naval Research Laboratory[24], suggests that, with appropriate R&D, this can happen.

Coatings that look like stainless steel. . . coatings for metal that emerge from high-speed applicators with patterns and designs permanently fixed in the coating, without the need for inks . . . self-healing coatings which would minimize—or perhaps actually prevent damage from handling abuse. . . solventless coatings. . . self-stratifying coatings[25] which would require only a single application to produce both primer and topcoat—and possibly even pretreatment, as well. . . coatings based on hyperbranched/dendritic polymers with advanced—possibly even *new*—properties. . . coatings that could be cured using high-energy, *visible light*. . . . What if we were able to apply the principles of combinatorial chemistry, currently being utilized with dazzling virtuosity by the pharmaceutical industry, to facilitate the rapid development of new coatings with undreamt-of properties? . . . . The list goes on and on, and these are not pipedreams—they are all possible. The real question is whether or not we are willing, as an industry, to provide the quality and quantity of research to take full advantage of these opportunities. Will we invest in new R&D approaches for the future, or will we attempt to "make do" with products and processes based upon incremental improvements (when available) to the same old technology?

Personally, I see no choice, and I am deeply concerned that, as an industry, we seem to be pondering this question as if there *is* a choice. At the end of the day, the answer is—the answer *must* be—that we are willing to tackle the future in the same way that we built our past: By setting our sights high, and by supporting each other in an industry-wide pursuit of new opportunities. It's our choice—let's make it a thoughtful, well-reasoned, *strategically sound* one.

REFERENCES

[1] Wismer, M. *American Paint and Coatings Journal*, **1984**, 68 (49), p.12.

[2] "Accelerated Corrosion Testing," Chess, J. A.; Hastings, M. R.; and Paolini, A. *J. Coatings Tech., 1977*, 49 (633), 55-61.

[3] "Accelerated Weathering: Science, Pseudo-science or Superstition?" Chess, J. A.; Cocuzzi, D. A.; Pilcher, G. R.; Van de Streek, G. N.; in Bauer, D. R. and Martin, J. W. *Service Life Prediction of Organic Coatings: A Systems Approach*, American Chemical Society/Oxford University Press, *1999*, 130-148.

[4] Q-Panel Lab Products, Product Bulletin LU-8160, *1994*.

[5] "Evaluation of Accelerated Weathering Tests for a Polyester-Urethane Coating Using Photoacoustic Infrared Spectroscopy," Bauer, D. R.; Paputa Peck, M. C.; Carter, R. O. *J. Coatings Tech., 1987*, 59, p.103.

[6] New Developments in Coil Coating Paints," Simmons, G. C. *ECCA Conference Transcript*, 17-22 May, *1996*.

[7] Gerlock, J. L., Ford Motor Company, Personal Communication, 1987.

[8] "Accelerated Weathering Test Development with Fluorescent UV-Condensation Devices," Fischer, R. M. SAE Technical Paper Series, #841022, *1984*, p.2.

[9] *Ibid*.

[10] *Ibid.*, p. 1.

[11] "Polyester Weatherability: Coupling Frontier Molecular Orbital Calculations of Oxidative Stability with Accelerated Testing," Sullivan, C. J., *J. Coatings Tech., 1995*, 67 (847). p.55.

[12] *Ibid.*, p. 56.

[13] ASTM Standard Method G53-95, *1995*, p. 7.

[14] Eastman Chemical Company, Publication N-335A, *1996*.

[15] "Coil at the Crossroads of the Millennia: Transcending Limits to Create New Opportunities," Pilcher, G. R. *European Coatings Journal*, April, *2001,* pp.134-143.

[16] These are composite figures which are drawn from multiple sources, including the Metal Construction Association (MCA), the Metal Roofing Alliance (MRA) and an independent report by the Freedonia Group.

[17] "Pushing Powder Further," Esposito, C. C. *Coatings World, 2000*, 5 (6), pp. 62-65.

[18] "UV-curable Powder Coatings: "A Variety of Applications," Zune, C. and Buysens, K. *European Coatings Journal, May, 2000*, p. 18-30.

[19] "Coating R&D: Where Is the Money Going?" Bailey, J. *Industrial Paint & Powder, 1999*, 75(2), 18-22.

[20] "Update on Powder Coatings for Coil Coating," Binda, P. H. G. *ECCA Conference Transcript,* 1-4 June, *1999*.

[21] "Corrosion Resistant Visible Light Curable Coatings, Part II," Mejeritski, Alexander; Marino, Thomas; Martin, Dustin. *RadTech 2000 Technical Proceedings, 2000*, 440-461.

[22] "Combined Galvanising/Painting Lines: Economic Advantages of the Concept," Delaunay, D. *ECCA Conference Transcript*, 1-4 June, *1999*.

[23] "Nanotechnology: The Next Big Thing," Schultz, W., *Chemical & Engineering News, 2000*, 78 (18), 41-47.

[24] "Clean Hulls Without Poisons: Devising and Testing Nontoxic Marine Coatings," Brady, R. F., Jr., *J. Coatings Technology, 2000*, 72 (900), 45-56.

[25] "Coatings Developments for the New Millennium," Rassing, J. *ECCA Conference Transcript*, 22-23 November, *1999*.

Macromol. Symp. 187, 17–21 (2002)

Powder and Waterborne Coatings 2000–2010 – Is Past Growth Sustainable?

New Technological Developments and the Impact on Future Markets, A World Overview

Franco Busato

Managing Partner of Irfab Chemical Consultants, Brussels, Belgium

Summary: Globally, the penetration of POWDER COATINGS in the Industrial Paints sector is 6%.The deepest penetration is in Europe with just over 9%, of which Italy leads, also worldwide, with the highest degree, 15%. Follow Asia-Pacific, North America. In ten years time, global penetration of powder coatings will reach the 10% level. Penetration of powder coatings is much different in the individual European countries. By the year 2010, penetration of WATERBORNE industrial coatings will reach a level of over 30%. The European Waterborne Industrial Coatings Industry was over 500,000 Tonnes in 1999 and will reach the one million Tonnes mark by the year 2010. Overall, the penetration of Waterborne Coatings in the Industrial Paints sector is currently around 20%. When considering individual market segments, penetration varies from 3% to 70%. Selected industrial coatings sectors, having low penetration of 3% in 2000, will reach up to 15% by the year 2010 at the expense of solvent borne paints and other non-environmentally-friendly systems. Main resin systems include: acrylics, epoxies, alkyds/polyesters, polyurethanes, PUDs, 2K-PU, alkyd emulsions and other. The paper will also cover an in-depth analysis of the inter-competition between solvent borne coatings, waterborne, powder coatings, radiation cureables. The increasing use of a combination of environment-friendly technologies such as *powders / WB: powders/radcure* (UV, EBC) and *WB/radcure* will be considered. Chemical intermediates and additive manufacturers have been developing novel raw materials both for waterborne paints and powder coatings.

Keywords: Waterborne coatings, powder coatings, markets, UV-cureable powders, UV-cureable waterborne coatings, paint technology forecast.

Introduction

Waterborne products, including both Architectural and Industrial paints, currently, account for an estimated 50% of the Paint and Coatings consumption in Western Europe,

   CCC 1022-1360/00/$ 17.50+.50/0

roughly estimated at 6.4 Million Tonnes. By the year 2010, Waterborne Paints and Coatings, will represent over 55% of total European paint consumption. Powder coatings and radiation-cureables will account for at least 11%, while in the same year, about one-third of all paints and coatings will still be solvent-based.

In contrast to early projections, paint technologies and markets have developed in different directions and at different rates. Penetration degree and growth of Powder Coatings and Waterbornes are very different in the various Regions. Resin, chemical intermediates and additive manufacturers have been developing novel raw materials for latter paint technologies. Reportedly, pigment manufacturers are lagging behind in the development of products for waterborne and powder coatings applications. Because of the variety and diversity of the paint technologies in the different regions, this provides tools for strategic decision making to identify untapped opportunities and explore new geographical markets.

Environmental considerations continue to be a major force behind technical innovation, exemplified by the search for technically viable coatings with little or no solvent content. Different research goals prevail in the three leading markets of the world.

Environmentalists, crying for more green, have elevated Powder and Waterborne Coatings to the forefront of paint technology development in the Industry. Is this growth sustainable over the next ten years, until the year 2010? Baring current economy conditions, what developments are occurring today and what are the trends driving powder and waterborne coating applications to 2010 in terms of technologies, markets, applications, raw materials and strategies?

POWDER COATINGS

Penetration degree of Powder Coatings

A differentiated development pattern of powder coatings compared to solvent-based, waterbornes and radiation-cureable coatings in the different regions of the world exists. Globally, the penetration of Powder Coatings in the Industrial Paints sector is 6%. The deepest penetration of powder coatings worldwide is in Europe with just over 9%, of which Italy leads, also worldwide, with the highest degree, 15%; Asia-Pacific follows with 5-6%, North America with 4-5%. In ten years time, global penetration of powder coatings will reach the 10% level. This means that there is still plenty of room for powder coatings to penetrate in the world paints arena.

The use of the combination of environment-friendly technologies such as *powders /WB, powders/radcure* (UV, EBC), is increasing.

The Trends

Globally, for the period 2000-2010, a yearly growth rate for powder coatings of about 7% is forecast. The hiccup in Asian demand of recent years is being taken over by growth, driven by the relentless demand for improving standards of living.

For newly industrialised countries in particular, powder coatings offer a clean and performing technology without the problems and consequences of solvent based coatings. The technology is easily accessible. One can consider investment in powder coatings is the technology of choice.

Powder Coatings Developments In The Asia-Pacific Region

There is a major trend towards the use of 'superdurable' polyesters for exterior applications. This is due to the high heat and humid climate which prevails in the Region. TGIC-free powders have only recently successfully started in Australia.

Japan is a forerunner in the application of powder coatings for blanks for white goods. In Japan has the largest demand for polyurethanes and is also developing UV curable powders. The newly and more industrialised countries such as China, Taiwan and Korea mainly require standard hybrids.

Japan, one of the most industrialised countries in the world, is still averse to the use of 'clean' technologies. The latter only represent a meagre 16 percent of the total Industrial Paints in Japan. In February 1998 a coatings care plan and promotion committee was established and a prototype guidance for paint producing companies is being developed, but on a voluntary basis.

Powder coatings and resin prices, particularly of polyesters, are significantly lower than those in Europe and the USA.

Like in Japan, in Korea dedicated coating lines exist for the powder coating of blanks for household articles. VOC regulation is expected to be applied within 3 years.

Major international companies have started buying companies or forming joint ventures.

WATERBORNE COATINGS

Trends and Growth rates

Preliminary analysis shows that waterborne paints and coatings as a whole, by the year 2010, will not grow as dramatically as Raw Material Manufacturers and the Paint Industry, overall, have anticipated. By the year 2010, penetration of waterborne industrial coatings will reach a level of 30%. The European Waterborne Industrial Coatings Industry was over 500,000 Tonnes in 1999 and will reach the one million Tonnes mark by the year 2010. The penetration of Waterborne Coatings in the Industrial Paints sector, overall, is currently around 20%. When considering individual market segments, penetration varies from 3% to 70%. Penetration of waterbornes in the General Metal industrial sector will double by the year 2010. Selected industrial coatings sectors, having low penetration of 3% in 2000, will reach up to 15% by the year 2010 at the expense of solvent borne paints and other non environment-friendly systems. This means there is still plenty of room for increased penetration of Waterborne Coatings in the world paints arena. The CAGR of waterborne coatings will outpace that of the overall industrial coatings industry by at least four times. Average growth rates for *waterborne* industrial paints vary, depending on the single end-use segments, from a minimum of 3% to over 15% per year.

Main resin systems present in waterbornes include: acrylics, epoxies, alkyds/polyesters, polyurethanes (PUDs, 2K-PU), alkyd emulsions and other.

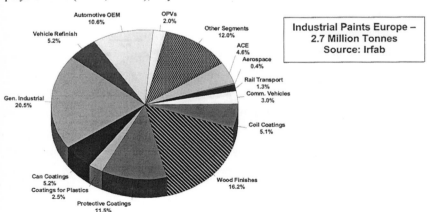

Participation of SMEs

It is important to stress that not only multinational companies, but also SMEs are heavily involved in developing global strategies to exploit renewed growth in the waterborne and powder coatings arena, as is clearly shown for Europe in Table 1.

**Table 1. SALES BREAKDOWN BY "PAINT TECHNOLOGY –
SMEs AND LARGE COMPANIES 2000 – Europe**

PAINT TECHNOLOGY	SMEs	% Of TOTAL
Conventional SB	58 %	67 %
High Solids SB	21	9,5
Waterborne	9	15
Powders	9	6.5
Radcure	3	2
TOTAL	100%	100%

Source: Irfab Chemical Consultants

Macromol. Symp. **187**, 23–34 (2002)

Coating System for Infra Structural Works

Andreas Heutink, B.Sc., Willem Bonestroo, Jo van Montfort B.Sc.

Ministry of Transport, Public Works and Water Management, "Rijkswaterstaat", Civil Engineering Division, P.O.Box 20.000, 3502 LA UTRECHT, The Netherlands

Summary: "Rijkswaterstaat" is a government organisation responsible for the construction, operation and maintenance of a large number of structures. The organisation consist of a general engineering division and a number of regional offices responsible for the Rijkswaterstaat objects in their particular region. Rijkswaterstaat spends between 40 to 50 K€ per year on coating and coating related activities. When the steel structures of the Eastern Scheldt barrier ($\pm$ 200,000 m^2 of steel) were coated in the mid eighties the applied high solid epoxy system developed extensive cracking. This resulted in numerous locations with pit corrosion. Investigations were carried out and it was concluded that the cracking of this specific high-solids epoxy system was caused by restrained shrinkage. This resulted in the build up of high internal mechanical stresses in the coating system. Problems with stress build up were also encountered on other projects, although less severe and mainly concentrated at edges and welded joints.

To prevent future problems, RWS has taken a number of initiatives such as the developed a procedure for the selection of coating systems. This selection procedure has proven to be useful for RWS in selecting coating systems for use on infra structural works.

The basic principle of the selection procedure of the RWS selection procedure is different from most other procedures. The RWS selection procedure uses the object to be coated as a starting point instead of the coating system. This is done by first evaluating all relevant aspects of the object which may influence the selection of a system. These aspects are then translated into so called "Functional requirements". Coating systems with properties that are in compliance with the performance requirements are gathered and through comparison of requirements and properties a selection is made.

The objects that, until now, have been coated with systems selected with the selection procedure have performed adequately. Systems that have been selected with the procedure to replace or repair those that have developed cracking have until now not shown any new problems.

At the present time more and more alternative tests and test procedures become available, most of them derived from ISO 12944. These will be studied closely and compared with the currently used methods on the level of functional requirements.

RWS is already actively involved in the search for alternative test methods and will remain to do so in the future. In view of this the Rijkswaterstaat selection procedure will remain in function.

Keywords: Cracking, e-modulus, fingerprinting, functional requirements, high-solid epoxy-based coating systems, integral life cycle, mechanical stress, tensile strength, restrained shrinkage, rijkswaterstaat selection procedure, shrinkage, structural data, WOM Whether-O-Meter test

   CCC 1022-1360/00/$ 17.50+.50/0

INTRODUCTION

"Rijkswaterstaat" is a government organisation responsible for the construction, operation and maintenance of a large number of structures. These include all the major flood barriers that have been constructed in the Netherlands as well as a large number of bridges, tunnels and locks. Most of the objects are designed for a service life of 100 to 200 years.

The organisation consists of a general engineering division and a number of regional offices responsible for the Rijkswaterstaat objects in their particular region. Rijkswaterstaat spends between 40 to 50.000.000,- euro per year on coating and coating related activities.

Until just a few years ago, the steel structures of flood barriers in the Netherlands were usually coated with tar-epoxy coating systems. The tar-epoxies performed very well, but the use of tar is now prohibited in the Netherlands. The coating industry and the Dutch government have furthermore agreed that the emission of volatile organic compounds must be reduced in the coming years. This has prompted a major effort on the part of the coating industry to develop new coating systems which can meet the requirements and are more environmentally friendly.

When the steel structures of the Eastern Scheldt barrier ($\pm$ 200,000 m^2 of steel) were coated, it was decided halfway through the project that the coal tar based system would be changed to a high solid epoxy system. Within a few years, the high solid epoxy system developed extensive cracking. This resulted in numerous locations with pit corrosion. At approximately the same time similar problems were noticed on other steel structures, although less severe and mainly concentrated at edges and welded joints.

Investigations have been carried out to establish the cause of the cracking of these high-solids epoxy-based coating systems.

The investigation showed that the cracking behaviour can be simulated using the accelerated Weather-O-meter test. It was concluded that the cracking of this specific high-solids epoxy system was caused by restrained shrinkage. The main cause of the shrinkage was due to evaporation of compounds during a stage, in which the coating system had already become fairly rigid. This resulted in the build up of high internal mechanical stresses in the coating system. Stresses were found to be highest in areas where transitions in coating thickness occur and changes are seen in the geometry of the structure. The level of stress and probability of cracking is high in these areas. Problems with stress build up were also encountered on other projects.

Since these defects have resulted in a significant amount of costly repair work it was decided that initiatives were needed to avoid further damage and costs in the future.

INITIATIVES

The initiatives taken by Rijkswaterstaat to improve on performance of protective coatings were:

- Design and implementation of a selection method for protective coating systems
- Improving contracts and specifications
- Development and implementation of maintenance strategies,
- Improving the standard of workmanship by taking part in an initiative for personal certification of painters, setting standards and implementing training for inspection personnel.

In this paper only the selection method for protective coating systems will be outlined and discussed.

BASIC PRICIPLE

The basic principle of the Rijkswaterstaat selection procedure is different from most other procedures. The RWS selection procedure uses the object to be coated as a starting point (instead of the coating system). This is done by first evaluating all relevant aspects of the object which may influence the selection of a system. These aspects are then translated into so called "Functional requirements". Coating systems with properties that are in compliance with the performance requirements are gathered and through comparison of requirements and properties a selection is made.

SELECTION PROCEDURE IN THEORY

The objective of the selection procedure is not to act as a "coating system select machine" but to support the selection process in a logical way so the most suitable coating system for the intended purpose can be selected. This selection is made in four steps. Selecting a coating system is trying to find a perfect balance between aspects such as environment, costs, lifecycle, application, maintenance strategy and product quality.

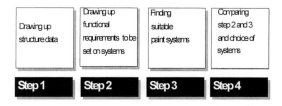

FOUR STEPS:

Step 1: Drawing up structural data

In step 1 all data relevant about the object to be coated is gathered. Typical data will include function of the object, its form (shape), type of exposure, esthetical aspects, substrate, maintenance, economic life time, etc. This data will be used in step 2 as the basis for the functional requirements of the protective coating to be selected..

Step 2: Drawing up functional requirements to be set on coating systems

Data of the object structure is translated into functional requirements for the coating system in step 2. The requirements that introduce a risk with respect to the selection of a coating system must be quantified or made quantifiable and measurable. Functional requirements will not only be test related but will also include requirements with respect to application properties, experience with a system, environmental considerations, etc.

Example:

If in step one it was found that the structure has a large number of very complicated details there is a significant risk of excessive film thickness being applied. This would be caused by a high degree of overlap during airless application. This aspect would then be translated into a functional requirement with respect to the tolerance in <u>maximum</u> allowed film thickness.

Step 3: Finding suitable coating systems

In step 3, matching functional properties are sought for the functional requirements drawn up in step 2. Data on the functional properties of protective systems must therefore be available. The data is generated through testing on the initiative of Rijkswaterstaat and/or is delivered by the coating manufacturers. The data does not only include test data but, even more important, data with respect to practical experience (references), application properties en environmental aspects.

Step 4: Selecting a system by comparing functional requirements and functional properties

In the fourth and final step, the functional requirements are compared with the functional properties of protective systems. The degree to which the properties match the requirements determines the selection of a protective system. Before considering whether a protective system meets the requirements, it must be determined whether the available data is complete. Incomplete data does not necessarily mean a system is unsuitable but introduces additional risks. Depending on the importance of a project and especially on the amount of indirect costs and risks in case of a premature coating failure, insufficient data may be accepted or additional data gathered for instance through testing or evaluation of additional references. Carrying out a more detailed investigation may result in new data, after which the information will have to be reconsidered.

SELECTION PROCEDURE IN PRACTICE

Step 1: Drawing up structure data

Since not all objects are completely unique, with respect to the aspects that influence the selection of a coating system, standard formats have been drawn up that cover the majority of aspects of the Rijkswaterstaat objects. These have been divided into the following three major categories;

- Atmospheric exposure with direct UV light, for instance topside of bridges.
- Atmospheric exposure without direct UV light for instance underside of bridges.
- Immersion service, for instance flood gates.

When a coating system has to be selected in step one an evaluation is made with respect to differences between the standard list of relevant aspects compared to the actual aspects of the object for which the system is to be selected.

Step 2: Drawing up functional requirements

The standard lists of aspects in step 1 have been translated into standard lists of performance requirements for the three mentioned categories. In case differences are found between the standard and an actual aspects during step one this may result in the addition of new performance requirements, altering of existing requirements or perhaps even omitting standard requirements.

Example:

In step one it was found that a structure had an existing coating system with micro cracking. This existing system had to serve as the substrate to which the new system would be applied and therefore differs from the standard Sa2 ½ blasted steel surface mentioned in the standard performance requirements for this structure.

The reason for applying a new coating system, in this particular case, was to prevent the micro cracking form developing into a much more serious type of cracking or even flaking. The anti corrosion properties would however still have to be realised by the existing system. (the substrate) Therefore there is no need for corrosion test data (salt spray test, immersion, etc.) and the functional requirements with respect to these were omitted. Given the nature of the problem additional requirements with respect to adhesion and mechanical behaviour on this specific substrate were however considered necessary. Additional functional requirements in the form of the standard WOM test have being carried out on a simulated cracked substrate (test panel) instead of the standard Sa2 ½ blasted steel panel were added to the lab test programme.

Step 3: finding suitable coating systems

Rijkswaterstaat has initiated a number of test programmes since 1994 to generate sufficient data with respect to the properties of various coating systems. Systems meeting the standard requirements of the three categories mentioned under step 1 are available for use on Rijkswaterstaat objects. They have been categorised under an internal standardisation system called "NBD standards".

Recently a programme has been finished in which nine systems of various manufacturers have been tested. There are ongoing discussions with the manufacturers, especially with respect to some of the performance tests required in the selection procedure. Eventually the manufactures will be responsible for generating and submitting the data in conformance with the selection procedure.

Step 4: Selecting a system by comparing functional requirements and functional properties

As stated before the degree to which the properties match the requirements determines the selection of a protective system. In case of a situation which, as far as functional requirements are concerned, is within the "standard" one of the approved systems from the relevant categories may be used. The selection is a more or less straightforward affair. In case of deviations and incomplete data the evaluation of the subsequent risk is carried out on a project level in close conjunction with the regional offices.

TESTS AND TEST PROCEDURES

The current accelerated weathering tests used in the selection procedure are as follows:

- TNO corrosion cabinet;
- Weather-O-Meter (WOM) test;
- Saltwater spray test ASTM B 117;
- Fresh water immersion test ISO 2812-2;
- Saltwater immersion test ISO 2812-1.

The tests are designed in such a way that the two main deteriorating mechanisms, namely deterioration of mechanical properties and corrosion, are tested. The TNO corrosion test is a cyclic corrosion test and the only test so far that uses artificial rainwater with a combined SO_2/CO_2 exposure.

The WOM test is used to artificially age coating systems by utilising artificial sunlight (Xenon) with frequent and significant changes in temperature. The test panels for the WOM test have a number of "V" shaped grooves that simulate an in- and outward edge in the substrate at which concentrations of internal mechanical stress in the coating system may develop, ultimately resulting in cracking of the system. The WOM test is not used as an attempt to try and create an artificial environment which accurately represents the actual environment but to measure the development of specific properties of a coating film during the ageing process. Important properties that are determined before and during the WOM test are loss of mass and volume, elasticity, tensile strength, elongation at break, etc. Each individual property or more likely a combination of altering properties may result in cracking of the coating film.

Test programmes

Since 1994 five test programmes have been completed in order to be able to select coating systems for the repair of specific coating problems on the Eastern Scheldt barrier, the Maeslant barrier and the Haringvliet barrier. In 2001 a general programme was initiated to test systems of various manufacturers several systems according to the three different standards of functional requirements. There are plans to initiate an additional programme before the end of this year.

Experience with test programmes

In contradiction to conventional (standardised tests) the test programmes have proven that the tests used are reliable in reproducing actual defects on a laboratory scale. By exposing the relevant coating systems to the present WOM cycle the problems with the cracking of coating systems on various projects such as the Eastern Scheldt barrier were reproduced. (see also figure 1 and 2)

Based upon the measurement of specific properties at intervals during the test the responsible mechanisms could be traced. Also the reference system that has been tested in all the programmes until now has shown a constant performance which proves the consistence of the test methods

Compared with conventional (standardised) test methods the Rijkswaterstaat test procedure has proven to be adequately discriminating for coating systems for infra structural works. As an example of this in table 1 two glass flake epoxy systems are compared. These systems were

genercally identical but were supplied by two different coating manufacturers. The performance
nonetheless differs considerable in some aspects of the test.

Figure 1. crack at depression in film as a
result of concentrated internal stress

Figure 2. surface cracking of coating

System A has failed in practice on a flood barrier. After two years of exposure unacceptable
cracking developed. Both systems meet the same generic description stated in part 5 "Protective
coating systems" of ISO 12944. When tested in accordance with the procedure under part 6 of
ISO 12944 both systems would have been considered suitable.

Table 1

	System A		System B	
	Before exposure 1500 hrs WOM	After exposure 1500 hrs WOM	Before exposure 1500 hrs WOM	After exposure 1500 hrs WOM
E-Modulus KN/mm²	0,980	2,904	2.144	2.021
Shrinkage %	-	2,5	-	0,3
Tensile strength in Mpa	16.1	36	31.1	27

Discussions with manufacturers

A number of accelerated weathering tests, some of them standardised such as ISO 12944 and ISO
20340-A, are being used at the moment. Manufacturers have in the past suggested alternatives for
the Rijkswaterstaat tests in the selection procedure. After careful evaluation by a national panel of
coating experts it was concluded that these alternatives are less suitable for Rijkswaterstaat. A
rather large difference exists between the number of test cycles in the selection procedure

compared with other cyclic tests. The WOM test is the only test which utilizes a Xenon gas discharge lamp. The light spectrum of the lamp comes closest to natural sunlight.

Systems that have failed the test programme and have also failed in practice will easily meet the requirements of ISO 12944 part 6. Considering this, as well as the substantial financial risks in case of premature failure of a coating system in relation to the costs of testing as such, the present test methods will be adhered to until alternative procedures have proven to be suitable. This suitability will not only be a matter of comparing the working principle of different test methods but will be decided by the extend to which alternative tests are able to generate comparable data with respect to the functional requirements. Alternative programmes could be validated by exposing coating systems that have already (preferable more than once) been tested in the present procedure to alternative tests and compare the data.

REFERENCES

Relevant references are considered the most important aspect in the selection of coating systems. During the most recent test programme manufacturers were asked to supply information with respect to projects were the system to be tested has already been successfully used. The amount of relevant results that have been received has however been somewhat disappointing. The supplied data is often insufficiently detailed and no more then a list of objects supposedly coated with the system at some point in the past. The current status of the system is not known nor is there information about the composition of the system (film thickness, number of coats, etc.) in relation to the system to be tested. Not all objects are readily available for inspection. Considering the importance of practical experience in the selection of a system this should be improved.

FINGERPRINTING

Since the performance of a coating system in practice or in a test is directly related to the composition of the individual products fingerprinting was introduced to guard against unnoticed or unannounced changes in the composition of products. Other end users as well as a number of manufacturers have recognised the importance of fingerprinting. Fingerprinting is also included in ISO 20340.

A fingerprint involves an Infra red and XRF analysis as well as the assessment of a number of basic properties such as mass, solids by volume and viscosity.

The purpose of the fingerprint is not to fully identify the products as such but to obtain general data with respect to the generic composition. This data can at a later stage be used to verify the composition of products to be used on an actual project.

Given the complex nature of the analysis and composition of some coating products the interpretation of comparative results can not be outlined in absolute grades or percentages but should be left to experts. It is important that the technical perimeters of the fingerprint are defined clearly to avoid misinterpretation. It is also important that, when using the fingerprint as a quality control tool during an actual coating job, a procedure is drawn up clearly outlining the sequence of actions to be taken as well as the responsibilities of all parties concerned.

SUMMARY AND CONCLUSIONS

Rijkswaterstaat has developed and used the selection procedure for coating systems since 1994. The procedure consists of four steps.

- Object data is translated into functional requirements for the coating system.
- The functional requirements include test data but also practical performance, environmental aspects and application properties.
- The functional properties are compared with properties of possibly suitable systems.
- Based upon the extend to which the properties match up with the functional requirements a selection is made.
- The object properties and functional requirements have been summarised in standards for three specific object categories.

The selection procedure has proven to be useful for Rijkswaterstaat in selecting coating systems for on infra structural works. The Dutch Directorate-General for Public Works and Water Management has no intention of permanently committing itself to the use of the present test methods. The test methods that have been adopted in the selection procedure have however proven to be reliable in reproducing actual defects on a laboratory scale and provide adequate and accurate information with respect to the functional requirements. The objects that, until now, have been coated with systems selected with the selection procedure have performed satisfactory. Systems that have been selected with the procedure to replace or repair those that have developed cracking have until now not shown any new problems. In view of this the selection procedure will remain in function.

At the present time more and more alternative tests and test procedures become available, most of them derived from ISO 12944. These will be studied closely and compared with the currently used methods on the level of functional requirements.

Rijkswaterstaat is already actively involved in the search for alternative test methods and will remain to do so in the future. The initiative should however be taken by the coating industry.

FUTURE

Outdoor exposure test

During the most recent test programme additional test panels for outdoor exposure have been prepared. These will be exposed to atmospheric and/or immersion conditions for a period of at least five years. The purpose of this is not to add another test to the programme but to validate the exposure of the lab tests. The change in properties of the exposed systems such as elasticity, corrosion from a scribe, tensile strength, etc. will be measured once a year. A visual assessment of the panels will be performed twice a year. By comparing this data with the data obtained during the lab testing a comparison can be made between the rate and nature of weathering in the lab test and actual outdoor exposure. This information will be used in the future development of the selection procedure.

Fingerprints

Recently there have been some discussion and misunderstanding about the way fingerprinting of products should be handled on project level. A Fingerprint procedure will therefore be developed that can be used as an integral part of a specification. This procedure should clearly state all the technical as well as procedural aspects of fingerprint identification.

Project references

Considering the importance of references in the selection of coating systems the possibility of generating a data system for references will be investigated. This system will be fed by inspection data generated by or under supervision of Rijkswaterstaat. Its purpose is to serve as feedback for assessment of coating system performance but more importantly for the future design of structures.

Modelling

Rijkswaterstaat demonstrated that the risk for unexpected and unacceptable failure of coating systems can be significantly reduced by introducing a selecting method based on functional requirements. However, this method is still largely based on exposure tests. The main general known disadvantages of testing is that it takes (to much) time. Predicting the actual service life solely based on tests is, despite the reliability of the Rijkswaterstaat procedure, almost impossible. At this moment Rijkswaterstaat is supporting the start of a new development to simulate the behaviour of a coating system by using a computer model. This could offer many advantages in drawing up more reliable maintenance strategies. The same model could, in theory, be used by coating manufacturers for the development of new products.

Integral life cycle

Considering the fact that most of Rijkswaterstaat objects are designed for a service life of 100 to 200 years there is a strong demand for drawing up reliable protection strategies. The service life of an organic protective coating system such as considered in this paper performs an average service life of 25 years.

Based on this fact new strategies for maintenance of large infra structures are developed in The Netherlands. Key feature in this new strategy is development of long term (100 years) programs. The complete life cycle from "cradle to grave" is considered. Integral Life Cycle Thinking is strongly related to drawing up functional requirements. In providing a durable protection of an infra structure not only traditional organic coatings will be selected. Alternative non coating solutions such as spray applied metals may prove to be a suitable solution for specific problem areas of a structure.

In designing new constructions considerations with respect to accessibility for coating, preventing the influence of micro climates at specific details and the selection of materials not requiring coatings are to be made as well.

Reference:

By Jan Bijen and Jo van Montfort, "Cracking of High-Solids epoxy coatings on Steel Structures in the Netherlands", Materials Performance vol. 38, no. 5, May 1999

Appendix:

Functional requirements and tests for selection procedure

Macromol. Symp. **187**, 35–42 (2002)

From Molecular Aspects of Delamination to New Polymeric Coatings

M. Rohwerder, M. Stratmann

Max-Planck-Institut für Eisenforschung, Max-Planck-Str. 1, D-40237 Düsseldorf, Germany, Email: stratmann@mpie.de

Summary: The delamination of organic coatings from reactive metal surfaces has been examined by different kinds of techniques. The electrochemical behavior of the delaminating system has been proven by the Scanning Kelvin Probe technique to be a combination of localized electrodes between the defect, delamination front and intact interface. A first idea of the extent of the delamination front could be gained. To achieve an even deeper insight in the characteristics of the delamination, Scanning Kelvin Probe Force Microscopy (SKPFM) has been used for in situ delamination experiments for the first time. This technique permits a high lateral resolution of the potential distribution and, at the same time, detection of the topographic changes of the substrate. An overview of the development of delamination research from macroscopic to submicroscopic resolution is given here.

Keywords: Kelvin probe; delamination; corrosion; coatings; cathodic debonding

Introduction

The delamination of polymers from reactive materials such as steel or galvanized steel has been investigated in the past by many techniques (1-4). Among others, the Scanning Kelvin Probe (SKP) has been used to analyze potential profiles along the delaminating interface. With this technique surface areas which are intact (anodic electrode potential) and those which are already delaminated (cathodic electrode potential) can be readily distinguished. Using these data, the kinetics of the delamination process is elucidated as a function of relevant parameters, e.g. ionic concentration of the electrolyte or the pre-treatment of the metal surface. Furthermore, principle reactions leading to delamination have been described using the SKP in combination with spectroscopic and gas exchange experiments (5-7).

However, little is known about the lateral extension of the reaction zone, which is defined as the

   CCC 1022-1360/00/$ 17.50+.50/0

transient area between the intact metal/polymer interface and the delaminated interface. Also kinetic details of the electrochemical and chemical reactions within this reaction zone are little understood. Therefore the lateral resolution of the electrochemical techniques being able to detect the local reaction zone is increased while the chemical und structural complexity of the interface is decreased, leading to improved applicability of sophisticated spectroscopic techniques. In this paper, first results which have been obtained in this research field are summarized.

EXPERIMENTAL

Only a short overview of the special experimental parameters of the new SKPFM technique is given here (8-9), since the number of different sample set-ups is high. For more detailed information on the other sample set-ups, see publications (5-6).

The SKPFM measurements have been carried out with a Dimension 3100 extended (Digital Instruments, Veeco) atomic force microscope. For the experiments n-doped silicon tapping mode tips were used. The measurements were carried out in a custom-made glass cell purged with humid air to control the relative humidity of the environment.

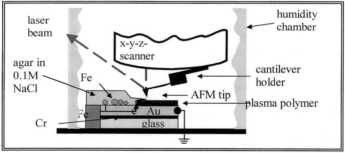

Figure 1:   Schematic sample set-up for model cathodic delamination experiments.

To achieve a reasonable local resolution in this study, a specific optimized flat sample set-up was preferred (Fig.1). Flame annealed gold was covered by an ultrathin plasma polymer film made from hexamethyldisilane and electrochemically coupled to iron to set the electrode potential of the polymer coated gold to $-0.4V_{SHE}$. 0.1M NaCl in agar was used as the electrolyte and applied onto the surface. The gold surface, which allows electron transfer but no ion transfer reactions, behaved similar to a passive iron electrode below the organic coating and served as the local cathode during delamination while the iron acted as the local anode.

RESULTS AND DISCUSSION

Scanning Kelvin Probe

Some information on the lateral extension of the delamination zone was obtained by comparing potential profiles as measured by the SKP with local de-adhesion measurements (Fig.2). These data have been obtained for galvanized steel (5). The local de-adhesion was measured by detecting the force necessary to locally remove the polymer film from the substrate. The SKP showed a continuous transition from a detached interface represented by negative potentials to an intact interface represented by anodic electrode potentials. The corresponding plot showing the mechanical adhesion clearly reveals 3 regions: intact interface (tensile force 0.3N/mm), delaminated interface (tensile force approx. 0) and an area in between (tensile force 0.15N/mm) which could be attributed to the reaction zone, since the interface was already weakened but adhesion was not yet lost.

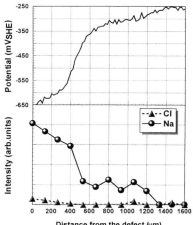

Figure 2: Galvanized steel. Top: Potential drop from delaminated (left) to intact interface (right). Bottom: Correlated tensile force. (5)

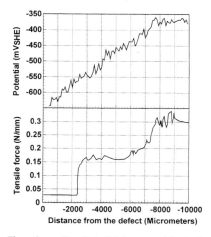

Figure 3: Top: Potential drop over delamination front. Bottom: Correlated ESCA line scans.

The information concerning the reaction zone of interest is even more precisely described, if the SKP potential profile is compared to the interfacial chemical analysis measured by ESCA (Fig.3) (5). Again the potential profile shows a continuous transition between negative and positive electrode potentials whereas the concentration profile, in particular of sodium ions, is quite interesting. The fact that the sodium concentration grossly exceeds the chloride concentration proves the existence of an electrochemical galvanic cell with the oxygen reduction taking place below the organic coating.

Between 0 and 400μm the coating was already delaminated, as seen by the low electrode potentials and the high sodium concentration. This points to a high rate of oxygen reduction. Between 600 and 1200μm the coating is still adherent (high potentials) but small amounts of sodium are already found at the interface. Obviously within this region oxygen is already reduced and the front of de-adhesion lags behind a front of oxygen reduction as marked by the incorporation of sodium. Therefore the zone between 600 and 1200μm could be identified as the crucial reaction zone. As the time necessary to shift the point of de-adhesion from 600 to 1200μm is in the order of 60min and the rate of oxygen reduction at the buried interface is in the range of $1\mu A/cm^2$, the charge necessary to electrochemically destroy the interface amounts to approximately $4mC/cm^2$. This corresponds to a consumption of 10 monolayers of oxygen.

Obviously, the SKP has a precision to detect unambiguously the size of the reaction zone for weak coatings. However, for better adherent coatings, the lateral extension of this zone is smaller and not easily detectable with the SKP. Moreover the delamination frontier appears to be even and any potential differences based on inhomogeneities on a mesoscopic scale is missing due to a lack of lateral resolution. This limit is overcome by the Scanning Kelvin Probe Force Microscopy (SKPFM).

Scanning Kelvin Probe Force Microscopy (SKPFM)

Set-up and principles of the SKPFM have been described elsewhere in detail (10-13). In particular, Frankel et al. have used the SKPFM technique extensively in order to analyze the corrosion of Al-based alloys (14-16). However, the delamination of organic coatings has not yet been analyzed with SKPFM. The special sample set-up already described in the experimental section successfully initiated cathodic delamination which proceeds similar to the cathodic delamination on iron. Fig.4 compares the AFM-topography a) with an SKPFM potential image b).

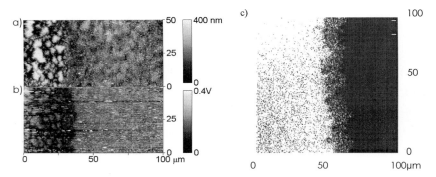

Figure 4:   a) Topography of a plasma polymer coated gold surface during delamination.
b) Correlated potential map. c) TOF-SIMS: Correlated sodium distribution.

Only the delamination front was visible which is situated at a distance approximately 5000μm from the defect. Delamination is seen in the AFM map as an elevated section and is therefore easily visible. The delamination front as deduced from the height profile is marked by the white dashed line. On the Voltapotential map the delamination is also clearly visible as a region of low potential similar to the conventional SKP results. However, the front of the negative electrode potential exceeds the delamination front by a constant distance of approximately 10μm.

The magnified cross-section of the potential map for a similar sample is shown in Fig.5. A potential transient is clearly visible over a distance of 7μm, which is rather comparable to the difference between the de-adhesion front and the sharp potential decline. Obviously this potential transient marks the reaction zone which leads to de-adhesion. The size of this reaction zone is well within the lateral resolution of the SKPFM technique and therefore lies easily within the detection limits. It is rather astonishing however, that the potential drop detected by SKPFM is

40

quite small in comparison to the potential drop as measured by the SKP on the same surface (600mV). The reason for this difference is a subject of present investigations.

The dimension of the reaction zone was further confirmed by TOF-SIMS measurements as shown in Fig.4c. SIMS results prove that only sodium was incorporated into the interface and a sharp transition is seen between the delaminated area (bright region) the intact area (dark region) and the reaction zone (medium brightness) the width of the latter being again approximately 10μm. In

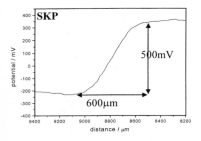
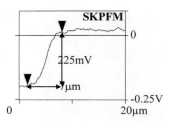

Figure 5:   Potential scan at the delamination front a plasma polymer coated Au surface a) Scanning Kelvin Probe b) Scanning Kelvin Probe Force Microscopy.

contrast to the continuous potential profile, the chemistry between the reaction zone and the delaminated zone changes abruptly for reasons not yet understood.

However, it is quite obvious that a well defined reaction zone exists with dimensions well beyond molecular dimensions. The chemistry and electrochemistry going on within this reaction zone is the subject of ongoing studies.

Delamination of molecularly well defined interfaces

First studies have been undertaken on model interfaces whose chemical and structural composition is well defined (Fig.6a). Cathodic delamination was also initiated on these samples and the TOF-SIMS analysis of the reaction zone now shows a fine structure which was not seen on the disordered plasma polymer/metal interfaces (Fig.6b).

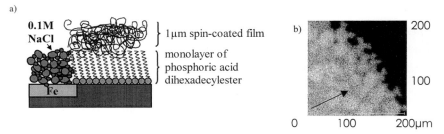

Figure 6:   a) Schematic sketch of the sample set-up. b) TOF-SIMS: Sodium distribution at the delamination front (delamination direction indicated by arrow).

Once again, only sodium was incorporated into the interface not by a homogeneous front but by local patches which grow together and form domain-like structures. The size of the structures which incorporated sodium is quite comparable to the domain size of the underlying Langmuir-Blodgett film. This could point to the fact that the local diffusion path and the local electroreduction of oxygen is determined by the local disorder of the metal/polymer interface.

Conclusion

In this study, Volta potential profiles have been measured for the first time on a delaminating interface with a local resolution in the sub-μm range. A well defined reaction zone of approximately 10μm has been observed, which marks the transition from an adherent to a delaminated interface. Within the reaction zone a preferential incorporation of sodium is observed which demonstrates, that within this zone oxygen reduction is the predominant electrochemical reaction leading to delamination. The oxygen reduction and the corresponding incorporation of ions may be determined by the order and supermolecular chemistry of the interface.

Acknowledgements

The Deutsche Forschungsgemeinschaft is gratefully acknowledged for financial support.

References

[1] D.M. Brewis, D. Briggs (Eds.), Industrial Adhesion Problems, Orbital Press, Oxford, **1985**.

[2] R.A: Dickie, F.L. Floyd (Eds.), Polymeric Materials for Corrosion Control, American Chemical Society, Washington, **1986**.

[3] H. Leidheiser Jr. (Ed.), Corrosion Control by Organic Coatings, NACE, Houston, Texas, **1981**.

[4] M. Rohwerder, M. Stratmann, MRS Bulletin, *24*, 43 **(1999)**.

[5] W. Fürbeth, M. Stratmann, Corrosion Science, *43*, 207 **(2001)**.

[6] W. Fürbeth, M. Stratmann, Corrosion Science, *43*, 229 **(2001)**.

[7] W. Schmidt, M. Stratmann, Corrosion Science, *40/8*, 1441 **(1998)**.

[8] E. Hornung, M. Rohwerder, M. Stratmann, GDCh-Monographie, *21*, 22 **(2000)**.

[9] E. Hornung, M. Rohwerder, M.Stratmann, Scanning Kelvin Probe Force Microscopy–Chances and Limitations for in situ Delamination Measurements Electrochem. Soc. Proc. PV 01-22 **(2001)** 618-622

[10] M. Nonnenmacher, M.P. O'Boyle, H.K. Wickramasinghe, Ultramicroscopy, *42-44*, 268 **(1992)**.

[11] H. Jacobs, P. Leuchtmann, O. Homan, A. Stemmer, J. Appl. Phys., *84/3*, 1168 **(1998)**.

[12] H. Jacobs, H. Knapp, A. Stemmer, Rev. Sci. Instrum., 70/3, 1756 **(1999)**.

[13] F. Robin, H. Jacobs, O. Homan, A. Stemmer, W. Bachtold, Appl. Phys. Lett. *76/20*, 2907 **(2000)**.

[14] P. Schmutz, G. Frankel, J. Electrochem. Soc., *145/7*, 2285 **(1998)**.

[15] P. Schmutz, G. Frankel, J. Electrochem. Soc., *145/7*, 2295 **(1998)**.

[16] V. Guillaumin, P. Schmutz, G.S. Frankel, J. Electrochem. Soc., *148/5*, B163 **(2001)**.

Macromol. Symp. 187, 43–51 (2002)

New Trends in the Chemical Pretreatment of Metal Surfaces

Joachim Heitbaum

Chemetall GmbH, Trakehnerstraße 3, 60487 Frankfurt a. M., Germany

Summary: The driving forces for the development of new chemical pretreatment products are both the need for more environmentally compatible processes and improved productivity at the customer. The first trend is met by introducing easily biodegradable surfactants, by avoiding waste and waste water by closed loop pretreatment systems and by substituting hazardous chemicals such as chromates. In the second trend, pretreatment steps are transferred from part lines upstream to the coil lines of sheet metal manufacturers thereby reducing assembling, and painting costs in automotive and white goods industries as well as in architectural application.

Introduction

Chemical pretreatment, although it is necessary within a production process, is in many cases an unliked part of the assembly line. It does not contribute to the outside appearance of the product and therefore cannot be used as a marketing argument. In addition, it often is hot, humid, produces waste and is a frequent cause of failures. Chemical pretreatment is, however, essential for the quality of the product because it fulfills the following tasks:

- it completely removes all residues from earlier production steps such as oils, fats, greases, sputter particles and grinding dust;
- in provides the so-called water break free surface;
- it constitutes a reprodusible and well defined surface condition capable of supporting all subsequent coating operations;
- it builds up a first coating layer usually called „conversion coating" which is essential for corrosion protection and paint adhesion.

This dilemma of being unwanted but necessary at the same time gives rise to the trends in chemical pretreatment we encounter at present: On the one hand, customers ask for environmentally compatible processes to improve their workers safety by avoiding poisonous chemicals and to reduce energy consumption and wastes to meet regulations. On the other

   CCC 1022-1360/00/$ 17.50+.50/0

hand, we observe a trend to transfer pretreatment steps from part lines to the sheet metal manufactures. Thus, precoated substrates are getting more and more in use thereby improving quality standards and reducing assembly costs.

Thus, pretreatment processes presently introduced in the market or upcoming as the next generation will at the same time reduce environmental risks and improve productivity.

Cleaning

The cleaning stage of a pretreatment line has to remove completely all oils and greases and to condition the surface for all subsequent pretreatment steps. Although, the use of biodegradable surfactants and bath maintenance procedures are state-of-the-art, a new situation has come up by the demand of the EU to introduce socalled „easily biodegradable surfactants" until 2004. These surfactants have to meet the OECD 301 regulation saying that 60 % of the surfactant has to be decomposed into CO_2 within 28 days. The goal is achieved by the development of a new generation of surfactants with a narrow range of EO groups. Fig. 1 shows EO distribution of both „old" and „new" surfactants.

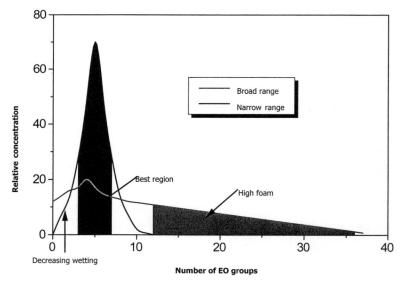

Figure 1.

These new raw materials result in

- better biodegradability,
- significantly better cleaning ability,
- less foaming even near to the cloud point.

Since synthesis of these new surfactants is more complicated, raw material costs are slightly higher which is partly overcome by better efficiency. But as a general rule it turns out that protection of the environment is not for free.

Heavy Metal Free Processes

The discussion carcinogenic chromates is more than ten years old and on one hand in some industries such as the European automotive industry chromates are probably going to be completely banned in 2003 due to the European Directive "End of life vehicle". For the passivation of zinc and the pretreatment of Aluminium, on the other hand chromic acid treatment still is standard.

As an alternative to yellow chromating of aluminium especially for architectural applications, a new process has been developed that is based on cerium chemistry. It has the interesting feature that it provides a yellow surface which is almost identical to the colour obtained by yellow chromating. Thus, the cerium treatment allows for the same easy quality inspection as the chromate: a uniform yellow coloured surface shows that the process is under control. Fig. 2 provides some test results indicating that the Cerate process has comparable performance to chromate coatings under paint. It must be stated, that bare corrosion protection is not as good as with the chromate coating, which means that Cerate technology is intended to be a pretreatment prior to paint. The customer can use the process in his existing chromating dip line with minor adjustments.

But it is not the chromate ion alone that is under discussion. Nickel compounds are rated as being possibly carcinogenic and other heavy metals give also rise to concern, if only in the waste water treatment system. Thus there is a tendency to avoid heavy metals completely if possible.

As a fist product of this kind we have recently introduced the so-called SAM-technology into the market. It again can be used on aluminium surfaces as a pretreatment prior to painting. The conversion coating in this case consists of monomolecular layer of self assembly molecules. With their head groups, the molecules chemisorb to the hydrated metal oxide layer whereas the tail group is bound to the paint. Thus, we achieve excellent paint adhesion and at the same time very good corrosion protection (Fig. 3). Reason for the corrosion resistance of the layer is the first densely packed layer of molecules which consist beside their end groups of alkane chains which cause the layer itself to be hydrophobic. Thus the layer immediately in contact with the metal acts as a barrier against the diffusion of water and ions.

Figure 2.   Cerate Process
Treatment: Cerate pretreatment under Polyester Powder Paint (Herberts Al-88-9010)

Test results	Cerate-Treatment	Chromating-Treatment
Filiform Test, 1008 h (DIN EN 3665)	U < 1	U < 1
Salt Spray Test CASS, 240 h (DIN 50021	U 0	U 0
Salt Spray Test ESS, 480 h (DIN 50 021)	U 0	U 0
Humidity Test KK, 240 h (DIN 50017)	Gt 0	Gt 0

Figure 3

SAM – PROCESS

Treatment: Alk. Cleaning, Acid Pickling, SAM-Treatment Painting; Substrate: AlMg1

Test Results	Two Layer Coil Coating Paint	One Layer PE-Powder Paint
Salt Spray Test ESS (4000 h)	< 1	< 1
Salt Spray Test CASS (1000 h)	< 1	1
Filiform Test (3000 h)	< 1	1
Outdoor Exposure (34 monts)	0	0
Cross Hatch Test (240 h constant humity)	-	Gt 0
T-Bend-Test T0/T1 [% paint adhesion]	≥ 95 / 100	-

Another group of products that are principally free of heavy metals are based on Silane chemistry. Fig. 4 gives an overlook on what products are available at present and where they can be used. The figure also summarizes test results obtained with silane coatings under paint.

Although some customers use these products already, the technology still under development and further improved products have to be worked out. Although there is still the risk of failure, we see the following advantages for the Silanes:

- heavy metal free,
- applicable to all standard metal surfaces,
- easy, if any, waste water treatment,
- can be used for passivation (without paint),
- paintable coatings

Figure 4. Commercial Products based on Silane Chemistry, Application and Test Results

| OXSILAN | Blank | Painted | Substrates | SST | | | Cross hatch |
				NSS 1008 h [mm]	ESS 1008 h [mm]	CASS 240 h [mm]	
AL-500	X	X	Al		< 1		-
			Z	< 2			0
			CRS	≤ 2			0
MM-702		X	Al		< 1	0	0
			Z	< 1			0
			CRS	≤ 2			0
MM-705		X	Al		< 1		
			Z	< 2			0
			AZ	< 1			0
			ZA				
			CRS	≤ 2			0

On the other hand, the silane chemistry is complex and hydrolyzed silanes tend to react not only with the surface in question but also with each other or impurities which somehow get into the pretreatment bath. Thus, stability of bathes and concentrates are critical. And although paintability was proven in lab test, there is still the question to what extent silanes will interfere with subsequent painting stages in a paint shop.

To sum up this chapter, we conclude that there are new pretreatments on the market or under investigation that can not only replace chromating but may also result in considerable cost savings by optimizing processes andavoiding or minimizing waste and waste water treatment.

Precoated Substrates

Thin organic coatings with coating thickness of 1 to 5 µm have the potential to change the world of chemical pretreatment considerably. These coatings are usually applied by roll coating at the sheet manufactures, already, galvanizing in a coil coatingdirect application in galvanizing lines are being discussed. Most variants need a preceeding pretreatment step but there are some that provide pretreatment and protective coating at the same time. Such coatings were first introduced into the market in the early 90's with what we call Permanent Coatings. The first generation was chromate containing and was designed for the corrosion protection and anti-fingerprint coating of Galvalume. It was not meant to be painted although panting was possible especially with powder paints. Thus, a number of applications were founed mostly in the general industry, where customers used such precoated materials – e. g. Galfan – for the simplified manufacturing of white goods and the like. Pretreatment at the final customer consists of simple cleaning, and the finished good is powder painted, only.

Meanwhile, a second generation of such Permanent Coatings is being introduced into the market. The new products are chromium-free and corrosion protection is achieved by the use of nanoparticles. Fig. 5 summarizes test results with these new coatings. They not only provide excellent corrosion protection but possibly also better edge protection than the old products.

Figure 5. Gardobond Permanent-Coating - Cr-free - Product Program

Gardobond	Solids content	Application System	Substrate	Planned bath make up		Corrosion protection properties			
				Polymer component	Additive	Dry film coating w.	Salt spray test	VDA cycle test	Humidity cycle test
PC 4615	30 weight %	Chem-coater	AZ	99,0 weight % PC 4615	1,0 weight % GBA H 7409	1,5 – 2,0 g/m²	<5 % face corrosion after 480h	<5 % face corrosion after 20 cycles	n.a.
PC 4617	30 weight %	Chem-coater	Z ZA ZE	95,0 weight % PC 4617	5,0 weight % GBA H 7410	1,5 – 2,0 g/m²	< 5 % face corrosion after 360h	< 5 % face corrosion after 10 cycles	n.a.
PC 4619	30 weight %	Chem-coater	CRS	95,0 weight % PC 4619	5,0 weight % GBA H 7411	1,5 – 2,0 g/m²	n.a.	n.a.	< 5 % face corrosion after 20 cycles

The next product group, which was tried to be established in the market were the so-called Dry Lubricants. These are organic coatings formulated with an high amount of lubricating waxes. They were intended as substitutes for oils and greases in press shops. But although some advantages could be found, Dry Lubricants have not been used extensivly so far. Their main problem is that they need to be removed in the first cleaning stage of a pretreatment line thus giving rise to an enormous load of polymers and waxes which have to be removed by filtration means and causing high amounts of waste. The idea of dry lubricating may come up again, however, when the organic coating can remain in the surface and is overpainted in paint lines. The Permanent Coatings mentioned above contain waxes, already, to support roll forming operations.

The next step is pretreatment primers and especially welding primers that are on the market for several years and which will be substituted by a second generation of products in the very near future. The new generation of products again is chromium-free and at present they are intended to be used in automotive industry.

The requirements for welding primers are summarized in Fig. 6. The products contain electrocially conductive pigments such as zinc particles or iron phospide. Fig. 7 shows the coating structure. These coatings are suitable for subsequent electro-painting.

Figure 6.   Requirements for Welding Primer

- Application via roll-coating in coil lines
- Good formability (ideally oil-free)
- Spot weldable ($\geq$ 800 Spots, Welding area: 1,5 kA)
- E-coat paintability
- Highest Corrosion protection; no red rust after:
    1. Generation: 10 Cycle VDA-Test (Flange corrosion)
    2. Generation: 20 Cycle VDA-Test (Flange corrosion)
- Adhesive Bonding (Shear strength >20 MPa)
- Usable for bake hardening steel (T< 170°C)

50

Figure 7.

Welding Primer

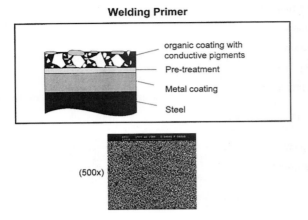

organic coating with
conductive pigments

Pre-treatment

Metal coating

Steel

(500x)

Fig. 8 summarizes the development of welding primers and also indicates what might come next: UV-cured welding primers. The advantage here is that even better cross linking is achieved and the coil manufacturer can apply the coating in the galvanizing line if he decides to invest in UV-lamps. Thus, he can avoid to transfer the coil to a coil coating line where at present the thermally cured welding primers are applied due to the necessary high curing temperature of 160 – 250 °C, depending on the product used.

Figure 8.

History of Welding Primers

Substrate	ZE	ZE	ZE Z	ZE Z	ZE Z
Pretreatment	Zn-ph, Cr CM Henkel	No-rinse, Cr CM Henkel	No-rinse, Cr-free CM Henkel 1456 PPG (Nupal)	No-rinse, Cr-free CM TP 10475 Henkel 1456	No-rinse, Cr-free CM TP 10475 Henkel 1456
Weldable Primer	1. Generation ~ 3 µm, 250°C pmt PPG Bonazinc	1. Generation ~ 3 µm, 250°C pmt PPG Bonazinc Henkel Granocoat	2. Generation ~ 6 µm, 160°C pmt PPG Bonazinc Henkel Granocoat CM/André BASF Akzo-Nobel	2+. Generation ~ 6 µm, 80°C pmt + UV CM/André	2++- Generation ~ 3 / 6 µm ? 160°C/250°C ? CM/André Henkel PPG BASF Akzo-Nobel

If such UV-curing stations are established, this will open the market for other type of products such as pretreatment primers that have excellent corrosion protection due to their highly cross linked organic matrix and that would have further advantages for the end user such as

- support of forming processes,
- coloured surfaces for easy visual inspection,
- excellent bare corrosion protection,
- paintability

It might even be possible to formulate coatings with self-healing capability. Such coatings would repair damages of a painted surface if they occur and moreover would provide an edge protection of a prepainted substrate thus opening up new potential for increased productivity at the end user.

Conclusion

As was shown in this paper, the suppliers of chemical pretreatment products strive to meet customers expections of environmentally more compatible products and of easy-to-handle and safe processes. At the same time, we understand ourselves as connecting link between metal producers, part manufacturers, and paint companies. Thus, we ensure that new pretreatment processes meet customer's demands.

Macromol. Symp. 187, 53–63 (2002)

Reactive Dispersions for Corrosion Inhibition

Hans-Juergen P. Adler , *Evelin Jaehne*, Axel Henke, Lu Yan, Andrij Pich*

Dresden University of Technology, Institute for Macromolecular Chemistry and Textile Chemistry, Mommsenstr. 4, D-01062 Dresden, Germany

Summary: Several concepts have been developed for the improvement of adhesion promotion and corrosion inhibition of reactive metal surfaces, like aluminum. The aim of these studies was to replace the present technical procedure for pretreatment of aluminum surfaces with chromium acid.

Therefore, the adsorption and organization process of mono- and bifunctional alkyl phosphonic acids and mono alkyl phosphoric acid esters has been investigated. The properties of the adsorbed SAM's were confirmed by industrial linked adhesion and corrosion tests. By functionalization of the terminal position with thiophene, molecules are prepared which can graft conducting polythiophene layers via surface polymerisation on silicon/metal substrates.
Also polymeric nano-particles with phosphoric acid groups form thin and smooth well-packed films on aluminum surfaces.
Stable core-shell polypyrrole (PPy) dispersions were prepared by using polystyrene (PST) or poly (styrene-co-butyl acrylate) (PST-co-BuA) as the core, and polypyrrole as the shell.

Keywords: SAM, microgel, adhesion promoter, corrosion inhibition

1   Introduction

Reactive metals like aluminum is one of the most used metallic materials in industry –from buildings over cars to aircrafts. Protecting reactive metals by covering their surface with organic coatings is a smart way to take advantage of mechanical properties of metals like aluminum while preventing them from corrosion and introducing requested surface properties in one step.

Mainly the treatment with chromic acid is applied to reduce the thickness of the natural oxide layer, to eliminate particles of other metals, which are present like iron, and to form a new layer of a mixed aluminum-chromium oxide inhibiting corrosion.

One topic of this work was the replacement of the chromating process on aluminum by thin organic adhesion promoter layers as an alternative pretreatment for the coating. Therefore, we used as reactive species self-assembly molecules. A strong and intact bond between coating

*corresponding author

  CCC 1022-1360/00/$ 17.50+.50/0

and metal can withstand electrochemical or chemical attacks from aggressive species found in the environment such as water, oxygen and pollutants.

Self-assembly molecules (SAM's) with two reactive groups, one to the metal surface and the other to the lacquer connected by e.g. an aliphatic spacer form such bonds, which can stand up to the attack of water and simultaneously inhibit any electrochemical reaction at the interface [1-3] (see figures 1a and 2). Those very stable bonds between reactive organic monomers and metal surfaces can be achieved, if the reaction conditions are optimized for the given metal/coating system.

Especially, phosphonic and phosphoric acid groups showed a strong effect on the aluminum surface [4, 5]. The formed chemical bond to the substrate inhibits any electrochemical reaction at the interface (figures 1, 2)

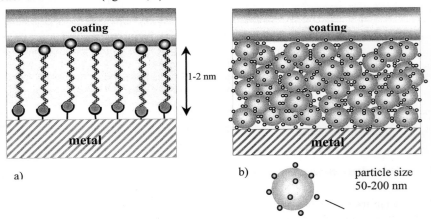

Figure 1: *Models of adsorbed layers of*

a) bifunctional molecules and   *b) microgels with functional groups*

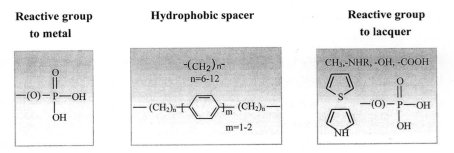

Figure 2:   *Structural concept for adhesion promoters*

Microgels are the basic of the second concept (see figure 1b). Polymeric core-shell particles are combinations of different mostly incompatible polymers and offer the opportunity for introducing functional groups on the surface of the particles. They were functionalized with phosphorous containing groups according to former works [4-6] and were tested as links between aluminum and a final coating.

Based on these investigations microgels with a conductive shell have been developed. In order to improve the processability of e.g., polypyrrole, a lot of efforts have been made to obtain conducting polymer-coated composites where the "core" consists of a non-conducting particulate material [7-10]. Nanometer-sized polystyrene (PST) or poly (styrene-co-butyl acrylate) (PST-co-BuA) core particles were synthesized by using three different kinds of monomer/polymer as the surfactant which can produce particles with different charge.

The applied SAM's have a thickness of few nanometers and the microgel layers of 50 to 200 nm. The characterization of these very thin layers is not easy. Special surface sensitive methods were used to investigate the adsorbed monolayers. With the help of X-ray photoelectron spectroscopy and FT-IR spectroscopy it was shown that the density and structure of the films strongly depends on the surface roughness, the pre-treatment of the metal substrate, the structure and concentration of the surfactant, and the adsorption time. Surface plasmon resonance spectroscopy (SPR) was used to follow the kinetics of the adsorption process. The effect of core particles on the morphology and conductivity of PPy containing composites have also been investigated [11].

2 Results

2.1 Monofunctional self-assembly layers

The adsorption of alkyl-phosphonic and phosphoric acid monoalkyl esters on aluminum surfaces is an acid-base reaction. The driving force is the formation of a surface salt (see further explanation [4]).

Surface Plasmon Resonance Spectroscopy (SPR) was used to examine the kinetics of the adsorption process. In figure 4 one can see a typical SPR kinetics curve from the adsorption of octadecyl phosphonic acid on chromium. The advantage of SPR spectroscopy is that the kinetics of the adsorption process is directly visible without destroying the surface.

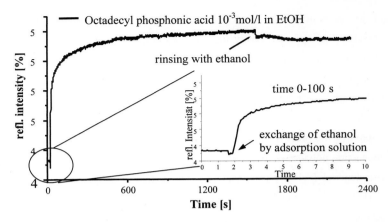

Figure 4: *Kinetically SPR measurement of the adsorption of octadecyl phosphonic acid on aluminum*

It is clearly seen from the curve that the adsorption process of phosphonic acids on aluminum starts very fast and reaches a plateau after some time (10 minutes to some hours). After rinsing with the used solvent some physisorbed species are removed and then the curve remains stable.

The organization and orientation processes of phosphonic acids also can be determined by angle-resolved X-ray photoelectron spectroscopy (ARXPS).

This method allows the quantitative detection of the element distribution on the substrate surface. These contributions change by varying the detector angles. Because of the limiting attenuation length of the electrons the depth information of XPS data is dependent on the varied angle α. A special region of the adsorbed layer can be seen at every detector angle. Though, at low angles (5-10°) the top of the adsorbed film is better detected than the deeper regions whereas at high angles (80-90°) one can get information from the substrate regions. Specific elements (marker elements) are watched at the different angles and from their distribution we have indications about the orientation and organization within the layer.

The thickness of the adsorbed layer was calculated from the experimental data and has a value of 2,21 nm. The theoretical C:P ratio is 12:1, but because of carbon contamination there could be detected an excess of C. Figure 8 shows the separation of adsorbed phosphonic (gray line) acid carbon and the carbon from the contamination (gray area). Now the layer thickness was about 1,1 nm.

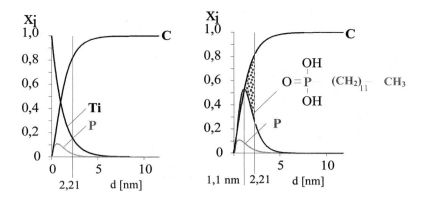

Figure 5:   *Atomic contribution in dependence of layer thickness*

The calculated size of the dodecyl phosphonic acid molecule was 1,536 nm. From the ARXPS measurements we can conclude that the adsorbed dodecyl phosphonic acid molecules are tilted at 44° from the surface normale.

2.2   Bifunctional compounds

To form a chemical composition between metal and organic coating it is necessary to tailor self assembling molecules with two functions; one to chemisorb on the metal and another to react with the organic material of the top layer. A chain length of 10 to 12 carbons seems to be ideal for the preparation of α,ω-fuctionalized alkyl phosphonic acids. As second functional groups were chosen the amino group, hydroxyl group, phosphonic- or phosphate. The synthesis of these compounds is described elsewhere [4, 12].

The objective attention of the investigation was to find out whether and how the molecules were orientated after adsorption onto the chromium surface. Therefore, angle resolved X-ray photoelectron spectroscopy data were taken (see figure 6). The spectra with small α give the most information of the top layer. With increasing angle α more information of the substrate were obtained. The ARXPS spectra of an 50 h adsorbed AlMg1 sample were taken from 5° to 75° between the sample surface and the detector. The phosphorous 2s peak and the nitrogen 1s peak were used for analysis (marker peaks). The shape of the intensity ratios proved that the phosphonic acid group is attached to the substrate and the amino group is orientated off the surface.

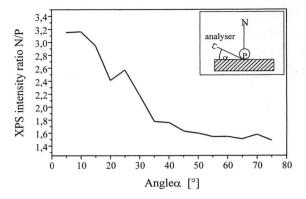

Figure 6: *ARXPS intensity ratio of P 2s and N 1s peaks of a 50 h adsorbed AlMg1 sample*

Model surface reactions were developed to give prove for the reactivity of the second reactive group. Aliphatic or aromatic isocyanates are cross-linking agents for many coating systems. Because the isocyanate-group is very reactive to hydroxyl and amino groups, we chose an alkyl-isocyanate for model reactions. The resulting urethanes and substituted ureas can be detected by their characteristic infrared bands [4].

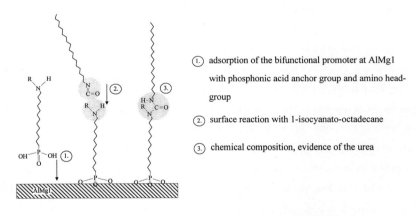

Figure 7:   *Model for surface reaction*

2.3   Surface reaction with thiophene-substituted phosphonic acids

Thiophene terminated self-assembly layers were synthesized as adhesion promoters to make polythiophene more compatible towards diverse metal substrates [13,14]. The surface polymerization can be undertaken either chemically or electrochemically. The principle of these reactions is shown in figure 8. $FeCl_3$ was used as an oxidative coupling reagent for the chemical polymerization, we used as. The terminal thiophene groups of the adhesion promoter take part in the polymerization process and a stable composite can be formed consisting of the conducting polymer, adhesion promoter and substrate.

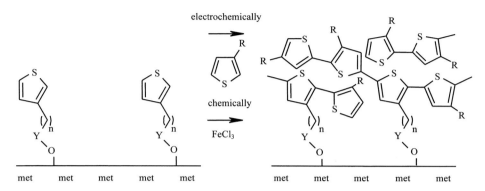

Figure 8:   *Principle of surface polymerization via an adhesion promoter*

For electro-polymerization, the electrodes were coated with the adhesion promoter and polymerised at the corresponding oxidizing potential.

The polymer thickness could be adjusted between 3 and 250 nm depending on the reaction conditions. The AFM image of polythiophene grafted onto a Si-wafer is shown in figure 9. The layer has a relatively homogeneous and dense surface with a shell-like structure. A roughness analysis gave values of $R_{ms} = 1{,}73$ nm and $R_a = 1{,}35$ nm.

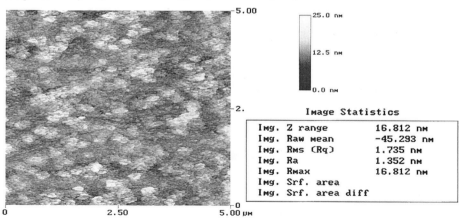

Figure 9:   *AFM Image of polythiophene grafted onto SiO_2*

2.4 Functionalized microgels

Polymeric nano-particles with phosphate groups and a core-shell-like structure were synthesized according to former publications [6] about self-assembling molecules based on phosphonate- and phosphate groups as reactive groups toward the aluminum surface. These particles can be used for corrosion inhibition of aluminum [15, 16].

According to figure. 10, functionalized particles were synthesized in a two-step emulsion polymerization. Cross-linked BuA/St particles were formed in a first step. In a second step a mixture of functionalized acrylate and BuA/St was added to the system.

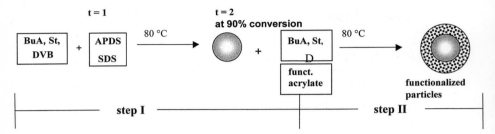

Figure: 10:   *Synthesis of functionalized composite particles*

Special P-functionalized acrylates were used to incorporate the phosphoric acid group into the particle surface.

For application, AlMg1-panels were dipped in phosphate-functionalized dispersions. Already after few seconds, particles were found on the AlMg1 surface. Time-dependent SEM pictures are presented for 160 nm particles in figure 11 [16].

Figure 11:   *SEM images of time-dependent SEM pictures for 160 nm particles*

Figure 12 shows that the film formation process begins within 10 s and is completed within 30 seconds in case of smaller particles. The transition of adsorbed microgel spheres to a closed film is clearly seen in the middle picture.

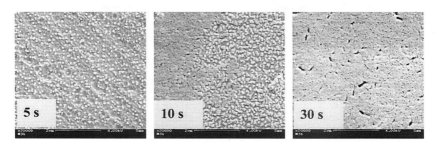

Figure 12: *SEM images of time-dependent SEM pictures for 60 nm particles*

Film formation of larger particles takes more time. The influence of particle size on film formation correlates with the results of corrosion tests.

For an application in corrosion protection, smaller particles (d~100 nm) seem to be more effective than bigger ones (d>150nm) due to better film formation properties during the adsorption process. This result correlates with the SEM pictures. Different testing methods exist controlling the corrosion protection of coatings on metal surfaces. For this work two usual methods were selected and performed by the CHEMETALL GmbH: the acetic acid salt spray-test (ASS-test, DIN 50021) and filiform test (DIN 50024). Therefore, the test panels were coated with SAM solution and a commercial clear coat, scratched and exposed to a mixture of NaCl-solution and acetic acid in a climate chamber for 1200h. The results were stated in mm infiltration. Infiltrations of < 1mm are the best values. The SAM treated samples have an excellent corrosion inhibition effect and better adhesion properties as the chromated surface.

The other industrial test results are summarized in table 1 with comparison with different coated samples.

Table 1: Industrial testing of coated AlMg1 panels with test panels

compound	ASS-Test DIN 50021 infiltration [mm]	Filiform-test DIN 50024 [mm]	T-Bend-Test Exfoliated area LPV 75 [%]
AlMg1	8	5	100
AlMg1 alkaline/acidic stained	4	4	85
AlMg1 phosphated	3	<1	10
AlMg1 chromated	<1	<1	25
Molecular adhesion promoter	<1	<1	<1
Microgels	<1	<1	<5

Polystyrene/Polypyrrole composites have been synthesized and characterized. Three different kinds of monomer/polymer, PM-MA, MEAK, and PEGMA have been used as the surfactant during the synthesis, which produce core particles with negative charge, positive charge and non-charge, respectively [11].

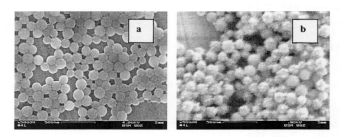

Fig. 13: *SEM images of (a) PST/MEAK core, (b) PST/MEAK/PPy composites Further details are published elsewhere [11]*

3 Discussion

Several methods have been developed to improve corrosion inhibition and adhesion promotion on reactive metals like aluminum.

The idea using self-assembly molecules as corrosion protection layers for chromium surfaces appears to be promising. The SAM's bearing a phosphonic acid head-group and polymer-reactive group were able to adsorb spontaneously onto the chromium surface. They formed orientated layers that could be proved by polarized external reflectance FT-IR measurements and X-ray photoelectron spectroscopy. Monofunctional phosphonic acids formed monolayers on the substrate. The chemical composition to organic coatings was realized with a second terminal functional group. This lacquer linking was confirmed by industrial adhesion- and corrosion tests. The effect of the bi-functional surfactants was equivalent to chromated panels due to corrosion inhibition and superiored due to adhesive strength.

Nanometer-sized composite PST/PPy particles with raspberry morphology were obtained when the core particles were stabilized by surfactant with charge, and with homogeneous morphology when non-charged surfactant was used. Stable polypyrrole dispersions can be prepared by using MEAK as the surfactant and Poly(ST-co-BuA) as the core, which give good ability of film formation. Conductivity of the composites was influenced strongly by the core material.

4   References

[1]   M. Volmer-Uebing, M. Stratmann, *Appl. Surf. Sci. 55* (**1992**) 19-35.
[2]   M. Stratmann, W. Fürbeth, G. Grundmeier, R. Lösch, C. Reinartz, in:
       Corrosion Mechanism in Theory and Practice, Chapter 11, M. Dekker; N.Y., **1994.**
[3]   M. Stratmann, *Advanced Materials 2* (**1990**) 191.
[4]   I. Maege, E. Jaehne, A. Henke, H.-J. P. Adler, M. Stratmann,
       Progr. Org. Coat. **1998**, *34*, 1.
[5]   E. Jaehne, A. Henke, H.-J. P. Adler; *Coating 6/***2000**, 218.
[6]   I. Maege, E. Jaehne, A. Henke, H.-J. P. Adler, C. Bram, C. Jung, M. Stratmann;
       Macromol. Symp. 126 (1997) **7-24.**
[7]   M. Omastova, J. Pavlinec, J. Pionteck, *Polymer*, 39 (**1998**), 6559.
[8]   D.B. Cairns and S.P. Armes, *Langmuir*, 15 (**1999**), 8052.
[9]   F. M. Huijs, J. Lang, *J. Appl. Polym. Sci.*, 79 (**2001**), 900.
[10]   F. Huijs, J. Lang, *Colloid Polym. Sci.*, 278 (**2000**), 746.
[11]   Lu, A. Pich, H. Adler, *Synthetic Metals*, submitted
[12]   R .D. Ramsier, P. N. Henriksen and A. N. Gent, *Surface Science 203*, **1988**, 72.
[13]   E. Jaehne, D. Ferse, H.-J. P. Adler, Ramya K., I. K. Varma, M. Wolter, N. Hebestreit,
       W. Plieth, *Macromol. Symp.* 164, **2001**,133-144.
[14]   E. Jaehne, D. Ferse, G. Busch, A. Singh, H.-J. P. Adler; *Designed Monomers &*
       Polymers, **2002** (in press).
[15]   A. Henke; *Farbe & Lack 106* 11/ **2000**, 60-70.
[16]   A. Henke, E. Jaehne, H.-J. P. Adler; *Macromol. Symp.* **2001**, *164*, 1-9.

Macromol. Symp. **187**, 65–75 (2002)

Novel Protective Coatings for Steel Based on a Combination of Self-Assembled Monolayers and Conducting Polymers

*U. Harm, R. Bürgler, W. Fürbeth, K.-M. Mangold, K. Jüttner**

Karl-Winnacker-Institut der DECHEMA e.V., Frankfurt am Main, Germany

Summary : Investigations for a new primer system for iron or low alloyed steel have led to first results. Several special phosphonic acids with thiophene derivatives as head groups have been synthesized. They form stable self-assembled monolayers (SAMs) on passivated iron by dipping the substrates into aqueous phosphonic acid solutions. SAM formation was validated by current potential curves and also by contact angle measurements, which showed an intensive hydrophobisation of the iron surface after the dipping process. Finally cyclovoltammetric (CV) experiments after SAM formation indicated the successful polymerisation of the immobilised thiophene derivatives.

Introduction

Several corrosion protection layers with conducting polymers for steel and other metals have been developed during the last few years.[1-3] Barrier or inhibitor effects, anodic protection or the mediation of oxygen reduction are proposed mechanisms for the anticorrosive effects observed with conducting polymers. [1]

Self-assembled monolayers (SAMs) heve been well known for more than a decade. Examples are thiols or other sulfur compounds on gold or silver which could become important for many potential applications, e.g. surface hydrophobisation, chemical or biological sensing and catalytical or optical devices. SAMs of phosphonic acids or phosphonates on passivated iron (also on Ti / TiO_2 or Al / Al_2O_3) have also been investigated in detail in the last years. [4-7] These investigations have shown that the phosphonic acid group is one of the best anchor groups for building stable SAMs on passivated iron. This was the starting point for the development of a new primer system for iron or low alloyed steel, which basically consists in SAM formation of special phosphonic acids on iron followed by polymerisation of the head groups to polythiophene chains. The new primer system promises to combine anticorrosive

CCC 1022-1360/00/$ 17.50+.50/0

66

effects (for example anodic corrosion protection) with improved adhesiveness for several top coats. After first investigations the practical application of the primer system appears to be feasible because the required SAM formation on passivated iron and later electropolymerisation of the head groups was achieved. Clearly, further investigations will have to be made before this process will be of industrial interest. Fig. 1 shows the procedure for the planned primer system.

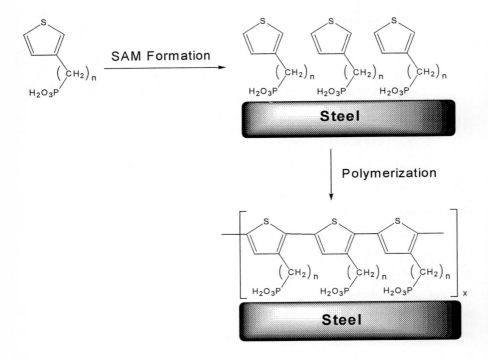

Figure 1 . Scheme of the procedure for the new primer system.

Synthesis of the phosphonic acids required

An overview of the structures of the synthesised phosphonic acids is given in Fig. 2.

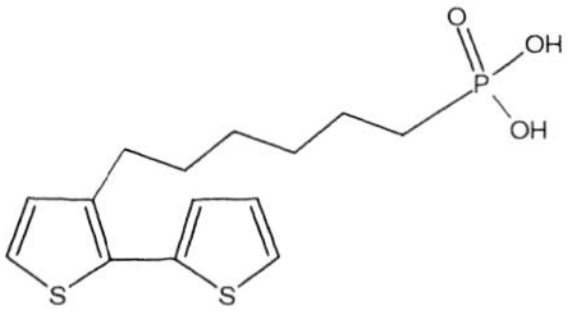

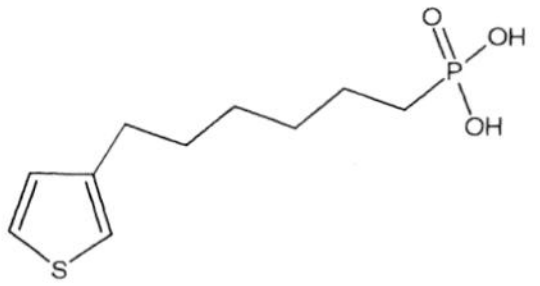

Bithienylhexanephosphonic acid (BTHPA) 7   Thienylhexanephosphonic acid (THPA) 6

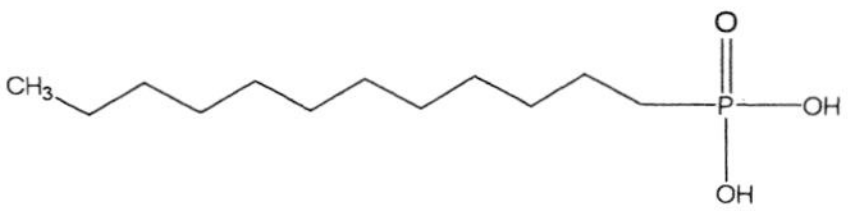

Dodecylphosphonic acid (DPA) 5

Figure 2. Overview of synthesized phosphonic acids used to date.

Synthesis of 6-(3-Thienyl)hexyl bromide **3** (the precursor of THPA **6**) was carried out by a method published by Bäuerle et. al. [8]. The reaction sequence is quite similar to the one of the corresponding 6-(3-Bithienyl)hexyl bromide **4** , which is shown below.

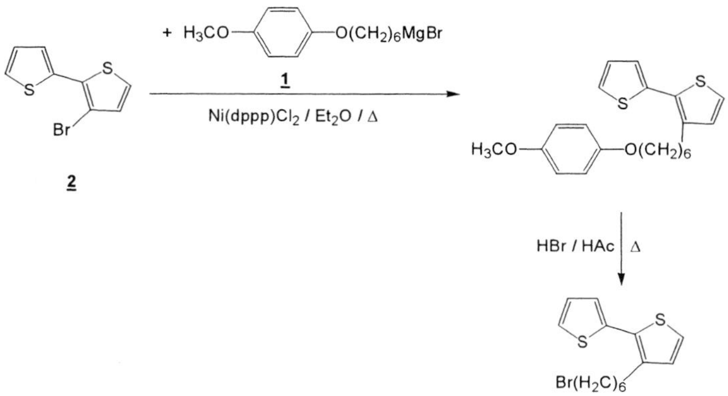

Figure 3. Synthetic route for preparing 6-(3-Bithienyl)hexyl bromide **4** .

3-bromo-2,2'-bithiophene **2** was prepared by a coupling reaction with 2-thienylmagnesiumbromide and 2,3-dibromothiophene according to literature descriptions.[9-10] Dodecylphosphonic acid (DPA) **5** was synthesized from 1-bromododecane and triethylphosphite by a standard procedure of the Arbuzov reaction. [11-12]

Thienylhexanephosphonic acid (THPA) **6** and Bithienylhexanephosphonic acid (BTHPA) **7** were prepared from the corresponding hexyl bromides **3** and **4** by a special variant of the Arbuzov reaction.[4+11] Instead of triethyl phosphite this method uses tris-trimethylsilyl phosphite as the agent, which allows more moderate reaction conditions.

p-Methoxyphenoxyhexylbromide 1

Into a 500 ml round bottom flask was placed 200 ml [1.30 mol] 1,6-dibromohexane (Merck) in 150 ml methanol. At 80^0 C under N_2 and with stirring was added dropwise a solution (over 1 h), which was prepared from 40.0 g [0.322 mol] p-methoxyphenol and 17.74 g [0.320 mol] KOH in 100 ml methanol. The reaction mixture was further stirred for 2 h at this temperature before it was allowed to reach room temperature. After removal of the methanol by evaporation, the resulting liquid was decanted from solid impurities and 100 ml chloroform and 200 ml cooled 5 % KOH was added to the oily liquid before extracting. The organic layer was washed with 200 ml 1 % $NaHCO_3$ solution and dried over sodium sulfate. After filtering through a 2 cm Na_2SO_4 layer, the chloroform and later the dibromhexane was removed by vacuum distillation. Recrystallisation of the residue from 400 ml 99 % methanol yielded 59.4 g [64.9 %] **1** as colourless crystals of melting point (mp) 52 - 54^0 C.

MS (relative intensity) : m/z = 286 (55 %) [M⁺] and m/z = 288 (57 %) [M⁺] ; m/z = 124 (100 %) $[C_7H_8O_2]^+$

3-Bromo-2,2'-bithiophene 2

A Grignard solution of 9.8 ml [102.8 mmol] 2-bromothiophene and 4.0 g [164.5 mmol] Mg in 150 ml ether was prepared (finally refluxing for 45 minutes under N_2). This Grignard solution was added dropwise under N_2 (over 1 h) to a stirred and cooled (about -5^0 C) suspension of 640 mg [0.78 mmol] Pd(dppf)Cl$_2$ / CH$_2$Cl$_2$ –Komplex (Aldrich) { dppf = 1,1'-Bis-diphenylphosphino-ferrocene }, 10.0 ml [88.2 mmol] 2,3-dibromothiophene (from Lancaster synthesis) and 100 ml dry diethyl ether. After stirring at 0° C for 3 h, 10 ml methanol was added (to end the reaction) and the mixture was filtered through a 3 cm Na_2SO_4 / silicagel double layer. After further eluating the layer with 30 ml dry diethyl etherthe solvent of the combined solutions was rotary evaporated. The residual oily liquid was purified by silica gel chromatography with hexane / CCl_4 (98:2) as the solvent to yield 19.0 g [87.9 %] 2 as slightly green oil.

Thin layer chromatography (DC) on silica gel 60 plates (Machery & Nagel) with hexane showed a pure product with a retention factor $R_f = 0.33$.

^{1}H-NMR [CDCl$_3$; 250 MHz] : δ_1 = [6.99 ppm ; d ; ^{3}J=6.2 Hz ; H-4 or H-5]

δ_2 = [7.05 ppm ; dd ; all ^{3}J=6.2 Hz ; H-4']; δ_3 = [7.15 ppm ; d ; ^{3}J=6.2 Hz ; H-4 or H-5]

δ_4 = [7.32 ppm ; dd ; ^{3}J=6.2 Hz and ^{4}J=1.25 Hz; H-5'/3']; δ_5 = [7.40 ppm ; dd ; ^{3}J=6.2 Hz and ^{4}J=1.25 Hz; H-3'/5']

6-(3-Bithienyl)hexylbromide 4

A solution of 4.30 g [0.015 mol] methoxyphenoxyhexyl bromide 1 in 40 ml dry diethyl ether was added slowly over 1 h under N_2 and with stirring to 0.52 g [0.021 mol] Mg (iodine etched) at 40^0 C and the mixture was refluxed for 5 h before it was allowed to reach room temperature. The Grignard solution was slowly transferred (over 1 h) via a double needle to an ice cooled suspension of 2.50 g [0.010 mol] 3-bromothiophene (Merck) and 50 mg [0.092 mmol] Ni(dppp)Cl$_2$ {dppp = 1.3-bis(diphenylphosphino)propane } (Aldrich) . After slowly coming to room temperature, the reaction mixture was refluxed for 20 h under N_2 . After an addition of ice, 200 ml 1M HCl and 50 ml diethyl ether, the mixture was extracted and the organic phase was washed with 1 % NaCl solution and with NaHCO$_3$ solution. Drying over Na$_2$SO$_4$, filtration through a 2 cm Na$_2$SO$_4$ / silica gel double layer and cold evaporation of the solvent led to a partly cristalline colourless product of the 3-methoxyphenoxyhexyl-bithiophene as intermediate, which seemed to be almost pure in thin layer chromatography on silica gel with CH$_2$Cl$_2$ as solvent ($R_f = 0.64$).

Without further purification, a solution, which was obtained by slowly mixing 13.6 ml [0.12 mol] 47 % HBr and 18.7 ml [0.20 mol] acetic anhydride (cooling necessary !) was added. The reaction mixture was stirred at 100^0 C for 20 h under N_2. Then 100 ml ice water, 200 ml 5 %

NaCl solution was added and after two-fold extraction with 100 ml diethyl ether, the combined organic phases were washed twice with 1 % NaCl solution and with 1% $NaHCO_3$ solution. After drying over Na_2SO_4 , hexane was added until hydrochinone precipitated. This suspension was filtered through a 2 cm Na_2SO_4 / silica gel double layer and the diethyl etherwas cold rotary evaporated. The residue was immediately purified by silica gel chromatography with cyclohexane to yield 1.80 g [5.47 mmol] (83 %) Bithienylhexyl bromide **4** as slightly yellow oil.

Thin layer chromatography on silica gel with cyclohexane indicated a pure compound with $R_f = 0.18$.

MS : $m/z = 328$ (100 %) [M^+] and $m/z = 330$ (92 %) [M^+] .

^{1}H-NMR [250 MHz ; CDCl$_3$] : $\delta_1 = 1.42$ ppm [m ; 4H ; H-4''/5''] ; $\delta_2 = 1.68$ ppm [m ; 2H ; H-3'']; $\delta_3 = 1.82$ ppm [m ; 2 H ; H-2'']; $\delta_4 = 2.75$ ppm [t ; ^3J=7.6 Hz ; 2H ; H-1'']; $\delta_5 = 3.37$ ppm [t ; ^3J=6.9 Hz ; 2H ; H-6'']; $\delta_6 = 6.92$ ppm [d ; ^3J=5.1 Hz ; 1H ; H-4]; $\delta_7 = 7.08$ ppm [m ; 4H ; H-3'/4']; $\delta_8 = 7.17$ ppm [d ; ^3J=5.2 Hz ; 1H ; H-5]; $\delta_9 = 7.29$ ppm [dd ; ^3J=5.0 Hz ; ^4J=1.3 Hz; 1H ; H-5']

6-(3-Bithienyl)hexanephosphonic acid (BTHPA) 7

A mixture of 2.01 g [6.11 mmol] Bithienylhexyl bromide **4** and 2.09 g [7.00 mmol] tris-trimethylsilylphosphite (Aldrich) was heated under argon (after removal of oxygen by using an ultrasonic bath) at 150^0 C for 4 h. Later 10 ml 80 % methanol were added at room temperature and the mixture was stirred for 15 h.

After the addition of 100 ml 1 M NaOH it was extracted twice with 50 ml tert. butylmethyl ether and the aquous layer was then acidified by adding 200 ml 5 M H_2SO_4. After twice extraction of this aquous solution with a mixture of 50 ml tert. butylmethylether / pentane (3 : 1) the combined organic phases were dried over Na_2SO_4 and this solution was filtered through a 2 cm Na_2SO_4 layer. Rotary evaporation of the solvents yielded 1.82 g [90.2 %] BTHPA **7** as a colourless highly viscous oil.

MS : m/z = 330.2 (100 %) [M^+].

^{1}H-NMR [250 MHz ; CDCl$_3$] : $\delta_1 = 1.35$ ppm [m ; 4H ; H-4''/5''] ; $\delta_2 = 1.65$ ppm [m ; 6H ;

H-2''/3''/6'']; δ_3 = 2.72 ppm [t ; ^3J=7.9 Hz ; 2H ; H-1''] ; δ_4 = 6.89 ppm [d ; ^3J=5.2 Hz ; 1H ; H-4]; δ_5 = 7.05 ppm [m ; 2H ; H-3'/4'] ; δ_6 = 7.13 ppm [d ; ^3J=5.2 Hz ; 1H ; H-5]; δ_7 = 7.27 ppm [dd ; ^3J=5.0 Hz ; ^4J=1.3 Hz; 1H ; H-5']; δ_8 = 9.70 ppm [s (breit) ; 2H ; -PO_3H_2]

Dodecylphosphonic acid (DPA) **5**

A mixture of 23.9 ml [0.10 mol] 1- bromododecane (Merck) and 18.9 ml [0.11 mol] triethyl phosphite (Aldrich) was heated (after removal of oxygen in an ultrasonic bath) under argon for 4 h at 160-190^0 C with continuous removal of the ethylbromide formed.

Now 100 ml 40 % HBr was added at room temperature and the mixture was refluxed under stirring for 3h (oil bath temperature about 140^0 C) with evaporation the hydrobromic acid at the end. On reaching room temperature a colourless solid was obtained, which was washed with pentane, recrystallised from hexane and from 0.1 M H_2SO_4 . Now 300 ml 0.1 M H_2SO_4 and 250 ml diethyl ether were added and after extraction the organic phase was dried over Na_2SO_4 .

Rotary evaporation of the solvent and new recrystallisation from hexane yielded 14.45 g [57.7 %] dodecylphosphonic acid **5** as colourless plates of a melting point (mp) 98-99^0 C .

Lit. [13] :   mp = 100.5-101.5^0 C (from hexane).

SAM formation of the phosphonic acids on iron

The first step in the formation of self-assembled monolayers (SAMs) of the new primer system with phosphonic acids is the passivation of the iron surface. Different methods of iron passivation have been described in the literature.[4-7] After several attempts with different passivation methods, a treatment with 10% HNO_3 (4 minutes at 25^0 C) [2] after grinding (paper with grain size 600) and degreasing with ethanol proved to be most effective.

SAM formation on the passivated iron was investigated after dipping the disc electrodes (armco iron) for 15 h in a 10^{-3} M aqueous solution of the sodium salts (1 : 1 - mixture of mono- and disodium salt) of the different phosphonic acids (pH 7.8 – 8.5). Before starting the electrochemical and contact angle measurements, the treated samples were stored dust-free in dry atmosphere to allow an optimal SAM orientation.

The anodic polarisation curves of the treated samples showed a significant decrease in current densities compared with those of untreated passivated iron (see Table 1). Contact angle measurements also impressively indicated the sucessful SAM formation of the phosphonic acids. The dipping process converts the hydrophilic surface of the passivated iron into a strong hydrophobic surface (see Table 1).

Table 1 shows the anodic current densities of the passivated iron disc electrodes ($A = 0.35$ cm^2) at 800 mV vs. Ag/AgCl in aerated 0.1 M NaClO$_4$ at a rotation rate of 1000 rpm before and after treatment with **DPA 5** and **BTHPA 7** solution.

Additionally the contact angles with water were specified. The values noted were the median of 4 - 8 measurements.

Table 1. Effect of phosphonic acid treatment of passivated iron, observed with the anodic current density and contact angle values.

	Passivated iron (untreated)	Passivated iron with DPA 5 treatment	Passivated iron with BTHPA 7 treatment
Anodic Current density at +800 mV vs. Ag/AgCl	6,14 µA / cm^2	2,56 µA / cm^2	0,37 µA / cm^2
Contact angle (water)	$18.5^0 \pm 4^0$	148.5^0 ($\pm 3^0$)	121.2^0 ($\pm 2^0$)

For comparison, the contact angle of unpassivated armco iron (after grinding and degreasing with ethanol as described) was 51.4^0 ($\pm 4^0$).

Electropolymerisation of immobilised bithiophene head groups on iron

Willicut & McCarley [14-15] described the electropolymerisation of pyrrol head groups after immobilisation of N-pyrrolyl-alkanethiols on a gold surface in propylene carbonate.

Now similar cyclovoltammetric experiments were carried out to electropolymerise the bithiophene head groups after SAM formation on passivated iron with **BTHPA 7** solution.

Anodic switching with potentials below 1.3 V (vs. Ag/AgCl) did not show any signs of electropolymerisation (see Fig. 4). After several further CV cycles with an increased switching potential of 1.4 V, where oxidation peaks appeared, reversable oxidation and reduction peaks could now be observed (see Fig. 5). They are characteristic for the formation of conducting polymers, e.g. polybithiophene.

Analogous experiments (anodic polarisations in CV cycles up to 1.7 V vs. Ag/AgCl) with thiophene head groups (iron samples, treated with **THPA** <u>6</u>) gave no indication of a polymerisation process. This can be explained by the higher oxidation potential of thiophene (head group) compared to bithiophene.

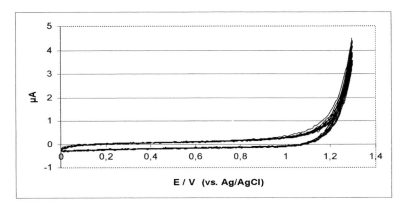

Figure 4. Cyclovoltammogramm of HNO₃ passivated armco iron after BTHPA <u>7</u> treatment in propylene carbonate / 0.1 M NBu₄ClO₄ (50 mV s^{-1} ; 1000 rpm rotation). CV cycles 1 – 5 with low maximum potential of 1.3 V .

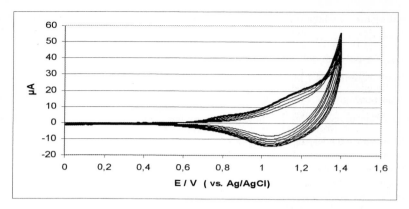

Figure 5. The electrode and system used above (Fig. 4) after 10 further CV cycles with a maximum potential of 1.4 V. The CV cycles 15 - 24 are shown with the increased maximum potential of 1.4 V vs. Ag/AgCl (50 mV s^{-1} ; 1000 rpm).

Discussion

Experimental results of both SAM formation of special phosphonic acids on passivated iron and of experiments to electropolymerise the bithiophene head groups of corresponding monolayers show that the new primer system planned (see Fig. 1) is feasible. Evidently, many further investigations are necessary before this new procedure will be of industrial interest. Furthermore direct structural information of the SAM's before and after polymerisation will be acquired by special surface analytical techniques, suchas AFM or XPS investigations.

The influence of other iron passivation methods with less surface roughness and more durable passivation layer on SAM formation and subsequent electropolymerisation will be studied in further experiments. Further phosphonic acids with bithiophene or other probably polymerisable head groups are also in development.

References

[1]   G. M. Spinks, A. J. Dominis, G.G. Wallace, D.E. Tallman, *J. Solid State Electrochem.* **2002**, *6*, 85.
[2]   S. P. Sitaram, J. O. Stoffer, T. J. O'Keefe, *J. Coatings Techn.* **1997**, *69*, 65.
[3]   C. A. Ferreira, S. Aeiyach, J. J. Aaron, P. C. Lacaze, *Electrochim. Acta* **1996**, *41*. 1801.

[4]   H.-J. Adler, E. Jähne, F. Simon, report from research project 287 / part B1, currently available on the internet at http://www.chm.tu-dresden.de/sfb/b/b1/b1.htm .

[5]   I. Mäge, E. Jähne, A. Henke, H.-J. Adler, C. Bram, C. Jung, M. Stratmann, *Macromol. Symp.*, **1997**, *126*, 7.

[6]   E. Kalmann, Electrochim. Acta **2001**, *46*, 3607.

[7]   U. Rammelt, P. T. Nguyen, W. Plieth, *GDCh monograph*, **2000**, *21*, 275   (ISBN: 3-924763-90-9).

[8]   P. Bäuerle, F. Würthner, S. Heid, *Angew. Chem.* **1990**, *102*, 414.

[9]   A. Carpita, R. Rossi, *Gazz. Chim. Ital.* **1985**, *115* (11-12), 575.

[10] N. Jayasuriya, J. Kagan, *Heterocycles* **1986**, *24*, 2901.

[11] Ph. D. Thesis Michael Wedel (University of Bremen) **2000**, 76.

[12] I. Maege, E. Jaehne, A. Henke, H.-J. P. Adler, C.Bram, C. Jung, M. Stratmann, *Progress in Organic Coatings* **1998**, *34*, 1.

[13] Beilstein, Handbook of Organic Chemistry, III. EW, *4*, 1785.

[14] R. J. Willicut, R. L. McCarley, *J. Am. Chem. Soc.* **1994**, *116*, 10823.

[15] R. J. Willicut, R. L. McCarley, *Adv. Mater.* **1995**, *7*, 759.

Macromol. Symp. 187, 77–85 (2002)

Pretreatments of Wood to Enhance the Performance of Outdoor Coatings

Dirk Lukowsky, Guido Hora

Fraunhofer Institute for Wood Research, Bienroder Weg 54 E;

38108 Braunschweig, Germany

Summary: The wet adhesion of water borne acrylic dispersions is a crucial factor on the performance of outdoor coatings on wood. Pine sapwood was treated with several methods for surface activation to increase the wet adhesion of water borne acrylic dispersions. The wet adhesion was measured by pull-off tests as well as with a modified cross-cut test. Atmospheric plasma, corona treatment and fluorination increased the wet adhesion of the coating which is attributed to the increasing polar portion of the surface free energy. Other ways of improving the wet adhesion are the addition of promotors, the use of primers and organisational improvements.

Keywords: coatings, wood; adhesion; dispersions; cold plasma

Introduction

The performance of outdoor coatings on wood is strongly influenced by the presence of liquid water that penetrates the substrate and the interface wood – coating. Changing moisture levels cause the wood to swell and shrink. These movements of the substrate may cause cracks in the coating as soon as the maximum in elasticity of the coating is exceeded. The deposition of water molecules in the interfacial zone reduces the bonding strength (wet adhesion) between the coating and the wood as well [1,2]. Especially, for water borne acrylic dispersions the wet adhesion is occasionally so critical that blistering, adhesive failure and other damages might occur (Figure 1).

The wet adhesion of coatings on wood depends on several factors. Therefore, an improvement can be achieved with several different methods that are highlighted in the paper.

CCC 1022-1360/00/$ 17.50+.50/0

Figure 1. Damages of coated wood due to the insufficient wet adhesion of the coating

Experimental

Scots Pine (*Pinus sylvestris*) sapwood was preatreated with several methods and afterwards coated with 3 layers of a transparent or stained waterborne acrylic dispersions. After distinct time intervals, the wet adhesion was measured by one of the following methods with at least 3 replicates for each parameter:

The cross-cut test was a modified test in relation to ASTM D 3359[3] – 93 and EN ISO 2409[4]. Eleven parallel cuts were performed under an angle of 0° and 90° to the grain direction of the wood – resulting in 100 squares of 2 mm x 2 mm. The cuts were made with sharp scalpel knives using either a template and cutting by hand or using an apparatus that produced 11 cuts at the same time.

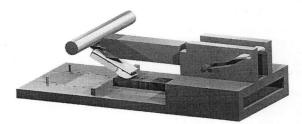

Figure 2. Aparatus for cutting 11 parallel cuts with scalpell blades at the same time.

Four layers of paper-towel were placed on the cut and intensively wetted with deionisised water. After two hours the towels were removed and the surface rubbed dry. 5 Minutes later a tape (Baiersdorf tesa® 4122) was rubbed firmly on the surface. The tape was then removed by hand at constant speed in an angle of about 60°. The classification of the cross-cut area was made according to EN ISO 2409 using the area of the flaking as classification system. A cross-cut classification of 0 represents a very good adhesion without flaking on the cross-cuts after the removal of the tape. A cross-cut classification of 5 represents a very poor adhesion with more than 65% flaking of the cross-cut area.

The pull-off strength was determined according to EN 311[5] fourteen days after the application of the coating. A circular groove of 0.2 mm depth with an inside diameter of 35.7 mm was milled into the surface and a metal dolly was glued in the center with a 2-k metal adhesive Metallon E 2082 (hardening time 12 hours) under pressure. With a laboratory syringe water was deposited into the groove and the pull-off strength was tested after 3 hours.

Results

Despite the inhomogeneity of wood and the inherent subjectiveness of the cross-cut test, both the cross-cut test and the pull-off test showed rather reliable correlation and reproducible values of the wet adhesion. The values correlated with the occurrence of adhesion problems in practice. When performing the cross-cut test on wood, very thin and sharp blades have to be used to avoid the squeezing of the substrate. It is also essential to wait 5 minutes after rubbing the wet surface dry before the application of the tape. Otherwise the adhesion of the tape may be too poor to achieve reliable results. Unfortunately the EN ISO 2409 classification of the flaking lacks of consistence as qualitative (kind of flaking) and quantitative parameters (area of flaking) have to be combined in the evaluation exercise without further specifications. Especially, as the image analysis of cross-cuts is one goal for the future, qualitative parameters should be removed from the standard or defined more precisely in relation to the flaking area.

The 60 seconds pretreatment of scots pine with 1% effective concentration of fluorine at 500 mb increased the wet adhesion of two acrylic dispersions (Figure 4).

The fluorine content in the wood was measured by Electron Probe Microanalysis (EPMA). At the very surface 0.25 atomic percent fluorine were present, in a depth of 0.4 mm following the grain direction about 0.1 atomic percent and in 1 mm depth the wood was fluorine free.

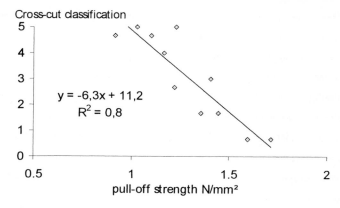

Figure 3. Correlation between the cross-cut test and the pull-off test

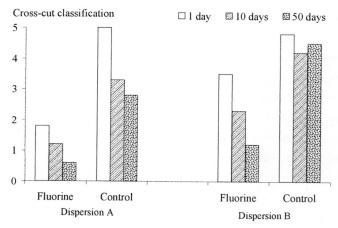

Figure 4. Cross-cut classification of two acrylic dispersions with and without a preatretment with fluorine in dependance of the timelag between coating and the cross-cut test. (Cross-cut classification of 0 = optimum adhesion)

The <u>corona-treatment</u> was performed with 100 V, 45 kHz, 100 μs pulses at a 2 kHz frequence and a movement of the substrate of 5 cm/sec. The cross-cut classification improved from 3,7 (untreated) to 0,7 after the corona treatment.

The <u>plasma treatment</u> was performed with a round nozzle under a distance of 3 cm and a movement of the substrate of 5 cm/sec. The results show in principle the suitability of the plasma pretreatment for acrylic dispersions. The wet adhesion of the two aqueous acrylate dispersions was improved by this treatment. Following the plasma treatment the wet adhesion after one day was not yet optimal, but improved gradually within the subsequent 9 days. Opposite to the observations made for the acrylate dispersions, the wet adhesion of the solvent based internal comparison product (ICP) according to EN 927-3[6] was reduced by the plasma treatment. The cause for the poor wet adhesion of the solvent-based ICP following the plasma pretreatment is not yet known. It was stated by microscopic investigations that the penetration of the ICP into the wood structure was not changed by the plasma pretreatment. It might be due to an incompatibility of the solvent or the functional groups of the ICP with the wood surface activated by the plasma treatment. In each case this is a clear reference to the fact that the plasma treatment has to be carefully fine-tuned to the coating material.

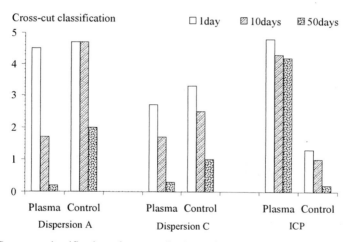

Figure 5. Cross-cut classification of two acrylic dispersions and ICP with and without a plasma preatretment in dependance of the timelag between coating and the cross-cut test.

The surface free energy of plasma treated pine and its respective disperse and polar portions where measured by the dynamic contact angle measurement according to Wilhelmy with water and methyleniodid. Except for the measurement immediately after the plasma treatment, both the surface free energy and its polar portion where increased by the pretreatment.

Table 1. Dependence of the surface energy and its disperse and polar portions at plasma-modified pine sapwood as a function of the time after the treatment compared with untreated samples

	Controls	Plasma treated				
		0 d	1 d	4 d	7 d	14 d
Surface tension N/mm²	45	37	52	51	47	52
Disperse portion N/mm²	40	24	44	40	39	42
Polar portion N/mm²	5	13	8	11	8	10

The wet adhesion can also be substantially improved by <u>adequate primers</u> (**Table 2**).It just turns out to be clear that the kind of primer is to be fine-tuned carefully with the selected final coating, since otherwise a degradation of the wet adhesion may occur. The temporal development of the wet adhesion proves that the positive effect of the primer is effective already briefly after the final coating.

Table 2. Cross-cut classifications of two coatings as a function of the pretreatment with different impregnations

	solvent [a]	solids [b]	Dispersion A		Dispersion B	
		%	cc1 [c]	cc50 [d]	cc1	cc50
Primer A	o	13	1.5	0.2	4.7	0.0
Primer B	o	15	4.7	1.2	5.0	1.7
Primer C	w	6	0.8	1.0	0.3	2.2
Primer D	w	18	0.7	0.0	0.5	0.0
Water	w	0	4.7	1.2	1.3	0.0
(Controls)						

[a] o= organic; w = water
[b] Solids content determined at 80 °C for 12 h
[c] Cross-cut classification one day after the coating
[d] Cross-cut classification 50 days after the coating

Numerous trials on the <u>time dependence of the wet adhesion</u> at room temperature after the drying of the film resulted in 3 typical features (Figure 6). Type A performs a poor wet adhesion without any mentionable change during the storage. Type B, the most common one, shows a rather poor wet adhesion after the drying of the film. Thus, within the first 10 days after coating a strong improvement occurs. Type C shows a good initial wet adhesion with little changes over the time.

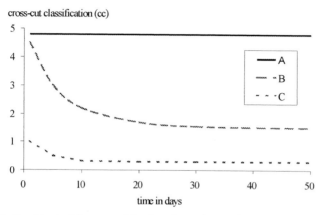

Figure 6. Idealised typical features of the time dependence of the wet adhesion of three aqueous acrylic dispersions

Discussion

The surface activation of wood is a promising approach to increase the wet adhesion of coatings. All three presented methods (corona, fluorination, plasma) are commonly used for the preatreatment of plastics to improve their coatablity. All three methods improved the wet adhesion of acrylic dispersions on wood, too. This can most likely be attributed to the increased polar portion of the surface free tension. As the corona technique is not appropriate for profiled timber, no further research was performed with this pretreatment technique. The fluorination of wood does have other disadvantages: The formerly halogen free wood is contaminated with fluorine and the price for the process is rather high for wood in practical dimensions. The vacuum steps, required in this process leads to an evaporation of terpenes from resinous softwoods, which might contaminate the reaction vessels.

For practical use, the pre-treatment with athmospheric plasma seems to be the most promising technique. A rough calculation of the preatreatment of one window unit resulted in approximately 0.3 Euro[7]. The continuous plasma treatment can be integrated into the existing production lines.

Nevertheless, in practice modern surface activation techniques will only be alternative, as long as other more conventional methods, listed in the following, for the improvement of the wet adhesion are not successful[7]:

- Reduction of the time lag between the last machining of the wooden surface and the first coating. (The wettability and the adhesion of coatings on resinous softwoods decreases rapidly after the last machining as a resin film develops on the surface[8])
- Increasing the time between the last coating and the exposure to weather
- Using an adequate functional primer[2]
- Improving the wet adhesion by adhesion promotors in the polymeric coating systems[1].

The optimum wet adhesion of coatings on wood depends on the expected moisture-conditions in use. Wooden constructions, that are regularly subjected to high moisture contents (fences, claddings) require a very high wet adhesion of the coatings. For wood, that should not become wet during its use (windows, doors) the optimum wet adhesion is more difficult to determine. On the one hand, a very good wet adhesion protects the coating from failure, especially when small defects are present. Little cracks, resulting from e.g. hail, will keep insignificant, as long as the wet adhesion is very high. On the other hand the wet adhesion of coatings should not exceed a certain value to allow an early visual detection of water that penetrates into the wood. From this point of view, it is advantageous if the coating shows blistering or flaking when the moisture content of the wood exceeds 30% for e.g. more than one week. If the wet adhesion is very high, the coating would appear defect free, even if the wood underneath is completely wet and e.g. decayed by fungi.

In any case, it is very important, that the wet adhesion has reached almost its final level when the wooden construction unit is installed especially in the cold months. As the development of the wet adhesion depends strongly on the temperature, it might take month to reach its final level at wintertime.

Acknowledgements

This work was sponsored by the Federal Ministry of Economics and Technology and the German Research Association for Surface Treatment (DFO)

References

[1] G. Hora (2002): Chemisch-physikalische Untersuchungen an wässrigen Holzaussenanstrichsystemen hinsichtlich Alterungsverhalten und Adhäsion. PhD-Thesis, Technical University Braunschweig.

[2] H. Turkulin; K. Richter; J. Sell (2000): Adhesion of waterborne acrylic and hybrid paint on wood treated with primers. Proceedings of the second woodcoatings congress, 23-25 October 2000, The Hague, Netherlands.

[3] ASTM-Standards, D 3359(1993): Standard test method for measuring adhesion by tape test.

[4] EN ISO 2409 (1994): Cross-cut test. Beuth Verlag GmbH, Berlin.

[5] EN 311(1992): Surface soundness of particleboards - Test methods. Beuth Verlag GmbH, Berlin.

[6] EN 927-3(2002): Natural weathering test. Beuth Verlag GmbH, Berlin.

[7] D. Lukowsky; G. Hora (2002): Final report 11739 N / 1 `Untersuchung der Ursachen einer verminderten Haftung von Anstrichfilmen auf Holzuntergründen im Außenbereich. DFO, Arnufstr. 25, 40545 Düsseldorf, Germany.

[8] R.M. Nussbaum (1999): Natural surface inactivation of Scots pine and Norway spruce evaluated by contact angle measurements. Holz als Roh- und Werkstoff 57, 419-424.

Macromol. Symp. **187**, 87–96 (2002)

Performance of Primers Containing Polyphosphate-Based Ion-Exchange Pigments for the Protection of Galvanised Steel

F.Deflorian[1], I.Felhosi[2], S. Rossi[1], L.Fedrizzi[3]*, P.L. Bonora[1]

[1]Department of Materials Engineering, University of Trento, Via Mesiano 77, 38100, Trento Italy, fax +39 0461 881977, e-mail defloria@ing.unitn.it

[2]Department of Surface Science and Corrosion Research, Chemical Research Center Hungarian Academy of Sciences, Budapest, Hungary

[3]Department ICMMPM, University of Rome "La Sapienza", Rome, Italy

Summary: The environment protection is pressing the replacement of chromate based pre-treatments and anti-corrosive pigments with environmental compatible coating systems, in particular in the coil industry, and many studies are being connected to this field.

A relatively new class of inhibitive pigments is the ion-exchange pigments. Commercial available ion exchange pigment based on calcium-exchanged silica gel are applied in wide range of coatings technology. The inhibition mechanism is related to the ion-exchange reaction and the deposition of a protective layer, but the mechanism is not yet completely known.

The aim of the present study is an electrochemical evaluation of corrosion protection performance of environmental friendly coating systems, whose primer is consisting of polyphosphate and ion-exchanger based anticorrosive pigments.

The results proved that the use of electrochemical techniques like electrochemical impedance spectroscopy is important in order to characterize and understand the corrosion protection mechanism and efficiency of organic coatings containing active pigments, as ion exchange pigments. Moreover it was shown that in the coatings the pigments have a significant inhibitive effect and they act as cathodic inhibitors.

Keywords: inhibitive pigments, organic coatings, electrochemical impedance spectroscopy, polyphosphate

Introduction

Nowadays the environment protection and the decrease of pollution are key subjects in different fields of research and development. The replacement of chromate based pre-treatments and anti-corrosive pigments with environmental compatible coating systems is also very important in the coil industry and many studies are being connected to this field [1]. The longest established low

CCC 1022-1360/00/$ 17.50+.50/0

toxicity anti-corrosive pigments are zinc phosphate, $Zn_3(PO_4)_2.2H_2O$ (in Europe) and barium metaborate (in the USA).

A relatively new class of inhibitive pigments are the ion-exchange pigments [2,3,4]. They are inorganic oxides with comparatively high surface area loaded with ionic corrosion inhibitors by ion-exchange with surface hydroxyl groups. The ion-exchangers are chosen for their acid and basic properties to provide cation- or anion exchangers, thus usually silica is used as a cation and alumina for anion support. The corrosion protection behaviour of pigments is controlled by the rate of ion-release process of exchanger [5,6,7]. The ion-exchange release mechanism allows efficient use of a comparatively low loading of inhibitive pigments.

The other advantage of this approach is to allow the use of otherwise highly soluble inhibitors. With the knowledge of kinetics of ion-release process within the primer it is possible to control the accessibility of inhibitors, and therefore to control its diffusion onto the metal surface.

Commercially available ion exchange pigment is Shieldex® [3] based on calcium-exchanged silica gel applied in the coatings technology. The inhibition mechanism is related to the ion-exchange reaction. Upon exposure of paint films to the aggressive environment, due to water uptake, exchange of calcium to hydrogen ions and release of calcium ions are taking place. Dissolution of polysilicate anions is also supposed [8,9]. The diffusion and deposition of both dissolved species onto the metal surface result in a formation of protective layer. Deposition of these species within pores of the coating is also presumed to improve its barrier function.

Polyphosphates form the base of third generation anticorrosive pigments. Polyphosphates are cathodic inhibitors, and protect most metals by precipitation mechanism; i.e., they form almost insoluble complexes with calcium, magnesium ions. The most important polyphosphates are tripolyphosphate and hexametaphosphate. Polyphosphates has low toxicity but they are alga nutrients. Due to their hydrolysis in acidic and alkaline conditions, they can easily transform to orthophosphates.

The aim of the present study was an electrochemical evaluation of corrosion protection performance of environmental friendly coating systems, whose primer was consisting of polyphosphate and ion-exchanger based anticorrosive pigments. The corrosion protection properties and mechanism of primers have been studied by electrochemical impedance spectroscopy.

Experimental

The studied materials were steel sheets with hot dipped zinc alloy coating (Zn + 1.5 ± 0.5 % Al) with thickness of 11 μm followed with fluotitanate pretreatment.

Sheets of galvanized steel were covered by coil coating technology with a 5 μm thick polyester primers cured with isocyanate containing inhibitive pigments (PVC 30 wt%) of:

CAPP – Calcium-Aluminium-polyphosphosilicahydrate: samples PC

ZAPP – Zinc-Aluminium-polyphosphatehydrate; samples PZ

The results of elemental analysis of inhibitive pigments are reported in Table 1.

**Table 1. Elemental analysis of the main
components of the inhibitive pigments**

CAPP	31% CaO	ZAPP	48% P_2O_5
	28% SiO_2		30% ZnO
	26% P_2O_5		12% Al_2O_3
	7% Al_2O_3		

The shape and the dimension of the pigments is very different as it is possible to see in figure 1: smaller in size and more regular for ZAPP (average dimension 2 μm), bigger and less regular for CAPP (dimensions 5-15 μm).

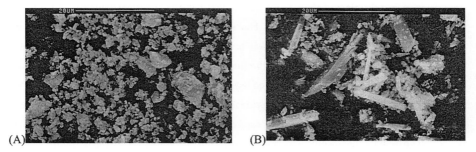

(A)　　　　　　　　　　　　　　(B)

Figure 1. Morphology of pigments ZAPP (A) and CAPP (B).

Polarisation curves were performed using a PAR 273 potentiostat at 0.5 mV/s of scan rate and the impedance measurements were performed using PAR 273 potentiostat and Solartron 1255 frequency response analyzer in the frequency range of 100 kHz – 1 mHz with a sine amplitude of 10 mV. The electrochemical cell was a three-electrode cell with an Ag/AgCl reference electrode

[+207 mV vs. SHE] and platinum counter electrode. All of the experiments were performed in 0.3 % Na_2SO_4 solution at room temperature. The studied surface area was 25 cm^2.

The morphology of the deposits was observed by Scanning Electron Microscopy and the chemical composition was evaluated by EDXS Analysis. In order to improve the chemical information, the zinc surface after exposure in the solution containing the pigments was analyzed by FTIR measurements (Nicolet DXL in reflectance).

Results and discussion

In order to characterize the effect of inhibitive pigments on anticorrosion performance of coatings, initially we studied galvanized steel samples covered with only primer without topcoat. In this case, the degradation process of coatings takes place in a relatively short time. By the use of electrochemical impedance spectroscopy technique, the complex electrochemical behaviour of coating system can be studied, e.g., the dielectric properties of coating, the corrosion properties of underlying substrate.

The impedance spectra of primer-coated substrate in 0.3 % Na_2SO_4 solution can be characterized mainly by three time constants. It means that in the Nyquist plot three different semicircles are visible. The high frequency one is related to the properties of the coating because the values of the impedance are independent on the applied potential and from the capacitance value just after immersion it is possible to calculate a value of the dielectric constant which is close to the theoretical one for the polymer matrix. The intermediate frequency loop is related to the electrochemical reaction on the metal surface because it is potential dependent and the value of the time constant is almost the same of the time constant of bare zinc corroding in the same solution. The cathodic reaction is the oxygen reduction and therefore it is possible to consider the low frequency tail caused by the diffusion process of oxygen molecule, which is the slowest, rate-determining process of corrosion. The proposed circuit, according with literature [10] is shown in figure 2. The high frequency elements describing the coating properties are the non-ideal coating capacitance (Q_{coat}), which is mathematically described by constant phase element (CPE) and the coating resistance (R_{por}) that is the ionic resistance through the coatings (through defects or pores and through the intact coating). The lower frequency part of impedance spectra is related to the corrosion process taking place on the metal surface: Q_{dl} is a non ideal capacitance of double layer and R_{ct} charge transfer resistance is related to the faradic reaction. Further, the lowest frequencies of impedance spectra is determined by the diffusion which can be modeled, for finite diffusion path, as a resistance in parallel with a constant phase element (R_{diff} and Q_{diff}) [11].

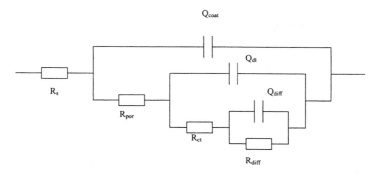

Figure 2. Equivalent electrical circuit for impedance spectra of primer coated metals.

Although the impedance spectra of corroding system is more complex, and likewise the complete electrochemical behaviour of the system contains more processes (for example zinc dissolution can involve two steps [12]), inclusion of further elements to the model is not required partly because these elements can be determine only with high relative error and for this reason its values are hardly uncertain, partly because neglecting these element from the model, the values of others elements are not significantly modified.

Figure 3 shows the change of evaluated parameters in time for the studied materials.

The coating capacitance (Q_{coat}) is increasing in time in both cases (PC and PZ), and the coating resistance (R_{por}) which is influenced by the presence of defects and conductivity through the pores, is decreasing in time. These changes refer to the pore formation and deterioration of primer due to the interaction with corrosive electrolyte. The general trend is the same, but clear differences are evident comparing the two kind of samples: the coating capacitance growth is higher in the case of PC samples. In this case the increase of Q_{coat} cannot be simply related to water uptake caused by diffusion of water molecules in the polymeric matrix, but there is also an important contribution due to the interaction of the electrolyte with the pigments which are soluble and can cause a further water uptake [13], even because the water can occupy the space becoming free by the pigment dissolution. It means that the coating capacitance change are due to two different phenomena: the water diffusion in the coating changing the average dielectric constant of the coating and the pigment solubilisation interacting with the polymeric matrix of the coating.

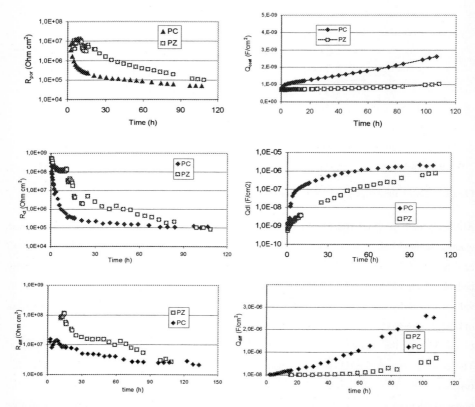

Figure 3. Coating resistance (R_{por}), coating capacitance (Q_{coat}), charge transfer resistance (R_{ct}), non-ideal double layer capacitance (Q_{dl}), diffusion resistance (R_{diff}) and diffusion capacitance (Q_{diff}) vs. the immersion time of polyester based primers pigmented with CAPP (PC) and ZAPP (PZ).

Also the coating resistance R_{por} is decreasing faster in the case of samples containing CAPP (PC), probably because the solubilisation of the pigments favours the ions penetration, but after about 120 hours the values become very similar. In accordance, the corrosion rate is increasing, which appears in the decrease of both charge transfer resistance (R_{ct}) of metal dissolution and diffusion resistance (R_{diff}). Also in this case, after an initial period in which the samples containing CAPP show a higher reactivity, after 5 days of immersion the values of R_{ct} and Q_{dl} are very similar. The only parameter which is becoming more and more different with the immersion time is the value of Q_{diff}. It is important to remember that we suppose that the species controlled by diffusion is oxygen. But with impedance we are measuring only the impedance of charged species and therefore R_{diff} is not the resistance to diffusion of oxygen, but the resistance to diffusion of ions. If the diffusion control is due to the presence of corrosion products it is well known that corrosion products can block the oxygen diffusion more that the ions diffusion. For this reason is more useful to look to the values of Q_{diff}, which are related to the presence of corrosion products inducing a capacitative contribution to the impedance. The values of Q_{diff} indicate that the diffusion processes are reduced in the case of the PC samples. This difference is probably due to the presence of more corrosion products under the coating in the case of PC primer acting as a partial oxygen barrier. This phenomenon is not so evident for the sample PZ because of the lower corrosion rate.

The deterioration process of primer is faster in the case of PC, which is pigmented with polyphosphate, calcium, silicon and aluminium (CAPP). The reason probably can be traced back to the difference in stability constants and solubility of Ca-polyphosphate and Zn-polyphosphate. The lower solubility of Zn-polyphosphates is well-kown in the literature [14]. This property can cause a lower dissolution rate of the pigments and also a higher tendency to re-precipitation on the metal surface.

In order to verify the accuracy of the proposed equivalent circuit model, and to get deeper knowledge on the anticorrosive mechanism of pigments, the polarization effect was studied after 2 weeks of immersion time. The primer-coated panels was polarized at first cathodically at -1100 mV vs. [Ag/AgCl], and then anodically at -800 mV.

During the cathodic polarization (which is lower of the free corrosion potential of the samples in the testing solution which is about -1000 mV) the low frequency resistance (R_{diff}) is decreasing, while the opposite occurs during the anodic polarization, which indicates, that the corrosion is controlled by the cathodic half-reaction (oxygen depolarization). On the contrary the values of R_{ct} do not change so much with the polarisation potential and the values remain lower than R_{diff} which therefore controls the total reaction.

After the cathodic polarization, the open circuit potential shifted to anodic direction in both cases (PZ and PC). Due to the polyphosphate, zinc and calcium ions are cathodic inhibitors; the enhanced adsorption should have contributed to the shift towards the cathodic direction. Thus the reason for anodic shift of corrosion potential and increase of polarization resistance seems to be the hydrolysis of polyphosphate to orthophosphate, which is anodic inhibitor and have good passivation effect. The hydrolysis of polyphosphate is probably activated by the increase of local pH near the metal/polymer interface due to the enhanced OH⁻ formation during oxygen depolarization reaction.

In order to have a confirmation of the cathodic inhibition effect of the studied pigments, some polarization curves in the same solution (0.3% of Na_2SO_4) saturated with the pigments were performed on bare zinc samples. The cathodic (figure 4a) and anodic (figure 4b) behavior is a clear indication of the cathodic inhibition, while ZAPP and CAPP pigments seems to have very similar electrochemical effect.

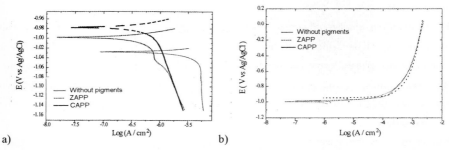

a)   b)

Figure 4. Cathodic (a) and anodic (b) behaviour of zinc in 0.3% Na_2SO_4 solution saturated with ZAPP and CAPP pigments.

Different is, on the contrary, the appearance of the zinc surface after the cathodic polarization: only on the sample immerse in the ZAPP solution it was possible to note a chemical conversion layer deposited on the surface (figure 5), looking like flat small crystal, while in the case of sample immersed in the CAPP solution it was impossible to note any difference in comparison with the zinc surface before immersion.

Also the EDXS measurements confirmed in the case of samples immersed in containing ZAPP solution the presence of P and Al (and a higher presence of oxygen) on the zinc surface while the sample immersed in the CAPP solution showed EDXS spectra equivalent to the spectra of new zinc surfaces.

In order to obtain more information on the chemical nature of the deposit in the ZAPP solution, a FTIR analysis was carried out and the results are in figure 6. The presence of peaks characteristic of phosphorous-organic compounds bonds (between 1000 and 1400 cm^{-1})are visible together with peaks which it is possible to relate to carbon-hydrogen bonds.

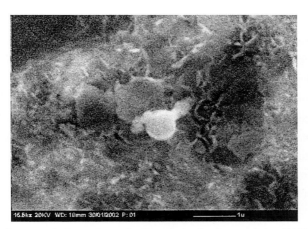

Figure 5. Zinc surface morphology after cathodic polarisation in containing ZAPP solution.

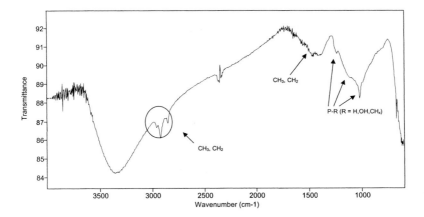

Figure 6. FTIR spectrum of Zn surface after cathodic polarization in the containing ZAPP solution.

From this characterization we can conclude that the ZAPP pigment is able to protect acting as a cathodic inhibitor by the deposition of a layer on zinc in alkaline conditions. The higher solubility of the CAPP pigments cause a faster primer degradation, moreover the deposition of a protective layer for this pigments is not proved, even if a cathodic inhibition activity was measured.

Conclusions

The use of electrochemical techniques, like electrochemical impedance spectroscopy, is useful in order to characterize and understand the corrosion protection mechanism and efficiency of organic coatings containing active pigments as ion exchange pigments.

For the studied materials it is possible to summarize the main results in the following points: In the primers the pigments have a significant inhibitive effect and they act as cathodic inhibitors. The pigment action results in a layer deposition on the metal surface reducing the corrosion rate. The higher solubility of CAPP causes a faster degradation of the corrosion protection properties of the primers containing this pigment in comparison with the primers containing the pigment ZAPP. Moreover the formation of a protective layer on the metal surface of the CAPP pigments was not proved.

References

[1] J. Sinko, Progress in Organic Coatings, 42 (2001), 267;

[2] B. P. F. Goldie, JOCCA, 9 (1988) 257

[3] Shieldex Product Inf., Grace, Worms, (D) (1989), http://www.gracedavison.com/products/coatings/shieldex.htm;

[4] L.W. Vasconcelos, I.C.P. Margarit, O.R. Mattos, F.L. Fragata, A.S.B. Sombra, Electroch. Acta, 43, (2001), 2291;

[5] R. D. Armstrong, S. Zhou, Corrosion Sci., 28 (1988) 1177;

[6] A. Amirudin, C. Barreau, R. Hellouin, D. Thierry, , Prog.Org.Coat., 25 (1995) 339;

[7] T. E. Fletcher, Proc. 11th Int. Corr. Congress, Florence, Ed AIM, Milan, vol. 2, (1990), 265,

[8] W.M. Liu, Werkstoffe und Korrosion, 49, 576, (1998);

[9] F. Deflorian, L. Fedrizzi, P. L. Bonora, Proc.of Symp. on Advances in Corrosion Protection by Organic Coatings III. (Electrochemical Society Proceedings) Vol 97-41, 45 (1997);

[10] E.P.M. van Westing, G. M. Ferrari and J.H.W. de Wit, Corros. Sci., 36 (1994) 1323;

[11] E.P.M. van Westing, PhD Thesis, University of Delft, 1992;

[12] C. Desluis, M. Duprat, C. Tournillon, Corros.Sci., 29 (1989) 13 ;

[13] L.Fedrizzi, F.Deflorian, G.Boni, P.L.Bonora, E.Pasini, Prog.Org.Coat., 29 (1996) 89;

[14] IUPAC Stability Constants of Metal-ion Complexes, Pergamon Press, 1979.

Macromol. Symp. 187, 97–107 (2002)

· Anticorrosive Pigments for Chemically and Thermally Resistant Coatings

Andréa Kalendová, Jaromír Šňupárek*

University of Pardubice, Faculty of Chemical Technology, Institute of Polymeric Materials, nám. Cs. Legií 565, 532 10 Pardubice, Czech Republic

Summary: The present state in the field of organic coatings requires such an anticorrosive pigment, which is adapted in its properties to the binder concerned and will contribute to the overall protection properties of the pigmented coating film at max. The paper presents as an example of thermally and chemically stable pigments the compounds based on mixed metal oxides. Other pigments exhibiting prospective results for the coating applications comprise the pigments of lamellar shape, which in addition to their alkalizing properties contribute to increasing the barrier and simultaneously inhibition mechanism by protective action of the coating.

Introduction

The conservation of metals in their pure form is possible only by their protection, which is secured by the chemical or electrochemical reactions of anticorrosive pigment with the metal alone or with the corrosion medium penetrating through the coating. The present-time task is devoted to investigating such anticorrosive pigments, which can be considered harmless from the environment protection point of view and, at the same time, highly efficient in their corrosion-inhibition action.[1,2] The toxic pigments could be replaced in practice with zinc phosphate $Zn_3(PO_4)_2.2H_2O$. At present there exist a series of anticorrosion pigments based on phosphate anions.[3] Also a phosphate-molybdate anion is possible, the applicable cations comprise Zn^{2+}, Al^{3+}, or Ca^{2+} and others.[4] A further possibility consists in using aluminum polytriphosphate as an anticorrosion pigment for water-dilutable and solvent-type coatings.[5] Aluminium polytriphosphate can be modified with Zn^{2+} ions, which results in an increase in the anticorrosive efficiency of the pigment concerned. The anticorrosive efficiency of primers can be secured by introducing the pigment acting on an ion-exchange principle.[6] This concerns the pigments consisting of silicate anions and Ca^{2+} cations.

CCC 1022-1360/00/$ 17.50+.50/0

Futher compounds distinguished by their anticorrosive effects are represented by barium metaborate, calcium, zinc, or strontium phosphosilicate, calcium borosilicate, and zinc molybdate.[7] Also the zinc and zinc lead(II) salts of organonitrogen compounds, organozinc compounds and calcium cyclotetraphosphate show an anticorrosive activity. A separate and rather problematic group of anticorrosive pigments consists of the so-called „core" pigments , the core of which is created by an inert carrier bearing an active component.[8] Examples of such pigments comprise iron (III) oxide, the surface of which is treated with zinc phosphate. [9] A possible solution is presented also by calcium carbonate, zinc oxide,[10] or calcium molybdate. As a replacement of standard anticorrosive pigments with the pigments of an equal activity but without harmful effects the spinel-type ferrites or the rutiles based on mixed metal oxides show to be prospective candidates.[11,12]

Effect of polymeric binders on the inhibitive efficiency of coatings

An appropriate formulation of the protective coating creates a general protection by increased life expectancy of the operational coating on the substrate. A key component determining the coating quality is the polymer matrix. No protective high-quality system can be obtained if no anticorrosive high-quality pigment is used in a combination with a high-quality binder. Properties of the binder, such as hydrolytic stability, chemical resistance[13], glass-transition temperature, cross linking degree, and polymer purity in certain limits are the parameters affecting principally the functional properties of coating.

Pigments in organic coating

For the behavior of anticorrosive coating from the corrosion protection point of view the properties of pigment component are determining items in addition to chemical and physical effects. It is not possible, however, to consider the pigment separately, but the pigment-binder system is always to be taken as a whole. The dispersing behavior of the system and the thickness of the film obtained are two remaining factors, which have a great significance from the application technique point of view. It is evident here that the quality of resulting film is affected by such factors as particle size distribution of individual pigments[14]. These parameters are then in a narrow connection to permeability for water and gases and through these characteristics also to the corrosion protection and efficiency of the film.[15] Pigmented organic coatings, which do not contain any corrosion-inhibitive matter (an anticorrosive pigment or organic corrosion inhibitor) can exhibit only the so-called barrier effect if attacked by corrosion-initiating species. This effect is, to a large extent, connected with the nature of chemical binder, the size and shape of filler (inert pigment) particles, the additive content and, eventually, with the fractions of individual components being present in the coating formulation.

Diffusion of corrosive medium through an organic coating

As it has been already described, each organic coating exhibits a definite permeability to corrosive substances as water, oxygen, Cl^-, SO_4^{2-}, and the like.[16] These substances penetrate to the substrate metal, at the surface of which the corrosion processes are initiated.[17] The corrosive medium components can penetrate through the film on several ways. From the morphology of surface film point of view the small-diameter and large-diameter pores can exist. The small pores having 1-5 nm in diameter or those of an even smaller size arise or decay at the movement of segments of macromolecular binder chains at temperatures above the glass polymer transition. This means that a single-phase system is concerned.[18] The temperature dependence of film penetrability reflects the temperature dependence of macromolecules mobility. If the corrosive substance penetrates these very low-diameter pores through the film, then the so-called activated diffusion is meant. The polymers, which have a high mobility of binder chains, such as elastomeres, are more susceptible to the activated diffusion than the polymer structures of high density. The system containing large diameter pores is designated as a two-phase one, and the shape of these pores does not depend on temperature. The permeability depends, on the contrary, on the properties of diffusing compound. The transport of diffusing substances depends on: the flexibility of macromolecular substances, the density of macromolecular network, the morphology of the structure formed, the glass-transition temperature (T_g) of polymer (binder), the penetrating-substance concentration, the molecular weight of penetrating substances. [13]

All these prerequisites do not consider the chemical reactions running at the diffusion through the film between the diffusing substance and a polymer forming the film structure and at all do not consider the reactions between the diffusing substances and the pigments and fillers contained in the film.[18]

From the corrosion of metal point of view, as it has been already stated, the acid medium is of primary significance, penetrating the protective film not only by physical diffusion, but thanks to the possibility of reaction with pigments, fillers and also binders by chemical diffusion or, maybe, by the physico-chemical art.[19]

Starting with these reasons the work was directed to following the penetration of acid solutions (hydrochloric acid, sulphuric acid, formic acid, acetic acid, and propionic acid) into a chemically resistant binder system - epoxy resin based coatings. Selection of these acids is given by the practical occurrence of hydrochloric and sulfuric acids in "heavy" corrosive atmospheres. This concerns the coatings exposed to atmospheres in chemical and power engineering related plants, but also to contaminated municipal and industrial atmospheres. Organic acids, formic, acetic, and propionic acids, belong among components encountered in foodstuff and also aggricultural plants. Expressive improvement of mechanical and chemical resistances of binders require comparatively low concentrations of fillers or pigments, as far as it is possible to create an appropriate structure of interface. The structure and properties of coatings depend, to a great extent, on the surface properties of pigment particles, on the composition of interface and also on the structure of polymer matrix. Interaction with pigment particles results always in a less or more expressive modification of binder.[20]

Mixed metal oxides as reactive pigments affecting the diffusion of corrosive substances

The spinels involve a large group of double metal oxides characterized by a general formula AB_2X_4 and a crystal structure similar to that of the natural spinel mineral, which is a magnesium aluminate ($MgAl_2O_4$). These mixed metal oxides are solid solutions or compounds consisting of two or more metal oxides.[21] The spinel lattices are rather stable, and the oxides showing this structure are characterized by high thermal and chemical stabilities and a high refractive index.

Thermostable spinel pigments

The outstanding effect of thermal stability of the spinel pigments can be utilized, when also other components of the protective coating exhibit a high resistance to the elevated temperature effects. Of the known polymers which have film-forming properties, the silicone resins are usable as binders for the coating compositions able to resist elevated temperatures. [22]

Experimental

The diffusion methods of investigating the anticorrosive pigments start with the existence of barrier efects and binding the aggresive compounds penetrating the organic coating. The diffusion of corrosively active compounds into the epoxy coating samples prepared was studied by a microscopy method starting with the second Fick law,[23] and independent calculation of the diffusion coefficient (D) was performed by means of the Einstein relation for D. As far as reactive pigments, i.e. the substances capable of chemical reactions with diffusing medium, are used, then the decceleration of diffusion of this medium through the film takes place. Microscopic investigation of the diffusion of aggresive acid medium into pigmented films consisted in measuring the depth of penetration of the medium. When a suitable form of test pieces was used the condition of diffusion in a single direction was satisfied, and the diffusions along the y and z axes could be disregarded.[24] The depth of diffusion can be measured microscopically as a function of time, the time being considered as the time of exposure of the samples to the corrosive medium being investigated.

Anticorrosive action of spinel pigments in coatings

Let us start with the assumption that the primary cause of corrosion at the metal substrate under the protective coating consists in the penetration of surrounding medium through the heterogeneous film created by a polymer matrix with pigment particles which are usually of inorganic nature. From this point of view we can appreciate the importance of the study of transport processes comprising water and corrosive acidic media into the coatings pigmented with spinel pigments. All the binders pigmented to a variable pigment volume concentration (PVC) with spinel like pigments showed a similar behavior for the penetration of acidic (1.0M hydrochloric acid (HCl), 0.5M sulphuric acid (H_2SO_4), and 1.0M acetic acid (CH_3COOH) medium through the coating. The curves obtained indicate that the diffusion of water through the coating layer dependend not only on the kind (defined by chemical composition) of the pigment, but also the pigment volume concentration in the system (Figure 1.). Titanium dioxide, which was used as a representative of inert pigments towards acid media, showed a rising coefficient of diffusion, which entails a higher degree of penetration of coating by the medium and, possibly, a reduction of the resistance of the system to corrosion as a whole. Figure1 depicts the influence of the pigment volume concentration (PVC) on the diffusion coefficient (D) for various spinel-type pigments. The curve/development dependence shows how the cations present in the spinel-crystal lattice affect the transport of acidic medium (0.1M HCl) into the epoxy resin-based coating film. The critical pigment volume concentration (CPVC) values of all the pigments tested lie in a range of 45-55 vol.% and it can be concluded that at such a pigment concentration a sharp growth can be observed for the diffusion of liquid medium. The PVC = 40 vol.% value is an optimum condition for the minimum diffusion of liquids into the coating films.

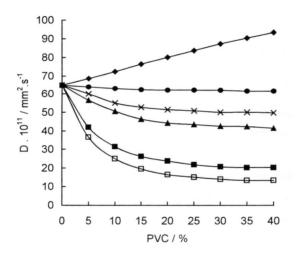

Figure 1. PVC dependence of the coefficient of diffusion (D) for 1.0M HCl into the coatings;

◆ = TiO_2, ● = $ZnAl_2O_5$, × = $TiMgO_5$, ▲ = $ZnFe_2O_4$, ■ = $ZnMgFe_2O_5$, □ = $MgFe_2O_4$.

Quite generally, the transport phenomena take part in all corrosion processes, and the appropriate pigmentation can reduce the diffusion of medium and thus prevent appearance of the primary corrosion cause. The capacity to neutralize the penetrating substances protects also, to a certain extent, the binder, by stopping chemical attacks on the molecular chain thereof. Based on the results obtained a model situation can be assumed, according to which at the surface of spinel pigment particles the salts (of appropriate acids) are formed.[23] This eliminates the chemical diffusion component. Also, an important role is here played by the degree of wetting of the pigment particle by the organic binder. An objective representation of the diffusion of corrosive media at the pigment/binder interface for both inert and reactive (spinel-type) pigment is given in Figure 2.

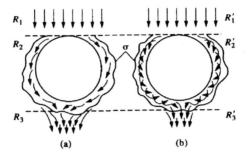

Figure 2. Scheme of diffusion of corrosive media at the pigment/binder interface; (a) inert pigment, (b) reactive pigment.

In the first stage the physical diffusion through the coating takes place, the diffusion running at a rate (R_1). At a moment when the diffusion field contacts an inert pigment tiranium dioxide (TiO_2) particle a change in the rate of diffusion takes place to a rate (R_2), which is determined by the degree of wetting (σ) of the pigment achieved during the dispersion process. The pigment particles act in this case as heterogeneties at a rate of diffusion $R_2 >> R_1$. If we consider the diffusion running only in one direction (along axis x) the rate of diffusion (R_3) will be higher than or equal to the rate R_1. If spinel-type pigments are used then in the phase of diffusion $R_1 = R_1'$. On contacting a reactive spinel and an acidic medium the reaction to the appropriate salt will take place, which also bind several water molecules. The particle surface becomes afterwards a coating consisting of these reaction products, which considerably reduces the diffusion process for the original medium. The rate of diffusion $R_2' << R_2$.

Preparation of the novel lamellar pigments

Due to the fact that the single layers cannot protect the metal effectively it is necessary to create a system consisting of several layers.[25] The top layer of coating system has to secure primarily the protection of coating system to the degradation by ultravioleght

(UV) radiation and in addition to it has an important anticorrosion function.[26] In this connection we speak on the barrier protection. An efficient barrier can be formed only from the binder, which is characterized by a low permeability for liquids and gases, and this function can be improved by using nonisometric, i.e. platelike or also flakelike pigments. The mechanism of active anticorrosion effects can be with barrier mechanism combined also in a single-layer system, which should comprise both the mechanisms.[27] The lamellar-pigment particle pigmented organic coatings are to be considered the composite materials. The matrix (binder) and also reinforcement (pigment) act with a synergic effect on the final physicomechanical properties. The strength, toughness and resistance to temperature and moisture liquid effects at the corrosion loading undergo raising. An important protection factor concerns also the adhesion properties[28] of the coatings formed. The adhesion of coating to the substrate, so as the adhesion of individual layers one to another are tightly conected with phenomena, such as osmotic blistering, peeling and cracking of the coating films. Improving the mechanical properties by using suitable pigments raises the stability of coating at changes in temperature and at mechanical loading. The connection with anticorrosion function of coating consists in a prerequisite that if no mechanical failure of the protective film takes place the corrosion of substrate does not run. The proper logic of anticorrosion protection starts with creating such a barrier by using the coatings to guarantee a tight insulation of the metal substrate from the environment[29]. The anticorrosion properties of these nonisometric pigments consist in a reduction of the velocity of diffusion of the corrosion medium in the direction to substrate metal (Figure 3.).

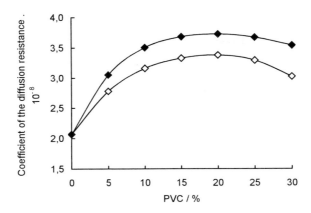

Figure 3. The dependence of coefficient of the resistance to diffusion of 1-C polyurethane coating on the concentration of surface treated iron muscovite; $\diamondsuit$ = iron muscovite, $\blacklozenge$ = lamellar zinc ferrite.

A disadvantage of natural mineral specularite consists primarily in the dark gray to black color and also the sedimentation in the binder resulting from a high specific mass. These disadvantageous negative properties are on the opposite not shown by the muscovite having been subjected to a surface pretreatment (hematite layer).[30]

Another type of such lamellar pigments, verified in a laboratory scale, goes out of the synthesis of primary starting reactants with lamellar particle shapes, which at a cautious calcination passes to a product showing no crystal system changes and thus no changes in their form. As an example zinc ferrite can be named for which as a starting ferric oxide –the specularite (a lamellar ferric oxide) with lamellar particle shapes (Figure 4.).

Figure 4. Morphology of lamellar particle of zincferrite (5. 000x).

Conclusion

The type of pigment used exhibits a considerable effect on increasing the chemical stability, resistance to the penetration of acid medium through the coating and thus also the corrosion protection of protected metal substrate. The results obtained, relating to the diffusion of corrosive medium through the coating pigmented with spinel pigments allow to take some general conclusions: the spinel pigments are active compounds capable to take an active part on the diffusion of corrosive medium through the film. This medium is mostly of acidic nature, or the gaseous substances are concerned. The spinel pigments actively protect the polymer film matrix by the reaction with diffusing medium to the degradative reactions causing the change of polymer structure due to effects of this difusing substance. Based on these conclusions it is possible to imagine the spinel pigments as reactive particles, on the surface of which at the diffusion of corrosive substances through the polymer film chemical reactions take place, which lead to the formation of less corrosive substances. The improvement of physicomechanical properties of the coatings is significant primarily in increasing the adhesion to the substrate material. The higher adhesion means a lower tendency to the formation of osmotic blisters and thus to the subcorrosion effects. The anticorrosion properties of these nonisometric pigments consist in a reduction of the velocity of diffusion of the corrosive medium in the direction to substrate metal. The lamellar surface treatment of muscovite by ferric oxide gives the pigments for the anticorrosive coatings acting as corrosion barriers.

References

[1]   M. Zubielewicz, E. Smieszek, T. Schauer, *Farbe und Lack* **1999**, 105, 136.
[2]   B. Amo, R. Romagnoli, V.F. Vetere, L.S. Hernández, *Prog. Org. Coat.* **1998,** 33, 28.
[3]   G. Adrian, A. Bittner, *J. Coat. Technol.* **1986,** 58, 59.
[4]   M. Takahashi, Polym. *Paint Col. J.* **1984,** 174, 281.
[5]   M. Deyá, V.F. Vetere, R. Romagnoli, B. Amo, *Pigment and Resin Technol.* **2001,** 30, 13.
[6]   B.P.E. Goldie, *Paint and Resin* **1985,** No.2, 16.
[7]   P. Kalenda, *Dyes and Pigments* **1993,** 23, 215.
[8]   P. Kalenda, *Kor. Ochr. Mater.* **1993,** 37, 5.
[9]   Z. Šrank, D. Jiráková, I. Charvát, B. Hájek, *Chem. Prum.* **1990**, 40, 90.

[10] W.C. Johnson, *J.Coat. Technol.* **1985**, 57, 67.
[11] A. Kalendová, *Prog. Org. Coat.* **2000**, 38, 199.
[12] A. Kalendová, *Pigment and Resin Technol.***1998**, 27, 225.
[13] P. Kalenda, *J. Appl. Pol. Sci.* **1992**, 45, 2235.
[14] P. Kresse, *Farbe und Lack* **1978**, 84, 156.
[15] B. Bettan, *Paint and Resin* **1986**, No.2, 16.
[16] O. Vogt, in: *"ACS Symposium Series 689, Organic Coatings for Corrosion Control"* ed., G.P. Bierwagen, Washington, DC, 1998, p.249.
[17] J.D. Keane, W. Wettach, W. Bosch, *J Paint Technol.* **1969**,41, 372.
[18]   P. Kalenda, A. Kalendová, *Dyes and Pigments* **1995**, 27, 305.
[19] P. Kalenda, *Pigment anr Resin Technol.* **2001**, 30, 150.
[20] W.C. Johnson, *J. Coat. Technol.* **1994**, 66, 47.
[21] A. Kalendová, *Anti-Corrosion Methods and Mater.* **1998**, 45, 344.
[22] A. Kalendová, *Pitture e Vernici Eur. Coat.* **2000**, 76, 57.
[23] A. Kalendová, J. Šňupárek, P. Kalenda, *"ACS Symposium Series 689, Organic Coatings for Corrosion Control"* ed., G.P. Bierwagen, Washington, DC, 1998, p.366.
[24] P. Kalenda, *Acta Mechanica Slovaca* **1999**, 3, 103.
[25] E. Carter, *Pigment and resin Technol.* **1986**, 15, 8.
[26] S. Wiktorek, E.G. Bradley, *JOCCA* **1986**, 69, 172.
[27] A. Kalendová, *Pigment and Resin Technol.***2000**, 29, 288.
[28] N. Sprecker, *JOCCA* **1983**, 66, 52.
[29] P. Antoš, *Kor. Ochr. Mater.* **1998**, 42, 58.
[30] P. Kalenda et al., *Prog. Org. Coat. J.* – in press.

*Macromol. Symp. **187**, 109–120 (2002)*

Corrosion Inhibited Metal Pigments

Authors: Alfried Kiehl*, Hans Brendel

Eckart GmbH & Co. KG, R&D Aluflake, 91235 Velden, Germany

Summary: Flake-shaped particles of aluminium are well known in the coatings and printing ink industry as "silver bronze pigments". For their use in waterborne coatings or outdoor applications, an effective corrosion protection of the highly reactive aluminium surfaces is required.
The traditional stabilization techniques for aluminium pigments are based on the addition of corrosion inhibitors or on chromate passivation. This publication presents new developments in the encapsulation of metallic pigments that are based on modern sol-gel techniques. All products are heavy metal-free and provide excellent applicational properties.

Keywords: effect pigments, inhibition, encapsulation, sol-gel, surface modification

Introduction

Flake-shaped particles of aluminium and copper-zinc alloys are an important group of lustre pigments. They are well-known in the coatings and printing ink industry as "silver bronze pigments" and "gold bronze pigments". In their applications these pigments act as small reflectors of light with a spectral response very similar to that of metallic silver or metallic gold.

This paper deals with aluminium pigments that imitate the optical appearance of metallic silver. The major problem in the practical use of these lamellar aluminium pigments is their exothermic reaction with water. Whenever this reaction takes place, the metallic gloss decreases significantly and hydrogen gas is evolved:

$$2\ Al\ +\ 6H_2O\ \rightarrow\ 2\ Al(OH)_3\ +\ 3H_2\uparrow$$

Therefore the highly reactive aluminium surfaces need an efficient corrosion protection whenever aluminium pigments are used in water based coatings or in outdoor applications. In waterborne automotive OEM coatings, the need of corrosion protection of the aluminium pigments is obvious.

     CCC 1022-1360/00/$ 17.50+.50/0

Other compliant coatings like powder or coil coatings have no water in their formulations. For economical reasons, outdoor applications made with these coatings are usually realised as one-coat paints. This means that the pigments in the single coating layer are exposed to all changes of the weather conditions. Since the colour of a coating shall not change for the whole lifetime (as a rule several years), the need of highly stable aluminium pigments becomes evident.

Methods of corrosion inhibition

There have been developed two basic concepts for the stabilization of aluminium pigments.[1, 2] One is the inhibition of the gassing reaction by the addition of suitable corrosion inhibitors (1). These inhibitors or their reaction products are able to adsorb on the active sides of the aluminium surface. Chromate treating leads to the formation of a conversion layer around the aluminium flakes.[3-5]

The second method for stabilization (2) is the full encapsulation of the aluminium pigments with a chemically inert, transparent layer. This protective layer is usually formed by sol/gel processes and can be either of organic or inorganic nature.[1, 6]

1. Inhibition

1.1. Phosphor-organic chemicals

Typical examples of suitable corrosion inhibitors for aluminium flakes are the organic derivatives of phosphoric and phosphoric acid. If necessary, the pH of these chemicals can be adjusted by the addition of suitable amines.

When the inhibitive treatment is completed, the gassing reaction of the aluminium pigments decreases significantly. Nevertheless the passivating layer cannot provide a perfect corrosion protection of the aluminium, but is permeable to some extent. This means that inhibited aluminium pigments show a delayed gassing reaction. More information about inhibited aluminium pigments is described in the literature and references therein.[1, 2, 6]

The phosphor-organic passivation of aluminium pigments provides an acceptable stability of the aluminium flakes in a lot of mild water based coatings. It is recommended for waterborne paint systems in which the pH does not exceed 8.0-8.5. The exact pH limit is depending on the binder/resin type and the choice of neutraliser. At higher pH-values (pH>8.0) and for most outdoor applications, more efficient corrosion protection is required.

1.2. Chromate treatment

A well-known technical process for the passivation of aluminium flakes is the chromate treatment. In this procedure the aluminium pigments are treated with a solution of chromic acid. The aluminium surface is oxidised rapidly and a dense, passivating conversion layer is formed all around the aluminium flakes. This layer is a complex combination of aluminium oxide, chromium oxides and water.[3-5]

Chromate-passivated aluminium pigments have an excellent gassing stability even in aggressive water based coating systems. If necessary, they are able to tolerate pH values up to 9.5. Chromate-treated aluminium pigments have a high performance in the humidity test. Furthermore they show low viscosities in aluminium pigment slurries, which are always the first step in the industrial production of waterborne metallic paints.

The following picture illustrates the gassing characteristics of untreated, phosphated and chromated aluminium pigments. It is evident that the chromated metallic pigments provide the best gassing performance and do not show continuous hydrogen development, even after months of storage in water based paint systems.

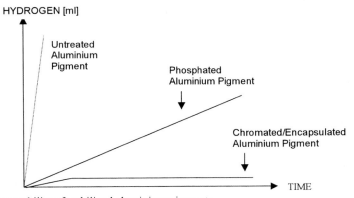

Figure 1.Gassing stability of stabilized aluminium pigments

Sol-gel encapsulated aluminium pigments provide a performance comparable to the chromated flakes. In the next chapter, this latest development will be discussed in detail.

2. Encapsulation

Inorganic coatings of the aluminium flakes with iron oxide, titanium dioxide or aluminium oxide provide unique coloured effect pigments.[1, 6-8] The oxide layers of these pigments are rather porous, for this reason often an additional stabilization is required.

The benchmark for stable aluminium pigments suitable for broad range of compliant coatings are the chromate passivated pigments.[5] Until now alternative, heavy metal-free stabilization methods like silica or polymer encapsulation have been applied successfully in some fields of application (e.g. powder coatings). But all these environmentally friendly stabilizations had considerable limitations in water based and outdoor applications.

This paper presents a new kind of sol-gel process that provides metallic pigments with a chemically stable, fully transparent silica encapsulation. The excellent application properties of these silica-encapsulated aluminium pigments in waterborne coatings will be described in detail.

Furthermore this paper will present a new type of aluminium pigment treated with a corrosion-resistant polymer encapsulation. This innovative technology has remarkable advantages in powder coatings for outdoor applications.

2.1. Silica Encapsulation

Synthesis

The process of silica encapsulation of aluminium pigments starts with the dispersion of the aluminium pigments of choice in an alcoholic solvent.[6] Any grade of untreated aluminium pigments can be selected as a starting material, preferably in paste form.

In the next step, tetraethoxy silane, water and a basic catalyst are added to the stirred suspension of aluminium pigments. The hydrolysis of the tetraalkoxy silane leads to the formation of silanole structures like $Si(OR)_3OH$, $Si(OR)_2(OH)_2$, $SiOR(OH)_3$ and finally $Si(OH)_4$. Under basic reaction conditions, all these silanole intermediates are able to undergo polycondensation reactions that finally lead to the formation of an insoluble SiO_2 network:

$$Si(OR)_4 + 4\,H_2O \xrightarrow{\text{Base}} Si(OH)_4 + 4\,ROH \quad \text{(Hydrolysis)}$$

$$Si(OH)_4 \xrightarrow{\text{Base}} SiO_2 + 2\,H_2O \quad \text{(Condensation)}$$

Many technical properties of effect pigments are related to their surface chemistry. It is a major advantage of the presented sol/gel chemistry to allow a controlled modification of the precipitated silica surfaces. The surface of the deposited layer can be modified by the addition of suitable bifunctional reagents like organosilanes.[9]

Controlled wetting properties

When the first reactive group of the bi-functional surface modifier has reacted with the silica surface of the pigment, the second functionality is still able to interact with the coating. The chemical nature of the second functional group directly affects the wetting properties and the orientation of the aluminium pigments in waterborne coatings. Under the curing conditions, the second functional group is able to react with the resin or other components contained in the coating. Since real chemical bonds can be created between the pigments and the coating, this process significantly improves the humidity resistance and the adhesion of the whole coating.[6,9] Last but not least the volume shrinkage of the coating during the curing process improves the planar orientation of the aluminium flakes.

If the surface treatment of the pigments is achieved with unpolar compounds like alkyl groups, a relatively poor wetting of the pigments in water based formulations results (Fig. 2). The visible effect is a "semi-leafing" with a parallel orientation of the aluminium flakes rather close to coating surface. This means an excellent brilliance of the application, but on the other hand the adhesion might be quite poor.

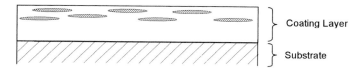

Figure 2. Poor wetting: "Semi-leafing"

If the surface modification is more polar (e.g. acrylic, amino, epoxy, ureido groups), the pigment wetting in waterborne coatings is improved. The orientation of the flakes in coating is still parallel but much more homogeneous in the whole layer (Fig. 3). Usually these conditions are the best compromise between a good pigment orientation and good adhesion properties.

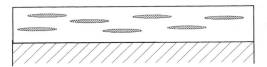

Figure 3. Wetting is OK: Flakes in parallel orientation all over the coating

When the polarity of the pigment surface is even higher, like with pure silica, the parallel orientation of the flakes in the coating starts becoming distorted (Fig. 4). There is a certain tendency of the flakes for precipitation and agglomeration. Consequently a very good pigment wetting in water often leads to a reduced brilliance and less hiding.

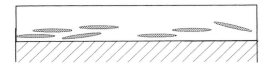

Figure 4. Very good wetting: Flakes tend to precipitate, orientation not strictly parallel

The wetting properties strongly affect the rheology of the pigment, too. A freshly precipitated silica surface contains many OH-groups leading to unacceptable high viscosity levels in most waterborne systems. The controlled end-capping of these OH-groups allows to adjust the rheological properties of the encapsulated aluminium pigments to the needs of a paint manufacturer.

Brilliant optics

When a coating with aluminium pigments is viewed face-on, it appears bright. If the application is turned sideways, the colour appears darker or in a different colour shade. This angle-dependant reflection of light is named "flop" or "two-tone effect". It is a characteristic criteria for all coatings with metallic effect pigments.

The metallic effect depends on the particle size of the flakes, the flake shape and on the wetting of the aluminium pigments in the coating system of choice. By means of the described sol-gel chemistry, pigments with tailor-made wetting properties can be manufactured.

Any passivation or coating of aluminium pigments has an effect on the optical appearance of the pigments, since the reflected light needs to pass the encapsulating layer. This means that the same basic aluminium pigment will show different optical appearances with different encapsulations. Table 1 shows the lightness differential ΔL of some sol-gel encapsulated aluminium pigments that are offered under the name Hydrolan. The ΔL values are calculated from the difference [L (silica passivation) – L (chromate passivation)] for each grade.

Table 1. ΔL values of silica-encapsulated aluminium pigments

Pigment	Particle Size (Cilas)			Lightness Differential ΔL		
	D10	D50	D90	25°	45°	75°
Hydrolan 2192	8	15	26	2,39	0,16	-0,95
Hydrolan 2156	10	17	28	2,38	-1,21	-1,56
Hydrolan 501	8	21	41	4,49	1,48	-1,58
Hydrolan 167	11	25	45	3,62	-1,29	-3,6
Hydrolan 214	17	34	57	3,54	-1	-2,23

All silica-encapsulated grades appear brighter viewed on face (25°) and darker viewed from the edges (75°). This means that the silica encapsulation yields aluminium pigments that do not only meet the optics of chromate passivated pigments, but appear slightly brighter and more "metallic".

Gassing stability

The test conditions for the gassing stability of aluminium pigments in water based coatings are described in detail in the literature.[6] Silica encapsulated aluminium pigments have an excellent gassing stability under these test conditions. Even in very aggressive paint formulations and at high pH values (pH>9) these pigments show no or a very small evolution of hydrogen gas. Their gassing stability is similar to the benchmark, the well-known heavy-metal passivated grades (chromate-passivation).

Controlled humidity resistance

All adhesion tests were performed with a commercial OEM coating system consisting of five layers: Phosphate primer (1), Cataphorethic Dip-Coat (2), Filler (3), Metallic Basecoat (4) and Clear Coat (5). The interaction of all these layers is responsible for the performance of the coating.

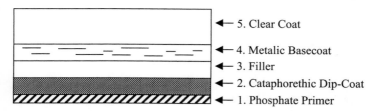

Figure 5. OEM coating

The standard test method for the humidity resistance of aluminium pigments is the condensation water test according to DIN 50017-KK (German standard). An OEM panel is exposed to high humidity conditions (100% with condensation) at 40°C for 240h.

After 10 days of humidity exposure, the adhesion of the coating is checked by the crosshatch test according to DIN ISO 2409. Table 2 clearly points out that the wetting properties of the silica encapsulated aluminium pigments can be controlled by the use of suitable surface modifications.

If silica encapsulated aluminium pigments without any surface modification are used, the humidity resistance of the coating is rather poor (GTC = 4-5). Water is able to penetrate layers with silica encapsulated pigments rather easily. Therefore blisters occur in the crosscut test.

When the silica surface is modified with groups of poor wetting properties (alkyl groups), the humidity resistance is improved, but still insufficient. The best results are obtained with an intermediate wet ability that can be obtained by the modification with amino or acrylic groups.

For automotive applications, the adhesion of a coating is also tested with an impact test with stone-chips or steel balls. An intermediate pigment wetting realised with amino or acrylic surface modifications showed the best performance in this test, too.

Table 2. Humidity resistance of modified pigments with sol-gel Encapsulation

Functionality	Cross Cut		
	1 min	1 h	24 h
Pure SiO_2	Gtc 5	Gtc 4	Gtc 4
Alkyl	Gtc 2	Gtc 2	Gtc1
Amino	Gtc 0	Gtc 0	Gtc 0
Acrylic	Gtc 0	Gtc 0	Gtc 0

Circulation stability

The circulation lines and pumps of the automotive industry can expose aluminium pigments to very high shear stress. This mechanical stress may result in deformations of the aluminium flakes leading to a significant colour shift of the coating. A suitable simulation of this effect can be done in a high-speed dissolver (20.000 rpm for 15 min) or a special kind of mixer called "Warring-Blendor". The silica encapsulation of Hydrolan provides an excellent mechanical stabilisation of the ductile aluminium core. This criteria is called "degrading resistance" and is superior to untreated aluminium pigments and many other passivated grades.[6]

Resistance to weathering

The weathering resistance of the silica-encapsulated pigments was tested under Florida conditions. In this harsh test OEM panels are exposed for two years to the weather conditions of south Florida (panel orientation 5° south). The new silica encapsulated pigments showed an excellent weather fastness under these conditions ($\Delta E < 1$ after two years).

2.2. Polymer Encapsulation

The encapsulation of aluminium pigments with an organic layer is described in the literature and references therein.[1, 2, 10] Recently a new process for polymer encapsulation has been developed. This technology is based on a modified polymerization process of acrylic monomers in the presence of aluminium flakes. In this paper, the encapsulated aluminium flakes manufactured by this process are called "New PCA". This improved polymerisation process yields aluminium pigments with an excellent performance in powder coatings for outdoor applications. All subsequent tests were carried out in a commercial polyester system with a pigmentation level of 4%.

Condensation test

The test conditions for the humidity test were in correspondence with DIN 50017-KK.

Table 3. Humidity resistance of powder coatings

Pigment	Encapsulation	First visible changes after
PCR 8154	Silica old	24h
Sillux 8154	Silica new (Sol gel)	168h
PCA 8154	Polyacrylic old	168h
New PCA 8154	Polyacrylic new	>1000h

The new type of polyacrylic encapsulation showed a superior humidity resistance. A humidity resistance of more than 1000h is considered to be sufficient for outdoor applications.

Kesternich test

The weathering conditions in an aggressive industrial atmosphere are often simulated by the Kesternich test (DIN EN ISO 6988). In this test, the powder-coated panel is exposed to humidity and SO_2 in several cycles.

Table 4. Kesternich test of powder coatings

Pigment	Particle Size D50	Encapsulation	First visible changes after
PCR 501	20μm	Silica old	7 cycles
PCA 501	21μm	Polyacrylic old	7 cycles
New PCA 501 ; V398	21μm	Polyacrylic new	11 cycles
New PCA 501, V535	21μm	Polyacrylic new	10 cycles
PCR 212	47μm	Silica old	12 cycles
PCA 212	53μm	Polyacrylic old	11 cycles
New PCA 212	54μm	Polyacrylic new	12 cycles

The improved polymer encapsulation has clear advantages for aluminium flakes of finer particle sizes. For coarse aluminium flakes, the chemical nature of the encapsulation seems to be of minor or no significance. Obviously the surface area of the aluminium pigments is more important for the results of the Kesternich test. Fine pigment particles have a much higher surface area, for instance the BET value of the new PCA 501 is around $6m^2/g$, whereas the values of the new PCA 212 are around 1 m^2/g.

Chemical resistance

For testing the chemical resistance of the pigments the powder coated panels are treated with drops of 1 M HCl, 10% HCl and 1 M NaOH. After 3 hours the effect of the corrosive agents on the pigment is visually examined.

The best samples of the New PCA showed no difference between the treated and untreated areas. Due to the short time of the treatment with chemicals, there occurs no significant deterioration of the base coat. Therefore this test is suitable for screening the chemical stability of corrosion inhibited aluminium pigments in powder coatings.

Table 5. Treatment of powder coatings with chemicals

Chemical Test with	10 % HCl	1 M HCl	1 M NaOH
Pigment	First changes after:		
PCA	30 min	1 h	5 min
PCR	30 min	1,5 h	5 min
New PCA	3 h	no change	no change

The chemical nature of the encapsulation proved to be the decisive criteria in this test, whereas the particle size distribution of the aluminium flakes was of minor importance.

It is obvious that the chemical resistance of the new PCA is far ahead of the resistance of standard silica (PCR)- or acryl ate (PCA)-encapsulated aluminium pigments. This opens up a wide scope of new applications.

Conclusions

The presented procedure of silica encapsulation provides a new class of aluminium pigments for waterborne coatings. These pigments have an excellent gassing stability, a very good mechanical stability and a high brilliance compared to the chromate-passivated types (flop). A patented surface-treatment allows to manufacture encapsulated aluminium pigments that are suitable for a broad scope of coating systems.

Furthermore there has been made remarkable progress in the polymer encapsulation of aluminium flakes. This new kind of aluminium pigments was especially developed for powder coatings that are used in corrosive media or in outdoor applications.

The examples in the article demonstrate that there are practically no limits for the use of aluminium pigments in coatings. Modern inhibition methods can provide a suitable stabilization of the aluminium flakes in almost any application.

References

[1]   W. Reisser, A. Fetz, E. Roth in "Waterborne Coatings and Additives", 180-189, The Royal Society of Chemistry, Cambridge 1995.
[2]   P. Wissling, Surface Coatings International, Vol.82, 335-339, 1999.
[3]   J. Ruf, "Organischer Metallschutz", 641-645, Vincentz Verlag, Hannover 1993.
[4]   K.-P. Müller, „Praktische Oberflächentechnik", 101-104, Vieweg Verlag, Wiesbaden 1999.
[5]   BASF, EP 0 259 592 (1987).
[6]   A. Kiehl, K. Greiwe; Progress in Organic Coatings, Elsevier, Vol. 37 (1999), 179-183.
[7]   Eckart, DE 195 20 312 (1995).
[8]   H. Birner, K. Greiwe, Coating 11 (1997), 432.
[9]   Patent pending, DE 198 20 112 A1 (1998).
[10] Eckart, EP 0 477 433 (1990).

*Macromol. Symp. **187**, 121–127 (2002)*   121

Investigation of Filiform Corrosion of Coated Aluminium

Andrea Rudolf, Wolf-Dieter Kaiser

Institut für Korrosionsschutz Dresden GmbH, Gostritzer Straße 61-63, 01217 Dresden, Germany

Summary: Filiform corrosion on different aluminium alloys, coated with a clear varnish has been studied. Image analysis and metallographics were used to characterize the corrosion. It has been shown, that filiform corrosion consists of a lateral propagation of filaments and an attack on the aluminium under the filaments. The extent of filiform corrosion is dependent on the sort of alloy and the pretreatment of metall surfaces.

Keywords: Aluminium alloys, filiform corrosion, pretreatment, image analysis, coating

Introduction

Filiform corrosion is a particular type of localized corrosion of coated aluminium and occurs by the following conditions:

- at a wet environment with relative humidity of 40-90 %

- at the presence of ionic substances, for instance chlorides

- at the presence of defects in the coating [1-8].

Filiform corrosion is characterized by a lateral propagation of filaments and an attack under the filaments on the substrat [9,10]. Figure 1 and 2 show two examples.

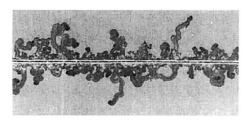

Figure 1. Lateral propagation of filaments of EN AW 2024, T42, coated with a 2K-PUR-AY coating

The aim of this work was to investigate the influence of the substrat material and the treatment of surfaces of the filiform corrosion of coated aluminium alloys. Different pretreatments of surfaces were used: alkaline degreasing, etching, yellow chromating, treatment with zirconiumfluoride

     CCC 1022-1360/00/$ 17.50+.50/0

and anodizing. It was used a clear coating material and many aluminium alloys used in industrial manufacturing.

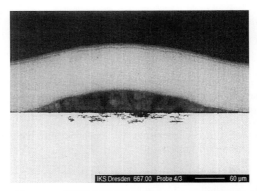

Figure 2. Attack under a filament of EN AW 6082, T4, coated with a polyester powder coating

Experimental

<u>Substrates</u>: Aluminium alloys used for this investigations are shown in table 1.

Table 1. Aluminium alloys

aluminium with high purity	Al99,999
EN AW 2XXX	EN AW 2017A, O EN AW 2017A, T42 EN AW 2024A, O EN AW 2024, T42 EN AW 2024, T62
EN AW 5XXX	EN AW 5005, H14 EN AW 5754, O EN AW 5754, F22 EN AW 5754, H22 EN AW 5182, O EN AW 5182, H19
EN AW 6XXX	EN AW 6060, extruded EN AW 6016, T4 EN AW 6082, T4 EN AW 6082, T651
EN AW 7XXX	EN AW 7020, T6 EN AW 7075, T76

Pretretment of metall surfaces [10]:

alkaline degreasing:	P3 almeco 20/HNO$_3$
etching:	P3 almeco 20/P3 almeco 40, NaOH/ HNO$_3$
yellow chromating:	P3 almeco 20/P3 almeco 40, NaOH/ HNO$_3$/Alodine C6100
Zr-F-Polymer:	P3 almeco 20/P3 almeco 40, NaOH/ HNO$_3$/Alodine 4830/31
anodizing:	P3 almeco 20/P3 almeco 40, NaOH/ HNO$_3$/H$_2$SO$_4$ (GS)

Rinsing processes were carried out, but they are not mentioned.

Coating:	2K-PUR-AY (clear varnish)

Loading:	Filiform corrosion was initiated by one-hour treatment with hydrochloric acid in according to DIN EN 3665 and following storage at 40°C and 82% relative humidity.
Exposure time:	2000 hours
Test procedure:	Image analysis of propagation of corrosion filaments and metallografical analysis of the attack on the substrat were carried out.

Results

Both phenomena - lateral propagation and appearence of corrosion within the substrat - are dependent on the pretreatment of metall surfaces and on the metal itself. The following figures 3 - 7 shows the lateral propagation of filaments.

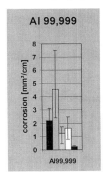

Figure 3. Al99,999

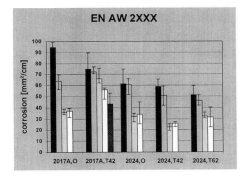

Figure 4. EN AW 2XXX

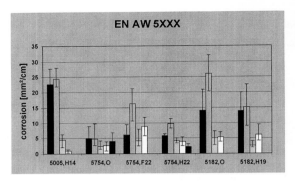

Figure 5. EN AW 5XXX

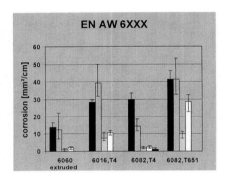

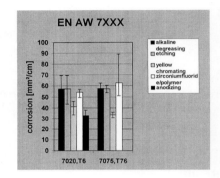

Figure 6. EN AW 6XXX   Figure 7. EN AW 7XXX

The formation of filiform corrosion is dependent on the pretreatment of metall surfaces. The best method to reduce filiform corrosion is to anodize, following by yellow chromating or treatment with zirconiumfluorides. Extensive filiform corrosion was also found when metal surface was only degreased or etched.

Lateral filiform corrosion and depth of the filiform corrosion attack are dependent on the metal. The chemical composition of the alloy plays an important role.

Alloys with copper and zinc have a high susceptibility to filiform corrosion. Aluminium with high purity shows a high filiform corrosion resistance.

The attack within the substrat has a special appearance. There are some alloys, they show intergranular attack (EN AW 2017, EN AW 6016), some alloys show roughening of the surface (EN AW 5005, EN AW 6006) and other show only even uniform corrosion (Al99,999). In some cases shallow pit building occurs (EN AW 5182, EN AW 7020). In the first case of intergranular attack precipitations on grain bonderies are important (table 2, figure 8-10).

Table 2.Examples of corrosion attack under the filaments

substrat	pretreatment of metall surfaces	attack under the filament	
		Appearance image of attack	max. depth [μm]
Al99,999	etching	uniform attack	<5
	yellow chromating		
	Zr-F-polymer		
	anodizing		
EN AW 2017A T42	etching	intergranular attack with grain disintegration	120
	yellow chromating		120
	Zr-F-polymer		130
	anodizing		140
EN AW 5182 H19	etching	shallow pit formation	35
	yellow chromating	intergranular attack and pitting	100
	Zr-F-polymer		100
EN AW 6016 T4	etching	intergranular with formation of pits	110
	yellow chromating		110
	Zr-F-polymer		130
EN AW 6060 extruded	etching	roughening	10
	yellow chromating		5
	Zr-F-polymer		10
EN AW 7020 T6	etching	roughening, shallow pit formation	25
	yellow chromating		25
	Zr-F-polymer		25
	anodizing		55
EN AW 7075 T76	etching	intergranular with grain disintegration	50
	yellow chromating		60
	Zr-F-polymer		55
	anodizing		80

126

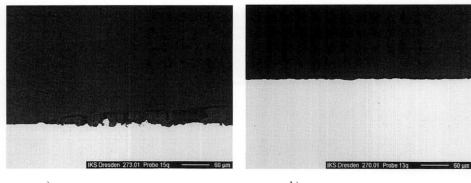

a) b)

Figure 8. Attack under the filament on EN AW 5005, H14
a: etching
b: Zr-F-polymer

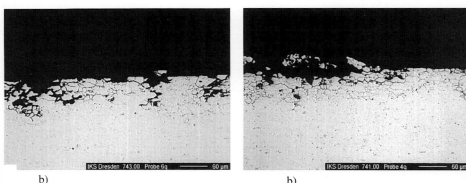

b) b)

Figure 9. Attack under the filament on EN AW 2017, T42
a: etching
b: Zr-F-polymer

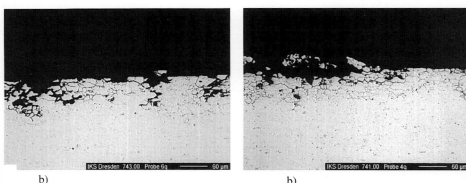

a) b)

Figure 10: Attack under the filament on EN AW 6016, T4
a: etching
b: Zr-F-polymer

Acknowledgements

This projekt was suported by the department of economy in Germany and the AiF (project number 11804B) . The autors are grateful to FPL, Stuttgart for performing image analysis.

References

[1]   Kaesche, H.: Werkstoffe und Korrosion 11(1959), S. 668 - 681
[2]   Ruggeri, R.T.; Beck, T. R.: Corrosion-Nace 39(1983), S. 452 - 465
[3]   Boutista, A.: Prog. Org. Coat. 28(1996), S. 49 - 58
[4]   Schmidt, W.; Stratmann, M.: Corrosion Science, 40(1998)8, S. 1441 - 1443
[5]   Haagen, H.; Rihm, K.-H.: Farbe + Lack 96(1990), S. 509 - 513
[6]   Haagen, H.; Gaszner, K.; Scheck, K.: Farbe + Lack 97(1991), S. 306 - 310
[7]   Pietschmann, J. E.: JOT (1993)11, S. 74 - 79
[8]   Defrancq, J. N.: Estal Eurocoat Symposion, Cannes 16.09.1994.
[9]   Rudolf, A.; Kaiser, W.-D.:   Aluminium 72(1996)10, S. 726 - 733
                                      Aluminium 72(1996)11, S. 832 - 835
[10] Rudolf, A., Kaiser, W.-D.: JOT 42(2002)4, S.98 – 103

Corrosion Processes below Organic Coatings

Sigunde Pietsch, Wolf-Dieter Kaiser

Institut for Corrosion Protection Dresden, Gostritzer Str. 61-63, 01217 Dresden, Germany

Summary: The protective mechanisms of paint systems of a 1-pack polyurethane- and an epoxy/2-pack polyurethane-coating system with zinc dust priming coats were investigated on blast-cleaned and on hand-cleaned steel substrates. The coated panels were exposed to the salt spray test and to a cyclic alternating test (VDA 621-415). The protective effect was assessed in determining adhesion, undermining at scratches, water uptake and the corrosion potential.
On blast cleaned steel substrates the adhesion of the investigated coating systems was not influenced by water uptake of the coatings. Scratches are especially cathodically protected. On hand-cleaned steel surfaces the rust layer between steel substrate and coating can participate in the corrosion process with rust reduction as cathodic partial reaction. The change of rust morphology is the reason for the loss of adhesion of coating. At scratches rust reduction takes also place at the edge of the defect which is independent from pigments of the base coating.

Keywords: adhesion, coatings, electrochemistry, zinc dust pigments, dielectric properties

Introduction

Corrosion processes below organic coatings are responsible for widespread damages of coated steel constructions. Normally these proccesses are electrochemical reactions. Therefore electrochemical methods are suitible for investigation of such processes. In this paper classical and electrochemical methods are used for investigation of the protective mechanisms of corrosion-protective paints.

Experimental

Paint systems

Paint systems of a moisture cured 1-pack polyurethane- and an epoxy/2-pack polyurethane-coating system with zinc dust priming coats were investigated on blast-cleaned (preparation grade Sa $2^1/_2$) and on hand-cleaned (preparation grade St 2) steel substrates (table 1).

CCC 1022-1360/00/$ 17.50+.50/0

Table 1. Paint systems

System-No.	Priming coat		Top coat	Paint system
	Binder	Pigment	Binder	Dry Film Thickness µm
1	1-pack polyurethane	zinc dust	1-pack polyurethane	200
2	epoxy	zinc dust	epoxy / 2-pack polyurethane	240

Exposure conditions

The coated panels were exposed to the neutral salt spray test (DIN ISO standard 7253) and to a cyclic alternating test consisting of periods of neutral salt spray, humid atmosphere and dry atmosphere (VDA standard Nr. 621-415).

Assessment of the protective effect

The protective effect was assessed in determining adhesion, undermining at scratches, wateruptake and the corrosion potential (table 2). All of these parameters were investigated in dependence of the exposure time.

Table 2. Paint system assessment

Parameter	Method	Standard, Literature
breaking strength (adhesion)	pull-off test	ISO 4624
undermining at scratches	measurement of the width of corrosion across the scratch	ISO 12944-6
water uptake	measurement of changes in dielectric constants by Impedance spectroscopy and calculation of water uptake using the empirical relation by Brasher	[1, 2]
potential	measurement of the potential of the steel/paint interface with the Scanning Kelvinprobe	[3, 4, 5, 6]

Results

The obtained results are shown in Fig. 1 to 7.

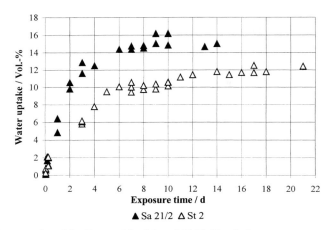

Figure 1. Water uptake of the System No. 1 in a 5 % NaCl solution

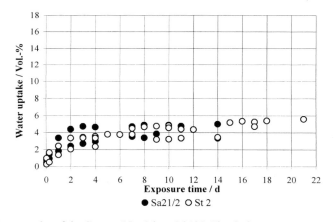

Figure 2. Water uptake of the System No. 2 in a 5 % NaCl solution

The water uptake in the saturated state of the system No.1 (Fig. 1) is 14 Vol.-% and of the system No. 2 (Fig. 2) 4 Vol.-%. In case of system No. 1 it seems to exist a dependence on surface preparation before painting.

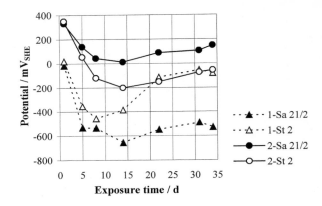

Figure 3. Potential of the steel/paint interface in dependence of the exposure time in humid atmospheres (rF>90 %) measured with the Scanning Kelvinprobe

On blast cleaned steel substrates (Sa $2^1/_2$) the potential (Fig. 3) of

- the coating system No. 1 is close to that of of a free corroding zinc/electrolyt interface,
- the coating system No. 2 is rather positive as the potential of a dry passivated metal surface.

On hand-cleaned (St 2) steel substrates the Fe^{3+}/Fe^{2+}-redox potential was observed for both coatings.

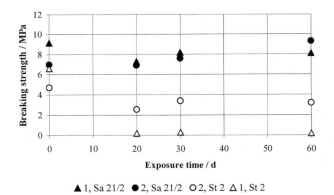

Figure 4. Adhesion of the paint systems exposed to the neutral salt spray test

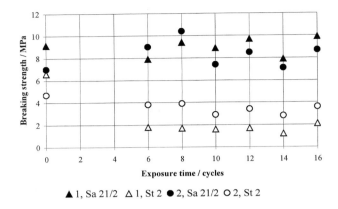

▲ 1, Sa 21/2 △ 1, St 2 ● 2, Sa 21/2 ○ 2, St 2

Figure 5. Adhesion of the paint systems exposed to the cyclic alternating test VDA 621-415

The adhesion of the coating systems No. 1 and No. 2 on blast cleaned steel substrates is not affected during the exposition in both tests (Fig. 4 and 5). On hand cleaned steel surfaces the adhesion of both coating systems decrease drastically. The break takes place in the rust layer. In the case of loading by neutral salt spray test the adhesion of coating system No. 1 is completly lost and by cyclic alternating test VDA 621-415 very much reduced. The adhesion of the coating system No. 2 ranges between 3 and 4 MPa in both tests.

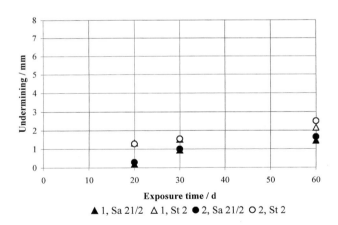

▲ 1, Sa 21/2 △ 1, St 2 ● 2, Sa 21/2 ○ 2, St 2

Figure 6. Undermining at scratches of the paint systems exposed to the neutral salt spray test

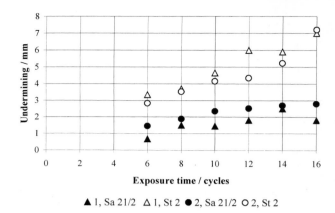

Figure 7. Undermining at scratches of the paint systems exposed to the cyclic alternating test VDA 621-415

The undermining at scratches of the investigated coating systems is higher on rusty then on blast cleaned steel surfaces (Fig: 6 and 7). The differences are more marked in the cyclic alternating test VDA 621-415 than in the neutral salt spray test.

Discussion and Conclusion

On blast cleaned steel substrates the adhesion of the investigated coating systems was not influenced by water uptake of the coatings, even in case of very high water uptake by moisture cured 1-pack polyurethane-coating system. Scratches are especially cathodically protected at their edges by anodic dissolution of zinc as part of the zinc dust primer.

On hand cleaned steel surfaces the rust layer between steel substrate and coating can participate in the corrosion process with rust reduction as cathodic partial reaction. In the presented paper the high water uptake of the moisture cured 1-pack polyurethane coating system enables rust reduction and thus, the change of rust morphology and the loss of adhesion of coating [7, 8]. At scratches rust reduction takes also place at the edge of the defect in form of a cathodic partial reaction which is independent from pigments of the base coating. This reduction is increased by the galvanic cell. In case of cyclic alternating loading reduced rust is reoxidized during dry phases and can participate again in the process during wetting [9, 10]. This is the reason for higher undermining at scratches in the cyclic alternating test (VDA standard 621-415).

References ·

[1]   F.M. Geenen, H.J.W. Lenderink: XX FATIPEC-Kongress Nizza 1990, 173-179
[2]   F.M. Geenen, J.H.W. de Wit: farbe + lack 98 (1992) 1, 9-13
[3]   A. Leng; H. Streckel, M. Stratmann: Corrosion Science 41 (1999) 547-578, 579-597, 599-620
[4]   M. Stratmann, A. Leng, W. Fürbeth, H. Streckel, H. Gehmecker, K.-H. Große-Brink-haus: Progress in Organic Coatings 27 (1996) 261-267
[5]   M. Stratmann, M. Wolpers, H. Streckel, R. Feser: Ber. Bundesges. Phys. Chem. 95(1991)11, S. 1365 – 1375
[6]   M. Stratmann; H. Streckel; R. Feser: farbe + lack 97(1991), S. 9-13
[7]   K. Hoffmann, M. Stratmann: Corrosion Science 34 (1993) 1625-1645
[8]   M. Stratmann, K. Hoffmann, J. Müller: Werkstoffe und Korrosion 42 (1991) 467-472
[9]   S. Pietsch, W.-D. Kaiser: FARBE&LACK 107 (2001) 6, S. 141-150
[10] S. Pietsch, W.-D. Kaiser, M. Stramann: Materials and Corrosion 53 (2002) 5, 299-305

Sugar Latexes as a New Type of Binder for Water-Based Paint and Coating

M. Al-Bagoury, B. Vymetalikova, Emile-Joseph Yaacoub*

Institute of Technical Chemistry, Department of Carbohydrate Technology,
Technical University of Braunschweig, Langer Kamp 5, D-38106 Braunschweig,
Germany; E-mail: e.yaacoub@tu-bs.de

Summary: Sugar latexes based on saccharid derivatives, such as 3-MDG, 1- or 3-MDF and ITDF, have been synthesized in batch and semi-continuous emulsion polymerization. The polymerizations were carried out at 60 or 70°C, initiated by potassium peroxodisulfate, (KPS), in the presence of either ionic or non-ionic surfactant. The effect of the type and concentration of the surfactant and the type of polymerization process on the colloidal and rheological properties was studied. It was found that the particles size increased with using a non-ionic surfactant. Monodisperse particles were obtained by using SDS below its CMC, and smaller polydisperse latexes were observed when the SDS conc. was above the CMC. The latexes exhibit different non-Newtonian flows depending on the solid content and on the additives.

Keywords: sugar monomers; emulsion polymerization; sugar latexes; rheology

Introduction

Water polymer dispersions based on renewable present a new type of binders for a wide range of applications especially for paint and coating. From point of view of environment protection, sustainable development and also new properties, we believe that these polymer dispersions present a high challenge for chemists and industries. Furthermore, they offer a serious alternative for limited petrochemical resource.

Low molecular carbohydrates such as Saccharose (from sugar beet), D-glucose (from starch) and D-fructose (from innulin), are available in a large industrial scale, low cost, exhibit high pure grades quality and permanent reproducibility. They are suitable as sources for sugar monomer synthesis. Figure 1 shows some examples of sugar (meth)acrylates such as 3-MDG, 1-MDF, 3-MDF and itaconic ester, ITDF.

CCC 1022-1360/00/$ 17.50+.50/0

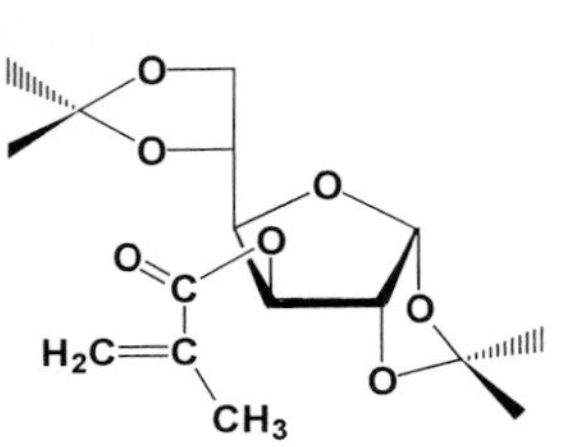

3-O-Methacryloyl-1,2:5,6-di-acetone-glucose
(3-MDG)

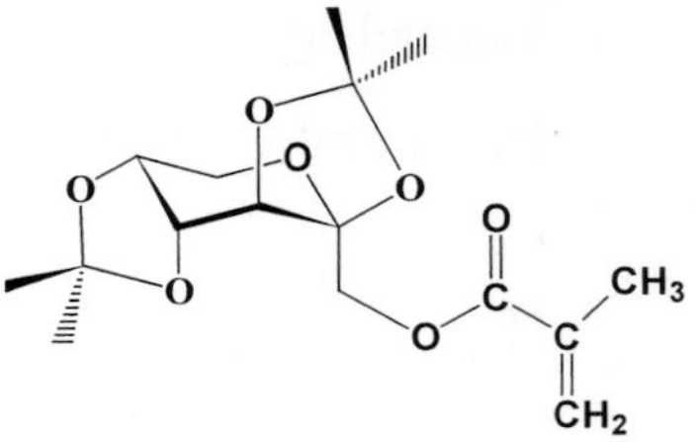

1-O-Methacryloyl-2,3:4,5-di-acetone-fructose
(1-MDF)

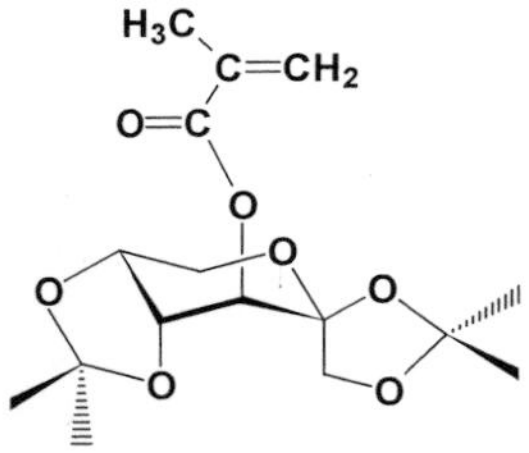

3-O-Methacryloyl-1,2:4,5-di-acetone-fructose
(3-MDF)

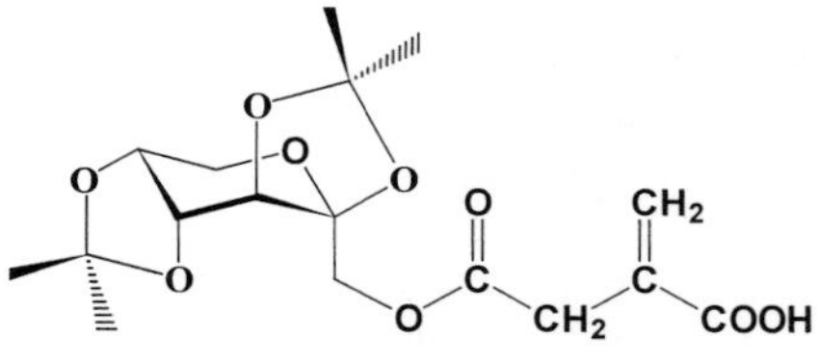

Itaconic- 4-(2,3:4,5-di-O-isopropylidene-ß-D-fructopyranose)ester
(ITDF)

Figure 1. Sugar monomers based on monosaccharides.

As comonomers for radical emulsion polymerization either the commercially available petrochemical monomers, such as butyl and 2-ethyl hexyl (meth)acrylate, or comonomer based on fats and oils, such as fatty alcohol (meth)acrylate or vinyl esters, could be choice for tailor made polymeric materials.

The use of such renewable based monomers in free radical emulsion polymerization, using various processes such us batch, semi-continuous or seed polymerization, lead to new tailor made polymer dispersions with various structures and morphologies. These nanoparticles exhibit interesting thermal, mechanical and rheological properties.[1-4] Example of copolymer latexes with different colloidal, chemical and physical properties are discussed.

Batch emulsion copolymerization

Batch emulsion copolymerization of 3-MDG and butyl acrylate (BA) was investigated at 70°C. First, the monomer reactivity ratios, $r_{(3-MDG)}$ and $r_{(BA)}$, were determined at 10 wt.-%

solid content, using 0.5wt.-% potassium persulfate (KPS), and 5 wt.-% sodium dodecyl sulfate (SDS). The emulsion copolymerizations were carried out using different mixture of 3-MDG and BA in feed. The reactions were quenched at low-conversion, the copolymer composition were determined from ^{1}H NMR and the monomer reactivity ratios were calculated. The values were found to be $r_{(3\text{-}MDG)} = 2.01$ and $r_{(BA)} = 0.54$.[3] In that case, 3-MDG is more reactive than the BA and consequently the polymer particles are richer with the harder segment (3-MDG) on the interior and the softer segment (BA) on the exterior. The Figure 2 shows the schematic structure of the copolymers.

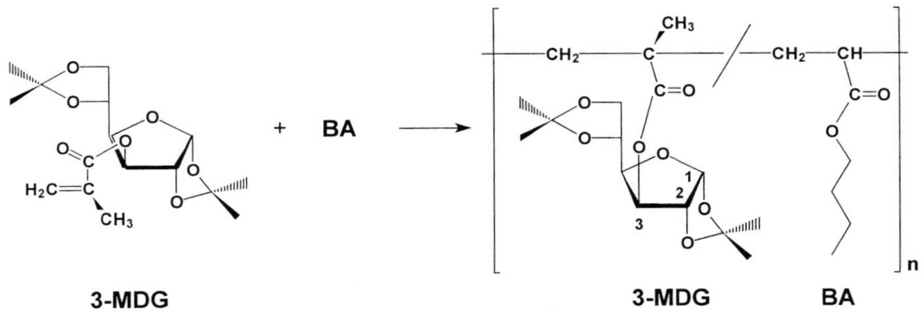

3-MDG + **BA** → **3-MDG** **BA**

Figure 2. Schematic presentation of 3-MDG/BA copolymer.

Sugar latexes were synthesized using other comonomers than alkyl (meth)acrylate. Itaconic esters derived from fatty alcohols and itaconic acid were used as comonomers. The influence of the aliphatic structure of fatty alcohol esters on the rate of polymerization was studied. The emulsion polymerization of 3-MDG and itaconic acid pentyl ester (ITSPE) was carried out at 70°C, at 15 wt.-% solid content and using fatty alcohol polyether sulfate as ionic surfactant (Disponil FES® 32 IS, Cognis). The reaction was followed by gas chromatography. Figure 3 shows the monomers and the total conversion versus time. It can be observed that the sugar monomer is more reactive than the itaconic ester. After 6h the polymerization is nearly complete. More than 99% yield was obtained.

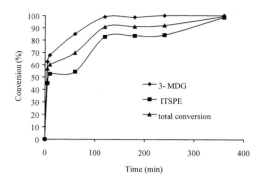

Figure 3. Time-conversion curve of batch emulsion copolymerization of 3-MDG/ITSPE.

Thermal behavior

Figure 4 shows the relationship between glass transition temperature (T_g) of (3-MDG/BA) copolymers and the sugar content in the corresponding copolymers. As expected, a small quantity of glucose units in the copolymer is sufficient to enhance obviously the glass temperature. Furthermore, from the plot it can also be observed that the glass temperature of the copolymers is strongly dependent on the polymer composition. The T_g increased nearly linearly with increasing the sugar moieties up to 100 mol-% in the copolymers Due to the rigidity of the sugar units as cyclic structure, it is though that saccharide moieties affect the flexibility of the polymeric chain and consequently enhance the T_g.

Semi-continuous emulsion copolymerization

Four experiments were carried out with different monomer addition rates (R_m) in order to study the effect of addition rate on the polymerization rate. The solid content was 20 wt.-% and SDS was used as an ionic surfactant with a concentration of 5 g/L. The reaction was carried out at 70°C under mild agitation. A two titrating pumps were used to feed separately the monomer mixture (3-MDG/BA; 35/65 mol%) and the initiator solution. The flow rates of these streams were in the range from 0.12 to 0.44 g/min and were kept constant during all the polymerization. The monomer feed was low enough to ensure monomer starved conditions.

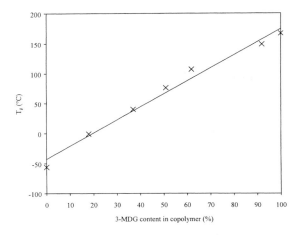

Figure 4. Effect of the sugar content in the (3-MDG/BA) copolymers on the glass transition temperature T_g.

Figure 5 shows the conversion-time curves at different monomer addition rate. The reaction rate in batch process is very high, more than 95% total conversion was reached after only 15 min. In semi-continuous reaction, the polymerization rate (R_p) is strongly depending on R_m. Under these conditions, R_p increases with increasing R_m. At the beginning of the addition of monomer, a finite time is required before the steady-state value of $[M]_p$ is reached. Initially, therefore, there is a gradual increase in R_p as the value of $[M]_p$ increase to its steady-state. R_p was calculated experimentally from the slopes of the nominally linear regions of the total conversion. It was found that the monomer addition rate (R_m) and the calculated reaction rate (R_p) are the same, which means all the reactions were in starved-feed condition.

Effect of the type and concentration of surfactant

The effect of the type and the amount of surfactant on colloidal characteristics, namely particle size (D) and particle size distributions (PSD) and rheological properties of the sugar latexes were investigated. Different anionic surfactants such as sodium dodecyl sulfate (SDS, Merck), sodium lauryl ether sulfate (SLES, Henkel), sodium alkylphenyl polyglycol ether sulfate (NOS10, Wittco) as well as the non ionic surfactant nonylphenyl polyglycol ether (HV25, Wittco) were used above their critical miccellar concentration (CMC). It was found that the surfactant type has a little influence on the reaction rate. Nevertheless, the number of particles and the particle size (D) are strongly related to the type of surfactant.

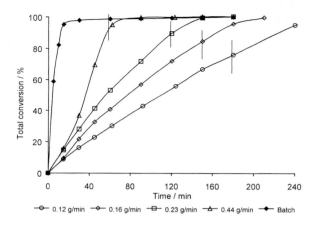

Figure 5. Effect of the monomer addition rate on the total conversion against time for batch and semi-continuous emulsion copolymerization of (3-MDG/BA) at 70°C.

By using anionic surfactants, the particles size slightly enhances during the reaction (Figure 6). The D are smaller by using ionic surfactants than that of the case of non ionic surfactant HV25. This is due to the art of particles stabilization, which could be ionically or sterically nature. The ionic stabilization of the formed particles results from the negatively head charge of the surfactant as well as from the sulfate charge, which arises from the initiator derived polymers end-group. This leads to the increasing of the surface charge density of the particles and allowed a high stability through greater mutual repulsion forces. In the case of HV25, in which the polymer particles are stabilized sterically, the D increased drastically during the polymerization reaction. This could be due to the poor stability of the polymer particles. These are not enough stabilized by the non ionic surfactant to counteract the increase in the ionic strength. It is also conceivable that this increase in particles size may cause the burial of more HV25 surfactant within the particles. As consequence coagulation takes place and the size of particles increases.

Furthermore, the concentration of the surfactant has a great effect on the particles size (D) and their distributions (PSD). Two sugar latexes have been synthesized by semi-continuous emulsion copolymerization, using SDS as ionic surfactant at two different concentrations, above and below the CMC. Figure 7 clearly shows this effect on the particle size and the PSD

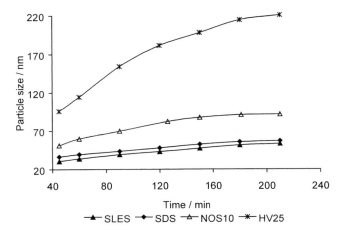

Figure 6. Effect of the surfactant type on the particle size (D) against time for semi-continuous polymerization of (3-MDG/BuA); SDS, SLES, NOS10 and HV25 were used as surfactant with a concentration 5 g/L.

measured by transmission electron microscopy (TEM). Below the CMC, monodisperse particles were obtained with D of 120 nm and a PSD of 0.04. Above the CMC, the particles are smaller (D= 61 nm) and polydisperse (PSD = 0.07).

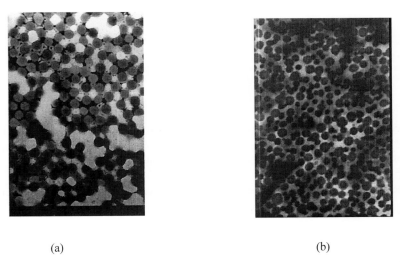

(a)   (b)

Figure 7. TEM of (MDG/BA) latex synthesized by semi-continuous emulsion polymerization; (a) [SDS]<CMC, D = 120 nm and PSD= 0,04 ; (b) [SDS]>CMC, D = 61 nm and PSD = 0,07.

Rheological properties

It is well known that the rheological properties of the latexes are strongly dependent on the particle size, the particle size distribution, the solid content and from other factors such as the functional groups on the particle surface. The effect of these parameters on the rheological behavior of latexes is of great practical importance. The flow behavior of (3-MDG/BA) latexes, synthesized by batch emulsion polymerization at 30 wt.-% solid content, with 5 wt.-% acrylic acid (AA), was investigated. The shear stress-shear rate curves of the sugar latexes at room temperature are shown in Figure 8 a and b. Both latexes exhibit non-Newtonian flow.

Before and after neutralization, the shear stress (τ) and shear rate ($\dot{\gamma}$) are not linearly related. Before neutralization, the curve flow shows that the shear stress increases by increasing the shear rate and the viscosity increases with increasing shear rate (Figure 8a). After neutralization, τ increases by increasing $\dot{\gamma}$ rate and the viscosity decreases with increasing shear rate (Figure 8b).

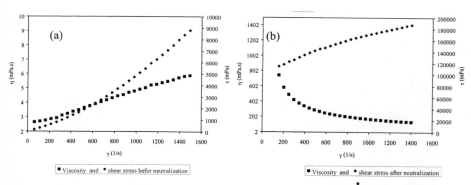

Figure 8. Evolution of shear stress (τ)-viscosity (η) versus shear rate ($\dot{\gamma}$), for a sugar latex at 25 wt.-% solid content. 3-MDG and BA were polymerized with 5wt.-% AA. Measurements were carried out at room temperature. (a) before neutralization; (b) after neutralization.

It can be assumed that in the first case, the behavior characterizes pseudoplastic flow or shear thinning and in the second case, a shear thickening flow (dilatent).

The flow behavior of sugar latexes at different solid contents (Ts) was investigated at room temperature. Figure 9 shows the different flow behaviors at 20 and 50 wt.-% solid content. At low solid content the sugar latex exhibits a Newtonian flow and the measured latex viscosity at the shear rate 1000 s^{-1} was 3.5 mPa.s. At higher solid content (50%), the sugar latex exhibits a non-Newtonian flow, e.g. shear thinning, in which the viscosity decreases with increasing shear rate. The constant, minimum value of the latex viscosity was 150 mPa.s at the shear rate of 1000 s^{-1}.

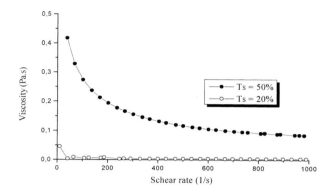

Figure 9. Evolution of the viscosity (η) versus shear rate ($\dot{\gamma}$), for two sugar latexes at 20 and 50 wt.-% solid content. Measurements were performed at room temperature.

Conclusions

It has been shown that the sugar monomer polymerized much faster than the BA and ITSPE. The glass transition temperature of the copolymers are linearly related to the monomer mixture in feed. The particle size and particle size distribution of the sugar latexes were strongly dependent on the type and concentration of the surfactant and on the type of polymerization process (batch or semi-continuous). The investigation of the rheological properties showed that the sugar latexes exhibit a non-Newtonian flow, e.g. shear thinning, at high solid content. Furthermore, two different non-Newtonian flows were observed depending on the pH of the latex solution when acrylic acid is used. Shear thickening flow and pseudoplastic flow or shear thinning were obtained before and after neutralization respectively.

References

[1] E.-J. Yaacoub, *German Patent*, DE 19945236.9 (**1999**).
[2] U. Koch, E.-J. Yaacoub, *Macromol. Chem. and Phys.*, **2002**, in print.
[3] U. Koch and E.-J. Yaacoub, *J. of Polym. Sc., Part A*, **2002**, submitted.
[4] M. Al-Bagoury, E.-J. Yaacoub, *Polymers*, **2002**, in print.

Macromol. Symp. 187, 147–154 (2002)

Forecasting and Optimisation of the Waterborne Anticorrosive Paint Composition with Non-Toxic Pigments

Małgorzata Zubielewicz, Elżbieta Kamińska*, Jacek Bordziłowski**,
Thadeus Schauer***

* Institute of Plastics and Paint Industry, Chorzowska 5044-100 Gliwice, Poland

** Technical University, Narutowicza 11, 80-952 Gdańsk, Poland

*** Forschungsinstitute fur Pigmente und Lacke, Allmandring 37, 70569 Stuttgart, Germany

Summary: The investigations of the waterborne anticorrosive composition containing non-toxic pigments and binders with different chemical nature have been carried out. The effect of binders, pigments, extenders and the level of pigmentation on the structure of coatings (Hg porosity test, DMTA) and on the protective properties (corrosion chambers, EIS, SVET) were tested. The test results show that protective properties of the coatings depend on the chemical composition of binders and the mechanism of film formation. The impact of pigments on protective characteristics of the coatings is connected with their chemical composition, physical shape and the level of pigmentation. The application of the statistical analysis of the results allows to define the weight of particular components of paints in shaping protective properties of the coatings and the selection of the optimum composition.

Keywords: waterborne paints; non-toxic pigments; coatings; barrier; electrochemical properties

Introduction

A lot of research has been carried out but the formulation of anticorrosive waterborne coatings containing non-toxic pigments is still problematic. There are many controversies on efficiency and the mode of performance of this kind of pigments. These problems come from the lack of the explicit laboratory method that would enable forecasting of the resistance of coatings and the differentiation of their protective capacity. [1]

The aim of the project was to determine the effect of main paint components – binders and

CCC 1022-1360/00/$ 17.50+.50/0

pigments/ extenders – on shaping of protective properties of waterborne coatings. The using of test methods, including both the physical structure and protective efficiency of coatings, has allowed to forecast protective properties and to optimise the paint composition.

Paint formulation

The components used in paint formulations are characterised in Table 1.

Table 1. Components and the level of pigmentation of paints

Binder	Urethane dispersion modified by fatty acids „A"	Alkyd emulsion (ok. 46% plant oils) „B"		Epoxy-ester dispersion „C"	Styrene-acrylic dispersion „D"
Pigment	Zinc phosphate „2"	Zinc calcium phosphate „3"	Zinc ferrite „4"	Micaceous iron oxide „5"	Jon-exchanged pigment „6"
Extender	CaCO$_3$ „a"			BaSO$_4$ „b"	
Λ (Reduced Critical Pigment Volume Concentration)	Λ=0,45 „I"		Λ=0,6 „II"		Λ=0,8 „III"

The paints contained the same amount of red iron and microtalc.

Adopted marking consists of four variable elements:

binder/ pigment extender/ Λ

for example: A/2a/I – urethane dispersion/ zinc phosphate, CaCO$_3$/ Λ=0,45

The coatings to be tested were formulated with possibly best packing of pigments and extenders which means their best compatibility in the area of the lowest CPVC identical with the efficiency of the anticorrosive performance in water dispersion.

Test methods

The following test methods were carried out for the paint compositions:

- Hg porosity test – at the apparatus constant of 21,63 μF / pF, max. pressure of 4,680 psi and wet angle Hg of 130 deg,
- dynamic-mechanical analysis (DMTA) – at the frequency of 1 Hz and the tension amplitude of 16 μm,
- salt spray test according to PN ISO 7253,
- *Prohesion* test: 1 h of spraying with electrolyte solution 0,5 g NaCl + 3,5 g $(NH_4)_2SO_4$/ l, temp. 25 ^{0}C/ 1 h of dry conditions, temp. 35 ^{0}C,
- humidity test according to ISO 6270,
- electrochemical impedance spectroscopy (EIS) at the amplitude of the measuring signal of 10 mV and the frequency of 0,01 Hz to 100 kHz, using as the electrolyte 3 % NaCl,
- scanning vibrating electrode technique (SVET) – at the frequency of 80 Hz and the amplitude of 50 μm, using as the electrolyte 10^{-3} M NaCl with complexing agent 10^{-3} M EDTA.

The paints were tested either as free film with the mean thickness of 70 μm or coatings with the mean thickness of 50 μm applied on steel plates of 70 x 150 mm.

Test results

Physical and mechanical properties of coatings, such as tightness, flexibility, homogeneity, often decide of their long-time durability, thus efficient corrosion protection. The type of binder and the way and level of pigmentation as well as the interaction of pigment/ pigment and pigment/ binder determine these properties in pigmented paint compositions. [2,3] It is especially important for waterborne paints in which the mechanism of acting of non-toxic anticorrosive pigments is still not fully recognised.[4]

The investigations of the effect of main paint components on the coating structure connected with anticorrosive properties have allowed both to explain the mechanism of pigment acting and to give guides to paint formulation.

150

The test results show that protective properties of the waterborne coatings depend on the chemical composition of binders and the mechanism of film formation reflected by cross-linking density and glass transition temperature T_g. Fatty-acid modified urethane dispersion (A) has the best barrier properties.

The impact of pigments on protective properties of the coatings is connected with their chemical composition (interaction with the binder), with their physical nature (size and particle size distribution) and the level of pigmentation (PVC/CPVC). [5-8]

The combining barrier- electrochemical acting has been found out for zinc calcium phosphate (3) and zinc ferrite (4). These pigments show better protective properties than others. Other tested pigments participate in barrier protection only.

Taking into account all the test results the binders can be ranked according to their decreasing protective properties as follows: A > B > C > D and the pigments: 4 > 3 > 2 = 6 > 5.

As an example, Figure 1 shows changes in impedance data depending on the kind of binder and Figure 2 - depending on the kind of pigment.

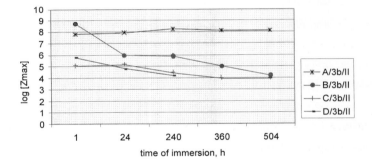

Figure 1. Changes of $[Z_{max}]$ in the time of immersion in 3% NaCl depending on the kind of binder.

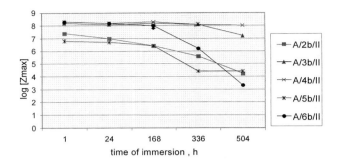

Figure 2. Changes of $[Z_{max}]$ in the time of immersion in 3% NaCl depending on the kind of pigment.

Forecasting and optimisation of the composition

Analysed data obtained from the test chambers comprised the factors of corrosion changes, such as blistering size (S) and density (D) and corrosion of substrate (C). These factors were determined in arbitrary range $0 - 5$ (0-without changes, 5 – maximum changes). The mean values, including all the factors, were calculated for each sample. They have been given in the range $0 - 100\%$ and marked respectively: Sm for salt spray test, Wm for water condensation test and Pm for *Prohesion* test.

The base of electrochemical data were the mean values of impedance modulus determined after 1 h immersion and total mean values.

The comparison of test results obtained by EIS, salt spray test, water condensation test and *Prohesion* test is given in Table 2.

Table 2. Comparison of test results

Symbol	Logarithm of impedance modulus Mean values		Unified factors of corrosion changes											
	after 1 h	total	Salt spray test				Humidity test				*Prohesion* test			
			S	D	C	Sm	S	D	C	Wm	S	D	C	Pm
A	7,83	6,27	2,6	3,0	2,0	50	2,6	4,3	0,0	47	0,4	0,5	0,7	10
B	8,10	6,47	4,0	4,5	3,0	77	3,5	4,0	0,0	50	2,0	2,5	2,8	49
C	5,30	4,61	3,5	4,3	4,2	80	2,6	2,9	0,0	37	3,4	2,2	2,8	56
D	5,64	4,72	4,0	4,3	4,2	84	1,3	1,7	0,0	20	3,0	3,2	3,6	66
2	6,53	5,36	3,6	4,0	3,6	75	2,8	3,8	0,0	44	1,5	2,0	2,6	41
3	7,08	6,09	3,3	3,7	2,5	63	2,5	3,6	0,0	41	2,1	2,0	2,3	43
4	7,03	6,43	2,1	2,4	1,1	37	2,4	4,2	0,0	44	0,8	0,5	1,1	16
5	5,75	5,02	3,7	4,1	4,3	80	2,8	3,1	0,0	39	1,8	2,3	2,5	43
6	7,02	5,68	3,1	4,1	2,5	64	1,4	2,1	0,0	23	1,8	0,9	0,9	24
a	7,78	7,11	2,4	3,0	1,5	46	2,8	4,7	0,0	49	0,3	0,3	0,7	9
b	6,59	5,48	3,4	3,8	3,2	69	2,3	3,2	0,0	37	1,8	1,9	2,2	39
I	7,31	6,49	3,0	3,5	2,5	60	2,5	4,0	0,0	43	1,1	1,4	1,8	28
II	6,06	5,18	3,5	4,0	3,5	73	2,3	2,9	0,0	35	2,1	1,9	2,2	42
III	6,75	5,16	3,3	3,6	2,8	65	2,4	3,2	0,0	37	1,8	1,7	2,0	37
Optimum	A(B)/3a/I A(B)/4a/I		A/4a/I				D/6b/II				A/4a/I			

From the data given in Table 2 it is evident, that the test results obtained by salt spray and *Prohesion* methods point clearly at paint formulation containing fatty acid modified urethane dispersion (A), zinc ferrite (4), $CaCO_3$ (a) with $\Lambda=0,45$. Furthermore electrochemical test results show, that the good corrosion protection can be obtained by paints basing on urethane dispersion (A) or alkyd emulsion (B) pigmented with zinc calcium phosphate (3) or zinc ferrite (4).

In case of humidity test the results are different from mentioned above. These results are not representative enough, because the most essential factor C – corrosion of substrate – has the value

0 and the assessment has been based on other factors only.

Taking into consideration the test results from salt spray, *Prohesion* and EIS methods the optimum paint composition should be as follows:

$$\boxed{\textbf{A/4a/I}}$$

The statistical analysis shows, that the salt spray, *Prohesion* and EIS methods are a good tool for the forecasting and optimisation of paint composition containing waterborne binders and non-toxic pigments. The more clearly results can be obtained in corrosion chambers. In the case of EIS it is very important to find more representative criterion for an analysis than a measurement after 1 h of immersion (i. e. time in deflection point of curve).

Conclusions

• The test results show that protective properties of the waterborne coatings depend on the chemical composition of binders and the mechanism of film formation reflected by cross-linking density and glass transition temperature T_g. Fatty-acid modified urethane dispersion has the best barrier properties and styrene acrylic dispersion – the worse protective efficiency.

• The impact of pigments on protective properties of the coatings is connected with their chemical composition (interaction with the binder), with their physical nature (tightness and homogeneity of coatings) and the level of pigmentation.

• Combining barrier-electrochemical acting has been found for zinc calcium phosphate and zinc ferrite. These pigments present better protective features. Other tested pigments participate in barrier protection. The more evident it is the more the coating is.

• The application of the statistical analysis of the test results allows to define the weight of particular components of paints in shaping protective properties of the coatings and to select the optimum composition as well as the correlation between the methods which have been used. Basing on the comparison of correlation coefficient *Prohesion* method has been chosen as the most compatible with electrochemical methods. The humidity chamber was eliminated because the test results do not completely reflect protective properties of the coatings.

References

[1]   G. R. Pilcher, J. Coatings Tech. **2001**, 72, 135
[2]   W. Bosch, W. Schlesing, M. Buhk, Europ. Coatings J. **2001**, 10, 60
[3]   D. Gysen, Ind. Minerals, June **2001**, 41
[4]   I. F. Veter, M. C. Deya, R. Romagnoli, B. del Almo, J. Coatings Tech. **2001**, 73, 57
[5]   C. J. Knauss: Am. Paint Coat. J. **1991**, 76, 99
[6]   A. J. Eickoff: Pitture e Vernici, **1998**, 12, 39
[7]   J. Austin: Paint Ink, **1993**, 6, 13
[8]   O. Leblanc: J. Oil Colour Chem. Assoc. **1991**, 74, 288

Macromol. Symp. 187, 155–164 (2002)

Application of Peroxide Macroinitiators in Core-Shell Technology for Coating Improvements

Tokarev V. S., Voronov S. A.*, Adler H.-J. P.[†], Datzuk V. V.*, Pich A. Z. [†], Shevchuk O. M.*, Myahkostupov M. V.**

[*] Institute of Chemistry and Chemical Technologies, Lviv Polytechnic National University
[†] Institute for Macromolecular Chemistry and Textile Chemistry, Dresden University of Technology

Summary: New approach to synthesis of core-shell latices is proposed in the presented work. The method is based on the peroxide macroinitiators utilization for formation of the seed particles with tethered to their surface peroxide moieties. As a result, initiation and chain-growing reactions of polymerization of the shell-forming monomer are localized precisely at the seed latex interface. That allows to diminish consumption of this monomer required for complete encapsulation of a core polymer. The advantages of this new method over the conventional core-shell technology are reflected in improving the properties of polymer coating derived from such latices.

Introduction

Latices with the core-shell particle structure are of great interest for various applications. Those of them constituted by a hard polymer core like polystyrene (PS) or polymethyl-methacrylate (PMMA) and a soft polymer shell, e. g. polybutylacrylate (PBA) or polybutyl-methacrylate (PBMA), have excellent film-forming properties [1, 2], therefore they might be used in diverse water-borne paint formulations. Another type of latices with soft corn and hard shell has been successfully applied as impact modifiers for rigid polymers [3, 4]. In any case latices with complex particle structure are synthesized via multistage sequential emulsion polymerization of two or more diverse by nature monomers and this technology can be briefly described as follows. At the outset, the first monomer is polymerized to form latex which is further served as seeds for successive polymerization of another kind of monomer forming the polymer shell on its particles. The complexity of this process, governed by many kinetic, thermodynamic and colloidal factors, is the main reason why formation of a true core-shell structure does not always succeed [4] and obtained particles thereby might have other distinct morphologies [1, 5, 6]. Among these factors the nature of the initiator used for performing

  CCC 1022-1360/00/$ 17.50+.50/0

second stage of polymerization is of high importance [7]. In the majority of cases better results were achieved with rather water-soluble initiators, e. g. potassium or ammonium persulfates, than with oleo-soluble ones.

Another approach to synthesis of core-shell latices is proposed in the presented work. This method intends using peroxide macroinitiators for formation of the seed particles with tethered to their surface peroxide moieties. As a result, initiation and chain-growing reactions of polymerization at the second and subsequent stages of the process are localized precisely at the seed latex interface. This permits to diminish the number of factors, which strongly affect shell formation, thereby this process becomes more stable and reproducible.

Peroxide macroinitiators (PMI) are oligomer products invented and intensively developed for two past decades in Lviv Polytechnic University [8]. Their synthesis is performed via the radical copolymerization of peroxide monomers with other various functional monomers; as a result the structure of PMI can be easily tailored in a wide range aimed at involving required functionalities what is achieved by choosing the proper initial functional monomers. It is worth mentioning that a number of monomers with diverse functional groups is technically available at the present time. Moreover, the polymer modification reaction involving these functionalities allows to widen the possibility for PMI design with diverse structures and required properties. On the other hand, many peroxide monomers with different thermal stability have been synthesized as well. Thus, this method provides incorporating of the hydroperoxide, perester, or ditertiary peroxide moieties into the structure of macroinitiators. In this way the precise control of performing the free radical processes is achievable due to generation of free radicals with defined rates under chosen conditions applying either thermal decompositions of PMI or their RedOx reactions.

Concerning the functional monomers, the peroxide copolymers of (meth) acrylic acid and maleic anhydride were found to have notable features, therefore they are most often used. Due to the presence of both the hydrophilic, e.g. acid, and hydrophobic peroxide moieties in macromolecules of PMI they demonstrate surface-active properties in water solutions [8, 9] thereby they can be used as reactive surfactants for emulsion polymerization. In the last case PMI serves simultaneously as a surfactant stabilizing monomer-polymer particles and as an emulsion polymerization initiator due to decomposition of its peroxide group [8, 10]. On the other hand, so far as PMI has high tendency to adsorption at the surface of dispersed phase and is capable of initiating radical processes, it might be a useful compound for the surface modification of polymers, particularly for creation of core-shell latices.

The presented work is targeted at demonstrating the achievements gained in improving the polymer coating properties when PMI are applied in core-shell technology.

Experimental

Materials

As <u>peroxide macroinitiator</u> we have used oligo(5-*tert*-butylperoxy-5-methyl-1-hexen-3-in-*co*-maleic anhydride) synthesized via radical copolymerisation of peroxide monomer with maleic anhydride in acetone in the presence of lauroyl peroxide as initiator at 60 °C as described elsewhere [9].

<u>Monomers:</u> styrene (Merck), butylacrylate (Aldrich), ethylacrylate (Aldrich), and methylmethacrylate (Aldrich) were used for latex synthesis after vacuum distillation.

<u>Other ingredients.</u> Ammonium persulfate (APS, Merck) was used as initiator of emulsion polymerization. Sodium lauryl sulfate (SLS, Merck) and Dowfax 2A1 (Dow Co) were used as surfactants for latex stabilization.

Procedure

<u>PMI immobilization on the seed particles.</u> Core-polymer latices have been synthesized by conventional methods [11] under a nitrogen blanket in a four-neck reactor equipped with a stirrer, reflux condenser, inlets for inert gas and feeding and outlet for taking probes during the process. The control of monomer conversion was performed via estimation of non-volatile residue as well as with the aid of liquid-gas chromatography. Water solution of approximately 12 wt % of PMI was prepared in a separate flask and then different amounts of this solution were added to core-polymer latex. The mixture was stirred firstly at ambient temperature for 30 min and then at 90 °C for ca. 6 hours, afterwards the reactor was cooled.

<u>Synthesis of core-shell latex</u> was performed in the same reactor. For that the second-stage monomer was added in a proper manner (either entirely in the beginning of the process, either by several portions or dropwise during the process). If required the components of RedOx system and additional amount of surfactant were charged as well. Thereafter the mixture was heated under continuous stirring. The monomer conversion was controlled as above.

<u>Particles sizes</u> were determined applying Malvern System 3000.

For obtaining the coating some amount of latex was poured on glass plate afterwards air-dried for 48 hours. The following assessments were performed to characterize the quality of latex films: adhesion was estimated in balls by conventional grid-cutting method; hardness was measured with the aid of pendulum apparatus and expressed in relative units (R.U.), whereby 100 corresponds to the glass hardness; opacity was calculated using the expression: $Opacity = \dfrac{D}{\Delta H}$, where D is optical density of the film and ΔH – its thickness (mm).

Results and discussions

The main idea described there consists in the following. Water solution of PMI is added to any ready-made latex and the mixture is heated at about 90 °C for several hours. As a result the PMI macromolecules are bonded to the surface of seed latex. Peroxide groups, immobilized in this way, serve further for initiating the graft-polymerization of vinyl monomer onto these particles, i.e. for forming the polymer particles with core-shell structure. This process is schematically shown below.

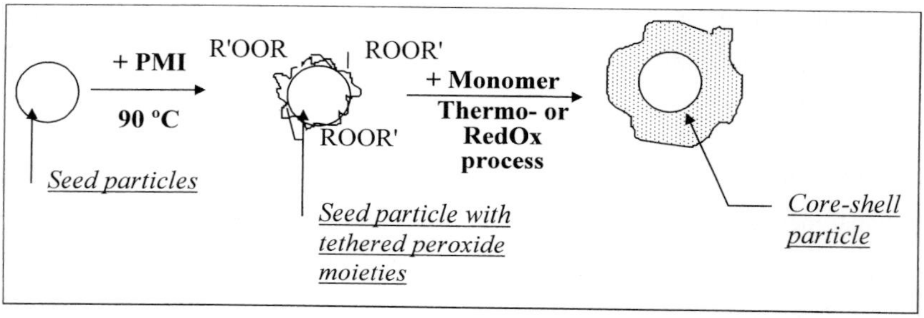

Scheme 1. Formation of core-shell latex particles applying PMI.

The process of PMI immobilization at seed latex particles is sufficiently complicated. It is suggested to occur via the radical mechanism and involve several elementary stages, among which the main ones are: adsorption of PMI on the particle surface; partial decomposition of PMI peroxide groups; activation of the polymer substrate surface via the chain transfer reactions; and chain termination reactions via recombination mechanism. These altogether leads to PMI coupling with the latex particle surface thereby the immobilization of the reactive peroxide groups at interface is achieved. The analysis performed with the use of different techniques (Thermo-Gravimetric Analysis, Differential Scanning Calorimetry and NMR-Spectroscopy of the latex polymer, potentiometric titration of the serum separated from frozen latex probes) proved almost complete bonding of PMI.

PS core latex particles modified with PMI in such a way have been used for synthesis of core-shell latices. Shell formation was achieved via the radical polymerization of BA initiated due to either thermal decomposition of peroxide moieties, immobilized at the core particles at 90 °C, or their decomposition facilitated by interaction with the reduction system (iron II sulfate –Chelaton B – Rongalit) at 40 °C. Figure 1 shows the kinetic curves of BA polymerization in the presence of PS core particles containing various PMI amounts.

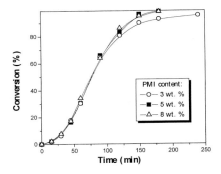 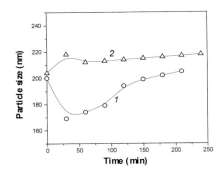

Figure 1. Kinetics of BA polymerization in the presence of PS seed particles modified with various amount of PMI.
PMI content is given in legend;
BA to PS ratio is of 3 : 10 by weight;
Temperature 90 °C.

Figure 2. Time dependence of the average particle size at BA polymerization in the presence of PS core latex initiated by 3 wt. % of PSA at 84 °C (1) and by 5 wt. % of PMI immobilized at 90 °C (2).
BA to PS ratio is of 1 : 3 by weight.

As it can be concluded from the presented data the process is not practically influenced by PMI content. The possible explanation for this phenomenon could be found in the theory of emulsion polymerization [11]. One of basic statements of this theory is an assumption that any latex particles may contain only one growing radical at any time. Because of small volume of the particles appearance two or more radicals there leads to their immediate interaction, i.e. to their termination. The main conclusion drawn from this theory is the statement that the total polymerization rate is independent on the rate of initiation reaction but is strongly influenced by the number of particles. This is the reason why increase in the amount of PMI tethered to seed particles does not effect the total rate of BA polymerization. On the other hand, this is also an evidence that BA polymerization occurs on the seed particles surface but does not occur in water phase. If the latter would take place, an increase in PMI concentration should brings about the nucleation of new polymer particles, i.e. enlarge their number in the system thereby accelerating the process, what is not observed in reality.

Although the PMI concentration has no effect on the rate of shell-polymerization, the core-shell latices, obtained with the use of various amount of immobilized PMI, differ in their aggregative stability, particularly against frosting (Table 1).

Table 1: Frosting effect on particle size of PS core – PBA shell latices

Amount of PMI immobilized on core particles (%)	Particle size (nm)	
	before frosting	after frosting
3.0	202	coagulation
5.0	203	292
8.0	208	254

Improved aggregative stability of latices modified by PMI has been observed in the previous investigations [12] and was attributed to enhancement of their electrosteric stabilization. It was supposed that immobilization of PMI macromolecules brings about an increase of charge density on the particle surface as well as appearance of the steric barrier hindering coagulation. It also results in some drop of the average particle size, presumably because of deterioration of their tendency to aggregation.

Comparative investigations of changes in the average particle size (Figure 2) were performed during BA shell-polymerization initiated by PMI tethered to the PS seed particles or by water-soluble initiator PSA in accordance with [13]. It is evident from Figure 2 that topochemistry of these processes is appreciably distinct.

Decomposition of PSA in aqueous phase brings about homogeneous nucleation of the number of small PBA particles at the initial stage of the process, what is indicated by sharp drop in the average particle size (Figure 2, curve 1). Because the adsorption layers of newly formed particles are strongly unsaturated they have strong tendency to coalescence, first of all at the surface of core particles which are essentially larger. Since this occurred the polymerization process takes place on the core-shell particle surface (particle size growth section of curve 1).

The topochemistry of shell formation involving PMI immobilized is essentially different. From the beginning of the process, BA polymerization occurs on the surface of PS particles, where the initiation sites are localized. Curve 2 in Figure 2 confirms the absence of homogeneous nucleation of PBA particles in this case, since the particle size is rising already at the initial stage of the process.

So, obtaining of core-shell latices with the use of PMI immobilized has some advantages over the conventional technique applying water-soluble initiators. Above all this concerns more effective utilization of shell-forming monomer. Because of its partial consumption in the process of homopolymer particle formation as described above, the core-shell latices obtained applying conventional technique demonstrate the film-forming properties at the weight ratio of core- to shell-polymers ranging from 1 : 1 to 1 : 9.

On the contrary, the proposed technique allows the localization of shell-monomer polymerization exactly at the surface of seed particles thereby diminishing its consumption required for complete covering core-polymer. Moreover, this is a case when the highest ratio of hard core- to soft shell-polymers, at which the latex possesses the film-forming properties, can be easily predicted on the basis of consideration the film formation process. Films derived from core-shell latices are typical heterogeneous materials, where the shell polymer serves as the continuous phase within which the core-polymer particles are distributed. The model of maximal packing predicts in the case of uniform spherical undistorted particles their highest volume partition to be of 0.67. In the case of polydisperse particles this value might however rise up to 0.77. In this model soft polymer just simply fills the space between those particles. On the other hand, one should bear in mind that the hard particles have to be completely encapsulated by soft polymer in order to achieve formation of a tight film, i.e. correction ought to be made on the thickness of shell layer. Assuming this thickness should be not less than 10…20 % relative to the core particle size, it is easy to estimate the required ratio to be in the range of 3 : 1 to 2 : 1.

Series of latices with various PS core to PBA shell ratios were synthesized applying both the conventional and new techniques in order to evaluate their efficiencies. With this purpose latex films of about 0.3 mm thickness, cast on the glass plates, were examined on adhesion, hardness and opacity (Table 2). For comparison there are given the properties of film derived from commercial vinyl-acrylic latex AC-264 widely used in water-borne paint formulation.

Table 2. Composition of latex dispersion and properties of film

No	Technique used for obtaining latex dispersion	Latex properties		Film properties		
		PS to PBA ratio	Particle size (nm)	Adhesion (ball)	Hardness (R.U.)	Opacity (mm^{-1})
1	New core-shell with PMI use	3 : 1	216	0	47.4	3.12
2	New core-shell with PMI use	2 : 1	210	0	41.6	2.69
3	Conventional core-shell with PSA use	3 : 1	198	2	–	–
4	Conventional core-shell with PSA use	1 : 1	208	0	24.8	2.87
5	Blend of homopolymer latices	1 : 1	201	2	31.4	2.99
6	Commercial vinyl-acrylate latex AC-264		206	0	15.7	0.17

Evident is the fact that the adhesion of core-shell latices obtained using PMI even at low content of PBA is similar to that of commercial latex. Applying conventional method for obtaining the core-shell latices a similar result is achieved at higher content of PBA. At least equal amounts of the core- and shell-polymers are required in this case. Plain blending of two homopolymer latices does not give acceptable results. Another distinct properties of the films derived from the newly synthesized core-shell latices is their essentially higher hardness (2…3 times) than that of the other latex films. This is an important result because low hardness usually attributed to latex coating is one of the main obstacles that hinders their wide application in paint formulations.

Film opacity is only one position where new latices lose to commercial product substantially. This is because the core-shell latices are capable of forming only heterophase films what is the result of thermodynamic incompatibility of the core- and shell-polymers. The high difference in their refractive indexes (n_D) causes light scattering on their phase boundaries, therefore such films cannot be as highly transparent as those formed by the latex with homophase particles.

There are two approaches to enhance transparency of the films formed by heterophase particles:

- choice of the core- and shell-polymers with as much as possible close or better with equal refractive indexes;
- creation of multilayered particles with a radial gradient of refractive index.

If the first of these approaches is clear, the second one requires some comments to be made. The graphic interpretation of the last approach is given in Scheme 2.

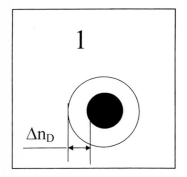 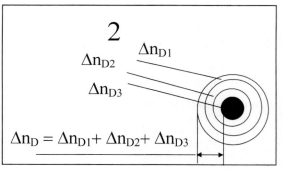

Scheme 2. Structure of plain core-shell (1) and multilayered (2) latex particles.

There every two neighboring layers do not differ in n_D as much therefore such particles as well as the material formed from them have a vague phase boundary, consequently slight scattering properties. In practice this might be realized via gradual changes in the content of copolymers forming each of these layers. Tables 3 and 4 show the receipts of multilayered core-shell latices synthesized with the use of PMI and film properties derived from them. There are appreciable gains in the transparency of the films while other characteristics are similar to those of previous latices (Table 2).

Table 3. Composition and properties of multilayered core-shell latices

		Core composition (%)				1st shell composition (%)			2^d shell composition (%)		Latex properties	
No	S	BA	BMA	MMA	PMI	S	BA	BMA	BA	BMA	NVR (%)[*]	Particle size (nm)
7	30	-	3	3	2	20		19	23	-	46.5	209
8	30	-	3	3	2	20		19	-	23	39.4	197
9	30	3	-	-	2	20	3	19	23	-	40.9	213
10	30	3	-	3	2	17	3	19	-	23	42.5	192
11	30	3	3	-	2	17	3	19	23	-	42.0	202

[*] – non-volatile residue.

Table 4. Properties of the films derived from multilayered core-shell latices

No	ΔH (mm)[*]	Adhesion (ball)	Hardness (R.U.)	Opacity (mm^{-1})
7	0.24	0	61	1.53
8	0.25	0	60	1.84
9	0.24	0	60	1.79
10	0.26	0	52	1.52
11	0.30	0	54	1.53

[*] – film thickness.

To prove the first approach to obtaining transparent heterophase coatings the core-shell latices have been synthesized with the use of PMMA as core-polymer ($n_D = 1.485$) and PBMA ($n_D = 1.483$) or some BMA copolymers with methacrylic acid (MAA) or ethyl acrylate (EA). The results are given in Table 5.

Table 5. Properties of the films derived from core-shell latices with MMA core-polymer

No	Shell-monomer composition (%)[*]	Latex properties		Film properties		
		NVR (%)	Particle size (nm)	Adhesion (ball)	Hardness (R.U.)	Opacity (mm^{-1})
12	BMA (50)	42.7	226	0	51.6	0.60
13	BMA (45); MAA (5)	51.8	224	0	50.8	0.42
14	EA (23); BMA (27)	40.0	208	0	41.3	0.14
15	MMA (48); EA (25); BMA (27)	41.1	212	1	34.0	0.26

[*] The monomer percentage is based on core-polymer weight.

Analyzing these data one can conclude that application of PMI in the core-shell technology allows to obtain the latices with the film-forming properties which are similar to those of vinyl-acrylate latex but exceeding the latter in the coating hardness.

Conclusions

As a result of the investigations performed a new approach to synthesis of core-shell latices has been developed. It is based on the peroxide macroinitiators utilization for formation of the seed particles with tethered to their surface peroxide moieties. As a result, initiation and chain-growing reactions of the shell-polymerization are localized precisely at the seed latex interface. This allows to diminish consumption of shell-forming monomer required for complete encapsulation of a core polymer thereby improving the properties of polymer coating derived from such latices.

References

1. Padget, J.C. J. *Coating Technology*, 1994, **839**, 89-105.
2. El Aasser, M.S.; Segall, I.; Dimonie, V.L. *Macromol. Symp.*, 1996, **101**, 517-521.
3. Okubo, M. *Macromol. Chem., Macromol.Symp.*, 1990, **35/36**, 307-325.
4. Okubo, M. et. al. *J. Appl. Polymer Sci.*, 1986, **31**, 1075-1082.
5. Goodwin, J. W.; Hearn, J.; Ho, C. C.; Ottewill, R. H. *Colloid Polym. Sci.*, 1974, **252**, 464-471.
6. Distler, D.; Kanig, G. *Colloid Polym. Sci.*, 1978, **256**, 1052-1060.
7. Rudin, A. *Macromol. Symp.*, 1995, **92**, 53-70.
8. Voronov, S.; Tokarev, V.; Petrovska, G. Heterofunctional Poliperoxides. Theoretical basis of their synthesis and application in compounds. Lviv: SU "LP", 1994, 85 p.
9. Voronov, S.; Tokarev, V.; Oduola, K.; and Lastukhin Yu. *J. Appl. Polym. Sci.* , 2000, **76**, 1217-1227.
10. Voronov, S.; Tokarev, V.; Datsyuk, V.; and Kozar, M. *Progr. Colloid Polym. Sci.*, 1996, **101**, 189-193.
11. Ivanchev, S. S.; Eliseeva, V. I..; Kuchanov, S. I.; Lebedev, A. V. Emulsion polymerization and its application in industry. editors. Moscow: Chemistry, 1976 (*In Russian).*
12. Datsyuk, V. S.; Tokarev, V. S.; Voronov, S. A.; Trotsenko, S. E.; Pich, A. Z. *Dopovidi NAN Ukrainy* (The Reports of National Academy Sciences of Ukraine), 1998, No **6**, 145-149.
13. Hidalgo, M.; Cavaille, J.Y., et. al. *Colloid Polym. Sci.*, 1992, **270**, 1208-1221.

Macromol. Symp. **187**, *165–175 (2002)*

Synthesis and Surface Properties of Microphase Separated or Nanostructured Coatings Based on Hybrid and Fluorinated Acrylic Copolymers

Valter Castelvetro[1,2], Antonella Manariti[1], Cinzia De Vita[1], Francesco Ciardelli[1,2] *

[1] Dipartimento di Chimica e Chimica Industriale, University of Pisa, Via Risorgimento 35, 56126 Pisa, Italy

[2] INSTM, Research Unit of Pisa, Italy

*Paper presented on behalf of AITIVA, Milano

Summary: Coating materials characterised by intrinsic inhomogeneity or nanostructured morphology can display unique interfacial (e.g. surface and adhesion) and bulk (e.g. mechanical, thermal) properties, when heterophasic or self-segregating components are obtained by suitable design of the constitutive copolymers' structure. With the purpose of obtaining intrinsically photostable low-surface energy coating materials characterised by good penetration into porous substrates and variable response of the adhesive and polymer-air interface, fluorinated acrylic-based copolymers and water-borne acrylic-organosilane hybrids have been considered. For the latter, dispersed phase polymerisation procedures based on combined emulsion copolymerisation and hydrolysis-polycondensation of organosilane precursors have been adopted.

Keywords: fluoropolymers, coatings, water repellency, latices, nanocomposites

1   Introduction

As described in a synthetic way in the following sections, in the past 10 years we have been involved in the synthesis and characterisation of fluorinated acrylic-based copolymers with variable structure, composition, and formulation (solvent or water-borne). Our attention has been mainly focussed on the understanding of the relationship between molecular structure and application properties as surface modifiers of porous substrates, achieved through tuning of polymer composition and comparative testing involving both fluorinated and parent unfluorinated products. The main application of fluorinated polymers is related to their surface properties and stability; accordingly, most of our testing, performed in collaboration with various research groups and with industrial companies interested in the improvement of surface properties of different materials, has concerned the evaluation of their coating properties as water-repellent materials with enhanced durability and, in a later development, with additional features such as adhesive and consolidating activity.

The early work, consisting in the screening of various macromolecular structures based on both commercial and newly synthesised fluorinated acrylates [1-3], was mainly aimed at providing new water-repellent protective coatings for porous materials (Section 2) [4-6]. The surface properties were therefore analysed, by studying the surface behaviour and dynamics in terms of response to contact with water of thin film cast on either porous or smooth substrates, and of LB multilayers (Section 3) [7]. Subsequently we have been involved in the preparation of water-borne formulations leading to either conventional [8] or nanostructured coatings; the latter are based on hybrid polymeric materials obtained by combined free-radical emulsion polymerisation of acrylates and sol-gel polycondensation (Section 4). The above aspects are presented here in a synthetic way, and their possible development discussed.

2 Fluorinated Acrylic-Based Polymers: Synthesis and Properties

In order to take advantage of the unique surface properties and improved chemical stability provided by the fluorocarbon structure, partially fluorinated polymers must be designed in such a way, that the advantageous complementary properties deriving from the unfluorinated and the fluorinated moieties be displayed while suppressing most of the respective points of weakness. Among the latter, the poor cohesive and adhesive properties of perfluorinated structures, and the inadequate water repellency and limited photooxidative stability of the unfluorinated ones.

To this purpose, the application procedure should be capable of contributing to the enrichment of fluorinated moieties at the polymer air interface at the lowest possible degree of fluorination, and the material itself allow self-assembling of perfluoroalkyl groups, leading to minimisation of surface energy and reduction of molecular mobility (that is, molecular dynamics). Indeed molecular dynamics at a hydrophobic surface can cause rapid decline of the surface properties upon contact with water.

In addition, the primary structure of the macromolecule and the structure of its repeat units should minimise the occurrence of hydrolytically or photooxidatively weak points, such as tertiary C-H bonds typical of polymers of vinyl derivatives, by reducing the number of such weak points to a minimum or through stabilisation induced by vicinal perfluoroalkyl substituents. This can not usually be achieved without restraining the conformational flexibility of the material, thereby worsening its thermal and mechanical properties; therefore a good balance must be devised.

Our research activity has been initially focussed on the investigation of the free radical polymerisation of (meth)acrylic esters bearing fluorine atoms or fluorinated groups either as substituents of the vinyl double bond or in the side chain [9]; copolymerisation of these monomers with other not homopropagating vinyl compounds such as α-olefins, alkenyl ethers, and maleic anhydride derivatives was also investigated. Depending on the location of the fluorinated carbons, the reactivity of these fluorinated acrylic esters with the above co-monomers can markedly change, and thus a macromolecule's primary structure and properties [10,11].

While perfluoroalkylethyl acrylates, such as 1H,1H,2H,2H-perfluorodecyl methacrylate (XFDMA), show a similar reactivity in the radical polymerization as their unfluorinated analogues, fluorine substitution at or closer to the vinyl bond as in methyl α–trifluoromethylacrylate (MTFMA), ethyl 2,2-difluoromethacrylate (EFMA), 1,1,1,3,3,3-hexafluoro-2-propyl α–fluoroacrylate (HFIFA), 1H,1H,2H,2H-perfluorooctyl α–fluoroacrylate (XOFA), 1,1,1,3,3,3-hexafluoro-2-propyl methacrylate (HFIMA) and 2,2,2-trifluoroethyl methacrylate (TFEMA) affects the electronic distribution in such a way that copolymerisation under conventional radical initiation with electron-rich vinyl monomers results in increasingly alternating structures in the order HFIFA ~ EFMA > MTFMA ~ HFIMA > IFA ~ TFEMA > XFDMA ~ MMA. Better control of comonomer sequence, and thus of nature of the co-units vicinal to the weakest bonds, can allow to modify the photooxidative pathway, although effective suppression of the photooxidative instability can only be achieved by using more intrinsically stable unfluorinated co-monomers such as short side-chain methacrylates.

Accelerated photoageing studies confirmed that the photooxidative instability of these structures is mainly due to the reactivity of the unfluorinated co-units, and is particularly high in the presence of tertiary C-H bonds not effectively stabilised by vicinal fluorinated carbons, or of long unfluorinated side chains, leading to increased mobility of the directly or indirectly photogenerated free radical species [12,13].

Accelerated photoageing experiments were routinely carried out with xenon lamp irradiation at $\lambda > 295$ nm, and the effects on macromolecular structure were followed by size exclusion chromatography (SEC), FTIR, and solubility tests. The photooxidative pathways of the MTFMA, HFIFA and EFMA copolymers with linear and branched alkyl vinyl ethers and butyl 2-propenyl ether (BPE) indicate that while photodegradation is still significant, the main mechanism can be quite different, with suppression of either crosslinking or chain scission mechanisms and selective cleavage of C-C main chain bonds only if the involved carbon atoms are not fluorinated (see scheme in Figure 1, based on FTIR and SEC observations). Representative examples are given in Figures 2,3: the copolymers of HFIFA and EFMA with butyl vinyl ether (BVE) and BPE display largely different tendency to primarily produce either insoluble gel (HFIFA/BPE) or cleaved low molecular weight volatile fractions (HFIFA/BVE and EFMA/BVE) [14] .

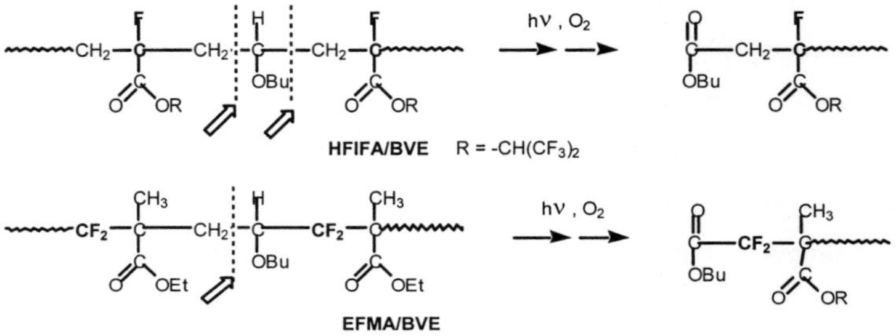

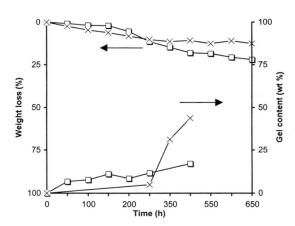

Figure 1. Photooxidative main chain cleavage at tertiary carbon: stabilisation of C-C bond by single carbon fluorination.

Figure 2. Weight loss and gel content of HFIFA/BVE (□) and EFMA/BVE (X) upon accelerated photoageing.

Crosslinking can even turn out to be beneficial for the polymer durability and the permanence of its surface properties. Indeed excellent results were obtained with an alternating copolymer of chlorotrifluoroethylene and vinyl ethers (Lumiflon™LF200), where reactive side chain groups provide the site for fast photo-crosslinking. However, crosslinking is not always a desirable feature, as in case of application of protective coatings onto the surface of artistic or monumental objects and buildings.

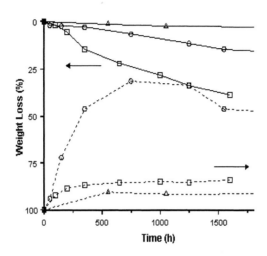

Figure 3. Effects of photoageing on HFIFA/BVE (□), HFIFA/BPE (O), and PHFIFA (Δ).

The coating efficacy on various stone substrates, evaluated in terms of reduction of the water pickup by capillary absorption, have confirmed the expected improved performance and better penetration of the fluorinated coatings as compared to the unfluorinated ones [4,15]. A generally reduced permeability to water vapour and, in the case of the copolymers with vinyl ethers, an increased photooxidative instability are undesirable features of these materials [16]. Furthermore, all coated surfaces undergo progressive decay of water repellency upon permanent contact with water. This is very fast for the XFDMA copolymer coatings, that is, those with the best initial performance, whereas the short side-chain TFEMA copolymers perform generally better at times exceeding 48 h. Such behaviour is related to a number of factors, as discussed in the following section.

3 Surface Properties and Stability

Static and dynamic contact angle studies have been carried out on thin films and on coated stones of different porosity, using both the sessile drop and the Wilhelmy balance method. Additional features such as the dynamics of hydrophobic recovery of the surface structure after swelling in water (see Table 1), and the true equilibrium contact angle [17] have also been determined, and related to the observed water repellent behaviour of the coatings on stone.

Table 1. Static contact angle of thin films before (θi) and after swelling up to 45 days in water.

Polymer [a]	Composition (mol %) [b]	F (wt %)	Tg (°C)	θi (deg)	θ vs. drying time 3 min	12 min	25 min
HFIMA/MA	66/34	40.6	48	105.7	103	105	105
HFIMA/EMA/MA	34/30/36	26.4	57	103.0	95	100	104
XFDMA/HFIMA/EMA/MA	11/12/41/36	30.0	27	110.5	94	102	104
IFA/BVE-1	75/25	11.7	52	99.1	93	86	92
IFA/BVE-2	66/34	10.8	41	92.2	90	92	91
HFIMA/TSPMA/BVE	56/8/36	34.0	29	101.0	64	71	70
HFIMA/TSPMA/MA	57/7/36	35.4	44	102.4	100	100	100

a) IFA = isopropyl α-fluoroacrylate, EM = ethyl methacrylate, MA = methyl acrylate, TSPMA=trimethoxysilylpropyl methacrylate
b) determined by ^{1}H-NMR

The still relatively scarce but already significant data, mainly collected from a series of copolymers of HFIMA, with one or more of XFDMA, TSPMA, BVE, EMA and MA, have provided useful information on the influence of the copolymer structure on the surface dynamics. Actually, surface energy and dynamics are determined by a complex combination of different contributions, such as:

i)   the chemical heterogeneity of the polymer surface, that is, the extent of maximum achievable surface coverage by fluorinated co-units based on copolymer composition and on purely geometric considerations (here, the microroughness of porous substrates plays also an important role: accurate measurement of the true equilibrium contact angle and of advancing and receding contact angles on both thin films and coated stone allows to quantify the effect of the roughness on the reduction of the coating efficacy, related to the receding contact angle, and to detect the presence of coating discontinuities [18]);

ii) the actual extent of the thermodynamically driven enrichment of fluorinated groups at the dry coating surface, reflecting the degree of incompatibility between unfluorinated and fluorinated components, and the favourable reduction of interfacial energy (the segregation of a highly fluorinated, poorly compatible fraction in a thin superficial layer reduces both the surface energy and the free energy associated with the existence of highly dispersed heterodomains with large associated interfacial area);

iii) the glass transition temperature, the presence and degree of crosslinking, and the existence of crystalline domains (as in the case of self-segregation of crystallisable long perfluorinated side chains in the XFDMA copolymers), affecting molecular mobility and therefore the kinetics of macromolecular rearrangement at the polymer-air interface, as observed upon replacement of the "hydrophobic" air with water.

As an example, the HFIMA/TSPMA/MA terpolymer bearing the reactive trimethoxysilane shows improved stability of its surface properties, as the consequence of post-application crosslinking and a comparatively higher Tg (Table 1).

In addition to the intrinsic surface properties of the coating material, penetration of the polymer in the bulk of the porous stone, film formation, film morphology and adhesion are important factors determining the ultimate properties of the coating as such and as the result of weathering.

Indeed, while the introduction of fluorine can improve the penetration of polymer solutions into the porous network [15], this can still be inadequate when consolidation or strengthening of the substrate is required; in addition, clogging of the pores with excessive film-forming material can unacceptably lower the water vapor permeability of the coated substrate. Therefore, particular attention must be payed to the choice of application procedure, product formulation (solvent-based or water dispersion), and amount of applied coating material.

4   Thin Films and Nanostructured Materials from Water Dispersions

Polymer blends, block copolymers or heterophasic organic-inorganic systems, such as siloxane-acrylate hybrids, offer new routes to tailoring the mechanical and surface properties of coatings.

In particular, aqueous phase polymerisation allows designing latices with nanosized heterophase morphology that can be maintained in the final material, while the reactivity of functional groups can be delayed and limited to within a polymer particle. Fine tuning of the polymer particle size, internal structure and composition can allow segregation of reactive groups within one particle's core, while the shell can provide film forming ability even in the presence of a high Tg or crosslinked core which, in turn, improves the mechanical properties of the final coating if good interfacial adhesion between immiscible components is achieved.

Following this approach, conventional and core-shell latex particles, with or without a fluorinated component in the shell, have been synthesised by semicontinuous emulsion polymerisation under "starved" feed conditions. Up to 4-8 wt% TSPMA was introduced in the copolymer to provide efficient coupling sites when the primary latex formation was followed by seeded condensation of various combinations of alkylalkoxysilane monomers and condensates. Control of pH, sequence of co-monomer addition, surfactants, temperature, and composition of the added organosilanes are crucial in order to improve the latex stability, control particle size and maintain adequate film-forming capacity. In fact, the reactivity of the hydrolytically unstable trialkoxysilane group can cause latex instability and poor quality of the resulting film, if the latex formulation does not allow to preserve a sufficient amount of unreacted silanol and alkoxysilane groups capable of forming bridging siloxane bonds upon latex application.

In Figure 4 are shown the TEM micrographs obtained from hybrid latices based on a MMA/BA/TSPM core polymer (Tg $\approx$ 10 °C, MMA=methyl methacrylate, BA= butyl acrylate), and a shell obtained starting from a 1/1 TEOS/DEDMS mixture (TEOS=tetraethoxysilane, DEDMS=diethoxydimethylsilane) fed to the latex in a second step. Controlled hydrolysis and polycondensation of the organosilane monomers onto the seed acrylic particles was achieved by allowing the silanes to equilibrate with the latex, and subsequently heating 5 h at 80° C in the presence of additional anionic surfactant and catalytic HCl in excess with respect to the ammonium carbonate buffer employed for the latex synthesis. The resulting hybrid latex, stable at pH=2 (close to the silanol isoelectric point, where its condensation rate is the lowest) or at pH 7, showed the expected core-shell structure (in Figure 4, the darker edges are due to the organosilane phase). The particle structure was maintained after film formation due to network formation within the shells, hindering coalescence between the acrylic particle cores; observation of a thin slice of the film obtained at room temperature (Figure 4b) reveals that the continuous phase in the hybrid film is the organosilane component, the overall morphology resembling that of a pomegranate. Heat treatment aimed at icreasing the conversion of the alkoxysilane and silanol groups into bridging siloxane bonds, by analogy with a classical sol-gel process, does not cause phase segregation, as shown in Figure 4c where the darker edges are indicative of an increasingly inorganic (silica-like) nature of the continuous phase.

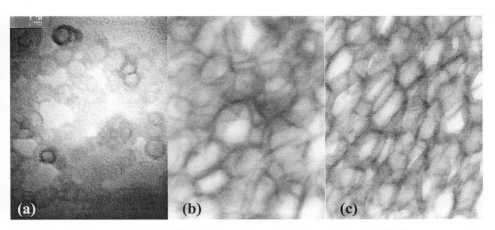

Figure 4. TEM images of core-shell hybrid particles and of the resulting hybrid film structure. (a) 50 nm acrylic core - 15 nm organosilane shell latex particles; (b) heterophasic nanostructure of a 50 nm thick section of the thin film cast at room temperature from a hybrid latex of 150 nm particle size and composition similar to the previous one; (c) heterophasic nanostructure after heat treatment at 120 °C.

The overall composition of the final hybrid latex, as well as the morphology, size and microstructure of the heterogeneous particles, are relevant features not only for the latex stability and its film-forming behaviour, but also for its penetration ability within a porous substrate and the adhesive and mechanical properties of the coating.

The present research is focussed at introducing additional features in these materials, such as fluorinated components in well defined portions of the structured latex particle. Among the goals of this approach the most interesting appear to be: the reduction of the amount of expensive fluorinated acrylic monomers required in order to maintain the same performance (surface properties of the coating), and the possibility to achieve nanosized control on the roughness of the film by suitable combination of high- and low-Tg components of the latex particles.

5   Future Perspectives

Based on the well established results of our previous investigations on fluorinated comonomer reactivity to form differently designed fluorocontaining multifunctional polymers, and on the surface properties and chemical stability of the resulting films, new heterophasic and hybrid systems with nanostructured morphology are presently under investigation. In particular water-borne dispersions, in addition to being obvious candidates for future environmentally friendly processes and applications, offer unique possibilities in terms of control of morphology and reactivity of a multicomponent heterogeneous material, as shown by the preliminary results presented here. However, a number of issues must be addressed and, in case of products to be employed as surface or bulk modifiers of porous substrates, are still far from a satisfactory answer. For water-borne polymers and hybrids the following are certainly among the most important:

1. effect of surfactants on surface properties, particularly when fluorinated polymers are employed to provide water repellency, or simply in order to reduce the sensitivity of the coating to water; this can require more sophisticated synthetic approaches, such as those involving the use of surfmers (surface-active co-monomers) or the design of surfactant-free polymerisation processes;

2. control on the particle size and, thus, of the ability of the latex particle to penetrate within a porous network or fill efficiently a void (e.g. with bimodal particle size distributions) if not only surface but also bulk properties of the substrate need to be modified;

3. film-forming, adhesion, and mechanical properties, all requiring good control of the delayed reactivity of alkoxysilane groups in the bulk of the acrylic polymer, and at the interphase and, if present, in the organosilane fraction.

To this purpose, the composition and morphology of the latex particles and of the resulting films, and the evolution upon ageing in different conditions (temperature, relative humidity), is under study by means of the described methods and of AFM microscopy, thermal analysis, and NMR (^{13}C and ^{29}Si solution, solid state CP-MAS, and spin diffusion) techniques.

Acknowledgements

Partial financial support for this research was provided by CNR - Progetto Finalizzato Beni Culturali, and from MIUR-PRIN co-financing program (2000-2002).

References

[1]   M. Aglietto, E. Passaglia, L. Montagnini di Mirabello, C. Botteghi, S. Paganelli, U. Matteoli, G. Menchi, *Macromol. Chem. Phys.*, **1995**, *196*, 2843.

[2]   F. Ciardelli, M. Aglietto, L. Montagnini di Mirabello, E. Passaglia, S. Giancristoforo, V. Castelvetro, G. Ruggeri, *Progr. Org. Coatings* **1997**, *32*, 43.

[3]   V. Castelvetro, L. Montagnini di Mirabello, M. Aglietto, E. Passaglia, *J. Polym. Sci. Polym. Chem. Ed.* **2001**, *39*, 32.

[4]   G. Alessandrini, M. Aglietto, V. Castelvetro, F. Ciardelli, R. Peruzzi, L. Toniolo, *J. Appl. Polym. Sci.* **2000**, *76*, 962.

[5]   V. Castelvetro, M. Aglietto, F. Ciardelli, O. Chiantore, M. Lazzari, L. Toniolo, *J. Coatings Technology* **2002**, *74* (928), 57.

[6]   L. Toniolo, T. Poli, V. Castelvetro, A. Manariti, O. Chiantore, M. Lazzari, *J. Cultural Heritage* (submitted).

[7]   N. Tirelli, O. Ahumada, U.W. Suter, H. Menzel, V. Castelvetro, *Macromol. Chem. Phys.* **1998**, *199*, 2425.

[8]   V. Castelvetro, F. Ciardelli, G. Francini, P. Baglioni, *Macromol. Mater. Eng.* **2000**, *278*, 6.

[9]   F. Ciardelli, M. Aglietto, L. Montagnini di Mirabello, E. Passaglia, G. Ruggeri, *Paints and Varnishes*, **1996**, *72* (3), 21.

[10] J.M. Bessière, A. El Bachiri, B. Boutevin, *J. Fluorine Chem.* **1992**, *56*, 295.

[11] T. Narita, T. Hagiwara, H. Hamana, T. Nara, *Makromol. Chem. Rapid Comm.* **1985**, *6*, 301.

[12] V. Castelvetro, M. Aglietto, F. Ciardelli, O. Chiantore, M. Lazzari, "Design of Fluorinated Acrylic-Based Polymers as Water Repellent, Intrinsically Photostable Coating Materials for Stone", in "Fluorinated Surfaces, Coatings, and Film", D. Castner and D. Grainger Eds., *ACS Symposium Series* **2001**, *787,* 129.

[13] M. Lazzari, O. Chiantore, V. Castelvetro, M. Aglietto, *Chem. Mater.* **2001**, *13*, 2843.

[14] O. Chiantore, M. Lazzari, V. Castelvetro, M. Aglietto, Conference papers, "Fluorine in Coatings IV", PRA Editions, Teddington, UK, 2001, Paper 15.

[15] F. Casadio, L. Toniolo "Micro-ATR study of the depth of penetration and distribution profile of water repellent polymers into porous stone materials", Proceedings of Vth Int. Conference of the Infrared and Raman Users Group, The Getty Conservation Institute, Los Angeles (USA), March 4-8, 2002. 62-6.

[16] O. Chiantore, M. Lazzari, V. Castelvetro, M. Aglietto, "Photochemical Stability of Partially Fluorinated Acrylic Protective Coatings. IV. Copolymers of 2,2,2-Trifluoroethyl Methacrylate and Methyl α-Trifluoromethyl Acrylate with Vinyl Ethers", *Polym. Degrad. Stab.* (submitted).

[17] C. Della Volpe, D. Maniglio, M. Morra, S. Siboni, *Oil Gas Sci. Techn.* **2001**, *56*, 9.

[18] M. Brugnara, C. Della Volpe, A. Penati, S. Siboni, V. Castelvetro, R. Peruzzi, L. Toniolo, Proceedings of the "6th Italian AIMAT Conference", Modena, Italy, 8-11 September 2002.

References

Macromol. Symp. **187**, 177–186 (2002)

Studies for a New Generation of Acrylic Binders for Exterior Wood Coatings

Roland Baumstark*, Franca Tiarks

* BASF Aktiengesellschaft, Development Architectural Coating Raw Materials, D-67056 Ludwigshafen, Germany
BASF Aktiengesellschaft, Polymer Research, Emulsion Polymers, D-67056 Ludwigshafen, Germany
(e-mail: roland.baumstark@basf-ag.de)

Summary: Modern acrylic binders for water-based exterior wood coatings should give films which are tack-free, hard and blocking resistant and at the same time very flexible to guarantee a long service life. This study shows that multiphase acrylic emulsions with controlled particle morphology give a means to overcome these contradictory requirements.
Binder parameters such as polarity, particle size and surface functionality as well as the type and quantity of surfactants used in the process also have a strong influence on fundamental wood coating properties, such as water protection, blushing resistance, viscosity, wet adhesion and durability.

Keywords: emulsion polymerization; wood coatings; particle morphology; polyacrylate, durability

Introduction

Modern binders developed for water-based exterior wood coatings have to fulfill a range of different application properties [1]. The most important ones are listed below:

- long-term weather stability

- permanent elasticity

- excellent blocking resistance and surface hardness

- good adhesion on wood, including adhesion under humid conditions (wet adhesion)

- very good blushing resistance in case of clear lacquers and stains

- good water repellency

- environmentally friendly

- ease of formulation.

        CCC 1022-1360/00/$ 17.50+.50/0

Most of these requirements are logical consequences of the sensitivity of wood as substrate. This esthetical, natural building material has to be protected against water, sunlight and microorganisms attack by the coating.[1]

Acrylic emulsions

Water-borne coatings based on straight acrylic binders are able to fulfill these demands by careful selection of the following basic binder parameters:

- the polarity of the copolymer
- the glass transition temperature (T_g)
- the surface functionality and crosslinking density
- the type and quantity of surfactants (or protective colloids)
- the particle size
- the polymer architecture
- and the particle morphology.

Copolymer composition and polarity

For the synthesis of acrylic binders normally standard soft acrylic monomers such as ethyl acrylate (EA), n-butyl acrylate (n-BA) or 2-ethyl hexyl acrylate (EHA) are used in combination with hard comonomers such as methyl methacrylate (MMA) or n-butyl methacrylate (n-BMA). Styrene is sometimes employed as an alternative for MMA. But higher styrene contents in the polymer and low level of pigmentation in the coating may lead to an increased photo-induced binder degradation. Therefore straight acrylic emulsions dominate the binder market for water-based wood stains and paints.

The binder polarity plays an important role concerning the water barrier properties of a coating. The more hydrophobic it is, the lower is the water swellability of the coating and the better is the protection of the wood substrate against moisture and fungi (see Fig. 1). Therefore in our study emulsions with the most hydrophobic soft monomer EHA in combination with MMA showed the best wood protection properties and the copolymers with the most hydrophilic acrylic monomer EA gave the worst results. N-BA/ MMA displayed an intermediate behavior. This was clearly seen in a series of weathered unpigmented stains on pine.

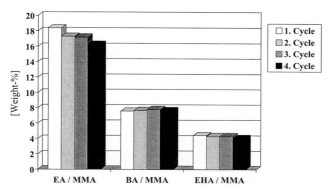

Figure 1. Water uptake of dispersion films with different copolymer composition after several wash/dry cycles.

By introducing special chemical functionality into the binder molecules, such as amino-, acetoacetoxy- or urea-groups the wet adhesion of the coatings to the wood can be significantly increased. This is a consequence of a polar and hydrogen bonding interaction of these functional groups with the wood constituents cellulose and lignin. This hinders flaking during weathering and helps to protect the wood substrate over a longer period.

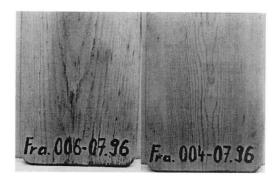

Figure 2. Influence on durability of a special wet adhesion promotor incoporated into the binder: outdoor weathering results (3 years, 45 ° direction south at Limburgerhof, South West Germany); a) binder without wet adhesion promotor, b) binder with special wet adhesion promotor.

The same is true for small particles, which have a higher tendency to penetrate into the complex wooden structure and therefore allow a better anchoring and adhesion of the coating. Nevertheless also the polymer architecture (molecular weight, branching, crosslinking density and gel content) plays a role in penetration and adhesion behavior.

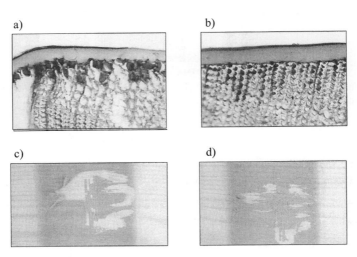

Figure 3. Influence of polymer architecture on penetration and wet-adhesion; a) standard binder; penetration only into first 2 wood cell layers, b) binder with special polymer architecture, penetration till 6 wood cell layers; c) and d) corresponding wet-adhesion tests by cross-cutting.

The T_g of the binder usually determines the balance between hardness and elasticity of the film. The higher the glass transition temperature, the lower the elasticity of the binder film and the better the blocking resistance and surface hardness of a coating.

Sufficient elasticity for the dimensionally unstable substrate wood can be realized by T_g values of the binder lower than 15 °C and preferably about 0 °C.[2] However such soft thermoplastic polyacrylates do not give sufficiently tack-free and blocking resistant films (especially at higher temperatures) when used to coat joinery articles. In other words, windows or doors coated with such materials may stick to their frames when exposed to pressure and higher temperature.

One strategy to enhance the blocking properties and hardness of a soft binder is the introduction of crosslinking.[3] But the established crosslinking systems for water-based paints are normally expensive, often cause pot-life and storage stability problems and sometimes suffer from toxicological concerns.[4] However their main disadvantage for exterior wood coatings is the fact that they also lower the coatings film flexibility and durability.

Particle morphology

A more attractive strategy for exterior wood coating binders is the creation of structured or multiphase acrylic particles by a stepwise semi-batch emulsion polymerization process.[5] This enables the presence of two or more different polymers simultaneously in one particle. Such dispersions that can be produced with non-toxic, standard acrylic monomers are often referred to as core/shell type dispersions.

The trick for exterior wood coating binders is the combination of a soft and a hard polymer phase in one acrylic particle.[6] The low T_g polymer (soft phase) is responsible for a good film elasticity and a sufficient coalescence with the presence of low or no solvents whereas the high T_g polymer (hard phase) provides a good blocking resistance and surface hardness of the films.

The hard phase acts as a nano-scale, transparent, organic filler and provides reinforcement of the soft film and at the same time a surface shielding of the soft-phase material. Both effects increase the blocking resistance. Stepwise emulsion polymerization allows not only the synthesis of core-shell particles, but other types of morphology for two step polymers like acorn, half-moon, strawberry, raspberry, octopus, mushroom, inverse core-shell, and inclusion structures are also possible.[7] In principle, thermodynamic and kinetic parameters determine the particle structure and the resulting application properties.[8] The most important polymerization process parameters for a two step, seeded emulsion polymerization process to create structured particles with a soft and a hard phase are the following: [9]

- the phase composition, which means the copolymer composition, the T_g and polarity
- the phase ratio
- the phase compatibility, which includes polarity but also grafting and crosslinking
- the polymerization sequence: soft phase first or hard phase first
- the molecular weight of the polymer phases, which is regulated e.g. by the quantity of initiator, chain transfer agents and the polymerization temperature.

With fixed phase composition the phase ratio, or weight ratio of soft to hard material, is one of the most important parameters to influence the particle structure and application properties (see example in Fig. 4).

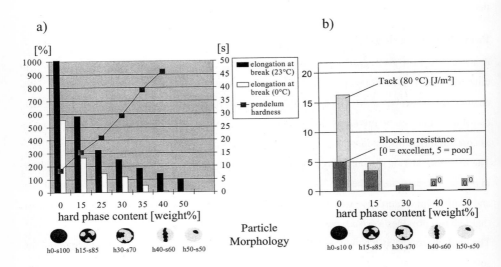

Figure 4. a) Elasticity and pendulum hardness and b) blocking resistance and tack as consequence of the phase ratio (hard phase content) and morphology.

The demand for wood coatings is to tailor the particle morphology to get the best compromise between blocking and elasticity.

Our studies showed that real core/shell particles are unsuitable for wood coating binders. Better properties are achieved with strawberry type or semi-engulfed (half-moon shaped) particles, where the hard phase material is located on the particle surface.

To hinder cracking of the coating on wood during weathering, especially on softer needle woods such as pine or spruce, the binder should fulfill minimum elongation at break values of approximately 100 % even at lower temperatures. This can only be realized with hard phase contents significantly lower then 40 weight % and a T_g of the soft polymer phase in the range of -25 °C to 0 °C.

In these cases the hard phase material, while acting as a transparent filler providing block resistance, does not detract from coating flexibility and elasticity as it is well distributed and oriented in the film because of grafting onto the soft phase during the polymerization.[10]

a)

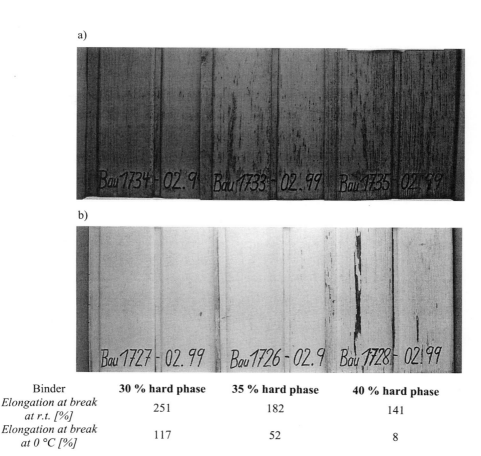

b)

Binder	30 % hard phase	35 % hard phase	40 % hard phase
Elongation at break at r.t. [%]	251	182	141
Elongation at break at 0 °C [%]	117	52	8

Figure 5. Influence of binder elasticity on durability of stains and paints: Outdoor weathering results (2 years, 45 ° direction south at Limburgerhof, South West Germany); a) stains, b) paints at pigment volume concentration 28 %.

Particle size and surfactants

Particle size for modern acrylic dispersions for wood coatings should also be relatively small, i.e. particle diameters significantly lower than 100 nm. This does not only enhance the penetration depth into the wood and therefore the coatings adhesion, but also has a positive influence on other essential properties such as the wet transparency, water protection and blushing resistance of stains. Another benefit of small sized dispersions is their huge internal surface which enhances their interaction with common associative thickeners of the Hydrophobic Alkali Swellable Emulsion (HASE)- or Hydrophobic Ethoxylated Urethanes (HEUR)- type. This, in combination with a low surfactant content allows a cost effective formulation of highly viscous coatings for spray application.

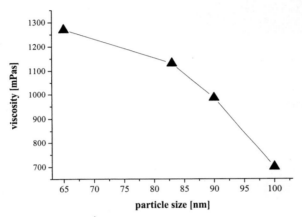

Figure 6. Viscosity (at 100 s^{-1}) of stains and particle size of binders with comparable surfactant content (all formulations with same HEUR-thickener type and amount).

Surfactants are necessary raw materials for the emulsion polymerization process. They control the particle size and give colloidal stability to the ready dispersions. But depending on their type (anionic or nonionic) and quantity they may increase the water sensitivity and blushing tendency of formulated water-based wood coatings and have a negative influence on the coatings adhesion under humid conditions (wet adhesion). They also hinder the development of viscosity in wood coating formulations by disturbing the hydrophobic interaction of associative thickeners with the acrylic particles. Therefore a careful selection of the type and quantity of surfactants is necessary. In Figure 7 an example for the influence of different surfactants on the viscosity of stains formulated with a HEUR-associative thickener is shown.

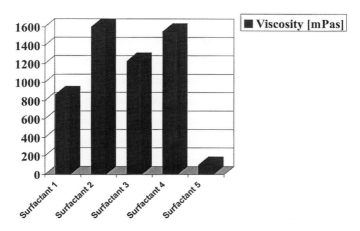

Figure 7. Influence of surfactants on the viscosity (at 100 s^{-1}) of stains formulated with a HEUR-thickener; binders with same copolymer composition produced with different surfactants.

Formulation

The best binder will fail if the overall coating formulation is not adjusted to accommodate it. For example the type and quantity of coalescent has to be optimized, to allow a crack-free and faultless film formation even under critical conditions. Another reason for using solvents, especially propylene glycol, is the extension of open time for brush application. But with modern binders possessing an optimized particle morphology and a sufficiently soft polymer phase, it is possible to obtain even low to zero volatile organic compounds containing coatings.

The thickener type and quantity always has to be optimized with respect to cost and water sensitivity. However, if there is no effective irradiation protection of the lignin in the wood substrate by introduction of UV-A-absorbers (e.g. of the benzotriazole type) or pigments (e.g. rutile or transparent ferrous oxide) in the formulation even with the best binder type the coating will fail after a short time period. The same is true if the layer thickness of the coating is not sufficient.

Conclusion

A modern binder for exterior wood coatings should be a fine-sized, tailor–made, acrylic emulsion with structured particles including soft and hard polymer fractions. This type of binder allows the formulation of wood coatings with a good balance of film forming ability, elasticity and durability, combined with sufficient blocking resistance and surface hardness.

In addition the binder has to be low in surfactant content and should be based on hydrophobic acrylic and methacrylic monomers. To strengthen the adhesion properties and weather stability a surface functionalization of the binder particles with a polar wet adhesion promoter is helpful.

However without an adjusted formulation, a sufficient layer thickness and an excellent UV-protection of the lignin in the wood by using UV-absorbers or pigments in the formulation, the coating with the best binder will fail during outdoor exposure.

References

[1] M. Schwartz, R. Baumstark, „ *Water-based Acrylates for Decorative Coatings* ", Vincentz-Verlag, Hannover 2001, Chapter 5, p.191ff.
[2] E. Schmid, *Applica*, **1998**, *105, No 3*, 10.
[3] R.D. Athey Jr., *Farbe und Lack*, **1989**, *95*, 475.
R. D. Athey, *European Coatings Journal*, **1996**, *816*, 569.
[4] E.S. Daniels, A. Klein, *Progress in Organic Coatings*, **1991**, *19*, 359.
M. Oaka, H. Ozawa, *Progress in Organic Coatings*, **1994**, *23*, 325.
B. G. Bufkin, J. R. Grawe, *Journal of Coatings Technol.* , **1998**, *50*, 41.
J. Feng, H. Pham, P. McDonald, M.A. Winnik, J.M. Geurts, H. Zirkzee, S. van Es, A. L. German, *Journal of Coatings Technol.*, **1998**, *70*, 57.
[5] D.I. Lee, *Makromol. Chem., Macromol. Symp.*, **1990**, *33*, 117.
[6] M.P.J. Heuts, R.A. le Febre, J.L.M. van Hilst, G.C. Overbeek, *"ACS-Symposium Series 648"* ,1996, Chapter 18, 271.
[7] M. Okubo, *Makromol. Chem., Macromol. Symp.*, 1990, *35/36*, 307.
I. Cho, K.W. Lee, *J. Appl. Polym. Sci.*, **1985**, *30*, 1903.
C.L. Zhao, J. Roser, W. Heckmann, A. Zosel, E. Wistuba, *"Proceedings of the 24th International Conference in Organic Coatings"*, 1998, 503.
S. Lee, A. Rudin, *J. Polym. Sci. : Part A: Polymer Chemistry*, **1992**, *30*, 865.
[8] S. Torza, S. Mason, *J. Colloid Interface Sci.*, **1970**, *33*, 67.
J. Berg, D. Sundberg, B. Kronberg, *Poly. Mat. Sci. Eng.*, **1986**, *54*, 367.
D. Sundberg, A.P. Casassa, J. Pantazopoulos, M.R. Muscato, B. Kronberg, J. Berg, *J. Appl. Polym. Sci.*, **1990**, *45*, 1425.
C.L. Winzor, D.C. Sundberg, *Polymer*, **1992**, *33, 18*, 3797.
[9] S. Lee, A. Rudin, *"ACS-Symposium Series 492"*, 1992, Chapter 15, 234.
A. Rudin, *Macromol. Symp.*, **1995**, *92*, 53.
Y.-C. Chen, V. L. Dimonie, O.L. Shaffer, M. S. El-Aasser, *Polymer International*, **1993**, *30*, 185.
[10] Y. Chevalier, M. Hidalgo, J.-Y. Cavaillé, B. Cabane, *"ACS-Symposium Series 648"*, 1996, Chapter 16, 244

*Macromol. Symp. **187**, 187–198 (2002)*

New Generation Decorative Paint Technology

Ad Overbeek*, Fred Bückmann, Ronald Tennebroek, Jan Bouman

NeoResins, P.O. Box 123, 5140 AC Waalwijk, The Netherlands

Summary: The growth of waterborne paints in the higher performance decorative paint segments (like in high gloss paints, trim paints and also in stains and varnishes) has been slowed down due to a lack of open time, rheology, film build and scratch resistance.
Over the years the improvement of the open time has been an intense area of research, but to date the typical paint rheology and brushing performance of a solvent based paint could not be achieved in a waterborne system. In this paper very low viscosity oligomers are presented as a solution for the open time problem. These oligomers remain liquid even after the evaporation of water and are self-crosslinking to achieve good final properties. The oligomers are combined with a dispersed polymer to reduce tack free times of the resulting coatings and to increase the rate of development of the properties. A model system is described where this effect is demonstrated.

1. Introduction

1.1 Definition of open time and wet edge time

Figure 1 schematically represents a drying paint on a substrate. In the literature [1], the open time of the paint has been defined as the period of time, during which a painter can make corrections to the freshly applied wet paint film without leaving brush marks. Wet edge time (also referred to as lapping time) has been defined as the period of time during which no edge marks are produced when a freshly applied paint film is lapped over a previously painted area. The edges of the paint film are thinner and therefore will dry more rapidly. If the edge of the paint is dry and is subsequently repainted with a new layer of paint, the dried edge will telegraph through the freshly applied paint layer and cause an irregularity at the paint surface. This is undesirable. Since the edges dry faster than the bulk of the paint the wet edge time is the most critical parameter determining how long the paint remains "open".

Wet edge / lapping time (edge)　　　　**Open time (bulk)**

Figure 1. *Wet edge time / lapping and open time.*

1.2 Open time, through drying and viscosity of paints and coatings

In Figure 2, three important viscosity based events are related to the speed of drying. Firstly, the paint requires a certain open time as discussed above (low viscosity). The next stage is inevitably a tacky paint film (medium viscosity), which can no longer be corrected but will easily be damaged. Preferably this stage is very short. Then a certain hardness and abrasion resistance will be achieved (high viscosity), and the paint film should become sandable if a second coat needs to be applied. At this stage, the paint film is not easily damaged any more. This final stage should preferably be reached as quickly as possible.

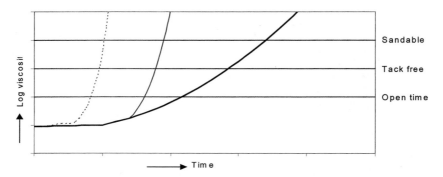

Figure 2. Drying profiles of different paints.

Conventional acrylic dispersions typically have a drying profile represented by the dotted line in Figure 2. They have a short open time, but are also rapidly achieving final properties, and become tack free in a short period of time. The bold straight line in Figure 2 represents conventional solvent based alkyd paints. Whereas they have a long open time, solvent based alkyds in many cases probably dry too slowly; tack free times are most usually many hours. Therefore, in a perfect system the open time would be much longer than for the current waterborne systems, but once this time has elapsed the final stage should be reached rapidly, like is depicted by the thin line in Figure 2. In other words, in the initial stages the drying behaviour of a solvent based alkyd is desired, while at the later stages the drying behaviour of a waterborne dispersion becomes more preferred.

1.3 Novel low viscosity oligomers

To achieve the desired improvement in open time and wet edge time the viscosity of the drying film needs to be low beyond the point after which most water has evaporated. The evaporation of water is too fast, even if techniques are used to delay this evaporation, or if very hydrophilic polymers are used. The desired very low viscosity after evaporation of water can only be achieved if low molecular weight oligomers, with preferably limited chain interactions (such as hydrogen bridging) are used. Of course small amounts of slowly evaporating solvents can be added to achieve even more control over the viscosity and rheology.

2. Results and Discussion

2.1 Experimental Section

2.1.1 Synthesis model system for a waterborne paint with good open time.

For the development of waterborne paint with a good open time and to illustrate the concept, we used a model binder system consisting of an oligomer dispersion and an acrylic dispersed polymer latex. The oligomer dispersion is prepared by dispersion of a carboxylic acid containing fatty acid functional urethane oligomer in water with the aid of a neutralising agent and a small amount of organic solvents. The urethane oligomer PU-1 is the product of the addition reaction between a diisocyanate and a polyol mixture consisting of a fatty acid polyol, dimethylol propionic acid, methoxy polyethylene glycol 750 and cyclohexane dimethanol [2]. A drier salt is used for autoxidative crosslinking of the fatty acid functionality of the urethane oligomer. The acrylic dispersed polymer latex P-1 is a BA/MMA/AA copolymer prepared by a standard emulsion polymerisation technique [3]. The oligomer polymer paint system is prepared by blending of the oligomer dispersion with the acrylic dispersed polymer latex and subsequent formulation of the binder to a paint system.

2.1.2 Test methods for wet edge open time, dust free and tack free time

To test the wet edge open time, the aqueous pigmented paint was applied to a test chart (18x24cm, form 8B - display, available from Leneta Company) using a wire rod, at a wet film thickness of 120μm. Wet edge open time tests were performed at fairly regular time intervals, the intervals between measurements decreasing towards the end of the run. The measurements were carried out at relative humidity levels of 50 +/- 5%, temperatures of 23 +/- 2°C and an airflow ≤ 0.1m/s.

The *wet edge open time* [4] was determined by brushing a virgin 25cm² edge area of the coated chart with a brush (Monoblock no 12, pure bristles/polyester 5408-12) carrying some more of the composition with a brush pressure of 100-150g during 30 seconds. In this time the brush was moved in one set of 5 times in the direction of the width of the substrate and 5 times in the direction of length of the substrate before visually assessing the coating.

Once the composition carried on the brush no longer formed a homogeneous layer with the coating on the substrate and/or a visible lap line could be seen the wet edge time was considered to be over. To test the wet edge time, the binders were evaluated in a pigmented paint formulation; see table 1:

Table 1. *A typical pigmented paint formulation*

	Ingredients		Part by weight
1.	Binder (based on PU-1 and P-1)		118.4
2.	Water		11.8
3.	Dehydran 1293	(defoamer trademark from Cognis)	0.8
4.	Disperbyk 181	(wetting agent trademark from Byk Chemie)	1.6
5.	TiO$_2$ RDIS	(Trademark from Finntitan)	45.8
6.	Dehydran 1293	(defoamer trademark from Cognis)	2.0
7.	NLS-210	(low shear thickener trademark from Hercules)	1.0
8.	Nuvis FX-1070	(high shear thickener trademark from Sasol Servo)	4.0
9.	Dapro 5005	(Drier salt trademark from Elementis Specialties)	0.15
		Total:	185.6

2.2 Phase inversion

The oligomers used for these new decorative coating systems are initially present as dispersed particles in the water phase. However, upon drying these particles will coalesce and eventually form the continuous phase after "phase inversion" has taken place. Since this newly formed continuous phase has a low viscosity and has the required rheological behaviour as also found in solvent based paints, the open time of the paint has been decoupled from the evaporation of water. As will be explained in more detail, the phase inversion point is a critical point in such a system. Certain aqueous binders, like for instance alkyd emulsions, are going from an oil in water to a water in oil dispersion during drying. Generally, a very high viscosity peak is observed during this phase inversion process, which is very undesirable when good open time and lapping are required. This can be observed in Figure 3, in which two solids-viscosity curves are presented.

In this graph a conventional oligomer is compared with the low viscosity oligomer PU-1. The latter has a significantly lower and more narrow phase inversion peak. As a consequence, it will require less force to brush through a paint film based on this novel oligomer at the phase inversion solids than it would be for a paint based on the conventional oligomer. With the conventional oligomer system a painter would experience severe tugging and a lack of flow of the paint.

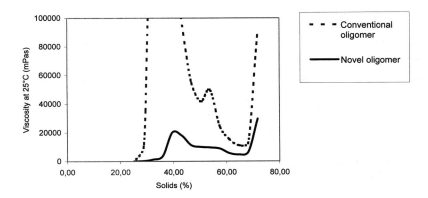

Figure 3. *Phase inversion peak of the low viscosity urethane oligomer PU-1 and of a conventional oligomer, in which the solids is decreased by diluting the oligomers with water.*

In Figure 4, the viscosity and conductivity of the urethane oligomer during phase inversion is measured, starting at low solids and increasing the solids level by blowing N_2 over the surface of the sample. If the conductivity is measured during drying a decrease in conductivity is found during the phase inversion, which is expected since water will no longer be the continuous phase, but will become a dispersed phase inside a continuous oligomer phase. A small increase in conductivity is found at around 50% solids, which is most likely caused by disappearance of the dispersed water droplets. The remaining water, and salts dissolved in this water, will now dissolve in the continuous phase causing the increased conductivity. From here on the conductivity will gradually decrease further since the dissolved water will continue to evaporate.

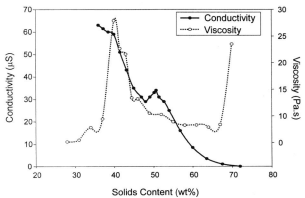

Figure 4. *Phase inversion of the urethane oligomer PU-1, starting at low solids; increase in solids is obtained by blowing N_2 over the sample surface.*

It is important to observe that the viscosity recovers to a very low value after the phase inversion, where in conventional waterborne systems the viscosity would remain infinitely high after the particles have come in contact with each other. This demonstrates that the concept of using low viscosity oligomers to control the rheology after evaporation of most of the water will indeed result in a significantly extended time during which the coating film will remain low in viscosity and hence "open". One of the reasons why the low and narrow phase inversion peak for PU-1 will not result in brush drag is that the solids increase due to water evaporation, as discussed earlier, will be dependent on the distance from a certain location in the coating to the coating surface. Therefore, the degree of phase inversion during drying will not be equal at all depths in the coating (See Figure 5) and especially for narrow phase inversion peaks this will result in a "smoothing" of the viscosity peak of the paint film. The painter will only experience the total force needed for brushing, so very thin layers with a solids close to phase inversion solids and with the concomitant high phase inversion viscosity, will not ruin the brushability, since the painter will only experience the average force needed to brush through the whole paint film. By strictly controlling the design of the oligomers, the undesired (high and broad) phase inversion peak viscosity could be further reduced and almost eliminated.

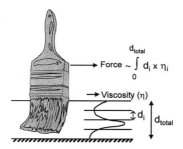

Figure 5. *Viscosity profile through the drying paint film.*

The phase inversion is also visualised in Figure 6 showing the appearance of the urethane oligomer PU-1 at different solids. It can clearly be seen from this Figure that the "phase inversion" followed by the dissolution of the water droplets, is completed at ca. 52% solids.

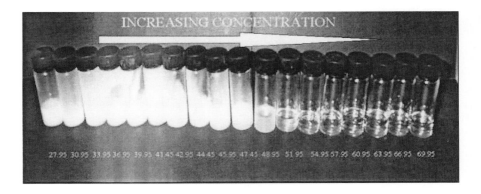

Figure 6. Appearance of the oligomers at increasing solids level.

During the drying process, the ratio of solvent to water will continue to increase (see Figure 7) since the evaporation rate of the solvents used is lower than that of water. Due to this increased solvent to water ratio the oligomer particle will "dissolve" in a solvent/water mixture after a critical solvent/water ratio is reached. If a less water miscible solvent is used,the "dissolution" of oligomer particles in the continuous phase will be delayed and a more conventional phase inversion from "oil in water" to "water in oil" will occur.

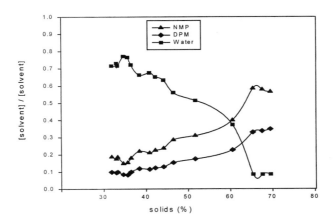

Figure 7. Fractional solvent content during drying of the urethane oligomer dispersion PU-1.

The rheology of the oligomer dispersion (see Figure 8) before phase inversion is (slightly) shear thinning like most waterborne dispersions. This can be expected since at this stage the oligomers are present as dispersed particles. The shear thinning behaviour progressively increases with increasing solids. At phase inversion and the associated viscosity maximum, the system is strongly shear thinning. After the phase inversion the system behaves Newtonian [5].

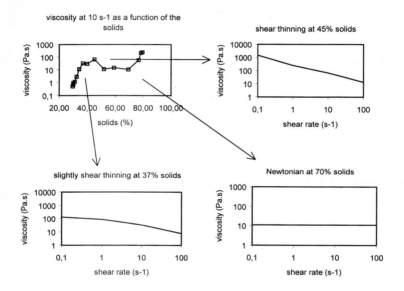

Figure 8. *Rheology profiles of the urethane oligomer dispersion PU-1 at various points on the solids/viscosity curve.*

2.3 Closing the paint

Obviously, if oligomers of low viscosity are used as the sole binder component, the resulting paint layer will remain open forever, which is clearly not desired. The final paint should have a reasonable dust and tack free time as well and should become sandable within a specific timeframe so that it can be repainted if required. Conventional solvent based alkyd paints close because:

1. the solvents are evaporating (alkyds are not that low in viscosity that they still flow at 100% solids).

2. alkyds are self-crosslinking by autoxidation (the crosslinking will build the molecular weight, and thereby the properties of the coating will improve).

Similarly, the closing of waterborne paints containing oligomers can be controlled by:

1. evaporation of water (and the solvent(s) that are optionally present).
2. self-crosslinking of the oligomers (to build the molecular weight and final properties).
3. combination with a dispersed polymer, of which it is known that good final properties can be achieved very rapidly.

The advantage of using small amounts of slow evaporating solvent(s) is that the viscosity is reduced even at relatively high oligomer molecular weights. When the oligomers have a higher initial molecular weight, less crosslinking will be required to achieve the desired properties, which will also occur faster.

The addition of a dispersed polymer of high molecular weight will certainly help to reach the desired end properties more rapidly (see Figure 9). If the dispersed polymer is designed to be stable as a dispersion in water and also stable as a dispersion in the oligomer phase after phase inversion, the final film will consist of a matrix of self-crosslinking oligomers with high molecular weight polymeric particles dispersed herein. Depending on the loading of the dispersed polymer in this system the polymer particles can even touch each other to form a bi-continuous film.

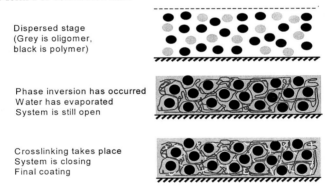

Dispersed stage
(Grey is oligomer,
black is polymer)

Phase inversion has occurred
Water has evaporated
System is still open

Crosslinking takes place
System is closing
Final coating

Figure 9. Film formation of the waterborne open time system.

But even if they do not touch the entanglement of the crosslinking oligomers with chains of the polymer inside the particles will cause a very rapid property development. The film formation process, as depicted in Figure 9, starts with the initial dispersion with dispersed particles of oligomer as well as dispersed polymer particles. Upon evaporation of water, phase inversion will occur and the continuous phase will now be the oligomer with polymeric particles being dispersed therein. At this stage the film is still "open". Due to the combined effect of molecular weight build up by self-crosslinking[20, 21]. and the continued evaporation of water and cosolvent, the wet paint will start to close and the final properties are building up rapidly. Using this approach the combination of the excellent final properties of acrylic dispersions can be combined with the strongly desired rheological behaviour of conventional solvent based paints.

2.4 Drying at different blend ratios of oligomer and polymer

A set of oligomer/polymer blend ratios were prepared to study the optimal ratio for a good wet edge open time with good through drying times. These blend systems were formulated into a paint (see Section 2.1) and checked for their drying characteristics. For reference purposes, a set of blend ratios of a conventional oligomer (alkyd emulsion) with the same acrylic dispersion was compared in the same paint formulation. The results are summarised in Figures 10 and 11.

As can be seen from Figure 10, good (wet edge) open times can be achieved by using PU-1 oligomer levels as low as 30% on binder solids, whereas the (wet edge) open time of alkyd emulsion based paints are short throughout the range of oligomer polymer ratios. From Figure 11 can be seen that acceptable dust and tack free times can be achieved by selection of the appropriate level of dispersed polymer in the paint. In conclusion, especially at oligomer levels between 20 and 60% an interesting combination of drying properties can be obtained. Below 20 % oligomer wet edge times become very short and above 60 %, tack free times get unacceptably long.

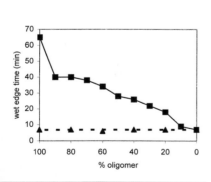

Figure 10. Wet edge open times of paint systems on Leneta test chart based on different ratios of the blend of the urethane oligomer PU-1 and the acrylic dispersed polymer P-1 (straight line), compared to similar blends of a commercial alkyd emulsion and P-1 (dashed line).

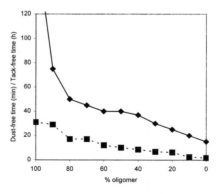

Figure 11. Dust free times (straight line) and tack free times (dashed line) of paint systems based on different ratios of the urethane oligomer PU-1and the acrylic dispersed polymer P-1.

2.5 Comparison of the new technology with commercial paints.

As illustrated already in some detail, there is a clear difference between paint systems based on solvent borne technology, waterborne latex technology, alkyd emulsion technology and this new oligomer polymer paint technology.

Benchmarking commercial products which are available in the decorative market resulted in the following overview in figure 12. For reference purposes, the results for open time of the new oligomer/polymer concept are compared to the formulated acrylic dispersion TX200, a formulated alkyd emulsion, four commercial products as well as a solvent based alkyd.

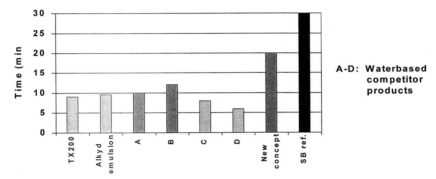

Figure 12. *An overview of the open time of waterbased paints compared to the new concept and a solvent based alkyd; all paints evaluated on Leneta test chart (non-absorbing).*

This figure clearly shows that still a big gap exists between open time of the current -state of the art- waterbased products and a solvent based alkyd. The new oligomer/polymer system is clearly outperforming all current waterbased paints available on the market with respect to open time.

3. Conclusion

A waterborne paint with an improved wet edge open time and a relatively short tack free time can now be obtained by the combination of a dispersion of a very low viscosity oligomer and a high Mw dispersed polymer. The lack of this piece of technology has retarded the changeover from solvent based to waterborne decorative coatings for the last twenty years. The expectation is warranted that we will see significant activity in the coatings industry in the coming years to explore the potential of this new technology for further improvement of the safety, health and environmental profile of decorative coatings.

4. Acknowledgements

We would like to thank Emilio Martin, Pablo Steenwinkel, Tom Annable John Padget, Jurgen Scheerder, Ludy van Hilst, Ad Ernst, Ankie van Gorkum, John Gerrard, Rob le Febre, Sjoerd Buil, Frank Nettenbreijers, Mourad Aabich, and Lex Donders for their contribution to the theoretical and practical work described in this paper.
Part of this article has been presented during the 2002 Athens Conference on Coatings Science & Technology.

5. References

1. Control of rheology of waterborne paints using associative thickeners. A. J. Reuvers, Proc. - Int. Conf. Org. Coat.: Waterborne, High Solids, Powder Coat., Institute of Materials Science, New Paltz, N. Y., 24th Edition, p. 337-356 (1998).
2. WO 2002/32980, Avecia.BV, Example 1.
3. WO 2002/32980, Avecia BV, Example 2.
4. WO 2002/32980, Avecia BV, page 26.
5. A.Overbeek, F. Bückmann, E. Martin, P. Steenwinkel, T. Annable, 2002 Athens Conference on Coatings Science and Technology, pp 199-216.

Macromol. Symp. 187, 199–206 (2002)

The Use of Nonionic Polymerizable Surfactants in Latexes and Paints

Christof Arz

Collano AG, CH-6203 Sempach-Station, Switzerland

Summary: In combination with anionic surfactants nonionic polymerizable surfactants lead to stable dispersions, which can be used as binders for paints. Generally the obtained results are worse than with a nonpolymerizable ethoxylated fatty alcohol. However the properties can be optimized through the right choice of the anionic surfactant. This further work is in progress.

Introduction

Emulsion polymers are often used as binders for paints, coatings, adhesives, etc. There are different ways for optimizing the properties. One possibility could be the use of polymerizable surfactants. 1959 a US-Patent mentioned the increase of the stability using polymerizable compounds with sulfonic acid groups [1]. These compounds are of anionic nature. About nonionic polymerizable surfactants there aren't so much datas in the recent literature. Ottewill and co-workers [2] investigated methoxy polyethylene glycol methacrylates in polystyrenes and found an increased electrolyte stability. This increase was only found after polymerization, blends from methoxy polyethylene glycol methacrylates with polystyrenes showed no effect. Yokota, Ichihara and Shin'ike used 1-Nonylphenoxy-2-polyoxyethylene-3-allyloxypropane to increase the water resistance of polymerfilms. Ferguson et al [4] could demonstrate, that the increase of the HLB-value from alkylethoxy acrylates increased the elecrolyte stability. Guyot and Co-workers [5] produced styrenes with hydrophobic carbonchains and hydrophilic polyethyleneoxid chains. They found, that longer polyethyleneoxid led to smaller particle sizes and better stabilities. The same researchgroup investigated also dialkyl maleates [6]. Gan et al [7] produced micro emulsions with ω-methoxy polyethyleneoxid-undecyl-α-methacrylate. Chern, Shi and Wu [8] found, comparing polymerizable NE-40 with nonpolymerizable NP-40, a larger particle size. They explained this result with the fact, that one part of the polymerizable surfactant is located into the

     CCC 1022-1360/00/$ 17.50+.50/0

particle and can not help to stabilize the particles. Reb et al [9] achieved, caused by a regular distribution of carboxylic groups, a high stability with polymerizable isophthalate surfactants. The problem of stabilization with polymerizable surfactants was investigated from Asua and co-workers [10] regarding to crotonates, methacrylates and maleates. The nonreactive crotonates led to unstable emulsions. The very reactive methacrylates showed a lot of coagulum. The maleates with a medium reactivity led to stable dispersions with increased water resistance and electrolyte stability. Guyot and co-workers [11] found the same results producing styrene or butyl acrylate latexes with polymerizable maleates.

In this article, the use of commercial available surfactants is investigated. 13 nonionic types were used to produce latexes and to formulate paints.

Experimental

Latexes were produced with classical methods for emulsion polymerization. To compare the results a nonpolymerizable ethoxylated fatty alcohol has been used. Table 1 contains a typical recipe.

Table 1. Polymer recipe

Reactor charge	Water	214,1
	30% Alkyl ether sulfate	0,9
	Polymerizable surfactant	0,5
Emulsion	Water	225,3
	30% Alkyl ether sulfate	29,9
	Polymerizable surfactant	16,5
	Sodium bicarbonate	0,9
	Vinyl acetate	100,5
	Methyl methacrylate	167,6
	Butyl acrylate	212,8
	Vinyl triethoxysilane	2,5
	Methacrylic acid	1,9
	Sodium bisulfite	0,5
Initiator 1	Potassium persulfate	0,4
	Water	5
Initiator 2	Potassium persulfate	0,7
	Water	20
	Total	1000

Coagulum, viscosity, particle size and water up take were determined for all latexes. After that paints were formulated and scrub resistance and pigment stability were measured.

Table 2 contains the different polymerizable surfactants:

Table 2. Polymerizable surfactants

Name	Chem. composition	Manufacturer
Emulsogen R 109	Vinyl polyalkylene glycol ether 10 EO	Clariant
Emulsogen R 307	Vinyl polyalkylene glycol ether 30 EO	Clariant
Sinnoester CPM 1	Dodecyl polyethyleneoxid maleat 17 EO	Cognis
Sinnoester CPM 2	Dodecyl polyethyleneoxid maleat 34 EO	Cognis
Sinnoester CPM 3	Dodecyl polyethyleneoxid maleat 43 EO	Cognis
Maxemul 5010	Alkenyl/carboxyfunctional hydrophobe with 25 EO	Uniqema
Maxemul 5011	Alkenyl/carboxyfunctional hydrophobe with 34 EO	Uniqema
PEM63P	Polyalkylene glycol methacrylate	Laporte
PPM63E	Polyalkylene glycol methacrylate	Laporte
MPEG 230 MA	Methoxy polyethylene glycol 230 methacrylate	Prochema
MPEG 400 MA	Methoxy polyethylene glycol 400 methacrylate	Prochema
MPEG 550 MA	Methoxy polyethylene glycol 550 methacrylate	Laporte
MPEG 750 MA	Methoxy polyethylene glycol 750 methacrylate	Röhm

Results

All surfactants led to stable latexes. Table 3 contains all technical datas. The solids of these latexes were about 50% and the pHs were about 5. The TG-values were measured with DSC-Spectroscopy. All values were in a range of about 11 °C.

Table 3. Technical datas of the vinyl acetate/acrylate-copolymers

	coagulum	viscosity	Water up take	Particle size	Pigment stability	Scrub resistance
	gr / 2000 gr Latex	mPa.s	%	nm		cycles
Ethoxylated fatty alcohol	0,07	128	9,3	129	No	850
Emulsogen R 109	0,1	80	12,1	176	No	550
Emulsogen R 307	0,1	64	10,9	197	No	500
Sinnoester CPM 1	0,23	420	13,1	142	No	650
Sinnoester CPM 2	2,44	65	17,7	211	No	370
Sinnoester CPM 3	4,68	64	21,2	237	Yes	500
Maxemul 5010	0,13	92	7,8	148	No	750
Maxemul 5011	0,1	136	8,3	144	No	630
MPEG 550 MA	2,44	136	16,2	147	No	370
MPEG 400 MA	0,1	224	9,5	152	No	450
MPEG 230 MA	0,1	408	7,9	136	No	430
Plex 6850-0	34,7	88	15,9	140	No	350
PEM63P	0,19	112	8,5	170	No	410
PPM63E	0,51	352	8,4	153	No	620

The formulated paint with the non polymerizable surfactant gave the best results. 5 polymerizable surfactants led to lower water up take and none showed a lower particle size as the non polymerizable surfactant. To explain these results tests were done with variations of the surfactant in the reactor charge. Figure 1 contains the viscosities of the latexes with methoxy polyethylene glycol methacrylate as the polymerizables surfactants.

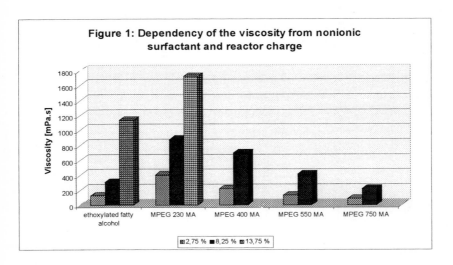

We see a strong increase of the viscosity with increasing quantities in the reactor charge. The three methoxy polyethylene glycol methacrylates with high molecular weight led to unstable latexes at the highest reactor charge. Interesting was the decrease of the viscosity with longer chain of ethylene oxid. The viscosity correlate to the particle size. Higher viscosity led to smaller particle size. Measuring the scrub resistance we receive totally different results, seen in figure 2.

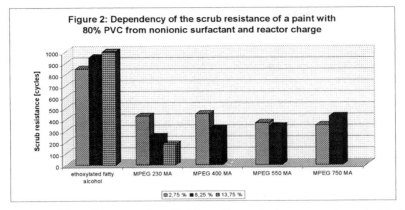

Figure 2: Dependency of the scrub resistance of a paint with 80% PVC from nonionic surfactant and reactor charge

Increasing quantities of the nonpolymerizable surfactant in the reactor charge led to higher scrub resistance. This is in contrast to what we see with the polymerizable methoxy polyethylene glycol methacrylates. A possible explanation could be that with the polymerizable surfactants one part of the surfactant is enclosed in the particle and can not take part at the stabilization of the pigments.

The Sinnoesters show another problem of polymerizable surfactants. Figure 3 contains the coagulum after polymerization.

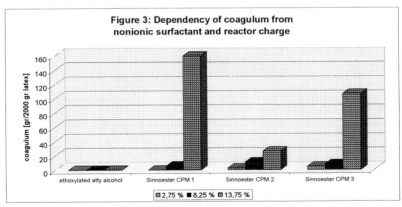

Figure 3: Dependency of coagulum from nonionic surfactant and reactor charge

Increasing the content of surfactant in the reactor charge the nonpolymerizable fatty alcohol shows that the content of coagulum remains stable. On the other hand we observe with the three Sinnoesters a strong increase of the coagulum content.

Measuring the water up take of the different polyalkyleneglycols, increasing contents in the reactor charge led to an increase or a decrease of the values, figure 4. These values correlate with the particle size. For a plausible explanation a further study is in progress.

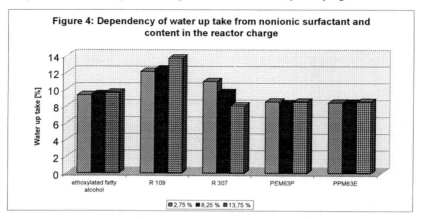

Figure 4: Dependency of water up take from nonionic surfactant and content in the reactor charge

In a second step four surfactants were tested with different monomer compositions. The TG-values of these different latexes remained constant. Figure 5 contains the viscosities of these latexes. We found very different values with no correlation to particle sice or scrub resistance. The terpolymer with Emulsogen R 307 as the polymerizable surfactant showed with 1050 cycles the highest value. Generally we can say that the polymerizable surfactants show with different monomer compositions worse results than with nonpolymerizable ethoxylated fatty alcohol.

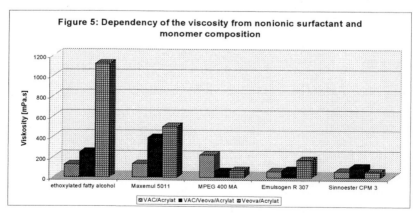

Figure 5: Dependency of the viscosity from nonionic surfactant and monomer composition

Measuring the water up take of dried films, we see a decrease of the values with increasing hydrophobicity of the monomers, figure 6. We also see smaller differences of the values. This means, increase of the hydrophobicity of the monomers lead to a smaller influence of the surfactant.

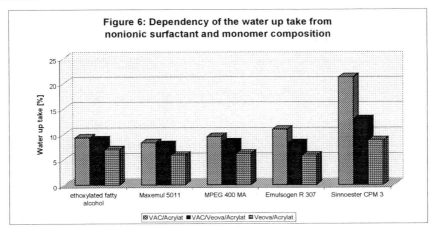

Figure 6: Dependency of the water up take from nonionic surfactant and monomer composition

In a third step we investigated the influence of the anionic surfactant. We made a combination of Maxemul 5011 with a sulfo succinate, an ethoxylated fatty alcohol sulfate and a dodecyl benzene sulfonate. Figure 7 contains the particle sizes.

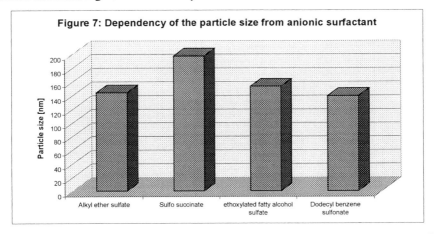

Figure 7: Dependency of the particle size from anionic surfactant

The results for the particle size analysis correlate with all other measured parameters. For example figure 8 contains the scrub resistance. It seems, that the polymerizable surfactant is strongly influenced by the anionic surfactant. This result was confirmed by Asua and co-workers [12] in a mathematical model of the polymerization with polymerizable surfactants. Studies continue to increase the scrub resistance of paints, formulated with polymerizable surfactants.

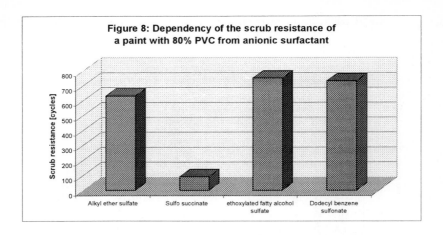

Figure 8: Dependency of the scrub resistance of a paint with 80% PVC from anionic surfactant

References

[1]   D. Sheetz, U.S. Pat. 2 714 499, **1959**
[2]   R.H. Ottewill, R. Satgurunathan, F.A. Waite, M.J. Westby; Br. Polym. J.; **1987**, 19, 435
       R.H. Ottewill, R. Satgurunathan; Colloid Polym. Sci.; **1988**, 266, 547
       R.H. Ottewill, R. Satgurunathan; Colloid Polym. Sci.; **1995**, 273, 379
[3]   K. Yokota, A. Ichihara, H. Shin'ike; Industrial applications of surfactants III ed. D.R. Karsa, **1992**, 29
[4]   P. Ferguson, D.C. Sherrington, A. Gough; Polymer; **1993**. 34, 3281
[5]   A. Filet, J. Guillot, T. Hmaide, A. Guyot; Polym. Adv. Technol.; **1995**, 6, 465
[6]   S. Abele, M. Sjöberg, T. Hamaide, A. Zicmanis, A. Guyot; Langmuir; **1997**, 13, 176
[7]   L.M. Gan, J. Liu, L.P. Poon, C.H. Chew, L.H. Gan; Polymer; **1997**, 38, 5339
[8]   C. Chern, Y. Shi, J. Wu; Polym. Int.; **1996**, 40, 129
[9]   P. Reb, K. Margarit-Puri, M. Klapper, K. Müellen; Macromolecules; **2000**, 33, 7718
[10]   M.J. Unzué, H.A.S. Schoonbrood, J.M. Asua, A.M. Goni, D.C. Sherrington, K. Stähler, K.-H. Goebel, K.
       Tauer, M. Sjöberg, K. Holmberg,; J. Appl. Polym. Sci.; **1997**, 66, 1803
[11]   O. Sindt, C. Gauthier, T. Hmaide, A. Guyot; J. Appl. Polym. Sci.; **2000**, 77, 2768
[12]   J.C. de la Cal, J. M. Asua; J. Polym. Sci., Part A: Polym. Chem.; **2001**, 39, 585

Binders for Exterior Coatings Exhibiting Low Soiling Tendency

Christian Meiners*[1], Ivan Cabrera[1], Thomas Fichtner[1], Rolf Kuropka[1]

Ralf Grottenmüller[2], Helmut Zingerle[2], Carlos Ibarrarán[3],

Jean-Yves Anquetil[4]

[1]Clariant GmbH, BU Emulsions, Frankfurt am Main, Germany

[2]Clariant GmbH, BU Textile Chemicals, Burgkirchen, Germany

[3]Clariant México S.A., 55540 Ecatepec, México

[4]Clariant France S.A., BU Emulsions, 60350 Trosly Breuil, France

Summary: Elastomeric coatings protect building facades and flat roofs from moisture and weather exposure. As a drawback, these coatings are prone to dirt-pickup due to the low glass transition temperature of the used polymeric binders. Strategies to overcome this enhanced soiling tendency are discussed, and the results of laboratory and outdoor soiling tests are compared. A novel method for the assessment of wet soiling tendency is presented.

Introduction

Elastomeric (wall) coatings are capable of bridging fissures in substrates up to about 2 mm width without necessity of prior treatment of the cracks with sealers. They are expected to withstand the dimensional changes caused by extension and contraction of the substrates. Furthermore, elastic coatings provide reasonable protection e.g. for flat roof constructions against water infiltration. The good protection can also be ascribed to the fact that by application of elastomeric coatings, further opening of the existing cavities by freezing water during the winter is avoided.

They should also impart protection against CO_2-passage through the applied film and against environmentally adverse influences. At the same time the coating should permit the ´breathing´ of the building: The potential of water vapour to pass the coating barrier must be given, otherwise moisture accumulating in the building might give rise to the growth of fungi.

Besides these technical requirements for elastomeric coatings, there exist also aesthetical ones, which have to be fulfilled in order to meet market´s expectations: The coatings

CCC 1022-1360/00/$ 17.50+.50/0

should not exhibit yellowing or other unwanted colour changes. Chalking and soiling should not occur either, since all these factors bring about substantial changes of the initial optical aspect of the paint film.

One potential drawback of these coatings is their inherent tendency for dirt-pickup due to the low glass transition temperature of the polymeric binders used. To provide the desired low-temperature elasticity needed for efficient gap bridging, the T_g of binders used for elastomeric coatings is usually in the range between -10 and $-40°C$.

Unlike conventional facade paints exhibiting higher glass transition temperature of the polymeric binder, elastomeric coatings contain polymers being always in a fluid-like state in the temperature range encountered in a temperate climatic zone (-10 to +35°C).

At the usual pigment volume concentrations of elastomeric paints, the soft continuous matrix of the binder shows poor resistance to indentation of e.g. dirt and dust particles.

To overcome this disadvantage, different concepts have been used to develop a technical solution to the problem.

Inter alia, polymer emulsions with high fluorine content have been studied.[1] Although these polymers impart hydrophobic and oleophobic properties to the coated surface, cost considerations exclude the general use of binders with high fluorine content in elastomeric coatings.

Also, silicones are widely used as anti-soiling agents. However, doubts about their effectiveness have been reported recently, since these additives did not prevent the surface deposition of dust particles in (styrene) acrylic-based facade coatings.[2] In this case, the soiling tendency may be ascribed to the hydrophobic coating surface, attracting the lipophilic fraction of the dust.

Another concept used to impart dirt repellence to elastomeric coatings is based on the observation, that soiling can be inhibited by superficial crosslinking reactions of the binder, triggered for example by UV-irradiation. Such a surface reticulation has successfully been accomplished by the use of photoinitiators like benzophenone. [3]

Further approaches to reduce soiling include microstructuring of surfaces[4], application of nanoparticle top-layers on coatings or thin films of hydrophobic silanes. Finally, the use of low molecular weight amphiphilic substances like perfluoroalkane carboxylic acid salts which migrate to the coating surface has been proposed for soiling reduction of coatings.[5]

Here we report on the soiling tendency of especially modified binders for elastomeric wall coatings (EWCs). The modified binders contain moieties of the type shown schematically in structure **I**[6]. Furthermore, an own newly developed system for laboratory wet soiling of samples will be described.

I

This method was established in our laboratory with the intention to better mimic the dirt-deposition processes occurring in outdoor weathering under changing humidity conditions, where continuous adsorption/desorption of particles from coating surfaces takes place. We believe that laboratory dry soiling tests are no suitable probe for the determination of soiling tendency, since the crucial role of the water in the soiling process is ignored: Water can transport dirt in form of aqueous dirt suspensions into film cavities, but also wash off dirt particles adhering to the surface. To mimic these transport mechanisms, we continuously pumped an aqueous dirt suspension over the test panel surfaces and evaluated the soiling after defined cycles.

Experimental

Paint manufacture:

Different pure acrylic (PA) and one styrene acrylic (SA) binders (see table 1) have been used to produce elastomeric wall coatings with pigment volume concentration 42%. Mill base has been prepared using a dissolver. The ready mill base was allowed to equilibrate for 24 h prior to addition of binder and final adjustment of paints (see table 2).

Dry soiling:

For dry soiling studies, coatings were directly applied to glass plates with a doctor blade (200 μm wet film thickness) and allowed to dry for 24 h at 23°C, 50% relative air humidity. Determination of the brightness L^* of the CIELAB system was accomplished with an Erichsen colorimeter model Nr. 526, using a white standard with $L^* = 94,33$. Then a mixture of 99.5% fly ash and 0.5% soot was applied to the substrates with a brush, and surplus particles were brushed off. The measurement of L^* was then repeated with the soiled substrates, and the difference to the initial L^*-value is named ΔL^*-value in this work. Soiling reduction in % is defined by the quotient of ΔL^*-values obtained for the coating containing the modified (ΔL^*mod) and the unmodified (ΔL^*) binder:

$$\% \text{ Soiling reduction} = [1-(\Delta L^*\text{mod}/\Delta L^*)] \times 100 \qquad (1)$$

Table 1. Tested aqueous polymer emulsions

Sample Nr.	Emulsion	Glass transition temperature [°C]	Solids content [%]
1	PA 1	-30	approx. 60
2	PA 2	-30	approx. 60
3	PA 3	-30	approx. 60
4	PA 4	-35	approx. 60
5	SA 1	-30	approx. 60
6	PA 1 mod 1	-30	approx. 60
7	PA 1 mod 2	-30	approx. 60
8	PA 1 mod 3 (comparison)	-30	approx. 60

Table 2. Preparation of EWC-coatings (PVC 42%)

	Component	Weight-parts	Comment
	Mill base:		
1	Water	100	Add components 2 – 6
2	Calgon N (10% aqu. solution)	5	in component 1 and mix
3	Coatex P 90	1,4	under stirring.
4	Foammaster 111 FA	2	Then shear the mixture for 15 min at 5000 rpm
5	Kronos L 2310	80	using a dissolver.
6	Durcal 2	380	Allow the mill base to equilibrate for 24 h.
7	Emulsion (s.c. about 60%)	382,3	Add components 7 and 8 at 500 rpm and stir for 5 min.
8	Ammonia (20% aqueous solution)	1	
	Final adjustment:		
9	Mergal K9	2	Add comp. 9 – 14 under stirring
10	Butyl diglycol	2	
11	Propylene glycol	10	
12	White spirit 17/18	5	
13	Coatex BR 100	7,5	dissolved in
14	water	22,5	
			Allow paint to equilibrate for at least 24 h.

Wet soiling:

Sample preparation was the same as for dry soiling, but as substrates Eterplan fibre cement panels (300 x 150 x 4 mm) have been used. Wet film thickness was 300 μm. Dried samples were fixed with the coated side up on a support, and the test panels formed an angle of 60° with the horizontal. Then a coating surface was continuously rinsed with a standardized aqueous dirt suspension by means of a peristaltic pump (500 ml/min) for 30 minutes per cycle. Dirt suspension dripping from the samples was collected in a petri dish equipped with a magnetic stirrer bar to keep dirt particles in suspension. After each 30 minutes-cycle, the panels were allowed to dry for 24h at 23°C/50% rel. air humidity. L^* and ΔL^* were determined after each cycle. Freshly prepared dirt suspension was used for each cycle. The dirt suspension was prepared as follows:

17g gas soot FW 200, 70g Japanese standard dust nr. 8 and 13g special pitch nr. 5 (Worlee) were weighed out in a 1000 ml powder bottle and 400 cm³ glass pearls were added. The mixture was homogenized for 24 h on a 'Rollenbock'. Then the glass pearls were removed by sieving. The resulting powder was further homogenized in a mortar. 1g of the resulting material, 1g of butyl glycol, and 998 g of water were mixed using a magnetic stirrer. The suspension was continuously stirred with a magnetic stirrer bar to avoid settling.

Outdoor exposure:

Eterplan panels were treated with a water-based primer and afterwards dried for 24 h. Then, twice about 400 g/m² (wet mass) paint was applied by brush. The first coating was allowed to dry 24 h prior to applying the next layer.

Sample panels were placed in a southward direction 45° inclined to the horizontal.

Initial brightness L^* of the panels was determined, then L^* was measured after 3 months.

Results and Discussion

Dry soiling:

The terms 'PA' for pure acrylics and 'SA' for styrene acrylics have been used in this work to name the tested emulsions.

As can be seen from table 3b, dry soiling tendency of acrylic polymer PA1 can be reduced by the polymer modification. However, the effect was rather moderate. As a comparative system the emulsion PA 1 was also modified by simple addition of a commercially available fluorinated additive (giving PA1 mod 3). In this case, however, the used additive had no functional group for reaction with the polymer backbone. Dry soiling was not reduced at all by addition of the fluorinated additive in this test. It has to be marked here that soiling tests reveal only qualitative tendencies and give no quantitative results.

Table 3a. Dry soiling results of EWCs containing emulsions PA 1 – SA 1, fly ash/soot mixture

Sample Nr.	Emulsion	ΔL^*
1	PA 1	32.3
2	PA 2	30.7
3	PA 3	30.7
4	PA 4	27.7
5	SA 1	25.1

Table 3b. Reduction of dry soiling by polymer modification, fly ash/soot mixture

Sample Nr.	Emulsion	Soiling reduction [%]
6	PA 1 mod 1	4.3
7	PA 1 mod 2	11
8	PA 1 mod 3 (comparison)	0

Wet soiling:

In contrast to the rather moderate dirt-repellent effect in dry soiling, a markedly reduced soiling of the corresponding paint films is observed when emulsions modified with the Clariant-technology are examined in the wet-soiling test. Soiling reductions between 30 and 50% have been observed (see PA 1 mod 1 and mod 2 in table 4b). The same trends as in dry soiling experiments are observed, but with much more pronounced differentiation. The coatings produced with unmodified binders do not exhibit the same ranking in soiling resistance as in the dry dirt-pickup test, emulsion SA 1 which has the lowest dry dirt-pickup, shows by far the highest wet soiling.

Table 4a. Wet soiling results of EWCs containing emulsions PA 1 – SA 1, 5 cycles

Sample Nr.	Emulsion	ΔL^*
1	PA 1	3.6
2	PA 2	3.1
3	PA 3	2.5
4	PA 4	2.6
5	SA 1	4.2

Table 4b. Reduction of wet soiling by polymer modification, 5 cycles

Sample Nr.	Emulsion	Soiling reduction [%]
6	PA 1 mod 1	47.2
7	PA 1 mod 2	32
8	PA 1 mod 3 (comparison)	5

Outdoor exposure:

In order to get quick results from the outdoor soiling tests, the test panels have been placed in the area of Mexico City, being faced with high dust immission. The TSP (total suspended particulate, corresponding to the dust fraction in air with particle sizes range from 0.1 to 100 µm) in Mexico City is roughly six times higher than the TSP values observed in the United States (about 300 µg/m³ mean value for Mexico City[7] compared to about 50 µg/m³ for US cities with 120 µg/m³ maximal immission concentration tolerated by the WHO).

As a result of the high concentration of suspended particulates in the air, soiling proceeds much faster than in cities with lower TSP: A parallel outdoor exposure soiling test in an industrial area of Frankfurt am Main, Germany, gave only about half as high reduction of L^* at equal exposure times and comparable test panel preparation. However, the general trends of dirt-pickup extent observed in outdoor soiling studies accomplished in Mexico and Frankfurt are comparable.

In the same way, Wagner reported on the high soiling tendency of facade coatings exposed to the atmosphere of Jakarta, Indonesia. After two months of outdoor weathering at 45° inclination against the horizontal, an optical differentiation of the test panels was not really possible any longer.[2]

Measurement of L^* after 3 months outdoor exposure in the area of Mexico City showed clearly that suitable binder modification can reduce the dirt deposition (table 5a,b). As in the case of the wet soiling ´quick test´ using an aqueous dirt suspension, outdoor weathering results suggest that modifications mod 1 and mod 2 of PA 1 are effective in terms of soiling prevention. They show a much better soiling resistance compared to the non-modified binder, which exhibit already reasonable dirt-repellence.

The styrene acrylic emulsion SA 1 and the unmodified pure acrylic emulsion PA 1 exhibit the highest dirt-pickup, whereas the pure acrylic binder PA 2 resists dust adsorption and indentation most.

Table 5a. Soiling after 3 months outdoor exposure in Mexico City

Sample Nr.	Emulsion	ΔL^*
1	PA 1	9.3
2	PA 2	6.3
3	PA 3	7.8
4	PA 4	8.2
5	SA 1	8.5

Table 5b. Soiling reduction by polymer modification, 3 months outdoor exposure in Mexico City

Sample Nr.	Emulsion	Soiling reduction [%]
6	PA 1 mod 1	43
7	PA 1 mod 2	13
8	PA 1 mod 3 (comparison)	2.2

Conclusion

The data show that the correlation between laboratory quick soiling tests and outdoor exposure is not reliable enough. However, the laboratory wet soiling test is a useful method to quickly determine the main trends in soiling behaviour. After such preliminary screening, the outdoor weathering of the selected samples is essential to get reliable results.

Exposure of test panels to areas with high TSP-concentrations in the air gives a differentiated picture of coatings' soiling resistance after few months.

The obtained results show that with the emulsions modified according to the Clariant-technology, binders are available which lead to elastomeric wall coatings with markedly enhanced soiling resistance.

References

[1] K. Nishiwaki, M. Katou, *Colloids and Surfaces* **1999**, *153*, 317.
[2] O. Wagner, *Farbe & Lack* **2001**, *107*, 105.
[3] US 3,320,198, T. B. Hill, D. Hill to E.I. du Pont de Nemours and Company, 1967.
[4] K. Tsujii, T. Yamamoto, T. Onda, S. Shibuichi, *Angew. Chem.* **1997**, *109,* 1042.
[5] US 4,208,496, R. A. Bergfeld, S. P. Patel, L. M. Schlickenrieder to N L Industries, 1980.
[6] Patent application 2001DE507, Clariant.
[7] Data from UNEP, United Nations Development Programme.

Macromol. Symp. 187, 215–224 (2002)

Modified Ureas: An Interesting Opportunity to Control Rheology of Liquid Coatings

János Hajas, Axel Woocker

BYK-Chemie GmbH, Abelstrasse 14 , D-46483 Wesel, Germany

Summary: Modified urea compounds can be used as very powerful liquid rheology additives in coatings. Their rheological impact in various coating systems is directly related to the specific chemical modifications that can be made to the polyurea molecule. Modifications evaluated were made using end groups of varying polarities such as low polarity alkyl groups, medium polarity segments, or highly polar structures. Analysis of rheological behavior in various types of coatings indicates the presence of two different thickening mechanisms (H-bonding between urea and urea groups of the additive, and the association of the additive with the binder). These additives result in coatings with exceptional resistance to settling during storage, along with good sag resistance. Some urea type additives used in combination with conventional rheology modifiers such as fumed silicas show synergistic effects. Advantages for coating producers and users will be shown using several practical application examples compared with conventional rheological additives.

Keywords: anti-settling, sag control, modified urea, paint additive, synergism, thixotropy, rheology additive

1. Modified urea rheology additives: chemistry and method of action

Modfied urea rheology additives are solutions of modified urea functionalities in the highly polar aprotic solvent N-methyl pyrrolidone (1-methyl-2-pyrrolidone). Their chemical structure is shown on Fig. 1. They can be produced by reacting monoalcohols with diisocyanates into monoadducts, which then can be reacted with diamines into diureas. Since the monoadducts still may contain traces of diisocyanates, besides the diureas also polyureas are formed in small amounts. However, the solubility properties of polyureas is worse that that of diureas, therefore the residual diisocyanate content of the monoadducts must be minimized in order to obtain the lowest possible levels of polyureas.

CCC 1022-1360/00/$ 17.50+.50/0

Three modified urea modifications have been developed: one with low polarity end blocks, another one with medium polarity end blocks and a third with highly polar ones. Shown below in Fig. 2 are the most important technical and physical data:

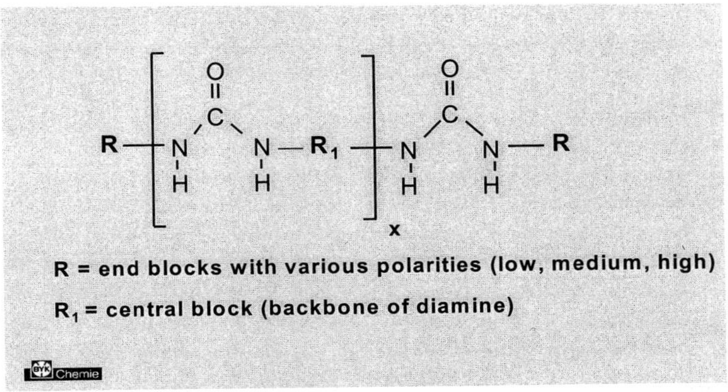

Figure 1.General structure of modified urea rheology additives

Figure 2. Modified urea thickeners

	R= Low polarity block	R= medium polarity block	R= highly polar block
Solvent	NMP (1-Methyl-2-pyrrolidone)	NMP (1-Methyl-2-pyrrolidone)	NMP (1-Methyl-2-pyrrolidone)
Solids (wt %)	Ca. 25	Ca. 52	Ca. 52
Density (g(cm3)	1,05	1,13	1,13
Visual appearance	Clear, low viscosity yellowish liquid	Clear, low viscosity yellowish liquid	Clear, low viscosity yellowish liquid
*Commercial name	BYK-411	BYK-410	BYK-420

When employed in binders, the active substance of these liquid rheology additives is selectively insoluble and separates out after incorporation — in a network of polyurea molecules in microcrystals, interacting via H-bonding with each other. The polyurea network demonstrates pronounced pseudoplastic properties, shear thinning and true thixotropy with an easily detectable yield point. The hydrogen-bonding structure is shown on Fig. 3.

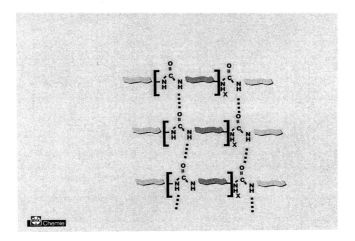

Figure 3. Hydrogen-bonding structure between modified urea molecules

Modified urea additives provide at lower shear rates a very strong viscosity increase, however, at elevated shear rates hardly any viscosity increase can be observed. This rheology is very favourable for most coating applications, since the pronounced increase of low shear viscosity improves the anti-settling behaviour as well as sag control characteristics, while the low viscosities at higher shear rates improve paint application properties. Low viscosities at higher shear rates also can help reduce solvent addition for spray applications, therefore the usage of polyurea additives fits very well within today's trends to formulate low-VOC systems.

The typical rheology behaviour of modified urea thickeners can be observed of Fig. 4, where the rheology of a high-solids long oil alkyd resin, modified with an organoclay, with hydrogenated castor oil and with the lower polarity modified urea thickener is displayed.

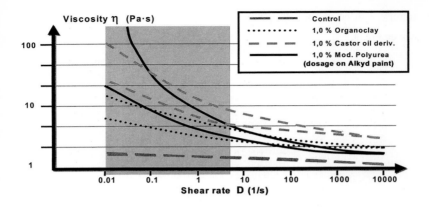

Figure 4. Viscosity characteristics of the modified urea thickener in comparison with organoclay and castor oil derivative

Additive level was in all cases 0.5 %. The very much different rheologies predict a stronger anti-settling and sag control behaviour with the modified urea thickener.

2. Rheology effects as function of modification

2.1 Characterization of modified ureas under various polarity conditions

It is very difficult to find a system, where all three modified urea componds can be characterized in a proper way. However, a suitable test system was found by using a thermoplastic acrylic resin which is soluble in aliphatic isoparaffinic solvents (Isopar H) and also in isopropyl alcohol. Resin solutions with 100, 75, 50, 25 and 0 % Isopar H and 0, 25, 50, 75 and 100 % isopropanol have been modified with each three modified urea compound. Viscosity measurements on these model systems display, that by using a higer polarity end group as modification, also the viscosity maxima of each product is shifted into the higher polarity domaine (Fig. 5).

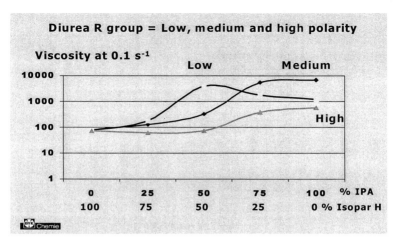

Figure 5. Influence of the system polarity on effectiveness of modified ureas with various polarities

However, in the case of the highly polar modification, the optimal polarity area can be reached only outside the range, when organic cosolven-water mixtures are used as solvent.

2.2 Low polarity urea thickener

The lower polarity new liquid thixotrope does not provide any remarkable thickening performance in pure solvents, but works well when binders – especially fatty acid modified ones – are employed. With increasing solids of the resin, modified with the urea thickener, the thickening effect will be exponentially improved [Fig.6].

The exponentially increasing sag control as function of resin solids content suggests that not only H-bonding interactions between polyurea molecules but also associative interactions with the resins may improve the rheology performance. Due to these pronounced associative interactions, the low polarity urea modification is ideal for higher solids applications but not for lower solids systems.

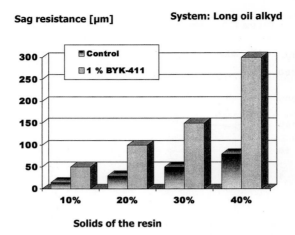

Sag resistance [μm]   **System: Long oil alkyd**

Solids of the resin

Figure 6. Sag resistance in a long oil alkyd paint, as a function of resin solids

According to lab test results, typical dosage levels of the non-polar modified urea thixotrope to improve anti-settling are 0.1-0.3 %. For sag control improvements, dosages above 0.3 % can be recommended.

2.3 Medium polarity urea thickener

This material builds up a thixotropic network in almost any kind of medium polarity solvents and solvent blends and works also in medium polarity resins or in their blends. The achieved rheology effect is almost independent from the binder content of the system. Just 2-3 % medium polarity polyurea thickener can "solidify" solvents such as xylene-butanol blends, resulting in gelly structures with a high yield point.

The best rheology performance can be achieved in typical medium polarity binders such as OH-functional polyurethanes, acrylic resins, polyesters, as well as in epoxies, vinyls, UV-binders, melamines and many other resins. Typical usege levels are 0.1-0.3 % for anti-settling, and 0.3-1 % for sag control.

2.4 Highly polar urea thickener

The high polarity urea modification provides best rheology effectiveness in aqueous formulations. Blends of water-soluble organic cosolvents with water can be modified with this additive, resulting in highly thixotropic properties. In heavily pigmented aqueous slurries and pigment concentrates – even without cosolvents – a very strong anti-settling effects can be achieved. Typical anti-settling effects of the highly polar urea modification are displayed on Fig. 7.

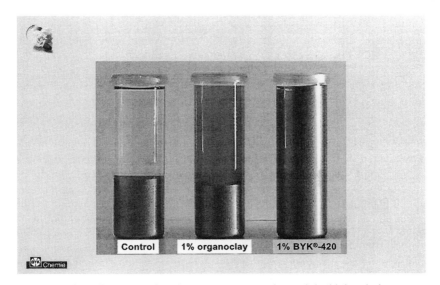

Figure 7. Anti-settling comparison between an organoclay and the high polarity modified urea in an aqueous pearlescent pigment slurry

3. Rheology Effects in Combinations with Other Rheology Additives

Since many other rheology additives also interact via H-bonding, it is also possible to combine the modified ureas with other rheology additives in order to achieve synergistic effects, and tailor-made rheology properties for certain applications. It has been found, that the medium polarity urea thickener shows very strong synergistic effects with fumed silicas, especially in solvent-free epoxy resin formulations. Thixotropy recovery properties can be optimized by using these combinations for maxium sag resistance in high-shear applications (airless spraying).

Thixotropy recovery properties by using fumed silica, modified medium polarity urea and combination of both in a solvent-free Bisphenol A-type liquid epoxy resin are displayed on Fig. 8. The viscosity improvement by using modified urea alone is better, than with fumed silica, but the network build-up after pre-shearing takes much longer time. In high-shear applications only the combinations of both additive, resulting in shorter recovery times and high viscosities, perform well for sag resistance.

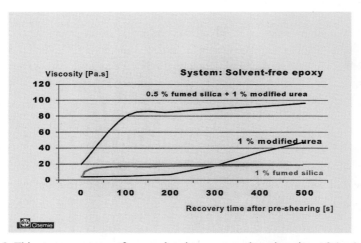

Figure 8. Thixotropy recovery after pre-shearing, measured as viscosity at 0.1 s-1 shear rate with modified urea, fumed silica and combination thereof

4. Incorporation

One of the greatest technical advantages of the modified urea additive chemistry is the fact that optimal rheology improvements (sag resistance, anti-settling) can be achieved just by post-addition to the finished paint. Neither grinding, nor heat activation or pre-gelling are necessary to achieve the best rheological performance.

The method of additive incorporation is also important. For optimal effectiveness and for homogeneous gel structure, a rapid and homogeneous incorporation of the additive is desirable, avoiding too high shear forces like pigment grinding conditions. Optimal is when using typical letdown conditions as used in the paint industry. An incorporation into the millbase is recommendable only if good anti-settling but no sag control is the target.

To test rheology properties, it is recommended to wait until the final viscosity build-up is completed, usually overnight.

5. Applications

The modified urea based liquid thixotropes can be recommended to improve anti-settling and sag control properties for a very wide range of applications.

The lower polarity modification is more often used in long and medium oil alkyd resin based systems, dissolved in mineral spirits. Applications include architectural coatings, marine coatings and also heavy-duty systems – almost everywhere where low polarity alkyd resins can be used. Applications in primers, under coats and topcoats are also possible.

The medium polarity version is successful in almost any type of medium polarity systems from coil coatings, wood coatings through general industrial coatings and automotive systems up to the protective and marine coating applications. In automotive base coats, special advantages can be achieved through the relatively slow thixotropy recovery of the polyurea network which can provide excellent anti-settling as well as very good metallic flake orientation and good levelling at the same time. Applications in printing inks and PVC plastisols also exist.

The high polarity modifications is mainly used as anti-settlig additive in aqueous pigment slurries and in pigment concentrates for architectural and industrial coating applications, also includung high quality systems such as automotive basecoats.

6. Advantages over conventional thixotropes

Compared to the commonly used rheology control additives, the new technology offers some unique advantages. The main advantage resides in the liquid form of the additive, which makes it much easier to handle. When replacing hydrogenated castor oil or polyamide modified thickeners by polyurea, other primary advantages are no requirement of heat activation for incorporation and the much lower heat sensitivity on application. Compared to organoclays and fumed silicas, the major advantage is the lower dosage and the very little or no influence on gloss. In certain cases, combinations of modified urea thixotropes with fumed silicas result in very strong synergistic effects (Fig.8).

7. Conclusion

Liquid rheology additives based on modified urea functionalities have been developed to solve settling problems and to improve sag control properties. Modifications for low polarity higher solids solventborne and solvent-free systems (typically alkyd resins), further for medium polarity low to high solids and solvent-free systems and for aqueous systems do exist, and provide unique advantages for the coatings formulators.

Macromol. Symp. **187**, 225–234 (2002)

Clearcoats Based on Maleimide/Vinyl Ether Combinations - Investigations into Their Properties and Curing Behaviour

Norbert Pietschmann

Institut fuer Lacke und Farben e.V. (iLF), Germany

Summary: UV curable maleimide/vinyl ether blends were discussed in literature during the last years. Results of basic research were published which mainly contained RTIR spectroscopy, differential photocalorimetry, and the mechanism of radical formation. Partially, these results indicated high reactivity and low sensitivity to oxygen inhibition. Therefore, UV curable clearcoats containing maleimides and vinyl ethers were formulated and UV cured in air atmosphere.In principle, it was possible to use equimolar maleimide/vinyl ether blends as photoinitiators for acrylates. However, their efficiency was inferior compared to commercial α cleavage photoinitiators. On the other hand, formulations based on certain maleimides and vinyl ether functional resins or/and reactive thinners could be cured without additional photoinitiators. The possibilities and limitations found are discussed.

1 Introduction

During the last years UV curable maleimide/vinyl ether blends (MI/VE) were an interesting new topic in the field of radiation curing. Possible mechanisms of radical formation based on H atom abstraction or electron transfer were discussed. Experimental investigations were presented which, e.g., presented results of RTIR spectroscopy and differential photocalorimetry. Partially, the results indicated high reactivity and low sensitivity to oxygen inhibition [1-10].

Two different pathways were proposed for the technical use of MI/VE systems:

i) The formulation of self-crosslinking coating vehicles containing MI and VE functionality which would not require external photoinitiators [2-4, 8, 10] and

ii) the use of MI/VE or other MI combinations as photoinitiators in acrylate formulations [3-5, 11, 12].

The goal of the study presented here was to check both possibilities with respect to curing behaviour and film properties. Thus, model formulations were prepared, applied on glass substrate, UV irradiated and tested.

                CCC 1022-1360/00/$ 17.50+.50/0

2 Experimental

Various MI derivatives were tested in the coating formulations. In this paper, the following compounds are mentioned:

- ADMI = bis-maleimide of versamin; liquid resin
- BMI = N-n-butyl maleinimide (Sigma-Aldrich)
- EMI = N-ethyl maleinimide (Sigma-Aldrich)
- tBMI = N-tert.-butyl maleinimide (Sigma-Aldrich)

Some other maleimides gave poor solubility in the coating formulations or other problems.

The maleimides were combined with vinyl ethers or VE blends, resp., in stoichiometric ratio (1:1). The general name "vinyl ether" is used here in spite of the fact that some of the compounds tested were other vinyl (or allyl) derivatives. The resins and reactive thinners considered in this paper were abbreviated as follows:

- BEDVE = butanedioic acid bis-4-[(ethenyloxy) butyl] ester
- BETVE = 1,2,4-benzentricarboxylic acid tris-[4-(ethenyloxy) butyl] ester
- CHVE = cyclohexane dimethanol divinyl ether
- DVE-2 = diethylene glycol divinyl ether
- DVE-3 = triethylene glycol divinyl ether
- HDDVE = hexane diol divinyl ether
- NVP = N-vinyl pyrrolidone
- PDVE = aromatic polyester divinyl ether resin
- PPVE = polyfunctional polyester vinyl ether resin
- UDVE = aliphatic urethane divinyl ether resin

The stoichiometric MI/VE mixtures were used either as self-crosslinking vehicles or as photoinitiators in acrylate formulations. The coating materials were hand-applied on glass (ca. 50 µm film thickness) and UV cured in air atmosphere. The following UV curing devices were used:

1. Beltron BE 20/III, two 80 W/cm electrode-powered radiators, first Ga-doped, second non-doped
2. Fusion LC 6/2, two 120 W/cm microwave-powered F300S units, first bulb Fe-doped (D bulb), second non-doped (H bulb).

The irradiation conditions considered in this paper are:

- one non-doped 80 W/cm Hg bulb, 5 m/min belt speed (abbreviation: 80/5)
- one non-doped 120 W/cm Hg bulb, 20 m/min belt speed (abbreviation: 120/20)

- all four radiators arranged in a series using both curing devices, belt speed 4 or 5 m/min in both curing units (abbreviation: all/4 or all/5)

The curing results were estimated by means of pendulum hardness (KÖNIG method) and scratch resistance (finger nail test). Furthermore, selected films were investigated by IR spectroscopy (ATR technique) and ERA (a radiometric method [13, 14]).

3 MI/VE photoinitiators in acrylate systems

Different commercial acrylate binder systems were used to investigate the photoinitiator efficiency of equimolar MI/VE blends. The example discussed here consisted of 70 % polyester acrylate oligomer and 30 % of an polyether acrylate reactive thinner with higher functionality (PPTTA). 97.5 weight parts of this mixture were completed with 2.5 parts of MI/VE to obtain UV curable clearcoats. Commercial α cleavage standard photoinitiators (i.e., HMPP, HCPK, and BDK) were used as reference.

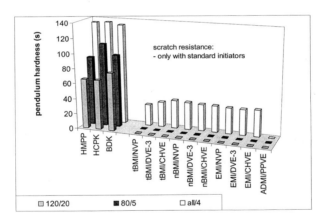

Figure 1. Curing results of polyester acrylate/polyether acrylate clearcoats (70:30) with 2.5 weight % of photoinitiator or MI/VE

As demonstrated in fig. 1 the MI/VE blends led to acrylate polymerization at very intense UV irradiation but they were not able to compete with the three standard initiators. Certain improvements were found after addition of tertiary amines, i.e. 3 % DMB (see fig. 2) or 5 % of a polyfunctional amine but the main conclusion was the same. This was also valid for higher photoinitiator or MI/VE concentrations (5 weight % photoinitiator, without and with 3-5 % amine).

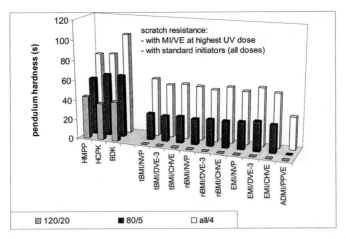

Figure 2. Curing results of polyester acrylate/polyether acrylate clearcoats (70:30) with 2.5 weight % of photoinitiator or MI/VE + 3 weight % DMB

Furthermore, other acrylate formulations containing an aromatic epoxy acrylate gave similar results.

4 MI/VE as binders and reactive thinners

4.1 Formulations with monofunctional maleimides

The following facts became clear during preliminary investigations on stoichiometric mixtures containing monofunctional MI components:

- EMI and BMI gave the highest curing efficiency.
- Many maleimides are solids and, hence, not all of them could be incorporated. Moreover, low-melting solids like EMI were incorporated at increased temperature but in some formulations they re-crystallized on storage.
- Neither a sole low-molecular mass vinyl ether (i.e., reactive thinner) nor an undiluted VE functional resin (PDVE, PPVE, or UDVE) gave optimum film properties in the combination with monofunctional MI.

  The achievement of lower viscosities and the incorporation of solid maleimides were additional reasons to blend the VE resins with VE reactive thinners.

Thus, the VE compounds were mixed in a weight ratio of 70 parts resin and 30 parts reactive thinner. The VE groups (sum of both vinyl ethers) and the MI amount gave a molar (stoichiometric) ratio 1:1. Fig. 3 shows that the curing results on glass, for a large part, were still insufficient. PDVE and PPVE (with varied reactive thinners) gave better results with respect to pendulum hardness compared to UDVE. PPVE, however, showed better scratch resistance than PDVE.

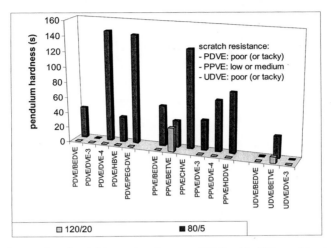

Figure 3. Curing results of stoichiometric mixtures of EMI with different blends of VE functional resins (70 parts by weight) and VE reactive thinners (30 parts by weight)

Fig. 4 demonstrates that improved hardness values (and good scratch resistance, too) could be obtained at an optimum content of the reactive thinner HDDVE. Similar results were also found using CHVE or DVE-3.

The addition of 5 % of a polyfunctional amine gave no significant improvement. In contrast, up to 5 % of additional liquid α cleavage initiator (HMPP) led to better hardness and scratch resistance (fig. 5). This was also supported by the crosslinking density investigated by ERA (fig. 6. Note: Lower M values indicate a higher crosslinking density [13, 14].).

Furthermore, within certain limits it was also possible to maintain sufficient curing properties at reduced MI amounts (fig. 7. The formulations considered here do not contain additional photoinitiators.).

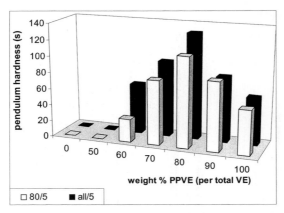

Figure 4. Pendulum hardness values of stoichiometric mixtures of EMI with varying blends of PPVE and HDDVE. The formulations do not contain additional photoinitiators.

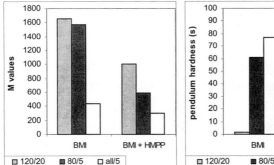 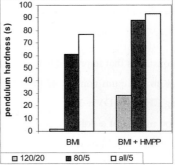

Figure 5.   Figure 6.

Pendulum hardness (fig. 5) and M values (from ERA; fig. 6) obtained from a stoichiometric mixture of BMI with a VE blend (70 parts PPVE + 30 parts CHVE).
Formulation without and with 5 percent additional α cleavage initiator (HMPP).

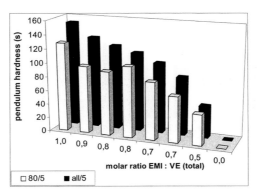

Figure 7. Pendulum hardness values of varying mixtures of EMI with a blend of PPVE (80 parts by weight) and HDDVE (20 parts by weight).

4.2   Formulations with ADMI

The higher-molecular bis-maleimide ADMI was combined with PPVE, PDVE and UDVE in stoichiometric ratio. Again, PPVE gave the best results with respect to scratch resistance after UV curing.

Fig. 8 shows that some combinations of PPVE with reactive thinners can give further improvements with respect to hardness. This is also supported by the results of IR/ATR spectroscopy (fig. 9) and ERA (fig. 10). Figs. 8-10 also demonstrate that the highest irradiation dose gives the most intense curing. However, the content of non-converted vinyl groups in the coating film could remain relatively high especially at the bottom side (fig. 9).

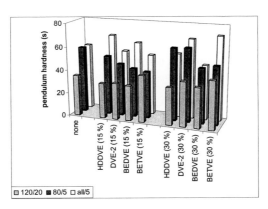

Figure 8. Pendulum hardness values of stoichiometric mixtures of ADMI with PPVE or with PPVE blends containing 15 % or 30 % of different VE reactive thinners.

232

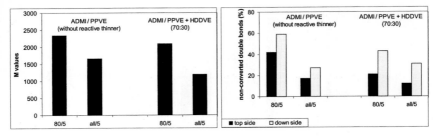

Figure 9.   Figure 10.

Non-converted vinyl groups at the top and down sides (fig. 9, from IR absorbance at 1616 cm^{-1} related to 1710 cm^{-1}) and M values (from ERA; fig. 10) obtained from stoichiometric mixtures of ADMI with pure PPVE or with a blend of 70 % PPVE and 30 % HDDVE.

5 Summary

Maleimide/vinyl ether (MI/VE) combinations were tested with respect to their UV curing properties in air atmosphere. Their use as photoinitiators for acrylate systems was possible. Compared with commercial Norrish type I initiators, however, the reactivity proved to be poor.

In contrast, MI/VE binder systems could be cured without additional photoinitiators. For that purpose EMI, BMI, or the bifunctional ADMI could be used. Useful results were obtained especially in combinations with a polyfunctional vinyl ether modified polyester resin. Further improvements were found when this resin had been diluted with VE reactive thinners.

Furthermore, within certain limits it was also possible to use a VE excess instead of the stoichiometric ratio.

However, during the experiments some limitations became visible:

- After several days or weeks changes in the properties of MI/VE combinations (and sometimes gelation) can occur. Therefore, the curing tests had to be carried out within 24 hours after mixing MI and VE. But on the other hand, no inhibitor (like MEHQ etc. which is common for acrylate binders) was added here.

- The reactivity in air atmosphere seems to be limited. After curing with a belt speed of 10-20 m/min (one UV lamp) this became visible both from the mechanical properties (empirical check of scratch resistance and through curing) and from the conversion of vinyl double bonds at the film top and bottom sides (compare fig. 9).

- Sometimes, irritation effects were observed which, probably, were due to the MI components. This could be caused by liquid formulations as well as by incompletely cured films on the substrate.
  Such problems were more pronounced for low-molecular types like EMI or BMI.

Thus, an improvement of the properties of MI/VE binder systems would require aliphatic MI types with further increased reactivity, higher functionality or/and molecular mass and, thus, lower irritation.

Acknowledgements

The results presented here were obtained during research supported by the Ministry of Economy and Technology of the Federal Republic of Germany.
The author is grateful to Mrs. R. Kupke and Mrs. E. Truhe for careful implementation of formulation, UV curing and testing as well as to other colleagues from the iLF Magdeburg for their co-operation and support in the spectroscopic and radiometric measurements.
Furthermore, thanks are given to the firms Allied Signal, BASF, Dexter-Quantum Materials, and ISP which kindly provided samples of raw materials.

References

[1] S. Jönsson, G. Doucet, G. Mattson, S. Clark, C. Miller, C.E. Hoyle, F Morel, and C. Decker,
RadTech Europe, Lyon 1997, Proceedings Academic Days, pp. 103-112
[2] C.E. Hoyle, S. Jönsson, S.C. Clark, C. Miller, and M. Shimose,
RadTech Europe, Lyon 1997, Proceedings Academic Days, pp. 115-120
[3] H. Andersson and A. Hult, J. Coatings Techn. 69, 02/1997, no. 865, pp. 91-95
[4] C. Decker, F. Morel, S. Jönsson, S.C. Clark and C.E. Hoyle,
RadTech Europe, Lyon 1997, Proceedings Academic Days, pp. 169-175
[5] S.C. Clark, S. Jönsson, and C.E.Hoyle, RadTech Europe, Lyon 1997,
Proceedings Academic Days, pp. 163-168
[6] C.E. Hoyle, S. Clark, C. Miller, J. Owens, J. Whitehead, S. Jönsson, P.E. Sundell,
M. Shimose, S. Katogi, F. Morel, and C. Decker,
RadTech Asia, Yokohama 1997, Proceedings G2-6, pp. 216-219
[7] S. Jönsson, G. Doucet, G. Mattson, S. Clark, C. Miller, C.E. Hoyle, F. Morel, and Decker, C.,
RadTech Asia, Yokohama 1997, Proceedings G2-15, pp. 255-258
[8] P. Kohli, A.B. Scranton and G.J. Blanchard, Macromolecules 31, 1998, pp. 5681-5689
[9] S. Jönsson et al., Pat. US 5446073 and EP 0618237

234

[10] C. Decker, C. Bianchi, F. Morel, S. Jönsson, and C. Hoyle,
      RadTech Europe, Berlin 1999, Proceedings, pp. 447-454
[11] C.E. Hoyle, C. Nguyen, A. Johnson, S.C. Clark, K. Viswanathan, C. Miller, C., S. Jöns-
      son,
      D. Hill, W. Zhao, and L. Shao, RadTech Europe, Berlin 1999, Proceedings, pp. 455-458
[12] S. Jönsson, K. Viswanathan, C.E. Hoyle, S.C. Clark, C. Miller, F Morel, and C. Decker,
      RadTech Europe, Berlin 1999, Proceedings, pp. 461-471
[13] S. Millow, Plaste und Kautschuk *26*, 1979, no. 1, pp. 44-48
[14] S. Millow and M. Osterhold, farbe + lack *98*, 1992, no. 9, pp.675-678

Macromol. Symp. **187**, 235–241 (2002)

New Epoxy-Siloxane Hybrid Binder for High Performance Coatings

Dr. Gerhard Reusmann

Tego Chemie Service GmbH, Goldschmidtstr. 100, 45127 Essen, Germany

Summary: The novel chemistry achieved by condensation reaction of an aliphatic epoxy with a polysiloxane results in an epoxy-siloxane hybrid binder whose unique physical characteristics allow its use as a durable binder for the protective coatings industry.
The new epoxy-siloxane coating system enables the zinc primer to be protected by a single topcoat. This results in reductions in application time, less overspray and a much simplified maintenance for corrosion protection. Furthermore the reduced number of coats and overspray results in about 70 % less solvent emission to the atmosphere.

Keywords: silicone-epoxy; corrosion protection; polysiloxanes; coatings; high performance polymers.

Introduction

Fighting corrosion can be an expensive and daunting problem [1]. Protecting objects such as oil platforms, bridges and storage tanks, ship decks, steel constructions, concrete walls and floors from corrosion requires high costs and investment of labor [2].

One of the best methods of protecting objects under varying conditions is though the use of protective coatings [3].

Coating systems traditionally used in the protective coatings industry rely almost entirely on organic binder systems [4]. Due to the variety of service conditions, a multiple coat system is most often required. A typical multiple coat system consists of a zinc-primer, a corrosion-resistant epoxy mid-coat and a weather-stable polyurethane coating [5].

New Silicone-Epoxy Hybrid Resins

A specific focus of the paint industry is the reduction of costs for corrosion protection. A special benefit as a result would be the combination of two coating layers in a single layer system (figure 1).

By exploring the inorganic silicone-based chemistry a siloxane hybrid polymer has been developed which combines the properties of organic and inorganic compounds in a new class of binders for protective coatings.

   CCC 1022-1360/00/$ 17.50+.50/0

236

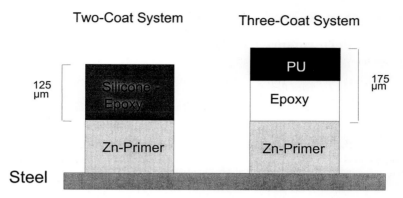

Figure 1: Profile of Anti-corrosion Coating Systems

Polysiloxanes exhibit excellent stability to heat and UV-radiation exposure due to the very stable [-(Si-O)$_n$-Si-] –backbone (figure 2) [7-13].

The idealized polysiloxane polymer structure is illustrated in figure 2:

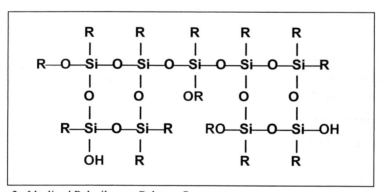

Figure 2: Idealized Polysiloxane Polymer Structure

By fine-tuning of the molecular weight, the degree of branching, the organic substituents R and the functionality, the resulting reactivity and the technological profile of the polymer can be adjusted [6].

The novel chemistry achieved by condensation reaction of an aliphatic epoxy with a polysiloxane results in an epoxy-siloxane hybrid binder [14] whose unique physical characteristics allow its use as a durable binder for the protective coatings industry.

Test Results of the Hybrid System as 2-K Corrosion Protection Coating

The quality and durability of the epoxy-siloxane hybrid coating is demonstrated in comparison with a three coat system of a traditional coating system in accelerated corrosion testing such as salt spray, water condensation exposure and UV exposure.

The new silicone-epoxy binder was tested by using the formulation in table 1:

Table 1. Formulation of a 2-component silicone-epoxy corrosion resistant coating

Component A p.b.w.	p.b.w.	Component B (Hardener)	
Pos. 1 Silicone-Epoxy Binder[1]	32.4	Pos. 11 AMEO[8]	16.0
Pos. 2 Tinuvin 1130[2] (1:1 in butanol)	2.0		
Pos. 3 Tinuvin 292[2] (1:1 in butanol)	1.0		
Pos. 4 Heliogen Blue L6901F[3]	1.6		
Pos. 5 Kronos 2160[4]	24.5		
Pos. 6 Talc AT extra[5]	2.0	1 Tego Chemie Service: Resin (SILIKOFTAL ED)	
Pos. 7 TEGO® Airex 900[6]	0.5	2 Ciba: UV absorber 3 BASF: pigment	
Pos. 8 Aerosil R8200[7]	1.0	4 Kronos: titanium dioxide 5 Norwegian Talc: Filler	
Pos. 9 Butyl Acetate	3.0	6 Tego Chemie Service: Deaerator 7 Degussa: Fumed Silica	
grinding in a bead-mill		8 Degussa: 3-Aminopropyl triethoxysilane	
Pos.10 Silicone-Epoxy Binder[1]	32.0		

The resulting application data are reported in table 2:

Table 2. Application data of the silicone-epoxy coating

Application data		
Pot life (25 °C)		4.5 h
Drying times (25 °C)	To touch	4 h
	Dry through	8 h
Adhesion to the primer	Cross hatch DIN 53 151	Gt 0
Hardness	Pencil hardness	F
	Pendulum hardness (König) DIN 53 157	86
Gloss	60° angle	87
Recoat time (25 °C)	Minimum	6 h
	Maximum	48 h

The new two coat system consisting of 75 microns of a zinc epoxy primer topcoated with 125 microns of an epoxy-siloxane hybrid coating (reference: 125 microns epoxy- + 50 microns polyurethane-coating) has been tested and passed the following performance tests (table 3):

Table 3. Comparison of the Test Results after 5000 h Weathering

Test	Zinc Dust/ Silicone-Epoxy 200 microns dried film thickness	Zinc Dust/ Epoxy + P.U 250 microns dried film thickness
Salt Spray Test – DIN 53 167 5,000 hours[1]	2	1 - 2
Humidity Test – DIN 50 021 5,000 hours[1]	1	1 - 2
QUV Weathering, 5,000 hours Lamp UV-B, Cycles 4 h/4 h Gloss Retention: Initial Gloss – 90 % (60°angle)	30%	10%
Colour Retention (delta E)	4.5	10.5
Chalking[1]	1	5

Note: Substrate: Bare Steel
        Surface Preparation: Sand Blasted to SA 2.5
        Drying/Ageing before Test: 10 Days Air Dry at Room Temperature		

[1] Rating: 1 (excellent) ... 3 (satisfactory) ... 6 (poor)

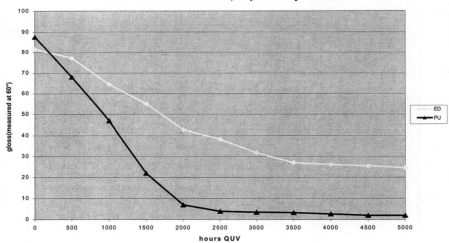

Figure 3: Comparison of the QUV-Test Results after 5000 h Weathering

Tests were carried out on sand blasted bare steel substrates. The coatings were air dried for ten days at room temperature before testing.

These tests in combination with the specified acceptance criteria are currently considered as severe performance test requirement for protective coating systems in the coatings industry.

Coating systems based on silicone-epoxy resins have been evaluated for gloss and colour retention through accelerated weathering via QUV testing. Gloss and colour retention are two of many factors that help to describe the weatherability of a coating and its ability to withstand weather-related effects such as sunlight, humidity, wind and temperature.

The qualitative comparison of gloss and colour retention showed that the epoxy-siloxane based coating outperformed the polyurethane based coating system.

Table 4. Chemical resistance after 24 hours and 7 days exposure

Chemicals	Zinc Dust/ Silicone-Epoxy 200 Microns Dried Film Thickness		Zinc Dust/ Epoxy + P.U 250 Microns Dried Film Thickness	
	24 h	7 days	24 h	7 days
Sodium hydroxyde (50 %)	2	3	2	5
Ammonium hydroxyde, conc.	2	3	3	4
Hydrochloric acid 1 molar	4	4	2	5
Sulfuric acid 1 molar	2	4	4	6
Nitric acid 1 molar	2	3	4	6
Citric acid 1 molar	1	2	2	3
Acidic acid 1 molar	2	5	2	4
Acetone	1	1	1	1
MIBK	1	1	1	2
Xylene	1	1	1	1
Butyl acetate	2	2	2	2
Ethyl alcohol	1	1	1	1
White spirit	1	1	1	1
Substrate: Bare Steel Surface Preparation: Sand Blasted to SA 2.5 Drying/Ageing before Test: 10 Days Air Dry at Room Temperature				

Rating: 1 (excellent) ... 3 (satisfactory) ... 6 (poor)

The high solids epoxy-siloxane coating can be used in various applications, including storage tank exteriors, offshore platforms, marine structures, bridges, exteriors of ships, structural steel, concrete walls and floors and the exteriors of railway coaches (figure 4).

♦ storage tank exteriors	♦ offshore platforms	♦ marine structures
♦ ship decks	♦ wind generators	♦ drilling rigs
♦ exteriors of ships	♦ structural steel	♦ bridges
♦ concrete walls and floors	♦ exteriors of railway coaches	

Figure 4. Applications for Silicone-Epoxy Based Coatings

In addition the epoxy-siloxane coating provides also excellent anti-graffiti and dirt-repellent properties. After the removal of the graffiti no residues of the graffiti remain and no change in gloss can be noticed (figure 5).

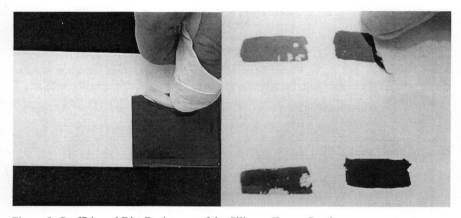

Figure 5: Graffiti- and Dirt-Resistance of the Silicone-Epoxy Coating

Correspondingly this new class of binders enable a new class of dirt resistant coatings [15]. In this respect the antifouling and anti-icing properties still have to be investigated.

Cost-effective and Enviromental-friendly System

This new epoxy-siloxane coating system enables the zinc primer to be protected by a single topcoat based on the innovative new polysiloxane hybrid binder, which combines the advantages of organic polymers and silicones in a single polymer.

This results in reductions in application time, less overspray and a much simplified maintenance for corrosion protection [16]. Furthermore the reduced number of coats and overspray results in about 70 % less solvent emission to the atmosphere.

Conclusion

The novel chemistry achieved by condensation reaction of an aliphatic epoxy with a polysiloxane results in an epoxy-siloxane hybrid binder whose unique physical characteristics allow its use as a durable binder for the protective coatings industry.

With the epoxy-siloxane hybrid resin, the advantages of the epoxy binder are combined with the strength of the polysiloxane providing a two component ambient curing thermosetting coating, which can be formulated to ultra high-solids and very low VOC.

The development of the aliphatic epoxy polysiloxane hybrid polymers has stimulated the formulation of a coating which can revolutionize the protective coatings industry by:

- Over 90% paint-solids, providing reduced VOC;

- high build application characteristics with standard application equipment;

- tolerance of high-humidity and/or low temperatures during curing;

- excellent colour and gloss retention by outperforming polyurethane topcoats;

- maximum corrosion resistance in a two coat system;

- excellent resistance to a wide variety of chemicals;

- cost-effective alternative to organic multi-coat systems;

- compliance with regulations regarding health, safety and environmental protection, reduction of solvent emissions, avoidance of health-damaging resins, crosslinkers or heavy-metal containing substances.

References

[1]   S. Sawant and A. Wagh, Corrosion Prevention and Control (1991), p. 75 - 77
[2]   C.H. Holl, conference CORROSION/86, Paper Nr. 31, Houston (Texas), NACE 1986.
[3]   C. Giudice and B. Del Amo, Corrosion Prevention and Control (1996), p. 43 - 47
[4]   K. Keijman, PCE, 7, 1996, p. 26-32.
[5]   G. Reusmann, Farbe & Lack 107, 2001, 11, p. 78 - 84
[6]   W. Noll, Chemie und Technologie der Silicone, Verlag Chemie, Weinheim (1968)
[7]   Tego Chemie Service GmbH: TEGO Journal, 1999, p. 109-115
[8]   Patents: JP 11106530, Deutsche Patentschrift 1520015, SILIKOFTAL® EW
[9]   Dow: US-Patent 5,952,439 and European Patent EP 0590954 B1
[10] US-Patents 5,618,860 and 5,804,161
[11] European Coating Journal 5, 1999, p. 64-67.
[12] European Coating Journal 4, 2001, p. 152-159.
[13] Patent-examples: EP 3385550, EP 283009, EP 153500 and WO 9844018
[14] Silicone-Epoxy Hybrid-Binder from Tego Chemie Service GmbH: SILIKOFTAL® ED
[15] G. Reusmann, Farbe & Lack 105, 1999, 8, p. 40-47.
[16] Surface Coatings Australia, Januar/Februar 1998, p. 24 - 27

Macromol. Symp. **187**, 243–247 (2002)

Epoxypolyesters as Film-Forming Materials

Piotr Penczek*, Zbigniew Bończa-Tomaszewski, Anna Bańkowska

Industrial Chemistry Research Institute, ul. Rydygiera 8, 01-793 Warsaw, Poland

Summary: A series of solid epoxypolyesters were prepared by the condensation of glycols, 1,2,3,6-tetrahydrophthalic anhydride and dicyclopentadiene followed by the epoxidation of the double bonds in the unsaturated polyester. The polyesters were characterized and then subjected to photoinitiated cationic polymerization using triarylsulfonium salt as the photoinitiator. It was demonstrated that the coatings prepared from these epoxypolyesters exhibited good hardness, impact strength and degree of drying. The straightforward synthesis and high reactivity in cationic UV curing make these epoxypolyesters attractive as film forming materials.

Keywords: unsaturated polyesters; epoxidation; epoxypolyesters; UV-curing; epoxypolyester film

Introduction

Epoxy resins based on bisphenol A[1-3] are the most important thermosetting polymeric materials. They have many excellent properties, such as high thermal stability, adhesion, mechanical and electrical properties and they are widely utilised in the field of coatings. However, it is well known that the commonly used epoxy resins are not resistant to UV light. The poor UV resistance of epoxy resins based on bisphenol A greatly limits their use as components of coatings for exterior applications.[4] Lots of efforts have been made to improve the UV resistance of the cured epoxy resins. One of the possible approaches consists in using epoxypolyesters that contain epoxy groups along and at the ends of the molecular chains.

This paper reports the synthesis and properties of a series of dicyclopentadiene modified epoxypolyesters, which were synthesised by epoxidation of double bonds in solid unsaturated polyesters.

The cationic photopolymerization and the mechanical properties of photocured epoxypolyester films were examined.

CCC 1022-1360/00/$ 17.50+.50/0

Synthesis of unsaturated polyesters

For the synthesis of solid unsaturated polyesters following materials were used: 1,2,3,6-tetrahydrophthalic anhydride (THPA, **1**), neopentyl glycol (NPG, **2**), 1,4-butanediol (BD, **3**), ethylene glycol (EG, **4**) and 4-cyclohexene-1,2-dimethanol (CHDM, **5**). For the purpose of increasing the crosslinking density, terminal dihydrodicyclopentadienyl ester groups were inserted into polyester chains by an addition of dicyclopentadiene (DCPD, **6**) to the carboxylic groups in the polyester.

The polyesters were obtained by two-step reaction (Scheme 1). Optimum content of DCPD in polyester composition (6% w/w) was settled referring to the analysis of properties of polyesters based on THPA, NPG and DCPD. The properties of polyesters are given in Table 1.

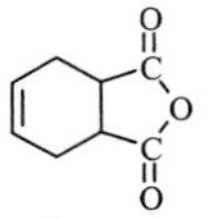

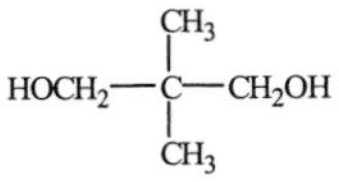

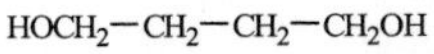

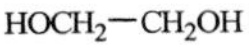

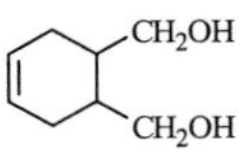

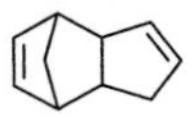

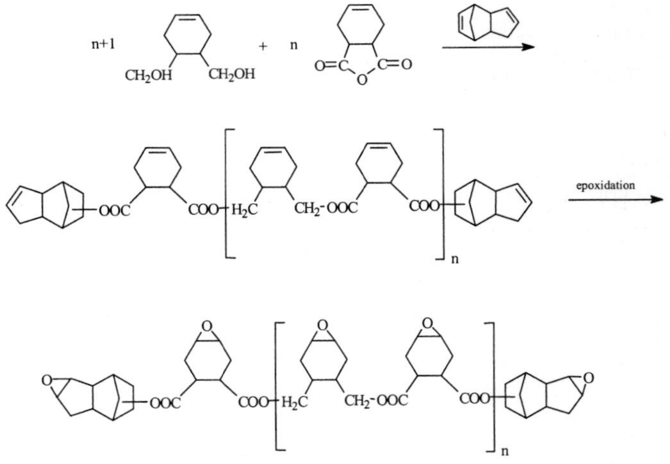

Scheme 1

The optimum properties: the highest softening point (102°C), high iodine number (83 g $I_2/100g$) and additional double bond derived from the cyclohexene ring were reached with the product that contained CHDM as a glycol component of the polyester (Table 1).

Table 1. Properties of unsaturated polyesters made of THPA, DCPD and various glycols

No.	Composition			Properties		
	THPA	Glycol	DCPD (6%)	AN	IN	SP
1	+	NPG	+	7.6	101	82
2	+	BD	+	4.0	90	83
3	+	EG	+	2.4	100	73
4	+	CHDM	+	3.9	83	102

AN, acid number (mg KOH/g)
IN, iodine number (g $I_2/100g$)
SP, softening point determined by the "ring and ball" method (°C)

Epoxidation of double bonds in unsaturated polyesters

Unsaturated polyesters were epoxidized with the following reagents: aqueous solution of peracetic acid[5,6] (PAA), a solution of PAA in organic solvents (ethyl acetate or dichloromethane) and m-chloroperoxybenzoic acid (MCPBA) (Table 2, 3).

Table 2. Epoxy numbers (mol/100g) of epoxypolyesters epoxidized with vario-us epoxidation agents

Epoxidation agent		A	B	C	D
Unsaturated polyester (Table 1) No.	1	0.33	0.06	0.07	0.17
	2	0.21	0.07	0.08	-
	3	0.21	0.05	-	0.21
	4	0.26	0.04	0.05	0.19

A: aqueous solution of PAA
B: solution of PAA in ethyl acetate
C: solution of PAA in dichloromethane
D: m-chloroperoxybenzoic acid

The highest yield was obtained from the reaction of polyesters with aqueous solution of PAA, which was prepared directly before epoxidation. PAA was obtained by the reaction of acetic acid with 60% aqueous solution of H_2O_2 in the presence of sulfuric acid, buffer salt (sodium acetate) and stabiliser (sodium polyphosphate).[6]

Table 3. Efficiency (%)* of epoxidation of unsaturated polyesters

Epoxidation agent	A	B	C	D
1	67	19	21	50
2	60	19	23	-
3	54	14	-	55
4	79	12	14	57

(Row labels, left side: Unsaturated polyester (Table 1) No.)

Efficiency of epoxidation was calculated referring to theoretical epoxy number, which is equal to iodine number given in mole $I_2/100g$.

Curing of epoxypolyesters

Epoxypolyesters (No. 1A and 4A) were chosen to UV curing because of the highest content of epoxy groups and the high efficiency of epoxidation of initial polyesters.

Solid epoxypolyesters with 0.5% cationic photoinitiator (50% solution of triarylsulfonium hexafluorophosphate **7** in propylene carbonate) were dissolved in a volatile solvent, spread on a glass plate and UV irradiated after evaporating of the solvent. The crosslinked films were smooth and of uniform gloss. They exhibited high surface hardness, good impact resistance and high degree of drying. Properties of the epoxypolyester films and obligatory standards for cured lacquer films are presented in Table 4.

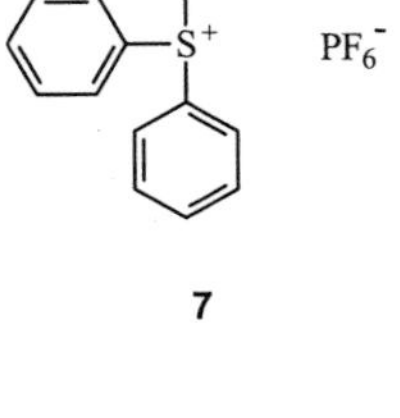

PF_6^-

7

Table 4. Properties of solid epoxypolyester films in comparison to obligatory standards for cured coating films

Parameter	No. of film 1a	No. of film 4a	Requested value	Obligatory standard
Degree of drying	6	7	min. 7	PN-79/C-81519, corresponding to DIN 53150-71
Impact resistance (Du Pont, cm)	45	50	min. 50	PN-54/C-81526, corresponding to ISO 6272:1999
Surface hardness	0.5	0.7	min. 0.5	PN-79/C-81530, corresponding to ISO 2815:2000

Conclusions

Four epoxypolyesters containing DCPD as terminal groups were synthesized and used as film forming materials. The mechanical properties of the cationic UV cured films were investigated. From the experimental results, following conclusions were drawn:

1. Epoxidation of double bonds in unsaturated polyesters with aqueous solution of peracetic acid was found to be the most effective method of preparing of epoxypolyesters.
2. The epoxypolyesters can be used as a solid film forming material while crosslinked by cationic photopolymerization. The film properties conform with obligatory standards for cured coating films.

Acknowledgements

The authors are grateful to the Polish Committee of Scientific Research for the support of this work (grant No. 3 T09B 086 16).

References

[1] Z. Brojer, Z. Hertz, P. Penczek: „Żywice epoksydowe" (Epoxy Resins, in Polish), WNT, Warsaw 1982, 3rd edition, p. 39.
[2] P. Penczek, J. Rejdych, D. Adamska-Rutkowska, Z. Bończa-Tomaszewski: *Chemik* (Gliwice, Poland), **1994**, *47*, 347.
[3] see [1], p. 489.
[4] W. Marquardt, H. Gempeler: *Farbe Lack* **1978**, *84*, 301.
[5] Z. Brojer, P. Penczek, S. Penczek: *Przemysł Chem.,* **1962**, *41*, 437, 684.
[6] A.E. Batog, I.P. Pet'ko, P. Penczek, *"Aliphatic-Cycloaliphatic Epoxy Compounds and Polymers"; Adv. Polym. Sci.,* **1999**, *144*, 49.

Climate Independent Painting: Is Infrared Heating a Solution for Professional Painters?

G. Jonkers, J.G. Nienhuis*, B. van de Velde*

Dutch Association of Professional Painters, Burg. Elzenlaan 321, 2282 MZ
Rijswijk, The Netherlands

* SHR Timber Research, Wildekamp 1c, 6704 AT, The Netherlands

Summary: Today's costs of maintenance of buildings are very high due to labour. Enlarging the period for painting outdoors and faster drying can reduce these costs. Therefore, the professional painters ask for paint products that should dry within 15 minutes. Infrared heating could be used to achieve these goals. Medium wave infrared heaters show accelerated drying of water borne coatings. Infrared heating results in fast drying, excellent film formation and anti-blocking properties. For solvent borne alkyds and high solids, faster tack-free times and through-dry times are achieved. Raw material suppliers, paint manufacturers and IR-equipment suppliers are challenged to develop new products and equipment to meet the needs of professional painters.

Keywords: infrared; drying; paint; water borne; solvent borne

Painters' problems

Today's costs of building maintenance, by painting windows, doors and cladding, are very high due to indirect labour activities. Interior and exterior maintenance should therefore be carried out with more efficiency. Normally the period from November until March cannot be used for outdoor painting. Outdoor painting is difficult due to humid and cold climate conditions. This has a negative effect on continuation and profitability of the companies. One of the possibilities of improvement is to enlarge the period for painting outdoors.

Earlier research

As a first step, in 1998 a research project, initiated by the Dutch Association of Professional Painters (Association), was carried out by the Dutch Research Institute TNO [1]. In this project, the impact of climate conditions on painting during winter times was investigated.

In the period November until March, there are 80 to 90 working days available. Due to climate conditions, only 50 – 65% of these days can be used, which leads to approximately 45 days. This is, until now, the main reason to stop outdoor painting in November.

CCC 1022-1360/00/$ 17.50+.50/0

Two assumptions were made to calculate the extra days that could be gained by painting during difficult climate conditions such as high humidity and/or low temperatures. First, some temperatures and relative humidity were found at which painting is not possible. The second assumption was that special paints are available to use during these conditions. Results are presented in table 1.

Table 1. Calculated number of available working days related to climate conditions.

Climate conditions *	1		2		3	
Temperature	10° C		5° C		1° C	
Relative humidity	80%		90%		95%	
Normal available days	45					
Extra days	Approx. 20%	9	Approx. 70%	31	Approx. 80%	36
Total days	54		76		81	

* Data from the Dutch Meteorological Institute (KNMI)

Forty paints (conventional alkyds, high solids and water borne paints) were tested under conditions as mentioned in table 1. Application features (viscosity), drying time, gloss reduction, chalking and cracking were checked. Some results on drying time and gloss are depicted in table 2.

Table 2. Tack-free time and gloss related to climate conditions.

	20° C/50% RH*		10° C/80% RH		5° C/90% RH		1° C/95% RH	
	drying time	gloss	drying time	Gloss	drying time	gloss	drying time	gloss
Conven. Alkyd	4.2 h	45	4.9 h	55	6.0 h	70	7.4 h	45
High Solid	4.9 h	60	6.0 h	45	7.8 h	30	9.5 h	5
Water borne	1.4 h	35	2.1 h	30	2.4 h	45	3.2 h	40

*The condition 20° C/50% RH is used as a reference.

The long tack-free time of conventional paints and high solids causes problems in closing windows. During application, the fast drying time of the water borne paints was a disadvantage.

The gloss of the high solids was reduced very strongly. Water borne paints showed a rather stable gloss.

The main conclusions were that outdoor painting could be possible in winter periods. However, the painters must be aware of negative effects of the performance of the paints due to application in wintertime. Especially drying time and film formation must be improved. This should result in new products with a longer durability.

Aim

The aim of the Association in this field is to stimulate development of paints (primers and topcoats) and drying techniques to reduce the drying time of paints to 15 minutes. The properties of the paint film should be at least identical to the conventionally dried paints. Experts concluded that a reduction of the drying time to 15 minutes is needed to have significant improvement in efficiency.

A desktop feasibility study [2] pointed out that infrared drying of paints could be very successful. In the automotive sector, it is already a well-known technique to enhance drying of automotive repair lacquers [3]. Some first experiments with infrared heating were carried out by SHR [4].

In this paper, the effect of medium wave length infrared emitters on drying of water borne acrylics and solvent borne conventional alkyd and high solid paints for the professional painter is presented [5]. This is the first step to increase the effective working time of the painter. To show the advantage of IR drying, experts of the Association calculated, based on the results, the effects of IR drying on the effective working time of the painter.

Experimental set-up

The experimental set-up is split into infrared drying of several paint types and economic calculations on time spending.

IR drying

Several types of infrared emitters can be used for paint drying. Available are short wave infrared light (wavelength around 1 μm), medium (wavelength 1.5-3.0 μm) and long wave (3 μm and higher). For painters' use, the best infrared emitter seems to be the medium wave length ones, due to the fact that they are robust and able to heat up the surface slowly to relatively low temperatures (not more than 100 °C). Furthermore, these emitters are available in 'ready to use' handhold equipment.

Subject of research in this paper were some commercial paint formulations, such as a water borne acrylic, solvent borne conventional alkyd, and a high solid paint. These paint types are available for the professional painter only. Specifications of the paint as stated by the manufacturer are given in table 3.

The water borne acrylic was tested in three formulations: with normal content of coalescent (100%), 50 % of coalescent and without coalescent to investigate the effect of higher drying temperatures on film formation.

Table 3. Drying times of tested paint, data provided by supplier.

	1	**2**	**3**
Paint type	Water borne	Solvent borne	Solvent borne
	Acrylic	Conventional alkyd	High solid alkyd
Tack free time (h)	1.5-2	6	5
Overcoat time (h)	4-5	18	24

First aim was to find out the optimal flash off time, surface heating temperature and drying time by infrared irradiation. Primed spruce and meranti panels were coated by brush application. The wet film thickness was about 80 – 100 µm. Meranti panels were chosen to evaluate blister formation on the pores during heating. Drying properties were evaluated by checking the tack-free time and nail hardness. The influence of colour on surface temperature was evaluated by application of paints in white (RAL1013) and darkgreen (RAL4823) colour and was measured with a pyrometer.

Secondly, the drying properties were evaluated by application of paint on glass strips with a BA applicator to wet film thickness of 80 – 100 µm. The paint was dried by using medium wave length IR light and putting the strips on a drying recorder (Brave Instruments type 5315). The drying process of the paints was evaluated with the drying recorder during 8 hours. A typical pattern of a drying process is shown in figure 1.

I	II	III	IV	V

Figure 1. Drying stages of paint: levelling (I), setting (II), surface-drying (III), through-drying (IV), dry (V).

Thirdly, film properties after infrared drying were evaluated by water uptake of coated spruce panels of 150x70x20 mm (method similar to EN 927-5). The panels were treated with the water borne acrylic dispersion according to the following scheme:

Reference 1	30 minutes drying of each layer at 21 °C and 55 % RH after application of primer and after application of topcoat
Reference 2	2 hours drying of each layer at 21 °C and 55 % RH after application of primer and after application of topcoat
Infrared	Drying with infrared after application of primer and after application of the topcoat. After each application 5 minutes flash off, 5 minutes drying at room temperature with ventilation, 2 minutes infrared heating at 60 °C. Total drying time of each layer was 30 minutes.

For water uptake, the panels have been floating on water for 72 hours. The coated surface was faced to the water; the other sides were sealed with two layers of a two component polyurethane coating. After water uptake, the adhesion was tested according to ASTM D 3359 A by making a double cross cut in the coating. The adhesion was tested by putting adhesion tape (3M Scotch Brand No. 800 Adhesion tape) on the cuts and pulling it of under an angle of 180° after one minute. Results were rated in classes. Adhesion class 5 indicates no adhesion, class 0 indicates total loss of adhesion.

Economic calculations

Experts of the Association carried out many time studies in the past, to be able to calculate the cost of labour. It turned out that approximately 80 – 85% of painting costs are labour costs. Costs for painting can be split into direct and indirect costs. Direct costs are defined as costs, which are directly related to the painting job. These costs contain activities like sanding, cleaning and actual painting. Indirect costs are defined as costs, which are made to realise the actual painting job. These activities are: (scaffolding), climbing, preparing the working place and materials, planning, appointments with owners to open doors and windows, instructions by management and cleaning the workplace. This results in a direct/indirect ratio of 57/43.

Results and discussion on IR drying

Water borne acrylic

First the results on drying parameters are presented. It appeared that a flash off time without heating and ventilation is important in obtaining good appearance of the coating. There is a difference between with and without infrared drying for water borne formulations with and without or with half of the amount of coalescent. It appeared that coalescent is necessary to obtain a coating film without orange peel. It turned out that the normal amount of coalescent is needed to obtain a smooth film.

A clear difference was observed for drying with and without ventilation. The water borne formulation could be dried till temperatures of 80 °C for 1 minute with 2-3 m/s air ventilation speed. Higher temperatures or longer drying times resulted in blisters. With ventilation it is possible to dry at higher temperatures. The results are presented in table 4 and are compared with the drying time of the reference (without ventilation and infrared heating).

Table 4. Results on IR drying for a water borne acrylic. + Indicates flash off or IR heating with ventilation

Water borne acrylic				
Substrate	Flash off time (min.)	IR heating time (min.)	Surface temperature (° C)	Results
S	-	-	-	**Reference, tack-free after 1.5 h**
S	2 / 10+	1+	100	Before IR already surface dry Blisters after 40 seconds IR irradiation After IR tack-free
S	5+	1+	60	After IR dust-free
S	5+	2+	60	After IR tack-free
S	5+	1+	80	After IR tack-free
S	5+	1/2+	80	After IR dust-free, almost tack-free
M	5	2	60	During application, appearance of blisters. During irradiation with IR, blisters increase without formation of more blisters

S stands for spruce, M indicates meranti

Solvent borne conventional alkyd

For the solvent borne alkyd, the surface could be heated up till a temperature of 100° C for 5 minutes. Higher temperatures or longer drying times resulted in blisters. Compared with the reference (without IR irradiation), initial drying was enhanced by IR heating. Through-drying time was the same.

During the tests, it appeared that air ventilation was needed to prevent blister formation at higher temperatures or longer drying times at high temperatures.

The results are given in table 5 and are compared with the drying time of the reference (without ventilation and infrared heating).

Table 5. Results on IR drying for a solvent borne conventional alkyd.

Flash off time (min.)	IR heating time (min.)	Surface temperature (° C)	Solvent borne conventional alkyd
			Results
-	-	-	**Reference, tack-free after 6 h**
10	5	60	After 2 h, no difference with reference
10	5	80	After 2 h, no difference with reference
10	5	100	After 2 h, no difference with reference

Solvent borne high solid alkyd

For the solvent borne high solid alkyd, a drying temperature of 100 °C with ventilation of 2-3 m/s can be reached. Without ventilation, blisters occur at a temperature of 80 °C. Compared with the reference, earlier dust-free times are obtained by IR irradiation. Through-drying is moderately influenced by IR heating; after 2 h the paint film is not tack-free. Results are shown in table 6 and are compared with the drying time of the reference (without ventilation and infrared heating).

Table 6. Results on IR drying for a solvent borne high solid alkyd.

Flash off time (min.)	IR heating time (min.)	Surface temperature ()	Solvent borne high solid alkyd
			Results
-	-	-	*Reference, tack-free after 5 h*
10	5	60	Earlier dust-free as reference
10	5	80	Earlier dust-free as reference and 5 min. at 60 ° C
10	5	100	During heating, blisters occur. After 30 min. the film is dust-free
10 / 10+	5+	100	No irregularities. After cooling down, film is dust-free. Seems to be through-dry earlier as reference
10+	5+	100	No irregularities. After cooling down, film is dust-free. Seems to be through-dry earlier as reference

+ Indicates flash off or IR heating with ventilation.
For flash off: 10/10+ indicate 10 minutes flash off without ventilation and 10 minutes with ventilation.

The results of drying on glass strips are depicted in table 8. Drying conditions before putting the glass strips on the drying recorder are depicted in table 7.

Table 7. Drying conditions before putting the glass strips on the drying recorder

	Flash off (min.)	Ventilation (m/s)	Temperature (°C)	Time (min.)
Water borne acrylic	5	2-3	60	2
Solvent borne alkyd	10	2-3	100	5
Solvent borne high solid alkyd	10	2-3	100	5

Table 8. Results of drying on glass strips

	Water. borne acrylic		Solvent borne alkyd		Solvent borne high solid alkyd	
	Ref.	IR	Ref.	IR	Ref.	IR
Tack free [min.]	145	10	70	15	70	10
Through dry [min.]	145	10	470	80	400	110

A remarkable increase in drying time was obtained by IR irradiation. Due to the fact that these results are obtained on the more heat-conductive substrate glass in stead of wood, differences in drying time occur. As a result of IR drying, the films do not show a levelling and setting period.

The results of water uptake are depicted in figure 2. It was noticed that the samples without IR heating were not dried enough. Therefore, no difference in water uptake is observed between reference 1 and the uncoated samples. A clear difference in water uptake (i.e. film formation) is observed between reference 1 (30 minutes drying without IR) and IR (2 minutes drying with IR). A 40 g/m^2 lower water uptake (i.e. a better film formation) is obtained for IR compared to reference 2 (2 hours drying). After water uptake, the coating on samples without IR irradiation is softer compared to the IR treated surfaces. Directly after water uptake, adhesion was measured. For all systems, the adhesion failed (class <3 according to ASTM D 3359). Two weeks after the water uptake, the adhesion was class 4 or 5 according to ASTM D 3359.

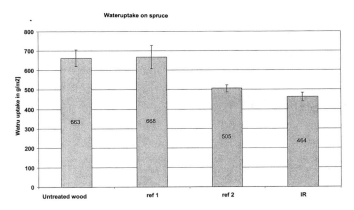

Figure 2. Water uptake in g/m2 for IR and non-IR dried painted spruce panels

Summarising the results, it was shown that infrared radiation is a powerful technique to shorten drying times of water borne acrylics to minutes. Water uptake as an indication for film formation showed a slight lower value for IR dried paint. Early good adhesion is not obtained; indicating that film formation is not completed after the short drying period.

The solvent borne types showed an earlier touch dry time. Although the through dry time was improved significantly, it was still in the order of hours. Differences in results on wood panels and glass strips could be explained by heat uptake of the glass thereby enhancing drying speed.

For all paint systems, air ventilation is necessary to obtain drying without blister formation. Air ventilation without IR heating is known to enhance drying properties too.

Results and discussion on economic calculations

The economical calculations are based on real time studies for maintenance of buildings. The results presented in this paragraph are an average of four time studies carried out during painting four different buildings. The impact of the use of infrared equipment is based on estimations, guided by the results of increased drying properties as described earlier. Experts in the field made the estimations and calculations. In general, it is recognised that IR heating is not a solution for all situations. For each project, a decision should be made whether it is a proper solution or not.

Using infrared equipment has only an effect on the indirect activities; the paint job itself does not change. The calculations were made for the following painting job, which is a typical maintenance scheme:

- cleaning and sanding,
- primering 50% of the surface,
- application of 1 primer layer,
- application of 1 topcoat layer.

The data available from the time studies and the estimations are depicted in table 9 and figure 3.

Table 9. Time spending during a painting job: direct and indirect activities with and without infrared drying.

		Normal	IR drying
Total direct activities		**56,6**	**65,5**
	cleaning, sanding, painting	56,6	65,3
	using IR-equipment	0,0	0,2
Total indirect activities		**43,4**	**34,5**
	climbing/scaffolding	3,7	1,9
	preparation	4,7	4,2
	waiting, due to occupants	1,2	0,8
	changing working place	3,8	1,5
	precautions due to rain	0,7	0,7
	repair and rework	7,8	7,2
	guidance/control	5,7	2,4
	cleaning working place	2,0	1,8
	personal hygiene/breaks	13,8	13,8
	handling IR-equipment	0,0	0,3
Total time spending		**100,0**	**100,0**
Index direct/indirect		**1,3**	**1,9**

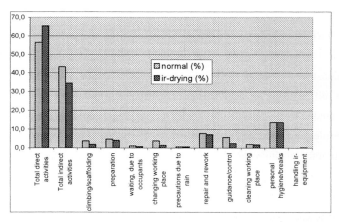

Figure 3. Time spending during a painting job: direct and indirect activities with and without infrared drying.

It is shown that the ratio direct/indirect changes from 1.3 to 1.9. This indicates an improvement of 46% in the ratio. With infrared heating, painters could stay longer on their working spot and finish the maintenance scheme in one time. Therefore, the indirect activities (such as climbing) will be done in a much more efficient way. Also time, spend to make appointments with occupants of the houses will be reduced.

However, using infrared drying probably even more important factors cannot be calculated such as:

- Faster closing of windows and doors,
- A smaller change of damaging of the paint work
- Less delay in finishing the paint job,
- Enlarged time of painting, because infrared drying could be less dependent on climate conditions.

By using infrared radiation, it is possible to create the proper climate conditions for a paint to dry. After application and drying of the primer layer, the topcoat can be applied immediately. Due to the fact that infrared radiation cures the paint layer, lower temperatures and higher relative humidity are believed to have less effect on the drying time. This indicates that the paint job could be done at temperatures of $1 - 5°$ C and at an RH of $80 - 95\%$. Herewith, the painting could be done during difficult climate conditions in wintertime. In the near future this kind of experiments will be made with newly developed and specially designed infrared equipment.

Conclusions

- Infrared radiation significantly reduces drying time of water borne acrylics
- Drying of solvent borne alkyds (conventional and high solids) is improved by infrared heating
- Air ventilation is needed to prevent blister formation.
- A significant reduction of indirect activities of painting is possible
- Improvements in drying results in less damaging and less delay of the paint film,
- Enlarging the period for painting in humid and cold climate conditions is realistic.

Future developments

After experimental work on laboratory scale, developments in the industry are needed. Developments on the paint formulation might result in even faster drying times. Portable infrared heating devices must be developed to enable the professional painter to cure the paint on the building site. Attention should be paid to the fact that these devices should be easy to work with and easy to handle.

In the Netherlands, the painters are convinced that these developments will be successful and offer new opportunities. They are optimistic about the preliminary results and are confident to start using IR heating within two years.

Therefore, the Dutch Association of Professional Painters challenges the raw material, the paint and the IR-equipment industry, to develop new products and equipment that meet the specification of completely dried paints within 15 minutes.

Organisations

The Dutch Association of Professional Painters (Mr. G. Jonkers B.Sc.) represents more than 6,000 companies with approximately 30,000 individual painters. The association stimulates innovations by sponsoring research programs. One of the goals is to introduce industrial techniques and high-tech applications in painters' daily business. The image of painters will change in this way to a more professional one.

SHR Timber Research (Mr. J. Nienhuis M.Sc., B. Van de Velde B.Sc.) is a research institute carrying out independent contract research for the wood working industry (joinery and furniture), its suppliers (such as coating industry) or its associated professions (such as professional painters). Specific knowledge of infrared heating of wood and curing of paints is used in this project.

References

[1]   'Schilderen het hele jaar door' (Painting in every season), TNO report BNG.98/003236-1/RC; March 11th 1998
[2]   'Haalbaarheidsonderzoek versnelde droging van bouwverven op hout' (Feasibility study fast drying of paints for wood), VAW990470.RAP; December 31st 1999
[3]   Modern coating and drying technology, Cohen and Gutoff, 1992 VCH publishers, New York
[4]   'Verkennend onderzoek naar de haalbaarheid van snelle droging door middel van infraroodstraling' (Preliminary study to feasibility of fast drying by infrared heating), SHR report 20.282, December 8th 2000
[5]   'Vervolgonderzoek versnelde droging verf met infraroodstraling' (Investigation of shorter drying times of paint by infrared irradiation), SHR report 1.263, October 19th 2001

New Developments in Catalysis

Werner J. Blank

King Industries, Inc. Norwalk Ct, 06897 Science Road USA

wblank@kingindustries.com, www.wernerblank.com

Summary: Environmental, safety and health concerns are major driving forces for the development of new coating systems, which in turn require catalysts with a different performance profile. One critical area for the development of new catalysts is the replacement of organotin compounds in polyurethanes with environmentally friendly catalysts, such as bismuth, aluminum and zirconium chelates. For applications in epoxies new catalysts for the epoxy-carboxyl reaction are also being developed. To gain the needed improved performance multiple cure mechanisms are being employed in coatings requiring dual action catalysts.

Introduction

New developments in the industrial coating market are driven by environmental regulations, worker safety, energy concerns and the need for improved performance. These problems are being addressed by the development of low volatile organic content (VOC) coating systems. Differences in regulations between countries have created a highly fragmented line of attack to these problems.

In isocyanate and blocked isocyanate crosslinked coatings organotin compounds are the primary catalysts used. New regulations require the replacement of these versatile catalysts with organotin-free candidates. High solids, solvent-free coatings, powder coatings and waterborne coatings have distinctly different application and cure characteristics, which are dissimilar from low solids coatings. Often existing chemistry cannot meet these application and performance requirements. A combination of different crosslinking chemistries is used to upgrade coating properties.

Coating Technologies under Development

System	Challenge
High solids, solvent-free	Viscosity/non-volatile, cure response, stability
Waterborne	Stability, co-solvents, reactivity, amine
Electrocoating	Co-solvents, cure temperature
Powder	Film thickness, reaction temperature, flow-leveling
UV/Electron beam	Pigments, complex shapes

Catalysts for Isocyanate Crosslinking

Dibutyltin dilaurate (DBTDL) is the typical catalyst used for aliphatic isocyanates in both 2K isocyanate and blocked isocyanate crosslinked coatings. Because of environmental concerns, however there is a need for alternate catalysts.

Table 1 shows the periodic table and the elements, which according to the literature and our own screening studies[1,2], are active catalysts for the isocyanate reaction.

Table 1. Periodic Table

1	1A	2A												3A	4A	5A	6A	7A	8A
2	3 Li	4 Be												5 B	6 C	7 N	8 O		
3	11 Na	12 Mg	3B	4B	5B	6B	7B	|------	8B	------|	1B	2B		13 Al	14 Si	15 P	16 S		
4	19 K	20 Ca		22 Ti	23 V	24 Cr	25 Mn	26 Fe	27 Co	28 Ni	29 Cu	30 Zn	31 Ga						
5		38 Sr		40 Zr				44 Ru	45 Rh						50 Sn	51 Sb	52 Te		
6		56 Ba	57 La	72 Hf		74 W	75 Re	76 Os				80 Hg			82 Pb	83 Bi	84 Po		

The metals were tested as the carboxylate salts, sulphonate salts or as dionate complexes.

We selected zirconium, bismuth and zinc compound for a detailed reaction study and compared their catalytic activity to DBTDL. The results are shown in Figure 1.

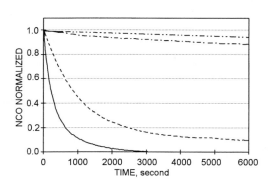

The disappearance of the IR isocyanate band at 2272 cm⁻¹ wave number is a simple tool to determine the reaction rate. Under the reaction conditions shown a zirconium dionate complex shows substantially higher catalytic activity on a metal basis in comparison to DBTDL, bismuth carboxylate and zinc octoate catalyst.

Figure 1. Reaction of 1,6-hexamethylene diisocyanate trimer with n-butanol (1.12 mol/l NCO) in xylene. Molar ratio NCO/OH 1/1. Catalyst 0.014 % metal on total reactant. Catalyst conc.: ----- DBTDL; —— Zr Chelate; — · — Bi carboxylate; — · · — · Zn octoate.

In the next experiments because of the low reaction rate with bismuth and zinc catalysts the level of catalyst was increased.

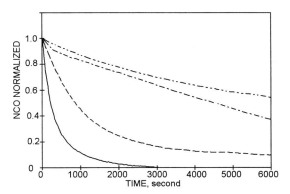

Figure 2. Reaction of 1,6-hexamethylene diisocyanate trimerwith n-butanol. (1.12 mol/l NCO) in xylene. Molar ratio NCO/OH 1/1. Catalyst conc.: ----DBTDL 0.014 % Sn; —— Zr Chelate 0.014 % Zr; - · - Bi carboxylate 0.13 % Bi; — · · — Zn octoate 0.27 % Zn on total reactants.

The results in Figure 2 illustrate, that under the reaction conditions shown both Zr chelate and DBTDL are faster catalysts than bismuth or zinc carboxylates. A faster catalyst is not necessarily superior in a coating formulation. Potlife, pigment absorption, application properties, selectivity of hydroxyl over water reaction and durability have to be considered.

We were also interested how reaction rates of catalysts vary with different functional groups. We used the catalyst concentration shown in Figure 2 for the additional experiments. As functional compounds, we selected the hydroxyl compounds shown in Table 2.

Table 2. Hydroxyl Compounds used in Reaction Rate Study

Compound	Characteristics
n-Butanol	Primary hydroxyl group, no steric hindrance
Isobutanol	Primary hydroxyl group, some steric hindrance
2-Butanol	Secondary hydroxyl group
2-Butoxyethanol	Primary β-hydroxylethyl ether group
2-Hydroxyethyl ester (HEE)	Primary hydroxyl, model for polyester
2-Hydroxyethyl carbamate (HEC)	Primary hydroxyl, polyurethane diol
Methoxtripropylene glycol (TPM)	Sec. hydroxyl, model for polypropylene glycol

Figure 3 shows the half-life of DBTDL, Zr Chelate, Bi carboxylate and Zn octoate as measured by the catalyzed reaction of 1,6-hexamethylenediisocyanate trimer with the alcohols shown in Table 2.

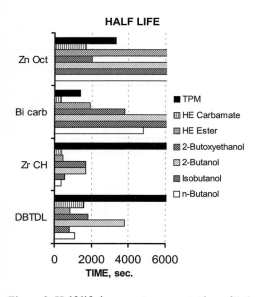

HALF LIFE

Legend:
- ■ TPM
- ▥ HE Carbamate
- ▨ HE Ester
- ▧ 2-Butoxyethanol
- ▨ 2-Butanol
- ▨ Isobutanol
- ☐ n-Butanol

TIME, sec.

Figure 3. Half-life isocyanate concentration of 1,6-hexamethylene diisocyanate trimer with alcohols. (1.12 mol/l NCO) in xylene. Molar ratio NCO/OH 1/1. Catalyst concentration: DBTDL 0.014 % Sn; Zr Chelate 0.014 % Zr; Bi carboxylate 0.13 % Bi; Zn octoate 0.27 % Zn on total reactants.

Half-life of the isocyanate concentration was calculated from the rate equation of isocyanate/time as determined by FT-IR. DBTDL catalyzes as expected, giving high reaction rates with primary alcohols. Even within the group of primary alcohols there are significantly different reaction rates. Secondary OH, as in 2-butanol, reacts substantially slower. TPM, a model for polypropylene glycol gives a very slow reaction rate. The Zr chelate gives substantially faster reaction with primary OH. 2-Butoxyethanol and 2-butanol are slower and TPM essentially deactivates the catalyst. The relative rates with bismuth are rather surprising. Neighboring groups have a substantial effect on reaction rate. Steric effects and secondary hydroxyl groups do not influence the reaction rate as much.

Conclusions for Isocyanate Catalysts

Many metal compounds can catalyze the reaction of isocyanates with hydroxyl groups. DBTDL is a very effective catalyst for the reaction of isocyanate with hydroxyl compounds. The notion of primary hydroxyls reacting faster with isocyanates is confirmed using DBTDL as a catalyst. But not all primary hydroxyl groups are equal in reaction rate. As shown in Figure 3, neighboring groups such as ether, ester or carbamate groups have a substantial effect on reaction rate.

Reaction rates with Zr chelate catalyst are substantially higher with all primary hydroxyl functional compounds with the exception of 2-butoxyethanol. Secondary hydroxyls are substantially slower. With TPM we see an almost complete inhibition of reaction.

With bismuth carboxylate catalyst the results are rather surprising, since neighboring groups have a substantial effect on reaction rate. The general rule of higher reaction rate with primary hydroxyl groups does not apply. Carbamate, ester and ether groups in the β position give superior reaction rates. Even secondary hydroxyl functional polyethers are faster reacting than primary hydroxyl compounds. This raises some interesting questions about the mechanism for the bismuth-catalyzed formulations. Generally it is been thought that bismuth functions as a Lewis acid catalyst and activates the isocyanate groups. Although zinc is a weak catalyst for the isocyanate reaction, we again see that neighboring groups affect catalysis. Ether and carbamate groups assist in increasing reaction rate, but ester groups have no effect.

Catalysis of Blocked Isocyanates

For many blocked isocyanates an elimination-addition mechanism is indicated[3] as shown in EQ. 1. The overall reaction rate depends on k_1, k_{-1} and k_2, k_{-2} and the volatility of the blocking agent, (Bl-H). Prior art suggests that catalysts only affect k_2 but not the deblocking reaction (k_1, k_{-1}). For a number of blocked isocyanates a displacement reaction[4,5,6] appears to be the predominate mechanism (Eq. 3). DBTDL is an excellent catalyst for both the isocyanate-hydroxyl and also for the displacement (transesterification) reaction. If indeed catalysis of the isocyanate-hydroxyl reaction determines the rate, there should be little difference in their catalytic activity between catalysts for 2K and blocked isocyanate coatings.

R-NHCOO-Bl $\underset{k_{-1}}{\overset{k_1}{\rightleftharpoons}}$ R-NCO + Bl-H   EQ (1)

R-NCO + R'-OH $\underset{k_{-2}}{\overset{k_2}{\rightleftharpoons}}$ R-NHCOOR'   EQ (2)

RNHCOOR' + R"OH $\underset{k_{-3}}{\overset{k_3}{\rightleftharpoons}}$ RNHCOOR" + R'OH ↑   EQ (3)

Comparative cure studies[7] of the catalysts used in Figure 3 and also Table 1 show that many catalysts that are effective in 2K formulations do not perform in blocked isocyanates. Deactivation of the catalyst[8] due to hydrolysis during cure and the loss of the ligands (carboxylate or dionate) are just some of the problems encountered during cure of blocked isocyanates. (Eq. 4), Table 4.

$$R\text{-COOMe}^{(a)} \quad + \quad H_2O \quad \longleftrightarrow \quad MeO\,(OH) \quad + \quad R\text{-COOH} \qquad EQ\,(4)$$
(a) Metal moiety

Table 3 shows the active catalysts for blocked isocyanates[9]. For the blocking agents ketoxime, 3,5-dimethyl-1H-pyrazole, e-caprolactam, phenol and malonate, a 1,6-hexamethylenediisocyanate trimer was used as isocyanate. For the alcohol[10] blocked isocyanate a polymethylene polyisocyanate was used. For these tests either the 2-ethylhexanoate, laurate or naphthenate salts of metals were used. 1,5-Diazobicyclo [4.3.0] non-5-ene (DBN) and 1,8-Diazabicyclo [5.4.0] undec-7-ene (DBU) were used as the amine catalysts. For malonate blocked isocyanates, which crosslinks via a transesterification reaction[11] no effective catalyst was found.

Table 3. Catalysis of Blocked Isocyanates

Blocking Agent	Temperature, °C	Catalyst
Ketoxime	130-150	Bi, Co, Cr (III), DBTDL, Zn, Ca,
3,5-Dimethylpyrazol	120-140	DBTDL, Bi, Co, Cr, Zn
Caprolactam	150-170	Zn, DBTDL, Co, Bi
Phenol	120-140	Zn, Mn, Bi, DBTDL, Ca, Cr (III)
Alcohol	150-170	Bi, Co, Zn, DBTDL,
Uretdione	170-200	Zn, Bi, DBTDL, Sn (II), DBN, DBU
Malonate	100-120	No effective catalyst

Table 4 shows the effect of ligands on the methylethylketone (MEK) resistance of a cationic electrocoating resin and a polymethylphenyl polyisocyanate blocked with 2-butoxy(2-ethoxyethyanol) and cured at different temperatures for 20 minutes. Metal concentration in the formulation was 0.25 % based on total binder. The incorporation method for certain metal catalysts was found important for catalyst effectiveness. Bismuth carboxylate catalysts are known to hydrolyze in water. We added the bismuth carboxylate catalyst to the solvent-free cationic resin/blocked isocyanate crosslinker and emulsified this system in water. We propose that the bismuth carboxylate catalyst hydrolyses in this formulation and forms nano particles of bismuth oxide. During heating to >120°C the bismuth carboxylate catalyst reforms.

Table 4. Bismuth Catalyst Effects of Ligands on Reaction Temperature (MEK double rubs)

Catalysts/ temperature of cure	180 °C	170 °C	165 °C
No catalyst	10	10	10
Dibutyltin dilaurate	63	30	10
Bismuth tris(2-ethylhexanoate)	150	70	20
Bismuth tris(isostearate)	200	150	20
Bismuth dimeric fatty acid	200	70	13
Bismuth tris(oleoyl sarcosine)	200	150	70

Epoxy Catalysts

Epoxy chemistry is important in low VOC coatings because of the absence of volatile reaction products and the excellent resistance properties that can be obtained. There is an interest in epoxy/carboxyl systems used in powder coatings and high solids coatings to improve the formulation stability, cure response and resistance properties. Traditional amine and phosphonium catalysts have deficiencies in yellowing, storage stability and resistance properties. We screened[12] a range of known amine catalysts, metal dionate complexes, carboxylate, phosphate and sulfonate salts in glycidylether and glycidylester/carboxyl formulations for cure response, stability and film properties. The metal catalysts evaluated were from the group of Al, Bi, Ca, Ce, Co, Cr(III), Fe, Mg, Mn, Ni, Sn(II), organotin, Ti, Zn and Zr compounds. Only Zn, Co, Mg, Sn(II) and Zr compounds showed activity. Generally the ligands had an effect on catalytic activity and the weaker acidic ligands such as dionate or carboxylate were superior to sulfonate. Zinc chelates and salts showed superior catalytic behavior and also room temperature stability. A new zinc chelate catalyst[13] was developed which gives improved stability, cure response and superior film properties. This catalyst is applicable for liquid and also powder coatings.

Epoxy resins can be homopolymerized as evidenced by UV cured coatings using cationic catalyst being in wide commercial use. The absence of functional groups which do not hydrogen bond produces reactants which are very low in viscosity[14]. This makes epoxy resins very attractive candidates for solvent-free and high solids coatings. We were interested in blocked superacid catalysts for thermal cure of epoxy resins[15] as an alternative to melamine crosslinked coatings. Thermally catalyzed cycloaliphatic epoxies are effective crosslinker for hydroxyl functional acrylic polymers and can be used to replace melamine crosslinkers in applications which cannot tolerate any formaldehyde emission.[16] The low viscosity of the system permits the formulation of 100 % reactive solvent free coatings.

Hybrid or Dual Function Catalysts

Crosslinking agents of different chemistries are being combined in coating formulations in order to improve performance and retain the low cost of existing formulations. Therefore, increasing emphasis is being placed on the development of catalyst for multiple curing mechanisms. As shown in Table 5 melamine formaldehyde crosslinked coatings can modified with isocyanates[17] and blocked isocyanates[18] to improve mechanical properties, acid resistance[19] and cure behavior. Isocyanate crosslinking can also be combined with the catalysis of siloxane reactions. DBTDL is an excellent catalyst for the siloxane and isocyanate reaction, but the level of catalyst required for siloxane reaction is several orders of magnitude larger than for the isocyanate reaction, leading to a heterogeneous network.

Table 5. Catalysis of Hybrid Crosslinked Coatings Polymer Hydroxyl Functional

Melamine Resin/Blocked Isocyanate [a]		Catalyst	Improvement
HMMM [b]		R-SO$_3$-Ester[20]	Acid etch resistance
		Zn Chelate/ Lewis acid[21]	Impact/stone chip
IMINO, Part Alkylated		R-SO$_3$-Ester	Acid etch resistance
		R-PO$_4$-Amine	Impact/stone chip
		Zn Chelate[22]	
HMMM	Siloxane [c]	R-SO$_3$-Ester	Acid etch resistance
Isocyanate	Siloxane	Al Chelate[23]	Scratch resistance
Isocyanate blocked [a]	Siloxane	Zn Chelate	Chemical resistance

[a] Blocked isocyanate; malonate, ketoxime, 3,5-dimethylpyrazol blocked aliphatic isocyanates; [b] Hexakis(methoxymethyl)melamine; [c] Alkoxysilanes

Conclusion

Catalysis plays a critical role in the development of new coating systems. To meet the requirements for solvent-free, high solids, waterborne and powder coatings, a variety of metal chelates are being developed which are environmentally acceptable and meet the cure requirements of the industry.

Any new catalyst is not a direct replacement for an existing product. The present know-how and experience does not necessarily apply to a new catalyst.

DBTDL as a catalyst shows increased reaction rates with primary hydroxyl groups over secondary hydroxyls. This information has been used in the design of polyols for polyurethane applications. With Zr catalysts this relationship still holds, although we find with some polyethers an extremely slow reaction.

With Bi based catalysts our conventional thinking of reaction rates requires a revision. ß-hydroxyalkyl ester, carbamate and ether groups show significantly higher reaction rates than sterically unhindered hydroxyls.

Thermally cationic catalyzed epoxy resins offers performance properties not attainable with other crosslinking systems. Low viscosity and high application solids make these systems candidates for zero VOC coatings.

Hybrid or dual crosslinked systems have achieved an important position in the automotive clearcoat market. Combinations of melamine resins with siloxanes or with blocked isocyanates are used to upgrade the acid resistance of melamine crosslinked coatings and still retain scratch resistance and lower cost.

Acknowledgement

I would like to thank King Industries Inc. for the permission to publish this work and for supporting the lifestyle I have been accustomed to. I also thank my coworkers in the R&D and TS department for their contributions. I also like to acknowledge the efforts of Dr. Leonard J. Calbo and Dr. Carl Seefried in proofreading this paper.

References

[1] Z. A. He, W. J. Blank, M. E. Picci " A Selective Catalyst for two component waterborne polyurethane coatings" Proceedings of the Twenty-Sixth Annual International Waterborne, High-Solids, and Powder Coatings Symposium, New Orleans, LA, 1999

[2] Communication with J. J. Florio, unpublished King Industries, Inc. Information

[3] S.P. Pappas, E.H.Urruti, Proceedings of the 13th International Waterborne, High-Solids, and Powder Coatings Symposium, New Orleans, LA, 1986

[4] Z.W.Wicks Jr., Prog. Org. Coat., 3 (1975); 73

[5] Z.W.Wicks Jr., Prog. Org. Coat., 9 (1981); 3

[6] Z.W. Wicks Jr. and B.W. Kosty, J. Coat. Technol.,49 (634) (1977); 77

[7] Private communication with Z. A. He and M. Piccie King Industries, Inc.

[8] US. 6,353,057 (2002), King Industries, Inc.,"Catalyzing cationic resin and blocked polyisocyanate with bismuth carboxylate". Invs.: Z. He; Zhiqiang

W.J.Blank, M.E. Picci;

[9] W.J. Blank, Z.A. He, M. E. Picci, (King Ind. Inc., Norwalk, CT 06852, USA). "Catalysis of blocked isocyanates with non-tin catalysts" Polym. Mater. Sci. Eng., 79, 399-400, 1998 ACS

[10] W. J. Blank, "Crosslinking with Polyurethanes", Polym. Mater. Sci. Eng., 63, 1990 ACS

[11] Z. W. Wicks, B. W. Kostyk, J. Coat. Tech. 49 No.634, (1977) 77.

[12] W. J. Blank, Z. A. He, M. E. Piccie, J. Coat. Tech. 74 No.926, (2002) 33-41

[13] US 6,335,304 (2002), King Industries, Inc., "Metal Salts of Phosphoric Acid Ester as Crosslinking Catalyst", invs.: Z. He; Zhiqiang W.J.Blank, M.E. Picci;

[14] Z. W. Wicks, G. F. Jacobs, I. C. Lin, E. H. Urruti, L. G. Fitzgerald, "Viscosity of Oligomer Solitions" J. Coat. Tech.., Vol. 57, No. 725 (1985), pp. 51-66.

[15] R. P. Subrayan, D. J. Miller, M. M. Emmet and W. J. Blank, "Catalysis of Thermally Curable High Solids Cycloaliphatic Epoxy Formulations", ACS PMSE Preprints Chicago Meeting (2001).

[16] King Industries, Inc. Technical literature "NACURE® Epoxy Catalysts"

[17] G..Teng, R.P. Subrayan, F.N. Jones, "Sag-Resistant High Solids Baked Coatings with Mixed Polyisocyanate and Melamine-Formaldehyde Crosslinkers," *Polym. Mater. Sci. Eng.*, 1997, *76*, 117-118.

[18] Ger. Offen. DE 4204518 A1 19 Aug 1993, (BASF Lacke und Farben A.-G., Germany). Invs: U. Roeckrath, G. Wigger, U. Poth.

[19] W.J. Blank, Z.A. He, E.T. Hessell, R.A Abramshe, (King Industries Inc, Norwalk, CT 06852, USA). "Melamine formaldehyde networks with improved chemical resistance.". Polym. Mater. Sci. Eng., 77, 391-392, 1997, ACS

[20] NACURE® 5414 catalyst, a product of King Industries, Inc. Norwalk, CT 06852 USA

[21] NACURE® XC-8212 catalyst, a product of King Industries, Inc. Norwalk, CT 06852 USA

[22] NACURE® XC-9206, XC-B219 catalyst, products of King Industries, Inc. Norwalk, CT 06852 USA

[23] NACURE® 5218 catalyst, a product of King Industries, Inc. Norwalk, CT 06852 USA

Macromol. Symp. *187,* *271–279 (2002)*   271

Tailor-made Crosslinkers for High Performance PUR Coatings - Hyperbranched Polyisocyanates

Bernd Bruchmann[1], Rainer Königer[2]*, Hans Renz[2]

[1]BASF-AG, Polymer Research, 67056 Ludwigshafen, Germany
[2]BASF-AG, Performance Chemicals for Coatings, 67056 Ludwigshafen, Germany

Summary: The introduction of Self-Crosslinking Isocyanates (**SCI**) as AB_2-building blocks makes it possible to synthesize hydroxyl- or isocyanate-group terminated dendrimers. Furthermore, inherent reactivity differences of non equivalent NCO-groups of specific diisocyanates (such as isophorone diisocyanate [IPDI] or toluylene diisocyanate [TDI]) can be used to build up hyperbranched polyurethane structures in a one pot synthesis without the need of arduous protection/deprotection steps. This synthetic approach allows the construction of tailor-made hyperbranched molecular architectures which are end-functionalized with either hydroxyl or isocyanate groups. These products were then tested as crosslinkers in 2-component coating formulations where they displayed better hardness than any other aliphatic isocyanate raw material.

Keywords: coatings; crosslinking; dendrimers; hyperbranched; isocyanates; polyurethanes; synthesis

Introduction

The development of Self-Crosslinking Isocyanates (**SCI**)[1] provided suitable molecules and functionalities for the synthesis of dendritic and hyperbranched structures based on polyurethane chemistry. SCI combine at least one isocyanate functionality with multiple hydroxyl or amine functional groups in one molecule. To obtain stable products, the hydroxyl or amine functionalities need to be capped with suitable protecting groups. These protecting groups are designed to react immediately with water to liberate the corresponding active hydrogen functionality which then can react with isocyanates (Figure 1). Originally, these SCI-systems were developed as reactive diluents for super high solids 2-component polyurethane coating systems. However, it quickly became apparent that SCI, which contain one isocyanate and two active hydrogen groups, have great potential to be used as building blocks for dendrimers and hyperbranched structures. SCI molecules can be regarded as capped AB_2-building blocks (Figure 1) which are necessary for the synthesis of dendrimers or hyperbranched polymers.

        CCC 1022-1360/00/$ 17.50+.50/0

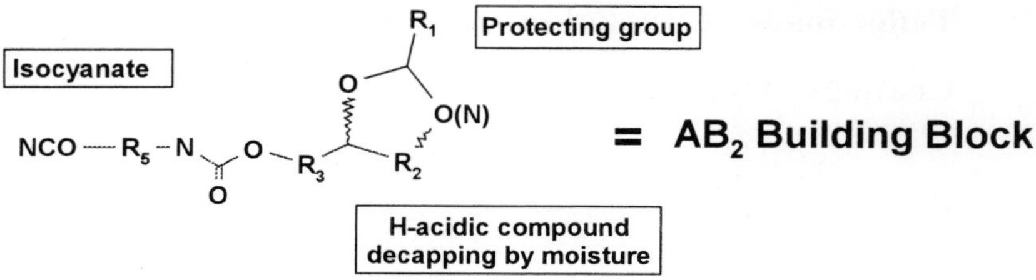

Figure 1. Schematic Representation of Self-Crosslinking Isocyanates (SCI)

Results and Discussion

Although a wide range of synthetic strategies have been developed for the construction of dendrimers[2] there are only few reports existing on the synthesis of dendrimers or hyperbranched polymers using isocyanate or urethane chemistry.[3-10]   Now SCIs open up new routes to simple, divergent syntheses of a variety of dendrimers.

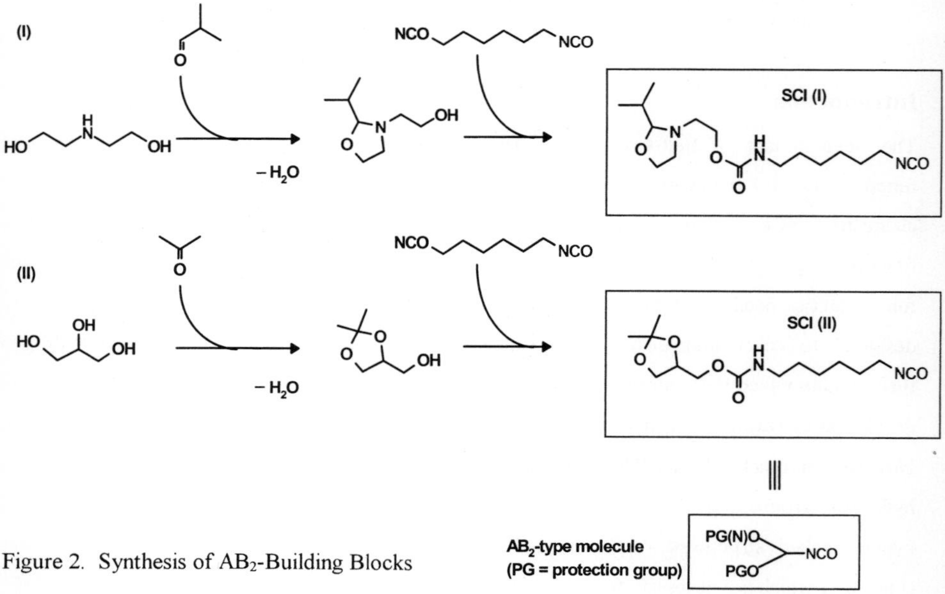

Figure 2. Synthesis of AB$_2$-Building Blocks

Two types of SCI were studied: (I) a 1:1 reaction product of hexamethylene diisocyanate with a hydroxyl functional oxazolidine and (II) a 1:1 reaction product of hexamethylene diisocyanate with a hydroxyl functional 1,3-dioxolane. Both, the oxazolidine and the dioxolane, are easily accessible by reaction of an aldehyde or ketone with diethanolamine or glycerol, respectively (Figure 2).

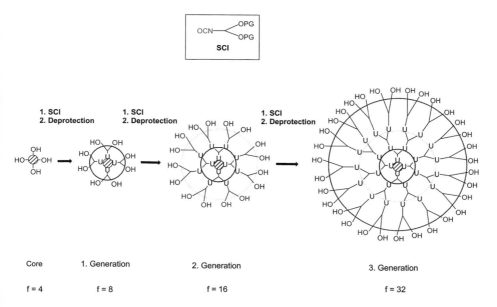

Figure 3. Synthesis of a Urethane Dendrimer

Starting from a multifunctional amine or alcohol as a core molecule the addition of e.g. SCI (II) will form the first generation dendrimer with a doubled functionality. Addition of water to deprotect and release the OH-groups sets the molecule up for the formation of the second generation dendrimer by adding again SCI. The repetition of these reaction steps allows the controlled synthesis of higher dendrimer generations (Figure 3).

Hyperbranched structures may be obtained more easily by simply adding water as well as a small amount of an alcohol to the SCI-system. The water will open the oxazolidine or dioxolane ring to yield the partners for an intermolecular reaction with the isocyanate group, while the amount of alcohol will regulate the average molecular mass by terminating the polymer/oligomer chain (Figure 4). Unfortunately, in this approach it is unavoidable that a certain amount of water may also react with the isocyanate group leading to side reactions. The choice of protecting group and, consequently, the speed of deprotection need to be chosen carefully to minimize this side reaction.

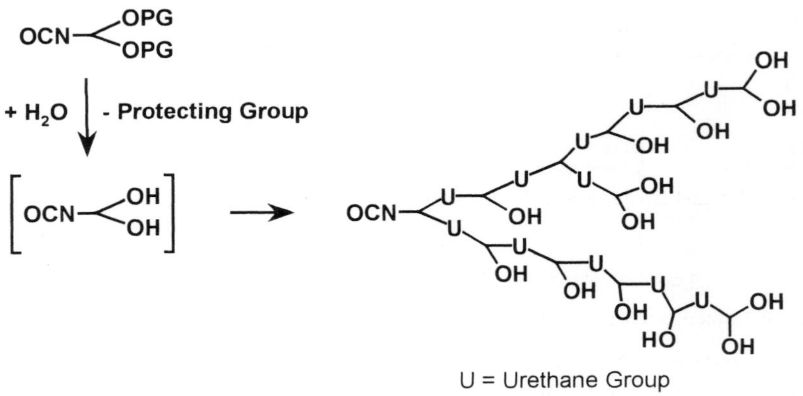

U = Urethane Group

Figure 4. One-Step Synthesis of Hyperbranched Structures

By employing modified self-crosslinking isocyanates this side reaction can be avoided. While the isocyanate group is blocked with an isocyanate blocking agent, e.g. a ketoxime, it is possible to deprotect the alcohol functionalities completely without risking a side reaction between isocyanate and water. Upon heating, the isocyanate group is regenerated and reacts to yield a hyperbranched structure (Figure 5).

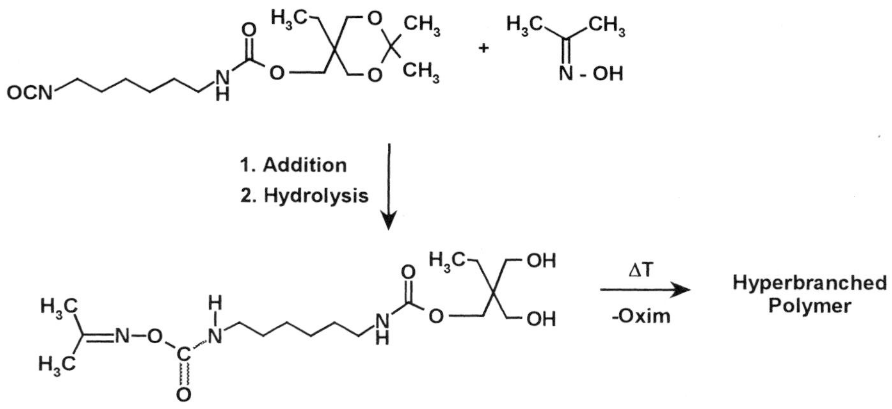

Figure 5. Blocked Isocyanates as Building Blocks for Hyperbranched Polymers

The syntheses described above for the production of hyperbranched structures have several attractive features such as the ready availability of the raw materials and the simplicity of the processes involved. However, protecting and deprotecting functional groups require additional process steps. Furthermore, the protecting groups which are employed require water for deprotection so that a rigorous drying step needs to follow every deprotection step in order to avoid side reactions. The deprotection, the removal of protecting group, and the removal of water limits the versatility of this approach. Consequently, a simpler process for the construction of hyperbranched structures was investigated.

By taking advantage of inherent reactivity differences in unsymmetrical isocyanates and in alcohols and aminoalcohols dendritic and hyperbranched structures may be assembled more satisfactorily. In the literature[10] the synthesis of 4-isocyanatomethyl-1-methyl-cyclohexyl-isocyanate has been reported.   This diisocyanate displays large reactivity differences between its two isocyanate functionalities which allow an effective kinetic differentiation. However, more readily available diisocyanates are also useful building blocks, despite having slightly lower selectivities between their isocyanate groups (Figure 6). The addition of multifunctional alcohols or amino alcohols in which the reactive groups show differing reactivities towards isocyanate groups allows the in-situ formation of AB$_2$-building blocks using the A$_2$B$_3$ approach. This approach allows the construction of a wide range of tailor-made, functionalized hyperbranched systems. A few representative examples were synthesized.

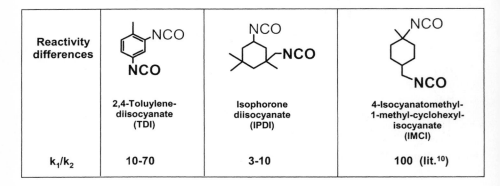

Reactivity differences	2,4-Toluylene-diisocyanate (TDI)	Isophorone diisocyanate (IPDI)	4-Isocyanatomethyl-1-methyl-cyclohexyl-isocyanate (IMCI)
k$_1$/k$_2$	10-70	3-10	100 (lit.[10])

Figure 6. Diisocyanates with Kinetically Non-Equivalent Isocyanate Groups

Starting from isophorone diisocyanate and diethanolamine an isocyanate-dialcohol compound may be produced in the first reaction step which can react in an intermolecular polyaddition to yield a hyperbranched structure which contains urea and urethane groups in the backbone and hydroxyl groups on the molecular surface (Figure 7). As described in a previous example, the addition of alcohol allows the control of the molecular weight in this system. Once the desired hyperbranched structure has been synthesized it can be modified to a polyisocyanate by further reaction with an excess of diisocyanate such as isophorone diisocyanate. Care needs to be taken to avoid side reactions in which the diisocyanate bridges two hyperbranched molecules. Due to the high functionalities this side reaction can quickly lead to crosslinking and gelation.

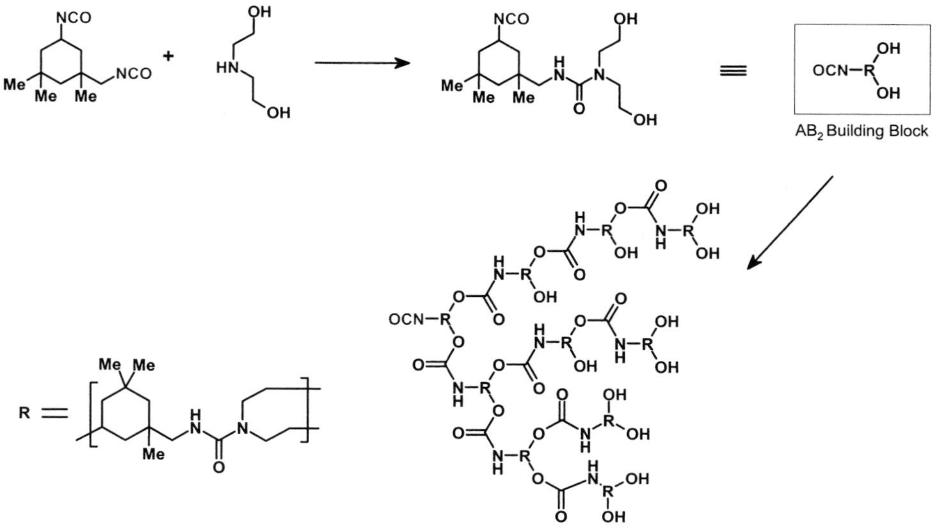

Figure 7. Synthesis of a Hyperbranched Polyalcohol

As an alternative to the NCO-functionalization of a polyol, a slightly different approach can be chosen to yield hyperbranched polyisocyanates. Starting from two molecules of diisocyanate and one molecule of a triol this $A_3(B_2)_2$-approach leads to an A_2B-type building block which bears two isocyanate and one hydroxyl group (Figure 8). In an intermolecular reaction this building block reacts to yield hyperbranched polyisocyanate structures. The addition of isocyanates allows an additional control of the molecular weight.

These hyperbranched polyisocyanates were tested as crosslinkers in conventional 2-component coating systems. A simple formulation containing isophorone diisocyanate trimeras crosslinker was employed; the isophorone diisocyanate trimer was then replaced by the hyperbranched polyisocyanate. The cured coatings containing these simple hyperbranched systems showed significant improvements regarding (pendulum) hardness, which were not achievable with any other known aliphatic isocyanate raw material.

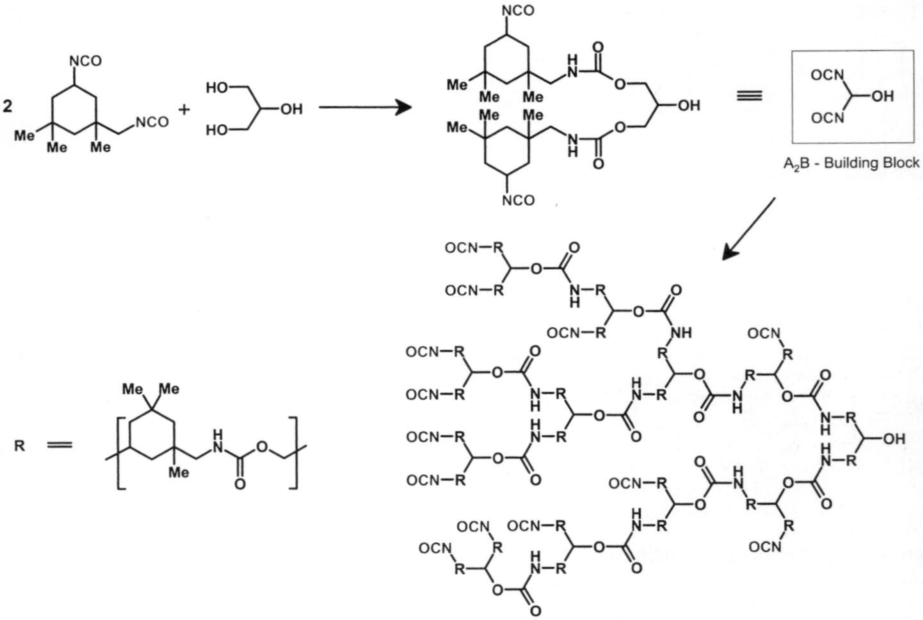

Figure 8. Synthesis of a Hyperbranched Polyisocyanate

Conclusion

Polyurethanes are an important polymer class for the coatings industry. Simple access to tailor-made, hyperbranched polyurethanes will open up new possibilities to improve properties of polyurethane coatings systems for matching customer demands. Hyperbranched polyisocyanates, for example, with their high NCO-functionality on the molecular surface represent a new class of crosslinkers for the formulation of coatings with improved properties. The synthetic approach exploits inherent reactivity differences of readily available, multifunctional raw materials to construct aliphatic as well as aromatic polymer backbones with isocyanate end-groups. Finally, all hyperbranched products are polymers according to the OECD and EU polymer definition.

Therefore, this new class of polyurethane polymers products fulfills the requirements for technical scale-up and commercialization. The unique properties of these products, especially with respect to the high hardness which they confer to a polymer film, may be of high interest to the coatings industry. Further studies on these hyperbranched polymers and their properties in coating systems are under investigation.

[1] H. Renz, B. Bruchmann, *Prog, Org. Coat.* **2001**, *43*, 32-40.
[2] *Top. Curr. Chem.* **1998**, *197* (Dendrimers) and references therein, by F. Vögtle (Ed.); *Top. Curr. Chem.* **2000**, *210* (Dendrimers II) and references therein, by F. Vögtle (Ed.); *Top. Curr. Chem.* **2001**, *212* (Dendrimers III) and references therein, by F. Vögtle (Ed.); *Top. Curr. Chem.* **2001**, *217* (Dendrimers IV) and references therein, by F. Vögtle, C.A. Schalley (Ed.).
[3] A. Kumar, S. Ramakrishnan, *J. Chem. Soc., Chem. Commun.* **1993**, 1453 – 1454.
[4] R. Spindler, J. M. J. Fréchet, *Macromolecules* **1993**, *26*, 4809.
[5] R. Spindler, J. M. J. Fréchet, *J. Chem. Soc. Perkin Trans. I*, **1993**, 913.
[6] G.R. Newkome, G.R. Baker, C.N. Moorefield, E. He, J. Epperson, C.D. Weis, *Polym. Mater. Sci. Eng.* **1997**, *77*, 65-66.
[7] S. Rannard, N. Davis, *Polym. Mater. Sci. Eng.* **1997**, *77*, 63-64.
[8] G.R. Newkome, C.D. Weis, C.N. Moorefield, G.R. Baker, B.J. Childs, J. Epperson, *J. Angew. Chem., Int. Ed. Engl.* **1998**, *37*, 307-310.
[9] R.M. Versteegen, R.P. Sijbesma, E.W. Meijer, *Papers presented at the meeting – Am. Chem. Soc., Div. Polym. Chem.* **1999**, *40 (2)*, 839 - 840.
[10] H.W.I. Peerlings, R.A.T.M. van Benthem, E.W. Meijer, *J. Polym. Sci., Part A: Polym. Chem.* **2001**, *39*, 3112-3120.

Macromol. Symp. 187, 281–292 (2002)

New Low Viscous Polyisocyanates for VOC Compliant Systems

Holger Mundstock, Dr. Raul Pires, Dr. Frank Richter, Dr. Jörg Schmitz*

Bayer AG, D-Leverkusen, Germany

Summary: Since 1961, aliphatic polyisocyanates based on HDI-biurets are used as light stable hardeners for 2-K polyurethane coatings. The market introduction of the HDI-isocyanurates or simply 'trimers' in the early 1980′s was driven by their lower viscosity and their better monomer stability. The most recent development in HDI-polyisocyanate research are the iminooxadiazinediones (asymmetric trimers). In this paper, a new product based on that technology with a very low viscosity (700 mPas@23°C) and a NCO-functionality of more than 3 is presented. Because of its low viscosity, this hardener can be easily incorporated into aqueous systems and is the solution for high and full solids coatings as well.

Keywords: aliphatic polyisocyanates, iminooxadiazinediones, asymmetric trimers, High Solids coatings, Waterborne 2K PUR coatings

Introduction

Since 1961, aliphatic polyisocyanates based on HDI-biurets are used as light stable hardeners for 2-K polyurethane coatings. The market introduction of the HDI-isocyanurates or simply 'trimers' in the early 1980′s was driven by their lower viscosity and their better monomer stability.

The most recent development in HDI-polyisocyanate research are the iminooxadiazinediones (asymmetric trimers). In this paper, a new product based on that technology with a very low viscosity (700 mPas@23°C) and a NCO-functionality of more than 3 is presented. Because of its low viscosity, this hardener can be easily incorporated into aqueous systems and is the solution for high and full solids coatings as well.

Today, 2-K polyurethane systems based on aliphatic polyisocyanates are the standard of the coatings industry for high quality applications such as automotive OEM, car refinish, plastics, furniture, commercial vehicle coatings, and general industrial applications [1].

However, after the discovery of polyurethanes in 1937 [2], the first polyurethane coatings have been made of aromatic hardeners such as TDI-adducts (Brand name 'Desmodur L', where L stands for 'Lack' German for: laquer, varnish, paint) and Polymer-MDI. They are, due to their low price and high reactivity, still the choice for a variety of applications especially for wood coatings and flooring (Fig. 1).

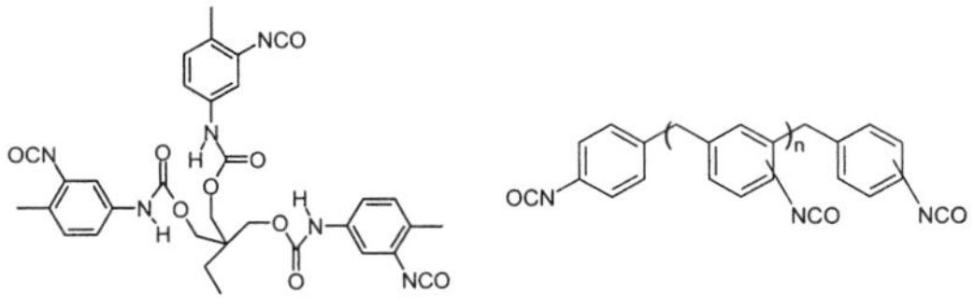

TDI-Adduct
Desmodur L

Polymer-MDI

L = Lack; German for: laquer, varnish, paint

Figure 1

   CCC 1022-1360/00/$ 17.50+.50/0

Unfortunately, high reactivity corresponds to a short potlife: in the case of aromatic polyisocyanates the potlife time can be in the range of only a few minutes. Additionally, and perhaps even worse, the light stabilty of coatings derived from aromatic isocyanates is limited.

This lead to the introduction of aliphatic polyisocyanate hardeners (Brand name 'Demodur N' where N stand for 'Nicht vergilbend'; German for "non yellowing") some forty years ago[3].

The first example was a biuret type, which could be made by the reaction of hexamethylene diisocyanate (HDI) with water, subsequent loss of carbon dioxide and addition of another molecule of HDI to the intermediate urea (Fig. 2).

Hexamethylene diisocyanate (HDI)

$\times 3, + H_2O, - CO_2$

HDI-Biuret (ideal structure)

Desmodur N

N = nicht vergilbend; German for: non yellowing

Figure

These hardeners quickly became the standard for aircraft and automotive refinish applications as they provide for several hours of potlife, not mentioning the excellent durability of the final coating.

Later on, the biuret-type of Desmodur N hardeners experienced support - and competition - by HDI isocyanurates or simply 'trimers' [4]. Their advantages over the biurets are a better monomer stability, and somewhat lower viscosity as well (Fig. 3). Both effects can be attributed to the absence of N-H-fragments the isocyanurate's backbone rendering the structure essentially inert towards reverse reaction (monomer formation) and intermolecular hydrogen bonding, respectively, the latter beeing one reason for a high viscosity of biurets versus trimers.

HDI-Trimer (Isocyanurate)

$OCN-(CH_2)_6$ ... $(CH_2)_6-NCO$

$(CH_2)_6-NCO$

=> lower viscosity
better monomer stability
vs. HDI-Biuret

Figure

The well established term 'trimer', however, only incompletely reflects the true nature of a polyisocyanate hardener. An HDI-isocyanurate, being undoubtedly the first stable product derived from the interaction three HDI molecules under the influence of a so called trimerization catalyst, contains three NCO groups which are essentially as reactive to further 'trimerization' as those of the starting monomer. This leads to the subsequent formation of higher molecular weight species, such as pentamers (containing four free NCO groups), heptamers (containing five free NCO groups), and so on. This reaction array is depicted in Figure beside a gel chromatogram of a typical commercial 'HDI-trimer' based hardener. Therefore, commercial 'trimer' based hardeners are more than NCO-trifunctional.

In addition, technical resins usually contain to some extend a number of species with structures like uretdiones, ureas, urethanes, and allophanates - to name the most common - that are different from the aming type; "isocyanurate" in the above mentioned case. They certainly do have some influence on ardener properties like NCO-content and -functionality (*vide infra*).

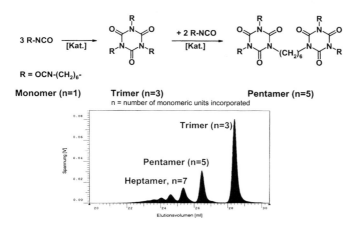

Figure 4

hese different, homologue isocyanurates exhibit different physical and chemical behavior. The lower the ontent of higher molecular weight oligomers in a given product, the lower the viscosity of that individual ardener, and the closer the NCO-functionality is to three.

his way, special hardeners for high solids coatings can be produced containing about 70 % of the real imer, here often referred to as the 'ideal structure'. These hardeners are very low viscosity, even in the osence of solvents (Table 1).

n the other hand, products containing a larger fraction of higher molecular weight species with a higher CO-functionality provide for faster drying films. Of course, these hardeners, are less well suited for the ormulation of high solids coatings due to their higher viscosity (Table 1).

Table 1. Influence of oligomer distribution on viscosity and drytime

HDI-Trimer	Viscosity @ 23°C [mPas]	NCO-functionality	Drytime	oligomer distribution		
				n=3	n=5	n>5
high viscous	15,000	~ 4.0	++	~ 30 %	~ 20 %	~ 50 %
Standard	3,500	~ 3.5	+	~ 50 %	~ 20 %	~ 30 %
ow viscous	1,200	~ 3.1	o	~ 70 %	~ 15 %	~ 15 %

Results

Polyisocyanates for High Solids

he search for VOC compliant solutions focuses on a number of different technologies, such as powder oatings, UV, waterborne and high solids, depending on the application. The interest in high solids systems olds especially true for clear coats in the automotive refinish sector.

The majority of the developments in recent years have been aimed at low viscous polyols, as well as reactiv
diluents. The goal was to minimize the content of volatile organic compounds in the formulated paint to le
than 420 g/l (for high solids) or even 250 g/l (for very high solids). Another option to achieve that
lowering the vicosity of the polyisocyanate hardener.

This can be done, as has already been outlined above, by using HDI-isocyanurate hardeners with a lo
content of higher molecular weight oligomers, or by employing HDI-uretdiones and/or allophanates. Bo
type of products exhibit considerably lower viscosities compared to isocyanurates (Fig. 5).

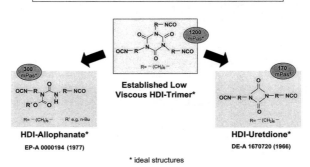

Figure

Contrary to isocyanurates with functionalities greater than three (*vide infra*), uretdiones and mono-alcoh
based allophanates are, independently of molecular weight, only NCO-difunctional. This leads to a low
crosslink density of the cured film resulting in inferior performance. As a rule, uretdione, as well
allophanate, based polyisocyanates are only used as reactive diluents in combination with isocyanurate
biuret based hardeners.

2. Low Viscous Asymmetric Trimers

At the end of the last century, the hitherto almost unknown iminooxadiazinediones - isomers of the we
known isocyanurates - have been investigated in more detail[5].

Structurally they differ from the latter only in one detail: one NCO-group is incorporated into the s
membered ring by opening the C=O double bond, instead of the usually occuring C=N bond scission. Th
detail renders the structure, in contrast to the flat and centrosymmetric isocyanurate ring, complete
asymmetric, and thus leads us to name them asymmetric trimers ("AST" for short, see Fig. 6).

In the course of our investigations, it has been discovered that HDI-ASTs are an almost ever prese
companion of their symmetrical counterparts. Their amount in the established products, however, is qui
low and can only be increased significantly by employing a new type of catalyst [5].

ASTs exhibit almost no difference in terms of reactivity, stability, and technological properties of cur
films, compared to isocyanurate based hardeners.

The color of the products is at present some 10 to 20 APHA units higher than that of the established trime
but a lot of improvement in that matter has been achieved over the last couple of years.

Moreover, the dilution stability, as well as the resistance of the hardener towards moisture, is considerab
better than those of the classical trimers.

Alternative Catalysts lead to Isomeric Trimers

Isocyanurate
(Symmetric Trimer, ST)

Iminooxadiazindione
(Asymmetric Trimer, AST)

Figure 6

Finally, the main advantage of HDI-ASTs is their dramatically lower viscosity over isocyanurates of comparable molecular weight. Therefore, ASTs are a promising approach to achieving lower viscosity, while maintaining at least NCO trifunctional hardeners, avoiding the use of difunctional products like uretdiones or allophanates (Fig. 7, Table 2).

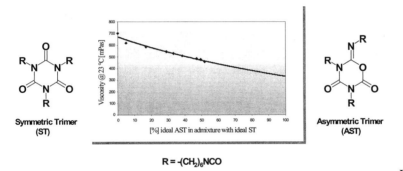

Symmetric Trimer
(ST)

Asymmetric Trimer
(AST)

R = -(CH₂)₆NCO

Figure 7

Table 2. Properties of various HDI-based hardeners

HDI-hardener	Viscosity @ 23°C [mPas]	NCO-functionality	NCO [%]	TSCA Status	Colour [APHA]	
					spec.	typical
Standard Symmetric Trimer (ST)	3,000	3.5	21.8	approved	<40	<15
Low Viscous Symmetric Trimer (LV ST)	1,200	3.1	23.0	approved	<40	<15
Very Low Viscous Asymmetric Trimer (VLV AST)	700	3.1	23.8	approved	<40	20 - 30
Allophanate	300	2.5	20.0	SNUR	<50	<30
Uretdione	170	2.5	21.8	approved	<80	<30

This paper will focus on a very low viscous HDI-AST that has recently been introduced to the market, from here on referred to as VLV AST. At a value of 700 mPas at 23°C, the viscosity of that new product equal the viscosity of the ideal HDI-isocyanurate. One has to keep in mind that the latter could only be isolated by rather cumbersome separation technique, like molecular distillation or extraction, and therefore would be very expensive to manufacture on a large scale[6].

The VLV AST enables the formulator to further decrease the VOC content in solvent borne high solids applications (see Chapter 4), while also showing utility in aqueous polyurethane systems due to its low viscosity, allowing it to be incorporated very easily (see Chapter 5).

3. Low Viscous Asymmetric Trimers in High Solids Applications

The VLV AST has been tested in automotive refinish clearcoats in comparison to other, established HDI trimers, all in combination with a high solids polyacrylate. No difference in mechanical (i.e., hardness and elasticity) or optical properties can be observed among these clearcoats. The results of weathering studies are identical as well (Fig. 8).

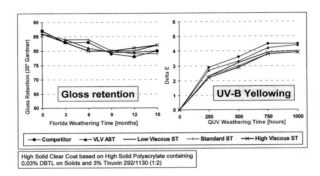

Figure 8

As can be seen from Fig. 9, employing the VLV AST in even an already optimized system, the solids content can be further increased by 1 to 3 % without sacrificing the properties of the cured film.

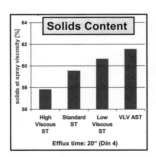 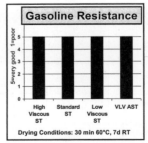

Figure 9

Due to the somewhat lower molecular weight of the VLV AST compared to conventional HDI-trimer hardeners at equal catalyst levels, the potlife of the formulation is increased. Additionally, the chemical, as well as physical drying of the paint, is slower (Fig. 10). This can be compensated by higher catalyst levels.

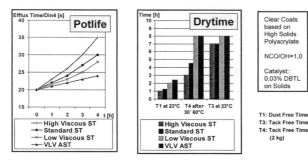

Figure 10

In order to answer the question of whether the NCO groups of an AST really are as reactive as those of an isocyanurate, 2-K high solids clear coats have been monitored for NCO decrease by IR spectroscopy while curing at room temperature (without forced drying) (Fig. 11).

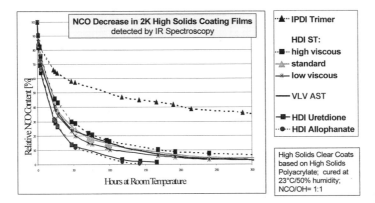

Figure 11

On one hand, Figure 11 clearly demonstrates the considerably lower reactivity of IPDI-based trimers over HDI-based products, which can be attributed to the presence of secondary NCO groups, as well as to a higher T_g of that system. On the other hand, a somewhat faster cure response has been detected for the low functionality derivatives, uretdione and allophanate, over the trimers. Among them, only negligible differences are observed, no matter if the polyisocyanates backbone is an isocyanurate or an AST.

As has been outlined above, the initial color of the VLV AST is slightly higher than that of the established products. This is, in our opinion, essentially due to the somewhat limited experience one has today manufacturing this new type of product. In addition, one has to take into account the shorter production periods, as the demand for these products is naturally not yet at the level of the established resins. However, the color stability of the VLV AST, no matter if a neat hardener, or a catalyzed, diluted solution is investigated, shows good comparisons to other hardeners (Fig. 12).

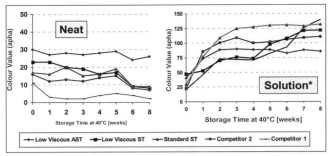

*: 60% Hardener in Butyl Acetate/Solvent Naphtha 100
+ 0,03% DBTL on Solids

Figure 1

The viscosity stability of the VLV AST is equal to, or exceeds, that of established products (Table 3).

Table 3. Viscosity increase of non-catalyzed, neat hardeners upon storage at 40°C

	Viscosity [mPas] at 23°C			Viscosity increase [%]
	Initial	after 4 weeks at 40°C	after 8 weeks at 40°C	
Standard ST	3000	3100	3270	11
Competitor 1	1330	1330	1420	7
Low Viscous ST	1200	1200	1300	8
Competitor 2	990	1100	1180	19
VLV AST	650	680	705	8

The VLV AST has also successfully been used for VOC reduction in 2-K PUR OEM applications, without deteriorating the technological properties of the clearcoats such as gloss, scratch and solvent resistance.

4. Low Viscous Asymmetric Trimers for Waterborne Applications

Waterborne systems, beside high solids, represent an interesting approach to coatings with a dramatically decreased solvent content, fulfilling today already the requirements of future VOC-legislation. Meanwhile waterborne systems have been developed for even high end applications.

Besides a proper choice of the polyol, the suitability of the corresponding polyisocyanate hardener is crucial for the performance of the whole system.

In order to make the incorporation of the hardener into the aqueous phase easier, specially designed, water emulsifiable aliphatic polyisocyanates have been developed. However, these products carry their hydrophilicity into the cured film, which results in a somewhat limited water and chemical resistance and deteriorates the corrosion protection properties.

These hydrophilically modified polyisocyanates are usually made of HDI- or IPDI-trimers that have been made water-dispersible by the internal or external incorporation of small amounts of hydrophilizing agents. The term "internal" stands for NCO-reactive agents, e.g. ROH-started polyethylene glycols (R = lower alkyl) which may form a urethane group upon reaction with the mother hardener (Fig. 13, top) or, more elegantly re further converted to an allophanate (Fig. 13, bottom). The term "external", on the contrary, stands for additives which do not react with the isocyanate groups of the hardener [7].

Hydrophilic HDI-Polyisocyanates

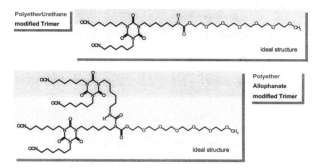

Figure 13

If the hardener has to be emulsified manually, e.g. by simple stirring with a mixing stick, hydrophilized products are indispensable. However, in many applications mechanical mixing devices like stirrer, dissolver, or even jet-stream-dispersers may be used, making the use of mixtures of hydrophilically modified and conventional hardeners possible.

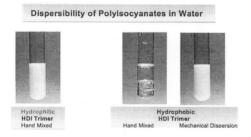

Figure 14

In some instances, the use of hydrophilically modified crosslinkers can be completely avoided (Fig. 14). Especially in the latter case, the viscosity of the hydrophobic hardener is important for good mixing results. This can be achieved by the use of solvents, but they, in turn, have a negative impact on the VOC level of the mixture. Therefore, low viscous polyisocyanates are being favored. Uretdiones and allophanates are being used in certain applications already, but the above mentioned lower NCO-functionality results here also in a lower crosslink density of the final film. The VLV AST represents an interesting alternative for waterborne applications, combining the advantages of low viscosity with a high NCO-functionality.

.1 Waterborne 2-K Industrial Topcoat

The use of hydrophobic polyisocyanates in waterborne applications, although providing for durable, chemically resistant coatings, has drawbacks in terms of gloss and other optical properties.

Due to its low viscosity, the VLV AST provides for a much better balance of chemical and optical film properties, no matter if used in combination with a hydrophilized hardener employing manual mixing, or as a single hardener and mechanical mixing with a dissolver as is exemplified in Figure 15 with an air-drying, 2-K PUR industrial topcoat.

The hardness of the film made from VLV AST exceeds that of those made from hydrophilic hardener (softening effect of the polyether tails) and with a low viscous, hydrophobic conventional trimer (film inhomogenity due to poorer mixing), respectively (Fig. 15, right). The flexibility of all three films, however is comparable (Fig. 15, left).

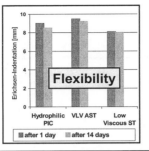
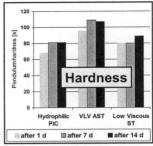

Waterborne White Top Coat based on Polyacrylate Secondary Emulsion
Mechanical Mixing of the Hardener (Dissolver, 2000 rpm)
NCO/OH= 1:1,5; 53% Solids (VOC: 140 g/l) Curing Conditions: 23°C/50% Humidity; 35 µ on Steel

Figure 1

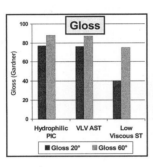
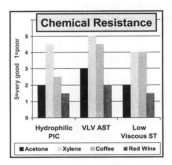

Figure 1

As expected, the gloss is highest for the hydrophilic hardener, but the VLV AST almost matches that resu (Fig. 16, left). On the other hand, the stability of the coatings to typical household chemicals is best in th case of VLV AST (fig. 16, right).

The same picture is obtained after storage in water. Adhesion and gloss retention both reach maximum leve when films are derived from the VLV AST (Fig. 17).

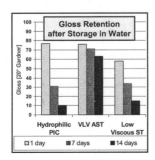
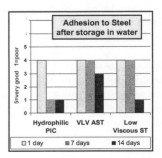

Figure 1

1.2 Waterborne 2-K PUR Automotive OEM Clearcoat

Conventional, solvent borne 2-K PUR coatings are standardly used for top quality OEM clearcoats today, combining an outstanding quality level of the coating in terms of weather and chemical resistance with excellent optical properties.

The development of waterborne systems with equally good properties is challenging, especially since the typical hydrophilized hardeners cannot be used for OEM applications due to their negative impact on the weatherability of the coating.

Meanwhile, waterborne 2-K clearcoats that equal the high quality level of the well established conventional 2-K solvent borne coatings are at the edge of a broad market introduction. These new systems are based on a combination of polyacrylate or polyurethane dispersions, respectively, which are crosslinked with a low viscous, hydrophobic hardener. The components are mixed prior to application using a jet stream disperser working at 50 bar (~ 725 psi) and contain 240 g/l VOC or less.

Applying the new VLV AST as a hardener in that kind of formulation allows for an even further VOC-reduction (from 240 to 120 g/l) and/or makes it possible to reduce the pressure of the disperser to 20 bar (~ 290 psi). At that pressure, even conventional gear pumps may be employed as they are used today in the recirculation lines at the automotive OEM sites.

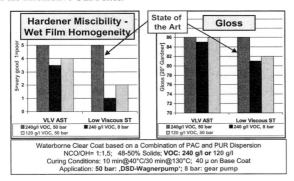

Figure 18

When using VLV AST instead of conventional low viscous isocyanurate based trimers, even at lower pressure, or at lower co-solvent content respectively, the miscibility of the components is improved, thus leading to a better homogeneity of the wet film and providing for higher gloss of the cured coating (Fig. 18). When using VLV AST as a hardener at either lower VOC or lower pressure, no deteriorations in terms of chemical or scratch resistance are observed (Fig. 19), compared to conventional trimer based coatings.

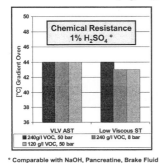

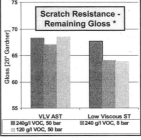

° Comparable with NaOH, Pancreatine, Brake Fluid   * after 10 car wash cycles

Figure 19

292

Conclusion

A new isocyanate hardener for 2-K polyurethane coatings based on HDI iminooxadiazinediones (asymmetri trimers, AST) has been developed. This product, herein referred to as VLV AST, combines the advantages c a very low viscosity (700 mPas/23°C) with a high NCO-functionality (≥ 3), thus making it an ideal candidat for promising future developments in the coatings industry aimed at low VOC or 'high solids' in th conventional, solventborne sector, and for waterborne applications as well.

The VLV AST is an interesting alternative to established hardeners in many applications such as: automotiv OEM, car refinish, coatings for commercial vehicles, plastics, furniture, and general industrial application up to thick film flooring and corrosion resistance.

The VOC content of high solids systems can be further reduced without sacrificing the mechanical, optica or chemical properties of the coating.

Waterborne coatings benefit from the low viscosity of the VLV AST, leading to a better mixing behavic compared to other hydrophobic, non-hydrophilized hardeners. Depending on the mixing techniqu (manually, mechanically, or jet-stream-dispersed) the amount of hydrophilic hardener in the formulation ca be diminished or removed entirely. That, in turn, leads to a significantly improved mechanical, optical, an chemical performance of the coating.

Acknowledgement

The authors are grateful to H. Brümmer, U. Freudenberg, T. Klimmasch, M. Mechtel, H. Neyers, R. Reye and N. Yuva who contributed to the success of the work discussed in that paper, to R. Halpaap and W Hovestadt for many helpful discussions, and to S. Grace for language polishing.

References

[1] Polyurethanes for Coatings, Ed. M. Bock, Vincentz Verlag 2001.
[2] DRP 728 981 (11.11.1937), see also: O. Bayer, *Angew. Chem. (A)*, **1947**, *59*, 257-288
[3] DE-A 1 101 394 (09.03.1961)
[4] DE-A 2 839 133 (08.09.1978)
[5] EP-A 0 798 299 (26.03.1996)
[6] R. T. Wojcik, *Modern Paint and Coatings*, **1993**, *83*, 39.
[7] R. Pires, H.-J. Laas, *European Coatings Journal*, **2001**, *11*, 16-22.

2 K Clearcoats Based on Silane and Urethane Technology

D. Hoppe, R. Lomölder, F. Plogmann, P. Speier

Degussa AG, D-45764 Marl, Germany

Summary: Despite significant advances since the late 80′s the balance between mar- and acid etch resistance of automotive clearcoats still is a subject of intense research.Among others silane- and PUR technology is used to produce clearcoats today. Generally speaking the silane based systems show better mar resistance whereas PUR based systems have advantages in acid etch resistance. We wish to show that a certain group of silane modified polyisocyanates can be used as crosslinkers for hybride systems exhibiting an improved balance between mar- and acid etch resistance.

1. INTRODUCTION

Until the late 80′s automotive clearcoats were based on melamine crosslinked acrylic polyols (TSA). Due to poor environmental etch resistance they were replaced to a significant extent by more acid resistant systems. Besides 2K PUR which is today the dominating technology used to attain high levels of acid etch resistance[1], 1K silane technology is another innovation that has also been used to improve acid etch during the 90′s. However in this latter case, silane modified acrylic resins are used in combination with melamine resins. Though displaying good mar resistance, a comparable level of acid etch resistance compared to a pure PUR system could not be reached[2] . It seemed interesting to investigate a hybrid crosslinking technology based on urethane and silane chemistry, because such a system hypothetically could improve the poorer mar resistance of a pure PUR while maintain a high degree of acid etch resistance.

2. STRATEGY FOR RESIN DESIGN

In the case of the PUR moiety of the hybrid technology it's unambiguous that this is introduced via a polyisocyanate component, typically based on isocyanate trimer structure reacting with the hydroxyl groups of a polyol. The crosslinking mechanism is relatively simple and well known. Silanes are capable of reacting with water (hydrolysis) and subsequent condensation or with hydroxyl groups of a polyol with heat activation (transesterification) [3] (fig. 1).

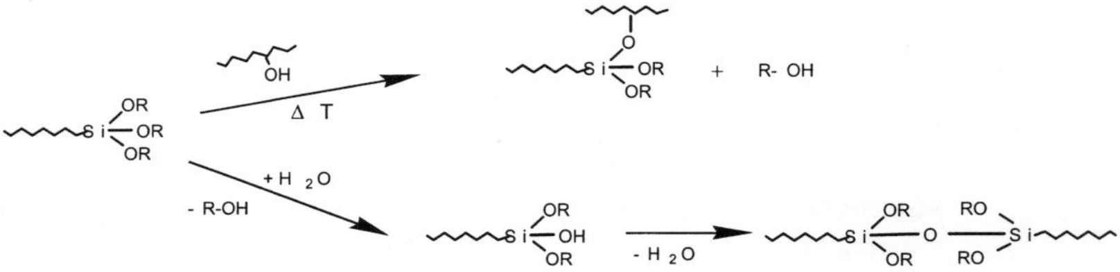

Figure 1. Crosslinking reactions of silanes in hybrid systems

There are several options for incorporating silane functionality into clearcoat resins. One would be a traditional approach, using silane functional (meth-)acrylic monomers[4] in acrylic resin synthesis. Such a resin should ideally contain both silane and hydroxyl functionality (fig. 2) in order to be capable for hybrid crosslinking.

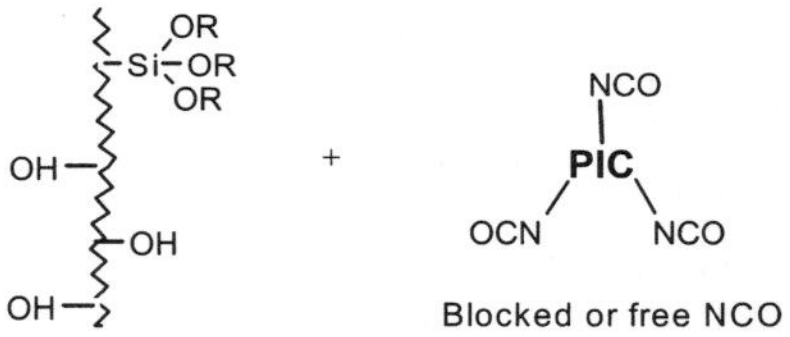

Figure 2. Option 1: Silane modification of acrylic polyol

Such a hydroxy- and silane functional acrylic resin would inherently have storage stability problems because of transesterification between OH- and silane functionality of the resin. Even if this could be avoided by the use of alcoholic solvents, the transesterification with the alcohol would lead to a different reactivity, because other substituents than methoxy are assumed to be less reactive. Furthermore, in the case of 2K systems, solvents with hydroxyl groups have to be avoided due to competitive reactions with the NCO groups.

The other option would be to incorporate the silane functionality into the polyisocyanate (fig. 3). The advantages of this strategy are evident: No need for stabilisation of a silane-OH-system avoiding all the problems mentioned before. Therefore, the option of choice was for us the modification of a polyisocyanate with a silane. This silane, in order to be linked to the polyisocyanate must bear a H-active functional group.

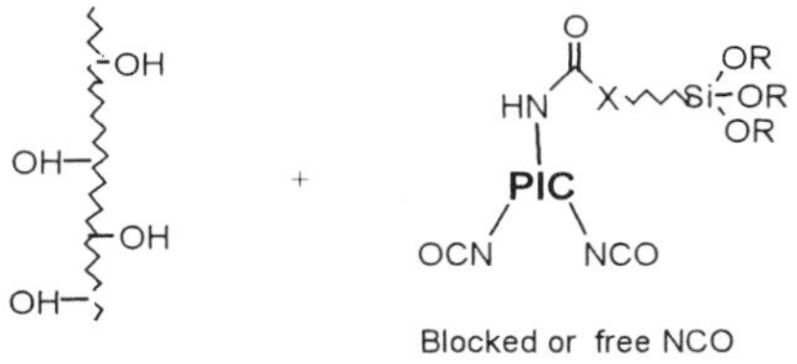

Figure 3. Option 2: Silane modification of the polyisocyanate

3. 1K SILANE-URETHANE TECHNOLOGY

In a first step the performance of this approach with regard to the etch and mar resistance should be studied. In order to have a clear view to this property and not to fight with a severe reactivity difference of free NCO and silane functionality we decided to investigate blocked polyisocyanates. Therefore, 1K-systems based on this novel hybrid technology were formulated. Due to the similar reactivity profiles compared to the silane functionality we decided to use MEKO as blocking agent.

Therefore, 40 % of the NCO groups of IPDI trimer were reacted with a H-active silane and blocked with MEKO.

In a serie of formulations this silane modified blocked derivative of IPDI trimer (Si-BPIC) was formulated with a low Tg acrylic polyol (Tg appr. −30°C, OH-Value 120 mg KOH/g). Such combinations are typically used to achieve hard but flexible coatings based on the high Tg polyisocyanate IPDI trimer. Clearcoats were formulated with a DBTDL catalyst level of 1 % calc. on resin. Formulating the silane modified polyisocyanate the stoichiometric level was varied in order to investigate the influence of additional silane crosslinking. As a control a straight PUR formula based on the MEKO blocked IPDI trimer was used, also at similar stoichiometric ratios (OH:NCO), curing cycle was 25 min. 140°C in all cases.

Mar resistance was checked in a wet rub test with an abrasive suspension under standardized conditions of high reproducibility, expressed as loss of gloss at a 20° angle of a cured clearcoat over a black basecoat. Afterwards reflow properties were tested. Conditions were 2 h at 40°C (reflow I) and additional 2 h at 60°C (reflow II). Results were expressed as a loss of gloss vs. the original unscratched panel.

Acid etch resistance was checked in a gradient oven test, employing 20 % sulfuric acid, listed are temperatures of first visible attack and damage of the coating.

Table 1. Data of cured films (clearcoats) by variation of stoichiometry

	BPIC (MEKO)			Si-BPIC 1 (MEKO/Silane)			
Stoichiometry (OH:NCO)	1:1	1:0.8	1:0.6	1:1	1:0.8	1:0.6	1:0.4
Diisocyanate content of total binder (pbw)	34	29	24	40	36	30	24
Dry film thickness [μm]	30	30	30	35	35	30	35
Pendulum hardness [König, sec]	163	144	90	178	165	157	108
Erichsen cupping [mm]	7	7.5	8.5	5	5	5.5	7.5
Impact resistance [in * Ibs]	> 80	70	> 80	40	40	50	> 80
Gasoline resistance [x1]	++	++	0/+	++	++	++	++
acid etch resistance (°C): first spot at	62	62	53	63	63	61	53
severe damage of clearcoat at	70	70	64	76	74	74	70

[x1]: ++ very good, + good, 0 moderate

Table 2. Scratch resistance and reflow properties of MEKO-blocked polyisocyanates

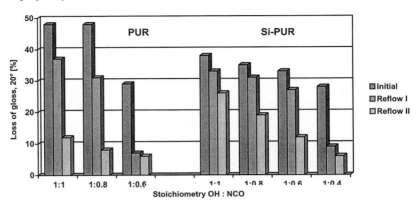

In the case of Si-BPIC hardness, elasticity and gasoline resistance are not affected down to a ratio of 1:0.6, while the straight PUR system exhibited poor hardness and chemical resistance already below a ratio of 1:0.8. This means, that this silane modified polyisocyanate can be formulated at stoichiometric levels of OH:NCO=1:0.6, the remaining OH-groups are obviously either crosslinked by the silane functionality via transesterification or the undercrosslinking is compensated via Si-O-Si-crosslinking. The relevant clearcoat properties considered in this study, i. e. the balance of mar and acid etch resistance were improved if the 1:0.6 crosslinking is compared with the 1:1 of the straight PUR. The expected good acid etch resistance of a straight PUR-crosslinking was maintained, initial scratch resistance was improved significantly, and the reflow properties were good.

The very good initial mar resistance of straight PUR at 1:0.6 and Si-PUR at 1:0.4 can be explained by a much lower hardness and higher elasticity. Such low levels of hardness are not acceptable for the targeted application. Therefore, these two formulations are not relevant in this study.

After this part of the study it can be concluded that with such silane modified blocked PIC a significant improvement of the balance of etch and mar resistance can be achieved, if the NCO:OH stoichiometry is adjusted properly.

4. TRANSFER INTO 2K TECHNOLOGY

Since in automotive clearcoat technology 1K PUR systems lost their importance in the late 90´s mainly due to competitive 1K technologies and because 2K PUR technology is still the state of the art for superior performing clear coats, the silane-polyisocyanate approach had to be transferred from 1K into 2K technology.

For the transfer from 1K to 2K technology a serious problem had to be solved: The reactivity variance, i.e. the strong difference in reactivity between silane and free isocyanate towards OH groups.The catalyst in the 1K system used for acceleration of silane and blocked isocyanate function at a level of 0.5-1% very likely could not be used in such a concentration in 2K systems. Such a DBTDL level would certainly raise the reactivity of the OH/NCO-reaction to such an extent that the pot life of the paint – a fact dominated by the PUR reaction - would be far too shortIn order to verify this assumption 2K-paints based on the silane modified but unblocked IPDI trimer and the aforementioned low Tg acrylic polyol were formulated at a stoichiometric ratio of OH:NCO = 1:0.6 (best ratio from 1K investigation), employing different DBTDL levels. Curing conditions were 25 min. 140°C. The corresponding data are shown in tab. 3 and tab. 4. As a reference the straight 2K PUR system at the optimum stoichiometric ratio of 1:1 was tested.

Table 3. Relevant mechanical and chemical resistance data of 2K systems

	PUR (control)	Si-PIC 1		
	0.05 % DBTDL	0.05 % DBTDL	0.1 % DBTDL	0.5 % DBTDL
Dry film thickness [μm]	30	30	30	40
Pendulum hardness [König, sec]	160	122	166	149
Erichsen cupping [mm]	7.5	8.0	7.5	7.0
Gasoline resistance[x1]	++	0+	++	++
acid etch resistance (°C) first spot at sev. damage of clearcoat at	56/72	57/74	56/71	61/78
gel time (h)	10	23	11	2.5

x[1]: ++ very good, + good, 0 moderate

Table 4. Influence of DBTDL catalyst level on scratch resistance and reflow

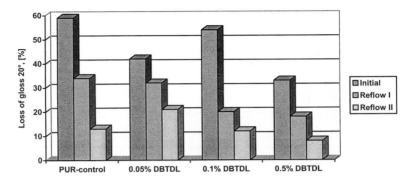

It can be seen that

a) as expected, 1K (tab. 1+2) and 2K-straight PUR systems show similar performance.

b) a DBTDL catalyst level of 0.05 %, suitable for typical 2K PUR technology is insufficient for silanes, because the development of hardness and scratch resistance are poorer.

c) a higher catalyst level of 0.1 % improves hardness and reflow properties.

d) the DBTDL catalyst level of 0.5 % is leading to the best results of the Si-modified 2K system with regard to acid etch and mar resistance and also reflow, but to an inacceptable short pot life of the paint.

Therefore we had to find a different catalyst approach. In order to transfer our approach from 1K to 2K there was a desire for a catalyst or combination of catalysts which activate the silane group without affecting the reactivity of the OH/NCO-combination. Due to the fact that silane groups are also activated by strong acids, we concluded that blocked sulfonic acids might be a very suitable approach. Such catalyst would be inactive at ambient temperature for both reactive groups, i. e. NCO and silane, but would become active with stoving after application of the paint.

Beside the very broad range of blocked sulfonic acids we tested a non ionically blocked sulfonic acid catalyst. This blocked catalyst becomes activate at temperatures 120°C. We used 5 % calc. on solid binder in combination with 0.05 % DBTL. The data conc. scratch and acid etch are listed in Tab. 5, comparing this formulation with 0.5 % DBTDL and the straight 2K PUR formulation.

Table 5. Influence of blocked sulfonic acid catalyst on the mar resistance

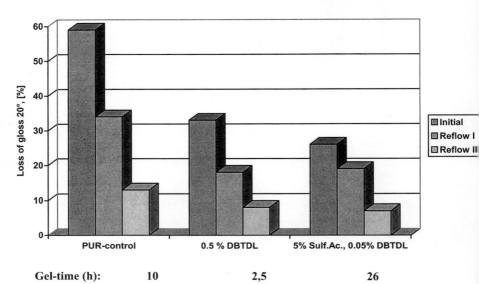

It is evident that the use of such type of sulfonic acid catalyst leads to the solution of the reactivity problem: While maintaining the good pot life of the control (straight 2K) it was possible to achieve the excellent mar resistance of the 0.5 % DBTDL-system or of the corresponding blocked system with 1 % of DBTDL catalyst.

5. 2K Blend Technolgy

In 2K PUR OEM clearcoats blends of HDI- and IPDI-trimer are commonly used to provide a good balance between chemical and scratch resistance[5),6)]. We should now like to discuss the performance of blends of silane modified trimers in 2K PUR OEM clearcoats.

In this investigation we used a high Tg acrylic polyol (Tg = 35°C; OH-number = 150 mg KOH/g) in order to achieve reasonable hardness with the HDI-trimer. In the case of the non silanated trimers the OH:NCO ratio was 1:1 and 0,05% of DBTDL were used as catalyst. In the case of the silanated trimers the OH:NCO ratio was 1:0.6 and 0,05% DBTDL and 5% DYNAPOL Catalyst 1203 were used.

From table 6 and 7 the following conclusions can be drawn:

- Blending of non silanated HDI- and IPDI-trimer results in an improved acid etch resistance and a poorer scratch resistance as expected (Clearcoats A and C)

- Using silanated HDI-trimer (B) and a silanated HDI- IPDI-trimer blend (D) instead of the non silanated products (A and C) a better scratch resistance is obtained in both cases. However, the improvement is more significant in the case of the blend (D) reaching almost the level of the silanated HDI-trimer (B).

- Among the clearcoats discussed here the silanated HDI- IPDI trimer (D) blend offers the best balance between chemical- and scratch resistance.

Table 6. Clearcoats Based on HDI or Blends of Crosslinkers

	HDI-Trimer	Si-PIC 2	Trimer blend HDI:IPDI=6:4	Blend Si-PIC HDI:IPDI=6:4
Clearcoat	A	B	C	D
Stoichiometry OH:NCO	1:1	1:0.6	1:1	1:0.6
Pendulum hardness [König, sec]	197	194	197	192
Acid etch resist. (°C) first spot at sev. damage of CC at	48 62	45 56	52 69	51 59

Table 7. Scratch Resistance and Reflow Properties

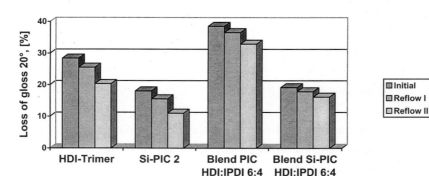

6. CONCLUSION

Silane modified (cyclo-) aliphatic polyisocyanates were investigated on their performance in automotive clear coat application. Both, blocked and unblocked polyisocyanates were part of the study. Due to the additional crosslinking capability of the silane functionality, understoichio-metric ratio of OH:NCO is recognized to be benefitial with regard to performance and costs. At a level of 40 % silanisation, OH:NCO should be appr. 1:0.6. 1K clear coats based on acrylic polyols and catalysed with 1 % DBTDL exhibited at such stoichiometric conditions a significant improvement of scratch resistance and/or reflow properties while nearly maintaining acid etch resistance.

A high tin catalyst level as used in 1K systems was not practical due to gel time limitations in 2K-systrems. Low tin concentration led to low reactivity of the silane moiety and poorer results with regard to mar resistance. Combinations of blocked sulfonic acid catalysts and low levels of tin resulted in a similar performance profile as 1K systems without sacrificing gel time. Therefore the significant improvement of the balance of etch and mar resistance by the use of partially silane modified polyisocyanates can also be transferred into the 2K-technology for automotive clearcoats.

Further improvements could be achieved by applying the blend technology used in 2K OEM clearcoats to the silanated (cyclo)-aliphatic diisocyanate trimers.

7. REFERENCES

1) M. Montemayor, Polym. Paint Col. J., 1999, 12-13
2) J. W. Holubka, P.J. Schmitz and Li-Feng Xu, J. Coat. Techn., 72, 77 (2000)
3) I. Hazan, API Congress, Dresden, Germany, Oct. 1999, communicated.
4) EP 549 643, WO 92/11327, WO 92/11328, US-P 5,225,248, DuPont
5) WO 93/05090, BASF Corp.
6) L. Kahl, R. Halpaap, Ch. Wamprecht, Industrie-Lackierbetrieb, 61, 30-34 (1993)

Macromol. Symp. 187, 305–315 (2002)

What Can Nano-Chemistry Offer to the Paint Industry?

I. Felhősi[1], J. Telegdi[1], K. Papp[1], E. Kálmán[1], Á. Stáhl[2], J. Bognár[2]

[1] Institute of Chemistry, Chemical Research Center, Hungarian Academy of Science, Budapest
[2] Research Institute for Paint Industry, Budapest

Summary: Iron surface was modified by organic, self assembled nano-layer of 1,7-diphosphono heptane. The self-assembling film formation and the self-healing process of the injured layers was monitored by electrochemical methods. The morphological changes of paint and lacquer layers which was due to different pre-treatments were monitored by surface analyzing techniques.

Keywords: surface pre-treatment, self-assemble layer, phosphonic acid, paint, electrodimpedance spectroscopy, atomic force microscopy, scanning electronmicroscopy

Introduction

Researchers make significant efforts to develop environmentally acceptable chemicals for surface pre-treatment in the paint industry where many toxic and acidic additives are used for coating. For the paint industry among the new materials the so-called self-assembled nano-layers could replace other toxic materials. The self-assembled molecules form an adhesive layer between the metal surface and the paint layer.

The self-assembly of disulfides on gold discovered in 1983 [1], and, soon after, of alkanethiols opened new perspectives in surface science. This is a spontaneous process that takes place on an appropriate substrate by immersion into a solution of surfactant-like molecules. This process consists of adsorption/chemisorption and results in self-organized, highly ordered mono-layers of dense, oriented and stable structure. Self-assembled monolayers (SAMs) on metal surfaces form organic interfaces with properties mainly controlled by the end-groups of the molecules. SAMs are excellent systems for the study of interfacial processes, of molecular structure – layer property relationship and of technical importance in fabrication of sensors, transducers and protective layers.

 CCC 1022-1360/00/$ 17.50+.50/0

306

The chemical surface modification of metals by self-assembly is a new encouraging method of corrosion protection [2], too, which is able to replace the previously applied environmentally not acceptable surface treatments. Because of the high attractive interaction among the alkyl side chains, SAMs can exceed the corrosion protection effect of commonly used inhibitor molecules. First successful attempts to apply self-assembly of alkanethiols for corrosion protection were done on iron [3-5] and copper [6-8]. The stable hydrophobic film depressed significantly the metal dissolution. The chemisorption of n-alkanethiols takes place only on metallic-state iron [9], which substantially limits its practical application. Practically more promising are those molecules, which preferably form adsorption layers on oxide- or hydroxide-covered metal surfaces. Not only alkanethiols but alkane-mono- and diphosphonates have been successfully applied to form well-oriented self-assembled surface layers on iron [10-13], aluminum [14-15] and zinc [14]. The self-assembly process of phosphono-functionalized molecules takes preferably place at metal surfaces covered with thin hydroxide layer. The self-assembled layer formation consists usually of two separate adsorption kinetics; a fast adsorption step controlled by the adsorption of molecules on metal surface, and by slow organization/orientation step determined by the intermolecular van der Waals interaction between alkyl chains, that depends on the length of molecule. In corrosion protection one of the most important application of self-assembly process is a pre-treatment of metal surfaces beneath organic coatings [15]. The immobilized molecules are able to act as adhesion promoters and corrosion inhibitors. They can also be applied as a temporary corrosion protection of metals. The advantage of the molecular self-assembly lays mostly in the wide variety of functional groups and length of the alkyl chain. Beside its practical importance, the self-assembling phenomenon on metal surfaces provides an excellent model system for studies of adsorption of organic compounds at atomic scale.

The aim of this study was to elaborate new chemicals (of nano-scale at least in one dimension), and to elucidate the processes and mechanisms of protective layer formation on iron surface in order to get stable layer that not only has good anti-corrosion activity but renders the metal surface more proper for coating. Experiments in the presence of alkane-phosphonates were presented in our previous communication [10]. In this paper some selected results obtained on protective layer formation in the presence of diphosphonates will be presented. Because of the limited water solubility of molecules with long alkyl chain, only shorter molecules were studied in aqueous solutions (n=5-12). The kinetics of layer formation, the mechanism of inhibition, and the self-healing properties of phosphonates will be presented through the example of 1,7-diphosphono-heptane. The influence of self assembled molecules (SAM) on the morphological properties of paint and lacquer layers will be demonstrated by scanning electronmicroscope (SEM) and by atomic force microscope (AFM).

Experimental

The protective layer formation was studied in aerated aqueous solution of 0.001 mol.dm^{-3} phosphonic acids (pH 7.0). Measuring electrodeimpedance spectroscopy (EIS) and open circuit potential (OCP) followed the layer formation of phosphonates. Electrochemical experiments were performed in a standard three-electrode cell of 500cm^3 volume at room temperature. The working electrode was Armco iron embedded into epoxy resin. The counter electrode was platinum net, and the reference electrode was a saturated calomel electrode (SCE).

Electrochemical impedance measurements were carried out in Solartron 1286 Potentiostat and 1250 Frequency Response Analyzer. Impedance spectra were measured at the open circuit potential in the frequency range of 30 kHz to 1mHz with ten points per decade. A sine wave with 10mV amplitude was used to perturb the system. Polarization curves was detected with the rate of 10 mV/min.

The influence of surface pre-treatment on adhesion, on the morphological properties of lacquer and paint layers were visualized on air by atomic force microscopy (AFM) and by scanning electronmicroscopy (SEM).

Results and Discussion

Surface treatment of iron by diphosphonates

The protective layer formation was monitored in situ in the phosphonate containing solution using impedance spectroscopy and measurement of the open circuit potential of Armco iron as a function of time. Our previous result [13] proved that the presence of electrolyte (e.g. sodium perchlorate) affects the adsorption of phosphonate and, interrupting the protective layer formation, the growth kinetics of alkane-phosphonates layers in perchlorate-free solution was followed.

When Armco iron was immersed in the 1,7-diphosphonoheptane containing solution the change of OCP in the first hour of immersion is shown in Figure.1. Measuring the continuous shift of OCP towards the anodic direction followed the initial transitivity stage. The change of corrosion potential is relatively slow; the maximum value around - 115 mV (vs. SCE) was reached after one day. This is the consequence of the adsorption of phosphono compounds that leads to the formation of a protective layer on the iron surface.

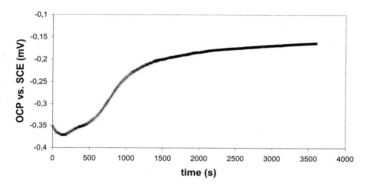

Figure 1. Change of OCP of Armco iron as a function of time in 10^{-3} M 1,7-diphosphono-heptane solution (pH 7.0)

The layer formation of diphosphonates on iron was monitored by impedance spectroscopy. After a 2-day-long immersion of iron in aqueous 1,7-diphosphono heptane solution the impedance spectrum shows the protective behavior of the adsorbed layer (Figure 2).

The effect of adsorption time on polarization resistance and corrosion potential of Armco iron in the 1,7-diphosphono-heptane solution is presented in Figure 3. With increasing immersion time the polarization resistance continuously increases. The polarization resistance is equivalent to the charge transfer resistance of iron dissolution that takes place at the uncovered metal surface, i.e. at pores and discontinuities within the film.

Therefore, the change of polarization resistance values gives direct information on the growth and quality of inhibitor layer.

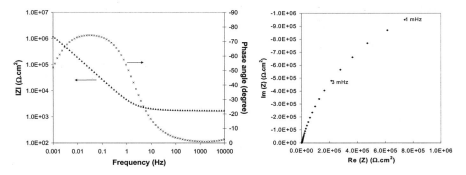

Figure 2. Typical impedance spectrum of Armco iron in phosphonate containing solutions
(a) Bode and (b) Nyquist plots of Armco iron in aqueous solution of 0.001 mol.dm^{-3} 1,7-diphosphono-heptane (pH=7) measured after 48 hours of immersion, and (c) Equivalent circuit modeling the impedance of modified iron. R_s: solution resistance, R_p: polarization resistance; Q_{dl}: CPE of the double layer

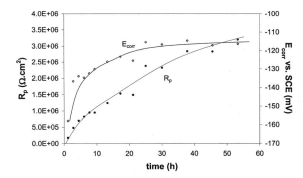

Figure 3. Time dependence of polarization resistance and open circuit potential of Armco iron in aqueous solutions of 0.001 mol.dm^{-3} 1,7-diphosphono-heptane (pH=7)

According to the results obtained by impedance studies, the iron dissolution is effectively hindered by the adsorbed phosphonate layer, which is supposed to have a continuous and dense structure. After few hours of immersion, significant protection could be achieved on bare iron. The fast adsorption of molecules is important in aqueous solution that can significantly decrease the disturbing iron dissolution process that runs parallel to the adsorption. Although, fast adsorption provides efficient protection of iron, a further increase in polarization resistance can be observed during a long immersion time. This observation refers to a time-dependent quality improvement of the film.

Due to the interaction of phosphono-groups with the solid surface, formation of a multi-molecular layer may also be considered. It was proved in our earlier SNIFTIRS studies [11] that thin multi-molecular adsorption layers of diphosphonic acids are formed on iron surface in aqueous solution. Although ordered structure and orientation of diphosphonates was observed in a certain degree, this was only qualitative information.

The composition of oxide layer influences the adsorption of phosphonates [16]. The adsorption of phosphonates takes place on aluminum surfaces, which is rich in OH-groups [17]. Presumably, the formation of thin iron oxide layer (that takes place at the first few minutes of immersion as shown in Figure 1 at the initial domain) ensures favored conditions for the protective layer formation of phosphonates.

All of the investigated diphosphonic acids inhibited the iron dissolution. The mechanism of inhibition was studied by polarization methods. Figure 4 shows anodic polarization curves of Armco iron in aqueous solution of 0.001 $mol.dm^{-3}$ 1,7-diphosphono-heptane measured after different adsorption time (30min, 1h and 3h).

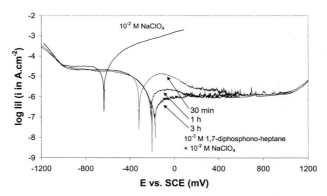

Figure 4. The effect of the adsorption time on the polarization curves of Armco iron in solution of 10^{-3} M 1,7-diphosphono-heptane + 10^{-2} M $NaClO_4$ (pH 7.0)

Significant decrease of anodic currents (which is related to the iron dissolution) in the active potential range was observed with increasing adsorption time, which suggests that the active iron dissolution was hindered by blocking of surface layer.

Film repairing properties of diphosphonates

A re-built protective layer formation can also be observed at defects of inhibitor film. This process is demonstrated on modified iron surface by using an artificial scratch by a sharp knife. Figure 5 shows the change of open circuit potential of modified iron electrode before and after an artificial scratch. Prior to scratching, a four-day-long surface modification was carried out by immersion of iron electrode into the aqueous solution of 1,7-diphosphono-heptane solution.

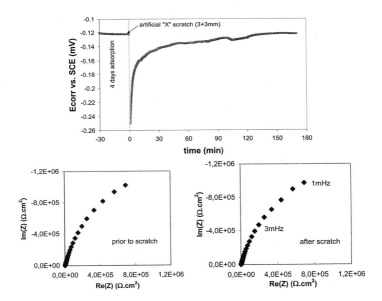

Figure 5. (a) Change of open circuit potential of Armco iron in 10^{-3} M 1,7-diphosphono-heptane before and after an artificial scratch; and impedance spectra measured (b) prior to, and (c) 3 hours after scratching

After replacing the scratched electrode into the solution of phosphonate, the open circuit potential first decreased which was the consequence of enhanced iron dissolution at the defect. But later, the open circuit potential increased showing the "healing" of the layer at the defect, which proved the repairing of the inhibitor film, the potential reached the initial value (−122 mV) within 3 hours. Impedance plots measured before (Figure 5b) and after the scratch (Figure 5c) shows actually the same electrochemical properties of modified iron.

Surface visualization by scanning microscopes

Assuming that the metal pre-treatment by self-assembling molecular layer can enhance the paint adhesion and improve the surface properties some measurements were done on lacquer and paint surface. The first question was how will be reflected the different coating treatment (e.g. heating) in the morphology of lacquer and paint surfaces. For the evaluation of surfaces atomic force microscope (AFM) and scanning electron microscope (SEM) were used. The influence of heating on the surface properties is demonstrated on Figure 6.

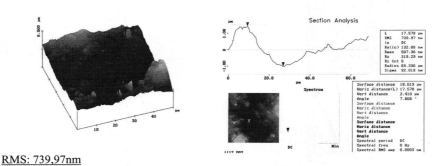

RMS: 739,97nm

Figure 6. AFM images and roughness factors of different paint and lacquer surfaces (continued next page)

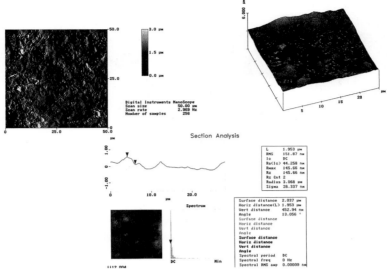

RMS: 152.87nm

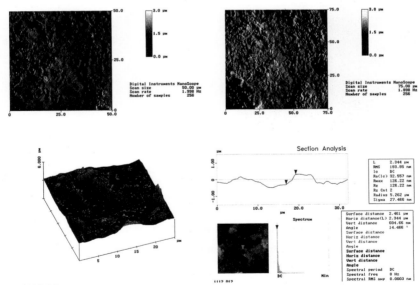

RMS:193.95nm

314

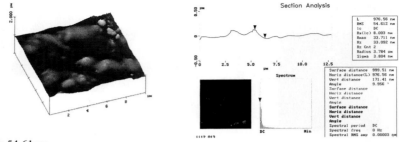

RMS: 54.61nm

Figure 6 (contd.). AFM images and roughness factors of different paint and lacquer surfaces

Not only perceptible the difference is but the roughness factor (RMS), too, reflects the smoothing of the layer as a consequence of the treatment. In other set of experiments the usefulness of the SEM technique was proved in evaluation of surface properties (Figure 7).

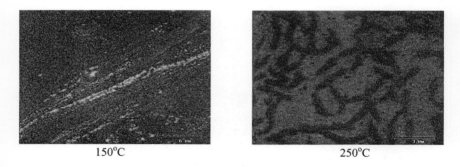

150°C   250°C

Figure 7. After heat treatment SEM images of paint surfaces

The influence of the self-assembled molecular layer on adhesion of paint and the surface roughness was visualized and measured by AFM. In accordance with electrochemical results the nano-structured surface layer improved the surface quality of coating.

Conclusion

Surface pre-treatment by 1,7-diphosphono heptane gave a self-organized, stable layer with ordered structure on oxidized iron surface. The pre-treatments of metal surfaces enhanced the smoothness and the adhesion of paints and lacquer layers.

Acknowledgements. This work was supported by the Hungarian Scientific Research Found (OTKA: T 037643, T 035122, and F 029709).

References

1. R. G. Nuzzo, D. L. Allara, *J. Am. Chem. Soc.,* **105,** 4481 (1983).
2. A. Ulman, *Chem. Rev,* **96,** 1533 (1996).
3. K. Nozawa, H. Hishihara, K. Aramaki, *Corros. Sci.,* **39,** 1625 (1997).
4. M. Volmer, M. Stratmann, H. Viefhaus, *Surf. Interf. Analysis,* **16,** 278 (1990).
5. G. Grundmeier, C. Reinartz, M. Rohwerder, M. Stratmann, *Electrochim. Acta,* **43,** 165 (1998).
6. K. Aramaki, M. Itoh, H. Nishihara, Proc. of 8th SEIC, Univ. of Ferrara, Ferrara, 1995, Vol 1, p.77.
7. P. E. Laibinis and G. M. Whitesides, *J. Am. Chem. Soc.,* **114,** 9022 (1992)
8. Y. Yamamoto, H. Nishihara and K. Aramaki, *J. Electrochem. Soc.,* **140,** 436 (1993)
9. M. Rohwerder, M. Stratmann, *MRS Bulletin,* 24, No. 7. 43 (1999)
10. I. Felhősi, J. Telegdi, G. Pálinkás, E. Kálmán, *Electrochim. Acta,* 47, 2335 (2002).
11. P. Póczik, I. Felhősi, J. Telegdi, M. Kalaji and E. Kálmán, *J. Serbian Chem. Soc.,* **66,** (11-12) 859-870 (2001).
12. I. Felhősi, E. Kálmán, *Mat. Sci. Forum,* in press.
13. I. Felhősi, E. Kálmán, P. Póczik, *Russ. J. Electrochem,* 2002, in press.
14. R. Feser and T. H. Schmidt-Hansberg, *Symp. Eurocorr '97,* Trondheim, **2,** 291 (1997).
15. I. Maege, E. Jaehne, A. Henke, H.-J. P. Adler, C. Bram, C. Jung, M. Stratmann, *Progr. Org. Coat.,* **34,** 1 (1998).
16. E. Kálmán, I. Felhősi, F. H. Kármán, I. Lukovits, J. Telegdi and G. Pálinkás, in *Corrosion and Environmental Degradation,* Vol 1, Ch. 9. (Ed. M. Schütze) Wiley-VCH, Weinheim, Germany (Series of Materials Science and Technology. A Comprehensive Treatment), 471-537 (2000).
17. M. Rohwerder, M. Stratmann, *MRS Bulletin,* 24, No. 7. 43 (1999)

Nano-Scaled Titanium Dioxide – Properties and Use in Coatings with Special Functionality

Dr. Jochen Winkler

Sachtleben, Duisburg, Germany

Summary: Nano-scaled titanium dioxides are used as UV-absorbers and to obtain certain color effects in pigmented coatings. For this they are generally required to have good phototresistivity. This is accomplished by doping of the crystal lattice with other elements than titania and by surface treatment with layers of inorganic substances. Currently the use of photoreactive nano-scaled titanias is being studied to formulate self cleaning and antimicrobial coatings.

Keywords: Nano-scaled titanium dioxide; inorganic UV-absorbers; frost effect; blue color shift; titanium dioxide photocatalyst

Introduction

Titanium dioxide white pigments are manufactured so that they scatter all wavelengths of the visible spectrum between 380 and 700 nm optimally. Contrarily, nano-scaled titanias that were originally developed with the aim to achieve transparent UV protection ideally scatter no visible light at all. In practice, there is , however, some selective scattering of blue light so that apart from the use as UV absorbers these materials are also found in metallic coatings and in other pigmented coatings where they lead to interesting color effects.

Titanium dioxide is itself also a photocatalyst. New research is focussing on the development of materials that will enable the formulation of self cleaning and/or anti microbial coatings.

The frost effect

When nano-titanias are put into metallic base coats these formulations obtain a viewing angle dependent appearance (1). When observed at a facial angle the coatings look yellowish whereas a slanting viewing angle brings forth a blue color tone.

Figure 1 helps to explain this phenomenon The incident light with all the wavelengths of the visible spectrum penetrates the coating. The red and green parts of the spectrum are hardly scattered so that they are reflected by the metallic pigments in an angle close to the incident angle. The blue part of the spectrum is preferentially scattered so that the reflected blue light leaves the paint film in a low angle.

    CCC 1022-1360/00/$ 17.50+.50/0

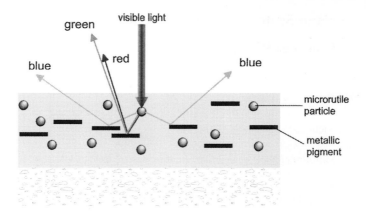

Figure 1. Schematic representation of scattering phenomena leading to the frost effect

Blue color shift

When combined with colored pigments nano-titanias will change the appearance of the colored pigments by selectively scattering blue light (2). Unlike the frost effect, this effect is independent from the viewing angle. Nano-titanias may be used to turn a red color into magenta or to formulate very clean blue colors in combination with copper phthalocyanines, for example. Although the effect resembles the well known "carbon black undertone" (CBU) effect of conventional titanium dioxide pigments (3) it is still quite different since nano-titanias have a low CBU when tested under standard conditions and, more important even, they lack the scattering power in the visible region of the spectrum so that there is no lightening effect on the coating.

Photoactivity of TiO$_2$

When UV light is absorbed by titania electrons from the valence band are hoisted into the conduction band of the semi conductor TiO$_2$ and so called excitons (separated charges) are formed. The electrons may migrate to the surface of the crystal where they can reduce species at the interface of the particle. Likewise, the positive holes remaining in the valence band may by successive electron transfers from their neighbors rearrange to the TiO$_2$ surface where they can oxidize adjacent molecules. Figure 2 depicts the generation of excitons and the type of reactions possible from then on.

This so called "chalking cycle" consisting of an oxidation reaction and a reduction reaction is well known (4) and in the presence of water and oxygen leads to a deterioration of the resin matrix surrounding the titania particles. Interestingly, this happens more pronounced on the illuminated sides of the pigment particles and less on the sides away from the light. At the end, the particles may even rest only on a pillar of resin (5). Since they are then no longer firmly attached they are easily wiped off the surface of a coating, thus causing the expression "chalking".

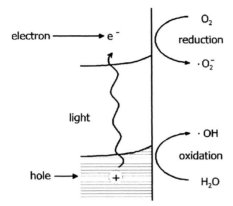

Reaction mechanism of TiO$_2$ photocatalysis

Figure 2.

In paint industry chalking is normally disliked although there are some exceptions such as road marking paints and also in some exterior wall paints. The reason is because chalking goes along with a self cleaning effect.

It is well known that titania in the crystal modification of anatase chalks more than rutile. This is surprising at first because the absorption edge of anatase (385nm or 3.29eV) is shifted more to the ultraviolet range than is the case for rutile (415nm or 3.05eV). Hence, under natural daylight conditions where far less photons are available below 385 nm than in the range between 385 and 415 nm less excitons will be formed in the case of anatase in comparison to rutile. Gesenhues (6) attributed the higher photocatalytic activity of anatase to the energy level of a positive hole in the valency band which, according to him, is 210mV lower than that of rutile. Therefore, the oxidation potential of a positive hole in an anatase crystal is higher than a comparable hole in a rutile crystal.

In order to suppress the photoactivity of rutile pigments the crystal lattices are doped with aluminum ions . The aluminum is introduced as AlCl$_3$ along with TiCl$_4$ into the burner in the chloride process. In the sulfate process, different aluminum chemicals are added prior to the calcination stage. The aluminum ions mostly take the place of titanium ions in the crystal lattice (7) and act as sites where exciton recombination preferentially takes place.

A further trick of the trade is to coat the particles with layers of inorganic matter. Although it is quite evident that this posttreatment is beneficial in decreasing the photoactivity the mechanism is unclear and the impact of a certain posttreatment is certainly not foreseeable. It is understood that the coverage should be 100% and that there is a certain threshold value where a further increase in thickness fails to further improve lightfastness or weatherability. The improvement with a certain kind of inorganic posttreatment depends to some extent on the core particle that is coated. For that reason, different pigment manufactureres have their own preferences when it comes to inorganic posttreatments. In the last fifty years more than two thousand patents dealing with inorganic surface treatments of titanium dioxide pigments have been filed (8)!

Nano-scaled titanias as UV absorbers

Since nano-titanias have a much higher surface area in comparison to pigment grades it is more difficult to suppress photoactivity. When durability is aspired the same steps like in TiO_2 pigment must be employed. As far as inorganic posttreatments are concerned their levels have to be raised, of course. Figure 3 shows an example for the effect of an aluminum oxide posttreatment on the photoactivity of a nano-titania with a particle diameter of approximately 20nm. The ordinate depicts a measure for photoactivity in a standard test. It is seen that in this particular case 12 weight% of inorganic posttreatment are necessary to inhibit photoactivity.

When manufactured accordingly, nano-scaled titanias in the rutile modification may be used as transparent UV-absorbers for substrate- as well as matrix-protection. In substrate protection the object that is to be protected from harmful UV radiation is covered with a clear coat containing about 0.5 to 1.5% pigment volume concentration (PVC) of nano-titania. An example for this is wood protection (9). As the Lambert Beer law is in effect, the PVC may be reduced if the film thickness is raised and vica versa. For matrix protection, that is when the nano-titania plays the role of protecting the resin itself, higher PVC levels are necessary.

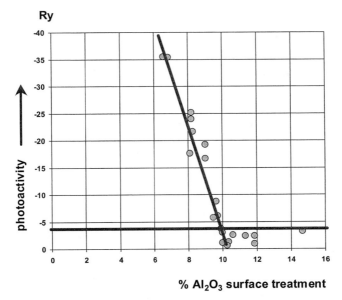

Figure 3. Effect of aluminium oxide posttreatment on the activity of a nano-titania with a particle diameter of ca. 20 nm.

Nano scaled titanias as photocatalysts

For photocatalytic applications the titania should either be in the anatase modification or otherwise x-ray amorphous. Naturally, when a highly active titania is aspired any measures that enhance its lightfastness or weather resistance should be avoided. A high specific surface area is beneficial and when the photocatalyst is to be evenly distributed in a certain matrix, dispersibility is an issue to keep in mind. Furthermore, the titania should have the tendency or at least the ability to adsorb whatever substance is to be degraded. In many cases, such as in water purification, this is a limiting factor. If at a given pH value the zeta potential of the photocatalyst has the same sign as the charge of the substance that is supposed to be mineralized no adsorption will occur and hence no catalytic oxidation will take place. The same problem can arise if during the course of photocatalytic oxidation charged intermediates are generated.

The quantum yield of photocatalytic processes is generally higher with low illumination intensities because mass transfer is inevitably the rate determining factor in these reactions.

322

Photocatalytically active coatings

In order to be photocatalytically active the titania catalyst surface must be accessible for the adsorbates. In a suspension reactor or in the case of titania ceramics this presupposition is fulfilled.

When trying to formulate photocatalytic coatings a way has to be found to incorporate the titania into the resin matrix without covering all of its surface with binder. This means that PVCs exceeding the critical PVC must be chosen, so glossy coatings cannot be formulated. Of course one has to compromise between film integrity and photocatalytic activity.

The next thing to consider is that the resin itself will inevitably be oxidized provided that it is vulnerable for this type of attack. By experience, any conventional organic resins will soon be destroyed and the coatings will loose their mechanical stability. The only exemptions are to some extent fluorinated polymers such as tetrafluoroethylene or polyvinylidenefluoride. It seems that the C-F bond is strong enough to withstand oxidation by titania photocatalysts. The drawback is that these polymers have extremely poor adhesion to substrates. Therefore, photoactive free films but unfortunately no coatings can be accomplished in this manner.

Figure 4. Lime sand stone with four differently pigmented silicate coatings. See text.

Not surprisingly, the best way to formulate photoactive, yet durable coatings is with silicate resins as binders. Figure 4 demonstrates the self cleaning properties of a formulation containing a nano-titania ("UV 100"; PVC 48%) in comparison to a formulation containing only clacium carbonate ("Org.", PVC 62%), one with a rutile ("R210", PVC 48%) and one with a standard pigment grade anatase ("LW-S",PVC 48%). A lime sanstone brick with the four formulations is shown after 6 months of outdoor weathering. There is an onset of growth of algea on all formulations except for the one containing the photocatalyst. The standard anatase grade formulation exhibits cracking and loss of adhesion as well. Since some titania photocatalysts are proven to have antimicrobial properties, these type of formulations may become important in the near future.

Last but not least, there is tremendous amount of development work focussing on the generation of transparent photoactive titania layers either by chemical vapor depositon or by sol gel methods. These layers show "superhydrophilicty" when illuminated with daylight, meaning that condensing moisture will not form droplets. Apart from that, these layers are self cleaning.

Conclusions

Standard applications of nano-titanias today are as UV-absorbers, and as effect pigments. In future, maybe a further use in photocatalytically active paint formulations or pure titania layers will become important. The possibility for this to occur seems promising.

References

1. EP 0270 472 B1; 29.07.1992
2. EP 0634 462 B1; 17.06.1994
3. "The optical theory of titanium dioxide", 11th. All-India Paint Conference; January 1983, Bombay, India, p. 171-175
4. H.G. Völz, G. Kämpf; A. Klaeren, farbe+lack 82 (1976) No.9, 805-810
5. G. Kämpf, W. Papanroth, H.G. Völz, G. Weber, XVI. FATIPEC Congress, 9.-14. May 1992, Liege, Berlgium, Congress Book III, p. 167
6. U. Gesenhues, farbe+lack 101 (1995) No. 1, 7-10
7. U. Gesenhues, T. Rentschler, J. Solid State Chem., 143 (1999) 210-218
8. R. Davies, G.A Schurr, P. Meenan; R.D. Nelson, H.E. Bergna; C.A.S. Brevett, R.H. Goldbaum, Adv. Mater. 10 (1998) No. 15, 1264
9. J. Winkler, W.-R. Karl, XXII: FATIPEC Congress, 10.-14. July 1996, Brussels, Belgium, Congress Book I, p. 195-215

Macromol. Symp. 187, 325–332 (2002)

Features of Reaction Amino-cyclocarbonate for Production of New Type Nonisocyanate Polyurethane Coatings

Oleg L. Figovsky,* Leonid D. Shapovalov**

*Israeli Research Center -Polymate Ltd., PO Box 73, Migdal HaEmek 10550, Israel

**Chemonol Ltd., PO Box 73, Migdal HaEmek 10550, Israel

Summary: We have studied different ways of preparing UV resistant oligomers with terminated cyclocarbonates, epoxy and amino groups. We have studied possibility of preparing HNIPU UV stable coatings. Linear and branched amino containing oligomers based on di- and tricyclocarbonates and primary diamines were investigated. It was found that oligomers should be used for curing epoxy-saturated resins, but since the residual quantity of amines this system can't be used for UV resistant coatings. The same problem was in system on the base of nonisocyanate epoxy-urethane oligomers cured by tertiary amines. Coating on the base of acrylic cyclocarbonates and oligomers with primary amines are similar to conventional polyurethane coatings on the base of acrylic hydroxyl containing oligomers and isocyanates, but mechanical properties and chemical resistance are better.

1. Introduction

Applications of polyurethane materials have significantly increased in comparison with some other thermosetting polymer materials. Conventional monolithic polyurethanes have good mechanical properties but they are porous and possess poor hydrolytic stability and insufficient permeability. The involvement of toxic components, such as isocyanates, in their fabrication process makes the production extremely toxic and dangerous.

Polyurethanes have weakness inherent to them due to their molecular composition. Within their polymer structure there are hydrolytically unstable bonds that make the material vulnerable to environmental degradation. By modifying the structure of the polymer, a new promising method of raising hydrolytic stability is introduced and readily displayed in nonisocyanate polyurethanes – a modified polyurethane material with lower permeability, increased resistance properties and safe fabrication processes.

CCC 1022-1360/00/$ 17.50+.50/0

Network nonisocyanate polyurethanes are formed as a result of the reaction between cyclocarbonate oligomers and primary amine oligomers [1, 2]. This reaction forms an intra-molecular hydrogen bond through the hydroxy group at the ß-carbon atom of the polyurethane chain as illustrated below:

$$R\text{-}O\text{-}CH_2\text{-}CH\text{-}CH_2 + H_2NR' \longrightarrow R\text{-}O\text{-}CH_2\text{-}CH\text{-}CH_2\text{-}O\text{-}C\text{-}NHR$$

Quantum-mechanical calculation, IR and NMR spectroscopic investigations have confirmed the stability of such a ring [2].

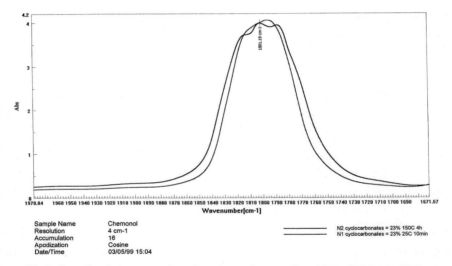

Sample Name	Chemonol
Resolution	4 cm-1
Accumulation	16
Apodization	Cosine
Date/Time	03/05/99 15:04

N2 cyclocarbonates = 23% 150C 4h
N1 cyclocarbonates = 23% 25C 10min

Fig 1. Reaction between cyclocarbonates and aromatic amines (Ethacure 100)

The blockage of carbonyl oxygen considerably lowers the susceptibility of the whole urethane group to hydrolysis. Moreover, materials containing intra-molecular hydrogen bonds display chemical resistance 1.5-2 times more as compared to materials of the similar chemical structure without such bonds [2].

Basic cyclocarbonate oligomers are formed by bubbling the carbon dioxide through epoxy liquid oligomers in the presence of a catalyst, and by interacting oligomeric chlorinehydrine ethers with carbonates of alkaline metals or oligomeric polyols with ester (chloride) of carbonic acid [2].

The reaction of aromatic diamines with cyclocarbonate and epoxy groups was studied by the authors, and a new method of synthesis of advanced cyclocarbonates is described below [7].

The data about the reaction between aromatic amines and cyclocarbonates are given on Fig. 1.

It was found out that up to 150^0C there was no reaction between cyclocarbonate group and aromatic amine group.

Nonisocyanate polyurethanes have superior properties over conventional polyurethanes. Their tensile strength and deformation are similar to those of conventional isocyanate polyurethane materials, but they display chemical resistance properties 30-50% higher, and have significantly reduced permeability: three to five times less than conventional isocyanate polyurethane materials [2]. Due to their unique molecular composition nonisocyanate polyurethanes have transformed into the material with practically no pores being insensitive to the moisture on the surfaces or on the fillers upon formation. As a result, nonisocyanate polyurethane materials exhibit superior resistance to degradation, making them optimum for their applications in crack resistant composite materials, chemically resistant coatings, sealants, glues, etc.

As nonisocyanate polyurethanes do not contain a highly toxic isocyanate, the material synthesis can be carried out in a safe and nontoxic process. Furthermore, nonisocyanate polyurethanes can be employed for the material hardening at room temperature [2].

Due to their superior structure and excellent resistance to degradation, NIPU is optimum for the numerous applications including crack resistant composite materials, chemically resistant coatings, sealants, glues, etc. Their outstanding properties are beneficial to many different industries [4,6]. Siloxane-polyurethane sealants fabricated from siloxane pentacyclocarbonate oligomers displayed the same high adhesion, elasticity and strength properties as conventional polyurethane sealants. However, their thermal stability and heat resistance properties proved to be twice as great [2].

Data about mechanical properties of filled nonisocyanate polyurethanes are given in table 1.

Due to the "data" it is possible to produce materials with different properties from elastic to rigid.

The coatings produced from nonisocyanate polyurethanes have excellent water-resistance properties equivalent to the conventional polyurethanes and epoxy coatings combined [2]. On addition of inorganic powdered substances during the fabrication process the chemical resistance of NIPU can be increased. These substances interact selectively with water and the aggressive medium, i.e., acids, alkalis, and salts transforming the system of high-strength hydrate complexes into durable inorganic adhesive cements. Since the substances interact with the medium, hydrate complexes (crystalhydrates) are always formed in the defects, i.e., micropores, microcracks of the material. This process eliminates the defects of the material and increases its strength.

Crystalhydrates form the volume and the contact-specific surface of an active additive that expands into a constant volume of the polymeric matrix. Due to the increased specific surface of the active filler and the material with viscous features formation, adhesion on the polymer-crystal boundary becomes stronger [5].Till now it was not investigated perspective way for synthesis of UV-stable nonisocyanate PU for coatings.

Table 1. Mechanical properties of filled nonisocyanate polyurethanes

Types of cyclocarbonate Oligomers	Functionality of cyclocarbonate Oligomer	*$\sigma t/\varepsilon p$ by applying primary amines				
		III				IV
		Functionality				
		2.92	3.47	4.31	5.08	2
I	~2	1.9/710	2.7/565	4.9/325	5.3/108	------------
	~3	8.8/220	13.7/114	15.3/90.5	15.4/63.5	26.6/88.5
	~5	21.3/72	21.8/50	18.7/31	17.2/22	20.3/48.5
II	2.45	8.2/315	17.3/98	23.9/47	2.6/19.6	30.8/20.5
	3.10	19.3/88	22.7/62	23.8/37	18.2/18.8	33.9/19.2
	3.95	19.9/82	26.1/42	25.4/28	17.1/14.3	34.2/18.3
	4.70	16.9/73	13.7/30.5	12.7/24.5	10.6/13.6	27.1/16.9

*σt- tensile strength, MPa; εp- elongation, %

I- aliphatic cyclocarbonate oligomer
II- aromatic cyclocarbonate oligomer
III- aliphatic polyamine
IV- Adduct of BPA epoxy resin and DETA

2. Results and Discussions

Cyclocarbonate olygomers were used for production adducts with polyamine olygomers. The following materials have been employed:

Laprolate 803-aliphatic cyclocarbonate oligomer (functionality~3) with polyoxypropylene chain (Russia);

Eponex 1510 (diglycidyl ether of hydrogenated BPA) Shell;

Polypox R-14 (neopentyl glycol diglycidyl ether);

Polypox R-19 (polypropylene glycol diglycidyl ether;)

Polypox R-20 (trimethylol propane triglycidyl ether);

UPPC (Germany)

and acrylic epoxy resin Setalux 17-1433 (Akzo Nobel, The Netherlands).

It were prepared different CC-carbonates by bubbling CO_2 with catalyst (tetrabutylammonium bromide) and their adducts with diamines:

Pentamethylenediamine (Du Pont)

Trimethylhexamethylenediamine

(Degussa)

Jeffamine EDR-148 (Huntsman)

Isophorondiamine (BASF)

3. Adducts Formation

Adducts have been manufactured by mixing cyclocarbonates with aliphatic and cycloalyphatic amines for 2-4 hours at $60-120^0C$. Synthesis of adducts is safe and easy: the control was provided by IR spectroscopy. When cyclocarbonate groups had disappeared, the reaction was finished.

Adducts formation process:

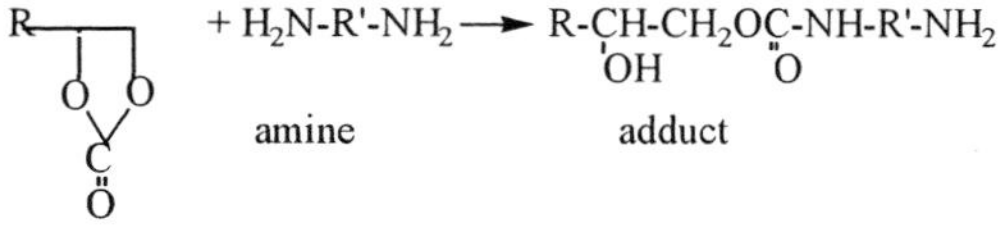

$$R-CH-CH_2-OC-NH-R'-NH_2$$

R—‖ + H_2N-R'-NH_2 → R-CH-CH$_2$OC-NH-R'-NH$_2$
   O O   OH   O
        amine   adduct
     ‖
     O

cyclocarbonate

These adducts were used for hybrid nonisocyanate polyurethanes (HNIPU) formation.

4. Analysis of Paint Properties

Linear and branched amino containing oligomers based on di- and tricyclocarbonates and primary diamines were investigated. It was found that oligomers should be used for curing epoxy-saturated resins, but since the residual quantity of amines this system can't be used for UV resistant coatings (yellowing). The same problem was in system on the base of nonisocyanate epoxy-urethane oligomers cured by tertiary amines.

It is possible to prepare only coatings with restricted UV stability by using adducts on the base of saturated cyclocarbonates and amines and saturated epoxy resins (Eponex 1510 and Polypox 14,16,18,19,20)

High light stability of acrylic resins is well known. We have synthesized cyclocarbonate on the base of acrylic epoxy resin Setalux 17-1433. The cyclocarbonate-terminated oligomers were synthesized by bubbling CO_2, in presence of catalyst at 140^0C, 10 bar.

Process of HNIPU curing:

-CH-CH$_2$-OCONH-R'NH$_2$ + ▽ ⟶ HNIPU
OH   O
adduct   epoxy oligomer

Process of acrylic NIPU curing:

```
      CH₃   CH₂
       |     |
-CH₂- C-CH₂-C-COOCH₂CHCH₂OCOO  +  NH₂-R'-NH₂
       |           |_____|
      COOR

                 ↓

      CH₃   CH₂
       |     |
- CH₂-C-CH₂-C-COOCH₂CH-CH₂OCONH₂-R'-NH-
       |           |
      COOR        OH
```

WHERE: R'- ALKYL, CYCLOALKYL

The main properties of the Epoxy based paint and the HNIPU and acrylic NIPU paint are listed in Tab. 2. According to the data, modification of epoxy materials leads to production materials with high mechanical properties (impact and abrasion resistance, elasticity, adhesion, etc).

On the base of saturated epoxy resin after 100hours UV test coatings slight yellow (Baroness 160, 70 w UV, 150 w IR)

Acrylic HNIPU coatings are much more light stable.

After 200hrs UV test coatings were very slight yellow.

By using of UV absorbers and HAL S it is possible to increase UV stability.

NIPU on the base of acrylic cyclocarbonate oligomer and aliphatic diamines were prepared by adding amines to acrylic cyclocarbonate. Curing at 110^0C during 2 hours. At RT it was impossible to prepare coating with advanced properties.

Coatings on the base of acrylic cyclocarbonates and primary amines with UV -absorber are similar to conventional polyurethane coatings on the base of acrylic hydroxyl containing oligomers and isocyanates, but mechanical properties and chemical resistance are better.

Typical properties of UV stable HNIPU coatings are given in the Table 2.

Table 2. Main properties of coatings

Properties	Aliphatic-Cycloaliphatic HNIPU	Acrylic HNIPU
NV, %	95-100	70-75
Pot life (doubling initial viscosity), hours	0.5-2	0.5-2
Drying time at 23^0C, hours	2-13	-
Curing time at 80^0C, days	> 0.5	-
Curing time at 23^0C, days	3-7	2hours @ 110^0C
Film appearance	clear smooth	clear smooth
Pencil hardness	2H	H
Elasticity, mm	1-3	1
Impact, kg cm	40-50	50
Adhesion mark ASTM D3359 mark	4B	4B
chemical resistance: H_2SO_4, 10% at 25^0C, days	> 20	>20
NaCl, 3% at at 25^0C, days	> 10	> 10
Abrasive resistance after 440 revolutions	0.3-0.4	-

5. Conclusions

Hybrid nonisocyanate oligomers based on saturated and primary amino containing oligomers epoxy resins can be used for manufacturing coatings with restricted UV stability. Nonisocyanate polyurethanes coatings on the base of acrylic cyclocarbonates and primary amines are similar to conventional polyurethane coatings but mechanical properties and chemical resistance are better.

References

1. Shapovalov L., Blank N., Tartakovsky A. Nonisocyanate Polyurethane for Protective Coatings. // Abstracts of the 2nd Conference of the Corrosion Forum - NACE - Israel, Tel-Aviv, Israel, 1996, No. 1.1.3.
2. Figovsky O. Improving the Protective Properties of Nonmetallic Corrosion-Resistant Materials and Coatings // Journal of Mendeleev Chemical Society, N.Y., USA, 1988, Vol. 33, No. 3, pp. 31-36.
3. Moshinsky L., Epoxy resins and hardeners. Areadia (book). Tel Aviv, Israel, 1995
4. Figovsky O., Shapovalov L., Blank N., Monolithic chemical resistant floor covering based on nonisocyanate polyurethanes. International Conference on Corrosion 1997 Report, EUROTECH Ltd. And Chemonol Ltd., Mumbai, India 1997
5. Figovsky O., Shapovalov L., Blank N., Nonisocyanate Polyurethane Materials. Polyurethanes World Congress 1997 Report, Poster No. 69, EUROTECH, Ltd. and Chemonol, Ltd. Amsterdam, The Netherlands, p.741, 1997.
6. Shapovalov L., Blank N., Tartakovsky A. Nonisocyanate Polyurethane for Protective Coatings. Abstracts of the 2nd Conference of the Corrosion Forum - NACE - Israel, Tel-Aviv, Israel, 1996, No. 1.1.3.
7. EP 1020457

Additives for UV-Curable Coatings and Inks

Dr. Sascha Oestreich*, Dipl.-Ing. Susanne Struck

Degussa AG, Tego Chemie Service GmbH, Goldschmidtstr. 100, 45127 Essen, Germany

Summary: UV-curable silicones are a highly specialised class of compounds that can be applied in various applications e.g. as additives in UV-curable inks and coatings or as release coatings on paper and plastic substrates.
Basically two classes of radiation curable silicones are available on the market today. Both, the free radical and the cationic curing process, offer each unique advantages to the customer. Applied as additives in UV-curable inks and coatings they offer several advantages such as improved wetting behaviour, scratch resistance, flow and levelling of the UV-curable inks and coatings. Additionally, the efficient cross-linking properties minimise the risk of migration.

Keywords: UV-curing; UV-curable coatings; UV-curable inks; cationic UV-curing; silicone acrylates; silicone polyether acrylates; cycloaliphatic epoxy silicones; additives

Introduction

High curing speeds, high cross-linking densities and the absence of organic solvents have made UV-curing a well established technology for all kinds of coating and ink applications. Today numerous UV-curable monomers and oligomers like polyether-, polyester-, epoxy-, polyacrylate- and urethane acrylates are available on the market. By the choice of raw materials - namely oligomers as binders and accompanying monomers, photoinitiators and synergists - the film properties such as hardness, flexibility, resistance and adhesion can be controlled in a very flexible way [1 – 4].

These basic components of an UV-curable formulation alone hardly ever create a coating or printing ink of acceptable quality. The pigment wetting properties of solvent-free formulations are rather poor. The removal of entrained air is difficult because of the short flash-off times before curing. The wetting of printed substrates, wood or film can be problematic and the surface of the cured formulation often is sensitive to scratches.

CCC 1022-1360/00/$ 17.50+.50/0

Where a few years ago the radiation curing technology was primarily used for clear systems, a disproportionate growth rate can be observed today for pigmented systems [5]. With increasing pigment content and the use of difficult-to-wet pigments the formulation of such highly pigmented systems becomes increasingly difficult. The market offers various UV-curable oligomers with improved pigment wetting properties. However, especially in UV flexographic ink with their high content of organic pigments or carbon black, acceptable rheology can hardly ever be achieved without the use of pigment dispersing additives. Therefore additive suppliers offer tailor-made solutions, often solvent-free and cross-linkable.

Additives for UV-Curable Coatings and Inks

To improve the deaeration of UV-curable formulations active ingredients with controlled incompatibility are used. Such additives are able to replace surfactants at the interface which stabilise the air bubbles. At the coating surface these bubbles burst wherever droplets of bubble-destabilising additives are present.

The balance between defoaming effectiveness and compatibility in a given system is always critical. Especially for clear non-pigmented systems special additives had to be developed which show high performance without creating strong turbidity or surface defects. Depending on the application and the specific requirements silicone-free additives (e.g. polyacrylate based) or organically modified siloxane (OMS) products are being used today.

Substrate wetting additives improve the wetting of various substrates. Such additives have to be very surface active. They reduce the surface tension of the given system significantly to improve the wetting properties. It is important that the effectiveness of these additives is also maintained in highly dynamic processes like printing. For standard UV-curable formulations these properties can be obtained by using low molecular weight silicone additives, organic surfactants or fluorinated surfactants.

Surface control additives are used to improve not only slip properties and scratch resistance, but also the flow and levelling of the coating. Like substrate wetting additives they are highly surface active but in contrast to these products they have a higher molecular weight and a strong tendency to migrate to the coating surface. UV-curable silicones form a thin lubricating film on top of the coating surface to improve the slip properties and the scratch resistance.

In many UV-curable formulations multifunctional additives are being used. Usually these products are organically modified siloxanes of medium molecular weight. They combine substrate wetting, flow and levelling with improved slip properties.

The enrichment of high molecular weight siloxane additives at the surface of the coating can be utilised to obtain even a release coating effect. To get this pronounced release effect siloxane-based additives with long siloxane backbones and relatively high incompatibility in a given system are being used. If the additive is cross-linkable this effect is permanent. Even pure UV-curable silicones are being used today. Typical applications for these compounds are release coatings for self adhesive label stock, tapes and personal hygiene products, like seals for diaper tapes. The silicone coating reduces the adhesion force to adhesives and so labels can easily be removed from the substrate for subsequent uses.

Today a tailor made additive design is necessary to meet customers' needs in the UV-curable coatings and inks area. Especially organically modified siloxanes with their wide variety of synthesis options allow a fine tuning of additive properties and performance in UV-curable formulations.

Synthesis of UV-Curable Silicones

It is important that surface active additives in UV-curable formulations are cross-linkable because the cross-linking effectively prevents the transfer of these additives to other surfaces and prevents extraction when in contact with liquids. Additionally, a more permanent effect is obtained.

The silicone backbone is a versatile building block and can be modified in various ways to design surface active additives for all kinds of applications. Today most UV-curable formulations are free-radically curable using unsaturated acrylate monomers and/or oligomers and so silicone acrylates are commonly used as surface active additives in UV-curable coatings and inks. Silicone acrylates can be synthesised by various condensation, addition and esterification reactions.

In 1983 the Goldschmidt AG introduced the first radiation curable silicone acrylates to the market. The first commercial silicone acrylates were based on condensation reactions of chlorosiloxanes with hydroxyalkyl acrylates. In addition to the salt precipitation additional drawbacks were the insufficient hydrolytical stability of the Si-O-C bonds and some remaining unreacted oligomeric acrylates which were introduced as by-products of the starting material.

A much better choice is the hydrosilylation of epoxy olefins and subsequent esterification of the epoxy group with acrylic acid. The hydrosilylation reaction leads to hydrolytically stable Si-C-bonds, no salt is generated as by-product and a broad variety of structural variations is easily accessible (Figure 1):

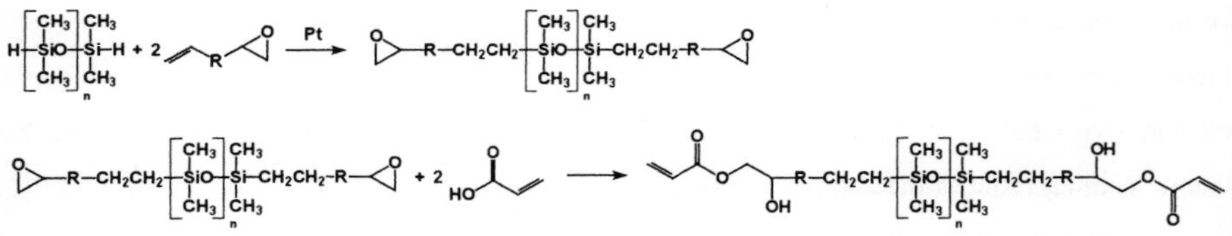

Figure 1. Esterification of epoxy silicones with acrylic acid

A further optimisation of that structural principle is the direct esterification of hydroxyalkyl silicones with acrylic acid (Figure 2):

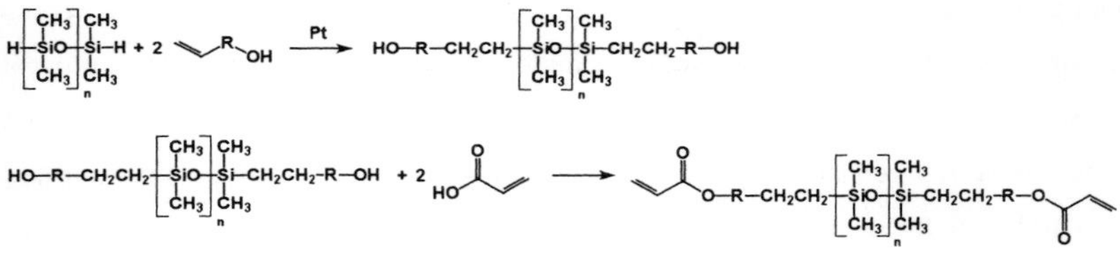

Figure 2. Direct esterification of hydroxyalkyl silicones with acrylic acid

Another very convenient route to synthesise silicone acrylates in a one step - one pot reaction is the direct hydrosilylation of multifunctional organic acrylates (Figure 3). This patented technology gives direct access to silicone acrylates with good release properties:

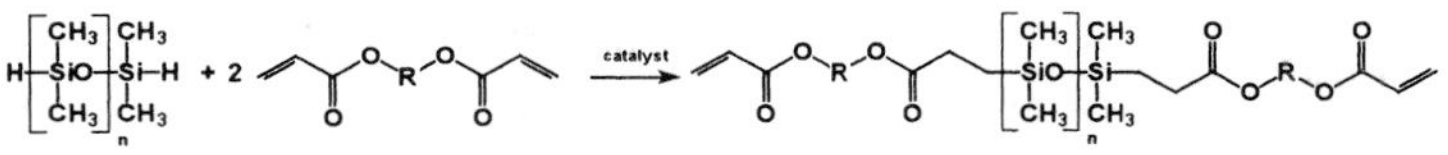

Figure 3. Direct hydrosilylation of difunctional acrylates

Recently, further progress has been made by using new technologies such as enzymatic catalysis [10]. The enzymatic catalysis, in contrast to the common acid catalysed acrylation, allows the production of low colour, highly transparent products free of undesired by-products.

Beside the free radical curing process also cationic UV-curable coatings and inks are being used today. In cationic UV-curable systems a strong acid is generated by UV-radiation which leads to a polymerisation of cycloaliphatic epoxy or vinyl ether groups. In contrast to the free radical polymerisation the cationic curing process is slower and it is not inhibited by oxygen. On the other hand the cationic polymerisation is inhibited by bases and is therefor substrate dependent.

Also for cationic UV-curable coatings and inks silicone additives are being used to improve properties like wetting behaviour, flow, levelling and release. The hydrosilylation of 4-vinylcyclohexene oxide leads to cationic UV-curable silicones (Figure 4):

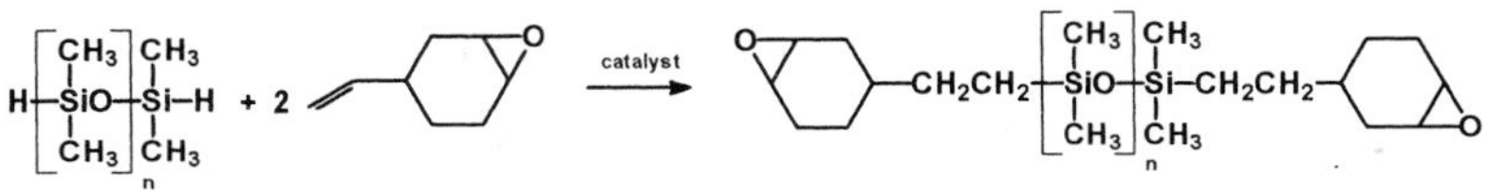

Figure 4. Hydrosilylation of 4-vinylcyclohexene oxide

In the case of cationic UV-curing also OH-functional silicone additives are able to cross-link during the curing process. However, to achieve a pronounced release effect cycloaliphatic epoxy silicones are required.

Results of Application Tests

Organically modified siloxanes can be adopted to the required property profiles by varying their siloxane chain length, the degree of modification and the selection of the modifying organic chain.

Longer siloxane backbones ensure stronger slip and release properties, while their specific incompatibility increases. A higher degree of modification improves the solubility, compatibility and recoatability, at the same time specific surface related effects such as release and anti-blocking properties decrease. By connecting polyethers to the siloxane backbone the polarity of the additives can be drastically increased, so that applications even in aqueous systems become possible.

Table 1 describes several silicone acrylates and their properties including their siloxane content.

Table 1. Silicone acrylate additives tested

product	siloxane content	functionality	solubility in monomers and varnishes	molecular weight	remark
Add. 1	low	5	very good	1500	
Add. 2		2	very good	3000	water soluble, silicone polyether acrylate
Add. 3		2	rather poor	1500	
Add. 4	very high	6	insoluble, dispersible	5000	

When comparing these additives in different UV-curable formulations, significant differences with regard to deaeration properties, flow and levelling improvement (cratering tendency), foam behaviour (Table 2), slip properties (Figure 5) and release effect (Figure 6) were observed.

Table 2. Flow and air release effect of silicone acrylate additives

product	varnish		pigmented coating		silk screen ink	
	Flow	air release	flow	air release	flow	air release
Add. 1	++	o	++	o	+	-
Add. 2	++	-	++	-	+	--
Add. 3	-	+	o	+	+	++
Add. 4	--	+	--	+	-	++

It can be seen from Table 2 that Additive 3 may cause turbidity, flow or levelling problems in sensitive clear coats. In silk screen inks, however, it combines flow and deaeration in a unique manner. In contrast to Additive 3, Additive 2 has a negative influence on the foaming properties of the silk screen ink, but is highly compatible in clear coatings and promotes flow and levelling.

With increasing siloxane content the risk of surface defects in sensitive systems is increasing; at the same time, however, the defoaming and deaerating properties grow stronger. Consequently it is not possible to recommend a certain structure as generally good or bad, judgements can only be made for specific formulations. Which additive finally finds its use in a certain application depends on the related specific requirements of the coating or printing ink.

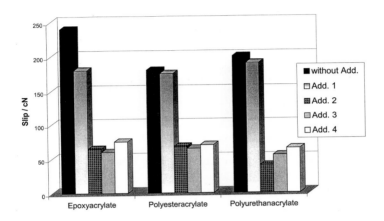

Figure 5. Slip properties of the silicone acrylate additives (0.1% addition level)

Another clear trend can be seen when comparing the slip properties in different systems. With increasing siloxane content the slip effect is increased. The slip effect is measured by pulling a standardised gliding object of defined weight across the cured formulation with defined speed and by measuring the required driving force. The lower the slip value, the more effective the additive [11].

Especially interesting is the difference in slip properties between Additive 1 and 2. Both additives are highly compatible in UV-cured formulations and both improve flow and levelling. Additive 1, however, only has a minimal influence on the slip properties while Additive 2 appears to be a highly effective slip additive. At addition levels of 0.1 % Additive 2 is able to reduce the gliding friction to about 25 % of the additive-free sample. This is possible because the increased siloxane content, which would normally cause incompatibility, is compensated by a polyether modification of the siloxane backbone.

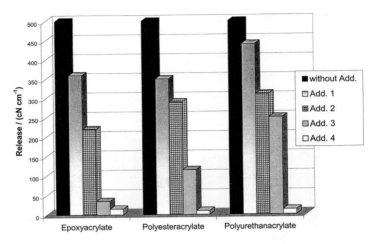

Figure 6. Release properties of the silicone acrylate additives (1% addition level)

Most extreme differences can be observed when measuring the release properties of the coatings. Here a special adhesive tape is applied to the cured coating under defined conditions. This tape is then peeled off at a constant speed and the required release force is being measured (Figure 7).

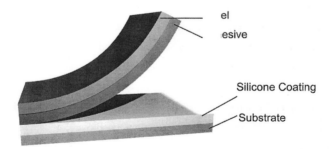

Figure 7. Release of labels from a coating containing silicone acrylates

Clearly Additive 4 is the most effective in this test [12]. It is a product with a very high siloxane content and it is highly incompatible. Its high functionality ensures that this silicone acrylate shows its dehesive effect not only on a short term basis, but also after storage and with repeated tape tests. In general the release force is increased with the density of acrylic groups along the silicone backbone.

By the selection of special cross-linkable additives the formulator can influence the surface properties and the foam properties of the coating or ink in an ideal way. Naturally, it is also possible to use additive combinations. The resulting variety of different property profiles is almost endless.

In most applications the multifunctional character of siloxane-based additives is especially appreciated. However, in some applications it may not be desired. Defoaming and deaerating properties of organically-modified polysiloxanes generally accompanied with a decrease in surface energy and slip resistance of the cured coating or ink. Problems with recoatability or overprintability may result. Whenever this has to be avoided, silicone-free additives are the products of choice. For such applications polyacrylate-based deaerators have been developed specifically for UV-curable systems. They combine good effectiveness with utmost compatibility and recoatability / reprintability and can even be used for curtain coating applications.

Conclusion

Today a tailor made additive design is necessary to be able to meet customers' needs in the UV-curable coatings and inks area. Especially organically modified siloxanes (OMS) with their high synthetical flexibility allow a fine tuning of additive properties and performance in UV-curable formulations.

Many properties of UV-curable formulations can be positively influenced by the addition of organically modified siloxanes. Acrylation of these additives provides higher permanency of release and slip effects, minimises the risk of migration and reduces the amount of extractable substances.

Siloxane-free additives, as well, can be optimised to meet the requirements of UV-curable formulations – they are used where modified siloxanes show drawbacks in special fields of application. It is possible to synthesise siloxane-free cross-linkable structures as well as silicone acrylates.

However, the most advantageous concepts can only be identified in close co-operation between raw material suppliers and formulators.

Acknowledgements

The authors would like to thank all the members of the global TEGO team for their assistance in preparing samples and for determining the performance of all additives described in this paper.

References

[1]   R. Schwalm, *Farbe & Lack* **2000**, *106*, Nr. 4, 58ff.
[2]   Ph. Barbeau, *RadTech Europe 99 Proc.*, p.123ff.
[3]   H.H. Bankowski, *RadTech Europe 99 Proc,*. p.131ff.
[4]   R. Königer, *RadTech Europe 99 Proc..* p.531ff.
[5]   J.P. Howard, *RadTech Europe 99 Proc.*,
[6]   DE-C-3841843, Goldschmidt AG
[7]   DE-C-3820294, Goldschmidt AG
[8]   EP-C-0336141, Goldschmidt AG
[9]   DE-19841559, Goldschmidt AG
[10] DE-19850507, Goldschmidt AG
[11] Tego Journal **1999**
[12]   EP-0761784, Goldschmidt AG

Macromol. Symp. 187, 343–355 (2002)   343

UV Stabilisation of Powder Clear Coats

Sevgi Zeren

Ciba Specialty Chemicals Inc., Klybeckstrasse 141, CH-4002 Basel, Switzerland

Summary: Powder coatings are gaining importance in exterior applications such as automotive and architectural finishes. The use of additives in powder coating formulations enhance the durability of the coating by reducing the effects of harmful UV light and providing high temperature stability during processing and curing. Customers' increased demands for better retention of aesthetical and mechanical properties has prompted the development of new additives dedicated to powder coatings.

This paper presents an overview of the performance of light stabilizers used in powder coatings.

Key words: powder clear coats; UV absorber; hindered amine light stabilizers; weathering

Introduction

verall development of the technology is mainly due to so called five E's associated to powder atings: Excellence of finish, Environmental friendliness, Economics, Energy savings, Ease of pplication. The latest technology advances in powder coatings stimulate further the growth by lowing outlet in new applications such as automotive clear coats or highly durable architectural nish. Powder clear coats are forward looking for the automotive industry as it is said by Johann haler from BMW [1]. Since 1996 these coatings are successfully used in series as clear top coat by is company. This development has been accessible thanks to new binders systems like olyacrylate used in combination with light stabilizers like UV Absorber (UVA) and hindered mine light stabilizers (HALS). Polyacrylate is a binder of choice for transparent finishes with eneral properties such as high transparency, excellent flow and acceptable durability. Light abilizers extend the life of such coatings to more than 5 years Florida exposure and meet the emand from the automotive industry [2, 3]. These high performance coatings match the quality of quid paints and so their use will grow more and more in the automotive industry for car bodies and omponents like wheel rims. In Europe the highest growth rates in powder coatings segments are uoted for polyacrylate powder coatings over the next five years [4]. The first large scale practical pplication of powder coating in general industry has been seen with bicycle frame [5].

    CCC 1022-1360/00/$ 17.50+.50/0

Currently the most common powder coatings used for exterior applications in industrial application and architectural finishes are based on polyester (PES) binder systems. One of the main restraints the development of these coatings are their limited durability where classical terephthalic acid PE resins are used. However in this area significant improvement have been achieved with the development of so called "super durable" polyesters. Lifetime of these super durable resins have been enhanced by use of appropriate light stabilizers. The aim of this paper is to show the prolonge life of powder clear coats via the use of light stabilizers. Artificial and outdoor weathering demonstrates the combined protective action of different UV absorbers (UVA) and hindered amir light stabilizers (HALS). To have a representative view of the market for powder clear coa (automotive and general exterior applications) four binder systems were chosen: two polyacrylate based on glycidylmethacrylate (GMA) and two based on polyester / triglycidylisocyanurate (PES TGIC).

Light stabilizers

Coatings used outdoor are exposed to detrimental processes initiated mainly by the UV portion of solar radiation, gaseous pollutants and acid rain. According to the intended end use, service life u to five or even ten years may be required. High performance coatings must be protected b stabilizers in order to withstand a harsh environment during its life time. The best protection for high quality clear coatings is obtained by using a combination of UVA and HALS [6-8]. The filte effect of the UVA protects the substrate mainly against color change and photochemical polyme degradation, which in turn leads to delamination and loss of mechanical properties. HALS preven surface degradation and protects the coating against loss of gloss and cracking. The four mo important UV absorber classes are benzophenone, oxalanilide, benzotriazole and hydroxyphenyl-s triazine, benzotriazoles are the most widely used for polymers and paints [9]. The mechanism of action of UVA is the absorption of UV-light and its rapid conversion into harmless energy e. heat. Absorbance follow the Lambert-Beer Law:

$$E = Abs = \varepsilon \cdot c \cdot d, \text{ where} \tag{1}$$

E = extinction; Abs = absorbance; ε = extinction coefficient [$L \cdot Mol^{-1} \cdot cm^{-1}$]; c = concentratio [$Mol \cdot L^{-1}$]; d = (film-, cell) thickness [cm].

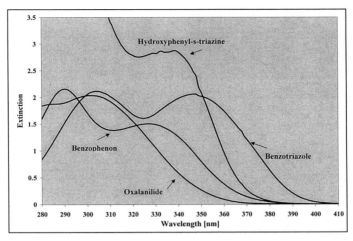

Figure 1: Absorption spectra of different UV absorber classes, c=1.4x10-4 mol/l in chloroform.

he initial extinction of UVA's is directly proportional to the extinction coefficient ε, the oncentration c and the film thickness of the clear coating.

he extinction coefficient ε is usually given at the wavelength corresponding to the maximum of bsorption of the product. Table 1 shows the ε of each UVA used in this work. The absorption rofile of the UVA's plays also an important role. Figure 1 shows the absorption spectra of the VA's participating in the study.

Table 1: Extinction coefficients ε of different UVA at maximum long wave absorption (λmax), c=1.4x10-4 mol/l in chloroform

UV Absorber	λ_{max} [nm]	ε [L · Mol^{-1} · cm^{-1}]
Hydroxyphenyl-s-triazine (UVA1)	339	~ 21700
Benzotriazole (UVA2)	346	~ 15600
Benzophenone (UVA3)	327	~ 10700
Oxalanilide (UVA4)	301	~ 14500

he strongest absorption in UVB (290-320nm) is achieved with hydroxyphenyl-s-triazine (UVA1), ollowed by oxalanilide (UVA4), benzotriazole (UVA2) and benzophenone (UVA3). The best overage in UVA (320-390 nm) is achieved with UVA 2 and UVA 1 whereas UVA 3 and 4 show a elatively low absorption in that part of the spectrum. An other parameter crucial in UV Absorber

selection is photo-stability, it is of paramount importance to keep in the film enough active produ
to ensure proper extinction. From that respect UVA1 and UVA2 have a definitive advantage wh
it comes to long exposure.

Sterically hindered amines (HALS) are almost exclusively derivatives of 2,2,6,
tetramethylpiperidine which is responsible for the stabilization effect of these compounds. T
"Desinov Cycle" (see figure 2) present a possible action mechanism for HALS. After activati
step conversion of >N-R to >NO* (nitroxyl radical) is trapping of radical build within the polym
matrix begin. As the reaction proceeds further with a peroxide radical the nitroxyl
radical is re-generated.

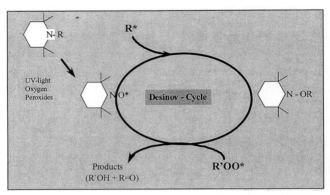

Figure 2: The 'Desinov Cycle'.

The substitution of the nitrogen atom in piperidine derivatives play an important role in HAI
efficacy. Experiments carried out in liquid clear coatings [10-11] have shown the effect of t
nitrogen substitution on the >NO* formation. These experiments have demonstrated that peroxi
radicals were formed right at the start of weathering. Their concentration start to decrease when
sufficient number of nitroxyl radicals is present in the coating. The faster the conversion of >N-R
<NO* the earlier the protection against free radicals start.

Unlike UV absorber, the efficiency of HALS does not depend on coating thickness and radic
scavenging take place throughout the film.

2. Experimental

All powder clear coats were mixed in a large kitchen cutter and extruded twice using a Prism
mm twin screw extruder, at 80-110°C barrel temperature.

Table 2: Basic formulations for GMA powder clear coat

Components	GMA 1	GMA 2
Synthacryl VSC 1436 / Additol VXL 1381	96.2	--
Almatex PD 7610 / Dodecanoid acid		94.5
Additol XL 490	0.5	1.5
Benzoin	0.3	--
Worlee Add 902	--	1
UVA	2	2
HALS	1	1
Sum	100	100
Curing	30' 140°C	30' 160°C

The materials were milled in a Retsch ultracentrifugation mill and sieved through a 125μm sieve.

The clear coats were sprayed onto panels pre-coated with silver metallic aqueous basecoat (30' 30°C) and cured in an electric oven (see table 2, 3). The dry film thickness after cure was approx. 60 mμ.

Table 3: Basic formulations for TGIC powder clear coat

Components	PES 1 / TGIC	PES 2 / TGIC
Uralac P 5000 / Araldit PT 810 (TGIC)	96.5	_
Crylcoat 2988 / Araldit PT 810 (TGIC)	_	96.5
Resiflow PV 88	1	1
Benzoin	0.5	0.5
UVA	1	1
HALS	1	1
Sum	100	100
Curing	15' 195°C	15' 195°C

In this work four formulations were used for evaluation of the different type of UVA and HALS. Detailed formulations are provided in table 2 and 3. GMA 1 is a system used for general industrial paints.

GMA 2 is especially suited for high performance automotive coatings. For TGIC system used in the architectural and general industry two types of polyester were chosen: a standard polyester as PES 1 and a super durable polyester as PES 2.

Light stabilizers tested were: hydroxyphenyl triazine (UVA 1), benzotriazole (UVA 2), benzophenone (UVA 3), oxalanilide (UVA 4), HALS 1 and HALS 2. The chemical formulae are shown in figure 3 and 4. All these UVA's and HALS are solid and have at 200°C less than 1% loss

of weight, this was determined by TGA (30°C – 300°C, heat rate = 10°C / min). GMA 1 wa
prepared with all UVA's in combination with HALS 1. GMA 2 was stabilized with UVA 1 an
UVA 2 in combination with HALS 2. Both TGIC powder clear coats were tested with UVA 1 ar
UVA 2 in combination with both HALS .

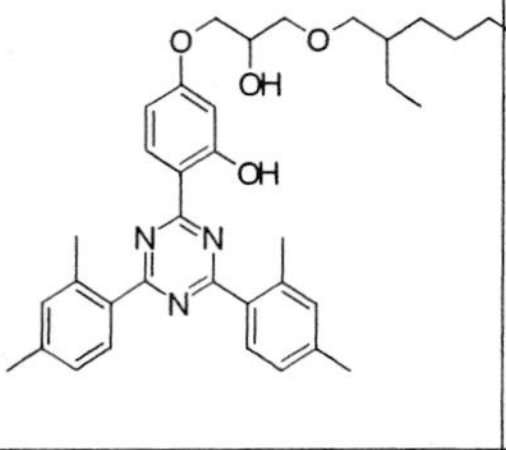 UVA 1	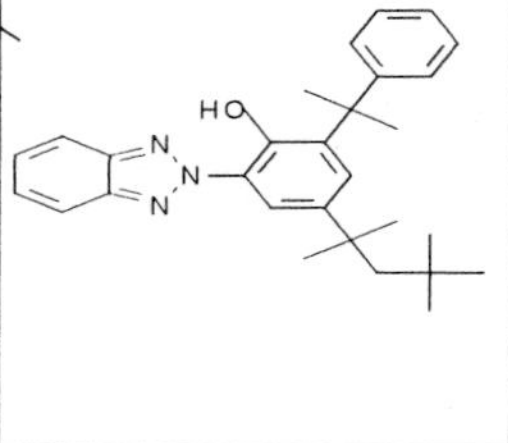 UVA 2	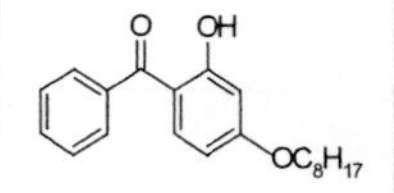 UVA 3	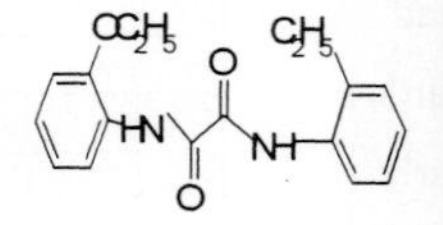 UVA 4
m. pt – 74°C	m. pt – 113°C	m. pt – 48°C	m. pt – 124°C

Figure 3: chemical structure of the different UVA used in this study.

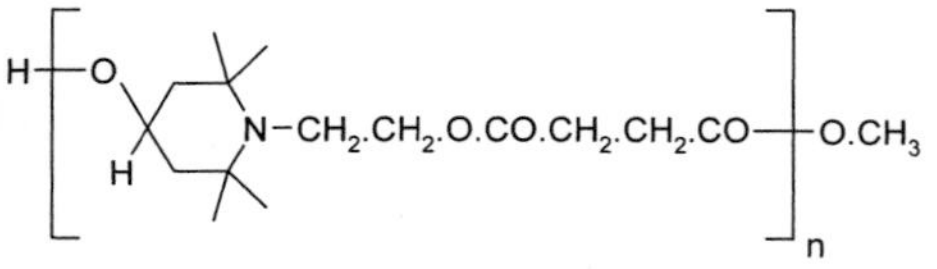 HALS 1	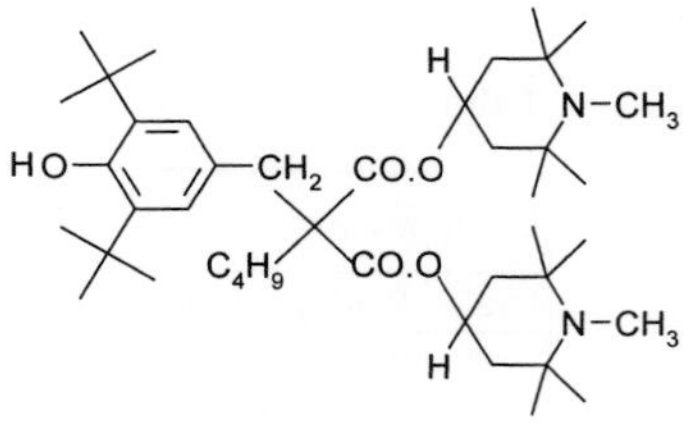 HALS 2
m. pt – 74°C	m. pt – 146°C

Figure 4: chemical structure of the different HALS used in this study.

Bake-over resistance was assessed by measurement of the Yellowness Index (YI), the higher the Y
measured is the higher the risk of yellowing of the coating during bake-over.

All powder clear coats were exposed to accelerated and natural weathering. The outdoor exposur
was carried out in South Florida (5° / black box, unheated). Accelerated weathering was carried ou
by UVCON (cycle: 8h light at 70°C; 4 h condensation at 50°C; 0,67 W / m² QUV-B radiation a
313 nm). Criteria used to assess coatings performance were: 20° gloss according to DIN 6753(
Yellowness Index (YI) according to ASTM D 1925.

Results

.1 GMA powder clear coats

olyacrylate powder coatings based on GMA combined with carboxylic or anhydride hardeners ₂ad to powder coatings with excellent flow and transparency. GMA 1 cured with an anhydrid ₂rdener is a common powder coating system used for industrial applications. GMA 2 cured with a ₂rboxyl hardener is a standard system targeted to the automotive industry.

₂ifferent types of UVA

ightstabilization with commercial UV absorbers like benzotriazole, benzophenone, oxalanilide ₂VA 2; 3; 4) and a newly commercialized hydroxyphenyl triazine (UVA 1) in combination with ₂ALS 1 were tested in a powder clear coat GMA 1. Natural and accelerated weathering results are ₂own in figure 5-6. In both exposure methods the hydroxyphenyl triazine UV-absorber gave the ₂st performance in term of gloss retention.

₂erformance gap between the different UVA classes is broader when it comes to colour retention ₂igure 7). Benzotriazole and hydroxyphenyl triazine show in this test clearly better performance

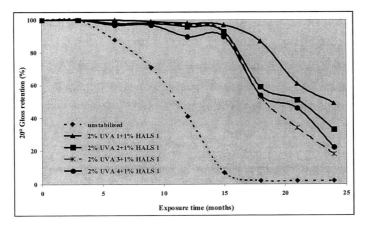

Figure 5: Florida exposure of GMA 1 in combination with different types of UVA.

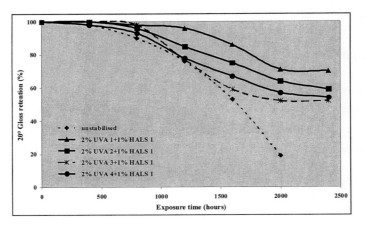

Figure 6: Accelerated weathering (UVCON) of stabilized and unstabilized GMA 1.

Performance gap between the different UVA classes is broader when it comes to colour retenti~~~ (figure 7). Benzotriazole and hydroxyphenyl triazine show in this test clearly better performan~~ compared to oxanilide and a benzophenone type. Limited differentiation in term of gloss retenti~~ and important gaps observed in term of color deviation confirm the contribution of HALS towa~~ stabilization of coating surface whereas the variable part (UVA) has a major impact on col~ stabilization.

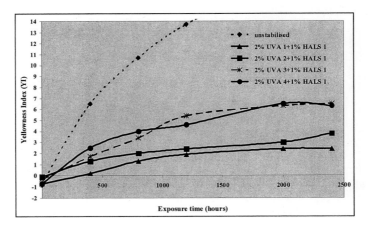

Figure 7: Color retention (UVCON) of stabilized and non-stabilized GMA 1.

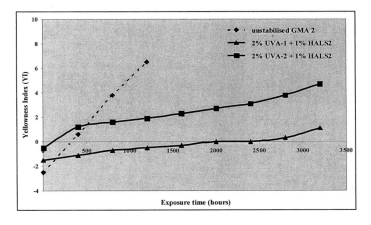

Figure 8: Color retention (UVCON) of stabilized and non-stabilized GMA 2.

1 GMA 2 HALS 1 could not be incorporated due to incompatibility with the binder. The lowest influence on initial color and the minimum YI development was achieved through the use of UVA 1 (figure 8). In this case the strong extinction of UVA 1 in UVB and slightly inferior absorbance in long UVA wavelength compared to UVA 2 is an advantage. A consequence is a lower initial color versus UVA 2 and better efficacy to filter the most damaging part of UV spectrum. GMA 1 and 2 without additives per-form both at relatively low level, worthwhile to mention that improvements reached by addition of light stabilizers may vary considerably from one binder system to the other s illustrated in figure 9.

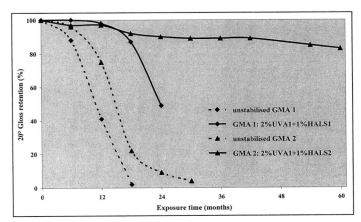

Figure 9: Comparison between GMA 1 and GMA 2 after Florida exposure.

Bake-over resistance

Figure 10 shows a comparison of the Yellowness Index between normal cured and baked-ove
powder clear coats of GMA 2 applied over silver metallic aqueous basecoat. In this system th
contribution of UVA 1 to the initial YI upon normal cure is minimum compared to UVA 2 .

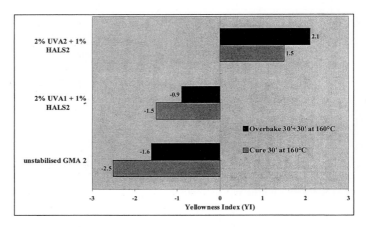

Figure 10: Influence of light stabilizer on thermal stability of GMA 2
powder clear coat upon silver metallic basecoat.

It is a definitive advantage for this high performance coating when applied over light colore
basecoat. This is due to the fact that the triazine UVA 1 is slightly blue shifted in comparison t
benzotriazole UVA 2. The color deviation in case of bake-over is reduced as HALS 2 contain
phenolic anti-oxidant moiety and prevents yellowing by auto-oxidation.

3.2 TGIC powder clear coats

TGIC powder coatings are still popular in architectural and general industrial paints used outdoo
In Europe these systems are labeled with 'T' (toxic) since 1998. There are several alternatives tha
can possibly replace TGIC [12]. TGIC and alternatives react with standard polyester resins (PES 1
based on terephthalic acid. Further development of these polyesters led to super durable polyeste
(PES 2) based on Isophthalic acid with better durability but weaker mechanical properties [13].

Different types of HALS and UVA

Table 4: Florida exposure of PES 1 in combination with different type of HALS and UVA

Stabilizer	20° Gloss after ... Months		
	0	12	21
unstabilised PES 1	99	46	
1% UVA 1	100	75	15
1% UVA 1 + 1% HALS 1	100	79	22
1% UVA 1 + 1% HALS 2	100	79	20
1% UVA 2	100	65	5
1% UVA 2 + 1% HALS 1	98	71	11
1% UVA 2 + 1% HALS 2	100	69	8

In both polyester systems PES 1 and PES 2 the two types of HALS were combined with UVA 1 and UVA 2 to assess the effect of different stabilizers on weathering.

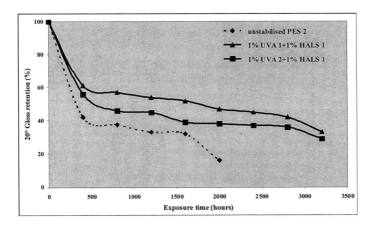

Figure 11: Accelerated weathering (UVCON) of stabilized and non-stabilized.

With the standard polyester system (PES 1) limited differentiation was seen in accelerated weathering between the different systems. Outdoor exposure data begins to show the effects of stabilization as demonstrated in table 4. UVA 1 show with and without combination of HALS better performance than UVA 2 (see table 4). However it is clear that limited improvement can be achieved through the use of additives in standard PES resins.

In the case of super durable PES resins all performances are clearly improved and UVA 1 outperform UVA 2 again (figure 11-12).

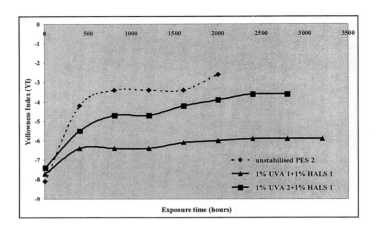

Figure 12: Color retention (UVCON) of stabilized and non-stabilized PES 2.

In addition the comparison between stabilized and un-stabilized coatings is particularly advantageous by the inclusion of additives (figure 13).

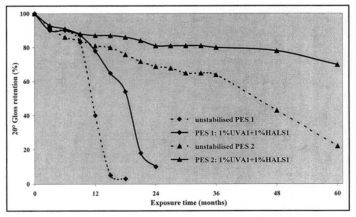

Figure 13: Comparison between PES 1 and PES 2 after Florida exposure.

4. Conclusion

The efficiency of light stabilizers in acrylates and super durable PES clear coat is confirmed in both artificial and outdoor exposure tests.

With the UV-absorber types the hydroxyphenyl triazine (UVA1) exhibits overall the best performance in term of gloss and color retention. Because of its ease of incorporation (low melting point), its slight contribution to the initial color and its great color retention performance this UV Absorber is dedicated to high quality powder coatings.

Additives have a valuable contribution provided that they are used in high performance binders!

References:

[1]   JOT 9/2001, p. 24
[2]   T. Türk , JOT 9/2001, p.16
[3]   W. Dohnt, JOT 9/2001, p. 28
[4]   S. Harris, PPCJ-November 2001, p. 37
[5]   R. Lüscher, JOT 2/2002, p. 22
[6]   A. Valet, F. Sitek, G. Berner, Farbe+Lack, 94 (1988), p. 734
[7]   A.Valet, Farbe+Lack, 96 (1990), p. 189
[8]   M. Holt, D. Bramer, Congressbook 13th Int. Conference on
        "Advance in the Stabilization and Degradation of Polymers" (1991), p. 23
[9]   A. Valet , "Light Stabilizers for Paints" (1997), p. 21
[10]   R. Bauer, J. Dean, L. Gerlock, Ind. Eng. Chem. Res. 27, 1988, p. 65
[11]   L. Gerlock et. al. Congressbook 11th Int. Conference on
        "Advance in the Stabilization and Degradation of Polymers" (1989), p. 25
[12]   M. Herszenhaut, Farbe+Lack, 107 (5/2001), p. 58
[13]   Y. Merck, „Super-durable polyesters: why?" In: Powder coatings, What's next?, Birmingham, 1999

Hydroplasticization Effect in Structured Latex Particles Film Formation

Jaromir Snuparek*, Bohuslav Kadrnka, Pavel Ritz

Institute of Polymeric Materials, University of Pardubice, 532 10 Pardubice, Czech Republic

and Otakar Quadrat

Institute of Macromolecular Chemistry CAS, 162 06 Prague, Czech Republic

Summary: Series of emulsion copolymers with structured particles were synthesized comprising copolymerized acrylic or methacrylic acid in the outer layer. All samples were based on particles, containing identical cores slightly crosslinked by allyl methacrylate and variable shells, weight ratio core/shell being constant 1/1. Each sample contained 10 wt% HEMA in the shell to achieve the film crosslinkability. In both series samples with different hardness and polarity (varible styrene/butyl acrylate ratio) of the shell layer were prepared. It was shown that the extent of particle swelling and hydroplasticization depends not only on the content of dissociated carboxylic groups, but also on the composition and crosslinking of the rest of polymer chain i.e. on its polarity and rigidity and on the origin of carboxylic groups. The effect of dissociated carboxylic groups on lowering the minimum film forming temperature was much more pronounced if the polymer chains were more polar.

Introduction

Polymer colloids in aqueous media form a very important group of polymers, which find a whole series of technical applications. In „Low VOC" or „Zero VOC" water-borne coating compositions, the aqueous copolymer dispersions operate as a binder component. A typical group of latex binders comprises products with a core-shell structure, which have different compositions of their cores and shells. The problems connected with utilization the aqueous polymer dispersions in coating compositions comprises the questions connected with film-forming properties of originally discrete polymer particles and with their rheological properties as well. Both these aspects are to be studied paralelly, especially in such systems, which have a hydrophilic particle surface layer and belong to the group of the so-called „alkali-swellable" polymer particles. Thanks to the hydrophilic nature of the surface layer and the dissociation of surface carboxylic groups, they can be, to a certain extent, plasticized with the surrounding water phase and can form films of sufficient quality even without any

© WILEY-VCH Verlag GmbH, 69469 Weinheim, 2002   CCC 1022-1360/00/$ 17.50+.50/0 addition of organic coalescence aids. The rheological behavior, due to a change in the hydrodynamic particle volumes after the alkalization, is usually of strongly pseudoplastic character. The important aspect connected with the utilization of structured polymer particles in the coating system is the ability of discrete polymer particles to bind the pigments. The film formation and, consequently, the formulation principles have to be investigated together with alkali-swelling behavior of particles and together with the effect of particle swelling onto latex rheology. This study was focused on the effect of hydrophobic styrene concentration in the shell layer copolymer onto particle swelling and hydroplasticization.

Experimental

Materials: Monomers: Styrene (S) - technical grade, Kaučuk Kralupy, butyl acrylate (BA) - technical grade, acrylic acid (AA) - technical grade, Eastman, Sokolov, 2 - hydroxyethyl methacrylate (HEMA), N,N'-methylene*bis*acrylamide (MBA), allyl methacrylate (AMA), methacrylic acid (MAA) - technical grade, Roehm, Darmstadt, Disponil AES 60 - sodium salt of alkyl aryl polyoxyethylene sulfate, Henkel, ammonium persulfate - Air Products. Synthesized latexes were neutralized using 2-amino 2-methyl propanol (AMP-95), ANGUS.

Polymerization recipe: Latexes were produced in a 2500 ml glass reactor by semi-continuous non-seeded emulsion polymerization under nitrogen atmosphere at temperature 80 °C. The reactor charge was put into the reactor and heated to the polymerization temperature. Then the monomer emulsion was fed into the stirred reactor at feeding rate about 10 ml/min. in three steps (1. seed preparation, 2. core preparation, 3. shell preparation). After that, during 2 hours of hold period the polymerization was completed.Two series of emulsion copolymers with structured particles were synthesized, the former comprising methacrylic acid, the latter acrylic acid in the outer layer. All samples were based on particles, containing identical cores slightly crosslinked by allyl methacrylate and variable shells, weight ratio core/shell being constant 1/1. Each sample contained 10 wt% HEMA in the shell to achieve film crosslinkability. In both series samples with different hardness and polarity (varible styrene/butyl acrylate ratio) of the shell layer were prepared.

Determination of the Minimum Film Forming Temperature (MFT): The minimum film forming temperature was measured using the MFT Tester (Synpo a.s. Pardubice). Minimum film temperature is defined to be the minimum temperature at which a film cast from dispersion becomes continuous and clear. The MFT Tester consists of a chromium coated copper slab in which a fixed temperature gradient is maintained by heating at one end (about 40 °C) and cooling at the other (about 0 °C). The polymeric emulsion was cast in a strip along this slab by a coating ruler with height of about 150 μm. The point at which the film becomes discontinuous when dry was observed, and this temperature was recorded as MFT. The slab is located in a box covered with a glass cover, in order to permit visual observation of the film during drying. Tests were carried out after a thermal equilibrium has been reached. Drying material was silica activated for 1 hour / 130 °C before the measurement.

Other analytical and testing methods used for characterization of latices and polymeric films: particle size - Coulter N4 Plus (Coulter Corp.), pH was measured using WTW 320 Ph-Meter, WTW G.m.b.H. Weilheim, Germany.

Results and Discussion

Swelling of the particle surface under alkaline conditions is important in the process of film formation form polymer dispersions. The alkali swollen latex particles are plasticized by the water phase this increases their viscosity and also enhances the ability of particles to undergo the process of coalescence. The extent of the particle hydroplastificization depends on the concentration of carboxylic groups an it is only effective at alkaline conditions. The effect of carboxylic groups content in styrene/butyl acrylate copolymer latex on changes in MFT at acidic and alkaline conditions is shown in Fig. 1.[1] The relationships document the increase in MFT with increasing content of acrylic acid due to increased Tg of copolymer. The particles need higher temperature to undergo the deformation necessary for a film formation *via* coalescence of particles. The sharp drop in MFT of samples with higher content of acrylic acid indicates their effective hydroplasticization under alkaline conditions. It is evident from the Fig. 1 that the minimum film formig temperature of the latex particles increased with incresing acrylic acid content as a consequence of increased T_g of the copolymer even at alkaline conditions. The MFT values of neutralized latexes, however, were lower than those of acidic ones. We investigated particles with shell layers based on butyl acrylate and styrene in different ratios, different concentrations of acrylic or methacrylic acid in the shell copolymer and different extent of the shell crosslinking by methylene-*bis*-acrylamide.[2, 3] All structured polymer particles contained a soft core based on butyl acrylate/styrene (wt. ratio

BA/S = 1.44) slightly crosslinked by allyl methacrylate and with a hard shell comprising 10 wt. % HEMA. The extent of particles hydroplasticization at alkaline conditions increased with

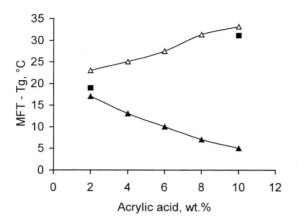

Figure 1. Effect of acrylic acid content in butyl acrylate/styrene/acrylic acid copolymer on T_g (■) and MFT in dependence on pH value. pH = 3 (▲), pH = 7,5 (△), wt. ratio butyl acrylate/styrene = 0.85 remained constant in all copolymers.[1]

lower concentration of styrene in the shell. The results were similar for samples with acrylic acid as well as methacrylic acid in the shell layer copolymer.

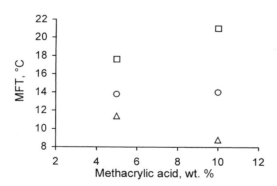

Figure 2. Effect of styrene and acrylic acid content in the BA/S/HEMA/AA/MBA shell layer copolymer on the minimum film forming temperature of neutralized structured core shell latexes. Shell composition BA/S/HEMA/MAA/MBA : HEMA = 10 wt. % , MBA = 1 wt. %, Styrene (△) = 35 wt. % , (○) = 40 wt. %, (□) = 45 wt. %.

As mentioned above, the extent of hydroplasticization depends not only on the content of dissocited carboxylic groups, but also on the composition of polymer chain and its polarity and rigidity. Thus, the hydroplasticization effect was not as high if the composition of butyl acrylate/styrene copolymer chain changed to higher content of styrene. This effect is illustrated in Figures 2 and 3. Here, the latexes comprising higher concentration of styrene in the shell layer of structured particles do not exhibit any drop in MFT as in the case of particles

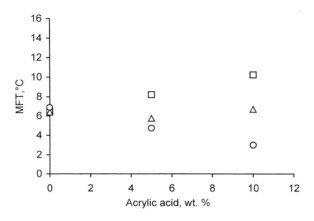

Figure 3. Effect of styrene and acrylic acid content in the BA/S/HEMA/AA/MBA shell layer copolymer on the minimum film forming temperature of neutralized structured core shell latexes. Shell composition BA/S/HEMA/AA/ : HEMA = 10 wt. % , Styrene (□) = 40 wt. %, (△) = 35 wt. % , (〇) = 30 wt. %.

It is also evident from Fig. 4 that crosslinking of the shell layer with methylene-*bis*-acrylamide was accompanied by the increase of MFT of neutralized samples to values close to those of acidic non-croslinked particles as the crosslinking impedes swelling and, consequently, the hydroplasticization of the particle surface. The effect of dissociated carboxylic groups on lowering the minimum film forming temperature was much more pronounced if the polymer chains were more polar and softer. Comparison of minimum film forming temperatures of structured polymer particles with variable content of styrene in the shells that were crosslinked with 1 % wt. methylene-*bis*-acrylamide was shown in Fig. 4.

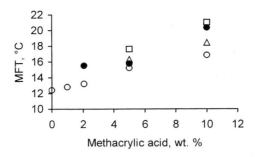

Figure 4: Effect of methacrylic acid and methylene-bis-acrylamide crosslinker content in the shell layer of structured core-shell particles on the minimum film forming temperature of neuralized (open points) and acidic (full points) latexes. Shell composition BA/S/HEMA/MAA/MBA : HEMA = 10 wt. % , S = 45 wt. %, MBA ($\bullet$O) = 0 wt%, ($\triangle$) = 0.5 wt. %, ($\square$) = 1 wt. %

Effect of surface layer hydroplasticization on the latex viscosity.

Surface carboxylic groups usually cause an increase in viscosity after neutralization of particles due to particle swelling and also due to the elecroviscous effect.[3] Capillary viscometry indicated that the dependences of viscosity of different alkalinized latexes on their particle concentration considerably differ.[4] It could be assumed that during alkalinization of originally acid latexes, due to electrostatic interactions of ionized carboxylic groups, the effective hydrodynamic volume of latex particles (intrinsic viscosity [η]) increased, which manifested in an increase of the latex viscosity. A significant difference was found in the effect of copolymerized carboxylic groups from acrylic acid and of those from methacrylic acid as a result of different monomeric acids polarity and, consequently, different concentration of surface and burried carboxylic groups. The latex viscosity and the intrinsic viscosity of diluted latexes indicates the extent of particle swelling, i.e. the extent of particle hydroplasticization. The difference between the effects of carboxylic groups from acrylic acid and those from methacrylic acid copolymerized in the shell layer of structured particles is shown in Figures 5 and 6.

The hydrodynamic volume of latex particles was characterized by their intrinsic viscosities [η] obtained by linear extrapolation of $\Phi/\ln \eta_r$ to zero volume fraction Φ of the latexes according to the Mooney equation[5] η_r = exp $\{[\eta]\Phi/(1-\Phi/\Phi_c)\}$, where Φ_c is the volume fraction of particles at which viscosity reaches an infinite value (the volume fraction at maximum packing). The relative viscosity $\eta_r = \eta/\eta_s$, where η is the viscosity of latexes and η_s is that of the dispersion medium, was measured using an Ostwald capillary viscometer.

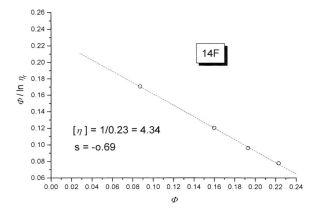

Figure 5 : Intrinsic viscosity of structured latex particles at pH = 8. Composition core//shell(wt. %): S/BA/MAA/AMA//S/BA/HEMA/MAA = 40/57.5/1.5/1//40/40/10/10. Mean particle diameter of acid particles 0.1 μm

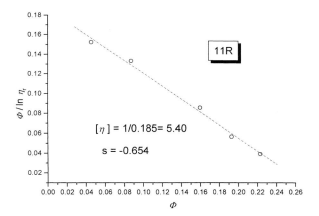

Figure 6 : Intrinsic viscosity of structured latex particles at pH = 8. Composition core//shell(wt. %): S/BA/AA/AMA//S/BA/HEMA/AA = 40/57.5/1.5/1//40/40/10/10. Mean particle diameter of acid particles 0.12 μm.

Similarly as in the case of minimum film forming temperature, the effect of styrene content in the shell layer on the particle surface hydroplasticization is also evident from viscosity measurements at higher solids content. Flow behavior of AMP-95 neutralized 40%solids latexes is shown in Figures 7 and 8. An increase in low-shear viscosity was found for latexes

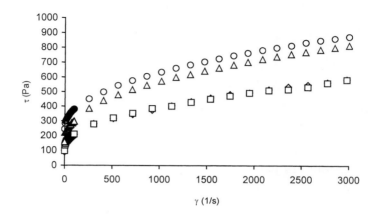

Figure 7: Effect of styrene content in the BA/S/HEMA/AA/MBA shell layer copolymer on the shear rate (γ) vs. shear stress (τ) relationship for neutralized latexes at 40 % solids. Shell composition BA/S/HEMA/AA/ : HEMA = 10 wt. %, AA = 10 wt. % and S = 45 wt. % ($\Diamond$), S = 40 wt. % ($\square$), S = 35 wt. % ($\triangle$) and S = 30 wt. % ($\bigcirc$).

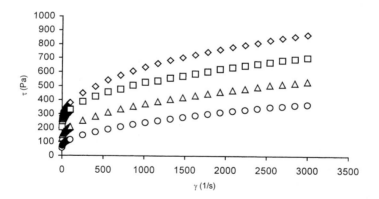

Figure 8: : Effect of MBA content in the BA/S/HEMA/AA/MBA shell layer copolymer on the shear rate (γ) vs. shear stress (τ) relationship for neutralized latexes at 40 % solids. Shell composition BA/S/HEMA/AA/ : HEMA = 10 wt. %, AA = 10 wt. %, S = 30 wt. %, MBA = 0 wt. % ($\Diamond$), MBA = 0.5 wt. %. ($\square$), MBA = 1.0 wt. % ($\triangle$) and MBA = 1.5 wt. % ($\bigcirc$).

containig 35 and 30 wt. % styrene in the shell layer. In the cases of styrene concentration 40 and 45 wt. % the the low-shear viscosity did not differe and were lower due to not so extensive swelling of polymer particles. The effect of swollen particles on the latex viscisity depends on the solids content, i.e. on the mutual particle distance. The higher solids, the higher effect of particle swelling on the rheology. Similarly, hindering of particle swelling by the particle surface crossllinking by MBA resulted in decrease in the low-shear latex viscosity at 40 wt. % solids as it shown in Fig. 8. [6]

Conclusion

The extent of hydroplasticization depends not only on the content of dissociated carboxylic groups, but also on the composition of polymer chain i.e. on its polarity and rigidity and on the origin of carboxylic groups. The effect of dissociated carboxylic groups on lowering the minimum film forming temperature was much more pronounced if the polymer chains were more polar. As the hydroplasticization is based on the particle surface swelling by water it also is accompanied by an increased latex viscosity. In latexes utilized as binders in water borne paints the functional groups, as they are e.g. carboxyl and hydroxyl ones, serve mainly as *loci* for crosslinking. Besides this, polar functional groups affect the most important application properties of water borne paint vehicles i.e. the film formation of the binder and the flow behavior of the paint.

Acknowledgement

The Ministry of Education of the Czech Republic (MSM 2531 00001), Grant Agency of the Czech Republic (No. 104/02/1360) and DuPont de Nemours are gratefully acknowledged for supporting this work.

References

[1]   D. Kubík, Diploma Thesis, University of Pardubice, 1995.
[2]   M. Faltejsková, Diploma Thesis, University of Pardubice, 2001.
[3]   Quadrat O.,Šňupárek J.: Progr. Org. Coat. **1990**, *18*, 207 .
[4]   Šňupárek J., Quadrat O., Horský J., Kaška M.: The Effect of Hydrophilic Non-ionogenic Comonomers on Flow Properties of Carboxylated Latexs *in* Polymer Colloids: Science and Technology of Latex Systems (E. S. Daniels, E. D. Sudol and M. S. El-Aasser, Eds.), ACS Symposium Series 801, American Chemical Society / Oxford University Press, Washington, D.C., Chapter 6, 71 – 79, 2001.
[5]   M. J. Mooney, *J. Colloid Sci.* **1951**, *6*, 162.
[6]   P. Ritz, Diploma Thesis, University of Pardubice, 2002.

Behaviour of Surface-Treated Mica and Other Pigments with Lamellar Particles in Anticorrosive Coatings

Andréa Kalendová[a]*, Petra Tamchynová[a] , Václav Štengl[b], Jan Šubrt[b]

(a) University of Pardubice, Faculty of Chemical Technology, Institute of Polymeric Materials, nám. Cs. Legií 565, 532 10 Pardubice, Czech Republic, (b) Institute of Inorganic Chemistry, Academy of Sciences of Czech Republic, Řež, Czech Republic

Summary: The paper deals with using lamellar pigments for anticorrosive barrier coatings. By depositing a ferric oxide layer on a muscovite particle a pigment is obtained, which being applied to coatings improves the mechanical properties thereof, resistance to UV radiation and acts as an anticorrosion barrier. The optimum concentration of lamellar surface-treated muscovite in the coatings amounts to 20 vol. %.

Introduction

Nonisometric lamellar pigments are used in anticorrosive coatings for a series of years. The most widely used lamellar pigment for this purpose is ferric mica.[1] From the chemical point of view it is a ferric oxide in crystalline lamellar structure (specularite). For the designation of natural-origin lamellar pigment the name „micaceous iron oxide" *(MIO)* has become popular in time. Specularite (also the so called iron mica) modified to the pigment form is characterized by a typical metal-gray color of sparking appearance. Practical experience and published papers concerning the application of *MIO* pigments to coatings destined to metal protection show the outstanding results.[2] The anticorrosive coatings pigmented with a MIO pigment show excellent barrier properties - they hinder the permeation of corrosive substances and water through the film, increase the adhesion of coating to the substrate, and the particles protect also binder to ultravioleght (UV) radiation. The *MIO* pigments are of natural origin and the deposits of specularite will be exhausted in the future and are also not rather broadened. In the first stage a path of preparing synthetic specularite (synthetic MIO pigment) was selected.[3] The synthetic *MIO* pigment preparation was performed at high temperatures and pressures in an autoclave. The pigment obtained is characterized by rather regular lamellar particles, color and chemical composition identical with the *MIO* pigment of natural origin.[4] When a synthetic *MIO* pigment is used, the anticorrosive barrier properties reach high values in the coatings. The economic evaluation is not favorable for the synthetic preparation of *MIO* pigment. With respect to this reason the synthetic pigment cannot compete with a cheeper starting material of natural origin. On the basis of preceding papers published in this research region a path of using mica was selected,[5] which has also the lamellar structure. Mica is chemically an aluminosilicate, occurring in large amounts in natural deposits and is more easily available than specularite. From a broad pallet of aluminosilicate minerals the potassium aluminosilicate $K_2O.3Al_2O_3.6SiO_2.2H_2O$, designated as muscovite is an especially appropriate compound for coatings. Muscovite has compared to *MIO* pigments an advantage consisting in a lower specific density (2.9 $g.cm^{-3}$) and thus a lower tendency to sedimentation in a liquid medium. Papers published already in the past indicate a lower anticorrosion barrier efficiency in the coatings in comparison to *MIO* pigments.[6] A contribution to solving this disproportion between two types of lamellar pigments consists in forming a ferric oxide layer on a muscovite particle. The chip of muscovite is a carrier and offers to a particle a lamellar shape. On the particle surface a thin ferric oxide layer (hematite) is precipitated, which nears a surface treated muscovite particle also from the properties point of view to a specularite particle - a *MIO* pigment.

Experimental part

Preparation of the surface-treated muscovite (Fe-muscovite)

The preparation of surface-treated muscovite (the so called surface-treated Fe-muscovite) was performed by a controlled hydrolysis with urea, at which a mixture of iron oxides-hydroxides is precipitated at the lamellar particle surface.[7] By annealing a strong chemical bond (Fe_2O_3-muscovite) is formed at the muscovite particle surfaces. The reaction mixture is heated to boiling and the change in pH value is continuously followed. The synthesis is completed after reaching pH = 8, which corresponds to a reaction time of 8 hours. The completion of reaction is indicated by ammonia developing in the reactor. The mixture is further kept in the reactor in moving at switched off heating. Then decantation, filtration, and drying at a temperature of 110 °C follow. The dry pigment was annealed at a temperature of 600 °C. Figure 1 shows the scanning electrone micrographs (SEM) of the sections of surface-treated Fe-muscovite. Tables 1-2 give physicochemical properties of pigments tested.

Preparation of coatings

To enable determining the most appropriate concentration of surface-treated muscovite in coatings the samples on a binder base of epoxyester resin were prepared. A concentration series of 5 to 30 vol. % pigments in the binder was prepared. The pigment component was replenished by titanium dioxide to an overall pigment volume concentration (PVC) of 60%. The formulation of model coating contains in addition to it bentonite to improve the rheological properties, and organic and inorganic corrosion inhibitors.

Cross section through the surface of particle (1.000x, Philips XL 30CP)

Morphology of the particle in detail (15.000x, Jeol 5600 LV)

Figure 1. Cross section and morphology of surface-treated Fe-muscovite.

Table 1. Characterization of lamellar pigments used.

Lamellar pigment	Composition	Density g/cm^3	pH[a]	Specific surface[b] m^2/g	Oil absorption[c] g/100g
Muscovite (white color)	Potassium aluminosilicate	2.92	9.51	5.83	28.0
Fe-muscovite (red color)	Potassium aluminosilicate Fe$_2$O$_3$ (5%)	3.31	7.23	34.28	30.0

[a] Determination of the pH value of water extracts for pigments (DIN ISO 787/9)
[b] Specific surface calculated by BET isotherm (*ASAP 2000*)
[c] Determination of oil absorption for pigments (DIN ISO 787/2)

Table 2. Characterization of lamellar pigments used.

Lamellar pigment	CPVC[a] (linseed oil)	Solubility in water[b]		Particle size distribution[c]		
		at 23 °C	at 100 °C	90%	50%	10%
		%	%	μm	μm	μm
Muscovite	53.04	0.291	0.314	22.51	11.85	2.46
Fe-Muscovite	47.95	0.185	0.266	26.36	11.56	1.57

[a] critical pigment volume concentration calculated by linseed oil consumption
[b] Determination of water soluble matter for pigments (CSN EN ISO 787/3)
[c] by laser beam difraction (*Coulter LS 100*)

Figure 2 shows morphology of lamellar pigments used (Jeol, JSM 5 600 LV).

Muscovite (4.500x)   Fe – muscovite (4.500x)

Figure 2. Morphology of lamellar pigments used.

Results

Effects of lamellar pigment concentrations on the coating properties

With the prepared coatings the development of coating hardnesses in dependence on the pigment concentration was followed. The measurements were performed by means of a pendulum instrument of the Perzos type (Pendulum damping test, ISO 1522). Figure 3 shows the dependence of coating hardnesses on the lamellar muscovite pigment concentrations (PVC). The dependences given in Figure 3 indicate that the coating hardnesses raise with the increasing concentration of lamellar muscovite in a range of 0 - 30 vol. %. The difference between the surface-treated muscovite and nontreated muscovite is minimal. Both the dependences show the same trend, and the difference can be caused by more porous structure of surface-treated muscovite. The porous structure appearing at the surface of a Fe_2O_3-mica particle consumes probably a higher amount of binder. The results indicate that the differences will appear at a concentration round PVC=20%. Practically the differences in coatings tested between both lamellar pigment types can be seen in the SEM photos (Fig. 4.).

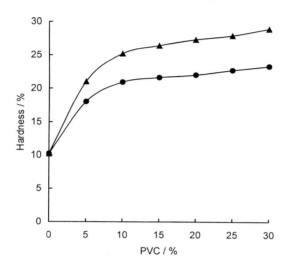

Figure 3. Dependence of the hardnesses of coatings pigmented with lamellar pigments on the pigment volume concentrations; ♦ = muscovite, ▲ = Fe-muscovite.

The lamellar mica particles positively affect the coating property, which is the coating adhesion to the substrate metal. Also affecting the cohesion of the coating alone is positive. Testing the coatings with all concentrations of both lamellar pigments was always connected with 100% cohesion fracture in the coating film.[8] Figure 5 quantifies the cohesion coating strengths in dependence on lamellar pigment amounts. As shown clearly by the dependence represented in Figure 5, the surface muscovite treatment by ferric oxide exhibits a positive effect on the cohesion in the coating. The surface treatment of lamellar particles contributes at a concentration of 20 vol. % in the coating with 0.5 MPa. At a pigment volume concentration of 30% the contribution to the cohesion strength equals 0.6 MPa. For the increase of film strength the Fe_2O_3 structure at the muscovite surface is responsible.

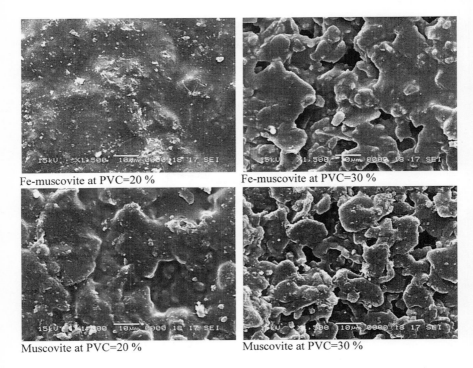

Fe-muscovite at PVC=20 %   Fe-muscovite at PVC=30 %

Muscovite at PVC=20 %   Muscovite at PVC=30 %

Figure 4. Structure of epoxyester coatings pigmented with lamellar pigments tested at a content of 20 and 30 vol. % (1.500x).

The corrosion testing results obtained with epoxyester coatings pigmented with surface-treated and untreated muscovite show the outstanding improvement of coating resistances to osmotic blistering. The appearance of blisters in the coating film is a reason of reduced adhesion of the film to substrate and local corrosion under the blister arch. The evaluation of blister sizes and frequencies was performed following the corrosion exposure by means of the ASTM D 714-87 Standard, and the obtained results were transformed to a numeric expression in a scale (0-100). The quantification method for evaluating the corrosion testing results was described already in the paper.[9,10]

Figure 6 brings the results of coating corrosion resistances after a 500 hour exposure to a salt spray chamber (Resistance to neutral salt spray, ISO 7253). The evaluation is directed to the appearance of osmotic blisters for the coatings with a fluctuating concentration of lamellar muscovite particles.

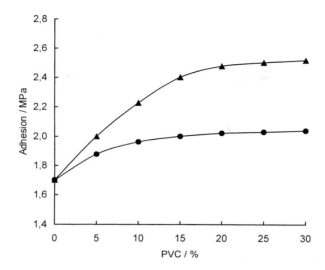

Figure 5. The cohesion component of the coating adhesion in dependence on the lamellar pigment concentrations; ◆ = muscovite, ▲ = Fe-muscovite.

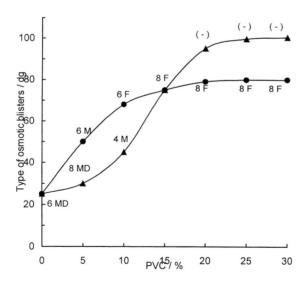

Figure 6. Manifestations of osmotic blisters in dependence on the lamellar pigment concentrations after a 500 hour exposure to a salt spray chamber (dg. 100 - without any blisters, dg. 0 - blisters of the type 2D); ◆ = muscovite, ▲ = Fe-muscovite.

As indicated in Figure 6, reproducing the dependence of osmotic blister appearances on the muscovite and Fe-muscovite concentrations, at a PVC value of 20% no changes are observed with both pigments anymore. The surface-treated Fe-muscovite in the coating at a PVC value of 20% hinders totally the appearance of blisters. At the same concentration of non-treated muscovite in the coating only a state can be reached, when the blisters of size number eight and frequency few (8F) type appear. With both the lamellar pigment types in the coating in a PVC range of 20 - 30% the blister type does not suffer a change. For muscovite without any surface treatment the blister type 8F is considered, with the surface-treated Fe-muscovite no blisters appeared at all. The overall anticorrosion efficiency of the coatings containing particles of both muscovite types is, as it was already described, considerably affected by the ability of these coatings to affect the appearance of osmotic blisters. Less affected appears to be the overall result of subcorroding under a coating, which manifests itself with the coatings containing a low amount of lamellar particles of both muscovites at a PVC of 5 - 10%. Also at higher particle concentrations there appears a state when no homogeneous coating is formed, but the pores enabling permeation of water or chloride solutions, and this results in subcorroding of steel under the coating. The effect of coating resistances to corrosion in an artificially prepared cut is really problematic for barrier pigments to be evaluated, as already from the principle of mechanism of anticorrosion action of lamellar particles this property is not afected in any way. The corrosion in cut values fluctuate round 2 mm without any respect to the concentration or type of muscovite in the coating film.

Figure 7 brings the dependences of overall anticorrosion coating efficiency on the muscovite and Fe-muscovite concentrations. In the evaluation the following factors were considered: osmotic blisters, subcorroding under the coating in surface, and subcorroding in the place of cut scribe. The results quite unambiguously indicate the positive effect of surface-treated muscovite on the anticorrosion barrier properties in the çoating. The tests with a natural non-treated muscovite have shown that the effect of particles in the coating is not of such a significance as on using the surface-treated Fe-muscovite.

Both pigments show as the most suitable concentration for the coatings with respect to anticorrosion properties a PVC of 20%.[9] For the comparison of observed results at a PVC of 0% (pure epoxyester binder) and a PVC of 20% (an optimum concentration) the growth of protection anticorrosion efficiency was shown:

- non-treated muscovite an 18% improvement
- treated Fe-muscovite a 39% improvement

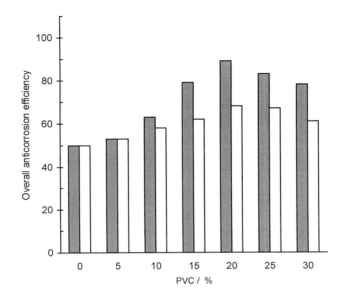

Figure 7. Dependence of the overall anticorrosion coating efficiencies on the muscovite and Fe-muscovite particle concentrations; ▓ = Fe-muscovite, ☐ = muscovite.

Conclusion

It was found that the surface muscovite treatment by controlled hydrolysis and precipitating a hematite layer on the lamellar natural-muscovite particles appropriately affects the properties of pigment particles. The surface treatment of muscovite by ferric oxide gives the pigments for the anticorrosive coatings acting as corrosion barriers. The surface-treated muscovite really advantageously causes a reduction of the formation of osmotic blisters in the coatings. From the anticorrosion efficiency point of view the surface-treated muscovite is more effective than the same muscovite type without any surface treatment. Also the mechanical coating properties show an advantageous effect of the surface-treated muscovite. A significant factor is connected with the improved cohesion coating strength. Also the film appearance and coating resistance prior to effects of the ultravioleght radiation which acts in the sense of degradation is better on using the treated Fe-muscovite. The optimum concentration of such lamellar anticorrosive pigment in the coating is a PVC value of 20%. At this concentration the best properties are reached in the coatings.

Acknowledgment

The work was performed with the support of the Ministry of Industry and Trade: FD-K/005 *"Application mica pigments with content ferric oxide into organic coating for ecological anticorrosive systems"*.

References

[1] E. Carter, *Pigment and Resin Technol.* **1990**, 15, 18.
[2] S. Wiktorek, E.G. Bradley, *JOCCA* **1986**, 69, 172.
[3] T.V. Kalinskaja, *Lakokras. Mater.* **1987**, No. 2, 4.
[4] A. Kalendová, P. Tamchynová,in: *5ᵗʰ International Conference „Inorganic Pigments and Binders"*, Ústí nad Labem, Czech Republic, **2001**, Book of papers pp. 154-158.
[5] N. Sprecker, *JOCCA* **1983**, 66, 52.
[6] A. Kalendová, *Pigment and Resin Technol,* **2000**, 29, 277.
[7] V. Štengl, S. Bakardjieva, J. Šubrt, in: *XXXII. International Conference KNH,* Pardubice, Czech Republic, **2001**, Book of papers pp. 83-86.
[8] A. Kalendová, *Pigment and Resin Techno.* - in press.
[9] P. Kalenda, *Dyes and Pigment* **1993**, 23, 215.
[10] P. Kalenda, *Acta Mechanica Slovaca* **1999**, 3, 103.

Macromol. Symp. 187, 377–386 (2002)

Possibilities of Affecting the Corrosion-Inhibition Efficiency of the Coatings by Means of the Zinc Powder Particle Sizes and Shapes

Andréa Kalendová[*], *Andrea Kukačková*

University of Pardubice, Faculty of Chemical Technology, Institute of Polymeric Materials, nám.Cs. legií 565, 532 10 Pardubice, Czech Republic

Summary: The paper deals with morphology and physicochemical aspects of zinc powder concerning the properties of coatings. The sizes or, sooner, the size zinc powder particle distribution manifest themselves especially in the mode of arrangement and affecting the electrochemical mechanism of action in the coating. The lamellar and isometric particle shapes affect by the pigmentation mode and height especially the barrier action and physicomechanical properties of the coating.

Introduction

The zinc metal particles have presented an anticorrosive pigment frequently used for the coating compositions destined to heavy corrosion protection of metals already for many years.[1]

The zinc metal was used in coating compositions in the year 1840, when zinc metal was applied for its high covering capacity without knowledge of its high anticorrosion efficiency. The zinc pigmented coatings were considered until recently for the coatings acting above all by the electrochemical mechanism, as the layer creating the so-called sacrificed electrode. The cathodic protection[2] of steel is secured by the time when the pores are sealed, and the electrochemical mechanisms passes to the barrier mechanism. The coatings contain in addition to zinc also binders, which are to package the zinc particles and thus to increase thus the resistance to such a degree that the electric conductivity drops down below a critical value, under which the coating cannot act in the sense of electrochemical protection. The electric conductivity of pigmented film is in connection with the concentration of zinc particles in coating composition binder. The highest electric conductivity is reached at a concentration of zinc particles in a range of 92-95%.[3] In applications to the corrosion protection[4,5], two types of zinc particle

CCC 1022-1360/00/$ 17.50+.50/0

shapes are used most often, namely the spherical (ball-shaped)[6] and lamellar (plate-like) shapes.[7]

Experimental

Preparation of the coatings pigmented with zinc

The study of zinc pigment effects occuring in the coatings requires to select a suitable binder unsaponifiable by the alkaline products of the electrochemical reactions running in the system. Such binders comprise two-component epoxy resins hardened by polyamines, a single-component polyurethane hardened by atmospheric moisture, and epoxy ester resins desiccating due to an oxidative process induced by atmospheric oxygen effects[9,10]. The coating formulations were prepared in such a way that the systems with volume content of spherical zinc, a pigment volume concentration (PVC) value = 0, 10, 20, 30, 40, 50, 60, critical pigment volume concentration (CPVC) value, and PVC value = 70% were obtained. The PVC = 0 denotes the nonpigmented binder - a transparent lacquer; the zinc content at CPVC relates to individual types of zinc pigment. On using the lamellar zinc a concentration, defined as PVC = 5, 10, 15, 20, 30, 40, and as $CPVC_{50.6} = 60\%$ was used.

Results

Pigments delivered by various zinc producers were used for the study of shape and primarily of the particle sizes of zinc pigments (Table 1. and Figure 1.). For further studies of the effect of zinc powder in anticorrosive coatings the zinc pigments, can be divided to two groups:

1. zinc with isometric - spherical particles
2. zinc with nonisometric - lamellar particles

Table 2 brings the selected results of mechanical tests performed in coatings with spherical zinc particles (Type 5) and lamellar zinc particles (Bend test, ISO 1519, and Cross cut test, ISO 2409).

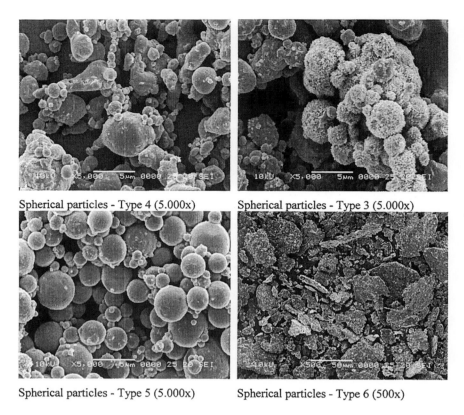

Spherical particles - Type 4 (5.000x)   Spherical particles - Type 3 (5.000x)

Spherical particles - Type 5 (5.000x)   Spherical particles - Type 6 (500x)

Figure 1. Morphology of particles of zinc powder tested.

The differences between the two types of zinc particles are more clearly evident in Figure 2 wherein the mass gains of the absorbent (silica gel) due to water vapor, passing through the coating film. The obtained dependences show that the lamellar zinc-pigmented coatings create a more efficient barrier for water vapor than the isometric spherical particles-pigmented coatings. The barrier created by the coating film to water vapor can be realized on the basis of a model shown in Figure 3 for the lamellar and spherical zinc particles. The photos are a result of the scanning electrone microscopy (SEM) coating analysis (Jeol, JSM 5 600 LV) of both types pigmented to a PVC=50%. The diffusion resistance value (μ) for the coatings pigmented to a PVC=50% is for he lamellar zinc (Type 6) = $7,21.10^8$, and diffusion resistance value (μ) for the spherical zinc (Type 5) = $1.9.10^8$.

Table 1. Characterization of types of zinc powder tested.

Type of zinc powder	Mean particle size[a] μm	Specific surface[b] m²/g	Characterization of zinc particles by SEM	CPVC %
Type 1 Spherical	5.0	0.240	Zinc particles under 1 μm are freely present, particles are not significantly covered by oxidation products	66.6
Type 2 Spherical	4.2	0.29	Zinc particles forming agglomerates, particles are partially covered by oxidation zinc products	65.9
Type 3 Spherical	2.9	1.427	The surfaces of many particles are covered by oxidation products, large representation of rather small particles forming agglomerates	58.1
Type 4 Spherical	3.9	0.519	Many zinc particles of irregular shape, overhelming content of particles smaller than 2μm	65.7
Type 5 Spherical	5.6	0.443	The pigment particles are of precisely spherical shape, the surface of particles is not oxidized	66.3
Type 6 Lamellar	25.6	2.079	A large representation of particles smaller than 1μm, and with the largest particles round 50 μm	50.6

[a] by laser beam diffraction *(Coulter LS 100)*
[b] specific surface calculated by BET isotherm *(ASAP 2000)*

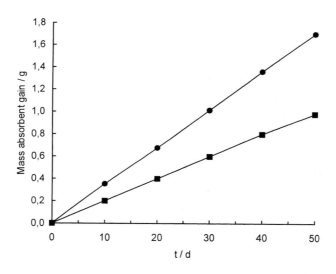

Figure 2. Dependence of the mass absorbent gain on time for the coatings pigmented to a PVC = 50%; ■ lamellar zinc – Type 6, ◆ spherical zinc – Type 5.

Table 2. Mechanical properties of coatings pigmented with lamellar and spherical zinc particles in dependence on the zinc volume concentration.

PVC	Bend test	Cross cut test
%	mm	Degree
Lamellar zinc particles (Type 6)		
0	< 4	0
5	< 4	0
10	4	0
15	4	0
20	4	0
30	5	0
40	5	0
CPVC	10	1
60	15	2
Spherical zinc particles (Type 5)		
0	< 4	0
30	< 4	0
40	< 4	0
50	< 4	1
60	6	2
CPVC	15	4
70	25	4

The Figure 4 brings graphical representations of the corrosion-testing results obtained in the salt spray test (ISO 7253) for formulated coatings (zinc powder Type 3). The dependence courses show clearly that the protective function of the coatings pigmented with zinc indicates improvements with the increases in concentration of zinc up to a respective PVC value (round 50 - 60 vol. %). After this concentration limit (CPVC$_{Zn}$ = 58 %) has been exceded the deterioration, i. e. a reduction of the anticorrosion protection of substrate metal takes place. The optimum values of anticorrosion efficiency are achieved at a concentration of zinc particles in the paint film of 50-55 vol. %. Provided the CPVC value for an epoxy-based zinc coating is known (CPVC$_{zinc\ type\ 3}$ = 58 %) a conclusion can be taken that the optimum protection is provided by a coating pigmented just below this value.

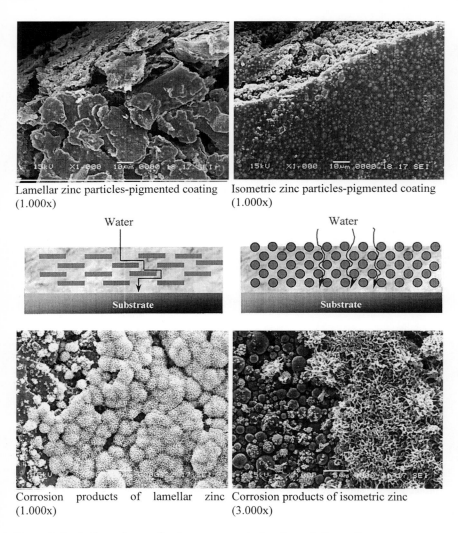

Lamellar zinc particles-pigmented coating (1.000x)

Isometric zinc particles-pigmented coating (1.000x)

Corrosion products of lamellar zinc (1.000x)

Corrosion products of isometric zinc (3.000x)

Figure 3. Ilustrative scheme of water vapor penetration through the coatings.

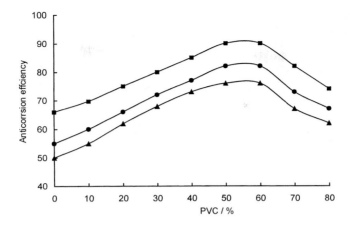

Figure 4. Anticorrosive efficiency of the coatings in dependence on the amount of zinc pigment (zinc type 3, an 1000-hour exposure in salt-spray chamber); ■ = epoxy resin, ● = polyurethane, ▲ = epoxyester.

In case of two-component (2-C) epoxy resin-based films containing zinc in a concentration range of 0-75 vol. % the SEM gave photos of the film surfaces and sections (Figure 5). This figure shows clearly that at the pigmentation by 15 vol. % of zinc powder the zinc particles (Type 3) are almost invisible, both in the film section and surface. At this concentration the protective action is in its essence only of barrier binder nature, and no other mechanisms of protection can take place. At a PVC value of 60 % the zinc concentration lies already above the CPVC value. The surface shows that the individual zinc particles are coated with the binder in shells, and at the same time among thus formed agglomerates the pores occur, which are of large sizes. The agglomerates are covered by the binder, here the electrochemical effect can partly manifest itself. At a zinc concentration of 75 vol. % a shortage of binder necessary for covering the zinc particles is observed, the particles become bare, and the coating is rather porous. In this case we cannot speak of any binder barrier protection against the diffusing substances, but the neutralization and electrochemical mechanisms manifest themselves really well.

384

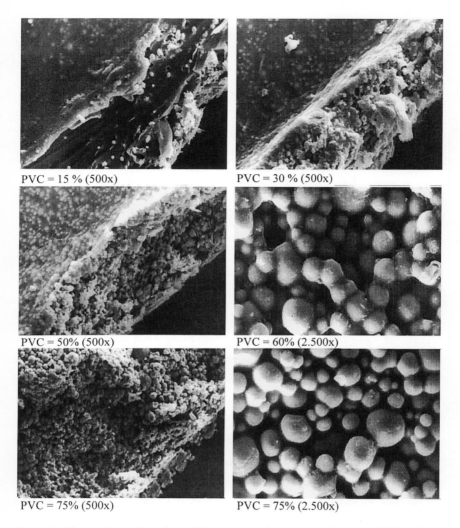

PVC = 15 % (500x) PVC = 30 % (500x)

PVC = 50% (500x) PVC = 60% (2.500x)

PVC = 75% (500x) PVC = 75% (2.500x)

Figure 5. The section and surface of the epoxy-resin based coatings at a PVC 15-75% zinc.

The results given in Figure 6 offer an idea of the anticorrosion efficiencies of coatings pigmented with zinc in an amount corresponding to the CPVC value (salt-spray test ISO 7253).

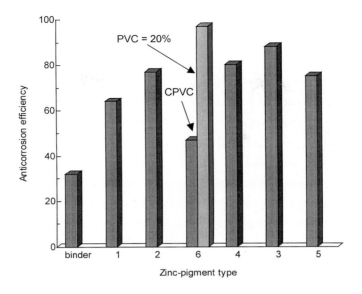

Figure 6. Effect of the zinc particle types on the overall anticorrosion coating efficiency following a 800 h exposure to a salt chamber medium.

The Figure 6 shows clearly that the spherical zinc particles (types 1, 2, 3, 4, and 5) in epoxyester coatings exhibit anticorrosion efficiencies in a range of 64-88%. Thus the difference between followed zinc types represents a 24% anticorrosion efficiency. The smaller zinc powder particles (Type 3), mean the higher anticorrosion efficiency in the coating. It can be concluded, that at an 800-hour exposure to salt chamber conditions the results of anticorrosion efficiency on using the lamellar zinc are almost by 40% worse than those obtained on using the spherical particles. The situation looks out quite differently at a lower concentration of lamellar zinc. At a PVC=20% the efficiency of lamellar zinc-pigmented coating reaches already 97%. If we compare the coating efficiency at a zero content of lamellar particles of an efficiency at a PVC=20% and at a PVC=CPVC=50% it is evident that already at low concentrations a rapid growth of anticorrosion efficiency takes part.

Conclusion

The action mechanism of zinc metal in coating films allows to conclude on the capacity to neutralize the acid corrosive medium penetrating through the film, in addition to the electrochemical protection, i.e. the capacity to create "sacrificed" electrode by shifting the potential to the region in which the corrosion of steel does not run. The alkaline products at the surface of zinc particles effectively restrict the diffusion of acidic components from the coating-film surface towards the substrate metal. This explanation is valid quite generally, as the protection effectiveness of such coatings depends on the concentration and location of pigment particles in the binder system concerned. The film will protect effectively the substrate as far as the optimum concentration of zinc pigment is used. It was found that the size of zinc particles considerably affected the anticorrosion properties of coatings. Better results were found on applying small zinc particles to coatings. Large spherical particles offer a lower anticorrosion efficiency. This phenomenon can be explained by easier filling free spaces among zinc spheres of smaller sizes. At larger particles the filling of pores by means of oxide is incomplete and the leakages can lead to more easy liquid and gas penetrations through paint film. At zinc powder of lamellar type the sealing of rather voluminous pores at CPVC is connected with problems, and the coatings exhibit a lower anticorrosion efficiency. The most appropriate concentration of lamellar zinc particles is round a PVC=20 vol %, whereat by up to one third higher anticorrosion properties in comparison with a series of isometric zinc particles can be reached.

References

[1]   R. Tulloch, *Amer. Paint J.* **1991**, 75,36.
[2]   M. Leclerq, *Eur. Coat. J.* **1991**, No.3, 106.
[3]   J. Ruf, in: „*Korrosion Schutz durch Lacke und Pigmente*" Verlag W.A.Colomb 1972.
[4]   S. Fuliu, R. Barajas, J.M. Bastidas, *J. Coat. Technol.* **1989**, 61, 63.
[5]   S.K. Verma, *Paintindia,* **1990**, 40, 31.
[6]   P. Kalenda, *Acta Mechanica Slovaca* **1999**, 3, 103.
[7]   A. Kalendová, *Pigment and Resin Technol.* – in press
[8]   P. Kalenda, *J. Appl. Pol. Sci.* **1992**, 45, 2235.
[9]   P. Kalenda, *Polym Paint Col. J.* **1991**, 11, 88.
[10] P. Kalenda, Sc*ientific Pap., Univ. Pardubice* **1990**, 54, 127.

Study of the Resistances of Organic Coatings to Filiform Corrosion

Petr Kalenda*, Michal Petrášek

University of Pardubice, Faculty of Chemical Technology, Institute of Polymeric Materials, nám. Cs. Legií 565, 532 10 Pardubice, Czech Republic

Summary: The filiform corrosion is a special-type atmospheric corrosion, which broadens below the organic coatings and is characterized by its manifestation in the form of fibres or filaments. An important factor for broadening the filiform corrosion involves also the barrier or chemical resistance of the organic coating. The paper deals with the modes of filiform-corrosion initiation and evaluation thereof under nonpigmented organic coatings based on various binder bases. Also the possibility of retardation of the filiform corrosion by means of zinc powder and the effects of lamellar pigment on the filament shapes are studied.

Introduction

The development and usage of new, steadily improved systems of surface protection is affected by a series of factors, among which the most important ones are the effects on environment and economics of surface treatment procedures.[1] Just the problems connected to the anticorrosive protection of metals lead us to the region connecting the corrosion engineering to the surface protection by organic coatings.[2] In the past a series of papers was devoted to investigating the corrosion processes of metals alone without any respect to this surface protection by organic coating films. On the contrary another series of papers was directed to the organic coating alone from the polymer matrix, pigment, and filler points of view, with respect to combination of individual components of coatings, the physical and chemical properties thereof.[3] It is evident that a future task will concern investigation of the corrosion processes running not at an unprotected metal, but in the system, where the metal will be fitted with organic protective film or with several layers of protective coats. A further shift in knowledge will be, therefore, conditioned by understanding and describing the processes running at the surface under the coating film.[4] At present it is possible to gather some knowledge to several points, concerning the corrosion processes under an organic coat.

CCC 1022-1360/00/$ 17.50+.50/0

Blistering

The blistering is one of the earliest signs of disturbing the protective coating functions.[5]. The blisterst can be defined as local regions, in which the coating loose its adhesion to metals substrate, where the water accumulation takes place and wherein the corrosion processes start to run.[6] The appearance of blisters on the coating can be explained by one of four basic possible mechanisms:[7]

a) Formation of blisters by a volume expansion originating in the swelling

b) Formation of blisters due to absorption or formation of gases

c) Osmotic formation of blisters

d) Formation of blisters due to the phase separation in the film formation

The corrosion mechanism in the blistering-attacked region consists in the action of locally cumulating water at the coating/substrate interface. The ferric ions are concentrated at the blister vaults, where the oxygen concentration has its highest value. In these locations also the cathodic reaction is running, and, on the oposite, the anodic reaction runs.

Flash corrosion

Under the term "flash corrosion" a really heavy problem is to be understood, which overwhelmingly concerns water-dilutable coatings. Defects of this type occur only in a case, when the coating film undergoing the drying-out process was subjected to a high relative moisture effect.[8,9] The film formation at aqueous latex coatings runs under the evaporation of water through particle coalescence, the deformation thereof and, at the end, through the diffusion of polymeric chains among individual particles.[10]

Rusting following the blasting operation

At the surface pretreatment of metals by blasting, when the blasting operation was peformed using steel grit there can appear scratchs or the particles can get jammed in the substrate metal. The unfavourable result can be hindered by a two-stage blasting operation or by a chemical cleaning of the metal surface.[11]

Anodic subcorroding

The anodic subcorroding represents the corrosion reactions running under the organic coatings, in which the driving force is exhibited by the anodic corrosion reaction.[12,13] The anodic subcorroding takes place when the steel substrate takes the role of anode, through the imposed potential. Without imposing the potential the protection of steel fails to a high degree through the cathodic subcorroding mechanism. The anodic corrosion running below the protective films has not been studied so intensely as the cathodic subcorroding processes.

Cathodic subcorroding

When the steel material fitted with a coating is permanently exposed to an electrolyte (as the case is with ships, pipelines, tanks for salt solutions, and the like) it possible to protect steel by imposing potential - the cathodic protection. An disadvantage of this protection is that in the vicinity of the place of coating damage the adhesion of coating to steel is mostly lost. This loosing of adhesion is usually designated as the cathodic delamination. The cathodic delamination takes place also in the case, where no potential was imposed. It is proven that the cathodic reaction leads to an increase of pH value under organic coating.[14]

Filiform corrosion

The filiform corrosion a specific manifestation of corrosion attack running under a coating film.[4] The filaments formed by the corrosion products have really various shapes, from the nodular ones to very fine sharply limited filaments. The filament width varies from 0.005 to 0.5 mm. The filaments can grow at a velocity of up to 1 mm/day. The conditions of formation of the filiform corrosion at the steel substrate and also at the galvanized steel substrate fitted with a coating material film rest on the steady exposure to the relative humidity of 70-85 % under simultaneous contamination of the substrate metal with soluble salts. At a really high value of humidity or at the exposure to a water medium the filament corrosion passes to the corrosion case of more general nature, and the thread character is lost.

Properties of filiform corrosion

The filiform corrosion[5] is a kind of electrochemical atmospheric corrosion, which broadens under an organic coating accross the metallic substrate surface in the form of orientedly growing filaments, which do not intersect each other anywhere. The filiform corrosion was also observed under oxidic layers covering the surface of appropriate metal. The steel, aluminum, and magnesium and the alloy substrates are attacked.[15-17] The corrosion appears at the so-called weak coating places. These places comprise namely unsufficient-thickness places of coating at the edges, local reduction of the coating thickness and mainly the mechanical damages of the coating, exposing the substrate part such as scratches or large pores. This kind of corrosion appears at specific conditions, which can be collected in the following manner:[4,18]

- Sufficiently high relative atmospheric humidity (65-95%) at laboratory temperature
- Sufficient permeability of water vapor through the coating
- Stimulation by impurities such as sulfur dioxide, sulfates, chlorides, acetates
- Presence of coating defects, as especially mechanical damages, pores, air bubbles, salt crystals, or dust particles
- Sufficient adhesion between the substrate and the coating. At insufficient adhesion the standard corrosion runs
- Presence of oxygen in the atmosphere

Other factors, such as temperature, substrate nature and pretreatment thereof, kind of impurity in the atmosphere, value of adhesion and permeability for water vapor, relative humidity value, thickness and kind of coating affect the nature and velocity of corrosion of that kind.

Description of filaments

The corroded surface is characterized by the form of filaments, which grow alongside by each other of a defective coating place, whereat they originated. The filament itself consists of a head and a body. The head is an active corrosion cell and moves at the surface and contains the concentrated electrolyte. A gellike amorfous membrane consisting of hydrated corrosion products separates the head from the body.[4] The trace itself contains the dried out rust. The filament head consists of a front and a backward part. The front side is an anode, whereas the backward part and head vicinity are of cathodic nature. The pH value at the filament front depends on the substrate kind and fluctuates from 1 to 4. The head rear is basic, and some sources report pH value up to 12. The filament widths vary approximately in a range from 0.1 to 1 mm and lengths from 2 to 30 mm and more.

Filament growth

A complete description of the filiform-corrosion mechanism is regrettably not known. Nevertheless, there are no doubts that the filiform corrosion starts with the appearance of an osmotic blister at the coating defect. The blister such formed (a corrosion cell) starts moving in time. It is not quite clear, wherein this movement has its cause. The possible explanation is the action of osmotic pressure in the head or transformation of corrosion products connected with changes in molar volumes. The head movement is, however, conditioned by a loss of the coating adhesion to the substrate in the direction of filament head movement. Literature reports two possible mechanisms of the coating delamination in case of this corrosion kind:[5,19] The so-called anodic and cathodic mechanisms. In both the cases the delamination is caused by hydroxyl ions appearing at the cathodes. The difference is that, which cathode is the main source of hydroxyl ions. The anodic mechanism ascribes the main role to the cathode situated in the rear head of the filament, the cathodic mechanism is ascribed to the cathodes in vicinity of that head. A series of authors propose the cathodic mechanism for the filiform corrosion at steel surface and the anodic one to the aluminum corrosion case.[4,20,21]

Experimental

The present work aimed at the initiation of filiform corrosion at comparing the stability of various organic coating kinds, applied to the substrates of varying composition (steel, aluminum) and pretreatments, to this corrosion kind. Further the effect of pigmentation of the coating, relative humidity, nature of initiation ions, and coating thickness on the filiform corrosion parameters was further sought.

Methods of initiation

In this work four kinds of artifitial initiation of filiform corrosion were tested. The metallic samples were fitted with a coating and following the drying out thereof (3 weeks); closely prior to initiation, fitted with a scratch down (cut) to the metal in a length of about 5 cm. With all the following methods the filiform corrosion was initiated by exposing the system to a medium containing the corrosion stimulators and followed by placing it in certain humid atmosphere.

Method with acetic acid (*Method 1 according CSN 673106*)

The metallic samples covered with an organic coating and fitted with cuts reaching down to the metal were submerged in an aqueous solution containing: sodium chloride (NaCl) 50 g/l, concentrated acetic acid 10 ml/l, and hydrogen peroxide 30% 5 ml/l for 1 hour (the aluminum ones for 6 h). After being removed from this solution they were placed in an atmosphere of relative humidity adjusted to 81% for 28 days at 23 °C.

Method with hydrochloric acid (*Method 2 according DIN 65472*)

The samples were placed in a medium of vapor above 32% hydrochloric acid (HCl) for an hour and afterwards exposed to an atmosphere of relative humidity of 80% and at 40 °C for 28 days.

Method with neutral salt spray fog (*Method 3 according CSN EN ISO 4623*)

The samples were exposed at first to the neutral salt fog (5%) medium for an hour and subsequently to the condensation chamber medium of relative humidity 100% and a temperature of 40 °C for 28 days.

Method of immersion in the solutions of natrium chloride, bromide, sulfate or thiosulfate as effect of the initiation ion kind (*Method 4*)

The samples were submerged in the solutions of natrium chloride (NaCl), bromide (NaBr), sulfate (Na_2SO_4) or thiosulfate at mass concentrations of 0.1 g salt/l for 30 s. After being removed they were dried with filter paper at places outside cuts, ventilated for 15 min and introduced into the chamber of a similar arrangement as in Method 1.

Description of the tests performed

The steel and aluminum panels were used as substrates. The steel was ground by an abrasive paper and degreased prior to application. With aluminum two pretreatment methods were used: on the one hand pickling in an acid medium and subsequent grinding, on the other pickling in a soda (Na_2CO_3) solution. Various kinds of organic coatings without any content on pigments (only binders alone or binders with a siccative content) were applied onto thus pretreated surfaces. Methods 1, 2, and 3 were used for the initiation. With the samples a scope of filiform corrosion (CSN EN ISO 4623) and coating thickness and with the binders the permeability values for water vapor (μ) and adhesion (Cross-cut test ISO 2409, Pull-of test for adhesion ISO 4624) from the substrates tested were evaluated. The aim consisted in comparing the efficiency of individual initiation methods and determining the effect of binder and substrate types and the substrate pretreatment effect on the manifestation of filiform corrosion. For studying the behavior of filiform corrosion under various organic coatings several tests were performed and the following binder types were tested: polystyrene, chlorinated rubber, two component (2-C)-solvent epoxy resin, 2 -C water-reducible epoxy resin, one component (1-C) polyurethane, epoxyester resin, alkyd resin, silicone resin, aqueous polyurethane emulsion, and aqueous alkyd emulsion.

Effect of relative humidity on the filiform-corrosion manifestation: the samples coated with one component (1-C) polyurethane were exposed to the initiation by the Method 2 and positioned in atmospheres of relative humidity 55, 63, 75, and 81% at 23 °C for 30 days. After ending the test the scope of filiform corrosion, the coating thickness and the maximum and minimum widths of filaments were determined (CSN EN ISO 4623).

Effect of the coating thickness on the filiform-corrosion manifestation: on to the samples the two component (2-C) epoxy resin was deposited in various thicknesses, and the filiform corrosion was initiated by the Method 2.

Effect of the zinc-powder pigmentation on the filiform-corrosion manifestation: paints containing zinc pigment were prepared from 2-C epoxy resin, one component 1-C polyurethane and epoxyester binder with a pigment volume concentration (PVC) value of 50, 55, and 60%. Filiform corrosion was initiated by the Method 1.

Effect of pigment showing a lamellar structure on the filiform-corrosion manifestation: for the pigmentation of 1-C polyurethane and epoxy resin the ferric mica was used and the pigment volume concentration (PVC) values were adjusted to 0, 10, and 20%. The filiform corrosion was initiated by the Method 2.

Effect of the initiation ion kind on the filiform corrosion scope: this test was performed at steel samples fitted with a 1-C polyurethane coating. The initiation was performed by the Method 4; as initiation ions sodium chloride, sodium bromide, sodium sulfate and sodium thiosulfate being used.

Figure 1. SEM photo of filament under coating based on 1-C polyurethane (35x).

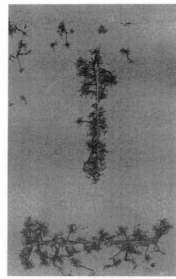

2-C epoxy resin based coat   2-C water-red. epoxy resin based coat

Figure 2.Photographic presentation of testing the filiform corrosion (steel, Method1).

Figure 1 presents scannin electrone microscopic (SEM) photo of filament under coating based on 1-C polyurethane. Figure 2 and Figure 3 give photo of filament under the coatings (based on 2-C epoxy resin, 2-C water-reducible epoxy resin, 1-C PUR , and silicone resin for steel substrates.

 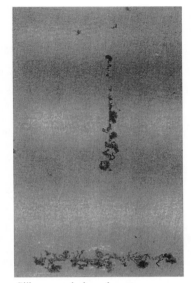

1-C PUR resin based coat   Silicone resin based coat

Figure 3. Photographic presentation of testing the filiform corrosion (steel, Method 1.

Conclusion

The binders were ordered by the surface size attacked by filiform corrosion. This order depends on the nature of substrate and the pretreatment thereof and further on the kind of initiation method used. The highest resistance values were observed in all cases of the substrates and the pretreatments thereof using by the silicone, polystyrene and chlorinated-rubber coatings. The relative humidity value showed the effects on the filiform corrosion parameters. With the growing relative-humidity value also the width and length of corroded filaments showed rising. Method 3 (positioning at 100% relative humidity) did not initiate this kind of corrosion, what confirmed the statement that at the relative humidity values higher above 96% this kind of corrosion passes to blistering.

In case of 2-C solvent epoxy resin the increasing coating thickness at steel surface entailed reduction of the filiform corrosion scope. This coating showed the highest barrier capacities of all tested binders. The coatings pigmented with zinc powder at steel and aluminum substrates showed an excellent resistances to filiform corrosion.

All the tested ions were capable of initiating the filiform corrosion. The ion kind affected as the scope of corroded surface so as the density of filaments along the section. The most effective initiator was sodium thiosulfate. Under the coatings of 1C polyurethane pigmented with ferric mica no filiform corrosion was observed. The surface was attacked by surface corrosion. In case of high solid epoxy resin the filiform corrosion was initiated and manifested the dependence of PVC value of ferric mica on the scope of corroded surface and filament density alongside the section.

[1] A. Kalendová, *Acta Mechanica Slovaca* **1999**, 3, 107.
[2] A. Kalendová, *Prog. Org.Coat.* **2000**, 38, 199.
[3] A. Kalendová, *Pitture e Vernici Eur. Coat.* **2000**, 76, 57.
[4] A. Bautista, , *Prog. Org.Coat.* **1996**, 28, 49.
[5] W. Funke, , *Prog. Org.Coat.* **1981**, 9, 29.
[6] P. Antoš, *Chem. Mag.* **1999**,9, 8.
[7] A. Kalendová, P. Antoš, *Koroze Ochr. Mater.* **1998**, 42, 55.
[8] P. Mošner, A. Kalendová, L. Koudelka, *Dyes and Pigments* **2000**, 45, 29.
[9] A. Kalendová, *Anti- Corrosion Met. Mater.* **1998**, 45, 344.
[10] A. Kalendová, *Pigment and Resin Technol.* **2000**, 29, 288.
[11] P. Antoš, A. Kalendová, *Koroze Ochr. Mater.* **1996**, 40, 70.
[12] A. Kalendová, *Sci. Pap. Univ. Pardubice* **1998**, Ser. A4, 137.
[13] A. Kalendová, *Pigment and Resin Technol.* **2000**, 29, 215.
[14] P. Antoš, O. Tokar, *Chem. Mag.* **2000**, 10, 8.
[15] M. Heinrich, H. Haagen, T. Schuler, *Farbe und Lack* **1994**, 100, 249.
[16] H. Haagen, K.H. Rihm, *Farbe und Lack* **1990**, 96, 509.
[17] H. Haagen, K. Gaszner, M. Heinrich, *Farbe und Lack* **1994**, 100, 177.
[18] E.L.J. Bancken, in: *6th Symposium on Paints*, 1997, Athens, Greece. p.281.
[19] A.W. Fangmeir, E. Kock, F. Vohwinkel, C.P. Brandt, in: ACS Symposium Book, Ser.689, *"Organic Coatings for Corrosion Control"*, edited by G.P. Bierwagen, 1998, p.238, Washington, DC. USA.
[20] P. Antoš, D. Beran, A. Kalendová, *Chem. Mag.* **1996**, 6 12.
[21] P. Antoš, M. Buchar, A. Kalendová, *Chem. Mag.* **1996**, 6 16.

Efficiency of Anticorrosive Pigments Based on Modified Phosphates

Petr Kalenda[a]*, Andrea Kalendová[a], Petr Mošner[b], Michal Poledno[a]

University of Pardubice, Faculty of Chemical Technology, [a] Institute of Polymeric Materials, [b] Department of General Inorganic Chemistry, nám. Cs. Legií 565, 532 10 Pardubice, Czech Republic

Summary: The paper deals with the study of modified phosphate compounds from their anticorrosion action point of view. The pigments start with zinc orthophosphate, the modification of which can give phosphosilicates, phosphomolybdates, or basic phosphates. In the region of cations the combination of zinc with calcium, strontium, barium, or aluminum are possible. The modified anticorrosive pigments differ in water solubility, inhibition efficiency of the aqueous extracts and anticorrosion actions in organic coatings. The various types of modified phosphates were studied. The effect of organic inhibitor in presence of anticorrosive inorganic pigments in coatings was determined.

Introduction

Zinc orthophosphate is presently the most widely used anticorrosive pigment for coatings determined to protecting steel to corrosion. Broadening this pigment in the region of organic coatings was strengthened by the ecological opinion regarding chromate and lead(II) anticorrosive pigments.[1] Zinc orthophosphate is, however, not a universal anticorrosive pigment and requires definite conditions.[2] The binder should contain carboxyl groups for securing the reaction between the pigment and the polymer chain, a result being the formation of an inhibition complex.[3] It is also important that zinc orthophosphate contains a definit amount of crystal water. Efforts to increase the anticorrosion efficiency of zinc orthophosphate lead to modifying this pigment by means of cations and also in the anionic part.[4] The modification can be applied also for adapting the pH value to an alkaline region.[5] Zinc orthophosphate in an untreated form exhibits a mild acidic nature in aqueous extracts.

The modified form is represented by double orthophosphates, in which one of the cations is represented by zinc.[6,7] This concerns primarily the Zn-Al, Zn-Ca, but also Zn-K, Zn-Fe, Zn-Ca-Sr or Ba-Ca and Sr-Ca combinations. In the anion part it is possible to combine the phosphate anion with further anticorrosively efficient anions, such as silicate, molybdate, or borate.[8,9] The modified phosphate pigments (as in the region of cations, so as in that of anions) can be modified by organic corrosion inhibitors in interest of increasing the anticorrosion efficiency.[10] The modified phosphate pigments can be exemplified by the following compounds: aluminum(III) zinc(II) phosphate, basic zinc phosphate, zinc phosphomolybdate, zinc calcium phosphomolybdate, zinc borophosphate. On combining the phosphate anion with silicates the combinations of cations giving the following pigments are used: zinc strontium phosphosilicate, calcium barium phosphosilicate, calcium strontium zinc phosphosilicate.[11]

As an organic inhibitor for modifying and increasing the anticorrosion efficiency of the pigment the following compounds can be used: zinc 5-nitroisophthalate, calcium 5-nitroisophthalate, calcium cyanurate, metal salts of dinonylnaphthalene sulfonic acids.[12]

Experimental part

Table 1 and Figure 1 give some characteristic properties and morphology of pigment particles tested.

Type A/2 (10. 000x)   Type A/6 (10. 000x)

Type C/2 (10. 000x)   Type B/1 (10. 000x)

Figure 1. Morphology of pigment particles tested (SEM, *Jeol 5600 LV*).

Table 1. Characterization of anticorrosive pigments tested.

Type	Chemical composition %	Loss on ignition[a] %	Oil absorp.[b] g	Density g/cm^3	Water abs. g	CPVC[c] %
		Phosphosilicate anticorrosive pigments				
A/1	ZnO 55%,SiO_2 24%, P_2O_5 21%	0.05	40.4	3.5	41.1	39.7
A/2	ZnO 35.5%,SiO_2 7.3%, PO_4 24.2%, $BaSO_4$ 29%, MgO 4%	0.09	30.1	3.3	34.3	48.4
A/3	SiO_2 39.7%, PO_4 8.8%, SrO 12.5%, CaO 39%	0.17	43.8	2.9	34.4	42.3
A4	SiO_2 41.1%, P_2O_5 7.1%, BaO 14.4%, CaO37.4%	0.14	36.5	3.0	41.0	46.2
A/5	SiO_2 45.7%, PO_4 5.4%, K_2O 2.6%, CaO 19.4%, Al_2O_3 11.9%, BaO 5.4%, Na_2O 3.9%, Sb_2O_3 2.4%, ZnO 1.4%, org. 1,9%	0.08	37.6	2.8	35.5	47.4
A/6	ZnO 42%, SrO 7%, P_2O_5 19%, CaO 16.5%, SiO_2 15.5%	0.11	39.1	3.1	43.0	43.4
		Anticorrosive phosphomolybdate pigments				
B/1	ZnO 55.8%, MoO_3 1.7%, PO_4 42.5%	0.16	24.2	3.5	30.3	52.3
		Anticorrosive phosphate pigments				
C/1	ZnO 40%,Al_2O_3 4.5%, PO_4 55.5%.	0.10	46.4	2.8	48.5	41.7
C/2	ZnO 52.5%,PO_4 47.5%	0.30	27.6	3.2	35.6	51.3

[a] Detertmination of loss on ignition at 200°C
[b] Determination of oil absorption for pigments (DIN ISO 787/2)
[c] critical pigment volume concentration calculated by linseed oil consumption

Preparation of coatings

The coatings were prepared by dispersing anticorrosive pigments in a styrene-acrylate binder. All the pigmented systems were formulated to a pigment volume concentration (PVC) value = 10 vol. %. The test coatings were prepared on standard steel panels. The coating film thicknesses fluctuated in a range of 75 - 80 μm in a dry paint film. The coatings were conditioned at standard conditions for 30 days.

Laboratory tests

The mass steel losses in aqueous pigment extracts

The steel panels were exposed to aqueous extracts prepared in 10% anticorrosive phosphate pigments in water. Following a 800 h exposure the panels were freed of the corrosion products (using etching solution) and gravimetrically the mass loss (K), the corrosion velocity (v_k), and the corrosion loss (X) calculated to blank experiment (corrosion in pure water) were determined as follows:

$$K = \frac{(m_1 - m_2).10000}{S} \quad \left[g.m^{-2} \right] \tag{1}$$

$$v_k = \frac{K}{t_1} \quad \left[g.m^{-2}.d \right] \tag{2}$$

$$X = \frac{K}{K_{H_2O}} . 100 \quad [\%] \tag{3}$$

m_1 mass of steel panel prior to exposure [g]
m_2 mass of steel panel after the exposure [g]
S surface area of steel panel [m²]
t_1 time of exposure [d]

Corrosion tests

Determination of the corrosion resistance in a condensation chamber

The test was performed according to ISO 6988 in a medium of 100% relative humidity and a temperature of 35°C. The coatings were exposed for 500, 1000, and 2100 h.

Determination of the corrosion resistance in a neutral salt fog

The testing principle consists in exposing the coatings to a fog of neutral 5% aqueous sodium chloride (NaCl) solution at a temperature of 35°C. The test starts with ISO 9227. The samples were exposed for 100 and 200 h.

Prohesion test

The method is a modification of salt-chamber test using the aqueous solution spray of 0.35% $(NH_4)_2SO_4$ and 0.05% NaCl. The samples were exposed also for 100 and 200 h.

Evaluation of the results

The overall evaluation of the anticorrosion protection of coatings was performed according to the already described method.[13] The result is expressed as a numerical value giving all the followed corrosion manifestations. The relation for the complete degree of corrosion (anticorrosion protection) after the salt chamber test (4) and to the condensation chamber test (5):

$$Overall\ rating\ =\ \frac{[Rating\ 1]\ +\ [Rating\ 2]\ +\ 2\ x\ [Rating\ 3]}{4} \tag{4}$$

$$Overall\ rating\ =\ \frac{[Rating\ 1]\ +\ 2\ x\ [Rating\ 3]}{3} \tag{5}$$

[Rating 1]   evaluation of the blistering degree with the coatings acording to ASTM D 714/87
[Rating 2]   evaluation of the failure in a cut with the coatings according to ASTM D 1654/92
[Rating 3]   evaluation of the corrosion degree with a steel substrate according to ASTM D 610/85

Results and discussion

Table 2. The properties of phosphate anticorrosive pigments.

Type	$pH^{a)}$	Conductivity $\mu S/cm$	Water-soluble matter$^{b)}$ %		K_m g/m^2	v_k $g/(m^2.h)$	X %
			at 20 °C	at 100 °C			
A/1	8.40	774	1.51	1.82	54.78	1.52	79.60
A/2	7.08	337	0.24	0.31	6.10	0.17	8.86
A/3	9.10	240	0.30	0.53	50.77	1.41	73.77
A/4	8.93	190	0.19	0.55	82.95	2.30	120.54
A/5	8.18	2270	2.51	2.41	3.44	0.10	4.99
A/6	7.19	97	0.10	0.16	49.77	1.38	72.33
B/1	6.89	154	0.14	0.16	65.09	1.81	94.59
C/1	6.28	75	0.08	0.11	55.89	1.64	89.63
C/2	6.44	94	0.06	0.06	54.96	1.49	79.87
Water	-	-	-	-	62.35	1.83	100.00

[a)] Determination of the pH value of water extracts for pigments (DIN ISO 787/9)
[b)] Determination of water soluble matter for pigments (CSN EN ISO 787/3, CSN EN ISO 787/9)

Table 2 shows that the anticorrosive pigments based on phosphosilicates (types A) have an alkaline extract at a pH value in a range of 7-9. Phosphomolybdate and phosphate pigments exhibit slightly acidic extracts with pH values of 6-7. The results show that at the corrosion steel losses in aqueous pigment extracts, the anticorrosion properties are positively affected by the organic inhibitors. The A/1 type pigment (zinc phosphosilicate) in comparison with an identical pigment A/2 (zinc phosphosilicate + organic inhibitor) differ in the velocity of steel corrosion running in aqueous extracts of these pigments. The A/1 type exhibits a corrosion velocity of 1.52 g/(m².h) compared to the A/2 type with a value of 0.17 g/(m².h). The inhibition efficiency of zinc phosphosilicate containing an organic inhibitor is thus by10 times higher than with zinc phosphosilicate alone.

The A/4 type pigment (barium-calcium phosphosilicate) and an A/5 type pigment (barium-calcium phosposilicate + organic inhibitor) exhibit similar differences. The organic corrosion inhibitor increases the conductivity of extract from a value of 190 μS/cm up to a value of 2270 μS/cm. Water-soluble compounds in both the pigments A/4 = 0.1% and A/5 = 2.51% give also a testimony of such a system, in which the organic inhibitor considerably increases the anticorrosion efficiency in an aqueous extract, but in the coating film the formation of osmotic blisters or an increase of film permeability to water and water vapor can take place. Comparing the anticorrosive pigments without any content of organic inhibitor (types: A/1, A/3, A/4, A/6, B/1, C/1, C/2) from the corrosion inhibition in an aqueous extract point of view shows clearly a comparatively nonexpressed anticorrosion effect. It is thus evident that the synergic effect accompanying the action of cations takes place. The anions in the followed anticorrosive pigments affect the inhibition properties of extracts in such a way that they create a series starting with the most efficient ones: phosphosilicates > phosphates > phosphomolybdates.

The results obtained after the exposure of coatings to a condensation chamber medium of 100% relative humidity do not give too expressive differences after 500 h between the anticorrosive pigments tested. 1000 h and 2100 h exposures to these conditions allow already to take conclusions from the state of coatings and steel corrosion under the coatings. If we consider the overall anticorrosion coating efficiencies in dependence on the solubilities of anticorrosive pigments in or the conductivities of the extracts thereof we find a striking agreement (Tab. 3).

The anticorrosively most efficient pigments comprise zinc-aluminum phosphate, zinc orthophosphate, calcium-strontium-zinc phosphosilicate, and zinc phosphomolybdate. It is surprising that always the compounds are concerned, which contain zinc as a cation. The effect of anion on the anticorrosion properties is of low demonstration ability.

Table 3. Order of anticorrosive pigment efficencies in a condensation chamber after 2100 h of exposure.

Pigment type	C/1	C/2	A/6	B/1	A/4	A/3	A/2	A/1	A/5
Overall anticorrosion efficiency	97	83	82	70	68	57	65	65	63
Conductivity of water extract μS/cm	75	94	97	154	190	240	337	774	2270

The contrary to following the corrosion inhibition in aqueous extracts whereat the better soluble pigments manifest themselves. A high pigment solubility or the content of water soluble compounds in the pigment lead to the increase in water penetration, when the pigments are used in coatings. In the most simple case the blisters and subsequently the corrosion of substrate metal appear just in the blister places. Having this in mind the pigments containing organic inhibitors do not manifest themselves in the test of exposure to a condensation chamber medium. These pigments are characterized by a higher water solubility.

In studying the inhibition effects of anticorrosive pigments it was necessary to derive a model which allows application in more porous binder matrix and which can demonstrate its inhibition properties. Owing to these reasons the coatings were subjected at first to 200 hour ultravioleght (UV) radiation of xenone arc. This exposure disturbs the coating surface, and at the same time, the hardening of binder takes place. The coating samples were after the irradiation transferred to a condensation chamber, salt spray chamber, and exposed for 500, 1000, and 2100 h. These modified test results give an image of that how a reaction of steel with pigment extract appears. The coatings are more permeable for water, and the amounts of water and water vapor having passed through the coating film are higher than at the preceding test in a condensation chamber. This corrosion test with a UV radiation also better reproduces the real atmospheric conditions which the coatings meet in practice.

Fig. 2 gives the results of corrosion tests performed with the coatings containing the pigments studied. A similar trend was observed in both test types without exposure to UV radiation and also after this exposure.

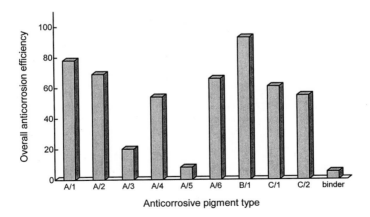

Figure 2. Overall anticorrosion coating efficiency in dependence on the type of anticorrosive pigment after a two-stage corrosion exposure for 200h to UV radiation and a subsequent 200h exposure to a salt chamber medium with the expression of blistering degrees.

The evaluation of anticorrosion efficiencies of the studied pigments was executed on the basis of all performed corrosion tests. Tab. 4 brings for every corrosion test the order of pigments in a series from the most efficient one to the pigments of low efficiency. The addition of anticorrosion efficiency values (in every test the max. value equals 100) for all five performed accelerated corrosion tests a value of ≤ 500 was obtained.

Table 4. The order of anticorrosive pigment efficiencies for all the corrosion tests performed.

Corrosion test	Pigment type / order of efficiency								
n_1 Condensation chamber 2100 h	C/1	C/2	A/6	B/1	A/4	A/3	A/2	A/1	A/5
n_2 200 h UV + condensation chamber 2100 h	A/6	A/4	B/1	A/2	A/5	A/1	C/2	A/3	C/1
n_3 Salt chamber 200 h	A/6	B/1	A/1	A/2	C/1	A/3	A/5	A/4	C/2
n_4 200 h UV + salt chamber 200 h	B/1	A/1	A/2	A/6	C/1	C/2	A/4	A/3	A/5
n_5 200 h UV + Prohesion test 200 h	A/1	B/1	A/4	A/2	C/2	C/1	A/5	A/3	A/6

Example of evaluation: Pigment A/1 has in the test n_1 the anticorrosion efficiency of or the test $n_2 = 43$, $n_3 = 81$, $n_4 = 78$, $n_5 = 83$.

$$\text{Anticorrosion efficiency} = \sum_{i=5} n_i = 350$$

The value 350 expresses the anticorrosion efficiency of pigment A/1 for all the performed corrosion tests. Table 5 reports the summary results for each anticorrosive pigment from the anticorrosion efficiency point of view in all the corrosion tests.

Table 5. Resulting anticorrosion efficiency values for the tested pigments.

Anticorrosive pigment type	A/1	A/2	A/3	A/4	A/5	A/6	B/1	C/1	C/2
Anticorrosion efficiency $\sum_{i=5} n_i$	350	338	198	313	214	349	389	261	263

Conclusion

In the conclusion we can say that the pigments based on modified phosphates are effective corrosion inhibitors in the coatings.

The results obtained allow to conclude:

- From the studied anticorrosive pigments the most effective one is zinc phosphomolybdate.

- Highly efficient pigments comprise also phosphosilicates containing zinc cation either alone or in combination with further elements such as strontium and calcium in their molecules.

- Zinc orthophosphate in comparison with modified phosphates exhibits only average anticorrosion properties.

- The combination of zinc and aluminum in orthophosphate brings no improvement of the anticorrosion properties.

- Phosphosilicates which do not contain zinc (barium-calcium and strontium-calcium phosphosilicates) are characterized by inferior anticorrosion properties to zinc orthophosphate.

- The effect of organic inhibitor in modified phosphate anticorrosive pigments is not too significant.

The order of anticorrosive pigments by the dropping efficiency was established:

Basic hydrated zinc phosphomolybdate

Zinc phosphosilicate

Calcium-strontium-zinc phosphosilicate

Zinc phosphosilicate + organic inhibitor

Barium-calcium phosphosilicate

Zinc orthophosphate

Zinc-aluminum phosphate

Barium-calcium phosphosilicate + organic inhibitor

Strontium-calcium phosposilicate

References

[1]   S. Krieg, *Pitture e Vernici* **1996**, No. 2, 18.
[2]   A. Kalendová, *Pigment and Resin Technology* – in press.
[3]   J. Ruff, in: „*Korrosion Schutz durch Lacke und Pigmente*", Verlag W.A.Colomb, 1972.
[4]   P. Antoš, A. Kalendová, P. Kalenda, *Farbe und Lack* **2001**, 107,.64.

[5]   M. Svoboda, *Koroze Ochr. Mater.* **1985**, 29, 51.

[6]   G. Adrian, A. Bittner, *J. Coat. Technol.* 1986, 58, 59.

[7]   M. Zubielewicz, E. Smieszek, T. Schauer, *Farbe und Lack*, **1999**, 105, 136.

[8]   M. Deyá, V.F. Vetere, R. Romagmoli, B.del Amo, *Pigment and Resin Technol.* **2001**, 30, 13.

[9]   P. Kalenda, *Dyes and Pigments* **1993**, 23, 215.

[10]   S.A. Hodgers, W.M. Uphues, M.T. Tran, *Pigment and Resin Technol.* **1998**, 27, 150.

[11]   P. Kalenda, *Acta Mechanica Slovaca* **1999**, 3, 103.

[12]   P. Kalenda, in: International Conference „*Inorganic Pigments and Binders*", Ústí nad Labem, Czech Republic,1999, p. 4.

[13]   A. Kalendová, *Pigment and Resin Technology* **1998**, 27, 225.

Macromol. Symp. 187, 407–416 (2002)

UV-Curing under Oxygen-Deficient Conditions

Barteld de Ruiter*, Robert M. Meertens

TNO Industrial Technology, Polymer Technology Department, P.O. Box 6235, 5600
HE Eindhoven, The Netherlands

Summary: Four different coating formulations have been formulated with different
amounts of photoinitiator and cured under different atmospheres (21, 5, 1 and 0% of
oxygen). Surface cure and through cure were assessed for each combination. It was
observed that the cure process for all formulations was strongly affected by the
atmospheric conditions, and that the formulations required substantially less
photoinitiator if cured under 5% of oxygen if compared to curing under air.
Calculations show that the lower price of the formulation due to the decreased
photoinitiator level may outweigh the additional equipment and running costs in
specific cases. Therefore, UV-inertisation is not only attractive from a product quality
point-of-view, but can also offer economic advantages.

Keywords: UV-curing; inertisation; oxygen level; photoinitiator concentration

Introduction

UV-curing represents a clean and fast technology for the curing of coatings for a vast amount of
applications. The technology can be considered as a standard technology in the printing industry,
in the wood coating industry and in the electronics industry.

A specific drawback of most UV-coatings is their sensitivity to oxygen during the cure step. The
problems concerning this oxygen sensitivity can generally be solved by modification of the
formulation, but this goes at the expense of the final coating properties. For a number of
applications these decreased properties pose no real problem, but for other potential applications
this intervention in the formulation makes UV coatings less attractive or even unsuitable.
Therefore, when a technically and economically attractive method could be found that counteracts
oxygen inhibition without damaging the final coating properties, this would certainly be of
interest for a large number of paint and coating appliers.

In recent years, UV curing under inert gas has attracted already considerable attention as one of
the possibilities to counteract oxygen inhibition. A lot of information is already available on the

                CCC 1022-1360/00/$ 17.50+.50/0

positive effects of curing under completely oxygen-free conditions, but only very limited data are available on UV-curing under partially inert conditions.

In the current investigation we have studied the effect of atmospheres with different oxygen concentrations on the cure behaviour of four different coating formulations and briefly evaluated its economical impact.

Brief theoretical background

An oxygen molecule is a biradical species, which, when dissolved in a formulation during a free-radical curing process, will interfere with the normal cure reaction in different ways. It is generally accepted that, under normal conditions, the main source for the inhibiting effect is the scavenging by oxygen of the growing free-radical chain end. Scavenging can be represented as:

Normal propagation (1): $R\cdot + M \rightarrow R\text{-}M\cdot$ (R = reactive chain end, M = monomer)

Oxygen scavenging (2): $R\cdot + O_2 \rightarrow R\text{-}O\text{-}O\cdot$ (R = reactive chain end, O_2 = oxygen)

The „normal" product $R\text{-}M\cdot$ has a reactivity very similar to the starting radical $R\cdot$, whereas the radical formed by reaction with oxygen, the peroxy radical $R\text{-}O\text{-}O\cdot$, is orders of magnitudes less reactive than $R\cdot$. Formation of peroxy radicals therefore effectively retards the polymerisation process.

The reactivity ratio of the two processes is given by the expression:

$$Rate\ (1)\ /\ Rate\ (2)\ =\ k(1)/k(2)\ \times\ [M]/[O_2] \qquad (1)$$

In this expression the rate constants $k(1)$ and $k(2)$ are roughly known; they depend only slightly on the exact structure of $R\cdot$ and M (in case of acrylates). The ratio $k(1)/k(2)$ is approximately 10^{-4}, which means that the the speed of the reaction with oxygen is approximately 10,000 times faster than the desired chain growth reaction. This means that, in order to have an appreciable polymerisation rate, the concentration ratio $[M]/[O_2]$ has to be large. The consequence is that, if carried out in air, the polymerisation will not effectively start before a large part (in radiation curable coatings typically >98%) of the dissolved oxygen in the coating has been removed. In practice, this leads to induction periods, a drop of the polymerisation rate and to incomplete conversion of double bonds in the surface layers of the coating, where diffusion of fresh oxygen is also a factor.

A lot of measures can be taken to overcome oxygen inhibition problems, including an increase of photoinitiator concentration, an increase of UV light intensity, addition of compounds that lower the solubility of oxygen in the coating, use of additives that act as oxygen diffusion barriers and the use of additives that react faster with oxygen than the growing chain ends and lead to relatively harmless side products (oxygen scavengers).

The most effective way to get rid of oxygen inhibition is the removal of oxygen from the system, i.e., the execution of the cure step in an oxygen-free atmosphere. However, such a procedure will increase both installation and running costs of the coating line. The main aim of this investigation is to find out if, and for which applications, the investments required for inerting the line can be compensated for by profits in the area of (decrease of) photoinitiator need.

Materials and methods

Base materials

In order to have complete control over the constituents of the formulations, we have used only very simple formulations based on a number of chemically different commercially available UV-resins. The following resin types were used:

A. A flexible aliphatic acrylated polyurethane (containing some HDDA)

B. A non-acrylic oligomer based on an aliphatic polyurethane and containing both maleate and vinyl ether functional groups, diluted with 33% of a special diluent of the same material type

C. A flexible aromatic acrylated polyester

D. A highly flexible acrylated epoxy (containing some TPGDA)

All materials were used as obtained. Resins A, C and D were further diluted with 33% of HDDA. All formulations had a viscosity in the range 2-10 Pa.s.

To these formulations the required amount of photoinitiator (PI) 2-hydroxy-2-methyl-1-phenylpropan-1-one (Darocur® 1173) was added in 0.05-5% concentrations (w/w) (see below).

Coatings were deposited on Bonder-type galvanised steel substrates with a coating knife with a slit of 40 μm, affording coatings in the thickness range 20-40 μm.

Methods

Coated substrates were placed in an aluminium frame covered with a 2-mm thick drawn HSQ synthetic quartz plate with a transparency of approx. 90% in the wavelength area 270-400 nm and of approx. 75% in the 230-270 nm area. The frame was supplied with a gas inlet and outlet. In order to carry out irradiations under a specific atmosphere, a specified gas mixture was administered to the frame, while an oxygen meter (Systech EC-90 MD MK II, useful for both % and ppm measurements) was attached to the outlet. Gas administration was continued until the measured oxygen level had stabilised and then continued for another 3 minutes. Both inlet and outlet were then closed and the frame was passed under the UV lamp. The frame was then opened and the samples investigated. Under the conditions used, temperature effects during irradiations can be considered to be of minor importance.

The gas mixtures (supplied by AGA Gas BV, Amsterdam) that were used in this investigation:

21% O_2: synthetic air

5% O_2: N_2/O_2 mixture (analysis: 5.12 ± 0.16%)

1% O_2: N_2/O_2 mixture (analysis: 0.993 ± 0.003%)

0% O_2: minimum nitrogen content 99.999% (AGA PLUS 5.0)

The samples were irradiated with a standard 100 W/cm Philips HOK UV lamp using two different intensities (110 and 190 mW/cm^2) and a conveyor speed of 4 m/min. The measured (International Light IL 390B Light Bug) UV-A doses were 360 and 700 mJ/cm^2, respectively.

After irradiation, through cure and surface cure of the samples were assessed. For through cure assessment, the standard MEK rub test was used: when 200 single rubs did not lead to any visible damage to the coating, the coating was considered well cured. Surface cure was assessed by pressing a finger on the irradiated surface and determining whether a print remained after removal or not (yes/no determination).

Experiments and results

UV cure experiments were performed inside the aluminium frame under different atmospheres and irradiation conditions and using different concentration of photoinitiator. After irradiation the two properties were assessed leading to results that are typified for one particular example (the formulation based on the acrylated polyester named C in the Materials and Methods section) in the following table.

Table 1. Example of assessment of properties of formulation based on an acrylated polyester of curing

Exp. #	% of O_2	UV-A dose mJ/cm^2	% of PI	Thickness (μm)	Tacky (yes/no)	# of MEK-rubs
B-1	21	360	5.0	43±8	no	>200
B-2	21	360	4.0	38±4	yes[a]	>200
B-3	21	360	3.5	39±5	yes[b]	>200
B-4	21	360	3.0	48±8	yes	>200
B-5	21	360	2.5	not meas.	yes[b]	>200
B-6	21	360	2.0	31±4	yes	>200
B-7	21	700	2.0	39±17	no?	>200
B-8	5	360	2.0	37±4	no	>200
B-9	5	360	1.5	not meas.	yes[a]	>200
B-10	5	700	1.5	40±12	no	>200
B-11	5	360	1.0	34±6	yes[a]	>200
B-12	5	700	1.0	29±7	no	120
B-13	1	360	1.0	44±5	no	>200
B-14	1	360	0.5	31±9	no	140
B-15	1	700	0.5	24±7	no	80
B-16	1	360	0.1	not meas.	yes	60
B-17	0	700	0.5	30±8	no	>200
B-18	0	360	0.1	61±2	no	75
B-19	0	700	0.1	47±15	no	100

[a] tackiness disappeared after 1 day
[b] only slightly

This specific example teaches us, that at high oxygen concentrations (21%) the surface cure forms the limiting factor. At the photoinitiator concentrations investigated, the through cure is good, as witnessed by the MEK-rub results, whereas at PI concentrations of 4% and lower the surface is always tacky. At low oxygen concentrations (0%), the through cure forms the limiting factor. At PI concentration down to 0.1% the surface is dry, whereas the MEK rub tests show increasing solvent sensitivity (no efficient through cure).

Similar tables have been composed for all different formulations. The following observations were made:

A. Formulation based on a flexible aliphatic acrylated polyurethane (containing some HDDA)

At 21% of oxygen concentration, no tack-free surface could be generated, not even at a PI level of 5%. By increasing the UV intensity from 360 to 700 mJ/cm2, the amount of PI required for a good surface cure is lowered from >5% to 3.5%.

At oxygen concentrations of 5%, the formulation can be cured to a dry coating at a level of approx. 2% of PI. Increasing the UV-dose leads to further decrease of required PI concentration (<1%).

At an oxygen concentration of 1%, the addition of <0.5% of PI is sufficient to realise a good surface cure and under pure nitrogen this amount decreases even to <0.1%. However, under these conditions through cure becomes the limiting factor: the coatings always seem to require ≥1% of PI in order to be fully cured.

B. Formulation based on a non-acrylic oligomer

These coatings are all relative thick due to rheology factors and this will have some consequences for the final values.

It appears that this formulation is also quite sensitive to the concentration of oxygen, present in the cure atmosphere. In air, only the formulation containing 5% of PI gives a completely dry coating. Decreasing the oxygen concentration to 5% leads to a significant improvement: 2% of PI suffices for a tack-free surface. Increase of the irradiation dose leads to further improvement. In an atmosphere with 1% of oxygen the amount of PI required for a tack-free surface drops further to approx. 0.5%, and in pure nitrogen an amount as low as 0.1% gives good surface cure.

Again, the MEK-rub tests show that the amount of photoinitiator required for a well-cured coating is ≥1%.

C. Formulation based on a flexible aromatic acrylated polyester

In air, 3.5% of PI already gives a dry surface, whereas 3% gives a tacky surface. Increase of UV-intensity enables the use of <2% of PI. Again, a decrease of the oxygen content to 5% leads to improvements concerning the photoinitiator requirements. At 360 mJ/cm2, 1% of PI is sufficient, at 700 mJ/cm^2, 0.5% even suffices. A 1% oxygen concentration leads to a further decrease: 0.5% and 0.1% of PI give dry surfaces at low and high intensities, respectively.

For these formulations, the MEK rub tests did not discriminate. Even at concentrations of 0.1% of PI, the maximum amount of MEK-rubs (200) did not lead to coating damage.

D. Formulation based on a highly flexible acrylated epoxy (containing some TPGDA)

This formulation shows the same over-all picture as the previous ones. Its sensitivity to oxygen in air can be best compared with formulation B. At the low UV dose, 5% of PI gives a good surface cure. The coating sharply reacts to a decrease of the oxygen content of the atmosphere to a 5% level: 1% of photoinitiator is already sufficient to cure its surface. Further decrease to 1% and 0% oxygen leads to limiting PI concentrations of 0.5 and 0.1%, respectively.

Formulations containing ≥1% of photoinitiator show good resistance against MEK.

Summary of relevant findings

The most relevant data are summarised in the following table. For each oxygen depletion step (starting point: air) it gives the attainable decrease in photoinitiator concentration for realising coatings with a dry surface and the concentration difference with the situation in air.

Table 2. Summary of relevant findings

form.	UV-A (mJ/cm^2)	O_2-level 21% → 5%	PI-diff.	O_2-level 21% → 1%	PI-diff.	O_2-level 21% → 0%	PI-diff.
A	360	PI >5 → 1.5%	>3.5%	PI >5 → 0.5%	>4.5%	PI >5 → <0.1%	>4.9%
B	360	PI 5 → 2%	3%	PI 5 → 0.5%	4.5%	PI 5 → <0.1%	>4.9%
C	360	PI 3.5 → 1%	2.5%	PI 3.5 → 0.5%	3%	PI 3.5 → <0.1%	>3.4%
D	360	PI 5 → 1%	4%	PI 5 → 0.5%	4.5%	PI 5 → <0.1%	>4.9%
A	700	PI 3.5 → <1%	>2.5%	PI 3.5 → <1%	>2.5%	PI 3.5 → <0.1%	>3.4%
B	700	PI 2 → <1%	>1%	PI 2 → <0.5%	>1.5%	PI 2 → <0.1%	>1.9%
C	700	PI 2 → <0.5%	>1.5%	PI 2 → <0.1%	>1.9%	PI 2 → <0.1%	>1.9%
D	700	PI 3 → <0.5%	>2.5%	no data	--	PI 3 → <0.1%	>2.9%

At UV-A irradiation doses of approx. 360 mJ/cm^2, the influence of inertisation is dramatic. Lowering the oxygen content of the atmosphere from 21% to 5% enables a reduction between 60 and 80% of the photoinitiator concentration. The absolute difference between the two concentrations ($[PI]_{21} - [PI]_5$) lies between 2.5 and 4%. Further reduction of the oxygen content leads to a further possible decrease: $[PI]_{21} - [PI]_1$ and $[PI]_{21} - [PI]_0$ lie between 2.5 and 5%. However, these very low oxygen concentrations tend to add only a limited amount to the possible photoinitiator reduction. Moreover, the photoinitiator levels that are possible for obtaining a tack-free surface at low oxygen concentrations are less meaningful, because it is clear that at these levels the through-cure of the coating will be endangered. We have found that for all investigated formulations (and using coating layers of 20-40 μm thick) the useful lower limit of photoinitiator concentration is close to 1%.

At higher doses, the relative effect of a decrease of oxygen concentration is similar to the one at lower doses. However, as the amount of photoinitiator for the 21%-case is significantly lower than for the low-intensity examples, the absolute decrease in photoinitiator concentration is significantly lower. In this case $[PI]_{21} - [PI]_5$ is generally close to 2%.

Discussion

Calculation of effect of photoinitiator reduction

The effect of the reduction of the oxygen level for the coating formulator can be easily calculated. If such a reduction leads to a possible lowering of the photoinitiator concentration of Δ_{PI}, then the price difference G between 1 kg of the two formulations concerned is given by:

$$G = [\Delta_{PI} (\$_{PI} - \$_{BF})] / 100 \qquad (2)$$

in which $\$_{PI}$ represents the price of 1 kg of photoinitiator, $\$_{BF}$ the price of 1 kg of base formulation without photoinitiator and in which Δ_{PI} is given in absolute percentage.

For a situation in which a base formulation costs 10 €/kg and a photoinitiator costs 30 €/kg and the PI concentration can be reduced from 4% in air to 1.5% in an atmosphere with 5% of oxygen, this affords a price reduction of G = 2.5 (30 − 10) / 100 = 0.50 €/kg formulation. This price reduction of the formulation can, in principle, be passed along by the paint producers to their customers, as this does not affect the profit margin.

A price reduction of this order of magnitude can be significant. If we assume, that a UV topcoat parquet lacquer station applies coatings of 1 m width at a thickness of 20 g/cm^2 at a speed of 40 m/min (productive period 1400 h/year), and that (partial) inertisation of the station would lead to a possible photoinitiator reduction from 4% (in air) to 1.5%, the following calculation can be made:

- Production: 40 x 60 x 1400 = 3,360,000 m^2/year
- Lacquer use: 3,360,000 x 0.02 = 67,200 kg/year
- Photoinitiator savings: 67,200 x 0.50 = 33,600 €/year.

This possible profit has to be compared with the additional costs that are connected with the use of an inerted production line, which are:

- Nitrogen (or other inert gas such as CO_2) consumption
- Rental or purchase of a inert gas installation (e.g., a liquid nitrogen tank + accessories)
- UV-line adaptations

Of these, the inert gas consumption generally is by far the largest cost factor. It is difficult to give a reliable estimation of these additional costs, as a lot depends on the situation on the spot. The price of liquid nitrogen is variable in time and differs from country to country. Also other factors such as the cheap availability of other inert gases (CO_2) will strongly affect the total picture.

However, the possible savings in photoinitiator use form a significant factor that contributes to the economic attractiveness of inertisation of UV-processes. Some application areas in which inertisation will be economically most attractive will include the following:

1. Of course, high throughput applications are interesting. The main reason for this is that the contribution of the investment costs becomes relatively small in the total cost calculation;

2. High-speed processes, where the residence time of the coating under the lamps is short. In these applications a good cure is generally realised by using relatively high amounts of photoinitiator and/or oxygen scavengers (other expensive solutions in the area of equipment are also possible). Inerting will significantly decrease the price of the formulation used, as the value of ΔPI in the G-formula given above will become relatively large;

3. In applications, where it is impossible or undesirable to use an oxygen scavenger. This can be the case in systems where the scavenger (mostly a tertiary amine) interferes with the cure chemistry or where high demands are exacted to certain properties of the product (e.g., related to smell, yellowing, and chemical resistance);

4. In applications, where an expensive photoinitiator is used. It is obvious that in that particular

case the G-value in the formula will be relatively large. Expensive photoinitiators are initiators that are specifically used in pigmented coatings (acylphosphine oxides and similar systems).

Conclusions

This investigation has shown that UV-curing under an oxygen-deficient atmosphere has many advantages. The most important outcomes are:

1. For „normal" UV-curable coatings, the effect of inertisation is not only noticed at oxygen levels in the 1-1000 ppm range, but even oxygen levels of 5% show already significant effects in comparison with air;

2. Curing under inerted conditions enables the formulator to use significantly less photoinitiator in his formulations. In general, this concentration can be decreased from 2-5% under air to 0.5-1.0% under inerted conditions, while maintaining the same cure speed;

3. The cost reduction due to a lower photoinitiator concentration may outweigh the extra costs necessary for running the UV-process under inert conditions (investments + running costs). Of course, the position of this „equilibrium" for a coating end user (coating applier), strongly depends on the willingness of his coating supplier to fully pass the cost reduction on to his customers and on the exact production conditions (location, production size, etc.);

4. Lower photoinitiator concentrations obviously lead to lower levels of odour and migration of initiator residues. In addition, the weatherability properties of the coatings will, in general, also be improved. This adds to the economic attractiveness of this process.

The inertisation conditions used in the current investigation (closed cell with excellent control over the atmosphere) are not fully comparable with the situation in an actual inerted UV-curing process, in which, e.g., dissolved oxygen will have much less time to leave the coating and oxygen can be dragged along with the fast-moving coating into the inerted chamber. However, it is reassuring that it was found that an oxygen content of 5% already has a markedly positive effect, which means that the levels of dissolved oxygen and mechanically introduced oxygen have to be quite high before they will actually seriously affect the observed results.

Macromol. Symp. **187**, 417–425 (2002)

Tribo Charging Powder Coatings

Zbigniew Bończa-Tomaszewski*, Piotr Penczek

* Industrial Chemistry Research Institute, Department of Polyesters and Epoxy Resins, ul. Rydygiera 8, 01-793 Warsaw, Poland

Summary: At present tribo-guns have captured a fairly large market share of the electrostatic spraying equipment used in the powder coating industry for powder paint applications. However, powder paint system based on carboxyl functional polyesters and some special powder paint formulations cannot be sufficiently charged with tribo guns to obtain a good deposition efficiency of the powder on the object. Attempts have been made to solve this problem through addition of special additives to the powder coatings premix or to the ready-to-use powder. Additives of this type have been proven not to be ideal because they can have a catalytic activity and thus effect the powder coatings properties.
We developed oligomeric additives that enhance the tribo charging of carboxylic polyester resins based powder coatings, without influencing to the kinetics of the curing process and other properties of the coating.

Keywords: tribo charging, corona charging, powder coatings, carbocylic polyesters

Introduction

In 1960, the first electrostatic spray experiments were carried out allowing the application of powder paint to metallic objects. With this process is possible to apply films with a thickness from 40 – 300 μm, in one single coat. There are three powder-charging techniques: contact, corona, tribo and charging [1].

Contact charging

In this process, conductive powder particles are propelled through the gun. When the particles come into contact with the electrode, they acquire the same charge as the electrode. Contact charging is effective, however, the requirement that the powder particles have to be conductive. Most powder coatings materials available are non-conductive, consequently only corona and tribo technologies are used.

  CCC 1022-1360/00/$ 17.50+.50/0

Corona charging

Corona charging is the most widely used of the powder charging techniques. From physicists' point of view corona is a form of plasma. Plasma is ordinary matter in such an energetic state that its atoms have all their electrons stripped away. Normally, a plasma is associated with temperatures of thousands degrees, but in the case corona means a cold plasma, not by thermal energy, but by strong electric field. The basis corona charging principle illustrated in Fig.1 Powder is pumped through feed hoses to a spray gun. A charging electrode in the gun is

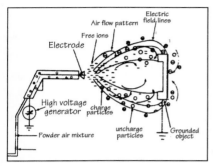

connected to a high-voltage generator, the surrounding air is ionized and free ions are formed in the front of the charging electrode. When powder particles are transported through this ion-cloud, partial adhesion of these ions on the particles takes place and the powder particles become charged. The polarity of the charging electrode can be reversed to create either positive or a negative charge on the powder particles. A negative charge is generally preferred because

Fig.1. *Principle of corona charging*

ions can be generated in greater numbers withless arcing. Because no electrical field exists in areas surrounded by earthed metal due to electrostatic forces, no deposition of powder will take place in these areas. This effect is commonly known as the Faraday Cage Effect. After spraying the final results will be a thin film with insufficient decorative and protective properties.

The second phenomenon that occurs with Corona charging is back-ionization (Fig. 2).

Back-ionization is caused by an electrostatic field in the deposited powder layer on the surface of an earthed object. During Corona charging only 0.5% are charging powder

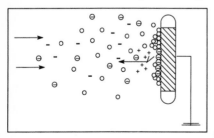

particles. The rest of these free ions move, together with the charger powder paint particles to the object and are incorporated in the powder layer. As layer build-up increases, field strength increases and leads to ionization of the air in the powder layer. The positive ions move in the direction opposite to the oncoming negatively charged

Fig. 2. *Back-ionization*

particles and ions. This results in neutralization of

powder particles, limiting further deposition of powder, creating disturbances in the powder layer, which after curing can results in a coating with a poor apperance.

Tribo Charging

The tribo or frictional charging technique does not make use of a high voltage-generating source, therefore no electrical field is generated between the gun and the contact object. The powder particles are charged by frictional contact between the powder particles and the material, of the body gun. The relative electronegativity of materials commonly used in tribo-charged system is shown Table 1 [2]

Table 1. Relative electronegativity of commonly used materials

MATERIAL	RELATIVE ELECTRONEGATVITY
POLYURETHANE	least electonegative
EPOXY	
POLYAMID	
POLYESTER	
PVC	
POLYPROPYLENE	
POLYETHYLENE	
PTFE (teflon)	Most electronegative

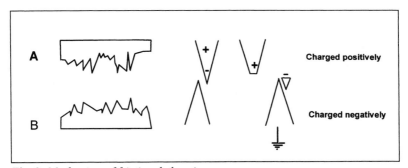

Fig. 3. *Mechanism of frictional charging*

The basic mechanism of tribo charging is illustrated in Fig.3. When the two materials A and B rub against each other little piece of material break off. If one of the materials is very electronegative and hard in comparison with the other material, as is the case for powder charging tribo system, the material transfer will cause a charge transfer. The application

420

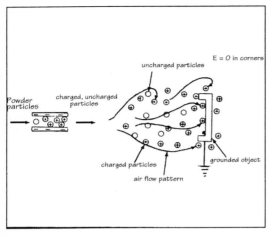

Fig. 4. *Principle of tribo charging*

process is illustrated in Fig. 4. A powder and air mixture enters a gun, where is passed through a tubular section. A conventional powder, such epoxy, polyester or hybrid, is passed through tube made of polytetra-fluoroethylene (PTFE, *Teflon*). As the powder particles collide with the walls of the tube, they pick up a positive charge by giving the electron to the tube, causing the tubes to become negatively charged. This negative charge is then passed to ground. Because is no external electric field in tribo applicators the movement of the particles from the gun to the object is primarily a results of aerodynamic forces. Thus little Faraday Cage-effect is generated, which is one of the greatest benefits that can attributed to tribo technology. Therefore this charging technique allows reach geometrically complex object to be powder homogeneously.

Due to the absence of free ions the Back-ionization does not occur for 10 to 20 seconds in tribo-charging system. Therefore it easier to obtain heavy or thick film with less porosity.

Stabiliation of generated positive charge

It should be noted that the tribo-charging could be improved if the powder is better suited to the stabilization of positive charge. The dielectric constant of PTFE (used in tribo guns as the frictional material) being very low ($\varepsilon = 2.1$), all substances with higher will get a positive charge when brought into intimate contact and separate from PTFE. However the systems used in powder coatings based on carboxy functional polyesters ($\varepsilon = 3$) have not been found to be appropriate for charge acceptance due to small differences in dielectric constant with Teflon [2].

There are several chemical structures that could accomplish of stabilization of positive charge on the powder paint particles. By analogy with what is used in xerographic toners, the addition of quaternary ammonium or phosphonium salts has been found to increase the

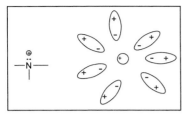

Fig. 5 *Stabilization of generated charge*

positive charging characteristic of the powder, improving transfer efficiency (Fig.5). Solution this type have been proven not to be ideal because they can also have a catalytic effect on the carboxy-epoxyreaction and thus have a negative influence on the flow appearance of the coating. Moreover, they a separated from the powder in the recycling process, as they differ from the specific weight, and thereby the powder loses its enhanced tribochargeability. Some other side effects of these compounds can be yellowing and poor corrosion resistant of the coating. To overcome these drawbacks, work is now in progress to find additives which enhances the tribocharging of polyester resins and subsequently of powder formulation of these without making concessions to the powder paint formulation. [3,4]

Research and development

Inorganic external additives [5]

Addition of the colloidal SiO_2 and Al_2O_3 due to their specific surface was used to improve the charging characteristic. The additive is incorporated into the powder coatings compositions by simply blending it in. The optimum levels are from 0.1 – 0.2% by weight. One of the great advantages of this invention is that permits much more precise adjustment of the tribo charging characteristics than would be possible with a "factory packaged" powder.

Nitrogen containing compounds as tribocharging additives [6]

As an electric charge-increasing agent of hybrid powders the following compounds were used:

- blocked isocynate
- guanamine
- benzotriazole
- anilid of the oxalic acid

Table 2 shows the example of the composition. The chargeability of this system is 2.8 μA [1]

---

[1] In order to obtain a proper triboelectric process, chargeability up to at least 2 μA will be necessary.

Table 2.

Component	weight parts
Esters Resin ER 8600*	90
Epikote 828 (Shell)	10
Adduct B 1530**	2
Modaflow (Monsanto)	0.7
Curezol C 172	0.3

   *   -Polyester resin M=3000, Acid number = 36
   **   -IPDI blocked with caprolactam

Azine and quaternary salts as tribocharging additives [7]

The following compounds enhanced tribo-chargeability without influence on the powder coating properties were described:

- triarylmetane derivatives
- azine derivatives
- tiaazine derivatives
- quaternary salts: e.g. $[PhCH_2N^+Bu_3]$ $[HO-C_{10}H_2SO_3^-]$

The additives are mixed with other components of powder system an extruded. In Table 3 is an example of formulation. The chargeability of this system is 2,1 μA.[2]

Table 3.

Component	weight parts
Alfatalt AN 721 (Hoechst)	399
Becopox EP 303 (Hoechst)	171
Kronos 2160 (TiO_2) (Kronos Titan)	295
Blanc Fixe F/$BaSO_4$ (Sachteleben Chemie)	100
Additol XL 496 (Hoechst)	30
Benzoin	5
C.I. Solvent Blue 124 (tribo agent)	0.5

---

[2] The chargability without tribo agent C.I. Solvent Blue 124 is 0.9 - 1.2 μA

Sterically hindered tertiary amine and amninoalcohol as tribocharging additives[8]

The use of nitrogen containing compounds can have a positive effect on the tribo-charging of powder paint, but also decreases the stability of powder coatings because of an undesired catalytic activity on the epoxy-carboxy reaction. Attempts have been made to solve this problem through addition sterically hindered tertiary amine or aminoalcohol is added during the preparation of polyester. An aminoalcohol containing two hydroxyl group copoly-merizes with polyester and, as results, diffusion of the additive from the coating is avoided. As additive the following compounds were used: N,N-diisobutyl-3-amino-2,4-dimethylpentane (1), N,N-dimethyl-3-amino-2,4-dimethylpentane(2), diisopropyloethanolo-amine, dimethylo-neopentanoloamine (3). The optimum levels of the additives are from 0.5 – 1% by weight. The chargeability are from 3.0-3.6 μA and the additives do not have an influence on other coating properties.

Oligomeric sterically hindered tertiary amine as tribocharging additive (ICRI invention) [9]

As tribocharging additive oligomeric sterically hindered tertiary amine modified with isophorone disocyanate (IPDI) was obtained.

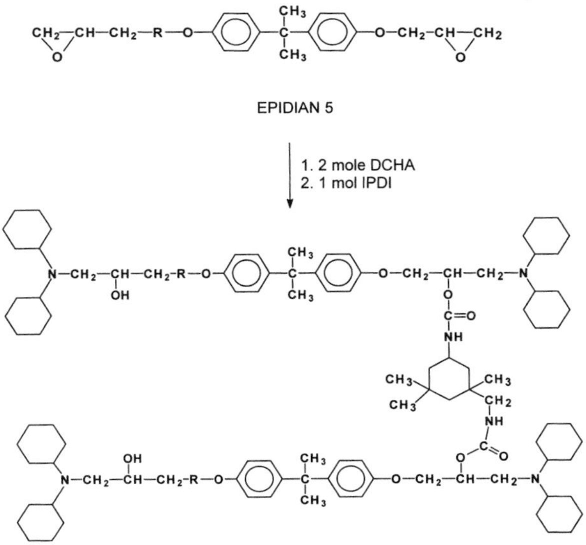

EPIDIAN 5

1. 2 mole DCHA
2. 1 mol IPDI

ELTRIBO

The additive (*Eltribo*) can be added to the cooling resin after synthesis, in the extruder, during the powder coating production process. The amount of Eltribo additive is 1-3% (wt) calculated on the carboxylic polyester resin. It enhanced tribochargeability to 2.5 mA without a negative influence on the coating properties. The example of the formulation is shown in Table 4.

Table 4.

Component	weight parts
Polyester resin *	38
Epidian 012 (ZCh Sarzyna)	23
TiO$_2$	40
Resiflow PV5 (Worlee)	0.7
Benzoin	0.5
Eltribo	3

URALAC 2450 or POLICEN 1100 (Polifarb Cieszyn-Wrocław S.A.), Acid Value 36 – 38

Summary

At present Tribo-charging technique have captured a fairly large market share of the electrostatic spraying equipment used in the powder coating industry for powder coating applications. The Tribo technique has certain advantages over the Corona technique such as the ability to create a fairly homogeneous powder layer on geometrically complex objects. This is due to the absence of a strong Faraday Cage-effect that occurs with the Corona Charging technique. However, the powder paint systems based on carboxy functional polyester cannot be sufficiently charged with tribo guns to obtain a good deposition efficiency of the powder on the object. Amines and organic salts enhance tribochargeability, however, these compounds also have an negative influence on other coating properties due to a catalytic effect on the carboxy-epoxy reaction. Some other side effects of these compounds can be yellowing and poor corrosion resistance of coating. This problem can be solve by using hindered tertiary amine or aminoalcohol as tribo additive or by modification of saturated polyester resins.

References

[1] Reddy, V., "Powder spray technology and their selection", *Plat. Surf. Finish.,* **76,** 34-38 (1989).
[2] Binda, P., "Polyester resin for tribo-charging powder coatings", *Polym. Paint Colour J.* **180,** 710-712 (1990). CA 115, 10915 (1992).
[3] Weigel K., "Beschichten mit außenbeständigen Pulverlaken im KPS-Verahren", *Metalloberfläche,* **40,** 491-4 (1986).

[4] Weigel K., "Triboabscheidbare Polyester-Epoxidharz-und Polyester-TGIC-Pulverlacke", *Metalloberfläche,* **42**, 179-81 (1988).
[5] EU Patent 300 818.
[6] Jap.Patent 91 00 775, CA **114**, 51984 (1991).
[7] EU Patent 315 084.
[8] EU Patent 371 528; Netherlands Patent 88 02 748, CA **114**, 83962 (1991).
[9] Polish Patent P-323017.

Macromol. Symp. **187**, 427–435 (2002)   427

New Materials with High π Conjugation by Reaction of 2-Oxazoline Containing Phenols with Polyamidines and Inorganic Base

Juraj Kronek[a]*, Jozef Luston[a], Frank Böhme[b]

[a]Polymer Institute, Slovak Academy of Sciences, Dubravska cesta 9, 842 36 Bratislava, Slovak Republic, e-mal: upolkron@savba.sk
[b]Institut für Polymerforschung Dresden e.V., Hohe Str. 6, 01069 Dresden, Germany

Summary: The non-covalent interactions between 2-oxazoline containing phenols and an aliphatic polyamidines as well as an inorganic base were studied. The reaction of a weak acid with a strong base results in the formation of the deprotonated species and subsequently in the formation of a new electronic structure. A bathochromic shift of the wavelength of the absorption maxima of the chromophores bounded to polyamidine was observed. Depending on the structure of the chromophore, the shift of the absorption maxima is 40-100 nm. The changes in photochemical behavior can be explained by the higher portion of quinoid structures in the conjugated π-system. The degree of deprotonation is dependent on the molar ratio of the chromophore and the polymeric base. Analogous results were obtained with an inorganic base.

Keywords: 2-Oxazoline, Phenol, Polyamidine, Ionic Complexes, Deprotonation, UV Spectroscopy

Introduction

It is well known that transformation of some molecules provides changes in their photochemical properties. This transformation can be achieved by the effect of light or heat, by the change of the molecule arrangement (geometry) or by interaction with certain reagents. One of the chemical methods can be a protolytical equilibrium reaction between a chromophore containing an acidic group and a base.

Low molar mass amidines are molecules which can interact covalently as well as non-covalently with a number of compounds[1]. Their pronounced basic behavior was emphasized[2]. Aliphatic polyamidines represent a basic polymer material which is able to form strong H-bonds with different H-donors or even to deprotonate them[2]. In the case, when an acidic compound contains a chromophore, changes in the electronic structure upon deprotonation are observed[3]. This effect can be utilized in several useful applications.

          CCC 1022-1360/00/$ 17.50+.50/0

Recently, ionic complexes of bis[(hydroxybenzylidene)cyclohexanone]s with aliphatic polyamidines were studied as active third-order non-linear optical materials[3,4].

Our effort was focused on the reaction of phenolic group containing 2-oxazolines with an organic polymeric base and a strong inorganic base. The non-covalent interactions between 2-oxazoline containing phenols and an aliphatic polyamidine and KOH in solutions were studied.

Experimental part

Materials

The starting compounds, as well as reagents and solvents, were used without further purification. Poly(1,10-decamethyleneacetamidine) (**1**) was prepared from 1,10-decanediamine and triethylorthoacetate according to Böhme et al[4]. 2-(4-Hydroxyphenyl)-2-oxazoline (**2**) (m.p. =208-210 °C) was prepared according to Saegusa[5]. 2-[4-(4'-Hydroxyphenyl)-azophenyl]-2-oxazoline (**3**) (m.p. = 236-238 °C) was prepared according to literature[6]. 2-(6-hydroxy-2-naphthyl)-2-oxazoline (**4**) (m.p. =198-200 °C), 2-(4'-hydroxy-4-biphenyl)-2-oxazoline (**5**) (m.p. = 238-240 °C) and 2-(6-hydroxy-3-pyridyl)-2-oxazoline (**6**) (m.p. = 180-185 °C) were prepared from the corresponding hydroxyacids according the literature[7].

Procedures and measurements

Methanolic solutions of the phenols **2-6** as well as the polyamidine **1** with a concentration of 10^{-3} mol/dm^3 were prepared. For the measurements, these solutions were mixed in appropriate molar ratios in such a way that the final concentration of the phenols was $5 \cdot 10^{-5}$ mol/dm^3. An analogous procedure was used for the preparation of mixtures of **2-6** and KOH in methanol. UV/Vis spectra were recorded with a Varian Cary 100 spectrometer.

Results and discussion

It has been mentioned earlier that the aliphatic polyamidine **1** with a pK$_a$ equal to 10.9 behaves like a strong organic base [2]. Its acid-base behavior is demonstrated in Scheme 1.

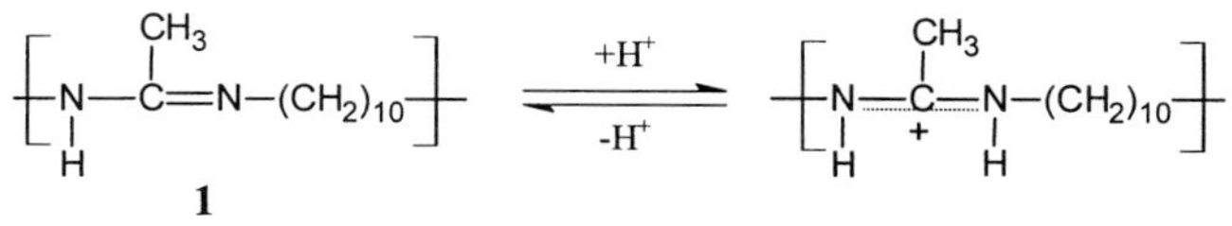

Scheme 1

It has been assumed that in a mixture with polyamidine **1** substituted phenol derivatives such as compounds **2 – 6** (Fig. 1) should act as proton donors. In order to know more about the extend of these interactions and its influence onto the optical properties of the mixtures UV/Vis spectroscopic investigations in methanolic solution were performed. For this, the solutions of the phenols were mixed with the polyamidine solution in different molar ratios.

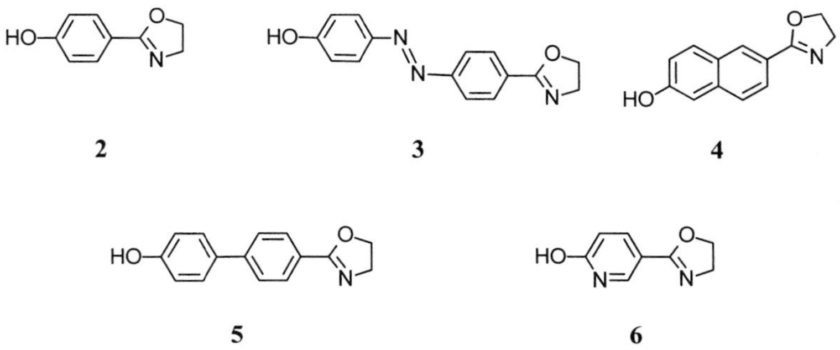

2   **3**   **4**

5   **6**

Fig. 1

UV/Vis spectra of the ionic complexes formed together with the dependencies of the particular absorbances on the molar ratio of the reaction components are shown in Figures 2 - 6. Wavelengths of the respective absorption maxima of the neutral and the deprotonated samples as well as the molar ratios of the polyamidine to the phenol at which almost complete deprotonation occurs are given in the Table 1.

430

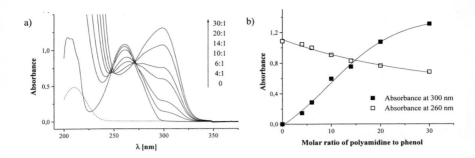

Fig. 2. UV spectra (a) of 2-(4-hydroxyphenyl)-2-oxazoline (2) with polyamidine 1 at different molar ratios of 1 to 2 (- - - pure 1) and (b) the dependence of the absorbance at 300 nm and 260 nm on the molar ratio of 1 to 2

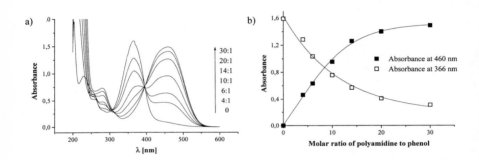

Fig. 3. UV spectra (a) of 2-[4-(4'-hydroxyphenyl)-azophenyl]-2-oxazoline (3) with polyamidine 1 at different molar ratios of 1 to 3 and (b) the dependence of the absorbance at 366 nm and 460 nm on the molar ratio 1 to 3

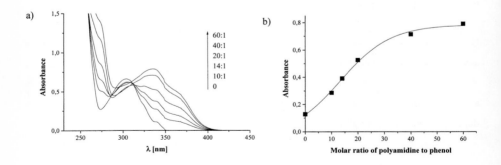

Fig. 4. UV spectra (a) of 2-(6-hydroxy-2-naphthyl)-2-oxazoline (4) with polyamidine 1 at different molar ratios of 1 to 4 and (b) the dependence of the absorbance at 337 nm on the molar ratio of 1 to 4

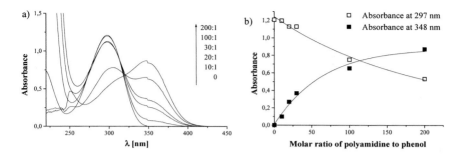

Fig. 5. UV spectra (a) of 2-(4'-hydroxy-4-biphenyl)-2-oxazoline (**5**) with polyamidine **1** at different molar ratios of **1** to **5** and (b) the dependence of the absorbance at 297 nm and 348 nm on the molar ratio of **1** to **5**

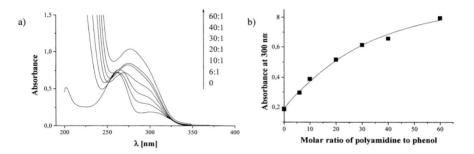

Fig. 6. UV spectra (a) of 2-(6-hydroxy-3-pyridyl)-2-oxazoline (**6**) with polyamidine **1** at different molar ratio of **1** to **6** and (b) the dependence of the absorbance at 300 nm on the molar ratio of **1** to **6**

Tab. 1. Reaction of polyamidine **1** with 2-(4-hydroxyphenyl)-2-oxazoline (**2**), 2-[4-(4'-hydroxyphenyl)-azophenyl]-2-oxazoline (**3**), 2-(6-hydroxy-2-naphthyl)-2-oxazoline (**4**), of 2-(4'-hydroxy-4-biphenyl)-2-oxazoline (**5**) and 2-(6-hydroxy-3-pyridyl)-2-oxazoline (**6**).

Phenol	λ_{max}[a]	λ_{max}[b]	$\Delta\lambda_{max}$[c]	Isosbestic point	Molar ratio[d]
	nm	nm	nm	nm	
2	260	300	40	248, 271	30
3	366	460	94	306, 396	30
4	303	337	34	287, 312	60
5	297	348	51	276, 320	200
6	261	300	39	-	60

[a] Wavelength of the absorption maxima of the origin chromophores
[b] Wavelength of the absorption maxima of the chromophores after reaction with polyamidine **1**
[c] Difference between wavelength of absorption maximum of original and deprotonated chromophore
[d] Molar ratio of polyamidine to phenol necessary for obtaining 100 % deprotonated chromophore

Figure 2a shows the spectral changes of the UV/Vis spectra of 2-(4-hydroxyphenyl)-2-oxazoline (2) at different molar ratios of polyamidine to phenol. It is seen that the absorption maximum at 260 nm decreases with increasing polyamidine concentration whereas the absorption maximum at 300 nm increases. This represents a bathochromic shift of the chromophor's absorption band about 40 nm upon deprotonation. In contrast to an inorganic base, a complete deprotonation of 2 is only reached at a big excess of polyamidine (Fig. 2b). A 30-fold excess of polyamidine is necessary to obtain nearly all molecules deprotonated. The reaction of 2-[4-(4'-hydroxyphenyl)-azophenyl]-2-oxazoline (3) is analogous to 2 (Fig. 3a). The new absorption maximum at 460 nm appears instead of the absorption band at 366 nm what represents a $\Delta\lambda$ of 94 nm. Complete deprotonation is reached at a molar ratio of the components equal to 30 (Fig. 3b). In this case, the reaction is accompanied with a color change. The orange color of the original chromophore changes to deep red after deprotonation. For the compounds 4 - 6, the shift in absorption maxima is in the range 35 to 50 nm (Figs. 4a-6a, Tab. 1) depending on the structure of the chromophore. For the compounds 4 and 6, complete deprotonation is achieved at a molar ratio of 60. The highest excess of base has to be added to derivative 5 which is completely deprotonated at a molar ratio of 200 (Figs. 4b-6b). In the case of derivatives 4 and 5 the reaction shifts the absorption maxima to the beginning of the visible part of the spectra.

The processes which proceed after mixing a phenol with the polyamidine can be described in the following way. In the first step, a hydrogen bond between the phenolic group of a phenol and an amidine moiety is formed (Scheme 2) which is in the next step followed by the deprotonation reaction of the phenolic group.

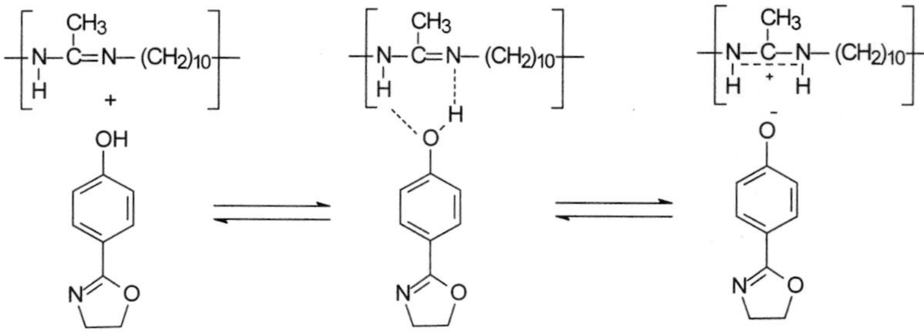

Scheme 2

The reactions are very fast what can be observed also visually by color changes during the reaction between polyamidine **1** and the azobenzene moiety containing phenol **3**.

All UV spectra show a strong bathochromic shift of the absorption maxima of the deprotonated chromophores. Obviously, their electronic structures are strongly influenced by deprotonation. Because of the π-conjugation, it is assumed that the negative charge is distributed over the whole molecule. One can distinguish between various resonance structures. For compound **3** the quinoid and the benzoid structure are shown in Scheme 3. The real structure of the molecule is somewhere in between.

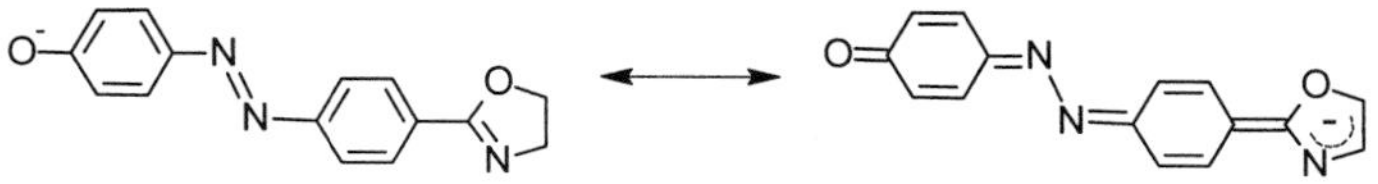

3

Scheme 3

Because of the strong bathochromic shift of the absorption maxima, a distinct influence of the chinoid structure can be expected.. This assumption is in accordance with the literature[4], where the interactions between bis[(hydroxy-benzylidene)cyclohexanone]s and aliphatic polyamidines bathochromically shifted the absorption maxima for about 100 nm. The structural changes were examined also by ^{13}C NMR spectroscopy. The authors observed positive and negative chemical shift effects Δδ when compared the signal position of the parent and the deprotonated forms. The strongest effect upon addition of a base was found for the carbon bearing the phenolic group where Δδ was higher than +12 ppm. Distinct negative chemical shift effects up to –6 ppm were observed at those positions were the negative charge can be stabilized. From these results, in analogy to our system shown in Schemes 2 and 3, delocalization of the negative charge was concluded.

For comparison, the interactions between phenols **2-6** with a strong inorganic base were studied as well. The results are illustrated in Figure 7 and Table 2. It is seen that also in the presence of an inorganic base a deprotonation reaction accompanied with the bathochromic shift of the absorption maxima proceeds. The shift of the absorption maxima is, as expected, the same as in presence of the polyamidine. This indicates that the processes occurring in the given systems are the same (Tab. 1 and 2). The difference appears in the amount of base that

434

has to be added to deprotonate the phenol derivatives completely. As shown in Table 2, complete deprotonation is already reached at a base chromophore ratio between 4 and 10.. This effect is the result of the higher basicity of KOH used in the deprotonation reaction.

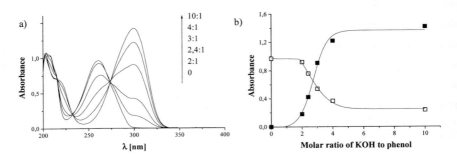

Fig. 7. UV spectra (a) of 2-(4-hydroxyphenyl)-2-oxazoline (2) mixed with KOH in different molar ratios to 2 and dependence of the absorbance at 300 nm and 200 nm on the molar ratio of 1 to 2 (b)

Tab. 2. Results of reaction of KOH with 2-(4-hydroxyphenyl)-2-oxazoline (2), 2-[4-(4'-hydroxyphenyl)-azophenyl]-2-oxazoline (3), 2-(6-hydroxy-2-naphthyl)-2-oxazoline (4), of 2-(4'-hydroxy-4-biphenyl)-2-oxazoline (5) and 2-(6-hydroxy-3-pyridyl)-2-oxazoline (6).

Phenol	λ_{max}[a]	λ_{max}[b]	$\Delta\lambda_{max}$[c]	Isosbestic point	Molar Ratio[d]
	nm	nm	nm	nm	
2	261	300	39	233, 274	4
3	366	460	94	300, 397	4
4	303	338	35	283, 314	6
5	297	349	52	258, 318	10
6	261	300	39	-	7

[a] Wavelength of the absorption maxima of original chromophores
[b] Wavelength of the absorption maxima of chromophores after reaction with KOH
[c] Difference between wavelength of absorption maximum of original and deprotonated chromophore
[d] Molar ratio of KOH to phenol necessary for obtaining 100 % deprotonated chromophore

Conclusions

The interactions of phenolic group containing 2-oxazoline derivatives with an inorganic base result in the deprotonation of the chromophores and subsequently in a change of their electronic structures which are characterized by higher portions of quinoid structures. The same effect is observed in the reaction of the chromophores with a polyamidine which has proven to be a strong polymeric base. Due to the different basicities of the bases used, different molar ratios of reactants are necessary to deprotonate the chromophores completely. These processes are accompanied by a bathochromic shift of the absorption maxima in the range of 40-100 nm. In several cases, the absorption maximum of the chromophores is shifted to the visible region. The interaction of the chromophore with a base provides new materials with higher π conjugation, which can be utilized for non-linear optical devices. From this point of view, the derivative **3** with the strongest conjugation seems to be the most promising.

Acknowledgements

The authors acknowledge the partial financial support of the Slovak grant agency VEGA (project No. 2/7038/20) and the fund for international cooperation of the German Federal Ministry for Education, Science, Research and Technology (project No. SLA-004-99).

References

[1] S. Patai, Z. Rappaport (Eds.), *The chemistry of amidines and imidates*, John Wiley and Sons, New York, 1991
[2] F. Böhme, C. Klinger, C. Bellmann, *Colloid Surf. A: Physicochem.Eng. Aspects* **2001**, 189, 21
[3] F. Böhme, L. Häussler, A. V. Tenkovtsev, A. V. Yakimansky, *Polym. Preprint*, **1999**, 40, 1140
[4] A. V. Tenkovtsev, A. V. Yakimansky, M. M. Dudkina, V. V. Lukoshin, H. Komber, L. Häussler, F. Böhme, *Macromolecules*, **2001** 34, 7100
[5] S. Kobayashi, T. Mizutani, T. Saegusa, *Makromol. Chem.*, **1984**, 185, 441
[6] J. Luston, J. Kronek, F. Böhme, H. Komber, *Macromol. Symp.*, **2001**, 164, 105
[7] J. Kronek, *PhD Thesis*, Bratislava, **2001**

Macromol. Symp. **187**, 437–445 (2002)

Waxes for Powder Coatings

Günther Heiling*

Clariant GmbH, 86368 Gersthofen, Ludwig-Hermann-Straße 100, Germany

Résumé: Dès les débuts, c'est-à-dire la fin des années 50, l'industrie des peintures en poudre a développée des produits et des procédés de fabrication présentant une alternative éprouvée aux peintures liquides. En principe, les peintures en poudre consistent d'un mélange de résines synthétiques, de pigments et d'additifs présentant des avantages économiques et écologiques par rapport à d'autres systèmes de revêtement. Les matières premières sont mélangées, extrudées et moulues en peintures en poudre.

Les cires jouent un rôle important comme additifs dans les peintures en poudre. Les différantes cires permettent d'influer favorablement aussi bien la fabrication, le stockage, la mise-en-oeuvre que les propriétés de la peinture en poudre appliquée. Les différantes cires et les avantages respectifs sont présentés à l'aide d'exemples pratiques.

Summary: Since the first beginnings at the end of the 50th, the powder coating technology has become a perfected alternative in product and procedure respect to liquid paints. Powder coatings are primarily a mixture of synthetic resins, pigments and additives offering economical and ecological advantages compared to other coating systems. The raw materials are mixed, extruded and ground to coating powders.

Waxes play an important part as additives in powder coatings. Production, storage and processing as well as the properties of the applied powder coating can be influenced positively by different waxes. Various waxes and their advantages are demonstrated by practical advantages.

Keywords: powder coatings; UV-systems; coating surface; compounds; extrusion; application

Since 1927, waxes are produced and finishes at the production site Gersthofen. The company originally belonging to Hoechst AG has become part of Clariant GmbH in 1997.

The term „wax" is a designation of merchandise origin for a number of naturally or synthetically produced substances having normally following properties:

 CCC 1022-1360/00/$ 17.50+.50/0

- workable at 20 °C
- firm to brittle hard
- coarse to fine crystallinity
- translucent to opaque but not glassy
- melts above 40 °C without decomposition
- relatively low viscosity even at slightly above the melting point
- strongly temperature-dependent consistency and solubility
- capable of being polished under light pressure

Since the first beginnings at the end of the 50th, the powder coating technology has become an economical and ecological alternative of liquid coating materials. With an annual growth rate of about 10 %, the world market demand increased from 270 thousand tons in 1992 to 590 thousand tons in 2000.

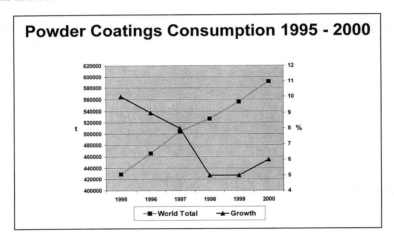

The most important areas are Europe, North America and Asia covering more than 90 % of the world market. The largest producers in Europe are Italy, Germany, France and England with 73 % of the total quantity produced.

Powder coatings are applied today successfully in different areas of industrial coating. A substitution of liquid coating materials is possible in many cases without any problem. Plastics painting and wood coating often cause problems because electric conductivity is not good. Examples for the use of powder coatings:

European Market Powder Coatings - 2000

Italy	80000
Germany	60000
Great Britain	36000
France	30000
Spain	25000
Switzerland	9000
Austria	8000
Netherlands	7000
Norway	6000
Belgium	5000
Sweden	4000
Portugal	2500
Greece	2500
Finland	2000
Ireland	1500
Denmark	1500

Traditional applications:

- Electric appliances (white material)
- General mechanical engineering
- Steel furniture
- Engineering of services and technical equipment of buildings
- Facade facing

New application fields:

- Motorcar accessories (e.g. wheel boss caps, wheel rims)
- Motorcar serial paints (transparent cover paints)
- Bicycles and motorcycles
- Coil Coating with thin-coat powder coatings

Future prospects:

- Plastics coating
- Wood and furniture paints
- UV cross-linked powder coating

Waxes play an important part as additives in powder coatings. Following properties can be influenced positively during production, storage and processing as well as on the powder coating applied:

Action of waxes in powder coatings

During manufacture and storage of coatings powder
– Improves the grinding characteristics and the shelf life because the coating particles do not agglomerate during storage or transport.

During extrusion and processing of coating powder
– Act as lubricant during extrusion thus reducing wear and extending the life and maintenance intervals of the equipment.
– Increase dispersibility and wettability of the pigments during extrusion. It will be achieved an higher output and at the same time power consumption will be reduced strongly.
– Improve the flow properties of the powder in pipe lines and spray equipment during application.
– Act as degassing agent and can replace proportionately benzoin.

Influence on powder coating surface
– Promoting flow, levelling and film formation. A smoother surface results.
– Increase the scratch resistance because objects which may scratch slip more easily off the smooth slippery surface.
– The resistance of the coating to mechanical stresses will be significantly increased. The reason is the higher smoothness of the surface, the hardness of the coatings film is unchanged.
– Achieve a fine to coarse texture, depending from the film thickness.
– Resistance to marking by metals and abrasion resistance are significantly improved by the reduction in coefficient of friction.
– Improves dirt- and water-repelling characteristics.
– Improves anti-blocking properties.
– Gloss reduction is possible.

In a hybrid powder coating we have tested a number of waxes and wax compounds of different chemical composition. Dosage for these tests was 1 % wax. Test formulation:

39.3 %   Alftalat AN 770 (Vianova Resins)

16.8 %   Beckopox EP 303 (Vianova Resins)

 3.0 %   Additol XL 493 (Vianova Resins)

 0.3 %   Benzoin (DSM Special Products)

 9.9 %   Blancfixe F (Sachtleben Chemie GmbH)

29.7 %   Kronos 2160 (Kronos Titan GmbH)

 1.0 %   Ceridust XXXX

Following properties were achieved with the different waxes:

PE wax powder

prolongation of service life of machinery improvement of the flowability of the powder; improvement of levelling

economical product: Licowax PE 520 Powder

Amidewax d_{50} = 6,5 µm

–   high melting point (140 °C)

–   degassing agent

–   entrapped air can escape

especially on porous substrates an partially replace Benzoin

Product: Ceridust 3910

Compound Amid / PE d50 = 7 µm

–   improvement of levelling

–   smoother surface

–   increased scratch resistance

–   better grindability

–   no lumping during storage and transportation

–   increased slip

–   an addition up to 2% supports matting effect

Product: Ceridust 9615 A

Synthetic hydrocarbon wax d50 = 6,5 µm

– reduction of gloss

– good levelling

– good scratch resistance

Product: Ceridust 2051

Compound wax d50 = 6,5 µm

– extremely high slip

– no gloss reduction

– haze is being minimized

Ceridust 3831 is a Clariant unique product

Compound PE / TF d50 = 6 µm

– smoother surface, slip

– excellent scratch resistance

– prevention of metal markings

– abrasion resistance

Product: Ceridust 3920 F

Compound PE / TF d50 = 13 µm

– fine texturing / structuring

– anti blocking

– excellent soil repelling

Product: Ceridust 3940 F

Polar waxes in particular can also clearly improve dispersibility of pigments as well as extruding ability of coating powders. This is shown by the example of a polyester primid system.

Montan wax d50 = 9 µm

– internal lubricant

– reduction of torque $\Rightarrow$ energy saving

– good wettability and dispersibility of pigments

Product: Ceridust 5551

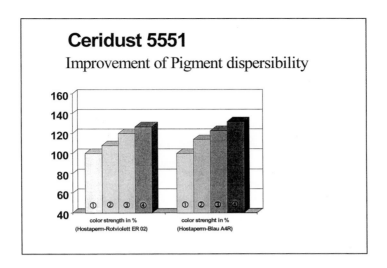

Besides the advantages of the use of waxes as additives for powder coatings, these explanations have also shown the versatile possibilities of the use of coating powders.

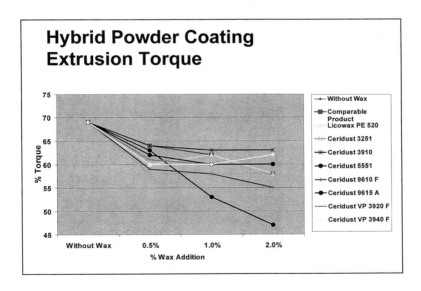

Due to the increasing environmental pollution by industrial paint and varnish plants resulting in solvent emissions, sewage contamination and volume of paint and varnish sludge, the importance of powder coatings as non-polluting alternative is constantly increasing. The production quantity in Germany has augmented continuously by about 6 % per annum during the past years.

New application possibilities, such as wood- or plastic coatings, can be realised more and more by formula optimisation. UV-reactive powder coatings are another direction of development. The potential of powder coatings to be achieved is estimated by experts to 30 % - 40 % of the entire demand of industrial paints and varnishes.

Waxes are an important part of powder coating formulations. They control a number of properties during production and processing of powder coatings and serve also to optimise the properties of powder coating. For UV-reactive powder coatings Clariant has marketed a new development under the designation **Ceridust TP 5091** which we would like to introduce to you in the following.

Literature:

Additive für Pulverlacke; G. Heiling, Lucas Meyer, Welt der Farben 01/99

Wachse für Pulverlacke; Dr. E. Krendlinger, Clariant Gersthofen,Welt der Farben 12/99

New developments in powder coatings; Dr. N.R. Kondekar, National Paints Factories, United Arab Emirates; Paintindia 03/2000

In particular I would like to thank Mr. Hans Dieter Nowicki – Clariant GmbH, Gersthofen – who has conducted all tests for this presentation.

Macromol. Symp. 187, 447–458 (2002)

How to Control Dirt Pick-Up of Exterior Coatings

Oliver Wagner*, Dr. Roland Baumstark

BASF AG, EDK/BA, 67056 Ludwigshafen, Germany

Summary : The composition of dirt and the tendency of coatings to pick up dirt are determined largely by the location. The factors affecting the tendency of coatings to pick up dirt are varied and complex. That is why the problem of dirt pick-up on exterior coatings can only be resolved satisfactorily by optimisation of the paint formulation as a whole (PVC, type of formulation, surface texture) and of all the raw materials (binders, pigments, fillers, additives).

Introduction

Exterior coatings are used to protect building materials from the weather and to give buildings a decorative finish and aesthetic appearance. That is why the contamination of newly coated exterior walls with dirt, which sometimes appears very quickly, is viewed as unsatisfactory by the end-users and requires improvements. To achieve these, however, we need to know more about the causes of this contamination and the composition of the dirt.

Types of dirt on buildings

Our own recent studies show that it is mainly deposits of soot, dust and inorganic crystallites on the surface of a coating which are responsible for the dirty or grey appearance of weathered coatings.

In studies with scanning electron microscopy (SEM) and atomic force microscopy (AFM) it was possible to detect both coarse dust particles (with a diameter of approx. 10 to > 50 µm) and also, for the first time, ultra-fine dust particles (with a diameter < 100 nm, see Figure 1) on weathered surfaces of exterior coatings.

          CCC 1022-1360/00/$ 17.50+.50/0

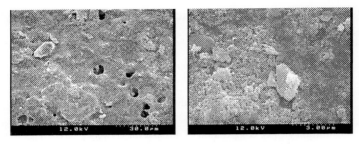

Figure 1a. SEM images of a very dirty exterior coating (PVC = 42 %)

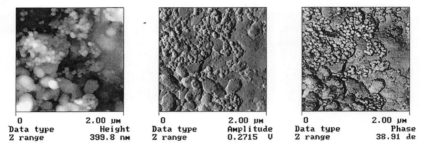

0	2.00 µM	0	2.00 µM	0	2.00 µM
Data type	Height	Data type	Amplitude	Data type	Phase
Z range	399.8 nm	Z range	0.2715 U	Z range	38.91 de

Figure 1b. AFM images (topography and phase contrast with finely dispersed (< 100 nm) deposits on a weathered exterior coating (PVC = 42 %)

At the same time, with regard to calcitic formulations, serious microscopic changes in the coating surface were observed – in the worst case after only one year of outdoor exposure (45°, south) – without any visible sign of chalking. Sometimes micropores and holes appear in the coating and particles of dirt can then become embedded in them (see Figure 2).

Figure 2. Particles of dirt embedded in holes; exterior paint PVC = 42 % after 1 year of outdoor weathering (45° south, Limburgerhof)

In addition, coatings can be affected by micro-organisms (algae, fungi, lichen and mosses) after prolonged weathering (starting after just one year, more evident after about 2 years of weathering at 45° south) and especially on the weather side of buildings. This can be kept under control within certain limits by using film preservatives, usually combined products consisting of algicides and fungicides (see Figure 3).

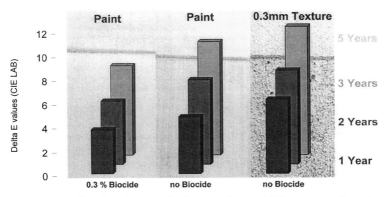

Figure 3. Results of outdoor weathering on paints and 0,3mm textured coatings with and without added biocide[1]

Plant growth tends to increase as the PVC increases and occurs on a larger scale with paints with a PVC exceeding the CPVC.

Overall, paints based on hydrophobic acrylate-styrene emulsions show a less marked tendency to plant growth than the more hydrophilic pure acrylate systems.

This is confirmed in Figure 4 by exterior paints modified with silicone resin. Owing to the styrene, which is a considerably more hydrophobic monomer, the acrylate-styrene emulsion absorbs less water than the pure acrylate and this can be seen in the differences in behaviour with regard to the swelling and shrinking of the coating film. As a result, the tendency to plant growth is reduced significantly.

After 5 years of subjection to outdoor weathering, however, it was impossible to detect any reduction in dirt as a result of the addition of the silicone resin emulsion, irrespective of the binder used.

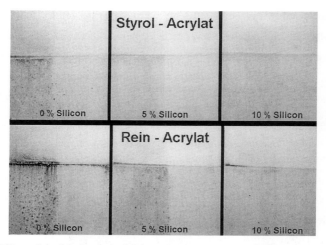

Figure 4. Effect of the binder and addition of silicone resin on plant growth in exterior paints after 5 years of outdoor weathering (45°, south, Limburgerhof)[1]. Left: dirty condition. Right: after washing.

The locational factor

The composition of the dirt depends very much on the location, as shown by studies by Nakaya[2] in Japan and confirmed by our own studies. In this connection, the proportion of carbon and the ratio of lipophilic dirt to hydrophilic dirt vary. In conurbations, organic dirt with a high proportion of carbon (e.g. diesel soot) predominates owing to the large amount of traffic and the many firing plants, whilst in rural areas there is mainly mineral, inorganic or biogenic material (parts of plants, pollen, spores etc.) and not much elementary carbon. Generally, however, significant amounts of inorganic particles containing Si, Fe, Al, K or Mg are always detectable on the surface of dirty exterior coatings, regardless of the location.

Nowadays, there is a very large amount of knowledge available in the literature regarding the sources, composition and particle size distribution of fine dusts in the environment[3].

Anthropogenic sources, such as road traffic, construction work and industrial activity, and combustion processes are responsible primarily for finely dispersed suspended particles (typically < 2.5 μm particle diameter, with some amounts down to particle diameters of less than 50 nm).

Natural processes, like the formation of marine aerosols, the swirling up of mineral dust or biological material by the wind (mainly with a particle diameter of 2.5 - 10 μm) and the formation of secondary aerosols from gaseous atmospheric pollutants (e.g. ammonium sulphate or ammonium nitrate) are other sources of fine dusts.

A study from Switzerland[4] shows that about one third of the suspended particulate comes from incomplete combustion processes (of which 50 % are soot particles), one third consists of secondary aerosols and the final third is made up of mineral particles from wind erosion, construction work, wearing away of roads etc..

Dust particles are generally spread around by air movement and rainwater or, in the case of the finest particles, by Brownian movement and thus come into contact with exterior coatings. And so the amount of suspended dust in the air at the weathering location has a determining effect on the dirtiness of the coating. That is why coatings in conurbations which have a high concentration of suspended dust become dirty much more quickly than in a rural environment with a lower overall dust concentration.

For an international comparison one has to turn to the so-called TSP values. TSP (Total Suspended Particulate) concentrations are the pollution figures for suspended dust in $\mu g/m^3$ and, by definition, comprise particles from 0.1 to 100μm (see Figure 5).

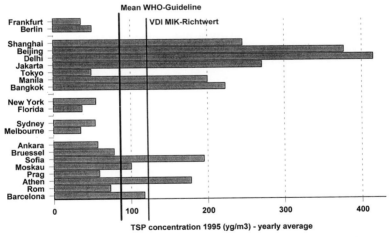

Figure 5. TSP concentrations (annual average) of various towns in 1995[1]

There are severe problems with dirt and atmospheric pollution in the developing and semi-developed countries of Asia, with figures of 200 $\mu g/m^3$ and higher, and South America (e.g. Mexico City with 410 $\mu g/m^3$). In industrial cities in China, such as Lanzhou (Gansu Province), TSP concentrations of 500 $\mu g/m^3$ and higher were measured, statistically, every second week in the first half of 2000. Cities like Peking also have ten times more SO_2 pollution in the atmosphere than, for example, Frankfurt. These differences make it clear that (formulation and weathering) results obtained in one region cannot necessarily be applied to other regions of the world.

Factors affecting the tendency to attract dirt

One problem with regard to the study of the factors affecting the dirtying of exterior coatings is the fact that, to date, no reliable laboratory dirt tests are available[1,5] and even outdoor weathering studies related to a particular site and kind of exposure (angle of inclination: 45°, 60° or 90°; and direction: south, north, west) provide conflicting results.

Based on many years of practical experience, it is nonetheless possible to make the following statements concerning the determining factors:

Apart from the polarity, resultant wettability, and swellability of the coating surface due to rain water, there are other factors, such as the thermoplasticity, surface tack, porosity and surface texture (roughness) of the coating, which have a crucial impact on its tendency to attract dirt.

Effect of the constituents of a formulation

The factors affecting the tendency of a paint formulation to pick up dirt are shown, together with an indication of their impact, in Table 1 and will be discussed below with reference to examples.

Table 1. Influencing factors on dirt pick-up of architectural coatings

Influencing factor	Impact
PVC	Strong
Type of formulation (emulsion, silicate or silicone paint)	weak – strong (to be explained in more detail below)
Binder	Strong
Solvent/plasticiser	Weak
Titanium dioxide	moderate – strong
Thickening agent	Moderate
Filler	Weak

The effect of the binder on the tendency of water-based coatings to pick up dirt is determined primarily by its polarity, glass transition temperature, morphology, cross-link density and thermoplasticity (see Figure 6, 7, 8) [1,6-10].

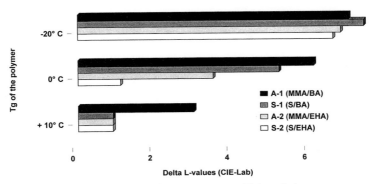

Figure 6. Effect of the glass transition temperature (Tg) and the monomers of the binders on the greying ΔL of exterior paints after 3 years of outdoor weathering (45°, south) in Limburgerhof[1]

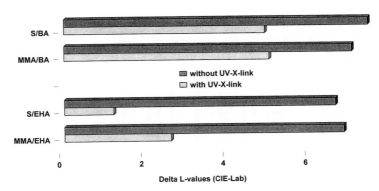

Figure 7. Effect of UV curing of the binders on the greying ΔL of pure acrylate and SA exterior paints after 3 years of outdoor weathering (45°, south) in Limburgerhof[1]

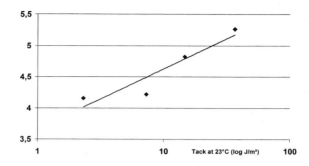

Figure 8. Correlation between the dirt pick-up ΔL of pure acrylate exterior paints (PVC=42%) and the tack of the emulsion film.

The film-forming aids and plasticisers (Figure 9) as well as the thickening agents (Figure 10), fillers and grades of titanium dioxide chosen (Figure 11)[7] are also formulation parameters which have an effect on the tendency of aqueous dispersion-based coatings to pick up dirt.

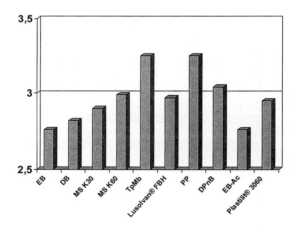

Figure 9. Effect of 2% of different film-forming aids and plasticisers on the dirt pick-up ΔL of exterior paints (PVC = 50%) after 1 year of outdoor weathering (Limburgerhof, 45°, south)

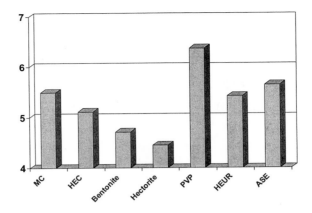

Figure 10. Effect of different thickening agents on the dirt pick-up ΔL of exterior paints (PVC = 42%) after 1 year of outdoor weathering (Limburgerhof, 45°, south)

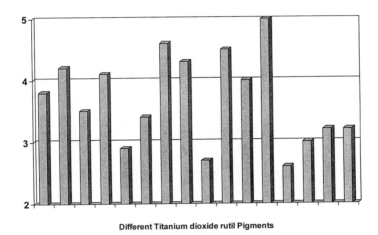

Different Titanium dioxide rutil Pigments

Figure 11. Effect of different grades of titanium dioxide on the greying ΔL of exterior paints (PVC = 45 %) after 2 years of outdoor weathering (45°, south, Limburgerhof)

In addition, the pigment volume concentration[1,10] plays a crucial role via the quantity of binder used (see Figure 12, 13), and the type of formulation chosen (emulsion paint, silicate paint, silicone resin paint – Figure 13) also has a determining effect.

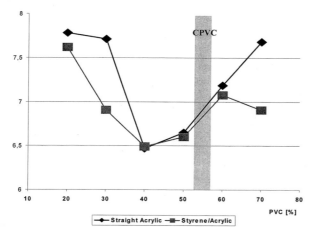

Figure 12. Effect of the PVC and the binder on the dirt pick-up ΔL of solvent-free and biocide-free exterior paints after 1 year of outdoor weathering (Limburgerhof, 45°, south)

It can be seen clearly from Figure 12 that dirt pick-up which is dependent on the PVC can be broken up into two areas with different causes. Below the CPVC the surface tack determines the amount of dirt pick-up, whereas above the CPVC the porosity encourages the depositing of biogenic material.

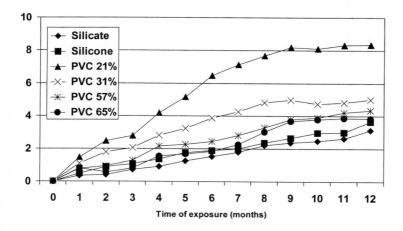

Figure 13. Effect of the PVC and the type of formulation on the dirt pick-up Δ L of exterior paints subjected to outdoor weathering, Indonesia Vertical

When the amount of dispersion in the various types of formulation (silicate, silicone and emulsion paint with PVC = 65%) is more or less the same, the tendency to pick up dirt is also approximately the same. Presumably, this is due primarily to the coating having similar surface tack values and porosity levels.

Conclusion

It has been shown, that as far as the binder is concerned, a high Tg, distinct underlying hydrophobic properties (styrene/acrylates are better than comparable pure acrylates) and curing (e.g. UV crosslinking) help to ensure that the coating formulations have a low tendency to pick up dirt. In addition, the grade of titanium dioxide chosen (manufacturing process and surface treatment) has a strong effect on the amount of dirt picked up by the coatings after weathering.

In general, as the PVC rises, up to the critical PVC (CPVC), the tendency to pick up dirt decreases and the differences between binders level off noticeably. Beyond the CPVC, however, the increasing susceptibility to contamination by micro-organisms has an increasingly detrimental effect on the aesthetic appearance of the coatings. The latter is due to the increasing porosity and reduced water resistance of the coatings as the PVC increases. High-boiling film-forming aids or even long-lasting plasticisers reduce the resistance to dirt pick-up, since they remain in the coating permanently and lower the glass transition temperature of the binder appreciably. The effect which the thickening system chosen to provide the rheological properties has on the tendency of the coating to pick up dirt is surprisingly strong.

References

[1] O. Wagner, Farbe & Lack, 107, 1/2001
[2] T. Nakaya; Progress in Organic Coatings 27, 1996.
[3] Literature on the sources, composition and particle size of fine dusts:
     a] G. Israel, A. Erdmann, J.Shen, W. Frenzel, E. Ulrich, 1992, VDI Fortschrittsberichte, Reihe 15: Umwelttechnik, VDI Verlag.
     c] M. Kleeman, J. Schauer, G. Cass, Environmental Science and Technology, 34, 2000.
     d] M. Kleeman, G. Cass, Atmospheric Environment 32, 1998.
     e] E. Bagda, Farbe & Lack, 108, 1/2002
[4] Ch.Hüglin, NFP41 Verkehr und Umwelt; EMPA Schweiz, Dübendorf vom 31.07.2000.
[5] H. Zeh, Weiß, Vortrag bei Fachgemeinschaft Kunstharzputze, 1999 in Dresden.
[6] R. Baumstark, C. Costa, M. Schwartz, Farbe & Lack 106, 10/2000.

458

[7]   M. Schwartz, Ch.-Le Zhao, Farbe & Lack 105, 9/1999.
[8]   O. Wagner, A. Smith, J. Coatings Technology, 68, No. 862, 1996
[9]   P. Pföhler, A. Zosel, R. Baumstark, Surface Coatings Australia, 1995, 18 - 23.
[10]   E. Bagda, Farbe & Lack, 108, 2/2002

Macromol. Symp. 187, 459–467 (2002)

Coatings with Self-Cleaning Properties

Martin Wulf*, Anja Wehling, Oliver Reis

Strategic Research, DuPont Performance Coatings, Wuppertal, Germany

Summary: The issue of self-cleaning significantly gained popularity due to the work of Barthlott and coworkers on the so called "Lotos-Effect®". They found out, that the cleanliness of the Lotos leaves originates from a combined effect of surface topography and hydrophobicity. The symbol of the beautiful Lotos flower as well as the fascination of surfaces being cleaned without any manual activity, simply by a rain shower, has since then stimulated the fantasy of many researchers. Our vision is to copy this mechanism from mother nature and to implement it into coating systems in such a way, that conventional application techniques, e.g. spray-coating, can be applied without the necessity of further process steps like e.g. soft lithography. Three different approaches will be presented in this paper. Roughness and contact angle measurements have been used to quantify the self-cleaning properties.

Keywords: coatings, structure, surfaces

Introduction

„Self-cleaning" is a prominent example for a desired property of a whole class of high-performance coatings, namely *functional coatings*. This term classifies systems, which posses besides the classical properties of a coating, i.e., decoration and protection, an additional functionality. Depending on the actual application, additional functionality of a coated surface can be manifold, e.g., soft-feel or soft-touch haptics for interior car parts, easy-to-clean (anti-graffiti) for storefronts or anti-fouling for ship bodies. Although the range of applications as well as the underlying chemistry is diverse, all these systems have in common a very specific benefit, which in each case provides an answer to an existing and well defined demand.

The issue of self-cleaning significantly gained popularity due to the work of Barthlott and coworkers on the so called "Lotos-Effect®" [1]. They found out, that the cleanliness of the Lotos leaves originates from a combined effect of surface topography and hydrophobicity. The symbol

CCC 1022-1360/00/$ 17.50+.50/0

of the beautiful Lotos flower as well as the fascination of surfaces being cleaned without any manual activity, simply by a rain shower, has since then stimulated the fantasy of many researchers. From a pure scientific point of view, however, the influence of roughness (also in combination with hydrophobicity) on water repellency was already known much earlier [2] and could also be understood on a sound thermodynamic basis [3, 4].

Our vision is to copy this mechanism from mother nature and to implement it into coating systems in such a way, that conventional application techniques, e.g. spray-coating, can be applied without the necessity of further process steps like e.g. soft lithography. The decisive challenge in reaching this target is of course a coating formulation which is capable of building up a suitable microstructure just on its own. In comparison, adjusting the hydrophobicity of the surface is much more straightforward. Three different approaches for coatings leading to microstructured surfaces will be elucidated. They are a) use of inorganic fillers, b) use of polymeric together with inorganic fillers and c) thixotropic textured coatings. These systems are compared to a conventional 1K automotive clearcoat.

Thermodynamics of contact angles on rough surfaces

From daily experience we know that a drop of water deposited onto a coated surface will form a sessile drop. The angle formed between the liquid-vapor interface and the liquid-solid interface at the solid-liquid-vapor three-phase contact line is conventionally defined as the *contact angle*. Despite the apparent simplicity of sessile drops on solid surfaces, contact angle phenomena are rather complex. Minimizing the overall free energy of a system consisting of a liquid in contact with a solid yields the Laplace equation of capillarity [5]

$$\gamma_{lv}\left(\frac{1}{R_1} + \frac{1}{R_2}\right) = \Delta\rho gz + c = \Delta P \tag{1}$$

and Young's equation [6]

$$\gamma_{lv}\cos\theta_e = \gamma_{sv} - \gamma_{sl} \tag{2}$$

where γ_{sv} is the solid-vapor interfacial tension, γ_{lv} is the liquid-vapor interfacial tension, γ_{sl} is the solid-liquid interfacial tension, θ_e is the equilibrium contact angle, R_1 and R_2 are the principal radii of curvature at a point of the liquid surface, $\Delta\rho$ is the density difference between the liquid and vapor phase, g is the acceleration due to gravity, z is the ordinate of a point of the liquid surface at which the principal radii of curvature are R_1 and R_2, c is a constant and ΔP is the capillary pressure or pressure of curvature.

While the Laplace equation essentially describes the shape of the liquid-vapor interface away from the solid-liquid and solid-vapor interfaces, the Young equation involves properties which are a function of the solid surface, i.e., γ_{sv} and γ_{sl}. The derivation of eq. (2) assumes that the solid surface in contact with the liquid is smooth, homogeneous, isotropic and non deformable.

On rough solid surfaces Wenzel [7] recognized that Young's equation may not be a universal equilibrium condition for the physical interaction between a solid and a liquid. He argued, essentially, that if the solid surface is rough, the interfacial tensions γ_{sv} and γ_{sl} should not be referred to the geometric area, but to the actual surface area. If we let

$$r = \frac{actual\ surface\ area}{geometric\ surface\ area} \tag{3}$$

this leads to the so called Wenzel equation

$$r(\gamma_{sv} - \gamma_{sl}) = \gamma_{lv}\cos\theta_W \tag{4}$$

where θ_W may be called the Wenzel contact angle. Equation (4) was also derived by Good [3].

In a thorough thermodynamic analysis Li and Neuman [4] have shown, that although θ_W is indeed the equilibrium contact angle θ_e, it is not accessible in experimental measurements. According to their analysis, a large number of metastable states of the free energy are existing for liquids on rough solid surfaces which correspond to the same number of metastable contact angles. Only the largest and the smallest contact angle can be observed experimentally. The former is referred to as the advancing contact angle (θ_a), while the latter is called receding contact angle (θ_r). The difference between θ_a and θ_r is the so called contact angle hysteresis, which may be conveniently measured by first advancing and then receding a liquid drop over a solid surface. Thus, for practical purpose, contact angle hysteresis measurements are best suited for quantification and evaluation of water repellency on structured surfaces.

The extraordinary water repellency, which can be observed on rough surfaces like e.g. lotos leaves, originates from a combined effect of surface structure and hydrophobicity, i.e. low surface tension of the solid, γ_{sv}. In this case one can observe an "air-cushion effect", if the water droplet "rests" on the peaks of the structures and is not able to penetrate into the valleys. Based on the theory of capillarity, i.e. taking into account eq. (1) and (2), it is possible to calculate the size of suitable surface structures for ideal geometries. This was done by Dettre et al. [2] for two cases. The first one has been a surface having holes of the shape of cylindrical capillaries, and the second one has been a surface consisting of vertically oriented cylinders. They derived the following equations for the capillaries

$$P = \frac{4\gamma_{lv}}{c}\cos\theta_a \tag{5}$$

and the cylinders, respectively.

$$P = \frac{\pi d\gamma_{lv}\cos\theta_a}{0.8660b^2 - (\pi/4)d^2} \tag{6}$$

Where P is the pressure, which is required to force water into the capillary (or between the cylinders, respectively), γ_{lv} is the liquid-vapor surface tension, c is the diameter of the capillaries, θ_a is the advancing contact angle, d is the diameter of the cylinders and b is the center-to-center distance between cylinders.

For this ideal geometries, Dettre et al. have calculated the maximum structure size for a self-cleaning wax surface, i.e., a very hydrophobic surface (comparable to the lotos leaves), of about 300 microns. Although the surface structures in practice will differ significantly from the situation described by eq. (5) and (6), the values calculated with these equations provide at least a valuable orientation and a good starting point for further experiments.

Materials and Methods

An overview of the different textured coating systems is given in table 1.

Table 1. Overview of textured coatings

#	Resin	Roughness due to	Hydrophobicity due to
1	OH-functional Polyacrylate Melamine X-linked	Inorganic particles	Fluor additive
2	OH-functional Polyacrylate Melamine X-linked	Inorganic and organic particles	Fluor additive
3	OH-functional Polyacrylate isocyanate X-linked	Thixotropic binder	Wax additive
4	OH-functional Polyacrylate Melamine X-linked	---	---

Roughness measurements

The roughness of the coated surfaces was measured with a Hommeltester T4000 equipped with a TKPK100 sensing device on an area of 2 mm x 2 mm with a velocity of 0.15 mm/s.

Contact angle hysteresis measurements

The contact angle hysteresis was measured with a G2 instrument from Kruess, using water as test liquid (LiChrosolv (Merck), water for chromatography). The needle of the syringe was kept in the sessile drop from above during the whole experiment. The drop was first slowly advanced over the surface by pushing additional water through the needle and then slowly receded by withdrawing the liquid. Simultaneously the contact angle was measured by aligning a tangent to the drop profile by means of the software of the G2 instrument.

Results

Roughness measurements

Fig. 1 shows the results of the roughness measurements. The surface of the conventional 1K clearcoat (#4) is, as expected, extremely smooth. Only a little weaviness with maximum heights well below 1 µm is present. In contrast, the surfaces of the other three systems are very rough. While the lateral structures of these three coatings are in the same order of magnitude (between 50 µm to 300 µm), the heights of the peaks are very different. As can be seen from the scale of the z-axis, the peak-to-valley height is about 10 times larger for systems #1 and #2 as compared to coating #3 (about 100 µm vs. about 10 µm).

Figure 1. Surface roughness of the four different coatings

Contact angle hysteresis measurements

A typical result of a contact angle hysteresis measurement is shown in Fig. 2 (system # 3). It can be seen that the water drop advances with a fairly constant contact angle of about 100 ° over the surface. After 50 measurements, the water has been withdrawn from the drop. While the drop is receding over the surface, the contact angle initially decreases by about 30° and reaches then a constant value of about 70°.

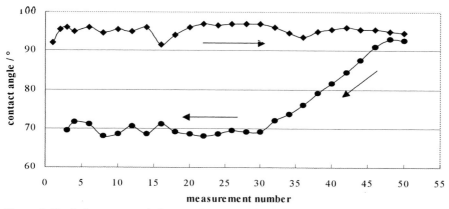

Figure 2. Typical contact angle hysteresis measurement (coating #3). The squares are advancing and the circles are receding contact angles.

The contact angle data of all coating systems are given in table 2 and typical images of receding contact angles are shown in Fig. 3.

Table 2. Results of contact angle hysteresis measurements

Coating #	θ_a [°]	θ_r [°]
1	140	140
2	140	130
3	102	70
4	86	20

(θ_a: advancing contact angle, θ_r: receding contact angle)

The behaviour of system #4 is typical for an automotive clearcoat without any self-cleaning properties: The advancing contact angle is relatively small (about 90 °) and the hysteresis is very pronounced. Due to the extremly low receding contact angle of about 20 °, traces of water remain on the surface. Dust or dirt will not completely be washed away by a rain shower.

Systems #1 and #2, in contrast, posses impressive self-cleaning properties, which are comparable to the leaves of the lotos flower. Both systems show extraordinary large advancing contact angles and virtually no hysteresis. Water drops take up dirt from the surface and roll off easily. System #3 shows an intermediate behavior: The advancing and especially the receding contact angle are considerably larger than on the standard (system #4), but smaller as for system #1 and #2.

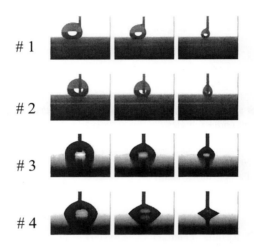

Figure 3. Receding contact angles for coatings systems #1-4.

Conclusions

It is possible to formulate coatings with extraordinary self-cleaning properties, which are applied simply with a spray gun. The results for system #1 and #2 have shown, that these surfaces are sufficiently hydrophobic and capable to build up a suitable microstructure. However, these structures are rather fragile. The mechanical stability is not yet sufficient for technical applications. Another approach was shown with system #3: This system has much better self-cleaning properties than the standard (#4) and the mechanical stability is ok.

Acknowlededegements

The authors thank Kornelia Armbruster and Michael Osterhold from the physics department of DuPont Performance Coatings for their support with the roughness and contact angle measurements.

References

[1] W. BARTHLOTT, C. NEINHUIS, *Planta* **202** (1997) 1
[2] R.H. DETTRE, R.E. JOHNSON (DuPont), US-Patent # 3354022 (1967)
[3] R.J. GOOD, *J. Am. Chem. Soc.* **74** (1952) 5041
[4] D. LI, A.W. NEUMANN, Thermodynamic Status of Contact Angles, in *Applied Surface Thermodynamics*, A.W. NEUMANN, J.K. SPELT (Eds.), Marcel Dekker Inc. (1996) 109-168
[5] P.S. DE LAPLACE, *Mechanique Celeste*, Supplement to Book 10, J.B.M. Duprat, Paris, 1808
[6] T. YOUNG, in Miscellaneous Works, Vol. 1, G. Peacock (Ed.), J. Murray, London, 1855
[7] R.N. WENZEL, *Ind. Eng. Chem.* **28** (1936) 988

Macromol. Symp. 187, 469–479 (2002)

The Effect of Fluoropolymer Architecture on the Exterior Weathering of Coatings

Kurt Wood*

ATOFINA Chemicals, Inc., Technical Polymers Research and Development

900 First Avenue, King of Prussia PA 19406 USA

Tel. +1 610 878 6914, Fax +1 610 878 6934, kurt.wood@atofina.com

Summary: Fluoropolymers of several different types (poly(vinylidene fluoride), alternating copolymers, perfluorinated ether polyols) are commercially available as coatings resins. While all these resins share certain chemical properties (e.g. hydrophobicity), the performance in the area of weathering depends crucially on details of the molecular structure. A number of recent studies have examined the mechanisms by which different fluorinated coatings weather, and have highlighted some ways in which the molecular structure affords photochemical protection that is not possible with conventional polyester or polyurethane coatings. The most weatherable fluoropolymer systems not only meet the most stringent worldwide industry specifications for high performance architectural and protective topcoats, but have demonstrated over thirty years chalk and fade resistance in south-facing Florida exposures.

Keywords: coatings; fluoropolymers; degradation; poly(vinylidene fluoride); photo oxidation

Introduction

High performance fluoropolymer resins for coatings applications have been commercially available for nearly forty years[i]. While it is sometimes convenient to consider fluoropolymers as a class, the differences between fluoropolymers are often more important than the similarities. It is in fact the unique molecular structure of each resin which leads to the special properties that render it useful.

Those fluoropolymers designed for use in exterior topcoats require specialized chemical and UV resistance properties. As with other coating properties, the performance in the area of weathering depends crucially on details of the resin molecular structure. Those details turn out to vary significantly among the different kinds of fluoropolymers proposed for use in highly weatherable coatings.

  CCC 1022-1360/00/$ 17.50+.50/0

A number of recent studies have examined the mechanisms by which different fluorinated coatings weather, and have highlighted some ways in which the molecular structure affords photochemical protection that is not possible with conventional polyester or polyurethane coatings. This paper will examine the effect of molecular architecture on exterior weatherability for four different types of fluoropolymer. A number of literature mechanistic studies of these materials will be reviewed, which are illustrative of degradation routes in commercial exterior coatings. Then, we present some new data from our laboratory, which illustrates particularly the kind of protective effects that can be obtained from thermoplastic PVDF coatings, leading to excellent color retention in South Florida exposure tests.

Important mechanisms of degradation of fluorinated coatings in the outdoors.

Among the mechanisms for coating degradation that have been identified as relevant for fluoropolymer coatings, are resin main chain or side chain oxidation, photooxidation of non-fluorinated crosslinks or adjuvant resins, hydrolytic or other non-photochemical attacks on crosslinks or the resin main chains, and effects associated with specific pigment grades.

A.   Resin main chain oxidation—example of FEVE resins.

Many conventional polyester or polyurethane type coatings are built up using diol or polyol type segments as an integral part of the molecular structure.   For purely aliphatic systems, which are normally preferable when outdoor weatherability is sought, coating network degradation can often begin at sites in the immediate vicinity of the linking groups (ether, ester or urethane) formed during prepolymer synthesis or through crosslinking reactions. Hydrogens in the alpha position relative to the diol oxygen are particularly susceptible to abstraction from any free radical species which may be in the coating[ii] (Figure 1). These kinds of hydrogen abstraction events can lead to the formation of hydroperoxides on the polymer backbone, which in turn lead to chain scission as well as the generation of new radical species that can start the degradation process in other locations.

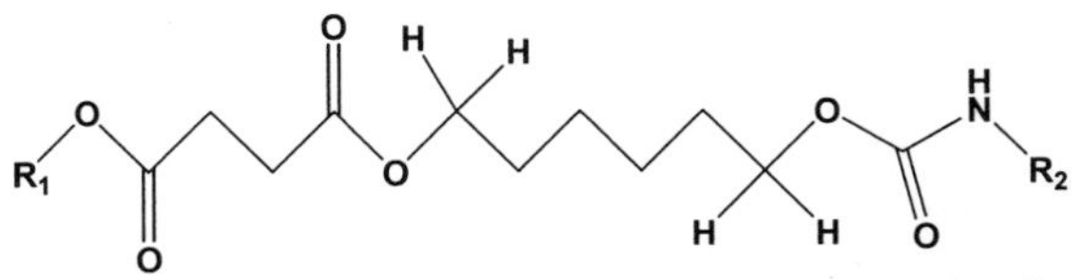

Figure 1. Generalized structure of an aliphatic polyester-urethane. The diol hydrogens in the alpha position relative to the ester and urethane oxygens are shown.

In so-called "FEVE" (fluorinated ethylene vinyl ether) resins, fluorinated monomers and vinyl ethers are copolymerized in an alternating fashion[iii]. (Figure 2) By using vinyl monomers containing pendant functional groups, e.g. hydroxyl groups, functional FEVE resins can be made and used with conventional crosslinkers like polyisocyanates and melamines, in 2-component crosslinked formulations. It has been reported in a series of studies by Gardette and co-workers[iv] that the hydrogen abstraction rate at the vinyl ether alpha position is greatly reduced relative to the rate for non-fluorinated polymers. This reduction is attributed to the electron withdrawing effects of the fluorines.

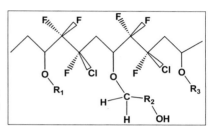

Figure 2. Schematic structure of FEVE resin made with chlorotrifluoroethylene as the fluorinated monomer.

This protective effect is limited to those atoms immediately adjacent to the fluorine. Urethane and ether linkages generated through crosslinking reactions are still subject to the same degradation processes as non-fluorinated thermoset coatings and would presumably need to be protected by other mechanisms, e.g. by using additives such as hindered amine light stabilizers (HALS).

B. Resin side chain oxidation—example of fluorinated acrylates.

In a few instances, the electronegativity of fluorine may actually destabilize neighboring C-H bonds, with respect to their susceptibility to hydrogen abstraction. Such an effect has been reported for some acrylic copolymers, made from monomers possessing fluorinated side chains[v]. The rate of hydroperoxide formation during irradiation, attributed to hydrogen abstraction at the side chain alpha position, was found to be greater when fluorinated side chains were present (fluorines in the alpha or beta position with respect to the hydrogen), compared to the case of a butyl group side chain (Figure 3). As with other kinds of acrylics, the decomposition of the hydroperoxides formed along the side chains can eventually lead to other kinds of more dramatic degradation (chain scission, polymer "unzipping", or crosslinking).

472

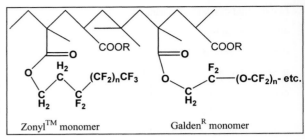

Figure 3. Fluorinated methacrylate monomers with side chains reported to have lower stability than butyl side chains (Reference 6).

C. Photooxidation of non-fluorinated crosslinks—example of perfluorinated polyethers

Another class of fluorinated resins which has become available in recent years, is based on perfluoroether oligomer ("ZDOL") diols:

$$HO-CH_2CF_2-O-(CF_2CF_2)_x-O-(CF_2)_y-O-CF_2CH_2-OH$$

Coatings made with these materials reportedly have very low surface energies, but the weatherability has been reported to be poor[vi]. This might be due in part to photochemical attack on unprotected secondary hydrogens at the alpha position relative to the non-fluorinated diol chain ends, as described in the previous section. However, Luda et al. have postulated a different degradation mechanism to explain the poor weatherability of this class of resins[vii]. They identify the non-fluorinated crosslinker region of the coating network as the locus of photodegradation effects. For instance, when IPDI trimer is used as the crosslinking agent, hydrogen abstraction at the position alpha to the urethane nitrogen is identified as the system's weak point (Figure 4):

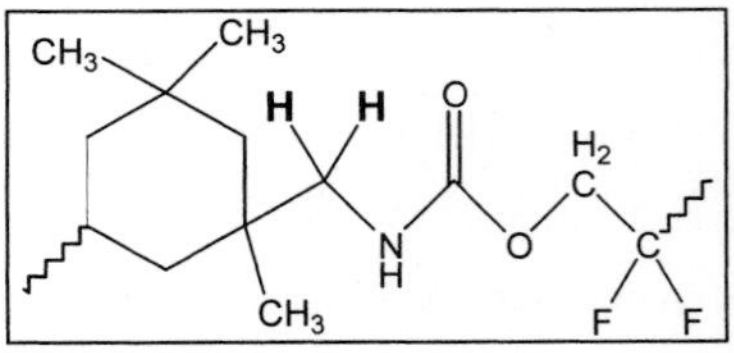

Figure 4: Hydrogens adjacent to nitrogen which are main attack points for IPDI-crosslinked perfluoroether urethane coatings, according to reference 8.

Regardless of the relative contribution of the two degradation mechanisms (i.e. hydrogen abstraction occurring at the aliphatic or fluorinated side of the crosslink), the use of HALS was found to significantly improve the weatherability of the ZDOL-based coating, as measured by the retention of film mechanical properties in accelerated tests.

Photodegradation of non-fluorinated crosslinked segments should in principle also occur with other classes of fluorinated polyols, e.g. FEVE resins, crosslinked with non-fluorinated isocyanates or melamines. In these systems as well, the use of HALS should theoretically improve the system weatherability.

D.   Photooxidation of non-fluorinated adjuvant resins—example of PVDF coatings.

The beneficial protective effect of the fluorine bond is most dramatically visible, however, in poly(vinylidene fluoride) (PVDF) resins. The structure of PVDF consists of alternating $-CF_2-$ and $-CH_2-$ units:

Figure 5. Structure of PVDF, or poly(vinylidene fluoride).

Poly(vinylidene fluoride)

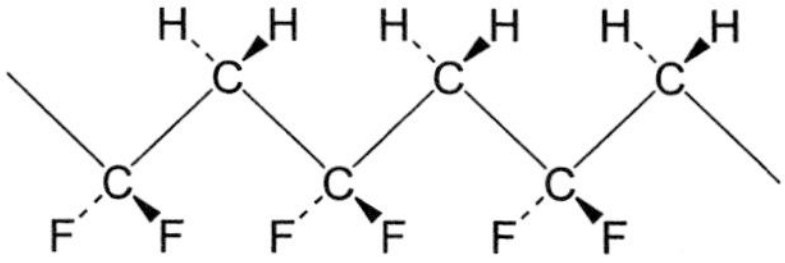

In the PVDF structure, every C-H bond is adjacent to four C-F bonds. The net result is a resin that is completely photochemically inert[viii], as well as electrochemically extremely stable. At the same time, the alternating structure gives the PVDF chain units a strong dipolar character. This confers excellent compatibility of the resin with a variety of other polymers such as poly(methyl methacrylate) (pMMA). It also means that low molecular weight ketones and esters can be used as active or latent solvents for the resin, so that it can be used easily in liquid coating formulations.

In fact PVDF coatings have been available commercially since the mid-1960s[ix]. Typical commercial formulations contain 70-80 % by weight PVDF in the coating binder, with the remainder of the binder being some sort of compatible acrylic such as pMMA. The acrylic resin is added because of the inertness of the PVDF molecule, in order to improve the pigment wetting and coating adhesion. Thus it functions as an "adjuvant" or helper resin. While the PVDF fluoropolymer is highly resistant to any kind of outdoor degradation, the acrylic component might be expected to be more susceptible to eventual photochemical attack. However, premium PVDF coatings contain no more than 30 weight percent acrylic on binder, limiting the impact of any degradation if it does occur.

The photooxidation of acrylic resins, including the acrylic resins in commercial PVDF coatings can be monitored in several different ways, but each of them has some limitations. The method that has been perhaps most widely used is to look directly for photooxidation products, such as carboxylic acid groups, in the infrared or Raman spectrum[x]. This method has the advantage of directly providing chemical information about the sample— and in some cases, more specifically, about the exposed surface of the sample. However, it can only be used to identify those chemical species that remain in the coating, and which have characteristic isolated vibrational bands.

Since low molecular weight degradation products may volatilize, a complementary method is simply to monitor the mass loss of the coating due to the exposure[xi]. While it does not provide chemical information, the weight loss method can be valuable especially in cases where little chemical change in the coating is apparent, for instance when the rate of coating erosion is at least as fast as the rate of chemical degradation. In one study in our laboratory of several weathered PVDF coatings, this was apparently the situation—only small changes were observed in the infrared spectra of a series of white coatings aged 10-15 years, yet they had suffered a substantial loss of gloss[12]. Unfortunately, the mass loss method could not be used in this particular case, because accurate initial weights for the samples were not available. This illustrates another disadvantage of the weight loss method: the experimenter must have the foresight to record initial weights, and to assure that samples do not suffer physical damage or accretions, which could affect the mass, during the entire duration of the test. Because of these limitations, mass loss measurements are best suited for laboratory accelerated tests, rather than outdoor tests that could last for years.

E.   Hydrolytic or other non-photochemical attacks on resins or crosslinks.

Besides photochemical oxidation processes involving, for instance, hydrogen abstraction, other chemical processes can occur which degrade fluoropolymers and particularly crosslinked networks. Such processes could include the hydrolysis of ether linkages formed from melamine crosslinking, or urethane linkages from isocyanate crosslinking. For a non-fluorinated polyester-melamine crosslinked system, the relative rates of photochemical and non-photochemical degradation have recently been quantitatively measured by Van Landringham et al.[xii], and the effect of purely hydrolytic effects was shown to be substantial.

Accelerants of hydrolysis and other non-photochemical processes can include low pH from the environment (acid rain), residual catalysts in the coating, and heat—such as might be generated when a dark colored paint sits directly in the sun on a warm day. Since many of the crosslinked coatings are based on low molecular weight resins, and depend crucially on crosslinks to achieve coating performance, the loss of crosslink density could have dramatic effects on coating properties. Of course, the degradation effects can also be retarded at least for a time, by having a high enough crosslink density to help exclude water and other agents from the bulk of the binder.

While many kinds of crosslinks can be acid sensitive, most fluoropolymers themselves used in coatings are inherently acid resistant. PVDF, for instance, is so acid resistant that the neat resin is used for CPI (chemical processing industry) applications involving strong mineral acids. PVDF is only attacked chemically under extremely strong basic conditions, which would not be expected in exterior coating environments.

In contrast to crosslinked coatings, thermoplastic coatings based on semi-crystalline resins-- such as commercial PVDF coatings-- should generally be highly resistant to the effects of hydrolytic degradation processes. Not only are the resins themselves inherently resistant to the loss of properties through hydrolysis, but the advantageous properties of the coating are enhanced through the crystalline associations of the component fluoropolymer[xiii]. As with thermoset crosslinks, the PVDF crystalline structures have excellent barrier properties, keeping water, oxygen and other destructive agents away from both the coating bulk and the substrate layers underneath the coating. At the same time, being non-covalent in nature, the crystalline structures have some limited ability to reform themselves, so that the loss of network structure is much less of a possibility.

F.   Pigment effects—example of PVDF coatings

For highly weatherable paints, it is essential that high performance pigments are used. These pigments should not only have exceptional inherent color stability throughout the coating application, curing, and weathering processes, but additionally should not promote any breakdown of the coating binder. For PVDF coatings, certain inorganic pigments and particularly mixed metal oxides have been used as pigments of choice for many years. For some formulations, masstone PVDF paints made with mixed metal oxide pigments have gone over thirty years with minimal color change.

To make light colored paints, rutile titanium dioxide is widely used because of its great hiding power, chemical stability, and cost effectiveness. However, for the most weatherable paints, grade selection is critical since TiO_2 crystals, like many other inorganic materials, are inherently photochemically active[xiv]. For coating binders with low outdoor weatherability, the addition of rutile TiO_2 almost always increases the weatherability of the coating, since the effect of UV absorption by the pigment generally outweighs the photocatalytic effects.

However, the more inherently weatherable the binder, the more important it is to carefully choose the grade of TiO$_2$ used in the coating[xv]. Figure 6 shows SEM micrographs of the surface of a white PVDF coating, made with a newer "universal grade" of TiO$_2$ said to have good weatherability, after two and five years Florida exposure. At the two year mark (left), some pitting of the coating around pigment particles can be observed. This kind of pitting was not observed for a control coating made with the standard recommended most weatherable grade. For the universal grade, this pitting was subsequently observed to lead to premature gloss loss in the coating, with a completely degraded coating surface after five years Florida exposure (right).

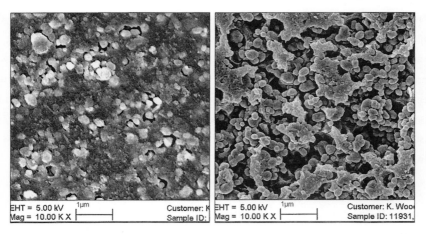

Figure 6. SEM micrographs of white PVDF coatings made with a "universal grade" TiO$_2$ pigment, after 24 and 57 months south Florida exposure.

Protective effects in PVDF coatings

As was mentioned previously, the crystalline structures in PVDF coatings are in a state of dynamic equilibrium. They have some limited ability to rearrange and reform themselves over time, due to the relatively low glass transition temperature (T$_g$) of the amorphous phase (pure PVDF has a T$_g$ of about –40 °C; since it is typically used with high T$_g$ acrylics, the blend T$_g$ is often around room temperature). This means that local stresses generated in the film can be relieved, and coating integrity maintained, much more easily than in a thermoset system where crosslinks enforce a degree of network rigidity. We believe that the ability of PVDF coatings to protect not only substrate materials, but also other components of the coating itself, can be attributed largely to the barrier properties generated by these labile crystalline structures.

One example of this protective effect can be noted in a series of panels, now fifteen years old, comparing PVDF coatings with color-matched polyester powder coatings. As might be expected, the non-fluorinated polyester coatings showed serious color fade and chalking, after just a few years Florida. The protective effect of the PVDF coatings is clearly seen in the same series, when comparing masstone paints of the same formulation, with and without an additional PVDF clearcoat. The degree of color fade in the PVDF color coat is reduced even more by having the PVDF topcoat over it—typical color change values are only about delta E =1—i.e. barely perceptible to the eye—after fifteen years (Figure 7).

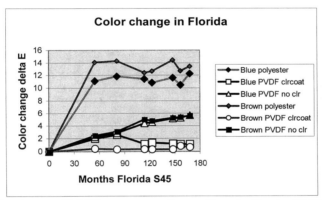

Figure 7. Color change of PVDF masstone coatings, with and without a PVDF clearcoat, in Florida. Results for color-matched polyester powder coatings are shown for comparison purposes.

Another study, which followed coating erosion during 15 year Arizona exposure by directly measuring the coating thickness, shows evidence for PVDF's ability to protect the acrylic component within the coating itself. Photodegradation of the acrylic inside the coating, with eventual volatilization, would be expected to lead to loss of coating thickness over time[xvi]. However, within the level of experimental uncertainty, a series of PVDF coatings showed <u>no</u> loss of film thickness during this extended period (see Table 1).

Conclusions

A number of highly fluorinated polymers are inherently highly weatherable, and in particular are resistant to hydrogen abstraction which leads to direct photooxidation of the fluoropolymer structure. However, in most commercial coating systems, non-fluorinated resin components are also present, being introduced either as crosslinkers or adjuvant resins. These non-fluorinated binder components can degrade according to mechanisms identified in non-fluorinated coating systems (polyester or acrylic based). It therefore remains important in all these systems to reduce the access to the bulk of water, molecular oxygen, and other destructive species. In crosslinked systems, some limited degree of protection for these less weatherable components may be provided by high crosslink density, until the integrity of the crosslinked network begins to be compromised.

By contrast, thermoplastic fluoropolymer systems such as PVDF coatings benefit from the semi-crystalline structure of the PVDF resin. This structure provides mechanical strength and barrier properties similar to conventional crosslinked coatings, but also has some ability to rejuvenate itself, so that the effects of any gradual damage to other coating components do not lead to catastrophic coating failure. As a result, protection is provided both for other components of the PVDF coating, and for the coating substrate. The most weatherable fluoropolymer systems not only meet the most stringent worldwide industry specifications for high performance architectural and protective topcoats, but have demonstrated over thirty years chalk and fade resistance in south-facing Florida exposures.

Table 1. Comparison of coating thickness for KYNAR 500® coatings exposed in Arizona for 15 years

Panel ID	Coating average thickness, in microns	
	Initial thickness	Final thickness
Control panel: no exposure	22.1 ± 1.5	20.3 ± 0.5
1997A	18.8 ± 1.8	18.5 ± 1.3
1998A	20.6 ± 1.0	20.3 ± 1.3
1999A	21.1 ± 1.5	20.1 ± 1.0
2002A	19.3 ± 1.0	18.3 ± 0.8
2003A	21.1 ± 2.0	19.0 ± 1.0

References

[i] R.A. Iezzi, Fluoropolymer Coatings for Architectural Applications, Modern Fluoropolymers (Wiley, 1997), p. 14.

[ii] J.-L. Gardette, B. Hailhot, F. Posada, A. Rivaton, and C. Wilhelm, Macromol. Symp. 143 (1999) 95-109.

[iii] T. Takakura, CTFE/Vinyl Ether Copolymers, Modern Fluoropolymers (Wiley, 1997), p. 557.

[iv] J.-L. Philippart, F. Posada and J.-L. Gardette, Polymer Degradation and Stability 53 (1996) 33-37; F. Posada and J.-L. Gardette, Polymer Degradation and Stability 70 (2000) 17-29.

[v] N.S. Allen, C.J. Regan, W.A.E. Dunk, R. McIntyre and B. Johnson, Polymer Degradation and Stability, 58 (1997) 149-157.

[vi] T. Temchtcko, "New Developments in Perfluoropolyester Resins Technology", Proceedings of 26th International Conference in Organic Coatings, July 2000, Athens, Greece (Institute of Materials Science, New Paltz, New York USA), 357-372.

[vii] M.P.Luda, G. Camino, E. Laurenti, S. Novelli, T. Temtchenko and S. Turri, Polymer Degradation and Stability 73 (2001) 387-392. See also F.X. Perrin, M. Irigoyen, E. Aragon and J.L. Vernet, Polymer Degradation and Stability 70 (2000) 469-475.

[viii] J.-L. Philippart, N. Siampiringue, A. Strassel and J. Lemaire, Makromol. Chem.. 190 (1989) 413-418.

[ix] The Pennsalt Corporation introduced the coatings grade KYNAR 500® PVDF resin in 1965.

[x] See, for instance, J.L. Gerlock, C.A.Smith, V.A. Cooper, T.G. Dusbiber and W.H. Weber, Poly. Degr. Stab. 62 (1998) 225-234; G. Ellis, M. Claybourn and S.E. Richards, Spectrochimcia Acta 46A(2) (1990) 227-241.

[xi] K.A. Wood, C. Cypcar and L. Hedhli, J. Fluorine Chem. 104 (2000) 63-71.

[xii] M.R. VanLandringham, N. Tinh, W.E. Byrd, and J.W. Martin, Journal of Coatings Technology 73(923) (2001) 43-50.

[xiii] S.R. Gaboury and K.A. Wood, "Tailoring Coating Properties Through Control of PVDF Copolymer Phase Behavior", Paper 17, Fluorine in Coatings IV Conference Proceedings (Paint Research Association, Teddington UK 2000).

[xiv] M.P. Diebold, Surface Coatings International 1995 (6) 250-256; 1995 (7) 294-299.

[xv] M.P. Diebold, "Unconventional Effects of TiO2 on Paint Durability", Proceedings of 5th Nurnberg Congress 371-389 (Paint Research Association, Teddington, UK).

[xvi] N.S. Allen, C.J. Regan, W.A.E.Dunk, E. McIntyre and B. Johnson, Polymer Degradationa and Stability 58 (1997) 149-157.

Preparation of Coatings via Cationic Photopolymerisation: Influence of Alcoholic Additives

Roberta Bongiovanni, Giulio Malucelli, Marco Sangermano, Aldo Priola*

Dipartimento di Scienza dei Materiali e Ingegneria Chimica Politecnico di Torino
c.so Duca degli Abruzzi 24, 10129 Torino Italy
e-mail: priola@fenice.polito.it

Summary: Coatings obtained through photopolymerisation of vinylethers, propenyl ethers and epoxy resins are described. The influence of alcohols and of comonomers bearing OH groups on the cationic mechanism of the curing process is discussed. The final properties of the coatings are evaluated and correlated to the structures of the additives.

Keywords: UV-curing, coatings, vinyl ethers, propenyl ethers, epoxides, alcohols

Introduction

The UV curing process is getting an increasing importance in the field of coatings: it allows to obtain a fast transformation of a liquid monomer into a solid film with tailored physico-chemical and mechanical properties. In the process, radical or cationic species are generated by the interaction of the UV light with a suitable photoinitiator, which induces the curing reaction of suitable reactive monomers and oligomers [1]: in the case of the cationic polymerisation, onium salts are used to generate very strong Brönsted acids upon photodecomposition [2]. The cationic photoinduced process presents some advantages compared to the radical one [3], in particular lack of inhibition by oxygen, low shrinkage, good mechanical properties of the UV cured materials and good adhesion properties to various substrates. Moreover, the monomers employed are generally characterized by being less toxic and irritant.

Different types of monomers and oligomers have been proposed and reported in the literature [4,5]. Among them the epoxides are largely available and show good properties after curing, but their reactivity is relatively low, especially when compared to the widely used acrylic resins [6]. The vinyl ethers on the contrary are rather fast, but their availability is limited   because the synthesis, which involves acetylene, is difficult. A good alternative is the use of propenyl ethers, as they are obtained by isomerisation of the allyl ethers and show good reactivity [7,8].

In order to modify the curing process of epoxide systems, the use of alcohols was first indicated by Penczek and Kubisa [9], then confirmed by Crivello [10,11]. The occurrence of a chain transfer reaction involving the OH groups causes the change of the kinetics of the process and of the properties of the cured networks. A further refinement of this research line is the use of monomers that incorporate the groups taking part to the chain transfer reaction: these products are called monofers [12].

In the frame of a work on the preparation of coatings via cationic UV curing we have investigated the influence of alcoholic additives on different types of resins, namely vinyl ether, propenyl ether and epoxy systems. In this paper both the kinetics of photopolymerisation and the evaluation of the properties of the obtained coatings are discussed.

Experimental

Materials: Reference resins

The vinyl ether monomers are: triethyleneglycol divinyl ether (DVE3), supplied by ISP Europe and diethylene glycol divinylether (DVE2), purchased from Aldrich. Their structures are:

$$CH_2=CH—O\left(CH_2CH_2—O\right)_3 CH=CH_2$$

$$CH_2=CH—O\left(CH_2CH_2—O\right)_2 CH=CH_2$$

The propenylether monomer is trimethylolpropane tripropenylether (TPE) synthesized in the laboratory of Perstorp according to the procedure described in [8]. The structure of TPE is:

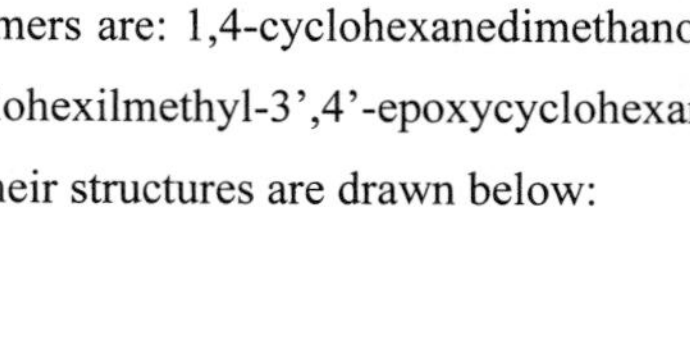

The epoxy difunctional monomers are: 1,4-cyclohexanedimethanol diglycidyl ether (DGE) from Aldrich and 3,4-epoxycyclohexilmethyl-3',4'-epoxycyclohexanecarboxylate (CE), kindly supplied by Union Carbide. Their structures are drawn below:

Materials: Alcoholic additives

Alcohols employed are: 2-propanol, n-butanol, 2-phenyl-2-propanol, bisphenolA, diethylenglycol, trimethylolpropane, hydroxy functionalized polybutadiene (PBOH, Mn = 1200; 55% 1,4 trans; 20% 1,4 cis; 25% 1,2; hydroxyl number 1.7 meq/g) all purchased from Aldrich. Monofers under investigation are:

o diethylene glycol monovinyl ether (HDVE2 from Aldrich)

o di propenylether trimethylolpropane (DPE) which was synthesised on purpose as reported elsewhere [8]. The structure is below. As a reference the acetic ester of it was used.

o epoxidized hydroxypolybutadiene (PBE), having the same structure of PBOH and containing about 20% of epoxidized double bonds (epoxide equivalent = 260), where the epoxy units are mainly of 1,4-trans type.

Materials: Initiator

In all the formulations triphenylsulfonium hexafluoroantimonate, from Union Carbide, was used as the cationic photoinitiator in a concentration equal to 2 wt%.

Methods

Film preparation

The photocurable mixtures were spread on a glass slide with a calibrated wire-wound applicator to obtain a thickness of about 100 μm.

The curing reaction was performed by UV irradiation with a medium pressure Hg lamp (Italquartz, Milano, Italy); the light intensity at the film surface was 120 W/m2. The irradiation was stopped when a constant reactive groups conversion was achieved, as determined by FTIR measurements (see below). The samples, after irradiation, were stored for one night and then put for 15 minutes in a closed system saturated with a 5% v/v water/ammonia solution vapors, in order to neutralize the acidic species.

Film characterization

The degree of the reactive groups conversion was calculated coating the mixture on a KBr disk, and measuring the ratio of the corresponding IR absorbance before and after UV exposure. The kinetic curves were obtained by measuring the decrease of the reactive group absorption band, after different irradiation time. A Genesis Series ATI Mattson (USA) spectrometer was used.

The Gel Content of the films was determined by measuring the weight loss after 20 hours treatment at room temperature with chloroform. DSC measurements were performed with a Mettler DSC30 (Switzerland) instrument, equipped with a low temperature probe.Dynamic mechanical thermal analyses (DMTA) were performed with a Rheometric Scientific MKIII (UK) instrument, at the frequency of 1 Hz in the tensile configuration.

Results and discussion

Curing of vinyl ethers

The kinetics curve of vinylethers such as DVE2 and DVE3 (in Figure 1 an example is reported) show that the rate of propagation of the photopolymerisation changes significantly in the presence of alcohols. The final conversion of double bonds increases, reaching even completeness in some cases, as reported in Table 1.

The alcoholic additive affects the polymerisation mechanism through a chain trasfer reaction as described previously [13,14]: FTIR analyses confirm this mechanism showing a decrease of the OH band at 3600 cm^{-1} and an increase of the ether band at 1100 cm^{-1}.

Considering the DVE3 system containing the HDVE2 monofer, we can assume that the same chain transfer reaction takes place together with the copolymerisation of the vinyl ether group present in the monofer. From the kinetic curve of Figure 1, it is evident that 5% of HDVE2 allows to increase the polymerisation rate and to obtain the complete conversion of the double bonds.

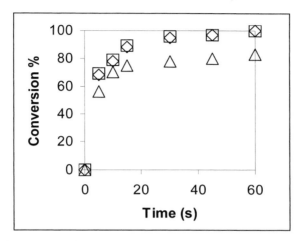

Figure .1 FT-IR kinetic curve of the pure DVE3 monomer (Δ) and in the presence of 5 wt% of HDVE2 (□) and 2-phenyl-2-propanol (◊).

The chain transfer makes the polyvinylether chain length decrease, so that free-dangling ends are introduced in the network. As a consequence, the network structure is more flexible and the mobility of the reactive species increases so that the polymerisation can be completed.

The flexibilisation is reflected by the lowering of the glass transition temperature of the cured films: as reported in Table 1, in the case of DVE3, the original Tg (58°C) is reduced down to 15°C by the addition of about 15 %mol of 1-butanol or 8 % mol of 2-phenyl-2-propanol. 2-phenyl-propanol is the most effective alcohol, probably due to its higher nucleophilicity [14]. At the same OH groups concentration, the Tg values are higher in the presence of HDVE2 with respect to the systems containing alcohols. These results can be attributed to the photopolymerisation of the HDVE2 double bonds, which increases the crosslinking density of the network.

Table 1: Effect of alcohols and hydroxylic monofers on the conversion and properties of vinylether systems (DVE3)

Type of additive	% mol#	η (%)	Gel content (%)	Tg (°C)
-	-	93	98	58
1-butanol	15.2	98	93	14
2-phenyl-2-propanol	8.3	100	92	15
HDVE2	14.2	100	96	30
	27.0	96	90	13
	49.5	96	71	-10

Moles of additives per moles of double bonds

Curing of propenyl ethers

The kinetics of the curing of TPE with and without alcohols or DPE are shown in Figure 2.

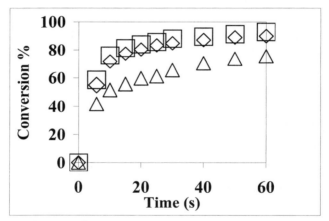

Figure 2. FT-IR kinetic curve of the pure TPE monomer (Δ) and in the presence of 5 wt% of DPE (□) and 2-phenyl-2-propanol (◊).

The kinetic rates are higher in the presence of 2-phenyl-2-propanol or DPE and the final conversions can reach 100%: for the TPE alone the yield never overcomes 80%. These results can be interpreted on the basis of a chain trasfer reaction involving the OH groups, as evidenced in the case of vinyl ethers systems. The FTIR spectra performed before and after the UV curing, confirm this mechanism.

A further evidence of this phenomenon comes from the investigation of the behaviour of the ester derivative of DPE. By adding the DPE acetate, there is no change in the polymerisation kinetics and the conversion is similar as in the case of the pure TPE [8].

In Table 2 some properties of the propenyl ether systems are reported.

Table 2: Effect of alcohols and hydroxylic monomers on the conversion and properties of propenylether systems (TPE)

Type of additive	% mol#	η (%)	Gel content (%)	Tg (°C)
Pure TPE	-	80	85	140
DPE	4	100	85	120
	2	96	83	125
1-butanol	6.5	93	83	108
2-phenyl-2-propanol	5.2	99	81	107

Moles of additives per moles of double bonds

The gel content decreases in the films containing additives, probably due to the formation of low M.W. branched products.

A further consequence of the presence of the additives is the clear decrease of the Tg: in the presence of DPE the decrease is less effective than in the presence of the alcohols, as obtained in the case of the vinyl ether systems.

Looking at the DMTA spectra of the TPE film without additive (Figure 3), one can see that the tanδ peak is quite broad and besides the maximum (from which the Tg value was estimated), a shoulder at lower temperature is present. There is therefore a hard domain, corresponding to a phase where the crosslinking is very high, together with a more flexible polymeric matrix. As the films are transparent, the heterogeneity of the networks are at submicron level. In the presence of additives, the DMTA thermograms are simpler (one signal in the tan δ curve), indicating that in this case homogeneous networks are formed (an example is reported in Figure 4).

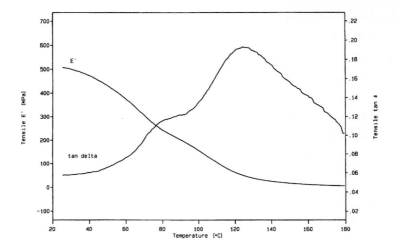

Figure 3. DMTA curves of TPE cured film.

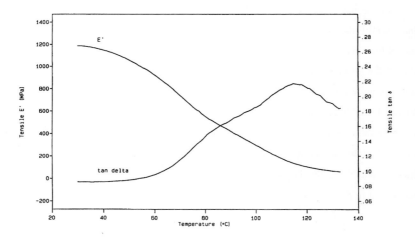

Figure 4. DMTA curves of TPE cured film containing 2-phenyl-2-propanol.

Curing of epoxy systems

The kinetics of the curing of DGE monomer in the presence of PBOH and PBE was investigated. The kinetic curve (Figure 5) of the DGE alone shows an initial high rate of polymerization, then a plateau at about 55% epoxy groups conversion.

In fact the propagation rate constant of the system is strongly affected by the molecular mobility of the growing chain and at high conversion the highly crosslinked network formed hinders the chains and stops the reaction. The addition of PBE in different amounts (ranging between 10 to 50 wt%) induces a clear increase of the curing rate and of the epoxy groups final conversion (Figure 5, Table 3). In the presence of 50 wt% of PBE the conversion is almost complete.

At the same time the gel content is very high (97-98 %), indicating that PBE is linked into the polymer network. Moreover the PBE contains also hydroxyl groups: in order to investigate the the chain transfer reaction, PBE was substituted, in the photocurable mixture, by PBOH. The results of Table 3 show an increase of the gel content to 98% and of the epoxy group conversion. We can assume that the hydroxyl groups interact with the growing chains, giving chain transfer reactions that increase the flexibility of the films so that the enhanced mobility permits a higher conversion. However this is lower than the final conversion achieved in the presence of PBE: this means that PBE is involved in the copolymerization with DGE through its epoxy groups[15].

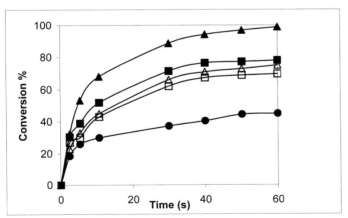

Figure 5 FT-IR kinetic curves of the pure DGE monomer (●) and in the presence of 20 (■) and 50 (▲) wt% of PBE or 20 (□) and 50 (Δ) wt% of PBOH.

Table 3: Properties of the UV cured films obtained from DGE-PBE and DGE-PBOH mixtures.

Sample	Gel Content %	η %	Tg °C
Pure DGE	90	56	53
DGE:PBOH 90:10 wt/wt	98	61	40
DGE:PBOH 80:20 wt/wt	97	70	39
DGE:PBOH 50:50 wt/wt	98	71	33
Pure PBE	95	48	30
DGE:PBE 90:10 wt/wt	98	68	47
DGE:PBE 80:20 wt/wt	97	80	43
DGE:PBE 70:30 wt/wt	98	85	38
DGE:PBE 50:50 wt/wt	98	99	32

The film properties are different depending on the type of reactive additive employed, PBE or PBOH. As reported in Table 3, by increasing the amount of PBE and PBOH in the mixture, a decrease of Tg is obtained. In particular the Tg values of the mixtures containing PBE are always higher than those obtained with the mixtures containing PBOH. These results can be attributed to a higher degree of crosslinking of the films obtained in the presence of PBE, due to the occurrence of the copolymerization reaction.

Similar results were obtained using the same additives and the CE as a monomer. In this case the ring opening polymerization proceeds more rapidly upon UV irradiation, but only 45% degree of conversion is achieved due to the vitrification effect. It increases in the presence of PBE, reaching completion in the presence of 50 wt% of the additive.

The gel content and the Tg values of the UV cured films obtained in the presence of PBE or PBOH additives, show the same behavior as obtained for DGE monomer, confirming the proposed reaction mechanism.

Table 4: Properties of the UV cured films obtained from CE/PBE and CE/PBOH mixtures.

Sample	Gel Content %	η %	Tg °C
Pure CE	88	45	214
CE:PBE 90:10 wt/wt	99	66	200
CE:PBE 80:20 wt/wt	98	77	190
CE:PBE 70:30 wt/wt	95	80	182
CE:PBE 50:50 wt/wt	97	100	172
CE:PBOH 90:10 wt/wt	97	56	200
CE:PBOH 80:20 wt/wt	98	65	184
CE:PBOH 50:50 wt/wt	97	73	171

In the case of the DGE system, the effect of the high molecular weight polibutadiene additives was compared with small alcohols, such as diethylenglycol, trimethylolpropane and bisphenol-A. The data of gel percent, conversion and thermal transitions are collected in Table 5: they indicate that in the presence of small alcohols a higher increase of conversion can be obtained.

Table 5: Effect of alcohols and hydroxylic monofers on the conversion and properties of epoxy systems (DGE)

Type of additive	% eq	η (%)	Gel content (%)	Tg (°C)
-	-	56	90	45
PBOH 50%	7	71	98	33
Bisphenol-A 5%	4.3	90	99	39
Trimethylolpropane 5%	11.2	82	98	40
Diethyleneglycol 5%	9.4	78	98	19

Conclusions

We investigated the behaviour of typical cationic UV-curable systems such as vinylethers, propenyl ethers and epoxy in the presence of alcoholic additives either in the form of alcohols or in the form of hydroxy comonomers. The presence of such additives changes the kinetics of photopolymerisation and the properties of the cured films. The results can be interpreted on the basis of a chain transfer reaction involving the OH groups which determines a flexibilisation of the network and increases the curing rate. In the presence of a monofer the flexibilisation is lower, but the curing rate remains clearly higher than the pure resin.

Acknowledgements

The financial support of Italian Miur, Prin 2001, is gratefully acknowledged. Thanks are due to Dr. Nicola Rehnberg of Perstorp for the fruitful cooperation on propenyl ether systems.

References

[1]   P. Fouassier, J.C. Rabek, in *"Radiation Curing in Polymer Science and Technology"*, Elsevier, London, 1993 Vol. I.
[2]   J.V. Crivello, in *"Photoinitiators for Free Radical Cationic and Anionic Photopolymerization"*, G. Bradley, Wiley, New York, 1998, 2nd ed. p. 329.
[3]   Y. Takimoto, in *"Radiation Curing in Polymer Science and Technology"*, J.P. Fouassier, J.C. Rabek Elsevier, London, 1993, Vol III p. 269.
[4]   S.P. Pappas, in *"UV Curing Science and Technology"*, Plenum Press, New York, 1992 p. 2.
[5]   C.G. Roffey, in *"Photopolymerization of Surface Coatings"*,(Wiley, New York, 1982 p. 74.
[6]   C. Decker, K. Moussa, *J. Polym. Sci. Polym. Chem.* **1992**, 30, 913.
[7]   J.V. Crivello, G.Lohden *J. Polym. Sci. Polym. Chem* **1996**, 34, 2051
[8]   M.Sangermano, G.Malucelli, R.Bongiovanni, A.Priola, U.Annby, N.Rehnberg *Polym.Int.* **2001**, 50, 998
[9]   S.Penczek, P.Kubisa, R.Szamanski *Macromol.Chem Macromol Symp.* **1986**, 3, 203
[10] J.V.Crivello, D.A.Conlon, D.R.Olson K.K. Webb J.Radiat. Curing **1986**, 13, 3
[11] J.V.Crivello, R.Acosta OrizJ.Polym.Sci.:Part A:Polym.Chem. **2002**, 40, 2298
[12] E.J.Goethals,   G.G.Trossaert,   P.J.Hartmann,   R.R.De   Clerq   *Polym.Prepr.   ACS Div.Polym.Chem.* **1993**, 34, 205
[13] M.Sangermano, G.MalucellI, R.Bongiovanni, ,A.Priola *Polymer Bullettin* **1999**, 42, 641
[14] M.Sangermano, G.Malucelli, F.Morel,C.Decker,A.Priola *Eur.Pol..J.* **1999**, 35, 636
[15] M.Sangermano, G.Malucelli, R.Bongiovanni, F.Peditto, A.Priola *J.Materials Science* in press

Macromol. Symp. 187, 493–502 (2002)

3-Dimensional Epoxy Binder Structures for Water Damp Permeable and Breathable Coating and Flooring Systems

Matthias Lohe, Mike Cook, Achim H. Klippstein*

Air Products & Chemicals Inc, c/o Kanaalweg 15, PO Box 3193

3502 GD, Utrecht, The Netherlands

Summary: Blisters and delamination are common problems in the epoxy and polyurethane flooring industry. Main reasons are osmotic effects and hydrostatic water pressure, because of the moister content of fresh concrete. Today's best recommendation is to wait 28 days, combined with perfect surface preparation. Even then, problems can't be excluded when following standard and conventional ways only. The patented technology for water based Emulsion Curing Agent technology enabled water-damp permeable Coating and Flooring systems. After completing the hardening process, this flooring system creates a highly durable micro porous structure, based on a 3D epoxy network. Due to the special nature of the matrix, only water vapour can pass through the system without liquid water penetrating through. This unique system is a perfect extension to conventional epoxy and polyurethane flooring systems. General performance properties, application examples and a description of the polymeric structure of the floor coatings are discussed herein.

Keywords: epoxy resin, polyamines, water based curing agents, crosslinking, microstructure, high performance polymers

Introduction

Concrete Coating and Flooring Systems

When a concrete substrate needs to be protected from environmental attack, there are many different types of coating and flooring technologies that can be used. Examples of these include polyurethanes, methyl methacrylates, polyesters and epoxies. Solvent-free epoxy systems are one of the most versatile technologies employed and are used in a wide variety of applications including primers, floor paints, self levelling, screed and mortar floors. Epoxy binder systems are typically two-component formulations consisting of a liquid epoxy resin and an amine functional curing agent. Epoxy resins are either based on bisphenol-A or bisphenol-F diglycidyl ethers. Their high viscosities require dilution with lower viscosity mono or difunctional reactive diluents.

CCC 1022-1360/00/$ 17.50+.50/0

In Europe the most commonly amine-curing agents for flooring applications are based on modified cycloaliphatic amines of which isophorone diamine (IPDA), meta-xylylenediamine (MXDA) and bis (p-amino cyclohexyl) methane, (PACM) are the most widely used building blocks. As stand alone curing agents, all of the above amines will only undergo partial cure when mixed with an epoxy resin at ambient temperature. The lack of full cure (B-staging), leads to brittle coatings with poor performance properties. In order to maximise the degree of cross-linking, thus overcoming B-staging, curing agents need to be formulated with non-reactive plasticizers, of which benzyl alcohol is most commonly used within the industry.

Waterborne Epoxy Curing Agent Systems

The first generation of epoxy waterborne curing agents were based on solvent free polyamides and were made water soluble by inclusion of a nonionic surfactant or through salting with an organic acid[1]. The high viscosity and water dilution profile of the polyamides allowed for low solids coating formulations only. Formulated systems tended to have short pot lives and coatings were often high in colour and slow to cure, particularly under condition of low temperature (<10°C) and high humidity (>80%). Higher performance products were subsequently developed and introduced into the market place[2] during the late 1980's and early[3] 1990's. Their chemistry was based on aliphatic amine adducts, which exhibit better colour stability and a higher solids content at the same viscosity compared to earlier polyamide products. The higher solids contents coupled with the fast reactivity inherent in aliphatic amine curing agent has lead to faster dry speed and through cure even under the adverse conditions mentioned above.

A common disadvantage of waterborne epoxy systems is that higher film build coatings, >500µm cannot be readily employed. Their use has been limited to sealers, primers and floor paints[4], where a wet film thickness is in the region 100-200 µm is usually applied. At significantly higher film build, such coating systems may encounter problems with entrapped water being present in the film after full cure. Entrapped water results in a downgrading of the performance properties, resulting in poor surface appearance, loss of gloss or soft films. Cracking ^(Figure 1) will be observed with typical waterborne curing agents used for self-levelling floor applications.

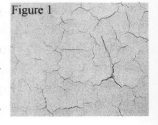
Figure 1

A novel waterborne curing agent technology has been developed[5] to overcome the limitations of previous waterborne epoxy systems. The new Emulsion Curing Agent (EMC) technology is based on a curing agent emulsion of an ultra high molecular weight aliphatic amine-epoxy adduct in water with high amine functionality and can be viewed as the 3rd generation waterborne curing agent for liquid epoxy resin.

This new technology incorporates three key design features for optimum performance:

(a)   Very high compatibility with liquid epoxy resin ensuring good emulsification

(b)   Optimised hydrophile-lipophile balance for good emulsion stability and

(c)   Removal of unreacted raw materials and by-products, which could reduce performance.

This new aliphatic amine emulsion technology results in coatings that dry significantly faster at ambient temperature. This is clearly demonstrated by comparing the Beckman-Koller thin film set times (phase I-IV), for three types of waterborne curing agent-epoxy coating systems. The aliphatic amine emulsion achieves a hard dry time (phase IV), within 6 hours compared to a hard dry time of 14 hours for a similar coating based on a waterborne polyamide curing agent. The new curing agent technology also exhibits a more effective dilution profile allowing the application of higher solids coatings at the same viscosity

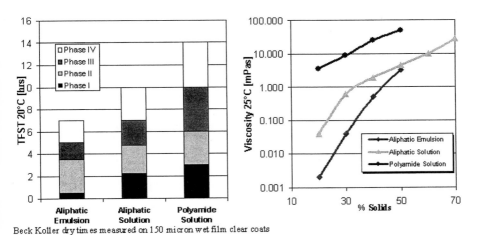

Beck Koller dry times measured on 150 micron wet film clear coats

Figure 2: Dilution Profiles and Drying Times of Waterborne Curing Agents

Waterborne Self-Levelling Floor Formulation

Ease of application combined with optimum mechanical and chemical protection as well as providing easy to clean surfaces has made solvent-free, self-levelling epoxy floors the first choice in concrete protection. Modified solvent free systems using cycloaliphatic curing agents are predominantly used for this type of application and can be viewed as the industry standard. Typically, such protective flooring systems are formulated with low filler to binder ratios (1,5 to 2,6:1), which are deemed necessary in order to ensure good flow and deaerating properties at the point of application.

In order to overturn the old belief that with waterborne technology higher film build coatings cannot be formulated, recent application studies focused on the development of a waterborne self levelling floor using the new generation waterborne curing agent. Development activities successfully lead to the starting point formulation as described in Table 1.

Table 1. Waterborne Self-Levelling Floor Formulation

A-Side:	Component	Supplier	Weight (%)
EmulsionCuring Agent	Epilink® 701	Air Products	11,00
Curing Agent	Anquamine® 401	Air Products	2,50
Defoamer	Byk-045	Byk Chemie	0,70
Pigment TiO_2	Kronos 2160	Kronos Titan	3,80
Diluent	Water	Local	9,10
Filler	Baryte Powder F	Sachtleben	36,00
Filler	Quartz Powder M6	Sibelco	18,00
Filler	Quartz Sand 0,1-0,3 mm	Local	18,50
Thixotropic Agent	Deuteron VT 819 (3% in Water)	Schöner GmbH	0,40
			100,00
B-Side			
Epoxy Resin	Epires® ER8	Air Products	10,00

The significant difference compared to a solvent free cycloaliphatic system is the fact that water can now be used as a diluent to achieve sufficient flow and deaeration. This allows for a reduction of the total binder content from approximately 30% in a conventional cycloaliphatic epoxy floor formulation, down to 15% in the waterborne system (Table 2).

This waterborne self-levelling floor contains approximately 15% by weight of water with theoretical volume solids of 70%. The composition would suggest a significant volume reduction to be observed upon evaporation of the water, eventually causing the floor to undergo severe shrinkage leading to cracking. However, upon visual inspection no defects are observed after the system has fully cured.

Table 2. Formulation Properties

	Waterborne Self-Leveller	Cycloaliphatic Self-Leveller
Binder content [%]	15	30
Filler : Binder	4,3 : 1	2,6 : 1
Volume Solids [%]	70,0	99,8
Water Content [wt%]	15,0	0,0

To quantify this observation an experiment was conducted to determine the degree of shrinkage. A specimen of the floor was prepared and cast into a mould of 2,5cm x 2,5cm x 25cm with two metal studs attached to each end. The distance between the studs was measured and used as a reference point. After 24 hrs the cured blocks were removed from the moulds and measurements were taken between the two studs every 24 hrs to determine the degree of shrinkage. The level of shrinkage after 14 days cure is the range of 1,3-1,5%, significantly lower than the maximum theoretical shrinkage of 30%. Most of the water is driven out during the very first day when the bulk part of the shrinkage occurs. Thereafter no significant change is observed as can be seen by the flat profile from day one onwards.

Comparative performance properties of waterborne and cycloaliphatic floors are summarised in Table 3. Flow properties as measured by flow out diameter from a cylindrical container are the same for the two systems. The handling time of the waterborne system at 45 minutes is within the same range as the cycloaliphatic system, as well as offering a comparable final Shore D hardness. Hardness after day one of the waterborne system is in excess of a Shore D value of 65, sufficient for foot traffic.

Bond strengths on shot blasted concrete coated with both the waterborne and solvent free epoxy self-levellers were determined by the dolly pull off method according to DIN 1048. Both systems exhibit bond strengths of >2 N/mm^2 (required minimum value), with a cohesive mode of failure occurring in the concrete substrate.

Abrasion resistance was measured in accordance with ASTM D 5178-91 utilising a Tabor abrasion tester fitted with wheel number C117. Similar weight losses within 1000 cycles are observed with both systems.

The surface finish of the waterborne floor is matte in comparison to the glossy cycloaliphatic finish. While it is the arguable which surface appearance is more attractive, the matte surface appearance of the waterborne floor proves to yield better scratch resistant. Application on uneven substrates also proves to better hide the application/surface differences.

It is anticipated, that the lower binder content in the waterborne floor formulation will affect the mechanical properties of the floor system. Experimental results confirm that the waterborne system exhibits lower compressive as well as flexural strength compared to a cycloaliphatic floor. Normal traffic areas typically require a compressive strength of 20MPa and the strength of the water based epoxy floor (40MPa) exceeds this basic requirement.

Table 3. Basic Performance & Mechanical Properties of Epoxy Self-Leveller Systems

	Waterborne	Cycloaliphatic
Flow out [cm]	15.9	16.5
Handling Time [min]	45	30
Hardness, 7d [Shore D]	80	85
Bond Strength [N/mm^2]	5	3
Abrasion resistance [mg/1000 cycles]	156	138
Surface appearance	Matte	Glossy
Compressive [MPa]	40	62
Tensile [MPa]	5	7
Flexural [MPa]	11	30

Water Vapour Permeability

Water vapour can readily pass through most concrete substrates due to small capillaries. However, when the concrete is coated with a conventional solvent free epoxy floor, the coating will act as a vapour seal and hinder water evaporation from the concrete. Typically, we are dealing with two different types of water transmission problems. One being the hydrostatic water pressure, which is a result of the differential between the highest elevation of a water column and the lowest physical point of a structure and the other being osmotic pressure.

Osmosis is defined as the spontaneous flow of a liquid through a semi-permeable membrane, from a dilute to a more concentrated solution. As a result, the liquid volume of the initially more concentrated solution increases until the hydrostatic pressure generated in its equilibrium with the osmotic pressure. The pressure generated by osmosis can greatly exceed other forces in concrete and ultimately yield in delamination of the floor coating. An estimate of the forces that can occur has been described applying the simplified description of osmotic pressure (p) according to Van 't Hoff (Equation 1)

$$p = Dc*R*T \hspace{4cm} \text{Equation 1}$$

Dc = difference in concentration of soluble salts in mol/l R = gas constant; T = temperature in Kelvin

Assuming a NaCl concentration of 1 mol/l yields a theoretical osmotic pressure of ~2,4MPa. This pressure or higher, at increased salt concentrations is sufficient to exceed the bond strength of the coating to the concrete substrate thus leading to disbondment of the coating.

Three requirements must be fulfilled for osmosis to take place:

(a)   Presence of water (liquid and/or vapour)

(b)   Presence of soluble salts

(c)   Presence of a semi-permeable membrane

All these conditions are typically found on concrete surfaces. Depending on the age of the concrete its moisture content can vary from about 4% when fully cured up to 18% in freshly prepared green concrete. The water content also depends on utilized water to cement ratio as well as cure conditions. The principal causes and effects of osmotic blistering for impermeable epoxy coatings on concrete has been extensively studied and reported by other authors[6]. It has been proposed that a higher concentration of soluble salts are present in the top section of the concrete that cannot migrate into the interior of the slab due to a more dense structure present in the bottom section. The difference in soluble salt concentration coupled with concrete permeability for inorganic salts between the top and the bottom section have been identified as major contributors to meet the osmotic cell conditions.

A comparison of how impermeable and permeable epoxy coatings affect blister formation is given in Figure 3.

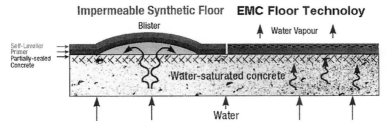

Figure 3: Osmotic Blistering

If pressures are allowed to accumulate and they are not dissipated or relieved by venting; developing pressures exceed the adhesive bond strength to the concrete, thus leading to disbondment from the substrate. Contrarily, permeable coatings are far less susceptible to the effects of these pressures than non-permeable coatings.

Two factors will influence the permeability of a coating. At pigment volume concentrations (PVC) higher than the critical pigment volume concentration (CPVC), the presence of capillary channels in the film causes the water vapour permeability to increase dramatically[7]. It has also been suggested to determine CPVC by measuring water vapour permeability[8]. The other contributing factor is a permeability provided by the nature of the polymer matrix itself. The lack of any substantial shrinkage in the external dimensions of the new waterborne floor as discussed earlier, suggests that the cured polymer matrix forms a micro porous structure. A study of water vapour permeability has therefore been conducted in accordance with DIN 52615 to calculate moisture resistance factor and ASTM E96-95 to calculate permeance comparing the conventional cycloaliphatic and the new waterborne floors[9].

The moisture resistance factor (m) indicates how many times greater the moisture resistance of the material is in comparison with the resistance of a motionless layer of air of the same thickness at the same temperature. The m-factor is a dimensionless quantity, hence is a material specific constant that allows direct comparison of two systems excluding the effect of coating thickness.

Results of the permeability study confirm the hypothesis of a water vapour permeable floor coating, (Table 4). The measured permeability of the waterborne floor is 30 times greater than that of the solvent free floor as expressed by the m-factor.

Coatings are classified as permeable when a U. S. permeability rating of greater than 3 perms* ($=1,7 \times 10^{-7}$ g/s/m^2/Pa) is achieved[10]. Standard concrete with a compressive strength of 21 MPa typically exhibits a permeance rating of 20-30 perms. ($=1,1-1,5 \times 10^{-6}$ g/s/m^2/Pa). With the new waterborne system a permeance of 11 perms is readily achieved, which is more than sufficient to comply with this definition of a permeable coating. Typically, cycloaliphatic floor systems do not exhibit the required permeability. This degree of permeability opens up the potential for the waterborne system to be used where osmotic pressure is a known problem.

Table 4. Water Vapour Permeability of Epoxy Self-Levelling Floors

	Waterborne	Cycloaliphatic
Film thickness [μm]	2,550	1,870
WVT [g/m^2/24hrs]	8,1	0,4
Permeance [perms]	11,7	0,09
Permeance [g/s/m^2/Pa]	$6,7 \times 10^{-7}$	$0,05 \times 10^{-7}$
μ-Factor	1,000	30,000

WVT = Water Vapour Transmission
1 perm = 1 grain/hr/ft^2/in Hg

The increased permeability as determined by the above test method is confirmed in scanning electron microscopy (SEM) images on floor castings of the two systems under investigation. The SEM images in Figure 4 show a magnification of 20.000 focused exclusively on the cured epoxy-amine binder system. A marked difference between the waterborne and the solvent free system is evident. A continuous structure with very few voids is observed with the cycloaliphatic thermoset whereas the waterborne system shows a sponge like appearance with voids and channels exhibiting a micro-porous structure that is permeable for water vapour. This analysis suggests that permeability is not only achieved due to a formulation beyond CPVC but that the underlying polymer structure created by the new waterborne curing agent is an integral part of this property. Work has focused on the development of an amine-curing agent that shows applicability in the area of permeable, water based epoxy flooring. This approach can provide the formulator with an additional tool to solve the osmotic blistering problem.

 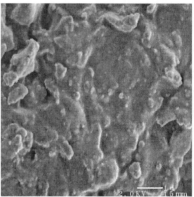

Waterborne Self-Leveller   Solvent Free Cycloaliphatic Self Leveller
Figure 4: Comparison of SEM Images (20,000 x Magnification)

Additionally, the control of the residual moisture in the concrete would not be viewed as critical, therefore reducing the waiting times prior to application of the protective coating. The epoxy system is also an alternative to the application of polymer modified concrete[11], where the new approach will offer advantages in the area of improved mechanical and chemical properties as well as an improved decorative surface appearance.

Conclusions

The new generation waterborne technology allows the formulation of waterborne high film build coatings with unique performance properties that are complementary to well established solvent free cycloaliphatic curing agent floors. The water vapour permeability of the waterborne self-levelling floor expands the range of applications for epoxy technology, allowing the formulator to apply floors in areas where osmotic blistering has been previously observed. For example, floors below grade, or floors where a damp proof membrane is absent or has started to fail.

Another new dimension in waterborne flooring technology is the development of a matte surface and the absence of amine carbamation. The latter property, allows floor coatings to be applied in colder and more humid conditions than currently possible with conventional solvent free epoxy systems. The additional advantages are high temperature resistance, impact resistance and good chemical resistance. The very low free amine content and the absents of non-reactive plasticisers enable a safer working environment.

References

[1]   Walker, F.H; Cook, M.I; ACS Symposium Series 663 Technology for Waterborne Coatings J. Edward Glass
Editor, 1997 The American Chemical Society

[2]   Godau C., Water based systems for surface protection

[3]   Akzo Chemicals, Epilink 660

[4]   Klippstein A., *Welt der Farbe 2/98*, Development of high performance coating under environmental aspects

[5]   Klippstein, A. WO 93/21250 Emulsion Curing Agent Technology

[6]   Pfaff, F.A; Gelfant, F.S; *J. Protective Coatings & Linings*, (1997), 52-64

[7]   Arcozzi, A; Arietti, R; Bongiovanni, R; Pocciola, M; Priola, A; *Surface Coatings International* 1995 (4), 140-143

[8]   Lyssy, G.H; Coating, 8, (1977), 204-205
Funke, W; Handloser, G; *Deutsche Farben Zeitschrift*, 27, (1973), 440

[9]   TNO Netherlands, private communication

[10]   Hall, C; O'Connor, S; *J. Protective Coatings & Linings*, 2, (1997), 86-102

[11]   Marohn H., Industrial Flooring Symposium 1991, Epoxy modified Mortar and Flooring Systems

Macromol. Symp. **187**, *503–513 (2002)*

High Scratch and Etch Resistance by Optimized Cross-linking Chemistry and Density

Dr. Carmen Flosbach

DuPont Performance Coatings, D-42271, Wuppertal, Germany

Summary: Target of development in the field of modern automotive clear coats is the combination of high etch and high scratch resistance.
For high etch resistance the cross-linking chemistry is the key factor. The property scratch resistance is mainly influenced by the cross-linking density and the flexibility of chains between netpoints. By combining these parameters high etch and scratch resistant clear coats can be formulated. As a representative example an optimized clear coat based on epoxy/acid cross-linking chemistry is discussed. The ways to achieve high cross-linking density are shown.

Keywords: etch resistance, scratch resistance, coatings, crosslinking, epoxy/acid

Introduction

In the recent years field damages of automotive clear coats have been observed which are caused by etchings from environmental pollution. Especially in aggressive atmosphere, clear coats based on conventional melamine cross-linking can show severe etching after short exposures. These damages have led to the development of etch resistant coatings by new cross-linking chemistries. With given polymer backbone chemistry, improved etch resistance can be achieved by delivery of cross-linking sites with better resistance against acidic attack. In figure 1 cross-linking mechanisms of clear coats are shown. These cross-linking reactions are leading to different cross-linking sites and have different potential in reachable cross-linking density. In acrylic/melamine the cross-linking sites are the weak point regarding etch resistance. On the other hand cross-linking density can be very high. Urethane structures which are generated in 2pack clear coats or by the use of blocked polyisocyanates show a much better chemical resistance. Steric hindrance of the urethane groups, increase in Tg of the coating and increase in hydrophobicity by the use of polyisocyanate hardeners based on isophorone diisocyanate increase this etch resistance even further. Cross-linking density is lower when compared to acrylic melamine systems. Other very resistant linkages are the siloxane linkage derived from silane condensation reaction and the ester

          CCC 1022-1360/00/$ 17.50+.50/0

504

linkage achieved by epoxy/acid addition. The ultimate in etch resistance and cross-linking density is achieved by a carbon-carbon-linkage which can be formed by radical polymerization processes like in UV-coatings.

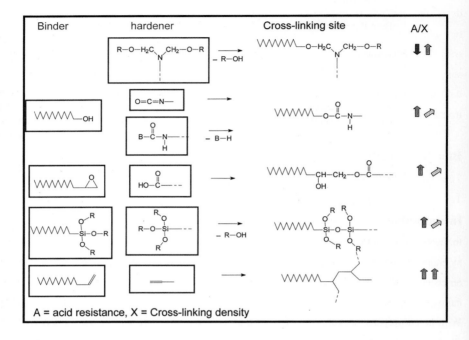

Figure 1. Different cross-linking mechanisms for automotive clear coats

Normally the increase in etch resistance is accompanied by reduced scratch resistance. Modern high performance binders combine high etch with high scratch resistance.

Testing methods

To check the etch resistance of clear coats, different artificial tests to simulate environmental influences are used. Among these artificial testing media besides sulfuric acid also a lot of different chemicals are used. Standard acrylic/melamine clear coats show severe damages after acidic attack. The clear coat is totally destroyed after a short exposure and often additionally an attack of the aluminum flakes of the base coat is observed.

The outdoor performance of clear coats normally is checked via outdoor exposure in aggressive atmosphere like e.g. in Jacksonville/Florida. Only three months of exposure are necessary to see severe damages in case of clear coats with poor etch resistance.

For the determination of scratch resistance also a lot of different test methods have been developed and established. The most common test in Europe is the bench top car wash simulation test with Amtec/Kistler device. Besides this simulation of real car washing stresses, a more scientific method has been developed with the nano scratch determination. In this test a single scratch is generated with a diamond indenter and increasing force. The plastic resistance (plastic deformation of the coating at a given force) and the fracture resistance (force at which an irrepairable fracture occurs) of the coating is measured in this experiment.

Binder chemistry

To achieve a high etch and scratch resistance three main parameters have to be fulfilled.

1. it is essential to generate cross-linking sites with good stability against chemical attack

2. a high cross-linking density has to be achieved

3. flexible chains between two net points have to be present.

If these parameters are adjusted in the right way, nearly every coating can be optimized according to superior etch/scratch performance.

As a representative example a 1pack etch resistant clear coat based on epoxy/acid cross-linking will be discussed in detail.

Figure 2 shows the basic cross-linking reaction. An epoxy functional resin is cross-linked via a carboxylic groups bearing hardener. During the addition reaction a hydroxyl group is generated which offers the opportunity of further cross-linking with auxiliary binders, like blocked polyisocyanates, melamine resins or transesterification hardeners.

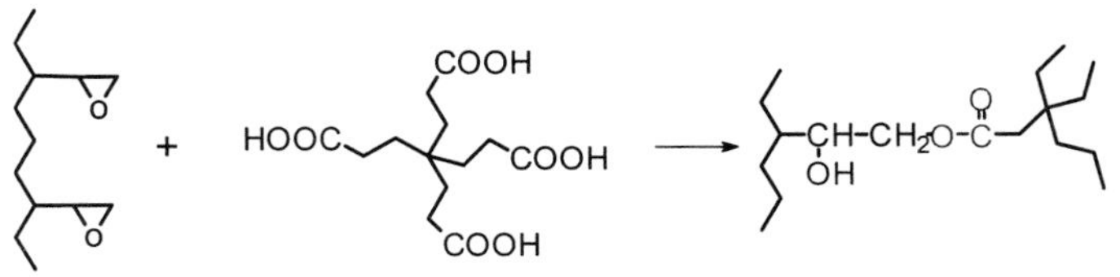

Figure 2. Basic cross-linking mechanism

To guarantee a good weatherability for clear coat applications aromatic epoxy resins based on bisphenol-A-diglycidylether can not be used. The binder class of choice are copolymers of glycidyl methacrylate with other appropriate co-monomers (Figure 3). The choice of the other co-monomers is mainly influenced by important features like hardness, hydrophobicity, flexibility, compatibility and flow behavior of the resulting main binder/ clear coat.

As a cross-linking agent a carboxy functional binder is used. Compared to acrylic resins acidic polyester resins show a higher reactivity and are the preferred binder class. The schematic structure of the acidic polyester hardener is shown in Figure 4. It is based on a highly branched polyester resin. The structure of this polyester resin is strongly influencing the mechanical properties of the cross-linked film. By using highly elastic hardeners with a high functionality, films with an excellent hardness/scratch resistance balance are obtained. An important structural feature is the partly addition of ε-caprolactone to the acidic groups. By this reaction the mobility of the acidic groups is increased and the reactivity is enhanced. On the other hand these flexibilising groups contribute also to the good scratch performance of the cross-linked film.

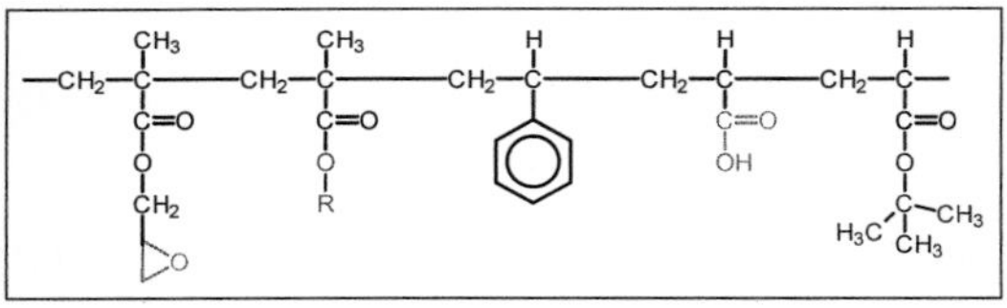

Figure 3.Schematic structure of epoxy acrylate

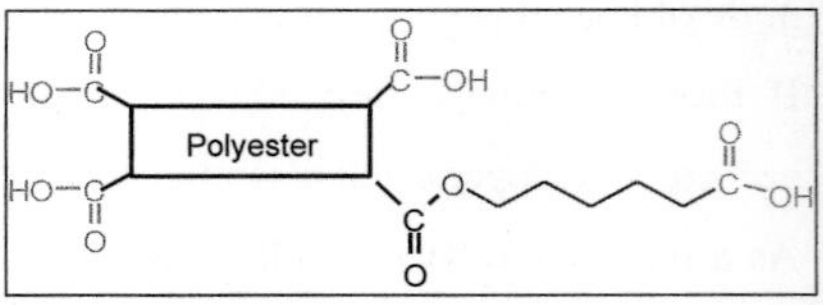

Figure 4. Schematic structure of acidic polyester

Cross-linking density

The reaction between epoxy and carboxy groups is an exothermic addition reaction. In Figure 5 the reaction of an epoxy functional and an acidic binder is observed via DSC. The highly exothermic reaction starts at approximately 110°C. In comparison the reaction of a butanonoxime blocked polyisocyanate (based on isophorone diisocyanate) with a hydroxyfunctional resin is in the first step an endothermic reaction starting at app. 130°C.

This high reactivity leads to increased cross-linking density. With this high reactivity it is possible to use binders with lower Tg and higher flexibility. Hardness of the final film is not mainly derived from the Tg of the binder but also from the Tg increase by cross-linking.

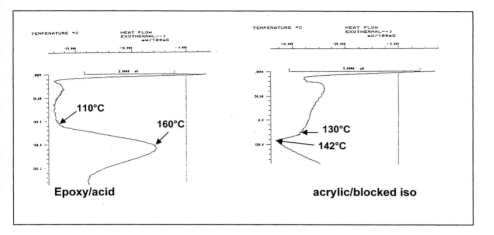

Figure 5. DSC measurements

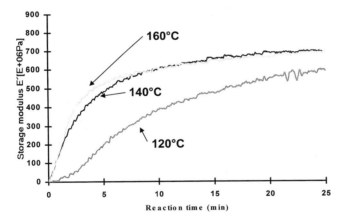

Figure 6. Determination of cross-linking density via DMA

The increase of cross-linking density in dependence of the baking temperatures and time can be measured by dynamic mechanical analysis. To perform the experiment the lacquer was applied on natural silk as a carrier. Figure 6 shows the results of this experiment. The cross-linking was monitored at 120°C, 140°C and 160°C over a period of 25 minutes, by determination of the storage modulus. The resulting curves show the increase of storage modulus which is a function of the cross-linking density during baking time. The higher the storage modulus the higher is the cross-linking density. The measured curves show, that the reaction at 120°C is slower than at 140° and 160°C, but after 25 minutes a similar level of cross-linking is reached. The reaction at 140°C and 160°C shows the same speed. After 10-15 minutes the final cross-linking density is reached. The baking at 160°C has no further advantages compared with the lower baking temperature [1].

The described experiments demonstrate the high reactivity of the epoxy/acid system. Despite the high reactivity of this system a good storage stability is realized.

Increase of cross-linking density under low bake conditions

For increased network density not only the high degree of branching of the acidic polyester hardener but also the structure of the main epoxy functional acrylic binder plays an important role. A further enhancement of the cross-linking density under low bake conditions can be achieved by different routes.

By incorporation of acrylic acid in the monomer mixture an internal addition reaction takes place. (Figure 7). During polymerization the acrylic acid reacts with the epoxy groups which leads to branched polymer chains, more net points in the final network and therefore increased cross-linking density.

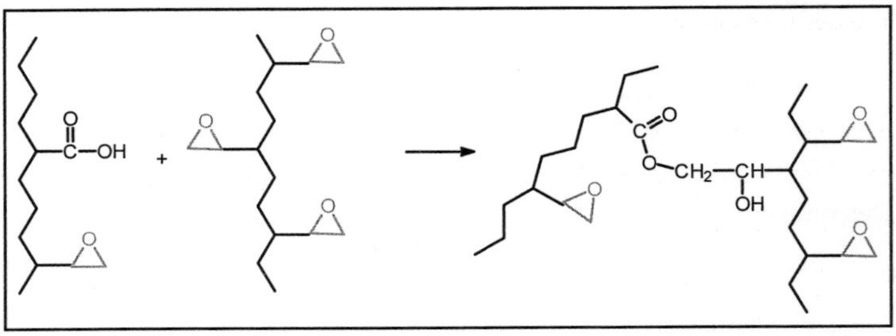

Figure 7. Internal branching during co-polymerization

The amount of branching is mainly depending on the amount of acid used in the polymerization. Figure 8 shows the differences in molecular weight in dependence of the amount of incorporated acrylic acid. The left column shows the result without acid, the middle column with an acid number of 3 (= 0,4% on monomer mixture) and the right one with an acid number of 5 (= 0.6% on monomer mixture). It is obvious, that by increasing the amount of acid the weight average of molecular weight increases. The number average of molecular weight stays on the same level. The dominant part of molecules shows no branching and therefore the viscosity of the resins at comparable solids contents is not much differing.

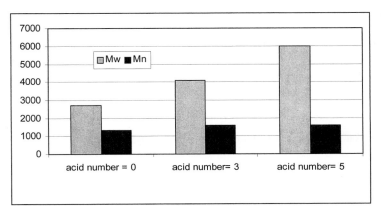

Figure 8. Dependence of molecular weight on acid number

Nevertheless this slight branching of the resins leads to a significant increase of cross-linking density in underbake conditions. The cross-linking density can be checked by determination of the xylene resistance of the film. During this test the baked film is penetrated for 10 minutes with xylene. 15 minutes after penetration the surface hardness of the film is measured by pendulum experiment (König). After 2h the measurement is repeated to control the regeneration of the film. The lower the cross-inking density is, the higher is the degree of decrease of hardness after xylene penetration.

Whereas the non-branched resin shows a strong decrease of hardness after xylene penetration, the branched resins show a very good xylene resistance.

Investigations on branched epoxy acrylates

For a better understanding of the process of branching several investigations were made. Besides the amount of acid in the monomer mixture also the speed of the different reactions, co-polymerization and addition is of importance. During synthesis of the binder different reactions will occur:

a) Copolymerisation of GMA

b) Copolymerisation of AA

c) Formation of a GMA/AA-adduct

d) Copolymerisation of the GMA/AA-adduct

e) Addition of an epoxyfunctional chain to a carboxyfunctional chain

To check which reaction dominates a comparative experiment was performed. If the reaction of GMA with AA is much faster than the co-polymerization, similar products would be achievable by substitution of AA against butanediol diacrylate (BDDA).

As can be seen in Figure 9, by using butanediol diacrylate both the weight and number average molecular are reduced dramatically. This observation leads to the conclusion that the co-polymerization is the faster reaction and the carboxy/epoxy addition takes place preferably only between already growing chains. Usually more than two chains are connected which leads to a broad molecular weight distribution.

The ratio of the reaction speed is depending on the reaction temperature.

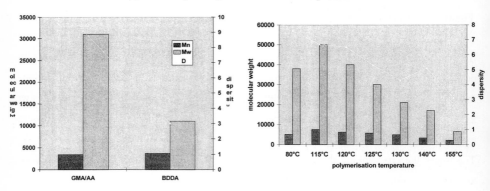

Figure 9. Use of BDDA versus GMA/AA   Figure 10. Influence of temperature

By elevating the reaction temperature from 80 to 155°C the molecular weight distribution is getting narrower (Figure 10). At higher temperatures the differences in reaction speed are smaller, so the weight average of molecular weight is reduced.

Incorporation of latent acidic groups

Another route to increased cross-linking density is the incorporation of latent acidic groups into the epoxy functional binder. The latent acidic groups are easily incorporated by co-polymerization of tertiary butylacrylate. This monomer has the attribute to decompose at elevated temperature especially in the presence of acid into isobutene and acrylic acid. During co-polymerization the acid content is too low to initiate the fragmentation. During baking in the presence of the highly acidic polyester the fragmentation takes place and additional carboxylic groups are generated, capable for additional cross-linking (Figure 11).

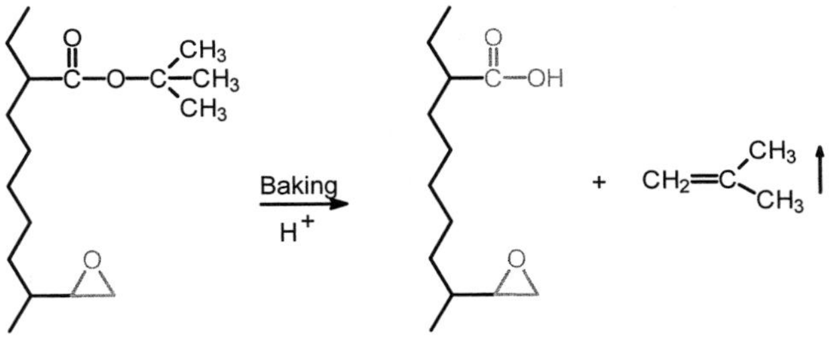

Figure 11. Latent acidic groups

The fragmentation of the tertiary butylester group can be easily observed via infrared spectroscopy. As can be seen in Figure 12 the absorption bands of the epoxy and the tertiary butylester groups decrease significantly during baking at 140°C.

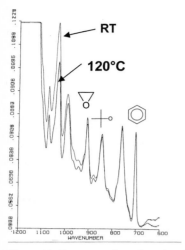 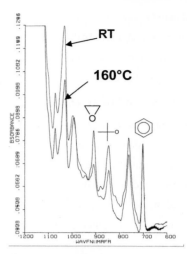

Figure 12. Fragmentation of tert.butylester group (Infrared Spectroscopy)

Etch and Scratch Performance

The superior performance of this optimized clear coat system has not only been proven by artificial etch tests but has also been demonstrated in outdoor exposure experiments in aggressive atmosphere. Whereas conventional 1K-TSA clears show significant etching in Jacksonville, the epoxy/acid clear coat hardly shows any damage. This clear coat shows the best performance in Jacksonville testing and can be compared with 2K clear coats.

Scratch resistance can be increased to up to 70% residual gloss after Amtec testing by implying the described techniques. Overall an excellent etch/scratch performance is achievable.

The concepts of this clear coat can be transferred to other cross-linking mechanisms. The sratch/etch performance of 2pack and 1pack clear coats can be increased significantly leading to new high performance systems with superior properties.

Conclusion

Mechanical properties like scratch resistance of a clear coat can be influenced by the cross-linking density and the flexibility of the coating. A combination of high cross-linking density with high flexibility of binder and/or hardener leads to increased scratch resistance. If these attributes are combined with a cross-linking reaction leading to cross-linking sites with high resistance against chemical attack the result is a high etch and scratch resistant coating.

References

[1]   W. Schlesing, M.Osterhold, H.Hustert, C.Flosbach, farbe+lack, <u>101</u> (1995) S.277-280

Macromol. Symp. **187**, 515–529 (2002)

Acid and Scratch Resistant Coatings for Melamine Based OEM Applications

Griet Uytterhoeven, Annie Fonzé and Hélène Petit

Resolution Performance Products (RPP),

Av. Jean Monnet 1, 1348 Ottignies-Louvain-la-Neuve, Belgium

Summary: Increasing the mar resistance of OEM clear coats has been one of the main R&D priorities of paint manufacturers over the last years. Reaching a good mar resistance level without compromising other coating properties such as acid resistance proves to be a major challenge. We investigated the use of branched glycidyl esters in melamine based OEM acrylic formulations. Our results show that systems based on this type of esters lead to good overall coating properties as well as a good balance between acid and scratch resistance.

Key-words: scratch and mar resistance, etch resistance, OEM, acrylic clear coats

Introduction

In the automotive industry increasing attention is given to the ability of coatings to retain their new car appearance by resistance to damage caused by weathering (e.g. acid rain) but also by resistance to chemical and physical damage such as scratches. There is a particular interest for OEM coatings providing both acid and scratch resistance as a combination of these properties is difficult to obtain.

It is well known that the acid etch resistance of polyacrylic-melamine (also called TSA-systems) is poorer than that of PolyUrethane Resin (PUR) coatings. The cause of this weakness is the hydrolytic degradation of the crosslinkages between the melamine resin and the polyacrylate [1]. The incorporation of Cardura® E10P, the glycidyl ester of Versatic® acid, is a way to minimise this problem [2]. This monomer imparts good acid resistance to coatings via steric protection of the crosslinks against hydrolytic degradation.

In order to obtain good mar resistance, a low glass transition temperature (T_g) of the coating is desirable [3]. As the incorporation of Cardura E10P in acrylics via the adduct with acrylic acid

CCC 1022-1360/00/$ 17.50+.50/0

contributes to a low T_g, we decided to study if the incorporation of Cardura E10P into acrylic resins would allow the design of melamine based clear coats with good acid as well as mar resistance. The structure performance study into these systems including the development of a simple and reliable test method to measure mar resistance are the subject of this paper.

Cardura E10P chemistry

Cardura E10P (CE10P) is the glycidyl ester of Versatic acid, a low viscosity, heavily branched, saturated C10 acid (see Figure 1)..

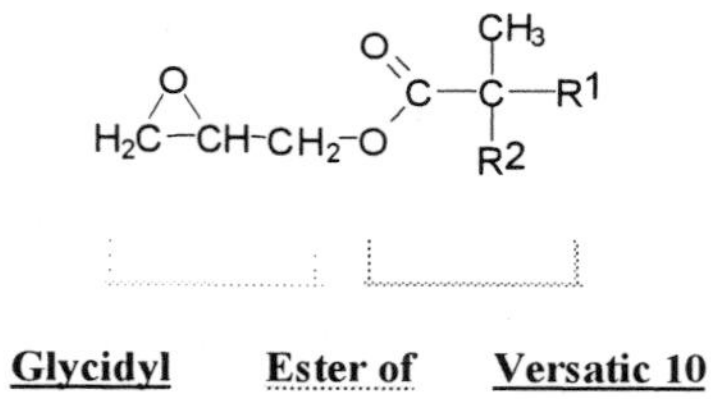

Glycidyl   **Ester of**   **Versatic 10**

Figure 1. Structure of Cardura E10P

It can be easily incorporated into acrylic resins by conversion into a radical-polymerisable monomer with unsaturated acids such as (meth)acrylic acid. CE10P based acrylic resins can be prepared in different ways. The potential advantage of Cardura in processing is however only fully exploited, if both the radical polymerisation and the esterification of Cardura are executed in one single step. The reaction between the epoxy group of the glycidyl ester and the acid group of (meth)acrylic acid takes place very rapidly under the reaction conditions used for the radical polymerisation of acrylic monomers. This allows Cardura-based acrylic resin production in a single step during which acrylic acid reacts with the epoxy group of the glycidyl ester and simultaneously polymerises with the other acrylic monomers. Upon esterification a hydroxyl group is generated, which can be used in the crosslinking process (Figure 2).

It has been demonstrated that the presence of a solvent can often be completely avoided by putting Cardura in the initial reactor charge, gradually esterifying the epoxy group with acrylic acid during the radical polymerisation. This is possible because of the low viscosity of Cardura-based acrylic resins at high solids contents and at the usual synthesis temperature. However, if desired, a wide variety of solvents can be used with Cardura.[4].

The performance benefits of the incorporation of Cardura, such as high solvent compatibility and low viscosity of paints as well as the impressive acid resistance and outdoor durability can be understood by taking a closer look at one of the possible resin intermediates, the acrylic monomer ACE (Figure 2).

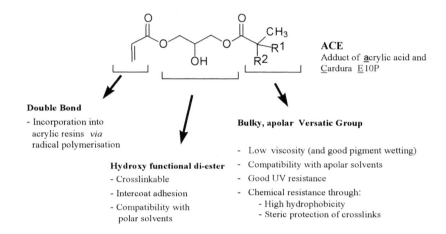

Figure 2. Structure and properties of the ACE adduct

The ACE monomer has a di-ester structure, which compatibilises the resin with polar and apolar paint components. The Versatic entity provides a high hydrophobicity and low viscosity due to the tertiary aliphatic structure. The low surface tension of the adduct also ensures a very good adhesion even on low surface energy substrates which are difficult to coat (e.g. plastics). In addition, a large T_g-region as well as changing co-polymerisation reactivity can be obtained by choosing different unsaturated acids to form the adduct. A comparison with T_g's of different acrylic monomers is given in Figure 3.

On December 14th 2001, Resolution Performance Products successfully started a new plant for the production of Cardura E10P. Cardura E10P is the same product as Cardura E10 except that it is produced at a higher level of purity. The higher purity finds expression in a higher Epoxy Group Content (EGC) and lower product colour

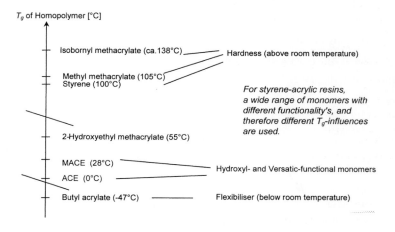

Figure 3. Glass transition temperature of the different acrylics

Scope of the study

The study carried out was limited to acrylic OEM clear coat finishes as, especially for these systems, acid etch and mar resistance are important performance features.

Resin composition

To study the effect of the incorporation of Cardura in polyacrylic-melamine systems, a range of acrylics resins with Cardura content between 0 and 40 wt% was made.

The hydroxy equivalent weight, which affects crosslink density and hence performance, was kept constant for all the resins at 425 g/mol. This corresponds with 4 wt% OH in the resin composition and an OH number of 132 mg KOH/g which is the recommended concentration to obtain good mar resistance [5][6]. The T_g of the various resins was varied between 0 and 25 °C. The different resin compositions are given in Table 1.

Table 1. Acrylic resin composition

Name:	Resin A	Resin B	Resin C	Resin D	Resin E	Resin F	Resin X
CE10P concentration [wt%]	20	20	20	20	20	40	0
Theoretical Fox T_g [°C]	24	15	15	15	0	0	15.8
Acrylic acid [wt%]	6	6	6	6	6	12.01	0.70
Hydroxy ethyl acrylate [wt%]	17.64	17.64	17.64	0	17.64	7.97	0
Hydroxy ethyl methacrylate [wt%]	0	0	0	19.8	0	0	31
Styrene [wt%]	30	30	30	27.8	26.7	18.45	30
Methyl methacrylate [wt%]	13.66	7.56	7.56	0	0	0	0
Butyl acrylate [wt%]	12.7	18.8	18.8	26.4	29.66	21.57	38.3

The resins were all prepared in the same way, i.e. at 140°C with 1.5 wt% di-tert. amyl peroxide (DTAP) as initiator. During the post cooking phase an extra 1 wt% initiator was added. This initiator was chosen over DTBP because we observed earlier that acrylic resins initiated with DTBP showed lower mar resistance than DTAP based ones. The reason for this is probably the broader molecular weight distribution obtained with the resins prepared with DTBP, as a result of the initiator ability to abstract hydrogen atoms and to form carbon-carbon crosslinks [7].

For the resins prepared with Cardura, the initial reactor charge consisted of the whole Cardura quantity together with 25 wt% of xylene. In the case of Cardura free resin formulation, only the solvent (xylene) was present as initial reactor charge. At the end of the reaction, the acrylic resins were cooled down and diluted with butyl acetate till a solids content of 60 wt%.

All resins had a similar low colour of about 20 Pt/Co and a low acid value of 4-6 mg KOH/g. The viscosity reducing effect of the incorporation of Cardura E10P becomes clear when comparing resins with similar Mw (Table 2).

Table 2. Viscosity cutting power of CE10P

	Resin X	Resin D	Resin F
CE10P content [%]	0	20	40
Mw [g/mole]	14400	11440	11300
Viscosity [mPa.s] at 22°C, 60 wt% solids	5330	1960	500

Clear coat formulation

Scratch resistance in 1K clear coat systems depends a lot on the melamine type. In our study Cymel 1158 (ex. Cytec Industries), a partially alkylated (butylated) high NH melamine resin with a low monomer content was chosen to cure the resins and to compare their coating performance. This product is recommended by Cytec Industries as a good crosslinker to obtain good overall coating properties. In addition, its low monomer content minimises the risk of self condensation which has a negative effect on coating performance [3]. Furthermore, this crosslinker has a high reactivity and consequently does not require the presence of an additional strong catalyst during curing. The carboxylic groups along the backbone of the acrylic resin are sufficient to catalyse the crosslinking reaction.

All the acrylic resins were cured with Cymel 1158 with a resin/crosslinker weight ratio of 70:30. This ratio was found to be the optimum in a small study carried out separately. The wet coatings were applied onto black base coated standard Q-panels with a wire-wrapped barcoater. After drying at room temperature for 15 minutes they were cured at 140°C for 30 minutes. The dry film thickness varied between 30 and 35 μm.

As reference system for the evaluation, a standard commercial acrylic resin, free of Cardura was included in the study (resin Y). It was cured in the same way as the experimental systems, not using any additives.

Mar resistance test method

There are 2 kinds of scratches, the fractured, irregular ones involving particle loss, and the smooth regular ones involving only elastic and plastic deformation. Mar resistance can be described as the ability of a coating to withstand permanent deformation against surface abrasions or scratches. The scratches of automotive top coats are mostly caused by car washing, and not necessarily by the nylon bristles but more by the dirt grit particles present on the coating : it is these scratches which reduce the gloss of the coating.. Most of these visible scratches belong to the category of the fractured scratches and therefore, we wanted to use a method which would simulate these damages. The literature places a lot of focus on methods to characterise the damage. These methods, which cover indentation methods, scanning electron microscopy and nano-scratch measurements, are rather complex and difficult to link to eye perception.

Therefore a simpler method which simulates car wash damage (combination of ASTM D2486 adapted with some recommendations given in reference [6]) was applied and fine-tuned. This method is based on gloss and ΔL* measurements.

Standard Q-panels were coated with a commercially available black waterborne base coat. Black was chosen as dark colours shows larger ΔL* values than light colours [8]. The clear coats were applied on the base coat as described above. Then, the panels were put for 4 days at 23°C before evaluation. The coatings were not evaluated directly after cure as it was found in an earlier study on two different acrylic resins, with different T_g and composition that the gloss after abrasion was influenced by the time the coatings were allowed to "rest" after cure.

As shown in Figure 4, highest gloss levels were obtained with coatings which were evaluated 4 days after cure. After 10 days there was no significant change anymore.

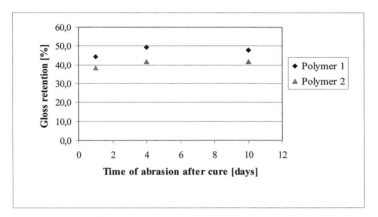

Figure 4. Experiment on determining full cure for melamine based coatings

After 4 days storage at 23°C, the panels were rinsed with water and carefully dried. The panels then were mounted on a glass plate in a wet scrub tester (Braive instruments). Subsequently 5 ml of an abrasive slurry, consisting of 6 wt% sand (Minex 4), 5 wt% of isopropanol and 89 wt% of deionized water was placed on the panel. The abrasive slurry was mixed 10 minutes before use.

Before use the nylon bristle brush (meeting the requirements as specified in ASTM 2486) was rinsed with water and then, the panels were scrubbed. Some extra weight was put on top of the brush so that the the total weight on the coating was 438g.

During the scrub test the brush was moved 10 times back- and forwards over the panel with a stroke frequency of 37 cycles per minute. The panel then was removed, rinsed with deionised water and carefully dried with soft paper. The damage was assessed by measuring the 20°gloss retention and also the ΔL*. Figure 5 shows that, there is a clear linear relation between ΔL* and the 20° gloss.

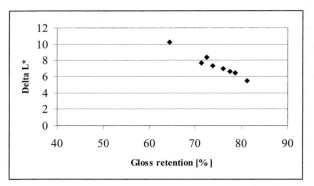

Figure 5. Relation between 20° gloss retention and ΔL* of the clear coats after abrasion

To determine the reliability of the test method, the level of variation in the mar resistance test was determined using the principle of the minimum discernible difference.

Determination of the minimum discernible difference in the wet scrub mar resistance test

For determination of the minimum discernible difference (MDD) in the wet scrub mar resistance test, the clear coat based on polymer 2 was applied onto 5 test panels which were subsequently subjected to the wet scrub mar resistance test as described above. Table 3 shows the results together with the calculated standard deviations. To allow a comparison of the test results, the MDD was calculated. For a single observation, the MDD is 4.52%. This means that if one panel per system would be evaluated, the results of two different systems would only be significantly different if their gloss retention differed more than 4.52%. Reducing the MDD is possible by increasing the number of test panels (n). For example if the number of test panels is increased to 4, the MDD decreases to 2.26%. This was considered as an acceptable level of variation and therefore always 4 clear coats were applied and tested for each system.

Table 3. Detection of the minimum discernible difference (MDD) in the wet brush test
WET SCRUB

Polymer 1	Average gloss before abrasion [%]	Average gloss after abrasion [%]	Average gloss retention [%]
Panel 1	89.1	40.8	45.8
Panel 2	89.6	38.0	42.4
Panel 3	89.3	38.1	42.6
Panel 4	89	38.7	43.5
Panel 5	89.4	37.2	41.6
Average:	**89.3**	**38.6**	**43.2**
Pooled standard deviation:	**0.24**	**1.36**	**1.61**
Standard error of average:	**0.10**	**0.68**	**0.81**
(standard deviation/ sqrt(n))			
95% conf interval based on s:	**0.18**	**1.87**	**2.06**
MDD for a single observation:	**0.67**	**3.81**	**4.52**
MDD for an average of 4 observations:	**0.33**	**1.91**	**2.26**

The recovery of our polyacrylic-melamine systems was measured to see if mainly fractured scratches were generated. This was done by remeasuring the gloss retention after one week storage at 23°C after abrasion and comparing the difference between the two gloss values obtained. The recovery was in general very small which demonstrates that almost all scratches were fractured. This is in line with other studies that showed that fractured scratches do not recover [1]. This shows that our test method is indeed adapted for measuring the permanent deformation caused by mar resistance. Another difference between fractured and plastic scratches is that fractured scratches are visible, independent of the incident light and observation direction [9]. Also this was the case for our test panels.

Determination of crosslink density and glass transition temperature T_g

From literature it is known that mar and acid etch resistance of coatings are influenced by their crosslink density and T_g and therefore a study into these parameters was done using dynamic mechanical analysis (DMA). The crosslink density is based on the tensile storage modulus, E' in the rubbery plateau region above T_g and is defined as v_e, i.e. the number of moles of elastically effective network chains per cubic centimetre of film. T_g was determined using the temperature at which E" is maximum. The free coating films were obtained by applying the clear coats on release paper from which they could easily be removed after cure. This technique is well documented in literature [10].

Discussion of results

The performance of Cardura E10P based acrylics

Melamine based coatings are primarily crosslinked via ether linkages, which are prone to hydrolysis upon exposure to acid rain.

Typical factors that improve the acid resistance of a coating are a high coating T_g, high crosslink density and high hydrophobicity as they minimise water absorption of the film. Also the incorporation of bulky hydrophobic groups is known to increase the acid resistance as such structures shield vulnerable linkages from acid attack.

In this context it can easily be understood that the incorporation of Cardura into the resin imparts improved acid resistance. To obtain the right acid resistance with Cardura free acrylic coatings, the coating T_g must be high. When using Cardura, this is no longer mandatory.

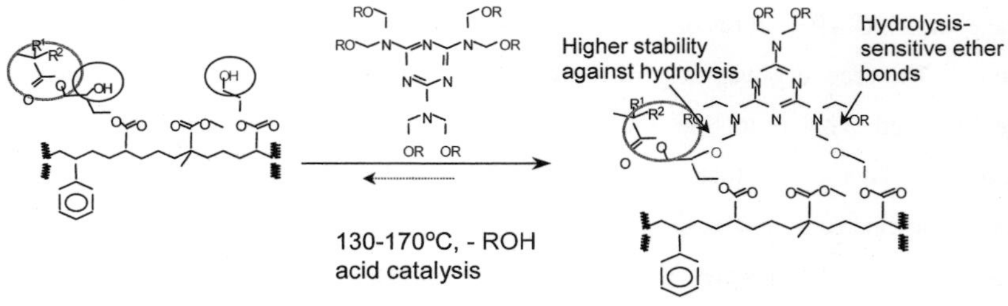

Figure 6. Protection of melamine ether bounds by the Versatile bulky group of CE10P

Mar resistance depends on a number of factors such as chemical composition of the resin, choice of crosslinker, crosslink density, additives, processing procedures but also coating T_g.

The possibility to decrease the coating T_g when using Cardura, without jeopardising acid resistance, is a real advantage as it might allow the design of coatings with good scratch and acid resistance.

To confirm this assumption concerning these anticipated benefits of incorporating Cardura, a range of acrylics all containing 20% Cardura E10P but with having varying theoretical Fox T_g's (in order to obtain different coating T_g's) were prepared. To obtain these differences in coating T_g's, the concentrations of BA and MMA in the resins were slightly varied. For resins D and E, also the styrene concentration was adjusted.

Within this series of resins, the crosslink density was almost the same (see Table 4).

Table 4. Crosslink density of the 20% CE10P made acrylic resins

Name :	Resin A	Resin B	Resin C	Resin D	Resin E
Cardura E10P concentration [wt%]	20	20	20	20	20
Crosslink density (mole/m^3)	1260	1310	1200	1300	1310

Figure 7 clearly shows the linear relation between the gloss retention and the coating T_g at constant crosslink density : the lower the coating T_g, the higher the gloss retention. The best scratch resistance, i.e. a gloss retention of 81%, is obtained with resin E, which has the lowest coating T_g. This is in line with previous results of the open literature publication [5]. The recovery of the coatings was also measured, but was not considered to be significant. More details on the overall coating characteristics of this resin E are given later.

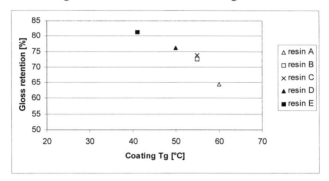

Figure 7. Gloss retention versus coating T_g for a series of 20% Cardura E10P containing resins

On the other hand, acid resistance normally decreases with decreasing coating T_g. However, Figure 8 clearly shows a stable performance in acid resistance, whatever the coating T_g. This is linked to the incorporation of 20% of Cardura E10P.

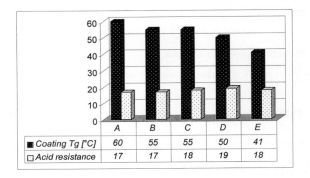

Figure 8. Acid resistance versus coating T_g for a series of 20% Cardura E10P containing resins

Figure 9 shows that, as expected, the coating hardness of the systems containing 20% CE10P increases with increasing coating T_g whilst lowering the coating T_g improves the flexibility of the coating. Ideally coatings should have both high hardness and high flexibility or at least a good balance between both. This balance is best with the resin E.

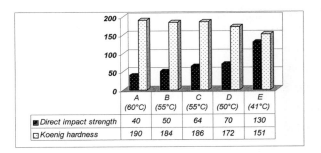

Figure 9. Hardness and flexibility for the 20% Cardura E10P containing clear coats with different coating T_g

The results discussed above concerned resins with a fixed Cardura content of 20 wt%. In the next step the effect of varying Cardura content on coating performance was studied. Therefore, 4 experimental acrylic resins, with 0%, 20% and 40 wt% Cardura were evaluated. A standard commercial resin for OEM automotive applications, which contained no Cardura (resin Y), was also used as benchmark.

Table 5. Coating characteristics of acrylic resins with different Cardura concentrations.

	Resin X	Resin Y	Resin E	Resin F
Cardura E10P concentration [%]	0	0	20	40
Final AV of the resin [mg KOH/g]	5.5	9.7	4.7	4.1
Crosslink density [mole/m^3]	1850	1630	1310	940
Coating T_g (by DMA) [°C]	64	44	41	30
Surface tension	34	38	<34	<34
Gloss retention [%]	78.7	71.4	81.3	77.5
Recovery [%]	2.8	5.5	2.7	1.7
Acid resistance	10	3	18	22
[0= poor,-30= excellent]				

Several remarks concerning the resins and coating properties listed in Table 5, can be made. Firstly, the obtained crosslink densities are somewhat different. This can most likely be explained by the differences in final acid value of the resins which plays a role on the crosslinking as the butoxymethyl melamine that is used, is catalysed by acid. Secondly, it is clear from the table that not only the crosslink density is important for obtaining a good mar resistance (expressed as gloss retention). The positive contribution of increasing the Cardura content at low T_g is clearly demonstrated. The two coatings based on the 0% CE10P resins are higher in crosslink density than those containing Cardura, but despite this higher crosslink density, their gloss retention after the scrub test is not as good as that of the resin E. This confirms that scratch resistance is not only influenced by the crosslink density, but by a combination of different parameters including the resin composition.

The analysis of the acid resistance results show the positive influence of incorporating CE10P in the resin: resins E and F clearly outperform the benchmarks in that respect. Resin E can be considered as an optimum if a good balance of acid and scratch is needed.

For commercial application in automotive OEM, coatings should in general impart not only a good acid and/or mar resistance, but also good mechanical properties. Spider charts are a way of simultaneously illustrating the various coating performance attributes relative to the assumed best achievable or desired level. The overall performance of the systems increases with increasing surface of spider chart.

528

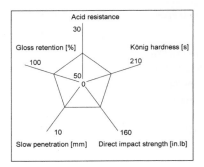

Figure 10. Legend of the spider chart on coating properties

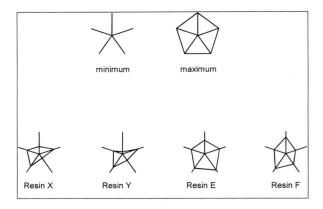

Figure 11. Spider chart on coating properties

Figure 11 shows the overall coating performance as a function of the Cardura content of the base acrylic resins. It clearly demonstrates that resin E imparts the best performance and provides unique properties, i.e. a combination of very good acid and mar resistance together with very good mechanical properties. This system outperforms the commercial benchmark as well as the Cardura free resin X.

Conclusions

As a first step of this study, we developed a simple and reliable method for testing the mar resistance of coatings. This method allowed us to assess the importance of incorporating Cardura E10P in OEM melamine-based systems. Our results show that Cardura is an excellent building block for the design of high quality resins as it allows to get coatings with a very good acid and mar resistance in combination with very good mechanical properties.

Acknowledgements

We are grateful to thank Gary Biebaut and Christian Ritter (analytical and statistical services, Monnet Centre, Louvain-La-Neuve, Belgium).

References

[1]   U. Meier-Westhues, U. Biskup, Ph. Yeske and M. Bock: "Polyurethane clearcoats with optimized resistance to scratching and chemicals", *Praxis-Forum meeting of the "Automobilkreis Spezial"*, Bad Nauheim (February 1999).

[2]   RPP: "Cardura E10P based acrylic polymers for high quality first finish automotive clear coats", *Product Bulletin, CM 4.2.1*.

[3]   B.V. Gregorovich and I. Hazan: "Environmental etch performance and scratch and mar of automotive clearcoats", *Progress in Organic Coatings, 24* (1994) 131-146.

[4]   RPP: "Solvent selection for acrylic polymers", *Product Bulletin, CM 4.1.1*.

[5]   Peter Betz and Angelika Bartelt: "Scratch resistant clear coats: development of new testing methods for improved coatings", *Progress in Organic Coatings, 22* (1993) 27-37.

[6]   J.L. Courter and E.A. Kamenetzky: "Micro- and nano- indentation and scratching for evaluating the mar resistance of automotive coatings", *PAINTINDIA* (1999) 115-128.

[7]   V. R. Kamath and J. D. Sargent Jr.: "Production of high solids acrylic coating resins with t-Amyl peroxides: A new way to meet VOC requirements", *PAINTINDIA* (1991).

[8]   K. Shibato, S. Beseche and S. Sato: "Studies on acid etch- and scratch resistance of clearcoats for automotive industry", *4th Asia-Pacific Conference* paper 4.

[9]   V. Jardret, B. N. Lucas and W. Oliver: "Scratch durability of automotive clear coatings: A quantitative, reliable and robust methodology", *Journal of Coatings Technology* (August 2000) Vol.72.

[10] L. Hill, K. Kozlowski: "Crosslink Density of high solids MF-cured coatings", *Journal of Coatings Technology*, 59 (751), 63-71 (1987).

Macromol. Symp. 187, 531–542 (2002)

Scratch Resistance and Weatherfastness
of UV-Curable Clearcoats

T. Jung, A. Valet

Ciba Specialty Chemicals Inc., Basle, Switzerland

Summary: UV-cuing has found an increasing number of industry allocations over the past decade due to its unique benefits, e.g. solvent free formulations, high cure speed and low temperature processing. In addition to these benefits two additional properties of uv-cured coatings are of today`s interest, especially in the automotive industry: scratch resistance and resistance against chemicals. One of the most important requirements for a broad use of uv-curable coatings in the coating industry is that coatings are stable against degradation caused by atmospheric influences since coatings for outdoor use are subject to especially harsh weathering conditions, e.g. uv-light, oxygen, moisture and air pollutants. This weathering leads to a degradation of the polymeric binder. Clearcoats containing photoinitiators based on bis-acylphosphinoxide ("BAPO") and a combination of hydroxyphenyl-s-triazine uv-absorber and a sterically hindered amine as a light stabilizer package show a very good curing behavior as well as an improved weatherfastness over a long period of time and a good scratch and chemical resistance.

Keywords: UV curable clearcoat, scratch resistance, durability

1. Introduction

Current state of the art clear coat technology involves heat to initiate cure. Such systems are for example 2P-Polyurethane, acrylic-melamine coatings etc. The UV-curing technology uses UV-light instead of heat to initiate the crosslink mechanism of the coating. The low temperature process and the possibility to formulate zero VOC (**V**olatile **O**rganic **C**ompounds) made the UV-technology successful over the last years.

In addition to these benefits two additional properties of UV-cured coatings are of today's interest, especially in the automotive industry: scratch resistance and resistance against chemicals. However, before this technology can be broader introduced to the car industry as well as to the general coating industry one major field where UV-curing has to show that it is able to compete with existing coating technologies is the area of outdoor applications [1].

CCC 1022-1360/00/$ 17.50+.50/0

2. Scratch and Chemical Resistance

The challenge for scratch and chemical resistant coatings seems to be rather simple: to combine the scratch resistance of a 2P-PUR with the chemical resistance like Teflon by keeping cost effectiveness [2]. Coating companies all over the world are currently developing coating systems, which come close to the target, however, there is room for further improvements. One possibility to solve the problems is the use of UV-cured systems [3]. From a cost effectiveness point of view, UV-cured coatings, however, should not only assessed by the raw material costs but also the possible cost savings of the whole coating process have to be taken into consideration.

In recent publications the comparison of the current used coating technologies were discussed with respect to scratch resistance or chemical resistance [18,19]. Clear coat technologies like HS-TSA, 2P-PUR or 1P water based, seam to have only one strong feature either they are chemical resistant or scratch resistant.

3. Light Stabilization upon Weathering

Coatings for outdoor use are subject to especially harsh weathering conditions, e.g. UV-light, oxygen, moisture and air pollutants. The absorption of UV-light by the polymer backbone or impurities leads to primary photochemical reactions resulting in a photo-oxidative degradation of the polymer. Reactions producing free radicals are particularly harmful. Due to the primary photo physical and photochemical processes, which take place during the interaction of harmful UV-light with absorbing molecules, the possible approaches to the stabilization of coatings can be radically reduced to two types of stabilizers [4]:

- UV-absorbers (UVA), filtering the harmful UV-light in the wavelength range 290-380 nm
- Radical scavengers (Hindered Amine Light Stabilizers = HALS)

Today a stabilizer package utilizing a combination of UV-absorbers and HALS is state of the art.

The filter effect of the UVA protects the substrate against color change and photochemical degradation, which would lead to delamination. However UVA`s do not function as "radical scavengers". Alternately, HALS does not absorb in the UV-area but they trap any radical formed during the outdoor use of a coating [7,8]. HALS protect the coating against loss of gloss and most importantly, against cracking. Figure 1 shows the liquid UV-absorbers and HALS used for this study. The concentrations of UV-absorbers and HALS in the different coating systems correspond to the level used in the European automotive industry.

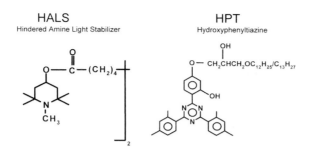

Figure 1. UV-absorber and HALS used in this study

The mode of action of UV-absorbers is the absorption of harmful UV-light and its rapid conversion into harmless energy [5,6], without being destroyed themselves. The four most important UV-absorber classes are the hydroxyphenyl-benzotriazoles (BTZ), hydroxyphenyl-s-triazines (HPT), hydroxy-benzophenones and oxalic anilides. From a technical point of view hydroxyphenyl-benzotriazoles and hydroxyphenyl-s-triazines are the most interesting UV-absorber classes [4].

4. Photoinitiators

The very beginning of the radical curing process of an UV-cured coating involves a photochemical reaction of a molecule within the coating by absorbing a photon, which leads to an excited state of the absorbing molecule. In a second step this molecule is cleaved leading to radicals, which are able to initiate the polymerization reaction. Those molecules are generally called *photo-initiators* and described in detail in the literature [23, 24, 25]. A bisacylphoshine-oxide (BAPO) and a α-hydroxy ketone (α-HK) were used for the present study and are shown in Figure 2.

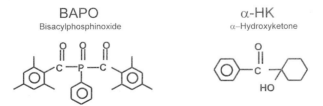

Figure 2. Photoinitiators used in this study

5. Results and Discussion

In this study a comparison of the currant coating technologies like 2P-PUR and HS-TSA with the UV-technology regarding durability scratch and chemical resistance is made. The UV-curing technology is one of the most desirable coating technologies for future use in automotive application. In the mean time not only a 100 % UV-Curable coating is discussed. With respect of the complex tree dimensional parts with shadow areas in the automotive application, which are difficult to cure, Dual Cure technology came in place. The Dual Cure system is working with two different cross-link mechanisms: the polyol / isocyanate being responsible for the curing in shadow areas. This mechanism is already used in 2P-PUR coatings today. The second mechanism is based on the radical induced polymerization of acrylate double bounds, which are part of the binder system.

The experiments comparing durability, scratch resistance, chemical resistance, were carried out using the following formulations.

- Commercial 2P-PUR OEM - stabilized with benzotriazole / HALS chemistry. Curing was performed at 130°C for 30 minutes
- Acrylic polyol cross linked with a fully methoxylated melamine resin (high solids thermosetting acrylic = HS TSA). Curing was performed at 130°C for 30 minutes
- Dual Cure coating - starting point formulation from Bayer AG/Germany
- UV-Cure model clear coat - aliph. urethanediacrylate / aliph. urethanetri/teraacrylate / TPGDA / TMPTA

5.1. Stabilization of UV-cured clearcoats

Figure 3 shows that the performance of an *unstabilized* UV-cured clearcoat (depending on the resin system used) can be comparable to commercially available clearcoats, e.g. High Solids (HS) TSA or HS 2P PUR.

The UV-cured coating shows a somewhat better gloss retention, while the conventional coatings show slightly better crack resistance. This led to the expectation that UV-cured clearcoats stabilized with UVA and HALS can also match the performance of stabilized conventional coating systems. Major concern about the use of UVAs in these coatings is their competition with the photoinitiators for UV-light, [1] that could potentially lead to a cure retardation of the coating. However, there are already many literature references to the use of UV-absorbers in UV- cured coatings [9-13], as well as new developments in the area of photoinitiators [14, 15] and UV-absorbers [16, 17] helping to overcome this issue. The chosen

HPT-UVAs have relatively weak absorption at 365 nm, thereby allowing efficient UV-curing to take place as shown in Figure 4. The spectra were taken in chloroform. The concentrations used correspond to an UVA/photoinitiator ratio of 2/3 and 2/5 (similar to a clearcoat stabilized with 2% UVA and UV-cured with 3% or 5% photoinitiator).

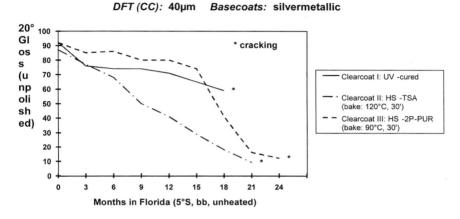

Figure 3. Comparison of gloss retention for conventional clearcoats versus a UV-Cured Clearcoat upon Florida exposure

If the PI contains additionally BAPO it shows a moderate absorption up to 430nm. This means that enough UV-light can be absorbed to impart a sufficient through-cure of the clearcoat.

Figure 5 shows a comparison of the influence of UVA and/or HALS on the performance of a clearcoat. Recent papers [22] reported that a minor influence of UVA on the performance of a UV-clearcoat. The results after 10 years Florida, however, show how important a UVA is regarding the overall stability of a UV cured clearcoat. The UVA/HALS combination shows a significantly improved performance compared to formulations stabilized with HALS or UVA. The proof is shown in Figure 6 which shows the 9 years results of a UV-cured model clearcoat over different basecoats compared to both a HS TSA and a HS-2P-PUR clearcoat over a silver metallic basecoat. No cracking occurred and the gloss retention is rather good. Although the basecoats described in Figure 6 are not the same, it can be concluded that the outdoor behavior of UV-cured systems is at least comparable to the behavior of the thermosetting systems.

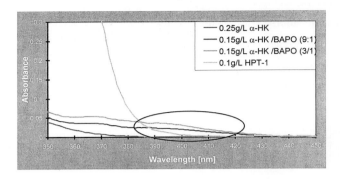

Figure 4. Absorption spectra of photoinitators and UV-absorbers

Under Xenon-WOM conditions (Figure 7) another UV-cured model clearcoat performs similar to UV cured clearcoat mentioned in Figure 6, which shows - as mentioned above - a very good gloss retention during 9 years Florida exposure.

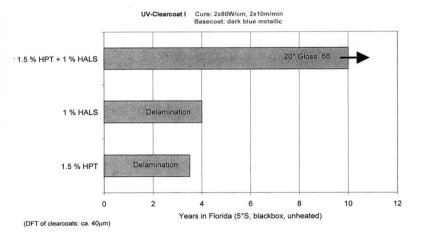

Figure 5. Comparison of the influence of UVA and/or HALS on the performance of a UV-curable clearcoat

Comparision of Florida Results: UV-Cured Model Clearcoat vs. Thermally Cured Clearcoats

Stabilization:
UV-Cured Clearcoat: 1.5% HPT/1% HALS
Thermally Cured Clearcoats: 2% HPT/1% HALS

UV-Clearcoat I   Cure: 2x80W/cm, 2x10m/min

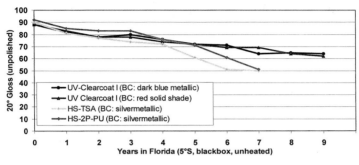

(DFT of clearcoats: ca. 40μm)

Figure 6. Comparison of UV-cured model clarcoat vs. thermally cured clearcoat after 9 years Florida exposure

Gloss Retention of UV-Cured Clearcoats (DFT: ca. 40μm) after Xenon-WOM Exposure

UV-Clearcoat I - Photoinitiator: 3% α-HK; Cure: 2x80W/cm, 2x10m/min
UV-Clearcoat II - Photointiator: 3% (α-HK/BAPO 7/1); Cure: 2x120W/cm, 5m/min

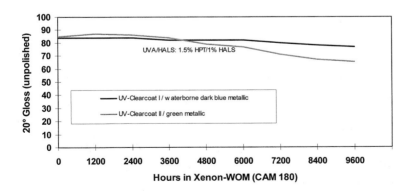

Figure 7. Comparison of two different UV-curable model clearcoats under Xenon Wom Cam 180 conditions

Figure 8 shows the performance of the UV-Cure model clearcoat a Dual Cure and a 2P-PUR clear coat. All coatings were applied over a black basecoat with a dry film thickness of 40μm. The UV-Cure II and the Dual Cure clear coat contain 2.7 % α-HK + 0.3 % BAPO + 1.5 % HPT + 1 % HALS. The concentration of photoinitiator and Light stabilizer are based on resin solids of both formulations.

The UV-cured model Clear coat was cured with two Mercury medium pressure lamps at 120 W/cm with a belt speed of 5 m/min. In case of the Dual Cure a 10 Min bake at 80°C on a heating plate as a thermal cure step and 2 mercury medium pressure lamps with a belt speed of 5 m/min was chosen to cure these coatings.

The UV-Cured model clear coat and the Dual Cure clear coat show by fare a better gloss retention compared to the 2P-PUR clear coat under the Xenon Womb Cam 180 weathering condition.

The crack resistance of the UV-Cured and Dual Cure clear coat is better than with the 2P-PUR clear coat. This result is also confirmed in UVCON weathering.

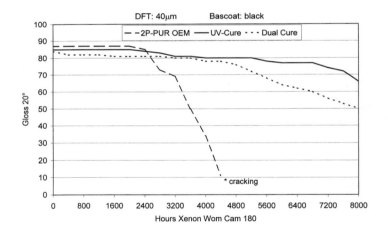

Figure 8. Performance of a UV-cured, Dual Cured clear coat and a 2P-PUR upon Xenon Wom Cam 180 weathering

Color change upon weathering is also a big concern of all paint and car manufactures. Using UV-technologies, including Dual Cure, the color matching is a major concern of the quality control.

Figure 9 show the yellowing behavior of the UV-cured model clear coat in comparison to a 2P-PUR clear coat. Both clear coats were applied over a white basecoat and Cured with the conditions mentioned above. The UV-cured coating shows somewhat better yellowing over time of weathering, while the conventional coating shows a slightly better yellowing after cure.

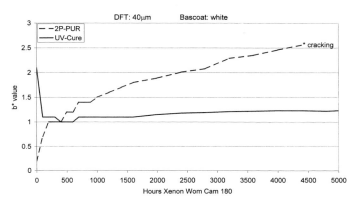

Figure 9. Yellowing behavior of a UV-curable clear coat vs. a 2P-PUR upon weathering.

5.2 Scratch resistance

For high performance coatings the durability is only one criteria to get approved. The Scratch resistance of the coating is also an important parameter. The Rota Hub method was developed by Bayer AG /Germany measuring the scratch resistance under dry conditions. The Gloss data were performed with a reflectometer form Byk-Gardner. The parameter under which the scratch tests were made, are the following: Speed in X and Y direction V_x / V_y = 15/15 mm/min; Speed of rotation ω = 5.0 U/s; Disc radius r= 35 mm; Rub material: paper.

The gloss 20° is determined before and after scratch. Figure 10 shows the Δ Gloss 20° data. The UV-Cure system show the smallest difference of Gloss followed by the Dual Cure clear coat. The Rota Hub scratch test has a big impact on Gloss of conventionally cured clear coat.

The results, which are performed with the Nano Scratch Tester for CSEM Instruments are shown in Figure 11. The figure displays according to Lin [20] the residual depth at 5mN in μm versus the critical Load. The critical load is the load in mN on the indentor at the weight the first crack is visible in the coating. The arrow shows the trend with an increase of scratch resistance. The increase in scratch resistance is in parallels with an increase in acrylate double bounds in the coating systems.

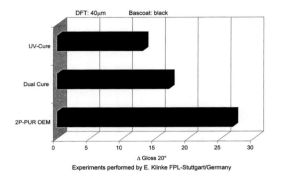

Figure 10. Scratch resistance of three different clear coats according to Rota Hub

Both methods Rota Hub and Nano Scratch Tester show the same ranking of the tested clear coat. Therefore a good correlation for both methods is shown.

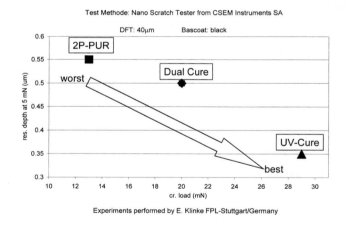

Figure 11. Scratch resistance measured with the Nano Scratch Tester of tree different clear coats according. Display according to Lin [20]

5.3 Chemical resistance

Coatings for outdoor use are subjects to especially harsh conditions. For example acid rain or birds shit. The coating used in automotive application should be stabile against these atmospheric and biological influences. Pancratine, soda and chloric acid were selected for the chemical resistance test. The experiment is performed in a gradient oven with a temperature range von from 40-75 °C. The chemical resistance was tested on two sets of panels. The first trial was carried out 4 days after cure, the second one 4 days after cure plus one week QCT (water temperature: 40°C) and one day recovery at room temperature.

Table 1. Chemical resistance of different clear coat technologies

		Pancreatine	NaOH	H_2SO_4
UV-Cure	After Cure	No impact	Strong impact	No impact
	After 1 week QCT	No impact	Strong impact	No impact
HS-TSA	After Cure	No impact	No impact	Some impact
	After 1 week QCT	Some impact	Some impact	Some impact
2P-PUR	After Cure	No impact	No impact	No impact
	After 1 week QCT	Some impact	Some impact	Some impact

Table 1 shows the results of the chemical resistance of three different coating technologies. The UV-Cured coating is sensitive to soda but stabile against Pancreatine and chloric acid. Even after the storage under wet condition the UV-cured coating show the same good results than before. The thermally cured coatings show instability against acids. Particularly after humidity influence the chemical resistance is weak.

6. Conclusion

The durability of UV-curable clearcoats can be significantly improved by using the correct light stabilizer package and an optimized photoinitiator combination. High performance photoinitiators combinations like BAPO and hydroxyphenyl-s-triazine UV-absorbers (in combination with HALS) are well suited to achieve both through-cure and weatherfastness of the clear coatings, which are at least comparable to those of thermally cured systems. UV-cured clearcoats show a significantly improved scratch resistance as well as chemical resistance. The overall performance characteristics of UV-cured coatings are also determined, as with other coating types, by the components used for the binder system.

7. References

[1] K. LAWSON, in "Catching New Light Under Loosened Regulations", Chemical Week, April 24, 1996

[2] M. MONTEMAYOR, Polym. Paint Col. J., Vol 189, 12 (1999)

[3] D.STRANGHÖNER,"Clear-Coat-Konzepte und Funktionsschichten in Entwicklung und Anwendung" Conference Proceedings, Bad Nauheim/Germany, June 14./15. 1999

[4] A. VALET, "Light Stabilizers for Paints", Vincentz-Verlag Hannover 1997

[5] H.J. HELLER, H.R. BLATTMANN, Pure and Applied Chem. 30, 145 (1972),

[6] H.E.A. KRAMER, Farbe + Lack, 92, 919 (1986)

[7] E.DENISOV, Int. Symp. on Degrad. and Stab. of Polymers, Brussels 1974, 137

[8] E.STEP, N.TURRO, P.KLEMCHUK, M.GANDE, Angew. Makr. Chem. 232, 65 (1995)

[9] K.J. O'HARA, Polym. Paint Col. J., Vol. 175, 776 (1985)

[10] L.GATECHAIR, H.EVERS, Mod. Paint and Coatings 76, 38 (1986)

[11] C.DECKER, K.MOUSSA, T.BENDAIKHA, J. Polym. Sci. 29, 739 (1991)

[12] A.VALET, D. WOSTRATZKY, RadTech Report, Nov./Dec. 1996, 18

[13] C.DECKER, K. ZAHOUILY, RadTech Europe 1999, Berlin Nov. 8-10, Conference Proceedings, 209

[14] W.RUTSCH et al., XXth Int. Conf. in Org. Coat. Sci. And Techn., Athens 1994, Congress Book, 467

[15] C. H. CHANG et. al., RadTech North America (1996) Vol. 2, 559

[16] A. VALET, Polym. Paint Col. J., Vol. 185, 31 (1995)

[17] USP 5.369.140 (Ciba Specialty Chemicals Inc.)

[18] U. Schulz, et. Al., Glänzend bestehen, Farbe & Lack 107 4/2001, P 179-191

[19] E. Frigge, Doppelt geschützt hält besser, Farbe & Lack 106 7/2000, P. 78-80

[20] L. Lin, G.S. Blackman and R. R. Matheson, A new Approach to characterize Scratch and Mar resistance of Automotive Coatings, Proceedings of the XXVth International Conference in Organic Coatings, Athens, 5. - 9.7.1999 P. 125-137

[21] A. VALET, T. JUNG, UV-curable Clearcoats – Ascratch Resistant and Weatherfast Alternative to Thermosetting Clearcoats, RadTech Europe (2001), 41

[22] M. E. Nichols, et. Al., Anticipating the Long-Term Weathering Behaviour of UV-curable automotive Clercoats, RadTech US (2002), Automotive Focus group

[23] W. RUTSC et. Al., Recent developments in photoinitiators, Prog. Org. Coatings,1996, 27, 227

[24] A. F. CUNNINGHAM, et. Al., Recent developments in radical photoinitiator chemistry, Chimia 1994, 48(9), 423-426

[25] H. F. GRUBER, Prog. Org. Coatings 1992, 17, 953

Macromol. Symp. **187,** 543–552 (2002)

CellFacts II – Single Cell Analysis in Real Time

Patrick Schwarzentruber

R&D Microbiology, Omya AG, Baslerstrasse 42, 4665 Oftringen, Switzerland

Summary: CellFacts II integrates electrical flow impedance and fluorescence to determine the number, size and fluorescence characteristics of individual cells in a conductive fluid. The instrument has been optimised to detect and enumerate viable and non-viable cells in fluid samples with varied particulate content, i.e. total viable counts, with discrimination of the physiological status of the individual cells.
The study shows the analysis of the physiological state of individual cells in a population, effectively in real-time, enabling the rapid determination of the effect of antimicrobial agents on these cells i.e. rapid determination and optimisation of antimicrobial agents in aqueous paint systems.

Keywords: CellFacts; fluorescence; impedance; physiological status; total viable count

Introduction

Traditional microbiological analyses for determination of the presence of low levels of contaminating microorganisms, such as Plate Count, Petrifilm or Easicult, are lengthy, often taking from 2 to 3 days to complete. In addition, there are various other factors to be considered, such as, for example, type of culture medium, partial pressure of oxygen (aerobic / anaerobic), selectivity, pH value and many more.[1-2]

A rapid method that could provide an accurate assessment of the number of microorganisms present would allow positive release of the paint product in the knowledge that it was free from effective contamination.

CellFacts II integrates electrical flow impedance and fluorescence to determine the number, size and fluorescence characteristics (e.g. viability, physiological status, speciation) of individual particles or cells in a conductive fluid. The instrument has been optimised to detect and enumerate viable and non-viable cells in fluid samples with varied particulate content, i.e. total viable counts, with discrimination of the physiological status of the individual cells (also discrimination between cells of different sizes, e.g. between yeast and bacteria).

    CCC 1022-1360/00/$ 17.50+.50/0

The following study shows the analysis of the physiological state of individual cells in a population, effectively in real-time, enabling the rapid determination of the effect of antimicrobial agents on these cells, i.e. rapid determination and optimisation of antimicrobial agents in aqueous paint systems. Since antimicrobial agents have to meet the new Biocidal Product Directive (BPD), and paint systems will be valued according to the directive RAL UZ12a (Blauer Engel) this level of proactive microbiological control becomes even more important.

The Operating Principle

CellFacts II uses patented technology integrating electrical flow impedance and fluorescence to determine the number, size and fluorescence characteristics. This is used to determine the viability, physiological status and speciation of individual particles or cells in a conductive fluid (Figure 1).

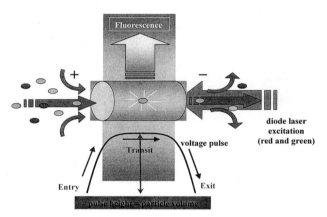

Figure 1. Schematic presentation of the CellFacts II principle. Particles (or cells), in a conductive fluid, pass through a 30 μm diameter orifice which has been laser etched in a 80 μm sapphire disk.

Two diode lasers (red and green) are focussed on the orifice. As a particle (or cell) enters the orifice it displaces its own volume of electrolyte and this generates a voltage pulse, the magnitude of which is directly proportional to the volume of the particle. This signal is analysed in considerable detail via the software and generates information on particle (or cell) size and the numbers within a given population. It may also be used to trigger fluorescence characterisation of the particle as it transits the orifice, i.e. the integration of impedance and fluorescence data. A key feature is the orifice block detection systems which ensure the instrument operates irrespective of the size heterogeneity of the particles on the sample side of the orifice and results in an extremely effective anti-blocking system.[3]

The instrument has a modular configuration (Figure 2) to meet the requirements of the broad spectrum of applicable market sectors.

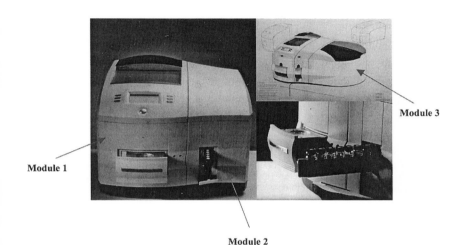

Figure 2. CellFacts II Modules:

Module 1   Analysis module, integrating both impedance and fluorescence measurements of individual particles or cells.

Module 2   Reagent addition module. This unit facilitates the addition of up to seven different reagents enabling multiple, complex, automated analyses to be undertaken for each sample.

Module 3   Shown here in the original concept drawing, is an automated sample-handling unit with sample mixing and bar code facilities.

Fluorescence: Characterising the Physiological Status of Cells

By following a protocol of minimal staining with dimeric cyanine nucleic acid dyes and lipophilic cationic dyes (membrane potential dyes, i.e. viability indication dyes) it is possible to characterise the physiological status of individual cells (Figure 3).[4-5]

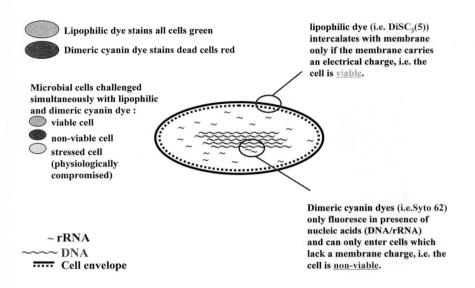

Figure 3. Staining to characterise cells.

Results

A study was carried out comparing the CellFacts II method, using a patented sample preparation technique which allows high solids suspensions to be analysed by removing preferentially microbial material, with traditional techniques, such as Plate Count, for evaluating the potential application in aqueous paint systems. The preservative properties of commercially available paint systems were also tested and different antimicrobial agents were compared against one another. Since antimicrobial agents have to meet the new Biocidal Product Directive (BPD), and paint systems will be valued according to the directive RAL UZ12a (Blauer Engel) this proactive evaluation methodology will become even more important.

Total Viable Count

On comparing the total viable count of all the samples tested (Table 1) it becomes clear that the results using the new technology CellFacts II tend to be somewhat higher than those of the traditional plate count method. This result was in fact expected, and can be explained due to the specific test conditions chosen for the traditional culture evaluation (nutrient, pH, temperature). This shows that not all cells are detected by the plate count method ! It is possible that cells are present in a system which can indeed have a negative impact on the properties of the product but that traditional methods cannot detect them. These cells could be successfully detected by the CellFacts II method, and, what is new, in real time.

Table 1. Total Viable Count of 16 Aqueous Paint Systems

Sample	PCA[a] (cfu/ml)[b]	CellFacts II (cfu/ml)
Interior Emulsion Paint 1A	< 100	< 100
Interior Emulsion Paint 1B	< 100	< 100
Exterior Emulsion Paint 1C	< 100	< 100
Exterior Emulsion Paint 1D	< 100	< 100
Interior Emulsion Paint 2	< 100	3.0×10^2
Exterior Emulsion Paint 2	< 100	< 100
Interior Emulsion Paint 3	2.0×10^2	1.3×10^3
Exterior Emulsion Paint 3	< 100	< 100
Exterior Emulsion Paint 4A	1.0×10^2	7.8×10^3
Interior Emulsion Paint 4B	< 100	3.5×10^3
Exterior Emulsion Paint 4C	< 100	< 100
Interior Emulsion Paint 4D	< 100	< 100
Interior Emulsion Paint 5A (without biocide)	< 100	1.8×10^3
Exterior Emulsion Paint 5B (without biocide)	7.0×10^2	6.7×10^3
Interior Emulsion Paint 5C (without biocide)	5.0×10^4	9.4×10^4
Exterior Emulsion Paint 5D (without biocide)	8.0×10^4	4.9×10^5

a) PCA = Plate Count Agar
b) cfu = cell forming unit

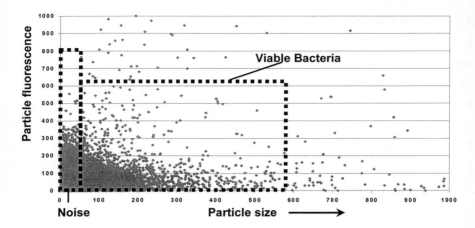

Figure 4. This graph clearly shows the area of the viable bacteria defined by fluorescence and particle size. The total viable count detected in the sample shown (Interior Emulsion Paint 4B) was 3.5 x 10³ cfu/ml.

Stability of Preservation

The samples in the series 1 - 4 were tested for their existing preservative properties. For this purpose, the samples were inoculated with a quantity of 5.0×10^4 bacteria per ml isolated from aqueous paint systems. After the subsequent storage of the samples for 24 hours at 30°C, the total viable count was determined by means of CellFacts II. Additional inoculations were made until the total viable count of a sample was $> 10^4$ cfu/ml (Table 2). The number of possible additional inoculations without viable contamination gives an indication of the quality of the preservative action in the product. The results show that significant differences can be found in the samples analysed. In this study the paints from series 1 could be identified as having the best preservative properties.

Table 2. Stability of Preservation of 12 Aqueous Paint Systems

Sample	TVC[a] before Inoculation	Number of Inoculations
Interior Emulsion Paint 1A	< 100	4
Interior Emulsion Paint 1B	< 100	4
Exterior Emulsion Paint 1C	< 100	3
Exterior Emulsion Paint 1D	< 100	4
Interior Emulsion Paint 2	3.0×10^2	1
Exterior Emulsion Paint 2	< 100	2
Interior Emulsion Paint 3	1.3×10^3	1
Exterior Emulsion Paint 3	< 100	2
Exterior Emulsion Paint 4A	7.8×10^3	1
Interior Emulsion Paint 4B	3.5×10^3	1
Exterior Emulsion Paint 4C	< 100	3
Interior Emulsion Paint 4D	< 100	2

a) TVC = total viable count

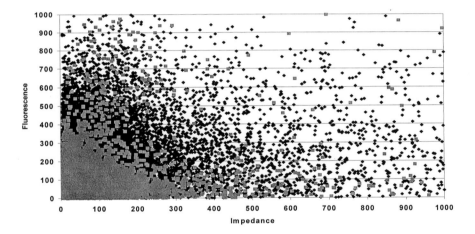

Figure 5. This graph shows the Interior Emulsion Paint, taken from series 3, before (violet) and after (blue) the inoculation amounting to approx. 5.0×10^4 cells/ml. The initial total viable count measured was 1.3×10^3 cfu/ml. After the inoculation and incubation for 24 hours at 30°C, however, the total viable count was 2.6×10^6 cfu/ml. The residue of the biocide in the sample was not sufficient to prevent a further contamination of viable microbial activity.

Performance of Different Antimicrobial Agents in Aqueous Paint Systems

In a further test the bactericidal properties of different active agents were compared by means of CellFacts II. The following substances were used in this study:

- 2-methyl-2H-isothiazole-3-one (MIT) / 5-chloro-2-methyl-2H-isothiazoline-3-one (CMIT) (CMIT:MIT = 3:1)
- 1,2-benzisothiazole-3(2H)-one (BIT)
- o-phenylphenol (OPP)
- 2-bromo-2-nitro-propane-1,3-diol (bronopol)

For this purpose, sterile aqueous paint samples were mixed with controlled quantities of the active agents, which are usually added in production, and stored for 24 hours at 30°C to achieve thorough mixing of the active agent. The samples were then inoculated with bacteria isolated from aqueous paint samples. The measurements with CellFacts II were made after 15 min., 30 min., 60 min., 2 hours, 4 hours, 6 hours and 8 hours. Both the decrease of the viable bacteria as well as the increase of the dead bacteria were measured.

Table 3. Performance of Different Antimicrobial Agents in Aqueous Paint Systems

Biocide	Time	% of cells detected as viable	% of cells detected as dead
MIT/CMIT	0 min	100	0
MIT/CMIT	15 min	97	3
MIT/CMIT	30 min	72	28
MIT/CMIT	60 min	16	84
MIT/CMIT	2 h	0	100
BIT	0 min	100	0
BIT	15 min	100	0
BIT	30 min	87	13
BIT	60 min	69	31
BIT	2 h	47	53
BIT	4 h	38	62
BIT	6 h	35	65
BIT	8 h	32	68
BIT [a]	24 h	32	68

a) To detect a possible deficiency in the action of the biocide, the samples were again measured 24 hours later.

Table 3. Performance of Different Antimicrobial Agents in Aqueous Paint Systems (contd)

Biocide	Time	% of cells detected as viable	% of cells detected as dead
OPP	0 min	100	0
OPP	15 min	37	63
OPP	30 min	0	100
Bronopol	0 min	100	0
Bronopol	15 min	98	2
Bronopol	30 min	82	18
Bronopol	60 min	65	35
Bronopol	2 h	30	70
Bronopol	4 h	9	91
Bronopol	6 h	0	100

Table 3 shows that BIT has a deficiency in its action against some of the bacteria used in this study, which is not necessarily recognised by the traditional methods due to their culture selectivity. Furthermore, it could be shown that the fastest bactericidal action is achieved with OPP, whereas with Bronopol the bactericidal effect only started after 6 hours.

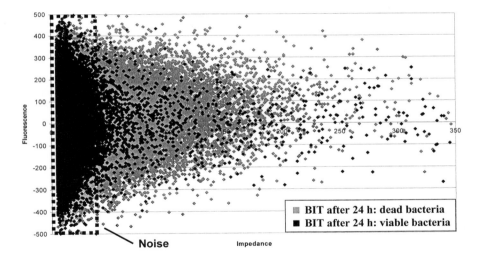

Figure 6. In this graph the deficiency in the action of BIT is illustrated. After 24 hours reaction time there are still viable bacteria. BIT is well-known for its deficiency in the action against some species of *Pseudomonas*[6].

Conclusion

CellFacts II has been shown to have a highly application-oriented functionality for both analytical flexibility and minimal operator intervention. The sensitivity and specificity of each particular analysis is given by the ability to determine the particle heterogeneity of the sample and the staining characteristics of the microbial population, which in turn is determined by the physiological status of the individual cells. Also discrimination between cells of different sizes, e.g. between yeast and bacteria, is possible.

The analysis of the physiological state of individual cells in a population, effectively in real-time, enables the rapid determination of the effect of antimicrobial agents on these cells, i.e. rapid determination and optimisation of antimicrobial agents.

References

[1]   Hall, G.H., Jones, J.G., Pickup, R.W., Simon, B.M., 1990. Methods to Study the Bacterial Ecology of Freshwater Environments. In Grigorova, R., Norris, J.R. (eds.), Methods in Microbiology, 22, pp. 181-209. London – San Diego – New York: Academic Press.

[2]   Fry, J.C., 1990. Direct Methods and Biomass Estimation. In Grigorova, R., Norris, J.R. (eds.), Methods in Microbiology, 22, pp. 41-85. London – San Diego – New York: Academic Press.

[3]   Schwarzentruber, P., Gane, P.A.C., 2002. Application of Microbiocides for the Storage Protection of Mineral Dispersions. In Paulus, W., Directory of Microbicides for the Protection of Materials and Processes. Dordrecht: Kluwer Academic Publishers.

[4]   Lloyd, D., Hayes, A.J., 1995. Vigour, Vitality and Viability of Microorganisms. FEMS Microbiol. Lett., 133, pp. 1-7.

[5]   Schwarzentruber, P., 2001. Microbiological Characterisation of $CaCO_3$ Slurries; Interim Report presented at the University of Warwick, Coventry, UK.

[6]   Paulus, W., 1993. Microbicides for the Protection of Materials – A Handbook. London: Chapman & Hall.

Macromol. Symp. **187**, 553–561 (2002)

Dual-Cure Processes: Towards Deformable Crosslinked Coatings

*Abdelkrim El-ghayoury[a,c], Chouaib Boukaftane[b,c], Barteld de Ruiter[*a]*
and Rob van der Linde[b]

a) TNO Industrial Technology, Department of Polymer Technology, De Rondom 1, P.O. Box 6235, 5600 HE Eindhoven, The Netherlands.

b) Eindhoven University of Technology, Coatings Technology Group, P.O. Box 513, 5600 MB Eindhoven, The Netherlands.

a) c) Dutch Polymer Institute, P.O. Box 902, 5600 AX Eindhoven, The Netherlands

Summary: Two dual-cure processes consisting of a UV-initiated radical polymerization followed by either a UV-induced cationic polymerisation, or a thermal addition reaction, were investigated. The feasibility of the processes was studied using an acrylate-oxetane monomer for the UV/UV combination, and an acrylated oligoester for the UV/Heat combination. It was shown by FTIR and Tg measurements, that both steps of each process could be performed efficiently and separately. This allowed the production of a deformable partially cured coating, whose cure can then be completed, leading to the required final properties. Furthermore, it was demonstrated that the increase of the functionality of the reactive diluent led to a decrease of the thermal crosslinking extent. This is probably due to the reduced mobility of the reactive species that is caused by an enhanced UV crosslinking taking place during the first step.

Keywords: dual-cure; coatings; acrylate; oxetane; polyester; crosslinking; UV-curing

Introduction

Photoinitiated polymerization or UV-curing of multifunctional monomers is widely utilized for rapidly producing highly crosslinked polymer materials.[1],[2] Among the advantages of this

---

* Corresponding author: Dr. Barteld de Ruiter: e-mail: B.deRuiter@ind.tno.nl

technology are the high cure speed, the reduced energy consumption and the very low organic emission. This allowed its rapid growth in a large variety of applications, in particular to achieve a fast drying of varnishes, printing inks and protective coatings, and a quick setting of adhesives and composites materials. [3],[4] A limitation of most photocrosslinkable systems, however, is the shrinkage occurring during the photopolymerization,[5] leading to a lack of adhesion on certain non porous substrates. Furthermore, the high crosslink density of the generated coatings often makes them very difficult to bend or to emboss, because of their brittleness. In this paper, we report on a two-step cure process that allows overcoming these problems.

The first process consists of the combination of two different UV steps based on different polymerization mechanisms. Few of such systems, combining (meth)acrylate with vinyl ethers[6],[7],[8],[9] or with epoxides[10],[11],[12] have been reported. However, in these studies the radical and the cationic steps were performed simultaneously. Moreover, only a few of them reported the efficient separation of the two steps such as compositions containing a photosensitive free radical reactive system and photosensitive cationic cure epoxy system. [13]

The second process consists in the combination of UV and thermal steps. Many of such systems have been reported where the formulations contain on one hand an acrylate function, and on the other hand a functionality having the ability to react at ambient temperature, like isocyanates or epoxy.[14] In addition to these systems, acrylates/amino resins were employed for radiation/heat technique that was more extensively studied.[15]

In our approach, the liquid formulation is applied on a flat substrate and then cured via a radical photo-initiated polymerization. This first step should lead to a tack-free, flexible and therefore deformable coating. In a latter stage, the coated sheet is shaped, and then the second cure, a cationic light-initiated or condensation/addition reaction, is performed to obtain a cross-linked hard coating, with improved physico-chemical properties. The feasibility of the dual-cure processes was investigated using acrylate-oxetane and acrylate-polyester systems for the UV/UV and the UV/Thermal combinations, respectively (Scheme 1).

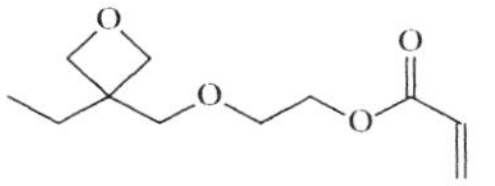

Acrylate-oxetane 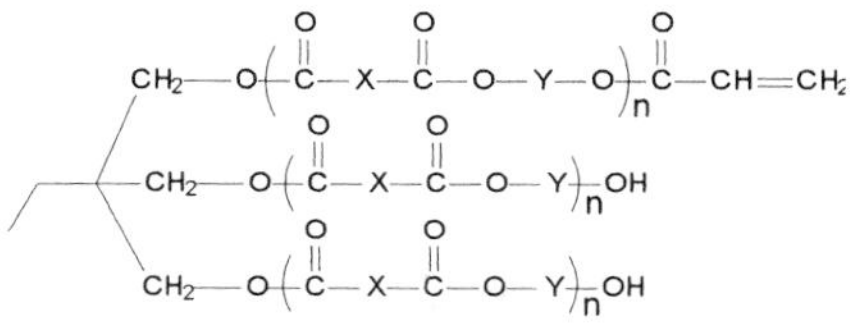   Acrylate-polyester

Scheme 1: Schematic representation of the compounds

Materials and Methods

Materials

1. Acrylate-oxetane monomer[†]

The acrylate-oxetane monomer was obtained after four steps reactions.

2. Acrylate-polyester

To have a binder bearing both UV and thermally reactive functionalities, one third of the hydroxy groups are converted into an acrylate function. Acrylation consists of adding dropwise at 0 °C in a dry medium, a solution of acryloyl chloride (ACl) (0.05 g/ml) to a solution of an oligoester obtained by direct catalysed esterification of trimethylolpropane (0.15 mole), neopentylglycol (1.21 mole), adipic acid (0.20 mole) and isophtalic acid (1 mole). For example, 0.5 g of ACl ($5.5 \ 10^{-3}$ mole) is added to a solution of 5g ($1.65 \ 10^{-2}$ mole OH) and 0.56 g of triethylamine ($6.0 \ 10^{-3}$ mole) in dichloromethane. After the reaction, the solution is washed with an aqueous solution of ammonium chloride and water to remove the produced salt, $(CH_3CH_2)_3NH^+Cl^-$. It is then dried with $MgSO_4$. The solvent and remaining TEA are removed under reduced pressure. The degree of acrylation of the polyester is 30% (NMR) and its molecular weight is 1970 g/mole (GPC).

3. Material mixtures used for the experiments

UV/UV dual-cure

93 % (w) Monomer

4 % (w) Irgacure 819 (Ciba Specialty Chemicals)

3 % (w) Cyracure UVI-6974 (Dow Chemicals)

Substrate: KBr plate

UV/Thermal dual-cure

39.6 % (w) Binder: Acrylated polyester

39.6 % (w) Thermal crosslinker: Blocked polyisocyanate (Desmodur BL3272 MPA, Bayer)

19.8 % (w) Reactive diluent (see text)

1% (w) Photoinitiator: Irgacure 819 (Ciba)

Methods

1. UV/UV process

For all experiments with the acrylate-oxetane monomer, 20-30 microns liquid films were prepared. A cut-off filter (allowing only wavelengths higher than 385 nm, where the radical photoinitiator is solely absorbing light) was combined with the lamp in the first step of the process in order to initiate only the radical polymerization. The second step was performed without the filter to allow the cationic photoinitiator to absorb light and start the second cure. The samples for FTIR and DSC (Perkin Elmer) measurements were applied on KBr plates and irradiated using a Uvicube curing system (Dr. Hönle), which was equipped with an F lamp (400 W). The samples were irradiated for 5 minutes and the UV and the visible doses were measured using an UV Power Puck radiometer. The measured doses were: Vis.= 10.6 mW/cm^2, UVA= 21 mW/cm^2 and with the filter: Vis. = 10.6 mW/cm^2. Inhibition of the radical polymerization by O$_2$ was prevented by exposing the samples to radiation either under a N$_2$ atmosphere or by applying a polypropylene foil on the liquid film.

2. UV/Thermal process

30-50 micron liquid films were irradiated under nitrogen using a Dr. Honle UV curing system equipped with an H lamp. At 20 cm height, the light intensity is (mW/cm^2) (UV Power Puck): Vis. =17; UVA= 24. The UV dose received by the coatings is 0.24 J/cm^2. The thermal curing was performed without vacuum or gas extraction. The coatings were cured at 120 °C for two hours.

3. FTIR and DSC (Tg) measurements

The dual-cure processes were studied by transmission FTIR using a Bio-Rad infrared spectrometer. Liquid films were applied on KBr plates using a wire bar coater. For the study of the UV/UV system, the cure was evaluated by following the decrease of the acrylate C=C (1630 cm^{-1}) and oxetane C-O-C (985 cm^{-1}) bands, after the UV induced radical and cationic steps, respectively. As for the UV/Heat system, the extent of the cure of the UV and thermal steps was assessed by monitoring the disappearance of the acrylate C=C (810 cm^{-1}) and OH (3450 cm^{-1}) bands, respectively. Tg measurements were carried out using a Perkin Elmer DSC.

Results and Discussion

UV/UV system

A two-step UV/UV coating process is possible by developing a monomer that combines two different functional groups. One of those groups can be polymerized via a radical mechanism (acrylate) and the second via a cationic mechanism (oxetane). In such a system, the free-radical step is to be carried out first, because during the cationic step, the photoinitiator undergoes photolysis through a radical reaction.[16] Therefore, the cationic photoinitiator would, if activated in the first step, initiate both the radical and the cationic polymerizations. The free-radical photoinitiator will not initiate the cationic step. The selected photoinitiators have different absorption spectra. The radical photoinitiator (Irgacure 819) is absorbing light between 200 and 420 nm. The cationic photoinitiator (Cyracure UVI-6974) is only absorbing light below 380 nm. Therefore, by using a filter that cuts off the light between 200 and 385 nm, we can selectively excite the radical photoinitiator during the first step and initiate only the free-radical photopolymerization.

The efficiency of the dual-cure process, that is the occurrence of subsequent UV-initiated radical and cationic polymerizations, was studied using the acrylate-oxetane monomer from scheme 1. The transmission FTIR spectra depicted in Figure 1 show that the first step, namely the radical cure, was performed successfully since the FTIR spectrum shows an almost complete disappearance of the acrylic C=C peak at 1630 cm^{-1}. This generated a soluble polymer with polyacrylic linear chains bearing oxetane side chains. When the second irradiation (without the filter) had been performed, the large decrease of the absorption of the oxetane C-O-C peak at 985 cm^{-1} indicated that the second cationic polymerization had also taken place. This latter allowed the formation of an insoluble 3D network. These findings could be confirmed by DSC measurements. The film obtained after the first step shows a glass transition temperature (Tg) of −16 °C. When the cationic polymerization step had been performed, the resulting cured film had a Tg of +34 °C. This large increase of the Tg observed between the first and second step is indicative of the formation of a highly crosslinked polymer film.

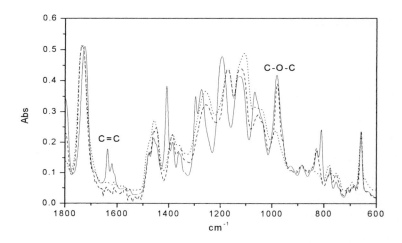

Figure 1. Dual-cure of acrylate-oxetane: Before cure (solid), step 1 (dash) and step 2 (dot); (UV dose (J/cm^2): 3.6 (step1), 12.6 (step2))

UV/Heat system

The formulation used for this study contains hexanedioldiacrylate (HDDA, UCB) as reactive diluent, besides the components mentioned previously. The coating obtained after the UV step is touch dry and has a Tg of 2 °C. Occurrence of the cure is confirmed by the disappearance of the acrylate double bond peak at 810 cm^{-1} (Figure 2.a). After the thermal step, the Tg of the coating increased by 35 °C, thus indicating that the second step was successful. This is also consistent with the large decrease of the OH band at 3550 cm^{-1} (Figure 2.b). These results show that the two steps of the dual-cure process can be performed separately and efficiently.

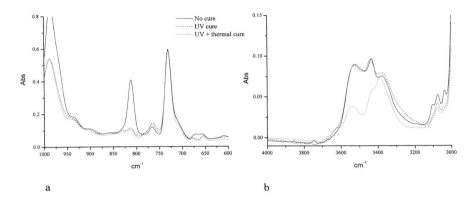

Figure 2. Study of the effectiveness of the UV/Heat process by FTIR

To study the influence of the reactive diluent on the second step of the process, three diluting acrylates having a functionality ranged between 1 and 3 were used. These monomers are isobornylacrylate (IBOA, UCB), hexanediol diacrylate and trimethylolpropane triacrylate (TMPTA, UCB). The extent of the thermal reaction involving OH and blocked NCO (OH/NCO=1) was evaluated by transmission FTIR. As shown by the FTIR spectra (Figure 3) of the three systems, higher functionality (IBOA<HDDA<TMPTA) of the reactive diluent, leads to lower conversion of the OH group, and therefore to a lower degree of thermal cross-linking. This behavior is probably due to the network generated by the UV step, since an enhancement of the functionality of the diluent leads to a denser UV generated network. This leads to a reduced diffusion of reactive species and thus to a lower conversion of the OH groups during the second step.

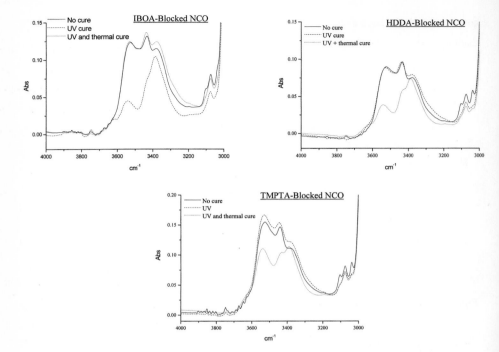

Figure 3. Effect of the reactive diluent functionality on the second step of the process

Conclusion

The two dual-cure processes combining on one hand a UV-initiated radical and on the other hand a cationic polymerization or a thermal addition reaction were shown to be effective in producing deformable cross-linked coatings in two separate steps. The efficiency of the second cure of the UV/Heat process was affected by the cross-linking density generated by the first cure.

References

† Full synthetic details will be given elsewhere.
[1] Roffey, C. C. *Photopolymerization of surface coatings*, Wiley, Chichester, **1982**.
[2] Pappas, S. P. *UV-Curing Science and Technology*, Vols.1 and 2, Technology Marketing Corporation, Stamford CT, **1978** and **1985**.
[3] Davidson, R. S. *Exploring the science technology and applications of UV and EB curing*.London: SITA   Technology, **1991**, 327.
[4] Decker, C.; In: Meijer, H. E. H. Ed., Materials Science and Technology, Weinheim: VCH Verlag, **1997**, *18*, 615.
[5] McGinnis, V. D.; Kah, A. Paint Research Institute, 8th Symposium, May **1978**.
[6] Vansteenkiste, S.; Matthijs, G.; Schacht, E.; De Schrijver, F. C.; Van Damme, N.; Vermeersch, J. *Macromolecules* **1999**, *32*, 55.
[7] Itoh, H.; Kameyama, A.; Nishikiubo, T. *J. Polym. Sci.: Part A: Polym. Chem.* **1996**, *34*, 217.
[8] Decker, C. *J. Macromol. Sci. – Pure Appl. Chem.* **1997**, *A34*, 605.
[9] Stohr, A.; Strohriegl, P. *Macromol. Chem. Phys.* **1998**, *199*, 751.
[10] Decker, C. *Polym. Int.* **1998**, *45*, 133.
[11] Vabrik , R.; Czajlik, I. ; Tury, G.; Rusznak, I.; Ille, A.; Vig, A.; *J. Appl. Polym. Sci.* **1998**, *68*, 111.
[12] Peeters, S.; In: Fouassier, J. P.; Rabek, J. F. Ed, Radiation Curing in Polymer Science and Technology, Vol. 3, Polymerisation Mechanisms, Elsevier Applied Science, **1993**.
[13] Rohm and Haas company, **1989**, EP 0 335 629 A2.
[14] A. Noomen, Sassenheim, Neth. Congr. Fatipec, **1984**, *17th (1)*, 255. A. Noomen, Sassenheim, Neth. Congr. Fatipec, 1980, *15th (1)*, 346. A. Noomen, *J. Oil Col. Chem. Assoc.*, **1989**, *64*, 347. S. Peeters, J M Loutz, M Philips, *Polymers Paints Colour Journal*, **1989**, *179*, 304.
[15] Boeckeler, R. H. *Radcure '86, Conf. Proc.*, 10th , **1986**, 16/1; US Patent 4 548 895 (Oct. 22, **1985**); US Patent 4 444 806 (Apr. 24, **1988**); US Patent 5 679 719 (Oct. 21, **1997**).
[16] Sasaki, H.; Kuriyama, A.; *J. Macromol. Sci.-Pure Appl. Chem.*, **1995**, *A32*, 1699.

Macromol. Symp. **187**, 563–571 (2002)

Physical-Chemical Conditions for Production of Combined Alkyd-Acrylic Dispersion

Eugene A. Indeikin, Olga A. Kulikova, Vladimir B. Manerov

Yaroslavl State Technical University, Yaroslavl Paint Research Institute,

Russia, Yaroslavl, 150044, Polushkina roshcha, 16

Summary: When preparing the mixing of acrylic polymer and copolymer dispersions with alkyd oligomer emulsions it is necessary to provide agregative stability of the combined dispersions. It was established that transfer of polymer particles mass of highly dispersed systems onto particles of less dispersed systems is seen during geterocoagulation of combined dispersions. Optimal physical-chemical and hydrodynamic conditions of the emulsification of alkyd oligomers with the required dispersion degree for producing the mixed systems were established. The combined dispersion stability is determined from the ratio of electrokinetic potentials of particles of polymers and oligomers being combined as well as ratio of their isoelectric points. The zone of dispersion incompatibility was established by method of microelectrophoresis.

Keywords: dispersions, mixing, oligomers, particle size distribution, incompatibility

Introduction

Combining of acrylic polymers dispersions and copolymers with emulsions of alkyd oligomers is used for purposeful regulation of materials' properties on their basis. It allows considerably to improve the adhesion to chalking and earlier painted surfaces, impregnating ability, to increase stability to the influence of the polluting substances, to improve water resistance of the coating. Thus properties, such as fast hardening, hardness of the coating and stability to UV -radiation, caused by presence of acrylic polymers in materials' structure remain. At combining it is necessary to maintain the aggregate stability, i.e. prevention of dispersion system heterocoagulation, both for stability of the material and for maintenance of important properties of the coating.

Heterocoagulation can be caused by various acid-basic character of the particles of the dispersion phase, different charges of the particles. In many cases it can be caused by various particle size distribution of combined dispersion systems.

   CCC 1022-1360/00/$ 17.50+.50/0

564

The study of heterocoagulation

By the method of polarized light diffusion is established [1,3], that at combining of dispersions, particles, strongly differing by the sizes, in the ratio of acrylic dispersion (AD): alkyd emulsion (AE) = 70:30, at the initial moment of combining the structure is characterized by a bimodal differential curve distribution (continuous line on the figure 1). In 24 hours the outstripping coagulation of alkyd emulsion is observed (dotted line), which is accompanied by reduction of a maximum appropriate to distribution of acrylic dispersion.

At reduction of a acrylic share in system the acceleration of the process of coagulation, down to complete disappearance of a maximum appropriate to distribution of acrylic dispersion is observed.

At combining of dispersion with smaller distinction at a rate of particles the distribution has an intermediate character in a kind of the greater affinity of two combined dispersions coagulation rates. In case of close on dispersion structure of systems coagulation is not observed.

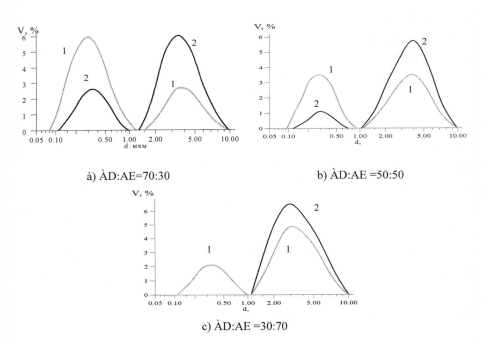

à) ÀD:AE=70:30   b) ÀD:AE =50:50

c) ÀD:AE =30:70

Figure 1. Differential curve distributions on the sizes of particles mixed dispersions (1 - at the moment of combining; 2 - after 24 hours).

In such a manner there was established, that in process of dispersion heterocoagulation the carry of mass of highly dispersed system polymeric particles on particles of system with smaller dispersity is observed, that defines the final distribution. The necessity of systems close on dispersion structure use for combining is shown.

The conditions of production

We investigate[4] influence of conditions of production on properties of emulsion of alkyd oligomer.

The choice of surfactant for production of the alkyd oligomer emulsions, used for combining with acrylic dispersion is caused as necessary efficiency of emulsifying action in view of the Bankroft law and concept of hydrophilic-lypophilic balance (HLB), and their influence on stability of acrylic dispersions. Thus minimization of surface-active substance in system is important for the declination of their influence on the coating properties. [5]

One of major parameters of the emulsification process is temperature, which influences as on HLB non-ionogenic surfactant according to the equation Griffin-Davis and viscosity of dispersion medium and formed dispersion phase. As have shown our researches [6], at use of anionic emulsifier the average size of the emulsion particles decreases with growth of temperature. Emulsifying ability of alkylphenolethoxylates in the relation to alkyd oligomer grows up to temperature of turbidity with growth of temperature. (Fig. 2).

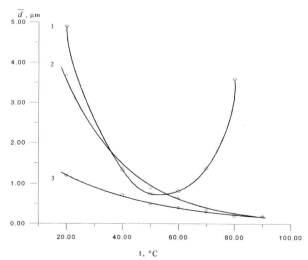

Figure 2. Dependence of the average size of particles on temperature of emulsification for octylphenolethoxylates: 1 - EO = 10; 2 - EO =30; 3 - EO = 40

Probably, it is explained to that the decrease of hydrophility is connected with a smaller degree of oxygen atoms solvation with a simple ether bonds, that causes the increase of the surfactant molecules hydrophobic sites solubility in emulsifying phase with orientation of polar parts into water medium, thus increasing hydrophility of alkyd oligomer.

For surfactant with a ethoxylation degree equal to 10 at achievement of the turbidity point at the expense of temperature increase the sharp decrease of emulsions' dispersity, connected with the beginning of its manipulation is observed. Despite of really high HLB meaning of these surfactants, the manipulation probably is connected with micelle formation of surfactant and with increase of their solubility in alkyd oligomer at the expense of ester, carboxylic and hydroxyl groups presence in it, which promotes surfactant dissolution in them with the meaning of HLB more than 7. The optimum of temperature is higher for emulsifiers with the greater degree of ethoxylation, that is connected to increase of the work, which is necessary for the carry of surfactant molecules from polar medium into non-polar medium with the growth of the ethylene oxide groups contents.

As have shown the results of research, the dependence of alkyd oligomer viscosity from temperature is adequately described by the Eiring equation.

$$\eta = \eta_0 \cdot \exp\frac{\varepsilon}{RT} \qquad (1),$$

where η - viscosity of alkyd oligomer (mPa $\cdot$ s),

η_0 - pre-exponential term which has the dimension of viscosity,

ε - - energy of viscous flow activation (J/mol);

R - universal gas constant, J mol^{-1} Ê$^{-1}$,

Ò - temperature, Ê.

On the basis of the carried out researches on influence of temperature on emulsifying ability of surfactant and rheological properties of alkyd oligomer there was chosen the optimum of emulsifying temperature, which allows to receive emulsions with high dispersity.

We have calculated the average energy of viscous flow activation, which is approximately identical to all considered alkyd oligomers and lies in the interval - 66-68 kJ/mol.

As it can be seen from figure 3, the growth of systems' dispersity with the increase of the surfactant concentration at its certain meaning is sharply slowed down. It is necessary to consider these concentration as expedient, because it is proved to be true also by data on adsorptive titration.

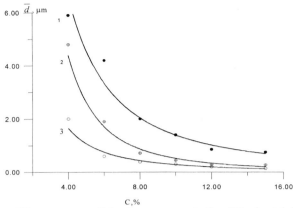

Figure 3. Dependence of the average particle size on concentration (C) of octylphenolethoxy-lates: 1 - EO = 10; 2 - EO =30; 3 - EO = 40

As have shown our researches, for production of alkyd oligomer emulsions with high dispersion the use of a concentration and temperature phases inversion method is expedient. Thus at the use of non-ionic emulsifier with the greater degree of ethoxylation the concentration inversion comes at the greater concentration of dispersion phase. (fig. 4). Temperature inversion is realized for alkylphenolethoxylates with a degree of ethoxylation equal to 10. ohm

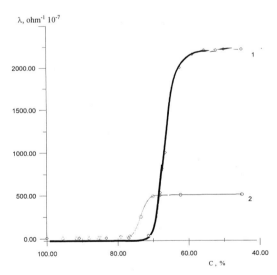

Figure 4. Dependence of electrical conductivity on concentration of dispersion phase for octylphenolethoxylates: 1- EO=10; 2- EO=40

The study of the dispersion phase rheological properties on the size of emulsion particles influence has shown, that dispersity is practically linearly depends on viscosity of alkyd oligomer and can be described by the equation:

$$d = k \cdot \frac{\eta_a}{\eta_c^2}, \quad (2)$$

Where k - factor which is directly proportional to the surface tension on the interface of the dispersion phase and dispersion medium and inversely proportional to the speed of shift.

Thus on the basis of the carried out researches there were chosen the optimum conditions for the alkyd oligomer emulsification, allowing to receive highly dispersed steady emulsions, which were used in a composition with acrylic dispersions.

Electrokinetic, optical and rheological properties

We investigate physical-chemical properties of combined alkyd-acrylic dispersions, and in particular there was investigated [7] the influence of dispersion media pH value on electrokinetic potential of dispersion phase particles and the meanings of dispersions isoelectric points were established, according to which the zones of dispersion incompatibility were determined.

There is shown, that the meaning of ζ-potentials for acrylic dispersions are determined by presence of anionic surfactant and for the dispersion of acrylate and vinyl acetate copolymer by ionization of carboxyl groups in micelles of polymer. There is determined, that ζ-potential of alkyd emulsion particles, caused depending on pH value by ionization of carboxyl groups or adsorption of neutralizator ions, has a negative mark in a wide range of pH value and does not depend on it dispersity and the type of surfactant.

As it can be seen from the figure 5, there are zones of incompatibility for the acrylic polymer dispersions and alkyd emulsion (**a** and **b**)- the zone of the electrophoretical mobility meanings opposite on a sign, and, hence, of ζ-potential. Besides, this zone is smaller for acrylate and vinyl acetate copolymer and alkyd emulsion, than at pure acrylic dispersion and alkyd emulsion.

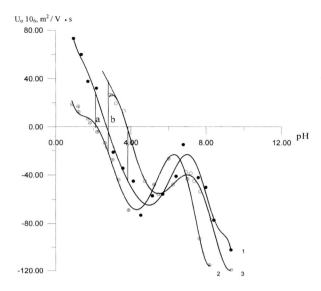

Figure 5. Dependence of electrophoretic mobility U_e of acrylic and alkyd emulsions on pH value 1- alkyd emulsion ;   2- acrylate and vinyl acetate copolymer dispersion; 3- acrylic dispersion.

At the formation of films from the dispersions the zones of incompatibility are equal with zones of increased optical density of polymeric coating (fig. 6).

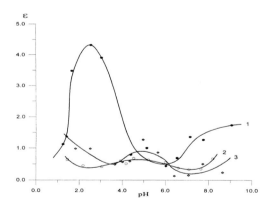

Figure 6. Dependence of film optical density (ε), generated from mixed dispersions from pH: 1- AD + AE ($\bar{d}$=2,6 µm); 2 - AD + AE ($\bar{d}$=0,5 µm); 3 - AD + AE ($\bar{d}$=0,2 µm).

In a case, when the zones of incompatibility are not present, the maxima of optical density are marked in zones ζ-potential negative meaning reduction for alkyd oligomer. Probably, it is connected with the increase of system concentration caused by evaporation of dispersion medium and results in coagulation. Besides, the particles of alkyd oligomer are the coagulation germs of the more highly dispersed acrylic dispersion particles.

The size of dispersion ζ-potential influences on rheological properties of dispersion. It was determined, that for the acrylic polymer and its copolymer with vinyl acetate the increase of viscosity with electrokinetic potential negative meaning growth is observed. Probably, it is connected with increase of diffusion layer thickness and adsorptive-solvative environment, i.e. with the increase of hydrodynamic volume. As against from polymer dispersions the viscosity of alkyd emulsion grows practically monotonously with the $\delta\acute{I}$ value growth, that is connected with increase of carboxyl groups ionization.

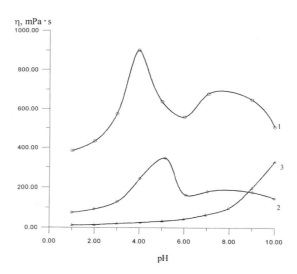

Figure 7. Dependence of dispersions' viscosity from pH:
1. - dispersion of acrylate and vinyl acetate copolymer;
2. - dispersion of acrylate and vinyl acetate copolymer + alkyd emulsion;
3. - alkyd emulsion.

For mixed dispersions as with the use of acrylic polymer and copolymer with vinyl acetate the character of viscosity dependence from $\delta\acute{I}$ is kept similar to character of their dependence in absence of alkyd oligomer in system structure (fig. 7).

Thus in the first case the constant absolute meanings of viscosity are kept, in the second case the downturn of acrylate and vinyl acetate copolymer dispersion viscosity is observed and at introduction of alkyd oligomer emulsion in it. Probably, it is connected with change of the particles charge and the thickness of diffusive environments because of adsorbing substances redistribution on interface of dispersion phases of a various chemical nature.

Conclusion

On the basis of the carried out researches it was determined, that in the process of dispersion heterocoagulation the carry of highly dispersed system polymeric particles weight on particles of the system with the smaller dispersity is observed, as it is defined with the final distribution of particles according to their sizes. The necessity of the systems close on dispersion structure use is shown. As a result of electrokinetic properties research, the zones of dispersions' incompatibility, described by opposite marks of electrokinetic potential are established. The dependence of alkyd oligomer emulsion dispersity from its' rheological properties is established and the optimal physical-chemical and hydrodynamic conditions of steady alkyd emulsion with necessary structure production are chosen.

References

[1]   V.V Verkholantsev, ECJ, **1997,** No.6 pp. 614-622
[2]   A. Hofland, Polym. Paint Col. J. 184, **1994,** No. 4350 pp. 261-263
[3]   O. A. Kulikova E. A. Indeikin, V. B. Manerov. ECJ, **1999,** No 10, pp. 44,45,48,50.
[4]   O. A. Kulikova E. A. Indeikin, V. B. Manerov. T. N. Shkumat., Lakokrasochnye mate-
       rial, **2000,** No. 1, pp. 10-12
[5]   F. T. Tadros and B. Vincentz, Emulsion Stability. Encyclopedia of Emulsion Technol-
       ogy. Becher P. (Ed.). Vol 1. - Marsel Dekker Inc., New York, **1983**
[6]   O. A. Kulikova E. A. Indeikin, V. B. Manerov, T. N. Shkumat.,., Lakokrasochnye mate-
       rial, **1998,** No. 6, pp. 4-6
[7]   O. A. Kulikova E. A. Indeikin, V. B. Manerov, N.B. Skopintseva, ECJ, **2000,** No.7-8

Macromol. Symp. **187**, 573–584 (2002)

Fumed Silica – Rheological Additive for Adhesives, Resins, and Paints

Herbert Barthel[a], Michael Dreyer[a], Torsten Gottschalk-Gaudig[a], Victor Litvinov[b], and Ekaterina Nikitina[c]

[a] Wacker-Chemie GmbH, D-84480 Burghausen, Germany

[b] DSM Research, NL-6160 MD Geleen, The Netherlands

[c] Institute of Applied Mechanics, RAS, Moscow, Russia

Summary: Fumed silica, a synthetic silicon dioxide, is a powerful rheological additive for resins and paints to introduce thixotropy or even a yield point. The rheological effectiveness of fumed silica is based on its ability to form percolating networks which immobilize large volumes of liquid. By a combination of advanced rheological experiments, spectroscopical investigations, and quantum chemical calculations it could be demonstrated that the formation and stability of the silica network is strongly influenced by particle-resin interactions. The results can be used to develop comprehensive models, which explain the rheological performance of different grades of fumed silica in different resins.

Keywords: fumed silica/ unsaturated polyester resins/ vinyl ester resins/ intermolecular interactions/ rheology

Introduction

Fumed silica is a synthetic amorphous silicon dioxide produced by hydrothermal hydrolysis of chlorosilanes in an oxygen-hydrogen flame. In this process, as a first step, SiO_2 molecules are formed which collide and react to nano-size proto particles, which by further collision in a second step form primary particles of around 10 nm in size. The flame process itself leads to the formation of smooth particle surfaces, which provides fumed silica with a high potential for surface interactions. [1, 2] At the high temperatures of the flame primary particles are not stable but are fused together to form space-filling aggregates. Leaving the flame, at lower temperatures, the silica aggregates stick together by physico-chemical forces building up large micron-sized agglomerates and finally fluffy flocks . [3, 4]

Fumed silica is widely used in industry as an efficient thickening agent providing shear thinning and thixotropy to liquid media like adhesives, resins, paints, and inks. Various parameters control the rheological performance of fumed silica: (a) The smoothness of the primary particle surfaces which provides a maximum contact area for various types of interactions like H-bonding and Van-der-Waals interactions of dipolar and dispersive character. (b) The space-filling structure of the aggregates with a mass fractal dimension of $D_m < 2.7$, leading to a fluffy structure of agglomerates, typically with a 'density' d = 50-100 g/l (remark: density of amorphous silicon dioxide d_{SiO2} = 2200 g/l), and agglomerate sizes > 1 μm. (c) The high physico-chemical interaction potential of the fumed silica surface is based on its reactive surface silanol groups (surface density 1.8 $SiOH/nm^2$), but is also due to its

CCC 1022-1360/00/$ 17.50+.50/0

polar Si-O bonds containing bulk. By surface modification, most commonly surface silylation, these interactions can be controlled precisely.

Particle interactions are the driving force for agglomerate and network formation, enabling fumed silica to form percolating networks in liquid media. [5-7] Basically, two kinds of networks are possible: Firstly, a network of fumed silica particles or aggregates originating from direct particle-particle-contacts and, secondly, a network based on polymer bridging where aggregates are interacted by polymers at least on two particles. Real systems may consist of both and mixed types. At rest or very low shear rates these networks are able to immobilize large volume fractions of liquids even at low fumed silica loading (< 5 wt%), resulting in very high viscosities or a yield point, respectively. Upon applying shear forces the network structure is reversibly destroyed and the apparent viscosity of the mixture decreases with increasing shear rate. When the shearing stops the system is able to recover the network structure. Fig. 1 depicts this process schematically.

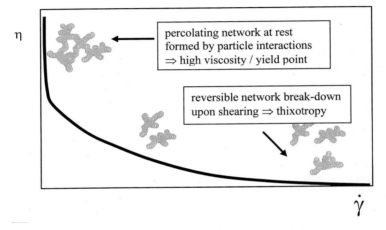

Figure 1. Dependence between shear rate, network structure, and viscosity.

Obviously, the rheological performance of fumed silica in adhesives, resins, and paints is mainly determined by the stability of the colloidal network. This raises the question which parameters influence the formation and stability of a colloidal silica network. In order to answer this question we have to understand a) the nature of interactions in silica/resin/solvent mixtures in terms of the interactions silica-silica, silica-resin, and silica-solvent, and b) how these interactions influence the rheological behavior of the mixtures. From these information we should be able to develop a comprehensive model that explains the rheological performance of different grades of fumed silica in different resin types.

Aim of this paper is to present comprehensive models of interactions comparing rheological results, analytical data from spectroscopic methods such as solid state NMR and IR, and quantum chemical calculations. Furthermore we demonstrate how these interactions influence the rheology of fumed silica in unsaturated polyester (UP) and vinyl ester (VE) resins.

Experimental

Two grades of fumed silica with a different degree of surface treatment and different polarity have been used: A non-treated, hydrophilic Wacker HDK® N20 (BET surface area 200 m²/g, 1.8 SiOH/nm² equivalent to 100 % residual SiOH) and a fully silylated hydrophobic Wacker

HDK® H18 (carbon content 4.5 %C, 15% residual SiOH), the latter is covered by a chemically grafted PDMS layer. Both silicas are products from Wacker-Chemie GmbH, Germany. The resins used are an unsaturated polyester resin Palatal P4 (UP resin), a co-condensate of a diol, maleic acid and orthophatalic acid, and the vinylester resin Atlac 590 (VE resin), a co-condensate of glycidine, methacrylic acid and bisphenol A. Both resins have a styrene content of 35% and were provided by DSM, NL.

Rheological studies have been performed using 3 techniques: (a) Measuring the step profile of the dynamic viscosity at a controlled but constant shear rate D at 1 s⁻¹, 10 s⁻¹ and 100 s⁻¹, each 120 s, respectively. (b) Recording the dynamic behavior and thixotropy using a shear rate controlled relaxation experiment at $D = 0.5$ s⁻¹ (conditioning), 500 s⁻¹ (shear thinning), and 0.5 s⁻¹ (relaxation). (c) Oscillation at a frequency of 1.6 s⁻¹ and a deformation sweep between 0.001 to 10. (d) Determination of the yield point from a log-log plot of deformation γ vs. shear stress τ. The yield point is defined as the stress τ where the corresponding γ - τ curve exhibits a deviation > 5% from a tangent defined by data points at low shear stresses. All samples contained 3%wt of fumed silica.

¹³C CPMAS spectra were recorded with cross polarization (CP) times of 0.5 and 8 ms for samples with a silica content of 4%.[8]

IR spectra were obtained from liquid samples using NaCl plates on a standard FT-IR spectrometer under compensation of the solvent signal.

Quantum chemical calculations were using cluster approach methods in a software Cluster Z1 and a modified PM3 parameter set.

Results and discussion

Rheological studies

It is well known that both, hydrophilic and highly hydrophobic fumed silica are efficient rheological additives for unsaturated polyesters. In more polar systems such as vinyl esters, however, only highly hydrophobic fumed silica is suitable. This observation is illustrated by the viscosity step profile of Wacker HDK N20 and H18 in Palatal P4 (UP) and Atlac 590 (VE), respectively, depicted in Fig. 2.

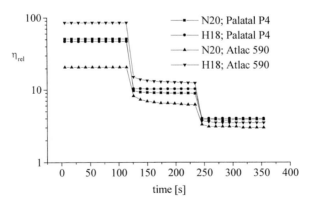

Figure 2. Relative viscosity in a shear rate step profile at D =1 s⁻¹, 10 s⁻¹ and 100 s⁻¹, each 120 s, respectively, of 4 resin systems: Wacker HDK N20 and HDK H18 in Palatal P4 (UP resin) and Atlac 590 (VE resin); 35%wt styrene; 3%wt fumed silica.

The step profile reveals that N20 and H18 dispersed in Palatal P4 exhibit an almost identical rheological behavior, whereas the apparent relative viscosities of N20 and H18 dispersed in Atlac 590 at low and moderate shear rates are distinctly different. To explain this behavior it is necessary to consider the polarity, functional groups, and chain length of the resins but also the surface properties of the fumed silica. All parameters together will influence the nature and strength of the colloidal forces, which govern the rheology of the mixtures.

Firstly we investigated the rheology of Wacker HDK N20 and H18 in pure styrene. Fig. 3a shows the relaxation test experiment, 3b) the deformation sweep, and 3c) the determination of the yield point.

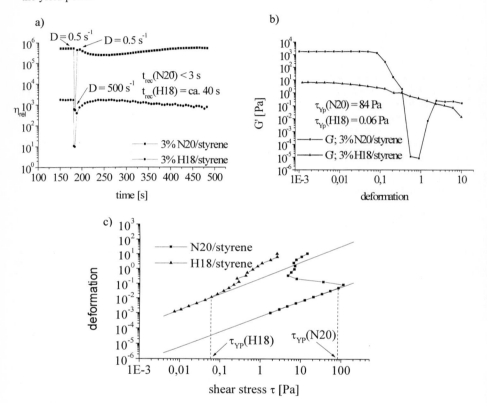

Figure 3: a) Controlled shear rate: relaxation experiment (profile D = 0.5 s^{-1}, 500 s^{-1}, and 0.5 s^{-1}) of N20 and H18 dispersed in styrene; b) oscillation: deformation sweep at a frequency of 1.6 s^{-1} and a deformation of 0.001 to 10 of N20 and H18 dispersed in styrene; c) determination of the yield point of N20 and H18 dispersed in styrene: data from deformation sweep.

Hydrophilic silica N20 forms an extremely stable and rigid network with a yield point of 84 Pa at 3%wt loading in pure styrene. Both experiments, relaxation and oscillation, reveal an instantaneous break down of the percolating structure upon exceeding the yield point, resulting in a low relative apparent viscosity η_{rel}, and a fast and complete recovering (< 3 s). However, under similar conditions, fully silylated silica exhibits a much weaker network with a yield point of around 0,06 Pa. The network is more elastic according to the deformation sweep experiment but recovers its structure after shear thinning markedly slower (ca. 40 s). The difference can be understood by H-bonding particle-particle interactions in the case of hydrophilic silica, which can recover the network structure almost instantaneously. For fully silylated silica we suggest a combination of hydrophobic interactions (interactions of hydrophobic particles in a hydrophilic environment are related to phase separation phenomena)[9] and entanglement of the grafted PDMS chains. This process of network forming can be seen as a kind of phase separation between the grafted PDMS and the styrene matrix. In this context it is important to know, that styrene is a worse-than θ-solvent for PDMS.[10] However, phase separation between compounds of comparable polarity are inherently slow. The occurrence of a yield point for H18 in styrene is a strong indication for a combination of hydrophobic interactions and chain entanglement. Systems flocculated exclusively by hydrophobic interactions are supposed to show no yield point.[11]

In the presence of resin oligomers the situation is changed distinctly. Mixtures of UP resin Palatal P4/N20/styrene exhibit a much weaker but more flexible network than N20/styrene mixtures indicated by a yield point of about 10 Pa and a slow degradation of the network structure by increasing the deformation in the oscillation experiment (Fig. 4b).

More interestingly this structure degradation occurs in a step-wise manner which can be explained with a polymer bridging of the resin molecules between the fumed silica particles. The relaxation time of the N20 network structure after shearing is dramatically increased in the presence of Palatal P4 (Fig. 4a).

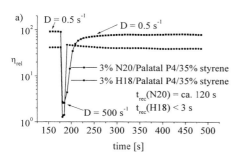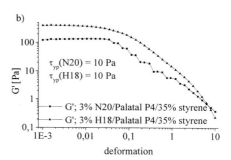

Figure 4. a) Controlled shear rate: relaxation experiment (profile D = 0.5 s⁻¹, 500 s⁻¹, and 0.5 s⁻¹) of N20 and H18 dispersed in Palatal P4/styrene; b) oscillation: deformation sweep at a frequency of 1.6 s⁻¹ and a deformation of 0.001 to 10 of N20 and H18 dispersed in Palatal P4/styrene.

This effect can be explained with a reversible adsorption of resin molecules to the freshly provided surfaces of silica particles produced by shearing down the cluster structure of the N20 network. The re-formation of the network requires at least a partial desorption of the resin molecules which is a slow and time consuming process due to the multi-point interaction of the resin chains with the silica surface.[12]

For H18 we observe an increase of the network stability (yield point 10 Pa) in the oscillation experiment and a decrease of the relaxation time to less than 3 s in the presence of Palatal P4. Both effects can be put down to the fact that the polarity of the mixture is enhanced by the resin oligomers which increases the interaction energy with respect to PDMS-PDMS entanglement and the phase separation between the grafted PDMS layer and the surrounding medium occurs faster. When hydrophilic fumed silica N20 is dispersed in a vinyl ester Atlac 590/styrene mixture the oscillation experiment reveals that the fumed silica is not able to build up a percolating network indicated by the lack of a yield point (Fig. 5b). This is also supported by the observation that the relaxation time of the system N20/Atlac 590/styrene is markedly longer than the time frame of the viscosity relaxation experiment of 500 s (Fig. 5a).

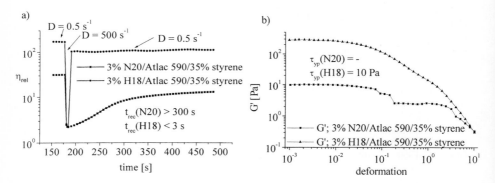

Figure 5. a) Controlled shear rate: relaxation experiment (profile $D = 0.5$ s^{-1}, 500 s^{-1}, and 0.5 s^{-1}) of N20 and H18 dispersed in Atlac 590/styrene; b) oscillation: deformation sweep at a frequency of 1.6 s^{-1} and a deformation of 0.001 to 10 of N20 and H18 dispersed in Atlac 590/styrene.

Vinyl esters resins are characterized by pending OH groups in the chain which are able to form strong H-bonds to the silanol groups of silica particles. Due to the strength of this interaction the adsorption of vinyl ester molecules at silica surfaces should be almost irreversible and result in a kind of steric stabilization of the silica particles. The formation of a silica network is suppressed and the achieved relative viscosities remain low.
In the case of H18/Atlac 590/styrene we observe a behavior in terms of network stability and relaxation which is comparable to the H18/Palatal P4/styrene system. This reveals that the net polarity of the medium is probably a more important driving force for the network stability of fully hydrophobic fumed silica than the specific chemical structure of the single components of the medium.

NMR and IR Spectroscopies
The results of our rheological study demonstrate that interactions between resin molecules and fumed silica particles significantly influence the network stability and its formation. In order to get a deeper understanding of the nature of such interactions a ^{13}C CPMAS study at different cross polarization times τ_{cp} from 0.5 to 8 ms has been performed. At short τ_{cp} spectra intensities are enhanced by ^{13}C resonances of the least mobile chain fragments. This study includes two different fumed silicas, hydrophilic Wacker HDK N20 and fully silylated Wacker HDK H18, with an unsaturated polyester resin Palatal P4 and a vinyl ester resin Atlac 590, respectively. Fig. 6 depicts the ^{13}C CPMAS spectra of N20 and H18, respectively, in Palatal P4/35% styrene at different cross polarization times τ_{cp}.

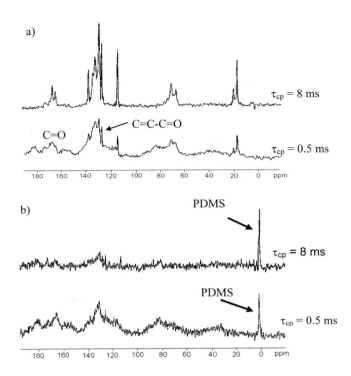

Figure 6. ^{13}C CPMAS spectra at different cross polarization times τ_{cp} of a) N20 in Palatal P4/35% styrene and b) H18 in Palatal P4/35% styrene.

The N20/Palatal P4 ^{13}C CPMAS spectra exhibit an enhanced intensity of the C=O and C=C signals at short τ_{cp} whereas Palatal P4 without added N20 shows no signals in the same experiment. This indicates that specific interactions of carbonyl and C=C-C=O groups of the resin oligomers with the fumed silica surface result in an immobilization of the resin molecules.[8, 13] This finding supports our interpretation of the rheological relaxation experiment, where we suggested that the increased relaxation times in the presence of Palatal P4 are related to the reversible adsorption of the resin oligomers at the silica surface after shear deformation.

The ^{13}C CPMAS spectra of H18 in Palatal P4/35 % styrene show an enhanced intensity of the grafted PDMS chains at short τ_{cp}. In a ^{13}C MAS experiment no signals for the PDMS chains of H18 could be detected. The fact that it was possible to detect the grafted PDMS by ^{13}C CPMAS indicates a strong immobilization of the chains, which is in agreement with our suggestion of H18 silica-silica network formation by a combination of hydrophobic interactions and chain entanglement. Both mechanisms are supposed to immobilize the PDMS chains in the grafted layer.

Further support for a specific interaction between the silanol groups of hydrophilic fumed silica and the C=O groups of resin molecules comes from IR spectroscopy, where additional to the carbonyl band of free Atlac 590 oligomers at 1724 cm^{-1} a second band at 1704 cm^{-1} appears under adsorption at the N20 surface (Fig. 7).

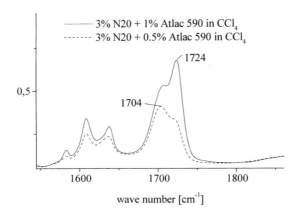

Figure 7. IR spectra of 3% N20 in CCl$_4$ after addition of 0.5 and 1% Atlac 590.

This indicates an interaction of the C=O function with the silica surface and particularly with the silanol groups of the silica.[14] At very low resin concentrations of less than 1.0 wt% the fraction of adsorbed oligomer is approximately 60% of the total resin, as seen from the IR intensities. Upon increasing the amount of resin the fraction of adsorbed resin oligomer remains small and does not exceed 5-8 wt% relative to the silica. The fact that only a small portion of the resin molecules is immobilized is in accordance with our model of thickening by hydrophilic fumed silica based on (a) direct particle-particle interactions, (b) polymer bridging, and (c) steric stabilization.

Quantum chemical calculations
Quantum chemical modeling is a suitable tool to elucidate the microscopic mechanisms of the adsorption processes of polymers on silica surfaces.[15] In the current study quantum chemical modeling has been used to quantitatively compare energies of interactions in the system silica and resin molecules.
For modeling of hydrophilic silica HDK N20 particles a hydroxylated silica cluster [SiO$_2$48-OH9],[15] containing 48 silicon dioxide units and 9 surface silanol groups, has been used; grafting two five-membered dimethylsiloxy (DMS) chains lop-wise (bonded at both ends) on it provided the model for the fully silylated silica HDK H18, silica cluster [SiO$_2$48-OH5-DMS$_5$2]. Two resin models have been simulated representing all typical functional groups of unsaturated polyester resins (UP) and vinyl ester resins (VE): sequence of UP model: methylether of 1,2-propanediol - maleic acid - 1,2-propanediol - orthophatalic acid - methylether of 1,2-propanediol; sequence of VE model: methacrylic acid - glycidine - bisphenol A - glycidine - methacrylic acid.
A special study has been dedicated to the nature of bonds in the system, in particular in the silica-resin system. The interaction energies of different kinds of H-bonds decrease in the order C-O-H···O(H)-Si > C=O···H-O-Si > C-(H)O···H-O-Si, which is summarized in Table 1. The oxygen atom of Si-OH bears a higher negative charge than that of C-OH, but the hydrogen of the latter is slightly more positively charged than in Si-OH; in consequence -C-OH···O(H)-Si is the strongest H-bond, but which occurs only in a VE system. Surprisingly, the carbonyl function also shows a rather high H-bond energy – linking an important interaction energy to carbonyl group immobilization as seen by ^{13}C CPMAS NMR and

carbonyl band red shift as observed by IR. In summary this indicates that a VE-like resin structure shows strong adsorption affinity towards hydrophilic fumed silica.

Additionally to the superior interaction energy of -C-OH···O(H)-Si vs. -C-(H)O···HO-Si steric considerations suggest that end-of-the-chain carbinols in a resin are forming stronger H-bonds with a silica surface than in-chain carbinols; however, this suggestion has to be verified in future studies.

Fig. 8 shows fully optimized structures of adsorption complexes of VE and UP resin models with the [SiO$_2$48-OH9] and [SiO$_2$48-OH5-DMS$_5$2] silica clusters.

a)   b)

c)   d)

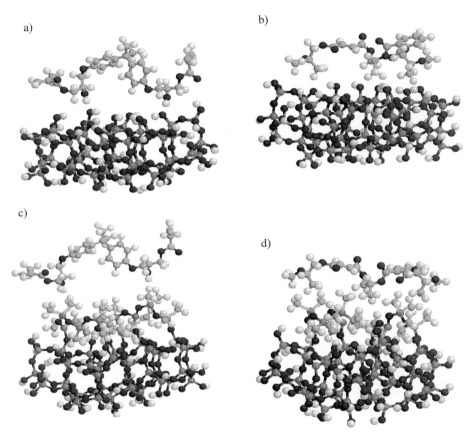

Figure 8. Fully optimized structures of the adsorption complexes of a) hydrophilic silica and vinyl ester [SiO$_2$48-OH9]/VE; b) hydrophilic silica and unsaturated polyester and [SiO$_2$48-OH9]/UP; c) hydrophobic silica and vinyl ester [SiO$_2$48-OH5-DMS$_5$2]/VE; d) hydrophobic silica and unsaturated polyester [SiO$_2$48-OH5-DMS$_5$2]/UP.

Calculated interactions energies of silica-silica, silica-resin, and resin-resin adsorption complexes are given in Fig. 9.

Table 1. Energies of specific interactions: silica model [$SiO_2$18-OH5][15] and VE model: methylether of glycidine - methacrylic acid.

System	energy of interaction [kcal/mol]	Bond distance [nm]
-C-H···O-Si(Si)	- 2.80	0.283
-C-(H)O···HO-Si	- 4.46	0.185
-C=O···HO-Si	- 5.60	0.183
-C-OH···O(H)-Si	- 9.51	0.177

VE system

Hydrophilic silica particles interact strongly with the VE molecule ([$SiO_2$48-OH9]/VE) due to H-bonds of the form -C-OH···O(H)-Si additional to H-bonds of the form C=O···H-O-Si and dispersion interactions – it is the most favorable interaction in all system combinations in terms of energy. As it is even stronger than the silica-silica interaction ([$SiO_2$48-OH9]/[$SiO_2$48-OH9]), strong adsorption is expected, leading to steric stabilization of the colloidal system by hampering direct particle-particle contacts. Rheologically, we would interpret this as low thickening efficiency – in deed, experiments show that N20 is not a stable thickener for VE resin systems. and resin molecule. The contact areas a = 0.82 nm^2 are identical (2-4 % deviation) for resin-resin and resin-silica interaction for both UP and VE, and hydrophilic and silylated silica, and therefore interaction energies are taken as directly comparable. Silica-silica contact areas are larger both by a factor of 3.2, the interaction energies are normalized therefore by the ratio of the contact areas.

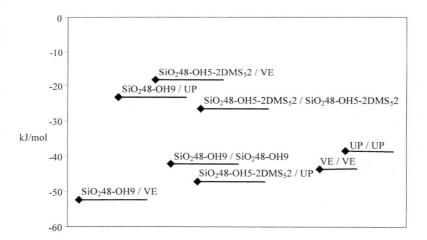

Figure 9. Calculated energies of intermolecular interactions in the system silica particles

The weakest interaction in the system is that of silylated silica with a VE molecule ([$SiO_2$48-DMS$_5$2]/VE). It is even weaker than the interaction of silylated silica with itself ([$SiO_2$48-DMS$_5$2]/[$SiO_2$48-DMS$_5$2]). The medium interaction energies of the latter surprises, as only nonspecific components are involved – however modeling makes evident a particle-to-particle entanglement of the 5-membered-DMS chains on the silica clusters: In fact, the additional immobilization of PDMS chains on H18 as seen by ^{13}C CPMAS NMR seems to be correlated with energies of interaction. The strongest complexes in the system VE and silylated silica are those of the VE resin molecules with each other. In consequence, the interface of highly silylated silica towards a VE phase is energetically less favorable: In order to optimize VE-VE contacts the silylated silica particle surfaces separate from the system by direct silylated silica-silylated silica particle contacts – a phenomenon which is well-know as 'hydrophobic interaction'.[9] However, real systems contain monostyrene, which is not or only very weakly adsorbed on a silica surface according to calculations and spectroscopic experiments. Hence, styrene is enhancing the effect of silylated silica separation from the VE-styrene phase. In fact, HDK H18 is an excellent and stable thickener and rheology control additive for all VE and epoxy resin-like systems.

UP system

The complex UP on hydrophilic silica is rather weak, but the interaction of two hydrophilic silica cluster is rather strong, even in competition to the resin-resin interaction, which are of comparable strength or weaker. Additionally, not depicted in Fig. 9, modeling shows that styrene is not adsorbed at all on hydrophilic silica. In consequence, following the interpretations given above, hydrophilic silica HDK N20 is an excellent and powerful thickener and thixotropic agent for UP resins, as strong particle-particle interactions occur. However, spectroscopic and rheological results indicate an adsorption of UP oligomers to hydrophilic silica surfaces. In accordance with our calculations this adsorption is only weak and reversible and therefore influences only the rate of gelation but not the network stability.

Surprisingly, the adsorption complex of UP molecule on the silylated silica cluster shows a distinctly higher interaction energy. Following the interpretations given above, this should result in wetting, adsorption, steric stabilization, and a weak colloidal network. However, as shown by the rheological data given above, HDK H18 is an excellent thickener for UP resins, too. It seems that in a system of real UP resins, the solvent styrene plays a key role: Being a bad or worse-than-θ-solvent for (the) PDMS (layer on silica) styrene is the driving force of hydrophobic interactions – in the case of silylated silica in UP resins the adsorption of UP molecules on the silylated silica surface seems not to hinder hydrophobic interaction: A reasonable assumption as the latter is not particle-particle contact but phase separation driven.

Conclusion

In order to develop a comprehensive model for the thickening behavior of different grades of fumed silicas advanced rheological and spectroscopic experiments as well as quantum chemical calculations have been performed. The results indicate that fundamentally different mechanism are responsible for the formation of colloidal networks of hydrophilic and fully silylated silica.

Hydrophilic silica forms a strong and rigid network by strong short-range H-bonds between silica particles. Addition of UP resin reduces the network stability but increases network flexibility and relaxation time after shearing. VE resins hamper the formation of a percolating network due to irreversible adsorption of OH end groups on the silica surface.

Fully silylated silica exhibits hydrophobic interactions between the PDMS covered surface and the solvent as well as entanglement of the PDMS layers. Addition of UP or VE resins increases the polarity of the medium. As a result, the mismatch between the matrix and the PDMS layer is increased which favors PDMS-PDMS interactions and therefore enhances the network stability and decreases the network relaxation time.

For both grades of silica the formation and strength of their percolating networks in terms of both particle-particle and particle-polymer-particle interactions is strongly influenced by the polarity of the surrounding medium and the nature of the resins. In particular, adsorption processes have been identified to play an important role for the rheological performance of hydrophilic fumed silica. From these results recommendations for the application of fumed silica in different types of resins have been derived, which are summarized in Table 2.

Table 2. Recommendations for the application of fumed silica in different resin types depending on their polarity and the potential to form strong specific interactions.

Resin oligomer	Polarity and type of interaction	Rheological additives
• Alkyd resins, • Saturated polyester resins	non-polar; no strong specific interactions	• hydrophilic fumed silica HDK N20
• unsaturated polyester resins	non-polar to medium polarity; no strong specific interactions	• hydrophilic fumed silica HDK N20 • fully silylated fumed silica HDK H18
• vinyl ester resins • melamine-polyesters	medium to highly polar; strong specific interactions	• fully silylated fumed silica HDK H18
• epoxy resins • polyurethane system • acrylate systems	polar; strong specific interactions	• fully silylated fumed silica HDK H18

References

[1] F. Achenbach, H. Barthel, H. Maginot, *Proc. Int. Symp. on Mineral and Organic Functional Fillers in Polymers (MOFFIS 93)* **1993**, 301.
[2] G. D. Ulrich, *Chem. Eng. News* **1984**, *62*, 22.
[3] H. Barthel, L. Roesch, J. Weis, *Vol. Organosilicon Chem. II,* (Ed.: J. Weis), VCH Weinheim, **1996**, pp. 761.
[4] H. Barthel, *Colloids Surf., A: Physicochemical and Engineering Aspects* **1995**, *101*, 217.
[5] D. Quemada, *Prog. Colloid Polym. Sci.* **1989**, *79*, 112.
[6] W. B. Russel, *J. Rheol.* **1980**, *24*, 287.
[7] T. F. Tadros, *Chem. Ind.* **1985**, *7*, 210.
[8] D. G. Cory, W. M. Ritchey, *Macromolecules* **1989**, *22*, 1611.
[9] J. N. Israelachvili, R. M. Pashley, *J. Colloid Interface Sci.* **1984**, *98*, 500.
[10] D. W. Van Krevelen, in *Properties of Polymers*, 3. ed. (Ed.: D. W. Van Krevelen), Elsevier, Amsterdam, Oxford, New York, Tokyo, **1990**, pp. 774.
[11] E. Killmann, J. Eisenlauer, *Eff. Polym. Dispersion Prop., [Proc. Int. Symp.]* **1981**, *66*, 36.
[12] Y. Otsubo, *Adv. Colloid Interface Sci* **1994**, *53*, 1.
[13] V. M. Litvinov, A. W. M. Braam, A. F. M. J. van der Ploeg, *Macromolecules* **2001**, *34*, 489.
[14] G. R. Joppien, K. Hamann, *J. Oil Colour Chem. Assoc.* **1977**, *60*, 412.
[15] E. Nikitina, V. Khavryutchenko, E. Sheka, H. Barthel, J. Weis, in *Organosilicon Chem. IV* (Ed.: J. Weis), Wiley-VCH Verlag GmbH: Weinheim, **2000**, pp. 745.

Macromol. Symp. **187**, 585–596 (2002)

A Model for the Drying Process During Film Formation in Waterborne Acrylic Coatings

Stefano Carrà[1]*, Deborah Pinoci[2], Sergio Carrà[2]

[1]Mapei S.p.A., Via Cafiero, 22, 20129 Milano (Italy)

[2]Dipartimento di Chimica Fisica Applicata, Politecnico di Milano, Via Mancinelli 7, 20131, Milano (Italy)

Summary: Establishing drying mechanisms during film formation in waterborne acrylic coatings is a technologically important problem, however complex, and still poorly understood. A model for the prediction of evaporation kinetics is proposed in this paper, where films are supposed to dry normally with respect to the film surface, and a drying front separates a top dry region from a bottom wet region. The model accounts for the competition between water evaporation and particle diffusion that determines the degree of vertical homogeneity, but also for the competition between water evaporation and particle deformation that ultimately establishes the rate-determining step in film formation processes. The model was validated by performing gravimetric water-loss experiments on latexes of acrylic polymers of various composition, various particle size and stabilizing systems, under different environmental temperatures and humidity, and various initial film thicknesses in order to evaluate the effect of the different factors that can in principle influence the film formation process.

Keywords: film formation; coatings; evaporation; acrylic latexes; mathematical modelling

Introduction

Establishing the mechanisms that govern drying during film formation is extremely important from a technological point of view for almost every coating application, but it is also a surprisingly complex and poorly understood problem, even if it has been frequently tackled in recent literature[1]. Film formation is generally described as a sequence of three steps: a stable colloidal dispersion is applied to a substrate; as water evaporates, polymer particles become more and more concentrated until they reach maximum packing (step I); when the forces accompanying drying exceed the viscoelastic resistance of particles, these start to deform to yield a mechanically

CCC 1022-1360/00/$ 17.50+.50/0

weak void-free film, where the original particles are still distinguishable (step II); finally reptation across interparticle boundaries renders particles undifferentiated and generates the entanglements that provide mechanical strength to the film. Water evaporation during film formation is more complicated than what it looks intuitively. The most fundamental variable related to evaporation mechanisms is the rate of evaporation itself, around which disagreement and controversy still linger in literature: Vandezande and Rudin[2] registered constant rates until particle concentrations around 93%, while Vanderhoff [3] observed decreasing rates at much lower concentrations and final rates consistent with diffusion of water vapor through a continuous polymer layer. As pointed out by Winnik[4], in order for a continuous layer to be able to restrain evaporation, polymer particles must be easily deformable. A brutal simplification in the description of drying phenomena would be to consider it homogeneous, meaning that water concentration remains uniform during the whole process. Dilute dispersions usually start drying homogeneously, but at a certain concentration dishomogeneities usually appear, as a consequence of several possible factors, among which the most significant are probably the presence of diffusional limitations that create particle concentration profiles from the beginning of the process, but also the possibility that, once maximum packing has been reached, the level of water might recede in the film, leaving a dry film layer on top, resting on a wet layer[5]. Much interest has also been drawn by the problem of drying fronts that can move normally to the film surface, but also propagate laterally[6]. In these conditions, and depending on the deformability of particles, the top of the film might reach a volume fraction of unity (no voids) before the rest of the film, sometimes before the rest of the film has even reached maximum packing. A useful criterion to understand the deformation behavior was proposed by Routh et al.[7], by comparing the evaporation time, nominally $H/\dot{E}$, where H is the initial film thickness and $\dot{E}$ the rate of evaporation, with the time scale for deformation $\tau = \eta_0 R_0 / \gamma_{wa}$, where η_0 is the viscosity of a melt of the latex polymer, R_0 is the particle radius and γ_{pw} the water-air interfacial tension. Temperature, or rather the difference $T - T_g$, where T_g is the gel transition temperature of the polymer, influences greatly η_0, and hence the overall deformation behavior.

There is a great applicative scope for the development of a model capable of describing the drying of water throughout the film formation process, at different environmental temperature and humidity, different coating thicknesses, and various latex properties (T_g, particle size distribution, stabilization type, eventually particle morphology), like the one proposed in this paper. Model results have been validated by the comparison with a series of experimental results.

Experimental

In this work the evaporation kinetics in film formation processes was investigated using several different acrylic latexes. These latexes were both laboratory prototypes and industrial products, with different monomeric composition and solid content, and various stabilizer types and amounts. The T_g of the different polymers was measured by Differential Scanning Calorimetry, on a Mettler Toledo TC 15 Thermoanalyser. Latex solid content was determined by gravimetry. Latex particle size distribution was estimated via light scattering using a LS-230 Coulter Particle Analyzer. Latex viscosity was determined using a Haake Viscotester VT5. Evaporation kinetics during film formation was determined adopting a gravimetric technique, registering on a scale connected on-line to a computer the loss of weight of a latex sample while this is drying on a glass plate. Borders were built on the glass plate by applying one or more layers of adhesive tape, in order to control the film thickness. Data were registered every 10 s. The control of the velocity of the air flow onto the sample was guaranteed by assembling a small wind tunnel where a series of small tubes convey air from a fan. It is important to check, using an anemometer, that all experiments are performed at a constant air velocity ($\cong 4$ m/s in the experiments described in this paper). The apparatus was located inside a climatic chamber where it was possible to vary environmental temperature and humidity, within certain limits (10 ÷ 40 °C, 30 ÷ 70% r.h.).

Results

Raw experimental data are given in terms of sample weight *vs.* time. Rates of evaporation can be calculated by numerical derivation. The solid content ϕ of the sample at every instant during the experiment can be calculated through a water mass balance, using the final solid content as a reference point. Experimental data are better compared in terms of the adimensional variable τ, defined as the time required to evaporate a certain amount of water from the latex divided by the time required to evaporate the same amount of water as pure water. Figure 1 shows a series of experimental τ vs. ϕ curves as obtained by drying an acrylic latex with a T_g of 20°C, with different environmental temperature, above and below the T_g. It is possible to observe that environmental temperature does not have any influence on the drying process until very high values of ϕ are reached. In any case the beginning of decreasing rate period is anticipated when drying is performed at high temperature, probably due to the stronger influence of diffusional limitations at higher rates of water evaporation. Evaporation hindrance increases when the temperature is set above the polymer T_g, due to the increased deformability of particles.

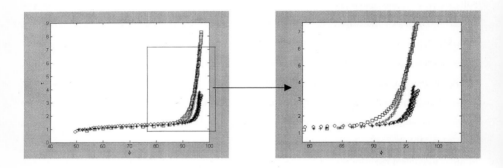

Figure 1. Experimental τ vs . ϕ curves (complete curved on the l.h.s, zoomed in on the r.h.s.) as obtained by drying an acrylic latex with a T_g of 20°C, with different environmental temperature (o: 10 °C, *: 20 °C, +: 30 °C, □: 40 °C).

Figure 2 shows experimental curves as obtained by drying an acrylic latex with a Tg of –39°C, by varying initial thickness. Results show that evaporation hindrance increases with initial thickness as an effect of diffusional limitations, and that the film with initial thickness equal to 1.5 requires a really long time to dry.

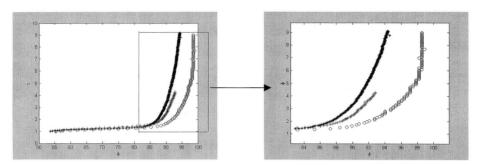

Figure 2. Experimental τ vs . ϕ curves (complete curved on the l.h.s, zoomed in on the r.h.s.) as obtained by drying an acrylic latex with a T_g of -39°C, with different initial thickness (o: 0.5 mm, + : 1 mm, *: 1.5 mm).

Figure 3 shows experimental curves as obtained by drying two different acrylic latexes with the same monomeric composition and a T_g of -39°C but different stabilizing system one prevalently anionic and the other prevalently non-ionic, where it is possible to notice that the former shows a lower evaporation hindrance, probably due to the fact that the electrostatic stabilization potential produced by the anionic emulsifier is more efficient in hindering and delaying particle deformation.

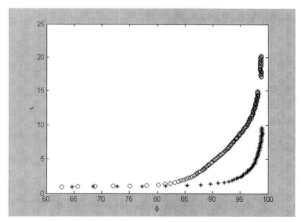

Figura 3. Experimental τ vs . ϕ curves as obtained by drying two different acrylic latexes with the same monomeric composition and a T_g of -39°C but different stabilizing system (*: prevalently anionic; °: prevalently non-ionic).

Figure 4 shows experimental curves as obtained by drying, at 20 °C, 50% of relative humidity and 1 mm of initial thickness, three different acrylic latexes. The latex at lower T_g displays the maximum evaporation hindrance, while the other two, that have similar T_g, almost tend to overlap, even if initial concentrations are very different.

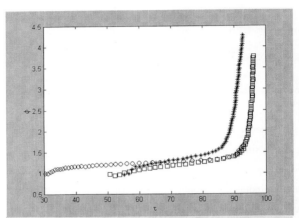

Figure 4. Experimental τ vs . ϕ curves as obtained by drying three different acrylic latexes (o: $T_g = 19.3$; o: $T_g = 20.0$; *; $T_g = -39.0$).

Mathematical Model

Effect of Diffusion Limitations

By developing the equations derived by Russel[8] to describe the diffusion we derived the following differential equation to describe vertical dishomogeneities during film formation:

$$\frac{\partial \phi}{\partial t} = D \frac{\partial^2 \phi}{\partial z^2} \tag{1}$$

relying on the hypothesis the sedimentation is neglectable.

Partial differential equation (1) can be solved using the method of the separation of variables, that, imposing the initial condition:

$$\phi(z,0) = \phi_0 \tag{2}$$

leads to:

$$\phi(z,t) = \phi_0 e^{\alpha z} e^{\alpha^2 D t} \tag{3}$$

Where α is the parameter introduced for the solution of the differential equation.

It is necessary to impose the condition that ϕ cannot surpass maximum packing fraction ϕ_{max}, i.e. 0.73 for monodisperse latexes, therefore:

$$\phi(z,t) = \phi_0 e^{\alpha z} e^{\alpha^2 Dt} \qquad \text{for } z < z*$$

$$\phi(z,t) = \phi_{max} \qquad \text{for } z \geq z* \tag{4}$$

where $z*$ is the coordinate at which the calculated ϕ is equal to ϕ_{max}.

The position of the film-air interface is defined by the variable h, that can be easily calculated by a water mass-balance:

$$-\frac{dh}{dt} = V_{ev}(t) \tag{5a}$$

with the initial condition:

$$h = h_0 \quad \text{at } t = 0 \tag{5b}$$

where V_{ev} is the rate of water evaporation, in m/s.

The mass balance for solid particles would satisfied by equating the average value of ϕ, as defined in equation (4):

$$\overline{\phi} = \frac{1}{h} \left\{ \frac{\phi^0 e^{\alpha^2 Dt}}{\alpha} \left[e^{\alpha z*} - 1 + 0.75(h - z*) \right] \right\} \tag{6}$$

with

$$\overline{\phi} = \frac{h^0 \phi^0}{h(t)} \tag{7}$$

By imposing this condition, we can calculate α. An example of calculation from the model developed above is given in Figure 5, where $\phi(z)$ profiles are shown at increasing values of t, for a set of parameters, $D = 1 \times 10^{-16}$ [m²/s], $h = 1 \times 10^{-3}$ [m], $U = 5 \times 10^{-8}$ [m/s], that corresponds to a value of the adimensional variable $HU/D = 2500 \gg 1$, that should lead to strong dishomogeneities in the film. Profiles are in fact very steep, and maximum packing at the film-air interface is reached when $\overline{\phi}$ is still in the order of 0.5.

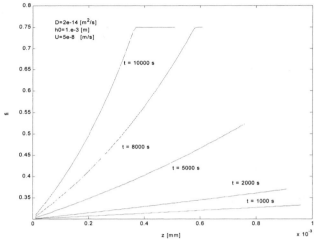

Figure 5. $\phi(z)$ profiles at increasing values of t, for a set of parameters, $D = 1\times10^{-10}$ [m^2/s]; $h_0 = 1\times10^{-3}$ [m]; $U = 5\times10^{-8}$ [m/s].

Effect of particle packing and deformation on evaporation kinetics

Evaporation is supposed to take place at a constant rate, approximatively equal to the rate evaporation of pure water, as long as the film-air interface is below maximum packing. Expressed in moles of water per unit area and time:

$$\widetilde{N} = k_g \left[P^0(T)\gamma - P_A \right] \tag{8}$$

where k_g is the mass transfer coefficient [moles/(s cm^2 atm)], $P^0(T)$ is the water vapor pressure at temperature T, P_A is the water partial pressure in air, γ is an activity coefficient to take into account possible effects of water soluble polymer, eventually present, onto vapor pressure.

At maximum packing, particles undergo deformation driven by capillary pressure, opposed by their elastic resistance. Described as a purely elastic phenomenon, deformation is supposed to be instantaneous. This is not the case if, more correctly, their viscoelastic character is taken into account.

At a certain point during the film formation process, as dictated by a water mass balance, the water-air interface starts receding inside the film, leaving a layer of dried deformed particles above. Even when in a dry state particles keep on deforming by dry sintering, driven by the polymer-air interfacial tension, and opposed by the viscous characteristic of the polymer. The situation at an instant t is represented in Figure 6:

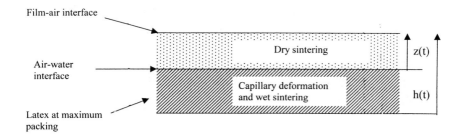

Figure 6. Schematic of inhomogeneous drying

The state of packing and deformation at a height z in the film can be defined by a variable $\varepsilon(z,\tau)$, that is essentially the degree of void, where τ is the time, relative to the instant at which the layer of particle at level z has reached maximum packing. Eckersley and Rudin[9] proposed a model where the contact radius of particles during stage II is the sum of contribution from the capillary and interfacial force.

$$a = a_{capillary} + a_{int\,erfacial} = \left(2.8R^2\sigma/G*\right)^{1/3} + \left(3\gamma Rt/2\pi\eta*\right) \qquad (9)$$

Contact radius and degree of void are related by a geometric relationship:

$$\varepsilon = 1 - \frac{4 + 2z^2 - 0.083z\left(3\xi^2 + 0.25z^2\right)}{1.333 + 4\left(1 - 0.5z\right)} \qquad (10)$$

where $\left(a/R\right) = \xi$ and $z = 2\left(1 - \sqrt{1-\xi^2}\right)$.

Therefore the instantaneous degree of void can be expressed as a sum, on an initial contribution that is complementary to the situation of maximum packing that identifies the condition at which deformation begins, of an instantaneous contribution due to capillary deformation, ε_c, and a time dependent contribution due to dry sintering:

$$\varepsilon(\varsigma,\tau)=\varepsilon_0+\varepsilon_c+\varepsilon(\tau) \tag{11}$$

Where $\varepsilon_0=1-\phi_{max}$. The dry film layer is therefore characterized by a profile of $\varepsilon(z)$, and forms an obstacle to the evaporation of water from the air-water interface below, that can be expressed in terms of a pressure drop ΔP. The rate of evaporation of water is therefore:

$$N=k_g\left(P^0\gamma-P_A-\Delta P\right) \tag{12}$$

The pressure drop in the dry layer can be described as it was in a porous bed:

$$\frac{dP}{dz}=-\frac{1}{K\psi[\varepsilon(t)]}u^{(g)} \tag{13}$$

where $u^{(g)}$ is the gas velocity $u^{(g)}=N/\widetilde{\rho}\varepsilon(t)$. Parameter K is defined as d_p^2/μ, d_p being the particle diameter and μ the latex viscosity. Several different functions have been proposed in the literature for function Ψ. We applied the classic Ergun's equation, according to which:

$$\Psi[\varepsilon(t)]=\frac{\varepsilon^3}{180(1-\varepsilon)} \tag{14}$$

Defining as $Z(t)$, the thickness of the dry state, it is:

$$dZ=u^{(l)}dt \tag{15}$$

where $u^{(l)}$ is the liquid velocity. Continuity equation states that:

$$u^{(l)}\varepsilon^0\widetilde{\rho}^{(l)}\Omega=u^{(g)}\varepsilon(t)\widetilde{\rho}^{(g)}\Omega \tag{16}$$

By integrating (11), and using (13) and (14), we derive:

$$\Delta P=-\int_0^Z\frac{u^{(g)}}{K\Psi[\varepsilon(t)]}dz=\frac{\widetilde{\rho}^{(g)}}{\widetilde{\rho}^{(l)}\varepsilon^0}\int_0^\tau\frac{u^{(g)^2}\varepsilon(t)}{K\Psi[\varepsilon(t)]}dt \tag{17}$$

Evaporation kinetics in the presence of a dry upper state is then described by the following equations:

$$u^{(g)}(\vartheta)=\frac{k_g}{\widetilde{\rho}^{(g)}}\frac{1}{\varepsilon(\vartheta)}\left[\left(P^0\gamma-P_A\right)-\frac{\widetilde{\rho}^{(g)}}{K\widetilde{\rho}^{(l)}\varepsilon^0}\int_0^\vartheta\frac{u^{(g)^2}\varepsilon(\tau)}{\Psi[\varepsilon(\tau)]}d\tau\right] \tag{18}$$

$$\frac{dZ}{dt}=\frac{k_g}{\widetilde{\rho}^{(l)}}\frac{1}{\varepsilon^0}\left[\left(P^0\gamma-P_A\right)-\frac{\widetilde{\rho}^{(g)}}{K\widetilde{\rho}^{(l)}\varepsilon^0}\int_0^t\frac{u^{(g)^2}(\vartheta)\varepsilon(\vartheta)}{\Psi[\varepsilon(\vartheta)]}d\vartheta\right] \tag{19}$$

Simulations and comparisons with experimental data

Two examples of comparisons between simulations and experimental data are given in the two Figure 7 and 8. In Figure 7 rate vs. time curves are shown for a acrylic latex latexes with a T_g of −39 °C and a 58% solid content at two different initial film thickness. In Figure 8 rate vs. time curves are shown for a latex of a polyvinyl acetate butyl acrylate copolymer with a Tg of 20°C and a 55% solid content at three different initial film thickness.

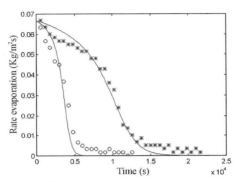

Figure 7. Rate vs. time curves for a acrylic latex latexes with a T_g of −39 °C and a 58% solid content at two different initial film thickness (o: 0.5 mm, *: 1mm).

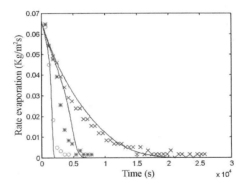

Figure 8. Rate vs. time curves are shown for a latex of a polyvinyl acetate butyl acrylate copolymer with a T_g of 20°C and a 55% solid content at three different initial film thickness (o: 0.2 mm; *: 0.5 mm; x: 1 mm)

The agreement between model predictions and experimental data is striking, but it has been obtained optimising somehow the values of G^* and η^*, in the range of values obtained by DMA, since the shortcomings of our model are somehow similar to those of Eckerling's and Rudin's[9], and the time scales of the deformation and flow processes are not exactly known. However our model takes explicitly into account the competition of particle deformation and water evaporation, and therefore it is able to discriminate if film forms under the effect of dry rather than dry mechanisms, and if a dry G^* should be used, rather then wet (wet values of G^* are lower, due to plasticization).

Conclusions

Establishing drying kinetics during film formation is a problem that is important as it is difficult. Results from a series of gravimetric experiments have been shown, demonstrating the effect on drying kinetics of environmental consitions and latex properties. A model to predict the rate of evaporation of water in a system where water evaporated normally to the film surface was proposed in this paper, showing excellent agreement with experimental results, but the estimation of some model parameters is critical, and should be further addressed in order to obtain a fully predictive model.

References

[1] Y. Holl, J. L. Keddie, P. J. Mc Donald, W. A. Winnik, in *"Film Formation of Coatings"*; T. Provder, M. W. Urban, Eds., ACS Symposium Series; American Chemical Society: Washington DC, Vol. 790, 2001, p. 2

[2] G. A. Vandezande, A. Rudin, *J. Coat..Tech.* **1996**, *68*, 63

[3] J. W. Vanderhoff, H.L. Tarkowski, M.C. Jenkins, E.B. Bradford, *J.Macrom. Chem.*, **1966**, *1*, 361

[4] M. A. Winnik, *Emulsion Polymerization and Emulsion Polymers*, Editors: P. A. Lovell, M. S. El-Aasser, John Wiley and Sons, 467 (1997)

[5] A.F. Routh, W.B. Russel, *AIChE J.*, **1998**, *44*, 2088

[6] M. A. Winnik, J. Feng, *J. Coat..Tech.*, **1996**, *68*, 39

[7] A.F. Routh, W.B. Russel, J. Tang, M. S. El-Aasser, , *J. Coat..Tech*, **2001**, *73*, 41

[8] W. B. Russel, D. A. Saville, W. R. Schowalter, *"Colloidal Dispersions"*, Cambridge University Press, 1992

[9] S.T. Eckersley, A. Rudin, in *"Film Formation in Waterborne Coatings"*, T. Provder, M. A. Winnik, M. W. Urban, Eds., ACS Symposium Series; American Chemical Society: Washington DC, Vol. 648, 1996, p. 2

Tailor-Made UV-Curable Powder Clear Coatings for Metal Applications

Dr. Andreas Wenning

Degussa AG, Coatings & Colorants, D-45764 Marl, Germany

E-mail: andreas.wenning@degussa.com

Summary: UV-curable powder coatings combine most of the benefits of conventional powder coatings together with the advantages of radiation-curable liquid coatings. This new coating process is not only environmentally friendly. It can also be used to coat substrates like wood, plastic, glass or metal at low temperatures within a short curing time. Several coatings based on binders like urethane vinylethers/unsaturated polyesters or methacrylates are developed for metals, wood or pvc flooring. This paper describes urethane acrylates as a new resin system for UV-curable powder clear coatings. The binders can be amorphous or (semi)crystalline. By combining both types it is possible to get tailor-made binders which exhibit an unique coating performance of excellent adhesion, good flexibility and high hardness on metal substrates. The appearance of the clear coatings can be varied from high to low translucent. In addition, the coatings show a good weatherability.

Introduction

Conventional thermosetting powder coatings are well-established since more than 30 years for many interior and exterior applications. They are accepted because of their outstanding balance between energy savings, no solvent emission and high film performance. Liquid UV coatings run with high line speeds because of the fast crosslinking reaction of the binder. However, they contain reactive diluents which are irritating and smell badly. Furthermore, these coatings have adhesion problems due to their high shrinkage during the crosslinking.

It seems reasonable to assume that a combination of both technologies, generating UV-curable powder coatings, should bring benefits. In deed, UV powders don't need irritating monomers, have low shrinkage during curing and can coat temperature sensitive substrates like thermoplastics, wood, paper and pre-assembled metals or alloys. In contrast to thermosetting powder coatings the melting and curing processes are separated from each other (see figure 1). So, very smooth films can be achieved. The performance of such films is good to excellent.

CCC 1022-1360/00/$ 17.50+.50/0

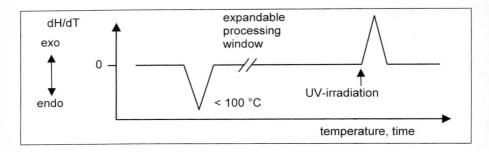

Figure 1. Melting and curing process of UV-curable powder coatings (DSC)

Therefore, UV-curable powder coatings will play an important role in the very near future of the coatings industry.

UV powder coatings are commercially available for several years. The binders used for these coatings differ chemically from the system introduced here.

Solid epoxy resins based on bisphenol-A are cured cationically. A photoinitiator generates, upon irradiation with UV light, a strong acid resulting in cationic polymerization of the epoxy groups [1]. This kind of polymerization exhibit low shrinkage, no inhibition by oxygen and can be post-cured thermally. Alcohols can added as chain extenders to influence mechanical properties and cure characteristics of the coatings. Lower polymerization speed and inhibition of the curing reaction by water are disadvantages compared to radically cured films.

Unsaturated polyesters combined with urethane acrylics or urethane vinylethers as hardeners have been designed recently [2,3]. These UV-curable coatings are cured by free radical polymerization.

Acrylated and/or methacrylated polyesters as special amorphous or (semi)crystalline binders are developed for the coating of metal, wood or pvc substrates [4]. The films exhibit excellent surface properties like chemical and scratch resistance. By using binder mixtures of amorphous and (semi)crystalline resins the flexibility of the cured films can be increased. However, the hardness of the coating doesn't remain on a high level.

New binders are needed which gives flexible and hard films. Urethane acrylates are a new class of binders for UV-curable powder coatings. They are cured by a free radical polymerization process.

New resins for UV-curable powder coatings

The performance of one amorphous and three (semi)crystalline urethane acrylates (UA) was studied (see table 1).

Table 1. Physical properties of urethane acrylate binders

Binder	Melting Point (°C)	Tg (°C)	Viscosity (Pa·s, 120°C)
Amorphous UA 1	80-82	49	270
Crystalline UA 1	77	/	1.3
Crystalline UA 2	68	/	1.8
Crystalline UA 3	110	/	1.0

Tg and viscosity can be adjusted by mixing the amorphous resin with a (semi)crystalline resin. The higher the amount of (semi)crystalline urethane acrylate in the mixture the lower the values of Tg and melting viscosity.

These binders are formulated to powder clear coatings (see table 2).

Table 2. Standard Formulation for UV powder clear coatings

Binder 1	Binder 2	Photoinitiator	Degassing agent	Flow agent
827 g Amorph. UA	146 g Cryst. UA	10 g AHK[1]	10 g	7 g
778 g Amorph. UA	195 g Cryst. UA	10 g AHK*	10 g	7 g

[1]AHK: α-hydroxyketone

The ratio of amorphous to (semi)crystalline resin is 85 : 15 or 80 :20, respectively.

Curing conditions

The clear coatings have been applied on steel substrates with a layer thickness of 80 ± 20 μm. By melting the powder coatings with infra red light within 100 s the temperature of the films raised to 125 °C maximum (see figure 2). After a cooling phase of about 60 s down to 95 °C the films have been cured with UV-light. We used a mercury vapour lamp of 120 W/cm. The UV-dose of 3000 mJ/cm^2, measured with the Power Puck, consists of the sum of UV-A to UV-V radiation regions.

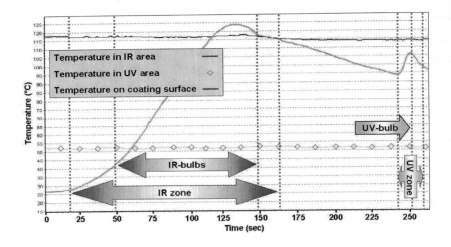

Figure 2.Typical heating and curing cycle of UV powder coatings on metals

Coating properties

UV powder coatings based on amorphous urethane acrylate resins have a high hardness (see table 3). But flexibility and smoothness of the coatings are poor. (Semi)crystalline urethane acrylate resins give very flexible but soft films.

Table 3. Film properties of UV powder clear coatings

Physical test methods	Amorphous UA	Crystalline UA
Pendulum hardness (s)	224	87
Cupping (mm)	2.0	> 10.0
Impact (d/i) (inch·lb)	40 / < 10	> 80 / > 80
Smoothness (1-10)	6	/

Mixtures of amorphous and (semi)crystalline were tested to overcome the insufficient coating performance. The mechanical properties could be adjusted depending on the structure and amount of the (semi)crystalline resin (see table 4). The smoothness got perfect. Although a soft (semi)crystalline resin was used the hardness values stayed on a very good level.

Looking deeper at the clear coatings based on mixtures of amorphous and (semi)crystalline urethane acrylate resins a different light transmittance behaviour was seen (see figure 3). Using the (semi)crystalline resin 1 clear coatings with a high transparency were obtained. The (semi)crystalline resin 2 decreased the transmission values of the clear coating. Clears with the (semi)crystalline urethane acrylate resin 3 absorped the visible light nearly completely.

So, by using the right (semi)crystalline resin we are able to produce clear coatings with a high to very low transparency.

Table 4. Coating performance of UV powder clear coatings based on resin mixtures

Binder mixture	Pendulum hardness (s)	Cupping (mm)	Impact (d/i) (inch·lb)	Smoothness (1-10)
85 % amorph. UA/ 15% cryst. UA 1	193	> 10.0	40 / < 10	9
80 % amorph. UA/ 20% cryst. UA 1	183	> 10.0	50 / 20	9 – 10
85 % amorph. UA/ 15% cryst. UA 2	203	> 10.0	70 / 40	9
80 % amorph. UA/ 20% cryst. UA 2	192	> 10.0	> 80 / > 80	9 – 10
85 % amorph. UA/ 15% cryst. UA 3	214	> 10.0	> 80 / 70	8
80 % amorph. UA/ 20% cryst. UA 3	204	> 10.0	> 80 / > 80	8

The translucency of the clear coatings was influenced by enhancing the molecular weight of the (semi)crystalline urethane acrylate resins (see figure 4). The light transmittance of the semi-transparent clear coatings based on the (semi)crystalline resin 3 was improved significantly. The UV powder clear coating with the high transparency got higher transmission values in the UV-V region. Even this little improvement striked one immediately.

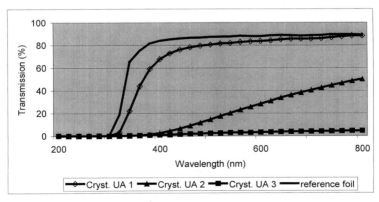

Figure 3. Transparency of UV powder clear coatings

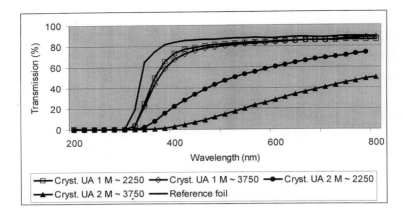

Figure 4. Increasing the transparency in UV powder clear coatings

The UV-curable powder coatings based on urethane acrylate resins showed a good weathering performance. The gloss retention in the weather-O-meter (WOM) was better than for urethane vinylether/unsaturated polyester UV powder coatings (see figure 5).

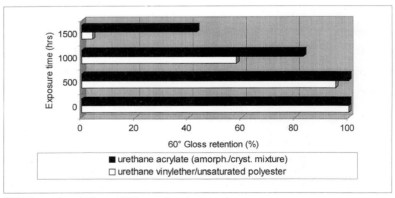

Figure 5. WOM weatherability of UV powder clear coatings

The QUV A resistance was excellent (see figure 6). There was nearly no drop in the gloss after 2800 h and an improvement in the yellowing behaviour comparing to the urethane vinylether/unsaturated polyester system.

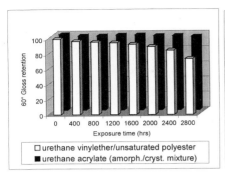 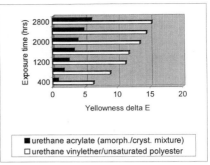

Figure 6. QUV A resistance of UV powder clear coatings

Conclusions

Urethane acrylate resins are suitable binders for UV-curable powder clear coatings. Perfect mechanical properties on metal substrates can be achieved by mixing amorphous and (semi)crystalline resins. The clear coatings with a good outdoor stability can be adjusted from high to low transparent.

The building-block principle enables customers to make tailor-made UV-curable powder coatings for their applications.

Acknowledgement

The author would like to thank E. Spyrou and G. Franzmann for their contributions to this work.

References

[1]   M. Reisinger, „UV-curable powder coatings for temperature sensitive substrates", RadTech Europe 99, Berlin, 628-633; Patent EP 0 667 381.
[2]   D. Fink, G. Brindöpke, „UV curing powder coatings for heatsensitive substrates", European Coatings Journal 9, 1995, 606; Patent EP 0702 040.
[3]   F.M. Witte „Radiation curable powder coatings", European Coatings Journal, 3, 1996, 115; Patent EP 0 636 669; Patent WO 99/14254.
[4]   D. Hammerton, K. Buysens, Y. Souris, „New UV powder systems for metal, wood and pvc", RadTech North America 2002, Indianapolis, 592-600; Patent EP 0 739 922; Patent EP 0 934 359.

Macromol. Symp. 187, 605–615 (2002)

UV Curable Electrodeposition Systems

Oliver Reis, Andreas Fieberg

DuPont Performance Coatings GmbH & Co. KG, Christbusch 25, 42285 Wuppertal, Germany

Summary: This article reviews the development of a UV curable electrodeposition system. Tailor-made acrylic functionalized polyurethane resins, which have been designed, are described and basic properties of the final e-coat system are shown. In addition several fundamental studies have been performed in order to analyze as well the homogeneity of the deposition as the efficiency of the UV curing process.

Keywords: coatings, irradiation, UV, electrodeposition, polyurethane

Introduction

UV-radiation curing has become a well-accepted technology which, because of its distinct advantages, has found manifold industrial applications, mainly as fast-drying protective coatings, printing inks and adhesives.[1] A liquid resin can be transformed almost instantly into a solid polymer material by simple exposure to UV light at ambient temperature, without emission of volatile compounds.

With respect to the application of coatings, electrodeposition is ecological and economical the most efficient method. The objective and therefore an innovative challenge was the development of a electrodeposition and UV-curing process, which combines the advantages of both techniques.

Urethane-acrylates have been selected as telechelic oligo- and polymers, because of their known overall balance of properties, in particular the superior resistance to abrasion and scratching, and the high impact and tensile strength of the cured polymer. As these compounds are not soluble in water, ionic groups had to be introduced into the polymer backbone to obtain stable dispersions.

Experimentals

Principles of anodic electrodeposition (AED)

Electrodeposition offers high corrosion protection, low cost, and compliance with

---

   CCC 1022-1360/00/$ 17.50+.50/0

environmental regulations. It is used for coating of articles of various sizes including steel building trusses, car bodies, furniture, appliances, toys, and details.[2, 3] The current success of electrodeposition is due to the water-dispersible, synthetic, electrodepositable macro ions as filmformers. The process is most frequently called electrocoating, though various names, such as electropainting, elpo, e/coat, etc, are in use. Electrodeposition combines many advantages of other painting methods with new and desirable features:

- Formation of films in highly recessed areas such as cavities, box sections, creases, and flanges results in excellent corrosion protection.
- Use of water as continuos phase virtually eliminates the fire hazard and environmental pollution and reduces the cost of control equipment.
- Low viscosity of the bath facilitates agitation and pumping and allows fast entry and drainage of workpieces.
- Freshly deposited coats are composed of nearly 95% non-volatile substances and therefore allow immediate gentle handling; there is no tendency to sag or wash off during cure.
- A second coat, usually a color coat, of waterborne or solvent-borne spray paint can be applied directly over the uncured electrocoat.
- Approximately 95% of the applied paint is utilized because the liquid paint, which adheres or fills the cavities of freshly coated pieces, is rinsed back by ultrafiltration into the coating tank.
- Overall savings, accounting for materials, labour, capital investment, energy, etc. are 20-50% compared to spray, electrostatic-spray, or dip-coat painting.

The plating cell contains a solution or dispersion of ions, which provide the electrical conductivity. Two electrodes, which have to be made of conductive materials like metal or graphite, are placed into the bath.

When the electrodes are being used in a direct-current system they are referred to as the *anode* and the *cathode*. The anode is the one, which is connected to the positive pole of the power source, and the cathode is the one, which is connected to the negative side of the power source.

Within the cell, the electrolyte is a solution or dispersion of positively and negatively charged ions – the electrolyte. When electricity is applied to the cell the electric field draws negatively charged ions towards the anode and the positively charged ions towards the cathode. Negatively charged ions are therefore known as, *anions* and positively charged ions as, *cations*. The current within the electrolyte solution is carried by these ions. The electrons can carry the current as far as the surface of electrode and no further. So at the surface of the electrode, which is immersed in the solution, there has to be some way of transferring the conduction process across the boundary of the electrode.

In the bulk of the solution, the composition is kept more or less homogenous by mechanical agitation. Because of the viscosity of the solution this agitation does not extend right up to the surface of the electrode, thin layer of solution exits over the surface of the electrode which is essentially free of convection. This is known as the *Nernst diffusion layer*. Any movement of reactants towards the electrode, or products away from it, is therefore either by *diffusion*, following Fick's laws, or *electrical migration*.

Drying and UV-curing

The coating is applied in several process steps. In the first step the conductive substrate is electrocoated using voltages of 100 – 200 V and bath temperatures between 25 and 30 °C (the determination of this application window is described below).

Then this electrocoagulated film is washed and pre-dried. It can be dried by using an air stream or additionally by using temperatures of 60 – 80°C for some minutes. The residual water is nearly quantitatively removed during this step.

The resulting film is UV cured using 2 passes on a Fusion UV-lamp system (Hg-emitter at max. 240 W/cm) at a belt speed of 5 m/min. For the UV curing a typical photo-initiator (α-hydroxyketone type) has been added to the formulation.

Basics in binder chemistry

Aqueous polyurethane dispersions have been gaining increasing importance in a wide-range of applications, due to their excellent properties.[4] These include:

• Adhesion to various substrates

• Resistance to chemicals, solvents and water

• Abrasion resistance

• Flexibility and toughness

• Fast dry

• Outstanding flexibility and impact resistance

• Low volatile organic content = less pollution

Conventional polyurethane resin systems usually contain a proportion of volatile organic solvents. At the end of the 1970s, the manufacturers of polyurethane resins developed processes that permitted the synthesis of low-solvent or solvent-free aqueous polyurethane dispersions. The increased use of polyurethane dispersions is attributed not only to the above-mentioned profile of properties but also to pressures from coating industry and environmental legislation for raw materials with a low organic-solvent content.

Central to the utilisation of polymers in aqueous media is the fact that certain polar functional groups are capable of conferring water solubility or water dispersibility to otherwise water-insoluble polyurethane. Best known are carboxylic acid groups, sulphonic acid groups, and tertiary amine groups. The concentration of such functional groups in the polymer highly influences its solubility or dispersibility in an aqueous environment. Thus, at high concentrations, the polymer may be water soluble, and at lower concentrations the polymer may be water dispersible, provided its molecular weight/viscosity is not excessive. At even lower concentrations, the polar group may be capable of providing charge or steric stabilisation to a dispersion of the polymer in water.

A range of synthetic routes is available for the preparation of waterborne polymer systems. The polymers can differ widely in terms of (1) the nature and concentration of the polar solubilizing group, (2) molecular weight, and (3) the hydrophobic/hydrophilic characteristics depending on the units in the polymer chain. As a result, aqueous polymer systems with a wide range of different morphological and physical characteristics can be obtained.

Special binders for UV-ED-Systems

Aqueous polyurethane dispersions (PUD's) are of special interest. They offer a good balance between mechanical properties like abrasion, hardness or tensile strength. The properties of polyurethanes can be modified in a wide range by varying the composition of the typical building blocks, such as polyols and polyisocyanates or chain extension chemistry. The most common way to synthesize polyurethanes is the formation of a medium molecular weight isocyanate terminated prepolymer by the reaction of polyols with polyisocyanates. In the case for use the polyurethanes for anodic electrodeposition, it is also necessary to incorporate acid functional groups into the polymer backbone. After the prepolymer is formed, functionalization and / or chain extension can take place to form the final polyurethane.

The main route for the synthesis in this work is described as follows:

A polyester polyol of hexanediol, neopentylglycol, adipic acid and isophthalic acid is reacted with dimethylolpropanoic acid and isophorone diisocyanate at 80°C for ca. 5 h until the isocyanate is consumed to form the polyurethane prepolymer. The acid number is adjusted to

30 and the NCO/OH ratio of 1 to 1.5. The primarily synthesis route is shown in figure 3:

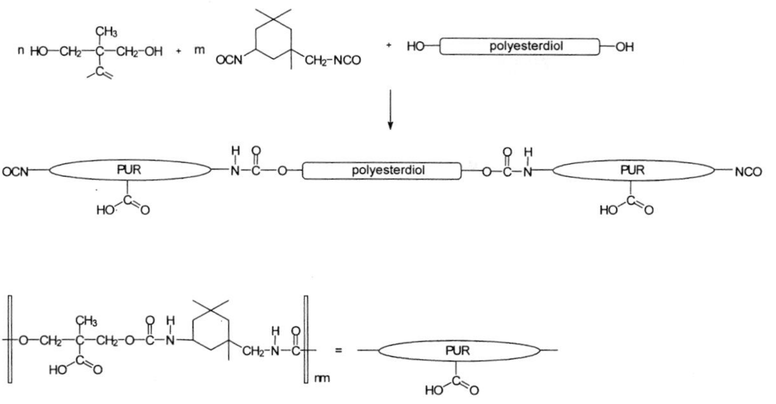

Figure 3. Schematically formation of a acid functional, isocyanate terminated polyurethane prepolymer.

Due to the need to prepare UV-curable resins, the above shown prepolymer is then functionalized with hydroxy ethylacrylate to give a acrylic functional polyurethane, shown in figure 4.

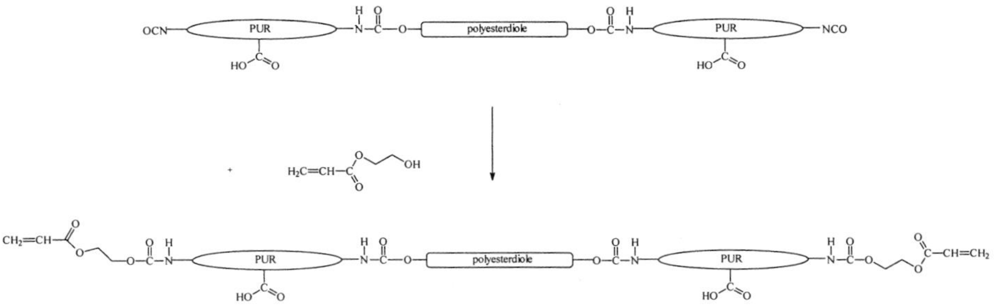

Figure 4. Acrylic functionalized polyurethane

Linear acrylic functional polyurethanes are very important to adjust overall mechanical properties, but with concideration to achieve also good solvent and / or chemical resistance, there is the need to use other UV-active cobinders to enhance crosslinking density and thereof chemical resistance of the final coating.

Results and discussion

Water content

In contrast to conventional electrodeposition coatings the UV system is not heated up to temperatures above 80°C during the whole application process. Although this offers more opportunities regarding the substrates which can be coated, it on the other hand becomes crucial to exactly control the water content of the deposited layer.

During the electrodeposition process most of the water is expelled from the coagulated coating film. However, it is well known, that the residual water content of the deposited film is still in the range of 5 – 10 % w/w. In a standard electrocoat this water is removed during the baking process very easily at temperatures well above 150 °C. In the new UV system this has to be ensured at temperatures of 60 – 80°C. If there is still water in the film when the fast UV curing takes place, it cannot evaporate but would be entrapped in the solid polymer network. The technological properties like hardness, gloss, resistance of the resulting coating film would significantly suffer. Consequently the water content of the deposited film has been determined as a function of the deposition parameters.

By means of a DoE the results have been collected and a response surface has been modeled. Figure 5 shows the water content of the deposited film as a function of T/°C and U/V. It can be clearly seen, that the water content increases significantly with increasing bath temperature and voltage.

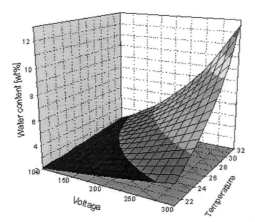

Figure 5. Water content of the deposited coating film as a function of bath temperature and voltage.

Based on these results, the application windows for the deposition process was defined. The usable voltage was limited to approximately 150 V, the bath temperature was varied from 26°C to 30°C. In this case it was shown, that after pre-drying for 10 min. at 60°C the residual water is quantitatively removed from the film. The UV cured coatings showed excellent technological properties.

Stability

A critical issue of electrodeposition coatings is the stability of the resulting bath materials. Since an e-coat bath in operation is refilled permanently, the dispersions have to be very stable.

In order to evaluate and improve the static stability of the system, a special test equipment LUMiFuge® which combines a centrifuge with an optical detection system was used.[5]

The homogeneous dispersion is placed into a measuring cell, which is then rotated with up to 3000 rotations per minute. This correlates to an accelaration of up to 1200 g , which significantly accelerates the migration of the particles within the dispersion.

The transmission is synchronously measured by an infrared flash along the whole length of the cell.

Zones of well distributed dispersions scatter and adsorb the light, so transmission is low. In contrast, any clarification allows more light to reach the CCD-lines: transmission raises.

The larger the particle size, the faster the demixing. The boundary between the light adsorbing sediment and the clear water phase can be detected as a function of time.

Due to this accelerated migration and fully automatic detection, the stability/instability of the dispersions could be assessed on a very short time frame. The sample is analyzed in user defined intervals over a whole measuring time of up to 24 hours. The corresponding transmission profiles are automatically collected and can be analyzed separately. Figure 6 shows an example for a typical transmission profile, which shows a creaming of the dispersed particles.

The red curves relate to the beginning of the test run. You can clearly see the air/dispersion interface, which is characterized by a strong drop of the transmission intensity. The well-dispersed material is scattering and adsorbing the test beam quantitatively. However, as a function of time, the dispersed particles start to cream. This can be detected by a shift of the boundary layer toward the top of the measuring cell (on the left hand side of the diagram). At the end of the test run (green transmission profile) you can see the floated material, which is located at the top of the cell, and an increased transmission through the rest of cell, which

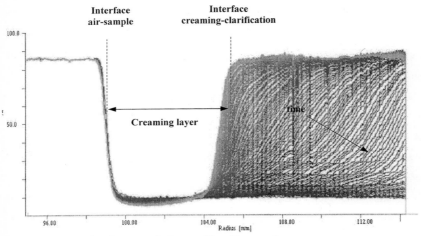

Figure 6. Example of a transmission profile with creaming.

relates to a clarification of the system.

In order to quantify these results, you can follow the clear/turbid interface as a function of time and thus get values for the sedimentation velocity (measured in μm/s). We have shown, that these data are correlating well with standard stability values determined in test cylinders.

The main reason for the improved stability was the adjustment of the compatibility of the two resins. In order to achieve this the type and amount of the hydrophilic diole component within the synthesis of the crosslinker was varied and optimized. The dispersions, which are based on these resins, have a small mean particle size of approximately 100 nm and thus show nearly no sedimentation.

Conversion of the unsaturated acryloyl groups

An advantage of the electrodeposition process is the early water-free coagulated film directly after the deposition, as already shown before. As a consequence only moderate heating is necessary to remove the residual water from the film and to proceed with the fast UV curing step. However, it is well known, that the conversion efficiency of UV curable systems strongly depend on the mobility of the polymers, which are engaged in the process.[7-9] If the viscosity of the polymeric matrix is very high it becomes more difficult to achieve a quantitative through-cure of the material, due to the immobility of the acrylic groups. Working at low temperatures could therefore possibly lead to an insufficient conversion of those groups. The influence of the substrate temperature on the conversion of the double bonds was investigated as a function of the applied UV dose. The extent of polymerization was calculated from the distinct band at 810 cm^{-1}, which can be assigned to the twisting mode of the acrylate double bonds.[6] Figure 7 shows the residual amount of double bonds as a function of the UV dose at two temperatures (room temperature and 80°C).

It can be clearly seen, that the substrate temperature has a significant influence on the conversion during the UV curing process. Whereas approximately 10% of the unsaturated double bonds remain unreacted at room temperature, a nearly quantitative conversion can be reached at a temperature of 80°C and a UV dose of 1 J/cm^2.

Therefore, in order to obtain cured films with good ageing and weathering stabilities, during this study the UV curing was done at increased temperatures (60 – 80°C) and with UV doses of > 1 J/cm^2.

Technological properties of the UV/EC coatings

The UV curable electrodeposition system, which has been developed within this project, can be in principal deposited on any conductive substrate. However, all tests have been performed on regular steel panels, which have been coated with a film thickness of approximately 20 microns. The films showed an excellent optical appearance and very good hardness. Critical issues like adhesion to the substrate and resistance against solvents (e.g. acetone) have been tested and compared to standard thermal cured e-coat systems (see Table 1).

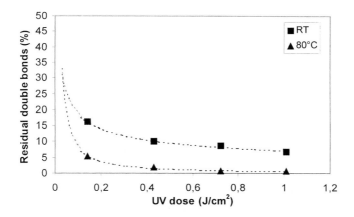

Figure 7. Relative amount of residual double bonds as a function of the applied UV dose.

Table 1. Technological properties of the UV/EC system compared to standard thermally cured electrodeposition systems

	UV/EC system	Classic thermal cured systems	
		Melamine	**Isocyanate**
Gloss @ 60°	>90	70-75	80-85
Cupping test (Erichson) (mm)	8,7	5,0	5,1
Acetone resistance 60s	Passed	Passed	Passed
Pendulum hardness (osc.)	145	110	130
Cross hatch test	0	1	0-1

The UV/EC system shows a superior gloss and hardness compared to thermally cured e-coat materials. In addition it still is not brittle, but shows an elasticity of 8,7 mm Erichson indentation. The material also passes the acetone and MEK tests, which proves the high crosslinking density resulting in a very good chemical resistance.

Conclusion

Waterbased polyurethanes have been identified as a powerful system for UV curable electrodeposition coatings. With respect to the mechanical and technological film properties a combination of two different resin components has been developed. In order to achieve an excellent performance of the final coating, the linear, acryl functional polyurethane is preferably used for adjusting the mechanical properties whereby the oligourethane is used to adjust the crosslinking density and therefore to increase the chemical resistance.

The resulting waterborne dispersions show very good stability and can be deposited very homogeneously. The coating can be applied on different conductive substrates and for example exhibits good adhesion on steel panels.

Dependent on the temperature during the curing process, a nearly 100% conversion of the double bonds can be realized.

References

[1] C. Decker, *Prog. Polym. Sci.* **1996**, *21*, 593.
[2] F. Beck, *Fundamentals of electrodeposition of paint* **1979**, BASF, 1 - 55.
[3] T. Brock, M. Groteklaes, P. Mischke, European Coatings Handbook, C. R. Vincentz Verlag,Hannover, 2000, 279 - 285.
[4] D. Dietrich, *Prog. Org. Coat.* **1981**, *9*, 281.
[5] L.U.M. Gesellschaft für Labor-, Umweltdiagnostik & Medizintechnick mbH, Berlin, Germany

[6]   T. Scherzer, C. Decker, *Vibr. Spectr.* **1999,** *19*, 385.
[7]   W. Reich, P. Enenkel, E. Keil, M. Lokai, K. Menzel, W. Schrof, *Proc. RadTech North America* *98*, Chicago **1998**, 258.
[8]   N. Pietschmann, *Proc. Radtech Europe 99,* Berlin **1999**, 785.
[9]   A. Tauber, T. Scherzer, R. Mehnert, *J. Coat. Techn.* **2000**, *72*, 51

Macromol. Symp. 187, 617–629 (2002)

20 Years of DPP Pigments – Future Perspectives

Olof Wallquist, R. Lenz

Ciba Specialty Chemicals Inc., P.O. Box, CH-4002 Basel, Switzerland

Summary: Diketopyrrolopyrroles (DPP), one of the major innovations in Pigment Chemistry in the last 20 years, have developed to an important class of organic pigments. Chemical transformations have been the door opener for most of the innovations. The insoluble nature of the DPP Pigments allows for solid-state phenomena to be investigated. Crystal modifications and solid solution formation have enhanced the scope of DPP pigments. Specific elaborations of the DPP chemistry have lead to interesting developments also for electronic applications like color filters for liquid crystal displays or for the second generation display technologies like the OLED involving electro luminescent materials.

Introduction

This paper will provide an overview on DPP (**d**iketo**p**yrrolo**p**yrrole) chemistry. In the last twenty years this new pigment class have been intensely studied by chemists at Ciba resulting in the commercialisation of many excellent products in various application areas. Milestones such as the finding of an efficient and general synthesis for DPPs and the elucidation of relationships between its structure and physical properties will be highlighted. Furthermore, the very interesting crystallographic properties of DPPs, including polymorphism and solid solutions, will be presented including two newly commercialised DPP pigments. Finally, factors controlling important properties in certain applications, such as particle size, heat stability and rheological behaviour, will be discussed and illustrated with typical examples within the DPP family.

Background

The DPP chromophore was mentioned for the first time in the literature in 1974, when Farnum et al. [1] briefly reported on the attempted synthesis of 2-azetinones via a modified Reformatsky reaction.

However, the proposed reaction failed to produce the target unsaturated β-lactam **1** *(Scheme 1)*. Instead, the authors isolated in unpredictably varying yields of 5-20% the unsubstituted DPP derivative **2**, in addition to other by-products. This unexpected product was described as "a highly insoluble, brilliant red crystalline compound" with a melting point beyond 350°C. These

     CCC 1022-1360/00/$ 17.50+.50/0

properties together with its structural features similar to many commonly known pigments, such as isoindolinones, indigos, and quinacridones, prompted chemists at Ciba in the early eighties to start an investigation of its performance as a pigment. The observed fastness properties were so encouraging that a series of new DPP derivatives were additionally synthesised for a rapid evaluation of the practical scope of these pigments and, at the same time, to seek appropriate patent protection [2]. A commercial synthesis was developed and the first industrial grade was commercialised in 1986.

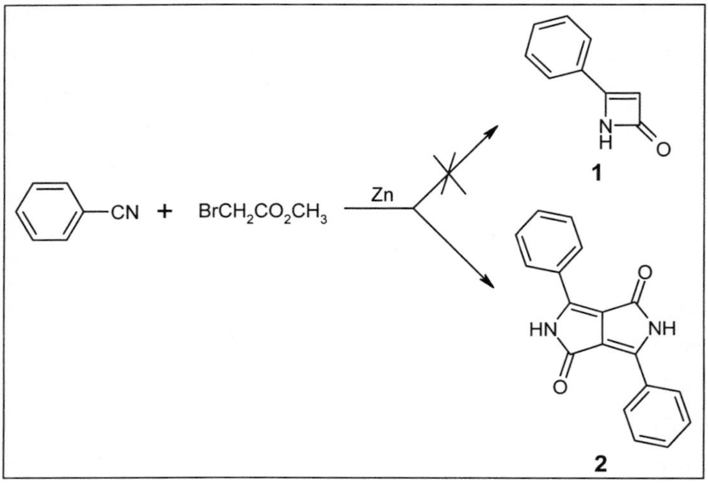

Scheme 1

Structure and Physical Properties

Inspection of several DPPs with varying substituents in the phenyl moiety, led to a few generalisations regarding its properties. First of all, a broad spectrum of shades ranging from orange-yellow via blue-red to violet could be accomplished by simply exchanging the substituents at the *para-* and *meta*-position of the two phenyl rings attached to the DPP chromophore unit. The following table shows the absorption maxima of such compounds measured in solution together with the ε_{max} values, as well as the absorption maxima in the solid state obtained by reflectance measurement on PVC-white reduction *(Table 1)*.

Table 1. Influence of substituents on shade and absorption (λ_{max}) of diaryl-DPP's

R	λ_{max} (NMP)	ε	Shade (PVC)
3-CF₃	509	~21'500	Yellowish Orange
3-CN	515	~20'500	Yellowish Orange
3-Cl	512	~27'000	Orange
4-t-Bu	511	~42'000	Orange
H	504	~33'000	Yellowish Red
4-Br	515	~35'000	Bluish Red
4-Ph	534		Bluish Red
4-NMe₂	554	~81'500	Violet-Blue

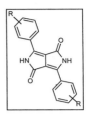

As with many other classes of pigments, all DPP compounds show a bathochromic shift of the maximum absorption in the solid state in comparison to maximum absorption in solution *(Scheme 2)*. This effect is due to strong intermolecular interactions in the crystal lattice such as hydrogen bonding, π-π- and van der Waals interaction. Detailed experimental [3] and theoretical [4] investigations on the influence of the intermolecular hydrogen bonding on the spectral properties of DPP in particular strongly indicate that this type of interaction is actually playing the most important role.

The solid state absorption maxima of DPP vary strongly depending on the nature and position of the substituent. In comparison to the unsubstituted DPP **2**, *meta*-substituted DPPs often show a hypsochromic shift and *para*-substituted DPP a bathochromic shift *(Table 1)*. The DPP pigments generally show bright shades, which are related to the high purity of the compounds and the sharp bands in solution as exemplified in *Scheme 2*. Most of the DPP pigments also display high colour strength, as shown by the relatively high ε_{max} values of the solution spectra.

The average level of light and weather fastness of DPP pigments is outstanding. Despite their low molecular weights, the diaryl-1,4-diketo-pyrrolo-pyrroles are highly insoluble and remarkably resistant to migration, even when carrying otherwise solubilising alkyl substituents, such as the *tert*-butyl group.

These excellent properties may be attributed, above all, to the presence of strong intermolecular bonding forces in the pigment solid state, as was also corroborated by x-ray structure analyses of individual members of this class of pigments exemplified below for the unsubstituted DPP **2** *(Schemes 3 and 4)*.

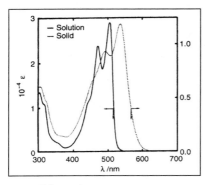

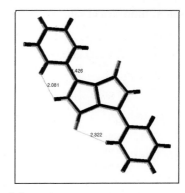

Scheme 2   *Scheme 3*

The DPP molecule is virtually planar, the two phenyl rings being twisted out of the pyrrolopyrrole plane by merely 7°. The distance between the two hydrogen atoms at N_2 and C_6 is 2.08Å, while that between the hydrogen atom at C_{10} and O_{11} is 2.32Å *(Scheme 3)*. For both cases, the sum of the corresponding van der Waals radii are substantially larger, 2.4Å and 2.6Å respectively, thus implying resonance interaction between the central unit and the phenyl rings. This appears to be confirmed by a C_3–C_5 bond length of 1.43Å, which is shorter, and hence possesses even more pronounced double bond character, than the bond (1.496Å) between the two aromatic rings of biphenyl. Such resonance interactions, amongst others, would explain the electronic influence of substituents on the earlier mentioned wavelength shifts of different DPP derivatives in solution *(see Table 1)*.

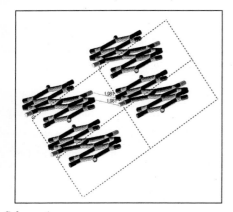

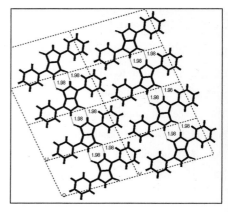

Scheme 4

The crystal packing along the a-axis is determined by π-π interactions between the layer of molecules. The interlayer distances (3.36Å - 3.54Å) are short enough to warrant significant overlap *(Scheme 4, left)*. Along the (a,b)-diagonal hydrogen bonding takes place. These bonds are formed between neighbouring lactam-N and carbonyl-O atoms leading to a one-dimensional tape-like structure *(Scheme 4, right)*. Intermolecular van der Waals contacts between the terminal phenyl groups determine the packing along the crystallographic c-axis.

Extension of Colour Space - Chemical Structure

The fact, that the absorption spectra of different DPP pigments in the solid state is varying significantly, despite that they have a very similar absorption spectra in solution, has allowed for the development of many new DPP Pigments covering a large colour area *(Scheme 5)*.

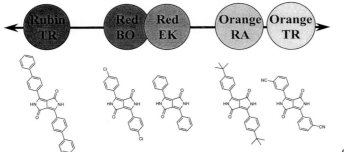

Scheme 5

Crystal Modifications

As with many other classes of pigments, polymorphism is also observed within the DPP family. Such polymorphs have the same chemical constitution but differ in their crystal modification which is usually characterised by powder X-ray diffraction. Some pigment polymorphs are almost indistinguishable in appearance whereas others differ very widely in their colouristics and other properties. A typical example for the latter case is C.I. Pigment Red 254 (p-chloro DPP) where a new crystal modification (β-form) has recently been identified [5] *(Scheme 6)*. In paint applications this β-form is significantly more yellowish compared to the thermodynamically more stable α-form.

Solid Solutions

Another crystallographic concept has proven to be very fruitful when applied to DPP chemistry: solid solutions [6]. Solid solutions can be obtained by combining two or more pigments of different constitutions by methods other than simple mechanical mixing. The resulting powder X-ray diffraction pattern is either the same as (or very similar to) that of one of the components, the so-called "host". In analogy to liquid solutions, one can therefore say that in solid solutions the "guest" molecules are dissolved into the crystal lattice of the "host". In some cases this can lead to a variation of the colouristic properties compared to a physical mixture of the individual pigments.

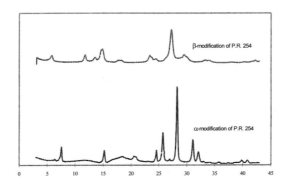

Scheme 6

The same is true for mixed crystals. Here, specific compositions of two or more pigments lead to a new unique diffraction pattern, different from either of the component pigments. In this case, loose analogy can be drawn to the formation of azeotropes in the liquid state. An example, where the concept of mixed crystals is advantageously exploited to extend the colouristic scope of DPP chemistry, is shown below *(Scheme 7)*. It has been found that equimolar amounts of the unsubstituted C.I. Pigment Red 255 and the *tert*-butyl substituted DPP C.I. Pigment Orange 73 can be combined and form an entirely new crystal lattice [7]. Compared to the parent pigments which exhibit a pure red and orange shade in paints or plastics, the resulting pigment shows a significant bathochromic shift leading to a bluish red hue. Interestingly, the new crystal lattice formed by the combination of the two DPPs is nearly identical to that of the corresponding asymmetrically substituted DPP derivative.

The very special conditions in a solid solution are illustrated by the fact that equimolar amounts of the unsubstituted C.I. Pigment Red 255 and the p-chloro substituted C.I. Pigment Red 254 form a mixed crystal, which in contrast to the above mentioned example exhibit a hypsochromic shift.

Solid solution formation can also be achieved by a "mixed synthesis" approach [8]. Instead of using just one aromatic nitrile in the DPP synthesis, two (or more) nitriles are used leading to ternary mixtures as depicted in *Scheme 8*. These mixtures consist of the two symmetric DPPs and the corresponding asymmetric derivative. The ratio of the three products strongly depend on the reactivity of the nitriles and the specific reaction conditions.

Scheme 7

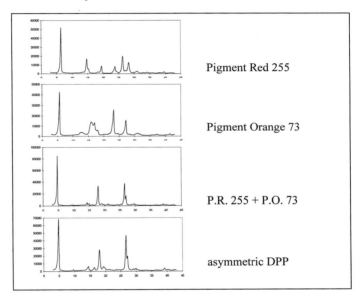

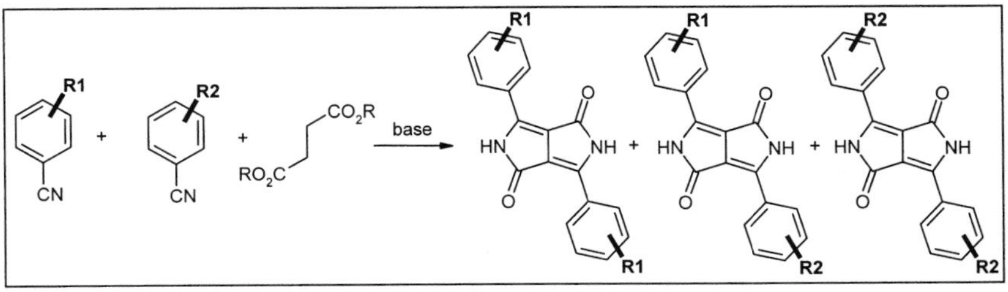

Scheme 8

An interesting example is the use of benzonitrile and p-chlorobenzonitrile in the DPP synthesis. As in the example discussed above, the resulting solid solution shows a hypsochromic shift in comparison to the corresponding symmetric DPPs (C.I. Pigment Red 255 and 254) providing evidence that this concept is capable of expanding the colour range in both directions *(Scheme 9)*. Again, the resulting X-ray diffractogram (XRD) is very similar to that of the corresponding asymmetric DPP.

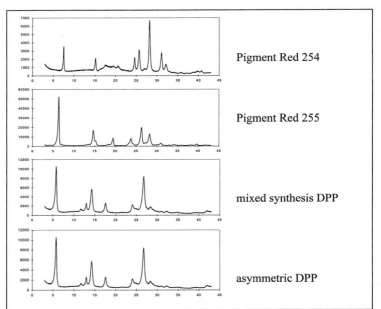

	Pigment Red 254
	Pigment Red 255
	mixed synthesis DPP
Scheme 9	asymmetric DPP

Extension of Colour Space – C.I. Pigment Orange 81

Through the combination of new DPP chemistry with manipulation of the solid state, a new very opaque and highly saturated orange DPP pigment has been developed and just introduced to the market. The new **C.I. Pigment Orange 81** combines a high opacity with an outstanding full shade colouristics. A comparison of the full shade chroma/hue as well as the opacity[1] is given in *Scheme 10*. The high opacity and the saturated shade consequentially renders the C.I. Pigment Orange 81 a very attractive value in use for coatings applications.

---

[1] Contrast dE (black over white) as a measure of the opacity – lower dE corresponds to higher opacity

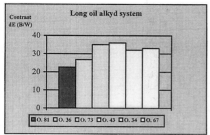
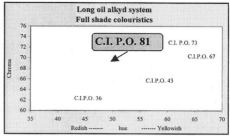

Scheme 10

Particle Size and Distribution Control

The crystal size of a pigment as well as the particle size distribution have an important effect on the tinctorial strength, hiding power, durability and viscosity, and can also affect the shade. It is therefore essential to efficiently control these parameters in order to obtain a pigment with the desired properties. Small particle sized pigments can be achieved by finishing methods such as milling or kneading of large particle sized crude pigments or by a (more economic) direct pigmentary approach. For DPP pigments several methods have been developed in order to condition the pigment particles *in situ*. The most important approach is to vary the conditions during the DPP synthesis by changing solvents, temperature, time or pH. C.I. Pigment Red 254 gives an example, illustrating to what extent the particle size of a pigment can be influenced. Its opaque form is obtained under elevated temperature and basic conditions, whereas by applying lower temperatures and non-basic conditions a very transparent version is formed. Transmission electron micrographs (TEMs) of these two pigment morphologies are shown in *Scheme 11*.

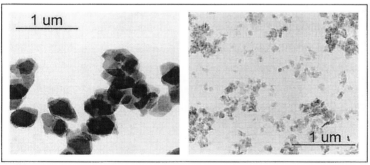

Scheme 11

626

The manipulation of the particle size towards very small particle size versions of DPP pigments opens up the possibilities of DPP pigments to be used in automotive metallic and effect colorations. One example is the C.I. Pigment Red 264.

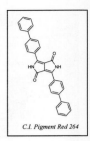

C.I. Pigment Red 264

Again the combination of a specific chemistry (constitution) and solid-state manipulations (particle size, size distribution etc.), have lead to the development of a strong medium transparent DPP with outstanding fastness properties in both organic and aqueous binder systems (see further p. 14).

Particle Surface Manipulations

Since the performance of a pigment is not only given by the chemistry and the solid state properties of the pigment itself but to a large extent also by the application in which the pigment is being used, the interaction between a pigment and a specific application media is of utmost importance. This interaction, however, depends to a large part on the surface characteristics of the pigment. In order to fully exploit the potential of a pigment in a given application, the surface of the pigment has to be optimised. Many surface treatment procedures have been developed [9]. Most of them have in common that they promote the formation of an adsorbed layer. Due to the intrinsic differences in the properties of the various application media, e.g. paints or plastics, individual surface treatments of a given pigment have to be found for each application or polymer system.

In paint application there is, for ecological and economical reasons, a trend towards higher pigment loadings. This poses a problem to pigment manufacturers since higher pigment concentrations in the paint inevitably results in higher viscosity of the whole paint system leading to difficulties in processing and inferior properties of the final paint such as reduced gloss. The rheological properties of the pigment have therefore to be improved by an appropriate treatment of the pigment surface. The result of such a treatment on C.I. Pigment Red 254 is shown below *(Scheme 12)*.

Although the untreated P.R. 254 already exhibits a relatively good rheological behaviour, its viscosity in AM-paint under high pigment concentrations could still be strongly reduced by adsorbing a DPP derivative bearing sulfonic acid groups onto its surface [11].

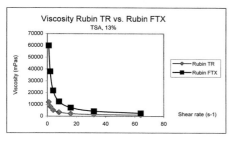

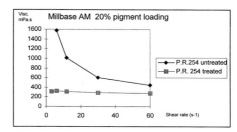

Scheme 12   *Scheme 13*

Transparent pigments, having a much larger surface area than opaque pigments, show higher viscosities in different binder systems leading to poor formulations both from a technical as well as from an economical and ecological point of view. Specific modifications of the surface of C.I. Pigment Red 264 have now lead to a new rheology improved version of C.I. Pigment Red 264 with strongly improved viscosity compared to the base pigment (*Scheme 13*).

New Applications

Utilisation of specific DPP chemistry to optimise and improve the physical and applicatory properties of DPP pigments has led to many new commercial products. Major objectives in the past have obviously been to find new and improved products for the coatings, inks and plastics industry. Some of the physical properties of DPP's, however, indicates, that the DPP structure could be of interest also for electronic applications. For example, the high chroma of the DPP pigments as illustrated by the steep transmission curve is not only an advantage for conventional applications but an essential requirement also for Colour Filter applications for LCD screens.

The red colour of the RGB system used for colour filter is not ideally positioned in the chromaticity diagram due to the spectral properties of the state of the art pigment. The steep transmission curve of many DPP pigments in combination with specific optimisations of the transparency, dispersibility and the rheological properties in the photopolymer have led to another new transparent DPP Pigment specifically developed for Colour Filter applications and now developing to the industry standard Red for Colour Filter applications (*Figure 14*).

628

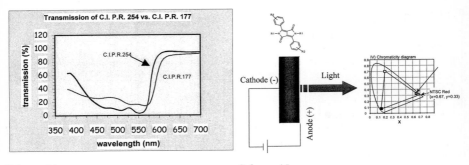

Scheme 14   *Scheme 15*

Another area currently being actively researched in many academic and industrial laboratories is organic light emmitting diodes (OLED). This technology is being viewed as one of the technologies which might partly replace the LCD's. The fact, that specifically substituted DPP's show fluorescence in the solid state makes them interesting for electroluminescence applications. Work on DPP's as elctroluminescent materials has been reported both in the direction of small molecules ([12]) as well as on polymeric DPP ([13]). In both cases, promising chromaticity and efficiencies have been found *(Scheme 15)*.

Conclusion and Outlook

Pigment chemistry is generally regarded to be a rather mature science, where industrial interests are mainly focused on the improvement of known products and on the development of new products within the established pigment classes. However, occasionally novel classes of pigments are discovered leading to new opportunities in many application areas, thus revitalising the pigment industry. Elaboration of the DPP chemistry in many directions has resulted in many innovative products for different conventional applications. Specific properties of this interesting chromophore have lately also allowed to find new applications in the electronic applications like for colour filters for LC display or potentially as electroluminescent materials for OLED applications.

References

[1] Farnum, D.G, Mehta, G., Moore, G.G.I., and Siegal, F.P., *Tetrahedron Lett.* **1974**, *29*, 2549-52.

[2] Iqbal, A. and Cassar, L. (Ciba-Geigy Ltd.), US Patent 4,415,685 (1983).

[3] Iqbal, A., Kirchmayr, R., Pfenninger, J., Rochat, A.C., and. Wallquist, O., *Bull. Soc. Chim. Belg.* **1988**, *97*, 615-643.

[4] Adashi, A. and Nakamura, S., *J. Phys. Chem.* **1994**, *98*, 1796-1801.

[5] Hao, Z., Schlöder, I., and Iqbal, A. (Ciba-Geigy Ltd.), EP 690,058 (1994).

[6] Whitaker, A. *J. Soc. Dyers Colour.* **1986**, *102*, 66.

[7] Hao, Z., Iqbal, A., Medinger, B., and Wallquist, O., (Ciba-Geigy Ltd.), EP 704,497 (1994).

[8] Iqbal, A., Pfenninger, J., Rochat, A.C., and Bäbler, F., (Ciba-Geigy Ltd.), EP 181,290 (1986).

[9] Hays, B.G., *Am. Ink Maker* **1984**, *30*, 270; *ibid.* **1990**, *10*, 13; *ibid.* **1990**, *11*, 28.

[10] Bugnon, Ph. and Herren, F., (Ciba-Geigy Ltd.), EP 466646 (1990).

[11] Jost, M., Iqbal, A. and Rochat, A., (Ciba-Geigy Ltd.), EP 224445 (1986).

[12] J. Otani et al., (Ciba Specialty Chemicals Inc.), EP 1087006 (2000).

[13] B. Tieke et.al., (Ciba Specialty Chemicals Inc.), EP 1078970 (2001).

Macromol. Symp. **187**, 631–640 (2002)

Pigments With Improved Properties

-Microreaction Technology as a New Approach for Synthesis of Pigments-

H. Kim, K. Saitmacher, L. Unverdorben, Ch. Wille

Clariant GmbH, Division Pigments & Additives, Pigments Technology, PT-PD,

Industriepark Höchst, Building G 835, D-65926 Frankfurt am Main, Germany

Summary: Clariant, as an important pigment producer forces the investigation of new pigments with improved qualities and properties to fulfill the rising tomorrow's demands of customers. For these reasons, new production ways like microreaction technology are included. This paper focuses on results obtained in manufacturing pigments in a lab-scale microreactor as well as in a microreactor pilot plant. Investigations of the diazotation, azo-coupling and laking steps of pigments have shown not only the principle feasibility of these reactions in laboratory microreactors but also significant improvement of coloristic properties. The microreactor pilot plant, realized by the concept of numbering-up instead of conventional scaling-up process, allowed more detailed investigations of the complete azo-pigments synthesis under production conditions.

Keywords: Pigments, coloristics, microreactor, numbering-up, pilot plant

1. Introduction

In the last two decades powerful processes have been developed for fabrication of three–dimensional microdevices from a wide variety of materials. Recently, microdevices have become highly interesting for chemical, pharmaceutical and biotechnical applications [1]. Nowadays so-called microreactors with channel dimensions in the sub-micrometer range are a state of the art tool for R&D [2-5].

Due to the microchannels, microdevices are normally restricted to gas/gas, liquid/liquid and gas/liquid reactions. Investigations on reactions done in microreactors involving or forming solid substances are quite rare. For example scientists at Institut für Microtechnik Mainz (IMM)[1] reported generation of micro- and nanoscale solid particles in IMM's interdigital micromixer like precipitation of copper oxalate. Although there is a strong bias that solids or pigment suspension will block microchannels, Clariant decided to check out the basic feasibility of pigments syntheses by using microreactors.

---

[1] Institut für Microtechnik Mainz, Carl-Zeiss-Straße 18-20, D-55129 Mainz, Germany.

     CCC 1022-1360/00/$ 17.50+.50/0

Microreactors, recognized by experts as a tool for process intensification by increasing mixing efficiency, have a high potential for application in the field of pigments technology. Producing pigments in microreactors may turn to profit to two respects. The main aspect is to fulfill customers demand on quality and properties for pigments of tomorrow. Intensified process conditions seem to allow yielding pigments with better application properties like coloristics (color strength, transparency, brightness and purity). Another aspect is to develop a continuous economic production technology for manufacturing of ordinary pigment volumes. The chemical industry's conventional approach of reducing production costs by economies of scale with larger plants could be challenged in future by another strategy. If industrial output via microreactors are demanded, several units have to be operated in parallel. This option makes possible to respond flexibly to changing demands by switching individual reactors on or off – in other words "production on demand" becomes feasible. Therefore, this work deals first with a principal feasibility study of producing azo-pigments in a laboratory microreactor. Additionally, the transfer of this technology to pilot plant scale to manufacture industrial quantities of azo-pigments with constant high quality is evaluated. For this, producing pigments in a pilot plant based on microreactors was investigated.

2. Evaluated microdevices

Figure 1 shows two CPC[2]-microreactor types used in this project. Upper right, you see a laboratory microreactor and at the bottom a microreactor that is mounted in a pilot plant. Letters (a) to (c) indicate in- and outlets for feed and products.

The microreactors are assembled of microstructured platelets mounted together by metallic bonding. Each platelet has numerous parallel channels in the sub-millimeter range with a special function such as

- dividing reactants into substreams and leading them into reaction channels,
- keeping reaction temperature within a desired range by cooling or heating,
- collecting product substreams and leading them out of the microreactor.

Based on a numbering-up concept [6] the microreactor pilot plant allows a higher output of several tons of pigment. This concept was realized by manifolding the reaction plane several times inside the microreactor and by operating several of those microreactors in parallel embedded in one housing. Additionally, the dimensions of the reaction channels have been slightly increased, but without changing the laminar flow conditions and especially without changing the mixing process. Adaptation of the channel geometries has also been done to maintain isothermal conditions for the reaction in respect to the increased product output.

---

[2] Cellular Process Chemistry Systems GmbH, Hanauer Landstraße 526 / G58, D-60343 Frankfurt am Main, Germany.

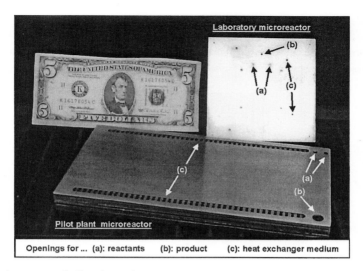

Figure 1. Laboratory and pilot plant microreactor

The laboratory and pilot plant microreactor have the same contacting principle of the reactants in common, shown in Figure 2.

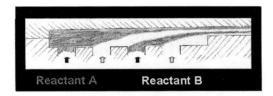

Figure 2. Sketch of microreactors contacting principle

634

Feeding the reactants onto each other leads to a multilamination. Mass transfer between these formed lamellas is due to molecular diffusion, which is an extremely fast process. To operate the microreactors, they are embedded in housings (see Figure 3) and equipped with pumps, heating bath and vessels (not shown as conventional laboratory equipment).

Figure 3. Housing and fittings of the laboratory microreactor

The Laboratory microreactor and the pilot plant allow investigating a continuous 3-step synthesis of azo-pigments – i.e. all three reaction steps like diazotation, coupling and laking are included – (see Figure 4). The key components of each reaction stage are microreactors embedded in one housing, several pumps and a corresponding number of vessels. The latter are for supplying educts and withdrawing product at the same time.

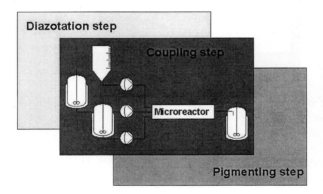

Figure 4. Flow sheet of the microreactor pilot plant experimental set-up

3. Pigments syntheses using lab-scale microreactor

To study syntheses of azo-pigments in microreactors with better properties two model pigments have been chosen, called model pigment 1 and 2, respectively. The first one is red coloured and the second one yellow coloured. Both pigments are synthesized by a 3-step reaction, i.e. diazotation, coupling and laking.

In the following the results of investigation of azo-coupling step of these two pigments are discussed. After reaction of a primary aromatic amine with sodium nitrite, the resulting diazo-solution /
-suspension was fed together with a coupling agent into the microreactor at reaction temperature. The coupling reaction itself is an exothermic electrophilic substitution of a diazonium compound ($Ar-N \equiv NY$) with a coupling component (RH). The produced free acid (HY) is buffered by additional feeding a solution of sodium hydroxide into the microreactor or by using an internal buffer.

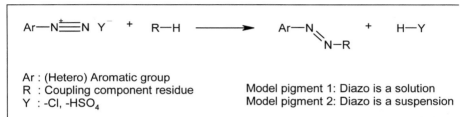

Ar : (Hetero) Aromatic group
R : Coupling component residue
Y : -Cl, -HSO$_4$

Model pigment 1: Diazo is a solution
Model pigment 2: Diazo is a suspension

Equation 1. Reaction scheme of the azo-coupling step [7]

Depending on the investigated pigment, flow rates of the raw materials were adjusted such that one of them could be analyzed in slight excess at the outlet of the microreactor. After coupling pigment suspension was worked up in a batch-wise comparable manner: separating the pigment from the yielded suspension, drying and milling. In the end, the pigment was tested and compared to the standard yielded in a batch process with respect to coloristic properties.
As there was no publication about producing pigments in microreactors, model pigment 1 was chosen for a first feasibility test. The reason for this procedure is, that in this case both reactants, i.e. diazo as well as coupling component, can be applied as solutions. This model pigment 1 is best to our knowledge the first pigment suspension produced inside a microreactor successfully. The product leaves the microreactor as deep red, highly viscous, almost pulpy suspension.
Table 1 shows comparison of the microreactor pigment with the standard of batch-wise produced pigment.

Table 1. Coloristic properties of microreactor model pigment 1 compared to the batch standard

	Microreactor Pigment 1
Color strength	119%
Brightness	5 steps glossier
Transparency	5 steps more transparent

As it can be taken out of table 1 microreactor model pigment 1 shows a tendency to more brightness as well as transparency compared to the standard. Higher transparency and an improved color strength of the microreactor pigment indicate smaller particle sizes, as will be shown later. The results indicated the possibilities for enhancements in product quality by using microreactors.

The next critical question to answer was, if one can also work with diazo-suspensions for coupling reactions in a microreactor. Therefore, model pigment 2 was chosen. Except the difference to feed now a diazo-suspension into the microreactor, the experimental procedures of producing model pigment 2 and 1, respectively, were the same. Also in this case, pigment suspension can be produced successfully: A diazo-suspension enters the microreactor and an intense yellow suspension leaves it. Furthermore, parameter studies show that coloristic properties of model pigment 2 yielded in the laboratory microreactor could be dramatically improved compared to the batch standard, i.e. conventionally produced pigment. Table 2 gives an idea about this.

Table 2. Coloristic properties of microreactor model pigment 2 compared to the batch standard

	Microreactor Pigment 2
Color strength	140%
Brightness	6 steps glossier
Transparency	6 steps more transparent

These laboratory microreactor experiments indicate process intensification e.g. optimum mixing of the raw materials when using microreactors for pigments production. Due to small dimensions of the reaction channels the pathway of one reactant into the lamella of the other reactant is pretty short.

Additionally, a high specific surface area of the multilaminated flow enables nearly isothermal process conditions without local overheating and, thus, suppressing secondary reactions such as decomposing of the diazo component. The latter explains the found high brightness of the microreactor pigment. Furthermore, like already indicated when investigating model pigment 1, a pigment with very high color strength as well as transparency was produced. These coloristic properties correspond to small particles and a narrow particle size distribution.

The comparison of the very first results of a totally new technology with the results of conventional technology after 100 years of optimization is very promising for the application of microreactors in the field of pigments synthesis. Furthermore, for conventional technology additional experimental work out is necessary when scaling-up the laboratory vessels to tank reactors. In contrast to this up-scale-struggle a huge advantage of microreaction technology is discussed in the scientific community for years: Numbering-up should avoid any risks in increasing production volumes. Unfortunately, experimental proof of this concept is rarely available. Therefore, the prerequisite for a successfull application of microdevices for industrial production of pigments is to prove that results of the laboratory microreactor are reproducible using a pilot plant based on microreactors.

4. Pigments syntheses using microreactor pilot plant

Azo-Pigments are traditionally produced in batch operations in large stirred vessels. After completing the reaction under defined reaction conditions, the resultant product suspension is further processed to produce the finished pigment. The advantages of batch processes in large stirred vessels – such as being able to change the production relatively rapidly – are however matched by some disadvantages in process control which often can have an effect on pigment quality. E.g. the formation of undesired by-products are due to reasons like hot-spots, non-uniform reactant concentration and backmixing.

In contrast to conventional scaling-up of batch process by enlargement of operation volume, microreaction technology achieves higher output by numbering-up of the operated microreactors [6]. Along this concept, there is no significant change of reaction channel geometry and, thus, of flow regime, although capacity was increased. Thanks to very small dimensions of the microchannels the stoichiometric ratio, temperature and residence time can be established very accurately and if necessary, adjusted till the optimal operation points are achieved. Therefore, no change of the achieved superior coloristic properties was expected when using identical parameters for pigments manufacturing in laboratory and pilot plant microreactors, respectively. This had to be proved and was finally shown for model pigment 2.

First, the pilot plant microreactor was designed and manufactured, in the second step a fully operable microreactor unit (including all peripheral equipment like pumps, vessels and piping to operate such a system in production environment) was built and in a third step the pigment synthesis was tested investigating model pigment 2. After bringing in operational use of the microreactor pilot plant process parameters and instrumentation (flowrates, sensor technology, reaction parameter control) in all three stages were optimized. Indeed, brightness and transparency of model pigment 2 produced in the microreactor pilot plant based on lab scale microreactors equal the values of the laboratory experiments (see table 2) and the color strength could even be further increased to 149%. This was primarily due to the used peripheral equipment. E.g. compared with the lab scale microreactor unit pumps were installed which are suitable for especially exact dosing of suspensions and for operating almost without pulsation. In this way, mixing efficiency as a the tool for process intensification could be enhanced at a very high level.

Figure 5 shows the coloristic results of model pigment 2 after running-up series and optimization of the continuous operation process. Now model pigment 2 can be produced with a constant quality at high level.

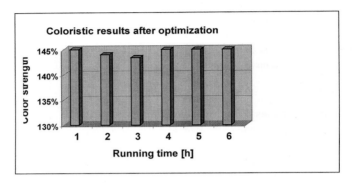

Figure 5. Coloristic results of model pigment 2 after optimization

Figure 6 shows a corresponding particle size distribution of the microreactor pilot plant pigment compared to the standard. Additionally, transmission electron microscope (TEM) pictures give a rough idea about a significant influence on pigment size of microreaction and batch pigments technology, respectively. Like indicated by coloristic properties, the particle size distribution of a microreactor pigment (1) is significant narrower than that one of the conventionally produced pigment (2) (standard deviation is $s = 1.5$ and $s = 2$, respectively). Also, the D_{50}-values differ by a factor of 6: Microreactor pigment has a D_{50}-value of 90 nm, the D_{50}-value of the standard batch pigment amounts nearly $D_{50} = 600$ nm.

In figure 7 the results of model pigment 2 yielded by numbering-up of microreactors are compared to data of the scaling-up phenomenon in batch operation. Increasing the reaction volume of batch syntheses color strength as well as color shade decreased at the same time. In contrast to this, there was no negative effect observed when realizing the numbering-up concept in the microreactor pilot plant. Therefore, nearly the same coloristic results were yielded under identical optimal operation parameters in the lab scale and pilot plant microreactor, respectively.

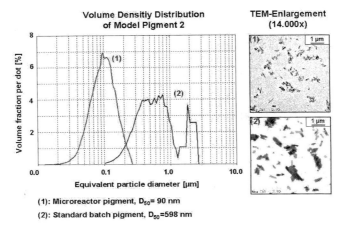

Figure 6. Comparison of model pigment 2 yielded in the pilot plant microreactor and batch standard pigment

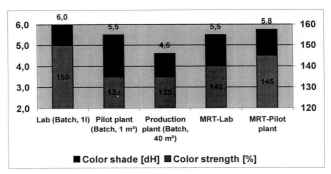

Figure 7. Coloristic results of model pigment 2 from scale-up operations compared with microreactor pigment

After reproducibility of laboratory experiments using the pilot plant is proved, there is at least one further hurdle to take. Regarding design and basic engineering of a pigment production plant, answers have to be found if fouling, coating or finally clogging of microdevices occur as well as how to minimize or to prevent these effects. Fouling of heat exchangers surfaces is still today an item of scientific investigations. Microdevices with its microstructures seem to be especially susceptible for fouling as well as coating. This would have a dramatic negative effect on performance of the microdevice. The advantages of enhanced mass and heat transfer as well as continuously operating would be lost.

Therefore, during some 100 h nonstop runs of the pilot plant the performance of the microreactor under production conditions was analyzed. And indeed, with increasing running time of the pilot plant a steady rise of the pressure loss in the microchannels can be observed. But by a special technique, partial removal of coating out of the microreactor and a drop down of the pressure loss to the level at the beginning of starting the reaction is ensured. Although this process management is still under investigation a high process stability is obtained by using this technique.

5. Conclusion and outlook

The successful production of two typical, commercially relevant azo-pigments in lab-scale microreactors was demonstrated. Significantly improved coloristic properties were found. A pilot plant scale microreactor unit, realizing the concept of numbering-up instead of conventional scaling-up process, was assembled. The first experiment series show the reproducibility of the results found in the lab scale microdevice. Effects like fouling and clogging, in the past regarded as major problems for operating microreactors especially with respect to handling of suspensions can be managed from our experience. Future investigations will firstly focus on development of new pigments with special properties produced in microreactors and secondly on development of automated microreator units for production in industrial scale.

References

[1]   W. Ehrfeld, V. Hessel, V. Haverkamp; *Ullmann's Encyclopedia of Industrial Chemistry: Microreactors*; **1999**, *6.Ed.*, Electronic Releases, WILEY-VCH, Weinheim.
[2]   O. Wörz, K. P. Jäckel, Th. Richter, A. Wolf, *„Microreactors, a new efficient tool for optimum reactor design"* in W. Ehrfeld, I.H. Rinard, R. S. Wegeng, (Eds.): *Process Miniaturization*: 2nd Int. Conference on Microreaction Technology, Topical Conference Preprints, AIChE, New York, USA **1998**,183-185.
[3]   *"Entwicklungszeiten einsparen"* in *Verfahrenstechnik* **2001**, *35, Nr. 6*, 28-29.
[4]   V. Autze, A. Kleemann, S. Oberbeck, *Nachrichten aus der Chemie* **2000**, *48*, Nr. 5.
[5]   Ch. Wille, V. Autze, H. Kim, S. Oberbeck, Th. Schwalbe, L. Unverdorben, *Progress in Transferring Microreactors from Lab into Production – An Example in the Field of Pigments Technology,* at IMRET 6, 6th International Conference on Microreaction Technology AIChE, New Orleams, USA **2002**.
[6]   W. Ehrfeld, V. Hessel, H. Löwe: *„Microreactors"*, **2000**, First edition, 1, 9, 71, 152, WILEY-VCH, Weinheim.
[7]   W. Herbst, K. Hunger: *„Industrial organic pigments – production, properties, applications"*; **1997**, 201, second completely revised edition, WILEY-VCH, Weinheim.

Macromol. Symp. 187, 641–650 (2002)

Impact of the Dispersion Quality of TiO$_2$ Pigments in Coatings on Their Optical Properties and Weathering Resistance

Dr. Peter Gross and Dr. Josef Schmelzer

KRONOS INTERNATIONAL, INC., Technical Service Department,

Peschstrasse 5, D-51373 Leverkusen, Germany

Summary: The dispersion status achievable with different dispersing equipment of an acrylic/MFA stoving enamel was monitored in the conventional way via the Hegman-gauge and gloss, gloss haze and tinting strength measurements. Then the same drawdowns of the cured coating were submitted to diffuse reflectance spectroscopy. Original and normalised remission curves were then compared and the distribution coefficients emerging from these measurements were correlated to the conventional methods for monitoring the dispersion process. The interpretation of the spectra with respect to the particle size, particle size distribution and the concentration of particles of a particular size was achieved on a qualitative and comparative way.

1. Introduction

TiO$_2$ pigment powder as delivered to customers consists of a mixture of primary particles, agglomerates and aggregates. During the dispersion process the primary particles and aggregates are merely distributed evenly in the liquid phase, whereas the agglomerates will be broken down to smaller particles to yield as an optimum purely primary particles and aggregates. Commonly, in coatings industry the quality, progress and efficiency of the dispersion process are monitored by the use of a grindometer (for the determination of the coarse particles) as well as by means of optical measurements like gloss, gloss haze and tinting strength. These methods give an overall and average account of the dispersion status in the coating and do not allow a detailed insight.

In the year 2000 J. Schmelzer[1] proved that the use of diffuse reflectance spectroscopy of TiO$_2$ containing coatings allows a deeper insight into the particle distribution of the pigment in the coating. In course of this investigation the dispersion status of an acrylic/MFA stoving enamel as achieved with different dispersing equipment was monitored in the conventional way via the grindometer and gloss, gloss haze and tinting strength measurements. Then the same

   CCC 1022-1360/00/$ 17.50+.50/0

drawdowns of the cured coating were submitted to diffuse reflectance spectroscopy. Original and normalised remission curves were compared with the results of the conventional methods. The interpretation of the spectra with respect to the particle size, particle size distribution and the concentration of particles of a particular size was achieved on a qualitative and comparative way. Based on an earlier work[2], two different TiO_2 pigments representing easy (Pigment A) and more difficult (Pigment B) dispersing behaviour were chosen for this investigation (table 1).

Table 1. Pigments under Investigation

	Inorganic Surface Treatment			
	SiO_2	Al_2O_3	ZrO_2	Production Process
Pigment A	0.1	3.0	0.4	Sulphate Process
Pigment B	2.8	3.5	---	Chloride Process

Pigments A and B differ not only by their surfaced treatment and ease of dispersion but also by the way of manufacturing (sulphate and chloride process) which may lead to consequences as described later.

2. Experimental

2.1 Monitoring the dispersing process with conventional means

In course of this investigation the dispersing behaviour of pearl mill, high speed stirrer (saw tooth blade) and Skandex® mixer were examined. Earlier laboratory studies proved that under the applied conditions the Skandex® mixer dispersion is the optimum. The dispersing conditions for the pearl mill were chosen with regards to the mildest conditions achieving a good dispersion of Pigment A. Table 2 gives a summary of the dispersing conditions.

Table 2. Dispersing conditions applied with the different dispersing tools

Skandex		60 minutes
High speed stirrer	Time	20 minutes
	Circumference speed	20 m s^{-1}
Pearl Mill	Volume of milling chamber:	125 cm^3
	RPM:	500
	Volume of pearls in milling chamber	80%
	Pearls	ceramic
	Diameter of pearls	1.6 mm
	Through put	$\sim 4.0 \text{ kg h}^{-1}$
	Temperature	23-25 °C
	Passes	maximum 4

Table 3 shows the results achieved with the different dispersing tools monitored with the conventional evaluation tools.

Table 3. Dispersion status of pigments A and B as achieved with the different dispersing tools

	Pigment	Gloss 20°	Gloss Haze	Hegman
High Speed Stirrer	A	80	16	7.25
	B	78	39	7.00
Pearl mill, 1. Pass	A	80	18	6.50
Pearl mill, 2. Pass	A	80	18	6.75
Pearl mill, 3. Pass	A	80	18	7.00
Pearl mill, 4. Pass	A	80	18	7.25
Pearl mill, 1. Pass	B	72	99	6.00
Pearl mill, 2. Pass	B	74	80	6.25
Pearl mill, 3. Pass	B	75	76	6.75
Pearl mill, 4. Pass[*3]	B	75	75	7.00
Skandex	A	80	16	7.25
	B	78	46	7.00

The results show that Pigment A is indeed easy to disperse in the coating system used. A perfect dispersion of this pigment can be achieved with all dispersing methods, even at circumference speeds as low as 5 m s^{-1} of the high speed stirrer. In the pearl mill dispersion a differentiation of the passes by optical measurements is not possible. Here only the grindometer readings show that there is a progress in dispersing. However, measurements show a readily achieved destruction of the gloss and gloss haze disturbing smaller agglomerates.

The dispersing behaviour of pigment B is noticeably different:

1. Perfect dispersion in all parameters monitored is achieved with the Skandex and the high speed stirrer only.

2. In the pearl mill dispersion process deficiencies in gloss and gloss haze remain even under much more vigorous conditions. The grindometer readings indicate perfect dispersion. Thus, whereas coarse agglomerates of Pigment B are broken down it appears as if the smaller agglomerates resist a break down.

---

[*] After 4 passes through the pearl mill noticeable greying of the white coating could be observed with Pigment B which resulted in a diminution of the Brightness L* from 98.1 to 97.9. This was not observed with pigment A, a sulphate process pigment with which the Brightness L* remained unchanged at 97.3. This demonstrates the known higher abrasivity of chloride process pigments compared to sulphate process pigments.

2.2 Diffuse reflectance measurements (remission)

The contribution of a TiO₂ containing coating to the remission at a specific wavelength is mainly due to particles of a specific size (approximately half of the wavelength). Thus a spectral comparison of TiO₂ containing coatings could reflect the details about the particle size distribution changes occurring during the dispersing process and/or give account of the differences in the dispersing behaviour of the pigments. Figures 1 and 2 show typical remission spectra in the visible light region yielded from white TiO₂-pigment containing coatings

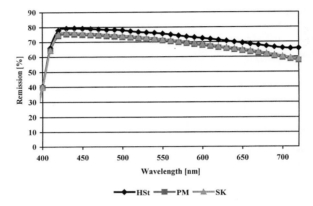

Figure 1. Remission spectra for Pigment A: Comparison Skandex mixer (Sk), pearl mill (PM) and high speed stirrer (HSt)

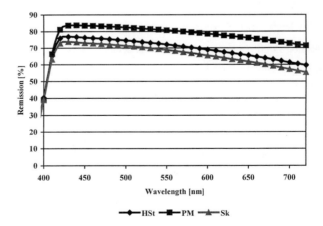

Figure 2. Remission spectra for Pigment B: Comparison Skandex, pearl mill and high speed stirrer

The intensities observed with a coating where the dispersion of the pigment was carried out with a Skandex mixer should be highest since this mixer yields the best dispersion. This is valid for coatings of identical thickness but for drawdowns as measured here the variation of coating thickness appears to be too large.

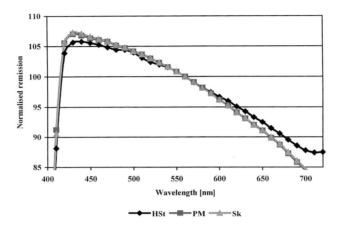

Figure 3. Normalised remission spectra for Pigment A: Comparison Skandex mixer, pearl mill and high speed stirrer

For these reasons normalisation of the spectra was carried out as described by Schmelzer[1] to allow a qualitative discussion of the particle size distribution. Normalisation was carried out by fixing the remission at 560 nm to be 100 % and calculating the "new" remission values at the other wavelengths. This procedure leads to the spectra as shown in figures 3 and 4.

In the following the term "smaller particles" refers to the pigment particles which account for the remission at wavelengths below 560 nm and the term "bigger particles" to the pigment particles accounting for the remission at wavelengths above 560 nm. In the same sense the terms "lower" and "higher" wavelengths are used.

The normalised spectra show that with the Skandex mixer the spectra exhibit highest remission at lower wavelengths and lowest remission at higher wavelengths. In terms of "particle size distribution" the Skandex mixer yields the highest portion of smaller, pigmentary particles. Thus, according to the remission spectra, too, the Skandex mixer is the most effective dispersing tool. For Pigment A the pearl mill dispersion leads to identical results as the Skandex mixer. With the high speed stirrer the portion of "bigger" particles is increased. This different dispersing behaviour is not observed with the conventional methods.

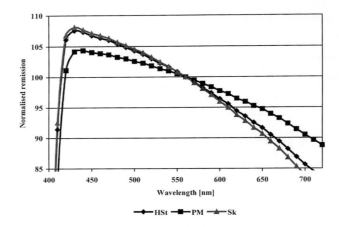

Figure 4. Normalised remission spectra for Pigment B: Comparison Skandex mixer, pearl mill and high speed stirrer

Dispersing with a high speed stirrer occurs to be more effective for Pigment B. Here the pearl mill dispersion is less effective which is in accordance with the "traditional" evaluation of a good dispersion. Figures 5 and 6 show the remission spectra of the progressing dispersion achieved after each pass through the pearl mill for pigment A and B respectively.

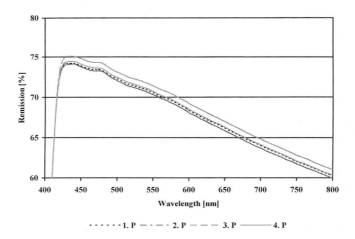

Figure 5. Remission spectra for Pigment A: pearl mill dispersion 1st to 4th pass

Uniformly, the remission level of the chloride grade pigment (B) is on a higher level compared to the sulphate grade pigment (A). This corresponds to normal optical measurements, where the chloride grade pigments exhibit a higher brightness L* (cf. footnote *, page

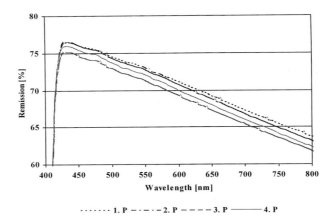

Figure 6. Remission spectra for Pigment B: pearl mill dispersion (1st to 4th pass)

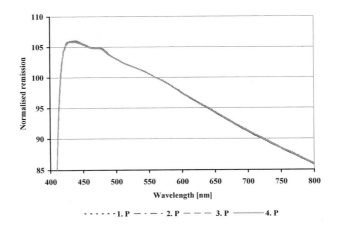

Figure 7. Normalised remission spectra for Pigment A: Comparison pearl mill dispersion (1st to 4th pass)

After normalisation of the spectra (figures 7 and 8) the spectra of the different passes of pigment A are identical. The different grindometer readings find no reflection in the spectra with regards to changing particle size distribution as a result of the progressing break down of coarse agglomerates.

The spectra of pigment B, however, differ noticeably. The biggest change is recorded between the 1st pass and the 2nd pass. Here the ratio bigger pigment particles to smaller pigment particles changes most. The changes between 2nd pass and 4th pass are only marginal. Again, this is found in the measured gloss and gloss haze values as listed in table 2, too. Here the biggest progress in dispersion was observed between pass 1 and pass 2.

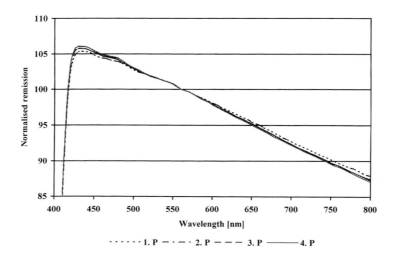

Figure 8. Normalised remission spectra for Pigment B: Comparison pearl mill dispersion (1st to 4th pass)

Distribution values and weathering

As seen before, dispersing pigment A with different machines gives different remission spectra although the values for gloss, gloss haze and grindometer do not indicate a difference. A good tool to quantify this difference in the remission spectra is the so called distribution values[4]. The distribution values are determined through the normalised remission curves. From those the ratio of the areas underneath the remission curve from 380 nm to 560 nm and from 560 nm to 720 nm is taken. For pigment A the ratios vary between 112 for the high speed stirrer to 115 for the pearl mill or Skandex mixer respectively.

In an earlier investigation coatings were prepared with pigment A exhibiting different dispersion values in its normalised remission spectra. The coatings were submitted to accelerated weathering. Figure 9 illustrates the correlation between distribution values of the coatings and the exposure time after which the deterioration of gloss to 50 % and the onset of chalking.

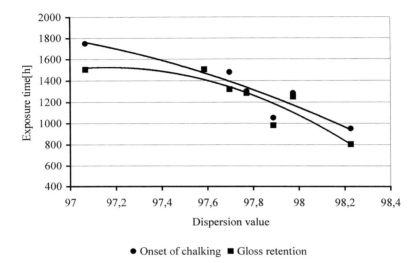

Onset of chalking ■ Gloss retention

The results give the impression that a good dispersion is not desirable, since the weathering results improve up to almost 50 % with a poorer dispersion status. This is expected since with a higher portion of fine particles the surface area of the TiO_2 increases considerably and with this its photocatalytic activity. However, a good dispersion leads to a reduction of the amount of TiO_2 pigment necessary for the desired UV-protection and the optical properties of the coating and is, therefore, counterbalancing the higher photocatalytic effect.

3. Summary

The dispersing process with different dispersing tools was monitored with the common methods like gloss, gloss haze and grindometer readings and compared to the corresponding remission spectra. Differences in the dispersing behaviour of the pigments could be linked to differences in the remission spectra. For pigment A, an easy to disperse pigment in the acrylic-melamine coating system the remission spectra proved, that the pearl mill is the adequate dispersing tool, whereas for pigment B the high speed stirrer proved to be superior. A correlation between remission spectra and gloss and gloss haze measurements were found, whereas the progress in the break down of coarse particles as monitored with the grindometer finds no obvious reflection in the spectra. The different action of the dispersing tools were revealed with pigment A where the traditional evaluation tools for a good dispersion give identical results. Whereas the dispersion of pigment A with a pearl mill gives identical results as the Skandex mixer with regards to the remission profiles (figure 3), the high speed stirrer dispersion results in a lower portion of "smaller" particles. Finally, the accelerated weathering correlates well to the distribution values and demonstrates that the weathering results can be influenced strongly by the dispersion status of the coating.

References

[1] J. Schmelzer, Farbe & Lack, 2/2001, 37-45, Mit Remissionskurven Pigmentverteilungen bestimmen

[2] P. Gross, APi Konferenz 2001: Untersuchung über das Dispergierverhalten verschiedener Titandioxid- pigmente mit unterschiedlichen Dispergieraggregaten

[3] B. Vielhaber-Kirsch, E. W. Lube, Abrasion characteristics of titanium dioxide pigments, KRONOS INFORMATION 6.30

[4] J. Schmelzer , KRONOS TITAN GmbH & Co. OHG, Charakterisierung der Verteilungsgüte von Weißpigmenten in einer Matrix, Offenlegungsschrift DE 10043038

Macromol. Symp. **187**, 651–661 (2002)

Pigment Encapsulation by Emulsion Polymerisation, Redespersible in Water

*Ph.Viala[1], E.Bourgeat-Lamy[1], A.Guyot[*1], P.Legrand[2], D.Lefebvre[2]*

1 CNRS-LCPP,CPE-Lyon, BP 2077,69616,VILLEURBANNE, France

2 Peintures JEFCO, MARSEILLE, France

* to whom correspondence should be adressed

Summary: Emulsion Polymerization was carried out in the presence of inorganic pigments such as TiO2, black FexOy, yellow FexOy, and red FexOy, and NP 30 as surfactants, and water soluble AZO compounds or KPS as initiator. Monomers with specific hydrogen bonding interaction must be used in the initial steps of polymerisation, methylmethacrylate and vinylacetate being the most convenient. Then a semi continuous feed of a mixture of monomers was carried out in starved conditions. In order to make the covered pigments water-redispersible a mixture of hydrophobic and hydrophilic monomers should be chosen with proper pH conditions. The amount of surfactant has to be chosen so that no agglomeration of the covered pigments take place. The covered pigments were dried upon lyophylisation, then formulation of powder paints was carried out using commercial powder binders and other additives. Good properties of the paints, such as brightness were obtained in that way.

Keywords: Pigments, Encapsulation, Emulsion Polymerisation, Acrylic copolymers, Titanium dioxide, Iron oxide

INTRODUCTION

Both patents [1-5] and open litterature papers [6-13] have been devoted to small particles encapsulation using emulsion polymerisation or related processes. Most often the obtained materials display a hybrid character, because the small particles are inorganic (silica, metal oxides, carbon black, etc ...) which are encapsulated by organic polymers (polystyrene, acrylic polymers, etc ....).

In the paint industry, one of the major problems is the dispersion of the pigments in either organic medium (solvent-based paints) or in aqueous medium (waterborne paints). Due to ecological reasons, this last category of paints is gaining more ane more interest. Encapsulation of the pigments with an organic polymeric coating offers an obvious possibility

CCC 1022-1360/00/$ 17.50+.50/0

of dispersion in an organic medium. Because emulsion polymerisation is carried out in water medium, it should be a good way for the dispersion of pigments in water medium.

Then several works have already been published in that direction[14-22]. One of the earlier work was done by Carris et al. [14,15,16]. These authors have used a TiO_2 pigment modified with an organotitanate unsaturated coupling agent Such modified pigment have strong bonds both with the pigment, through the alcoxyde group able to react with the surface OH groups of the pigment, and the unsaturated group able to participate in the radical polymerisation as a monomer, or ,possibly as a transfer agent. This work was continued[17,18], through copolymerisation of styrene or Methylmethacrylate (MMA) with the titanate coupling agent, which was followed by conductrimetry. By this way, it was possible to follow and detect the various steps of the process, initiation of the polymerisation , limited flocculation etc... the same monomers were used by Haga et al.[19] who have shown the importance of the pH, and other parameters, such as the nature of the initiator ; wh en the pigment and the growing polymer have opposite charges, a thick polymer shell is formed around the pigment particle, which cannot be extracted. They have studied also the thermal properties of that shell. A different approach has been used by the team of Templeton-Knight [20,21]. These authors did carried out the emulsion polymerisation, using ultrasonic stirring, at least during the first steps of the process ; then, the thermal decomposition of the persulfate initiator is strongly accelerated, and the polymerisation rate is also increased ; another effect is the disruption of the TiO_2 aggregates, so resulting in a much better encapsulation. Much more recently, in a series of 3 papers, Erdem et al.[22]have discussed the encapsulation of Titanium dioxide in miniemulsion polymerisation of styrene. They used both hydrophilic and hydrophobic pigments . Both pigment particles have a nominal primary diameter of 29 nm. In the first paper,a series of block-copolymer surfactants were tested, and one of them, polybutene-succinimide pentamine (OLOA 370) was retained and studied in more details. Using this dispersant, very good and stable dispersions were obtained in cyclohexane after sonification of the mixture with SDS as surfactant,which allows a very extensive de-agglomeration of the titanium dioxide particles. The smallest particle size were obtained from hydrophilic pigment (40-45 nm) and 1-2% of OLOA, which is almost totally retained at the particle surface and cannot be desorbed again. The second paper describe the transposition of the study to the styrene miniemulsion, while the third paper deals with the polymerisation. Here hexadecane and a small amount of polystyrene were used as hydrophobe to stabilize the miniemulsion. Density gradient column was used to study the encapsulation efficiency, allowing to estimate the distribution of the number of pigment particles by polymer particle. The maximum encapsulation efficiency was shown to be 83% for TiO_2 and 73% for polystyrene. The largest particle size was 209nm containing up to 22 pigment particles in the case of the hydrophilic particles. The hydrophobic lead to poorer encapsulation, with efficiencies around 60% for both polystyrene and pigment. It seems very difficult to obtain a 100% encapsulation efficiency, due to many causes, such as homogeneous nucleation, some remaining large aggregates of pigment and less than 100% of the droplets nucleated.

In the present paper, we describe a rather simple encapsulation procedure of commercial pigments (white TiO_2,,red, yellow or black iron oxides) based on emulsion polymerisation of acrylic monomers. An important feature of the work was to obtain an hydrophilic enough shell of copolymers, so that, after drying , the pigment can be dispersed again in water, upon gentle stirring. Then it should be considered as a component of powder paints. Indeed the powder paints present definite advantages as compared with the conventionnal water-borne paints, being stored in paper bags, and not needing biocides additives, and being able to support long term storage without ageing problems. Most of the work done about the development of redispersible powder paints has been devoted to the problem of binders[23]. Conventionnal binders, but chiefly those from vinylacetate copolymers, can be used ; theredispersion need to use hydrophilic protecting colloids, mainly polyvinylalcohol. The latter, make easy the coalescence of the binder, and then the particles have to be protected against that effect, and in most cases this is done through addition of charges such as montmorillonite.

In our knowledge much less efforts have been done to develop a redispersible pigment.

EXPERIMENTAL

*Materials :*The pigments are Titanium dioxide RXL from Tioxide, and iron oxides from Bayer : Red OXR 110 M, Yellow OXJ 915, and Black OXN318 M. They have been characterized by elemental analysis, Infra red spectroscopy, surface titration of OH groups, and particle size distribution by quasielastic light scattering.

The monomers, methylmethacrylate acrylic acid, methacrylic acid and others monomers (from Aldrich) were used as received, as well as the surfactants and initiators, all from Aldrich, except the azo compounds from Wako. Deionized water was used all along the experiments.

Pigment dispersion : The reactor was first filled with deionized water and the the surfactants were added, most often a mixture of Sodium dodecylsulfate (SDS) and ethoxylated nonylphenol (NP 40) with about 40 ethyleneoxide units. Finally the pigment was introduced under vigourous stirring. The dispersions were characterized by their pH and their isoelectric point, using a Zetasizer of Malvern, and finally by their average particle size and their particle size distribution, using either a sedimentometer (Br)ookhaven BL DCP Particle sizer) or a granulometer laser (Coulter LS 130) . The sedimentometer allows to determine the T 50 in the cumulative particle size distribution, which is the size corresponding to 50% of the sedimented weight , while the granulometer shows a differential distribution curve in the range between 0.1 and 800 µm.

Encapsulation through emulsion polymerisation : To the dispersion of pigment a slightly watersoluble monomer , such as methylmethacrylate (MMA) is added is added in amount lower than the limit of solubility. Then the temperature of the reactor is established and a semicontinuous addition of the mixture of monomer is added at a slow rate corresponding to starved conditions. Finally there is a cooking step of 2 hours to eliminate most of the residual monomer. After encapsulation the particle size and size distribution are measured , using the granulometer Coulter LS 130.

Water redispersion : Before redispersion, the product is dried through lyophilisation. The coloured powder obtained in the following conditions : congelation –40°C , sublimation – 30°C, final vacuum drying 40°C. The powder is finally grinded and stored in paper bags . For redispersion, 100g of dried powder are pourred in 400 ml of water and gently stirred to obtain a concentrate pigment paste.

<u>*Paint tests :*</u> In order to have a good estimate of the pigment dispersion, it is introduced in a varnish formulation given in Table 1.

Table 1. Varnish formulation

The pigment powder18.2 g) is first dispersed in 46.1 g of water containing 0.74 g of an anti-foam agent under stirring 5 min (250 rpm)
Then the following items are added : acrylic resin (mowlith DN 777) 127g, bactericide 0.36g, texanol (coalescence agent) 1.6 g, White spirit (coqlescence agent) 2.4 g, Thickener (viscotex) 2.4 g and ammonia 1g
A final mixing step of 10 minutes is carried out.

.When the dispersion seems good enough with only a few agglomerates, three items are measured : gloss at 60°, gloss at 85° and contrast ratio R.

Further estimate involve some comparison between a reference formulation of external paint and an experimental one containing the encapsulated pigment.

RESULTS AND DISCUSSION

The elemental analysis of the pigments is reported in Table 2, together with the number of surface OH groups. And indication of their acido-basic character (for instance, the ζpotential of the titanium dioxide is -21 at pH 7). The pigment dispersion state is strongly dependent on the surfactant system associated with it, and the best results obtained are shown in Table 3.

Table 2. Pigment analysis

Pigment	Ti%,Si%,Al %	C %	N oh $\mu mol/m^2$	Area m^2/g	character
TiO_2 RXL	45.2,4.8,2.3	<0.2	3.5	26	basic
Red,αFe_2O_3		<0.2		14.4	acidic
Yellow $\alpha FeOOH$		<0.2	20.0	20.4	acidic
Black,γFe_2O_3		<0.2	14.9	14.3	acidic

Table 3. Pigment dispersions in water

Pigment	Surfactant system, g	Particle size, nm	Poly.index
TiO_2 RXL	SDS 1.6 – NP40 1	330	1.4
Red,αFe_2O_3	SDS 1.6 – NP40 1	290	1.2
Yellow, $\alpha FeOOH$	SDS 1.6 – NP40 1	1080	1.4
Black,γFe_2O_3	SDS 1.6 – NP 40 1	3500	2

The coloured pigments are more difficult to be dispersed and their particle size remains rather big and polydisperse, except for the red pigment.A typical picture from Scanning Electron Microscope(SEM) is shown in figure 1.

In a first step, the encapsulation of the Titanium dioxide pigment from emulsion polymerisation has been carried out , using methylmethacrylate as monomer only. A variety of initiators have been tested and the bests were found to be Potassium persulfate (KPS) and a watersoluble azo VA86 , HOCH$_2$CH$_2$NHCO C(CH$_3$)$_2$N=N C(CH$_3$)$_2$CONH CH$_2$CH$_2$OH. Both initiators lead to high conversions within five hours before the cooking step.

Figure 1. SEM image of Titanium dioxide not yet encapsulated

The amount of MMA added in the semi batch process was calculated in order to cover the pigment particles with a homogeneous shell of 50 nm thick. However, it turns that the particle size increases from 330 to 830 (VA86) or 900 (KPS) nm, and the powder cannot be redispersed. A typical picture from SEM is shown in figure 2.

Figure 2. SEM image of Titanium dioxide encapsulated with emulsion copolymerisation of a mixture 3/1 of MMA and MAA, not neutralized

It was decided to add to the MMA a watersoluble monomer. A few methacrylic polyethyleneoxide(PEO) macromonomers were tested first. Promising results, in terms of redispersibility and particle size were obtained when the number of EO units was limited. A variety of watersoluble monomers were tested as well, and the best results have been obtained, using Methacrylic acid or acrylic acids in rather high amounts (25% versus MMA) . Typical data are reported in Table 4.

Table 4. Data of encapsulation of Pigment RXL using carboxylic acid monomers

Comonomer	% evrsus MMA	Varnish Test	Particle size μ	Poly. index
Methacrylic acid	5	A	3.7	1.4
	12.5	RG	2.5	1.1
	25	G	5.1	1.01
Acrylic acid	5	A	6.1	1.2
	12.5	RG	3.8	1.5
	25	RG	3.8	1.1
AHPSulfonate	5	B	0.43	3.8
V pyrolydone	5	B	2.05	1.7

A= acceptable ,B=bad, RG=rather good, G=good
The data of particle size and polydispersity index are obtained after grinding the powder

These products showed still a rather big particle size, even if the distribution is not very large as shown in figure 3.

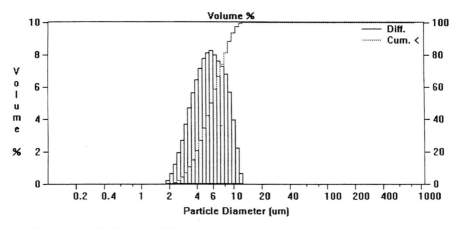

Figure 3.Particle size distribution of Titanium dioxide pigment encapsulated from emulsion copolymerisation of a mixture unneutralized of MMA and MAA (3/1)

More detailed study did show that floculation took place after about 30% conversion. In order to avoid that floculation, it was tempted to neutralized, at least partially , the carboxylic groups, so providing additionnal electrostatic stabilization to improved the stabilization given by the surfactants.Then,it can be seen in figure 4 that either the floculation do not take place or lead to agglomerate which can be destroyed upon grinding.. Corresponding data are reported in Table 5 ; showing results of the test of addition in the varnish formulation.

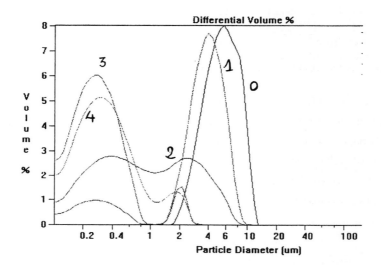

Figure 4. Particle size distribution of Titanium dioxide pigment encapsulated from emulsion copolymerisation of a mixture   of MMA and MAA (3/1)neutralized partially through addition of soda

While the data reported in Table 4 indicate the a rather large amount of carboxylic monomer is needed, Methacrylic acir being the best monomer and that the other monomers failed to give acceptable results in terms of paint tests, those reported in table 5 show that partial neutralisation of the acid groups lead to a definite improvement of the properties of the pigment powder in the paint formulations . However, it is clear that there are optimal neutralisation conditions which seemed to correspond to 2 g of added soda .The best quality seems to involve,in terms of gloss and contrast ratio, not the smaller size, but a polydisperse distribution of small size particles with 2 populations

Table 5. Data of encapsulation of Pigment RXL using partilly neutralized methacrylic acid

	%MAA	Soda g	Size μ	P.Index	N pop	V Test	G 60°	G 85°	R
Ref.	30	0	5.5	1.1	1	G	5.9	17.4	82
1	30	1	3.6	1.1	1	G	10.6	55.2	86.2
2	30	2	0.83	1.8	2	VG	26.9	68.5	90.6
3	30	3	0.33	1.5	2	G	15.6	34	85.4
4	30	4	0.28	1.5	1	G	26.6	23.2	83.2
5	30	10	0.52	2.3	2	RG	16.2	22.2	85.5

V.Test = quality of the varnish : G=good, RG=rather good, VG= very good
G 60° = gloss at 60° , G 85°= gloss at 85°, R= contrast ratio
P index = polydispersity index , N pop= number of particle populations

The encapsulated RXL titanium dioxide pigment was used also in paint formulations with a variety of commercial charges with good sucess. A high contrast ratio can be reached with less Titanium, but the abrasion resistance in wet conditions is slightly lower, but still better than powdered pigments recently introduced in the market.

Similar studies have been carried out with coloured pigments with good sucess. Some results with the three coloured pigments mentionned in Table 2 and 3 are reported in Table 6.

Table 6. Data for iron oxides encapsulation

Run	Soda g	Size μ	P index	N pop	V Test	G 60°	G 85°	R
1 Red	0				A	9.9	28.1	99.4
2 Red	4 VA86	0.77	1.3	2	VG	20.1	51.3	99.3
3 Red	4 KPS	1.25	1.4	2	VG	29.8	65.5	99.9
4 Black	4	2.8	1.4	2	VG	26.7	63.7	99.9
5Yellow	10	0.54	1.3	1	A	11.5	12.7	90.5
6Yellow	5	0.44	1.1	1	VG	36.7	62.1	97.6
7 Red	4 V501	2.65	1.2	2	G	25.2	46.1	99.9

All the samples have the same composition : MMA/ MAA = 3/1
P index = polydispersity index , N pop = Number of particle populations
V.Test = quality of the varnish : G=good, RG=rather good, VG= very good, A=acceptable
G 60° = gloss at 60° , G 85°= gloss at 85°, R= contrast ratio

Most of the trials have been very successfull, chiefly concerning contrast ratios, which are maximum in most cases. With the red pigment a cationic initiator lead to poorer results than the more classical initiators. Very good results for paints are obtained even if the particle size is big, as it is in the case of the black pigment ; in this case it seems that the polymerisation process helps to break the agglomerates of the initial dispersion, because, as shown upon comparing the data of Table 3 and 6, the final size is smaller than the initial one. It is possible that this feature can be a general one, where there is an equilibrium between agglomeration and dispersion during the encapsulation process. Such feature is illustrated in figure 5, where one can see the particle distribution corresponding to run 2 where a small population of bigger particles can be observed. Run 5 with the yellow pigment shows again that the best results are not obtained when the degree of neutralisation of the carboxylic groups in the copolymer is too high .

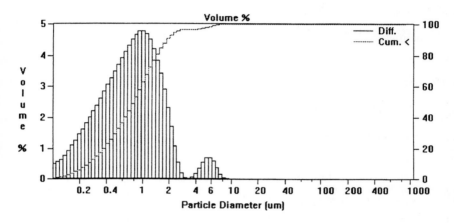

Figure 5.: Particle size distribution of Hematite red pigment encapsulated from emulsion copolymerisation of a mixture of MMA and MAA (3/1) partially neutralized upon addition of 0,1,2,3 or 4 g of soda.

CONCLUSION

Emulsion copolymerisation of MMA associated with a hydrophilic monomer such as MAA carried out in the presence of a mineral pigment (TiO_2 or Iron oxides of various colors) allows to obtain, upon partial neutralization of the carboxylic acid with soda, a pigment dispersion, which can be dried, grinded in powderpowder and then easily redispersed in water. The product can be used as a component of the formulation of a powdered paint, which are known to be a real progress from the ecological point of view, because they can be stored simply in paper bags for long time without biocide components, have no VOC and can be used leaving a minimum of residues which must be destroyed. At variance with other products recently introduced in the market, these powdered pigments affect only hardly the performances of the paints in the presence of humidity.

We think, however that progress should be made chiefly through optimisation of the process, for instance upon better analysis of the particle size distribution during the polymerisation, to obtain the best compromise between the dispersion and the trend of the particles to agglomerate. Another progress should be to extend the encapsulation to the organic pigments, and promising results have be obtained in our laboratory, using miniemulsion polymerisation (ref.24).

Finally, another kind of progress should be to combine, in the same powder both the pigment and the binder. We have begun to work in that direction, and have obtained just a few promising results, but a lot of work is still needed to get success in this connection.

References

1) US Pat.3,133,893, 1964, P.Newman, Cyanamid
2) US Pat.3,544,500,1970,D.Wilfrid,J.Osmond, ICI
3) US Pat. 4,421,660,1983,J.Solc, N.Halna,M.Mich , Dow
4) US Pat. 4,680,200,1987,J.Solc, M.Mich , Dow
5) US Pat. 4,608,401,1986,M.Martin, Union Carbide
6) T.Yamaguchi,T.Ono,Angew.Makromol.Chem.,*32,*177,1973 and *53,*65,1976
7) M.and K.Arai, J.Polymer Sci., Polymer Chem Ed., *20*,1021,1982
8) N.Konno,K.Shimizu, J.Polymer Sci., Polymer Chem Ed.,*23*,223,1987
9) M.Hasegawa, K.Arai, J.Polymer Sci., Polymer Chem Ed.,*25,*3117,3121,1987
10) P.Godard,J.P.Mercier, Double liaison, *387-388,*19,1988
11) K.Nagai,Y.Onishi, J.Applied Polymer Sci, *38*,2183,1989
12) J.S.Park, E.Ruckenstein, Polymer, *31*,175,1990
13) Ph.Espiard, E.Bourgeat-Lamy, A.Guyot, Polymer, *36,*4383,4391,4397,1995
14) C.H.M.Caris,L.P.M.Van Elven, A.M.Van Herk, A.L.German, FATIPEC 19, 1988 and 20, 1990
15) C.H.M.Caris,L.P.M.Van Elven, A.M.Van Herk, A.L.German, Brit.Polymer J., *21*,133,1989
16) C.H.M.Caris,R.P.M.Kuijspers, A.M.
17) R.Q.F.Janssen,A.M.Van Herk, A.L.German, FATIPEC 22,*1*,104,1994
18) R.Q.F.Janssen,A.M.Van Herk, A.L.German,J.Oil Colour Chem.Assoc.,*11*,455,1993
19) Y.Haga, T.Watanabe, Angew.Makromol.Chem.,*188*,73,1991 and *189*,23,1991
20) J.P.Lorimer, R.Templeton-Knight, Colloid Polymer Sci.,*269*,392,1991
21) J.Pal,S.Baydal, R.Templeton-Knight,J.Chem Soc. Faraday Soc.,*87*,991,1991
22) B. Erdem , E.D., Sudol ,V.L.Dimonie ,M.S. El- Aasser , J.Polymer Sci. Polymer Chem., *39*,4419,4431,4441,2000
23) Baumgartl, Zeh,Buelow, Proceedings EUROCOAT 96, Genoa,Sept. 1996
24) S.Lelu,C.Novat,C.Graillat,A.Guyot, E.Bourgeat-Lami, Polymer Internat. (Submitted)

Macromol. Symp. **187**, 663–675 (2002)

The Influence of Selected Synthetic Aluminium Silicates on Physicochemical Properties of Emulsion Paints

Andreas Feller, Hans-Dieter Christian

Degussa AG, Business Unit "Advanced Fillers & Pigments", Hanau, Germany

Summary: Synthetic aluminium silicates (AS) consist of amorphous particles with diameters between 5 and 10 µm. They are suited as spacers between the dispersed titanium dioxide pigment particles in emulsion paints, thereby providing an excellent distribution of the pigment. This effect makes a partial replacement up to 50% of titanium dioxide pigments by AS possible. In general the whiteness and the contrast ratio of an AS containing emulsion paints are improved, as well as the scrub resistance in particular cases. The usage of AS also has a favourable influence on the chroma and the brilliance of coloured systems. All these positive effects are primarily controlled by the particle size, porosity, and oil adsorption of the AS. For this reason synthetic aluminium silicates are not just ordinary fillers, but real functional pigment extenders.

Keywords: coatings; silicas; aluminium-silicate; emulsion-paint; whiteness

1. Introduction and History

Precipitated silicas and silicates first entered the coatings and printing inks industries on a sizeable scale around 50 years ago. Synthetic silicas and silicates are now firmly established as matting agents, thickeners, high-grade fillers and pigment extenders. From 1970 to 1999, worldwide production of precipitated silicas and silicates increased from 400,000 tonnes in to 1,100,000 tonnes. In emulsion paints and other decorative coatings, synthetic aluminium silicates are used as a partial replacement for titanium dioxide pigments.

2. Preparation of Synthetic Aluminium Silicates

Starting materials used in the production of precipitated aluminium silicates (AS) are solutions of alkali metal silicates, preferentially sodium silicate, from which amorphous aluminium silicate is precipitated by adding sulphuric acid and aluminium sulphate. Calcium silicate is obtained by using calcium sulphate for the precipitation instead of aluminium sulphate.

CCC 1022-1360/00/$ 17.50+.50/0

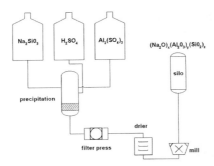

Figure 1. Industrial Production Process of Precipitated Silicas

The following schematic equation (not stoichiometrically balanced) illustrates the precipitation process:

$$Na_2SiO_3 + Al_2(SO_4)_3 + H_2SO_4 \rightarrow (Na_2O)_x (Al_2O_3)_y (SiO_2)_z + Na_2SO_4 + H_2O$$

By varying the major precipitation parameters such as temperature, pH, electrolyte concentration and time it is possible to provide AS with different morphology. Through the appropriate selection of precipitation conditions, AS with specific surface areas in the range from 50-200 m^2/g can be produced. The silica suspension obtained in the precipitation process is transferred to the filter press, where the salts formed during precipitation are washed out, and as much of the water as possible is removed. The resulting material is dried and ground in special mills.

3. The Structure of Synthetic Aluminium Silicates

The precipitation process results in the formation of fine, amorphous primary silica particles as determined by x-ray, which associate with other particles to form aggregates and agglomerates. Additional precipitation causes increased agglomerate formation, through the interaction of monomeric or oligomeric silica particles. Within these agglomerates the original primary particles, with an average diameter of about 15-50 nm, remain identifiable.

Figure 2 shows the transmission electron micrographs of four different synthetic aluminium silicates. The aggregates and agglomerates are evident as concretions with an approximate diameter from 0.1 to 1 μm. Synthetic aluminium silicates differ from natural products in several respects:

1. The specific surface area of synthetic products is higher than in natural products.

2. Synthetic products are more uniform.

3. Synthetic products are amorphous whereas natural products containing silica may be crystalline or partly crystalline.

4. Synthetic products possess a greater whiteness than natural products, even when the latter have been cleaned.

Even between the synthesized aluminium silicates, slight coloristic differences are observed. The various products can be distinguished simply by the eye as a result of their different whitenesses. These differences can be attributed to the iron content and the associated yellowing of the material.

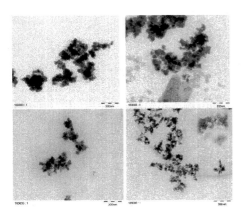

Figure 2. 4 TEM micrographs of different aluminium silicates (on a scale of 100,000:1)

4. Physicochemical Data of Synthetic Aluminium Silicates

Table 1. Physicochemical data

	SAMPLE A	SAMPLE B	SAMPLE C	SAMPLE D	SAMPLE E
Specific surface m²/g	80	111	51	67	64
Tamped density g/l [1]	288	397	386	228	227
Loss on drying 2 h/105°C % [2]	4.7	8.3	8.0	8.9	7.9
pH, 5% [3]	9.9	7.5	10.4	10.8	10.7
DBP adsorption [4]	160	107	150	160	173
Particle size d50%	7.1	4.7	8.9	5.4	9.1
Particle size d5%	15.2	12.6	20.3	12.2	22.7

	SAMPLE A	SAMPLE B	SAMPLE C	SAMPLE D	SAMPLE E
Standard colour value Y [5]	97.6	98.0	98.2	98.2	97.9
Whiteness (Berger)	96.6	96.7	95.4	95.4	95.4
Fe content	168 ppm	172 ppm	106 ppm	185 ppm	214 ppm

[1] (DIN ISO 787-11), [2] (DIN ISO 787-2), [3] (DIN ISO 787-9), (original material)[4], (DIN 53163)[5]

5. Experimental Section

5.1 Formulas

Table 2. Test formulations for replacement of titanium dioxide pigments by synthetic aluminium silicates

Formulation for an interior emulsion paint	R1 mass %	R2 mass %	R3 mass %	R4 mass %
Water	18.00	18.00	18.00	18.00
Cellulose ether, Tylose MH 6000	0.30	0.30	0.30	0.30
Sodium hydroxide 10%	0.10	0.10	0.10	0.10
Defoamer, Tego 8050	0.20	0.20	0.20	0.20
Dispersant, Dispex N 40	0.20	0.20	0.20	0.20
Dispersant, Calgon N 10%	0.70	0.70	0.70	0.70
Aluminium silicate (samples A to E)		2.5	5.00	7.50
Titanium dioxide (standard, medium and highly treated)	10.00	7.5	5.00	2.50
Talc, Luzenac 00C	7.00	7.00	7.00	7.00
Talc, Naintsch ASE	3.00	3.00	3.00	3.00
Omyacarb 5 GU; $CaCO_3$ d50 = 5 µm	25.00	25.00	25.00	25.00
Omyalite 90; $CaCO_3$ d50= 1 µm	7.00	7.00	7.00	7.00
Sty/Acr or VAc/E binder	16.35	16.90	17.40	17.90
Thickener, Acrysol RM 8	0.10	0.10	0.10	0.10
Defoamer, Tego LA-E511	0.20	0.20	0.20	0.20
Preservative, Actizide AS	0.10	0.10	0.10	0.10
Water	11.85	11.30	10.80	10.30
Total	100.00	100.00	100.00	100.00
PVC	70.0	70.0	70.0	70.0

5.2 Experimental Model

Producers of Emulsion paints want to produce high-quality products as economically as possible. This is one of the reasons why some of the titanium dioxide pigment is replaced by synthetic aluminium silicates. Accordingly, five commercial pigment extenders were compared in two industry standard formulations against three commercial titanium dioxide pigments. These titanium dioxide pigments received different aftertreatments (standard titanium dioxide pigment/low treated, TiO_2 content (ISO 591) $\geq$ 94%; special emulsion-paint titanium dioxide pigment/medium treated, TiO_2 content (ISO 591) $\geq$ 88%; special emulsion-paint titanium dioxide pigment/highly treated, TiO_2 content (ISO 591) $\geq$ 82%). All of the formulations were prepared in three different pigment volume concentrations (PVC = 70%, 75% and 80%) by varying the amount of binder. The experiments were conducted with a styrene-acrylate binder (Sty/Acr) and with a binder based on vinyl acetate-ethylene (VAc/E).

5.3 Experimental Procedure

All of the investigations were conducted in emission free and solvent free interior emulsion paint, whose formulation is given in Table 2. The emulsion paints were manufactured under conventional conditions using a laboratory dissolver whose disc was set at a rotational speed of 7 m/s (2 000 rpm). After the dispersing process, the binder was added with slow stirring. After ageing for one day, the paints were applied using a bar coater (200 and 400 μm slot height) to contrast cards (Byk-Gardner, No. 2801) for optical measurements and to Leneta test panels (P121-10N) to determine the wet abrasion resistance, in accordance with DIN ISO 11998.

The optical properties such as hiding power, whiteness and residual gloss at 85° (sheen) were measured after drying at 23°C and 50% relative humidity (RH) for 24 hours. The various wet abrasion resistance tests were conducted after storage at 23°C and 50% RH for 28 days.

5.4 Results

5.4.1. Hiding Power, Whiteness and Wet-abrasion Resistance

The hiding power is the ability of a dispersion coating to mask large differences in colour and lightness in the substrate. The hiding power should be as high as possible in order to achieve this even at low dry film thickness. The extent of the hiding power can be determined by way of the contrast ratio, comparing the lightness of a white paint on a black substrate with its lightness on a white substrate, according to DIN 53778-3. The hiding power depends on the difference in the refractive indices of the materials used (binder, pigment, filler). The higher this difference, the higher the hiding power. Titanium dioxide pigment provides the largest difference and therefore is the critical factor for the hiding power.

Besides the refractive index, particle size distribution, pigment volume concentration and the degree of dispersion of the pigment, has an effect on hiding power. Effective dispersion of the white pigment and the fillers in the binder is important for its effective deployment. Specifically this effect is achieved by using particularly fine, precipitated synthetic aluminium silicates. These silicates undergo optimum arrangement between the dispersed particles of titanium dioxide and exert, so to say, a spacer effect between the pigment particles. Consequently the precipitated synthetic silicates can optimize the maximum amount of titanium dioxide white pigment in the paint, which is necessary to enhance the hiding power and the whiteness of the formulation.

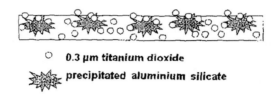

○ **0.3 μm titanium dioxide**

precipitated aluminium silicate

Figure 3. The spacer effect of precipitated synthetic aluminium silicates increases hiding power and whiteness

The fineness of the precipitated aluminium silicate particles gives them a high oil adsorption, thereby lowering the critical pigment volume concentration of the formulation and raising the porosity of the coating system. The refractive indices of calcium carbonate ($n = 1.55$) and aluminium silicate ($n = 1.46$) are quite similar. Both indices lie below the limit that applies to pigments of $n > 1.7$ (DIN 55943 and 55945). Nevertheless, up to 50% of the titanium dioxide pigment present can be replaced by the aluminium silicate without having a negative impact on the contrast ratio (Figures 4a to 4c) or whiteness (Figures 5a to 5b) of the formulation. In fact, in most cases both parameters are improved.

Figures 4a to c. Contrast Ratio (CR) as a function of aluminium silicate content (calculated on TiO_2) and PVC / VAc/E binder

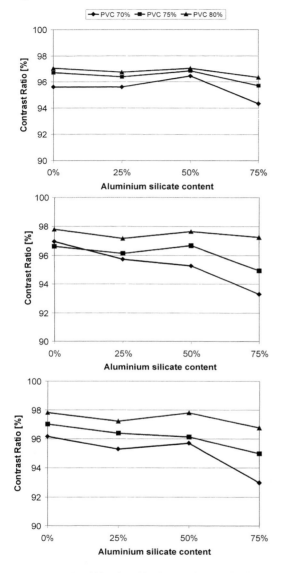

4a) Low treated TiO_2 (high gloss grade)

The formulation based on VAc/E possesses a CR of 95.61% at a PVC of 70%. If 25% of the TiO_2 is replaced by AS (Sample A), the CR remains approximately the same (95.59%). If the AS content is raised to 50%, a maximum of 96.45% is achieved. Only at an AS content of 75% does the CR fall off significantly. The effect on CR is similar at a PVC of 75 or 80%.

4b) Medium treated TiO_2

In this case a slight drop in CR from 96.94% to 95.74% or 95.29% (replacement of 25% or 50% of TiO2, respectively) occurs only at a PVC of 70%. At the higher PVCs, the CR remains almost constant at 25% substitution, and reaches a maximum at 50% substitution.

4c) Highly treated TiO_2

Despite high surface treatment of the TiO_2 used, again up to 50% of the pigment can be replaced by AS (maximum again at PVC of 70% and 80%) without a significant reduction in the CR (at 70% PVC from 96.17% via 95.30% to 95.71%).

A good filler should be devoid of any substantial inherent coloration, that it, should possess high whiteness, in order not to disrupt the colouring effect of the pigment. Synthetic fillers such as the aluminium silicates perform much more effectively here than natural products. Unlike the latter, they are even capable of raising the whiteness of a formulation, even where there has been a substantial reduction in the level of white pigment, as shown in Figures 5a and b.

Figures 5a and b: Whiteness as a function of aluminium silicate content (calculated on TiO_2) and PVC / VAc/E-binder

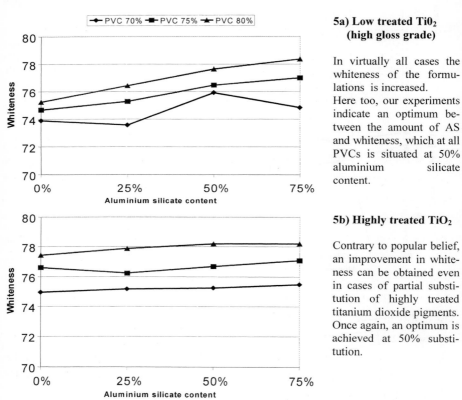

5a) Low treated TiO_2 (high gloss grade)

In virtually all cases the whiteness of the formulations is increased.
Here too, our experiments indicate an optimum between the amount of AS and whiteness, which at all PVCs is situated at 50% aluminium silicate content.

5b) Highly treated TiO_2

Contrary to popular belief, an improvement in whiteness can be obtained even in cases of partial substitution of highly treated titanium dioxide pigments. Once again, an optimum is achieved at 50% substitution.

The results of these experiments clearly show that a synthetic aluminium silicate is more than just a common filler, which always of course influences the physical and coloristic properties of a paint, but also serves to increase the volume and hence reduce the costs of a paint formulation.

Aluminium silicates function as an extender, reducing the amount of titanium dioxid pigment, and therefore the cost of the total paint formulation.

In previous experiments (same formulation based on VAc/E, 60% titanium dioxide substitution, PVC = 70%) we compared the AS (Sample A) with two natural calcined aluminium silicates (CC1/CC2) typical for the interior emulsion paint sector. In terms of all their coloristic properties these products were inferior to the synthetic aluminium silicate.

Table 3. Comparison - synthetic AS/natural AS (CC1/CC2)

	STANDARD TiO_2	SAMPLE A	CC1	CC2
Whiteness	74.14	75.8	73.56	74.79
Contrast Ratio	96.93	96.79	95.67	95.65
Particle Size d_{50} [μm]		5	4	3
Sheen (85°)	2.4	2.9	3.1	3.6

It can be seen in Table 3, that the synthetic AS performs better in all coloristic respects than the "natural products". Owing to the somewhat larger particle size of the synthetic AS, the increase in sheen is smaller. However, the sheen can be taken back to the original value with both types of AS. Coarser fillers also bring about a reduction in the binder demand; this should be beneficial to the wet abrasion resistance of the formulation.

The high porosity and oil adsorption of synthetic aluminium silicates reduce the wet abrasion resistance of the formulation. However, the extent of this effect is dependent on a variety of factors such as PVC, titanium dioxide pigment, aluminium silicate content, binder, wetting and dispersing agents. Therefore a standard titanium dioxide pigment or a medium surface-treated one often performs better in terms of wet abrasion resistance than a highly surface-treated titanium dioxide. The investigations also show that the wet abrasion resistance at a PVC of 70% is relatively independent of the aluminium silicate content. However, this does change as the PVC of the formulation goes up. Interestingly, we found that the effect of the binder on wet abrasion resistance can be greater than that of the aluminium silicate content!

Taking into account all of the experimental results and considering the overall properties of the formulation (whiteness, contrast ratio and wet abrasion resistance), additional investigations were conducted using the following combinations: aluminium silicate content = 50%, Sty/Acr binder or VAc/E binder, standard (low treated) or highly treated TiO_2, PVC = 75%

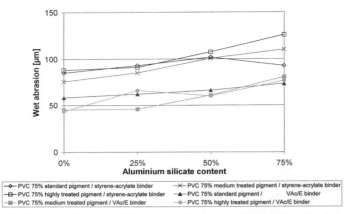

Figure 6. Wet abrasion as a function of aluminium silicate content, titanium dioxide pigment and binder type

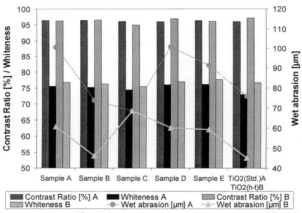

Figure 7. Whiteness, contrast ratio and wet abrasion resistance by ISO 11998 of a formulation based on

A: sty/acr, PVC = 75%, AS content 50%, standard TiO_2 / TiO_2 (Std.) A

B: VAc/E, PVC = 75%, AS content 50%, highly treated TiO_2 / TiO_2 (h-t)

On the basis of these preliminary experiments, the five pigment extenders were tested in the two stated combinations. Essentially, the results from the preceding experiments were confirmed. In formulations based on a standard titanium dioxide pigment, the whiteness was increased in every case and the contrast ratio was almost always improved. In terms of wet abrasion resistance, there are sharp differences depending on the AS used. Particularly, those AS types which do not have the best optical properties still have a favourable effect here and in some cases a very favourable effect, on the wet abrasion resistance which is even improved in comparison with a pure titanium dioxide pigment formulation.

A marked reduction in wet abrasion resistance, on the other hand occurs for the AS which do top the table. This effect is due to the different porosities of the AS types investigated. AS types having a relatively low porosity often equaling the original performance level of the straight titanium dioxide formulation, and in some cases the wet abrasion resistance is even improved by the use of low-porosity AS.

In terms of the partial substitution of highly treated titanium dioxide pigment by the five AS the previous results were confirmed. In the majority of cases the whiteness is increased by using AS. The contrast ratio of the pure titanium dioxide formulation is virtually achieved by all the AS. Again, it should be noted that the effect on the wet abrasion resistance of the titanium dioxide/binder combination used is greater than the effect of the aluminium silicate. Any negative effect of the AS can be more than compensated by choosing a different binder or by using a different type of titanium dioxide pigment.

5.4.2. Colouring Emulsion-Paints containing Aluminium Silicate

For the colouring tests we used a formulation (see Formulas) with a PVC of 70% based on Sty/Acr. In each case 50% of the standard titanium dioxide pigment was replaced by the tested aluminium silicates. The paint was subsequently coloured with 3% by weight of a cobalt blue pigment paste (Colanyl Oxide Blue Co100, pigment content: 60%). In all of the formulations based on aluminium silicate, the chromaticity of the coloured emulsion paints was raised, so that they appear more blue. Moreover, the AS-based coatings exhibit a higher "cleanness" of shade.

The chromaticity or chroma describes the interaction of the colour values a* and b* and is a measure of the brightness of the coating. A direct relationship exists between the AS pore volume size and the increase in chroma, i.e. a smaller pore volume results in a higher increase in chroma. This phenomenon can be correlated with the spacer effect of the AS. Aluminium silicates with a small pore volume are less efficient in maximizing the titanium dioxide pigment distribution in the paint, since they have far fewer large pores at their surface where the larger titanium dioxide particle could attach. In this case the small cobalt blue pigment particles have a much better chance of finding a site on the AS surface where they can anchor themselves.

Therefore, the number of titanium dioxide particles on the surface is decreased and the blue pigment is maximized, resulting in a coating, that appears bluer and darker → hue. This can be seen in Figure 8 for samples B and C. As seen in Table 1, the tamped densities of samples B and C are higher by about 150 g/l than the tamped densities of the other products A, D and E.

Table 4. Coloristic data for coloured interior wall paints

	L*	Δ L*	a*	Δ a*	b*	Δ b*	Δ E*	C*	h*
Sample TiO₂	77.72		-8.26		-21.70			23.22	249.17
Sample A	77.65	-0.07	-8.04	0.22	-22.26	-0.56	0.61	23.67	250.14
Sample B	76.96	-0.76	-8.25	0.01	-22.88	-1.18	1.40	24.32	250.18
Sample C	76.90	-0.82	-8.30	-0.04	-22.90	-1.20	1.45	24.36	250.08
Sample D	77.96	0.24	-7.98	0.28	-21.98	-0.28	0.46	23.38	250.06
Sample E	77.59	-0.13	-8.10	0.16	-22.31	-0.61	0.64	23.73	250.05

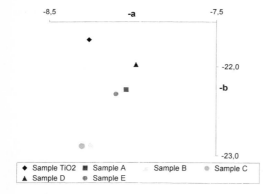

Comparing the Δ a* and Δ b* values for the samples A, D and E, it can be seen that there are no significant differences in relation to the original hue of the pure titanium dioxide formulation (Δ E = 0.46/0.64). Also, small deviations in shade can easily be corrected by minor modifications to the formulation

Figure 8: a*/b* values of the various AS-based formulations in comparison with the pure titanium dioxide formulation / coloured with 3% cobalt blue pigment paste

6. Summary

Synthetic aluminium silicates (AS) undergo optimum arrangement between the dispersed titanium dioxide pigment particles, optimizing their distribution in the paint, which improves the hiding power and the whiteness of the formulation. As a result it is possible to make a considerable reduction of the titanium dioxide pigment in the formulation of an interior or exterior paint. Our experiments showed that a 50% substitution is particularly advantageous in terms of whiteness and hiding power. These advantageous effects of the synthetic aluminium silicate, which lead to cost reductions in the formulation of paints, are also observed when highly treated titanium dioxide pigments are substituted. Synthetic aluminium silicates do not just improve the dispersion of the white titanium dioxide pigment; rather, their use leads to a general improvement in the distribution of inorganic pigments within coloured emulsion paints. It was also found that the chromaticity and colour brilliance of coloured emulsion paints with AS are increased, with both influences being dependent on the porosity of the various AS grades. This also applies to the wet abrasion resistance, which as a result of AS with relatively low porosity in many cases reaches the original level of the pure titanium dioxide pigment formulation, and in some cases is even improved. AS of high porosity have an unfavourable effect on the wet abrasion resistance. The experiments conducted have shown, that this disadvantage can be more than compensated for an appropriate choice of the the titanium dioxide pigment and the binder. Due to the described positive effects, synthetic aluminium silicates are no ordinary fillers, but can be acknowledged as "functional-pigment-extenders".

Acknowledgements

The authors thank Dipl.-Physiker E. Klinke (FPL-Stuttgart) and following staff members of the Degussa AG: Dr. P. Albers, Dr. R. Rausch, Mr. S. Heeg, Mr. N. Herrmann, Mr. H. Seikel and Mr. K. Thomas

References

Mischke, Peter/Brock, Thomas/Groteklaes, Michael, Lehrbuch der Lacktechnologie, 2. Auflage 2000, Vincentz-Verlag, Hannover

Schwartz, Manfred/Baumstark, Roland, Dispersionen für Bautenfarben, 1. Auflage 2000, Vincentz-Verlag, Hannover

Technical Bulletin Pigments No. 34, "Sipernat 820 A for Emulsion and Decorative Paints", Degussa AG, Business Unit "Advanced Fillers & Pigments", 2002

Technical Bulletin Pigments, "Precipitated silicas and silicates", Degussa AG, Business Unit "Advanced Fillers & Pigments", 2001

Macromol. Symp. ***187**, 677–682 (2002)*

Lotus-*Effect*® - Surfaces

Edwin Nun, Markus Oles and Bernhard Schleich

Degussa AG, D-45764 Marl, Germany

Summary: Surfaces are often characterized with phrases like "easy to clean", "dirt repellent", "dirt resistant", "self cleaning" or "**Lotus**-Effect®".Every one of those phrases is used to describe a behavior of surfaces – similar to each other but still different. This article is providing the definition of the **Lotus**-Effect®, techniques to manufacture self cleaning surfaces and methods to characterize them as well. How to generate a self cleaning surface depends on the substrate and the use later on. It can be as easy as a spray on but on the other hand as complicated as a three step process. Self cleaning surfaces are defined by four parameters – contact angle, roll-off angle, hystereses and C.I.E-Lab Δ-L value.

Introduction

Lotus-Effect® surfaces are hydrophobic nano-structured surfaces being cleaned by moving water! Surfaces are often characterized with phrases like "easy to clean", "dirt repellent", "dirt resistant", "self-cleaning" or "**Lotus**-Effect®". Each one of these expressions is used to describe a behavior of a surface – similar to each other but still different.

"Easy to clean" surfaces are smooth and hydrophobic and well known since decades. As indicated, it is not difficult to remove soil from such surfaces. Wiping with a wet cloth will clean them.

"Dirt repellent" suggests a surface that repels dirt. As of today no such surface is known.

"Dirt resistant" – it is harder to soil these surfaces but not impossible. This expression provides no hint how to clean.

"**Lotus**-Effect® " and "self-cleaning" are used synonymously. Some dirt might soil the surface. But a human cleaning activity is not necessary. Just expose the surface to precipitation and watch how water does the work.

Self-cleaning surfaces is the topic this lecture deals with.

The basic requirements for **Lotus**-Effect® surfaces are a combination of hydrophobicity and structure. As early as 1982 Abramzon [1] described that the structure of a Lotus leaf's surface and the hydrophobicity together create water drops having contact angles of 150 degrees. What he did not noticed at all was the self cleaning behavior. This property was also missed by SEKISUI

CCC 1022-1360/00/$ 17.50+.50/0

CHEMICAL CO. LTD who filed a Japanese patent application in 1994 [2]. In 1997 Barthlott[3] described the Lotus-leaf's surface self cleaning behavior. The following illustration shows a surface of a lotus leaf (Fig. 1).

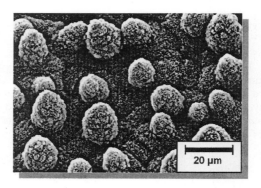

Figure 1. SEM image of a Lotus leaf (Nelumbo nucifera)

Other plants, too, force water to build out droplets rolling off at the slightest inclination [3]. Widely known is this behavior on cabbage, kohlrabi, nasturtium or gingko.

Not surprisingly each of these plants has a unique surface patterns. In common the plant leaves have a very hydrophobic top-layer and a structured surface.

As conclusion it can be stated that there is not only one possible surface structure for self-cleaning surfaces, there are a wide variety!

Results and Discussion

This wide variety of possible self cleaning surfaces opens up several possibilities to produce them.

The first approach was copying Lotus leaf's surface. But soon it was found that due to isolated structures the mechanical ability to withstand mechanical stress was very poor. Therefore the design was changed. A better mechanical stability should result by connecting the hydrophobic "knobs" to each other like it is realized in egg-shapes. In addition, the influence of the structure's size was evaluated. It was found that the self-cleaning behavior is not limited to the structure size found at Lotus plants. Even repeating structures from 250 nm and below are fulfilling the requirements for **Lotus-**Effect® surfaces.

Figure 2. SEM image of self-cleaning surfaces with different structure size

Alone the knowledge of working structures is not sufficient to manufacture **Lotus-**Effect® surfaces. Know how about the embossing tool, the embossing process and how to create a permanent hydrophobic surface are some of the keys being successful.

Currently it is in favor to emboss in a hydrophobic lacquer system under simultaneous cure. Most critical is the lacquer system that must have excellent adhesion to a carrier and must be nano-structured and hydrophobic after cure.

An other approach to create self cleaning surfaces is to tether nano-structured hydrophobic particles onto surfaces, accordingly (Fig. 2).

Bonding nano-structured hydrophobic particles onto any surface is a difficult task. It is to assure that the bounding is permanent. Usually, surfaces are more or less hydrophilic. Thus hydrophobic particles and common plastic surfaces are rather incompatible. Bridging these conflicting behaviors is the task for a coating system.

We developed this coating system with an excellent adhesion to most of the common polymer surfaces and showing affinity to hydrophobic nano-structured particles as well. Due to this unique combination of properties **Lotus-**Effect® surfaces on plastics are accessible. Even complicated 3d-shapes can be equipped.

Figure 3: SEM image of self-cleaning Plexiglas® (PMMA) surface with nano-structured hydrophobic particles.

The image above illustrates how polymer surfaces are covered with hydrophobic nano-scale particles. The very dense layer of particles covers the bonding resin completely. On this very surface the contact angle to water is above 150 ° and the roll-off angle below 2 °.

The ability to withstand mechanical stress is much higher than mother nature's model. It can be touched and brushed as well without diminishing self-cleaning performance.

Yet, it is not simply a paint. A standard lacquer system with the particles suspended in would fill the nano-structured particle's surface. And, as on Lotus plant's leaves too, losing structure is equal to losing self-cleaning performance.

Embossing hydrophobic lacquer or tethering hydrophobic nano-structured particles are leading to durable self-cleaning surfaces. A third kind of **Lotus-**Effect® surface is for non-permanent possible use.

An application for a European patent [4] reveals formulations of nano-structured particles mixed with binders to generate self-cleaning surfaces just by spraying onto surfaces. Depending on binders those surfaces are neither permanent nor non-permanent.

We developed a spray for non-permanent self-cleaning surfaces. No binder that will remain on a surface is needed to fix nano-structured particles onto nearly any surface. Just making use of some physical properties manufactures smart self-cleaning surfaces.

Lotus-Effect[®] surfaces are hydrophobic nano-structured surfaces being cleaned by moving water!

"**Lotus**-Effect[®] " and "self-cleaning" are used synonymously. Some dirt might soil the surface. But a human cleaning activity is not necessary. Just expose the surface to precipitation and watch how water does the work.

- So it is stated at the beginning of this lecture.

What actually is the definition for self-cleaning or **Lotus**-Effect[®] ?

There is no ASTM, DIN or ISO standard to characterize self-cleaning performance. Practically four essential analytical parameters were found. Only if all of them together meet the requirements a surface will be characterized as self-cleaning surface

- Contact angle to water: Young's equation for contact angles is related to the hydrophobicity of a surface. [5] Even very hydrophobic smooth surfaces have contact angles up to approximately 120° only. But with the right nano-structures on hydrophobic surfaces the contact angles increase. With a structured surface the question of baseline (cos θ) in Young's equation becomes indistinct. For practical purposes the angle is measured as it appears with having the surface on a scale pretending to be smooth.

  The first parameter that is necessary for self-cleaning surfaces is a contact angle above 140°.

- Hystereses of advancing and retreating angle. The advancing angle is the contact angle of water while increasing the volume of a drop that is pinned to a surface by a syringe. Decreasing the volume of the very same drop of water results in the retreating angle. The difference between advancing and retreating angle is called hysteresis.

  The second parameter, the hystereses, must be below 10°.

- The third geometrical characterization is the roll-off angle. A water droplet of 60 µl volume is placed on a horizontal surface. Tilting the surface and measuring of the angle when the drop begins to move gives the roll-off angle.

  The third parameter, the roll off angle, must be below 10°.

- The proof of self-cleaning behavior is performed in a so called "soil test". The surface in question will be soiled under standardized conditions with Printex 60® (Carbon black). Later on water mist precipitates the surface mounted in an angle of 45°. Immediately the carbon black starts to be washed off the surface. Typically, this procedure is repeated for 60 times. Afterwards the ΔL value is measured (CIE-L*a*b*) and it is defined as the difference of the L-values before soiling took place and after self-cleaning occurred,

  The forth parameter ΔL should not exceed 10.

Applications

Finally, some possible applications for the different types of **Lotus**-Effect® surfaces are presented.

Embossed self-cleaning surfaces are very useful on various overlaminates and protective films. **Lotus**-Effect® surfaces show pristine appearance even during nasty weather. Other fields of application are in traffic guidance and signs, window frames and solar cells.

Anchored particles can be used to manufacture self-cleaning surfaces on awnings, tarpaulins, wood, appliances or in bathrooms.

Non-permanent self-cleaning surfaces are preferably used to grand cleanness and dryness during a temporary outside storage or to protect furniture while transported on open trailers.

Of course there are much more applications.

References

1)   A. A. Abramzon, Khimia i Zhizu ("Chemistry & Life"), 1982, 11, 38 – 40
2)   JP 07328532 A to SEKISUI CHEM. CO. LTD
          3)   W. Barthlott and C. Neinhuis, Annals of Botany, 79, 667 – 677, 1997
4)   EP 1153987 A2 to BASF
5)   T. Young, Phil. Trans. R. Soc. London, 95, 65 (1805)

Macromol. Symp. 187, 683–693 (2002)

Hyperbranched Polymers as a Novel Class of Pigment Dispersants

F.O.H. Pirrung, E.M. Loen and A. Noordam*

EFKA Additives B.V., Innovatielaan 11, 8466 SN Nijehaske, The Netherlands

Summary: Hyperbranched polymers form a novel class of materials that are employed as components of resin formulations. They are appreciated for their low intrinsic viscosities, which is ascribed to their spherical shape. It was envisaged to devise pigment dispersants with hyperbranched moieties as structural elements. Commercially available hyperbranched polymers with polyester, polyesteramide and polyethylene imine backbones were chemically modified to develop a range of disperants with core- and shell-type pigment anchoring mechanisms. Both the shell- and core-type anchoring principles generally can be used for pigment dispersion and stabilisation at a low viscosity level.

Keywords: dendrimers, pigment dispersants, viscosity, coatings, polymers

Introduction:

Hyperbranched polymers belong to a class of polymers characterized by densely branched structures and a large number of reactive groups. Typical is the tree-like structure, which is obtained by polymerisation of *ABx*-type monomers.[1] Currently, the importance of hyperbranched structures has been identified by researchers in various industries and the polymers are being investigated in a broad range of coating applications.[2] This mainly was stimulated by the fact that several types of this polymer class have been made commercially available recently.[3,4]

The large number of functional groups on these polymers has a significant influence on the final physical and chemical properties. This allows various possibilities to derivatise (part of) these groups to obtain a certain desired property, such as reactivity, viscosity, compatibility or solubility and the use in different applications with their specific requirements. For example, the end groups can be transformed into unsaturated fatty acid esters or amides (for alkyd systems), acrylate- (radiation curing) or epoxy-terminated (for 2-pack systems), grafted with other polymers to obtain compatibilizers, equipped with ionic groups (for water solubility) or in particular with non-reactive groups for viscosity, polarity and compatibility adjustment.[5]

As a preliminary achievement, it has been recognized that hyperbranched structures show a non-

CCC 1022-1360/00/$ 17.50+.50/0

Newtonian relationship between viscosity and molecular weight, and these rheological properties (low viscosity at high molecular weight due to reduced chance of chain entanglement) are combined with other characteristics, such as reactivity, chemical resistance, mechanical properties. For coating applications, if used as a binder component, this can mean a lower solvent content with advantageous film properties, which is highly interesting for environmental reasons.[4]

Hyperbranched Polymers

Classical examples of hyperbranched structures, are those which are built up from monomers of the AB_2-type and are called dendrimers, *dendron* being the Greek work for tree. The first commercial available dendrimers, polypropylene imines and polyamido amines were monodisperse polymers, meaning that costly stepwise build up and purification steps were required, limiting their use to high added value products like medicinal applications. [6] Recently, one-pot polyester types *(PES)* are commercially available, which consist of a multifunctional core from which branches extend to give a highly branched inherent structure with a large number of hydroxyl groups.[3] The AB_2-monomer is 2,2-dimethylolpropionic acid. The polymers are described by exponential growth and defined as polydisperse hyperbranched polymers of which the molecular weight is controlled by stoichiometry without the use of protective groups. The geometric structure is sphere-like consisting of three different regions: the *iniator core* (polyfunctional *Bx* starter), the *intermediate layer* (tree-structure by monomers AB_2) and the *shell* (reactive groups at the peripheral surface). The term *generation* is used to describe the size of the polymer and represents one repetitive step during the synthesis. The interior ester functions are shielded by the dense tree-structure, showing a higher hydrolytic and chemical stability.[5]

A second class of polymers are hyperbranched polyesteramides *(PEA)* as possible starting materials for this study. They have been developed and commericialized at the end of the 1990's, being polymers with a large number of multifunctional groups.[4] Besides hydroxyl and carboxyl-groups, hyperbranched PEA can be modified with a broad range of functional groups such as tertiary amines, alkyl chains and unsaturated groups. The synthesis takes place as a one-pot reaction in which monomers such as cyclic anhydrides (*e.g.* phthalic anhydride, succinic anhydride and others) are reacted with diisopropylamine. A rearrangement of the reactive species leads to a complex branched polycondensation product.[7] Variations in the ratio between the two starting materials leads to a molecular weight (MW) range of hyperbranched PEA. The obtained polymers can be further reacted with monocarboxylic acids, cyclic anhydrides, dialkyl amines and ethyleneoxide grafted chains. Investigated application fields for these polymers are crosslinkers for (powder) coatings, toner resins, high-solid air-dying alkyds, polyolefin dyeing and as surfactants.[4] In these applications the high chemical and mechanical resistance is pointed out.

Another interesting starting point are polyethylene imines *(PEI)*. These products are polymeric amines based on homopolymers of ethyleneimine (aziridine).[8] The geometry is branched and spherical with a well defined ratio of primary, secondary and tertiary amine functions, where statistically this ratio is about 1:1:1. They can be reprensented by the partial structural formula - $(CH_2\text{-}CH_2\text{-}NH)_n$- with $n = 10 - 10^6$.

Depending on their molecular weights, they vary in viscosity. Their affinity for polar surfaces (adhesion) can be explained by the branched nature with many anchoring points, and especially the adhesion to pigment surfaces has been highlighted. The pure polymer is described as pigment dispersant and flocculant in one compound.[8] At amounts below 1% on solid pigment, or if high molecular weight PEI are used, flocculation is observed; at higher amounts or with low molecular derivatives, stable dispersions are obtained. The amine functions react readily with ketones and aldehydes, furthermore, amidisation reactions with a wide range of carboxylic acids can be carried out. It is known that PEI can be reacted with polyesters, polyetherketones, polyolefines etc. Some of these functionalised grades have been commercialised as dispersants, however generally they show a poor solubility and a tendency to crystallisation under use conditions.[9]

Dispersing Agents

In the study described below, it was envisaged to use hyperbranched polymers as basic building block for the design of polymeric dispersing agents, and to overcome at the same time mentioned disadvantages. In the EFKA technology, polymeric or high molecular weight dispersants are used to wet and stabilise dispersions of inorganic and organic pigments in pigment concentrates or by direct grinding.[10] This principle provides dispersions with a long term stable viscosity at high pigment loading and constant colour strength, without reflocculation and sedimentation of the solid pigment particles. For this purpose EFKA provides polymers based on two kinds of patented chemistry, namely acrylic copolymers and polyurethanes.[11, 12] In a simplyfied view both systems are based upon a linear backbone, from which functional side chains are pending. These functional groups consist of compatibility-enhancing moieties, such as polyester or polyether chains and pigment affinic groups, so-called anchoring groups, which are designed to adsorp onto the pigment surface, therefore enveloping the pigment particle with the polymeric dispersant. Compatibility enhancing chains in this representation extend into the liquid grinding medium and interact positively with the binder molecules and the solvent. Their main action is to induce the steric stabilisation mechanism, which prevents the penetration of the chains of neighbouring polymer/pigment complexes and subsequently their mutual agglomeration.[13] In addition, carboxylic acid groups can be connected to the backbone, which render a possibility to make the polymer water soluble after neutralisation.

These polymeric dispersants have proven to be efficient and are well established additives for the paint industry, however there exists some room for improvement, like in state of the art automotive paints. They lack an optimum low viscosity at a high pigment loading with certain organic pigments and show occasionally an undesired build-up of a high yield value (thixotropic behaviour, which can be reversibly broken down by low shear forces like stirring and shaking). This phenomenon can be explained by the specific polymeric structure, being linear and showing average MW's of 5,000 to 25,000 $g \cdot mol^{-1}$ in a distribution that includes MW's as high as 100,000 $g \cdot mol^{-1}$. Such linear polymers with pending polar groups can interact inter- and intramolecularly or with functional groups of binder molecules by polar forces which prevent Newtonian flow of the paste. We have identified two ways to investigate the solution to this phenomenon:

- by applying controlled polymerisation techniques to obtain much narrower molecular weight distributions and defined polymeric architectures.
- by using hyperbranched building blocks to avoid linear polymer geometries.

The first approach is considered outside the socpe the current study, whereas the latter principle will be presented in this paper. It was anticipated that hyperbranched systems, due to their characteristics of offering low viscosities at high molecular weights together with the possibility to functionalise the outer shell with the appropriate chemistry, can offer the solution to overcome viscosity build-up.

In the general concept, hyperbranched polymers would be reacted with side chains that provide solubility and specific anchoring moieties to ensure adsorption of the polymer to the pigment surface. Both dendrimeric PES and hyperbranched PEA, as well as PEI fulfill the requirements for further functionalisations.

Concept based on Dendrimeric Polyesters *(PES)*

In a first approach, dendrimeric PES were investigated, having 2, 3 and 4 generations, representing MW's of ca. 2000, 3500 and 5000 g·mol^{-1} with respectively 16, 32 and 64 reactive hydroxyl groups. Due to the extreme polar structure of the outer shell, these solid polymers dissolve readily in water and methanol at ambient or slightly elevated temperatures, but both solvents are unsuitable starting points for many organic chemical reactions. It was found that melting the solids (> 40 °C) in a liquid reactive system (such as cyclic esters and fatty acids), would render an adduct, which remains liquid at room temperature and dissolves in common coating solvents like aromatics, esters and ketones. Furthermore, the extension of the outer shell with one molecule of *e.g.* caprolactone would increase the molecular weight and extend the spherical structure of the dispersant. It was expected that the body of the dendrimer would take part in the required steric stabilisation of the pigment dispersion and steric side chains could be shorter (< 1000 g·mol^{-1}) than in the classical linear systems.[12]

Table 1. Dispersants based on dendritic polyesters (PES).

sample	MW of PES (g·mol^{-1})	hydroxy-car-boxylic acid [a]	capping agent [b]	aromatic group [c]	tert. amine [d]	approx. MW (g·mol^{-1})
4	3500	32 Cl	16 UFA	10 BA	-	12900
5	3500	16 CL	16 UFA	10 BA	-	11000
6	3500	16 CL	16 UFA	-	16 TA	16200
7	3500	16 CL	22 UFA	-	10 TA	15500
8	3500	16 CL	16 UFA	6 BA	10 TA	14500
9	3500	64 CL	-	-	10 TA	14900
10	3500	16 RA	16 UFA	-	15 TA	18800
11	2000	16 CL	6 UFA	-	5 TA	7500
12	2000	16 CL	6 UFA	4 PAH	4 TA	7700
13	2000	16 CL	6 UFA	-	5 TA	7500
14	5000	64 CL	-	-	15 TA	18400

[a] CL= caprolactone and RA = ricinoleic acid; [b] UFA = unsaturated fatty acid;
[c] BA = benzoic acid and PAH = phthalic acid anhydride; [d] TA = isocyanate-functionalised tertiary amine

In a typical example, PES (table 1, sample **4**) was reacted with an equimolar amount of caprolactone, then 16 terminal hydroxyl groups (50%) being extended with fatty acids leading directly to a compatibilising chain. Then 10 parts of benzoic acid were reacted. The remainder hydroxy groups were not functionalised. The thus obtained colourless polymer was diluted with solvents to ca. 50% and tested as a dispersant in an oil-free polyester based pigment concentrate with a standard pigment. As a reference a commercial linear polyacrylate dispersant and two competitive commercially available products were used. The pigment concentrate was measured on viscosity and subsequently applied with melamine on a sheet and cured at 130 °C, for determining gloss, seeding and transparency.

In a similar way, more samples were prepared (see table 1). The variation consisted of the amount of caprolactone initially applied to elongate the core dendrimer (12-hydroxy unsaturated fatty acid, ricinoleic acid, can also be used), the amount of fatty acid to cap some of the hydroxy groups, the utilisation of aromatic systems such as benzoic acid and phthalic acid anhydride as possible pigment affinic groups, or isocyanate-functionalided tertiary amines to enhance the anchoring efficiency, the structure of which is not disclosed for patent filing reasons.[14]

Table 2. Formulation for dispersant testing.

Components for **I** and **II** [a]	Amount (g)
Oil-free polyester	10.0
Dispersant	3.2 (at 50% solids)
Alkylbenzene	28.8
Irganzin DPP Red BTR	8.0
Glass beads	100
Total (**I**)	150 [b]
Above pigment concentrate (**I**)	2.50
Oil-free polyester	3.91
Melamine	1.67
Alkylbenzene	1.89
Silicone levelling agent	0.03
Total of (**II**)	10.0 [c]

[a] I = pigment concentrate and II = pour-out formulation
[b] grinding 1.5 h (Scandex) to < 5 μm ; c) mixed (dispermat) and poured out on plastic sheet, cured 0.5 h at 130 °C.

The above samples were selected by their gloss values at 20° to be above 70, and having no to only slight haze. The pastes were measured by their viscosity at different speeds with a Brookfield apparatus at different speeds (5 and 50 rounds per minute, *rpm*). The paste formulation that was used is based on an oil-free polyester as grinding medium, the pigment Irgazin Red BTR (16%), the dispersant at a level of 20% active material on pigment and solvent. An example of a typical formulation is given (table 2). Irgazin DPP Red BTR was chosen as a model pigment. It is based on diketopyrrolo pyrrole chemistry, and has a small particle size (BET = 15 $m^2 \cdot g^{-1}$) and known to easily build up viscosity after grinding.[15]

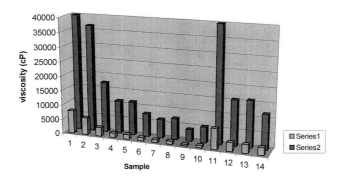

Figure 1. Viscosity measurements of selected pigment pastes. Series 1 and 2 measured at 50 rpm and 5 rpm respectively (Brookfield, spindle 4).

Figure 1 shows that dendrimers based on a PES with 3500 $g \cdot mol^{-1}$ (samples **4** to **10**) show lower viscosities in the pigment paste as compared to the reference linear polyacrylate (**1**) and two other competitive products (**2** and **3**). A decrease of the size of the dendrimer (2000 $g \cdot mol^{-1}$, **11** to **13**) gives a rise in viscosity. The chart also clearly shows that tertiary amine anchoring groups are more effective in viscosity reduction (**6** to **10**) as compared to the same polymer structure with merely aromatic anchors (**4** and **5**), this implying a stronger attachment of the dispersant to the pigment surface and therefore better dispersion.

Concept based on Hyperbranched Polyesteramides *(PEA)*

To investigate hyperbranched PEA, the results of the above dendritic polyesters were used for functionalisation. As a model structure, a PEA based on the condensate of succinic anhydride and diisopropanolamine having 8 hydroxyl groups throughout the branched structure with ca. 1200 $g \cdot mol^{-1}$ was chosen. This is lower than the used dendrimers and most similar to a PES with 2000 g·mol⁻¹, however with only half of the reactive groups.

The core polymer of the PEA-type, which shows low solubility and a solid state, was melted and reacted with caprolactone, followed by esterification of 3 hydroxy-groups with unsaturated fatty acids. The resulting polymer was treated with respectively 5, 3 and 1 equivalents of isocyanate-funtionalised tertiary amine, leading to three samples (table 3, **15** to **17**), which were submitted to the same application test with pigment Irgazin DPP Red BTR as described in table 2.

Table 3. Dispersants based on hyperbranched PEA.

sample	MW of PEA (g·mol⁻¹)	hydroxy-carboxylic acid [a]	capping agent [b]	aromatic group	tert. amine [c]	approx. MW (g·mol⁻¹)
15	2000	8 CL	3 UFA	-	5 TA	5800
16	2000	8 CL	3 UFA	-	3 TA	5000
17	2000	8 CL	3 UFA	-	1 TA	4100

[a] CL= caprolactone [b] UFA = unsaturated fatty acid [c] TA = isocyanate-functionalised teritary amine

All three samples showed an unexpected higher viscosity than any of the reference samples and competitive materials. Nevertheless, the gloss values of the three samples were high, at levels of 80 to 90 at 20°, showing good dispersion of the pigment. The undefined branched structure of this building block might be the reason of higher viscosities due to formation of linear structures. In addition, the low MW of this commerical product might be not sufficient for effective separation of the dispersed pigment particles, as it was the case with PES.

Concept based on Hyperbranched Polyethylene Imines *(PEI)*

The special feature of this concept is, that here the core of the polymer, the branched amine structure can be seen as the anchoring moiety, leading to the simplyfication that merely compatibility enhancing steric groups need to be connected to the core.

It appeared practical to block 50% of the available -NH- and -NH₂ groups of the PEI with MW's of respectively 800, 1300 and 2000 g·mol⁻¹ with unsaturated fatty acids. The remainder 50% of amino groups was treated with caprolactone, such that chains of 8 to 12 monomer units were connected to each available amine functionality. To avoid any viscosity build-up by terminal hydroxyl groups of the polyester chain, they were subsequently esterified with lauric acid. The thus obtained products (**18** to **20**) were first submitted to application testing (see table 2) for a range of pigments to determine which of the three PEI cores would be the most suited for further investigation. Table 4 shows that the increase of the size of the anchoring core of the hyperbranched dispersant leads to higher gloss values and lower viscosities in the pigment pastes. The PEI with MW of 2000 g·mol^{-1} (**20**) lead to better results than an comparisive linear polyacrylate dispersant (**1**).

Table 4. Performance results of PEI-based dispersants in pigment concentrates.

sample	**18**	**19**	**20**	**1** (ref.) [a]
MW of dispersant (g·mol^{-1})	10600	17600	26500	25000
(MW of PEI) (g·mol^{-1})	(800)	(1300)	(2000)	(n.a.)
Pigment used in formulation				
Heliogen Blue L 7072 D	49 [c]	81	98	92
with EFKA-6745 [b]	0 [d]	++	+++	0
Heliogen Blue L 7072 D	74	77	87	85
	-	-	+	0
Irgazin Red 2030	22	35	55	66
	+	+	++	0
Irgazin DPP Red BTR	62	80	92	72
	--	-	+	0
Colour Black FW 200	94	93	92	92
	0	+	++	0
Kronos 2310	82	90	85	89
	0	-	+	0

[a] EFKA-4401; [b] synergist for phthalocyanine pigments; [c] gloss (20°); [d] relative viscosity (reference set at 0).

Analysis of Results

From the above results, the PEI core appears to be the most promising building block for dispersant design. This can be visualised by comparing the geometrical structures of the resulting dispersants. Both PES and PEA lead to shell-type dispersants, while the PEI give core-type structures. In the first case, the anchoring groups are equally distributed over the surface of the ideally spherical shaped polymer, preventing an efficient pigment attachment of all the anchors, while core-types would experience a firm attachment by the core itself, avoiding the build-up of physical structures between the polar anchors and functional moieties of resins and others in the formulation. Generally it appears, that hyperbranched dispersants need sufficient MW (above ca. 15,000 g·mol^{-1}) in order to reach low viscosity levels. At lower MW's the volume of the polymer might not be sufficient for particle separation and as a consequence mutual interactions may prevent effective viscosity reduction. Further studies and optimization of the PEI dispersant architecture are required to determine the influence of the length and the amount of the steric side chains. The current results have provided insight in the structure-performance relationship of dispersants on the basis of which we are currently developing a new generation of polymeric dispersants, based on a new chemistry, which show an enhanced level of perfomance with regard to compatibility, a reduced tendency to crystallise and being liquid at room temperature.

Summary and Conclusion

It has been shown that the use of hyperbranched cores as carrier for the functional groups of a polymeric dispersant is a useful tool to improve the rheologic behaviour of polymeric dispersants in the described application fields. This is explained by the spherical shape of the polymers, leading to reduced interactions with polar groups in the liquid medium. The availability of numerous functional groups on the hyperbranched building blocks allows functionalisation of the core through versatile methods with desired moieties, such as pigment affinic groups and steric side chains.

References

[1] C. J. Hawker, J. M. J. Frechét, *J. Am. Chem. Soc.*, **1990**, *112*, 7638., G. R. Newkome, *Aldr. Chim. Acta*, **1992**, *2*, 31.

[2] B. Pettersson, *Pigment & Resin Technology*, **1996**, *25*, 4., M. Johansson, A. Hult, *J. Coat. Tech.*, **1995**, *67*, 849.

[3] SE 468 771 (1992), Perstorp Specialty Chemicals, invs.: A. Hult, M. Johansson, E. Malström, K. Sörensen.

[4] D. Muscat, R. A. T. M. van Benthem, in: *"Topics in Current Chemistry"*, Vol. 212, Springer Verlag, Berlin & Heidelberg, **2001**, 41.

[5] B. Pettersson, *Boltorn Dendritic Polymers as Thermoplastic Additives*, Perstorp Specialty Chemicals; B. Pettersson, *Hyperbranched Polymers – Unique Design Tools for Multi Property Control in Resins and Coatings*, Perstorp Specialty Chemicals.

[6] M. Fischer, F. Vögtle, *Angew. Chem. Int. Ed.*, **1999**, *38*, 885.

[7] D. Stanssens, R. Hermanns, H. Wories, *Progr. Org. Coatings*, **1993**, *22*, 379.

[8] Epomin Product Range, *Product Information*, **2001**; Shinwoo Advanced Materials Co., Ltd.; Lupasol Product Range, *Preliminary Technical Information*, **1996,** BASF AG.

[9] US 4,645,611 (1987), Imperial Chemical Industries Plc, invs.: F. Campbell, P. K. Davies, J. D. Schofield; US 4,861,380, Imperial Chemical Industries Plc, invs.: F. Campbell, J. M. Geary, J. D. Schofield.

[10] F. O. H. Pirrung, P. H. Quednau, C. Auschra, *Chimia*, **2002**, *submitted for publication*.

[11] EP 311 157 (1987), EFKA Chemicals BV, invs.: W. A. Wulff, P. H. Quednau.

[12] US 5,882,393 (1999), EFKA Additives BV, invs.: P. H. Quednau, F. O. H. Pirrung.

[13] J. H. Bieleman, in: *"Lackadditive"*, Ed. J.H. Bieleman, Wiley-VCH Verlag-GmbH, Weinheim, **1998**, p. 67., D. H. Napper, *"Polymeric Stabilisation of Colloidal Dispersions"*, Academic Press, London, **1983**.

[14] E. M. Loen, F. O. H. Pirrung, **2001,** *EP application.*

[15] O. Wallquist, in: *"High Performance Pigments"*, Ed. H. M. Smith, Wiley-VCH Verlag-GmbH, Weinheim, **2002**, p. 159ff.

Macromol. Symp. **187**, 695–705 (2002)

Stabilization of Carbon Black with Ionic-Hydrophobic Polyelectrolytes

Ch. Schaller [A,B], *K. Dirnberger* [A], *Th. Schauer* [B], *C.D. Eisenbach* [A, B]

[A] Institute of Applied Macromolecular Chemistry, University of Stuttgart,

Pfaffenwaldring 37, D-70569 Stuttgart, Germany

[B] Research Institute for Pigments and Coatings, Allmandring 37,

D-70569 Stuttgart, Germany

Summary: The mode of action and efficiency of amphipolar copolymers of different molecular architectures with copolymers consisting of ionic (acrylic acid) and hydrophobic (styrene) constitutional units and sequences (segments, blocks, grafts) as stabilizers of carbon black dispersion are discussed. The pigment-copolymer interactions were investigated by the ESA method as well as lumifuge and rheological measurements. The results indicate that there are distinctive differences in the stabilization behavior depending on the molecular architecture with better results being achieved for block and graft copolymers.

Introduction

The high potential of amphipolar copolymers as stabilizers for particulate systems has been shown in recent years especially as far as emulsions and dispersions are concerned [1, 2]. One promising area of application of copolymers is the stabilization of pigments. Using amphipolar copolyelectrolytes, both electrostatic and steric effects can be combined [3, 4] by the so-called electrosteric stabilization [5, 6]. In order to obtain an optimal stabilization of the particulate system one can take advantage of the variable macromolecular architecture as given in random, block or graft copolymers and different structural elements of ionic, hydrophilic and hydrophobic nature [7]. Despite the wide-spread application of these polymeric stabilizers, the general understanding of the interactions with the pigment surface (thermodynamics and kinetics of the adsorption) as well as their conformation in the adsorption layer is still limited. These interactions are influenced by the nature of the pigment, the matrix as well as the molecular architecture, the charge density and charge distribution within the polymeric stabilizer molecules.

In earlier papers we have reported about the potential of ionic-hydrophobic acrylic acid (AA)-

CCC 1022-1360/00/$ 17.50+.50/0

styrene (S) based copolymers as polymeric stabilizers and their pigment-polymer interaction in aqueous TiO_2 dispersion. Poly(acrylic acid) (PAA) and poly(styrene-*b*-acrylic acid) block (PS-*b*-PAA) and poly(acrylic acid-*g*-styrene) graft copolymers (PAA-*g*-PS) showed distinct differences in their interactions with the TiO_2 particles. It was found that PAA adsorbs via H-bonding and complexation forming trains at high pH values and coils or loops at lower pH values. Because of the possibility of intermolecular hydrophobic interactions, the amphipolar block and graft copolymers exhibit, in addition to the ionic interactions with the charged particle, so-called "solloid" and bilayer formation on the particle surface; this results in a better stabilization and long time performance as compared to PAA and is indicative of a higher efficiency of the electrosteric in comparison to electrostatic stabilization [8].

In this paper we describe the pigment-polymer interactions and the potential of these amphipolar copolyelectrolytes as stabilizers for aqueous carbon black dispersion. Carbon black is an organic pigment that is increasingly applied in automotive top coats or inks. Carbon black interacts preferably with the hydrophobic segments of a stabilizer [9]. In order to explore the potential of amphipolar copolyelectrolytes as polymeric stabilizers in general and to elucidate how the stabilization of carbon black dispersions [11] is effected by the architecture of the copolymer, well defined acrylic acid (AA)/styrene (S) model copolyelectrolytes with different molecular architecture were synthesized and investigated for the interactions with the particle surface as well as their stabilizing efficiency. The ratio of ionic hydrophilic and hydrophobic constituents was kept constant in order to be able to evaluate the influence of the polymer architecture on the stabilization of pigment dispersions. The structures of the investigated acrylic acid (AA)-styrene (S) macromolecules are shown in Fig. 1.

Experimental

Reagents

Tert-butyl acrylate (TBA), acrylic acid (AA), styrene (S) and *sec*-butyllithium were obtained from Fluka Chemie AG, Switzerland. Other chemicals needed for the synthesis were supplied by Merck, Germany. All chemicals were purified before use as described elsewhere [10]. The used carbon black FW 285 was provided by Degussa, Germany. FW 285 was characterized by an a particle size of d_{50}=11 nm and a specific surface of 350 m^2/g according to BET [11].

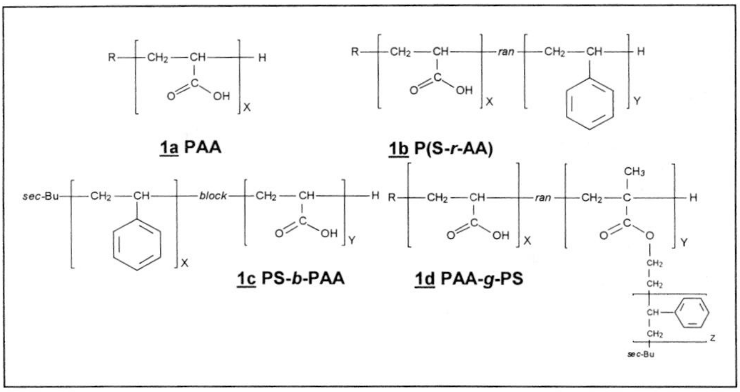

Figure 1. Molecular structure of the investigated poly(acrylic acid) (PAA; 1a), random acrylic acid-styrene copolymer (P(S-*r*-AA); 1b), poly(styrene-*block*-acrylic acid) (PS-*b*-PAA; 1c) and poly(acrylic acid-*graft*-styrene) (PAA-*g*-PS; 1d) (R: benzylmercaptane end group (BnS-))

Co/Polymer Synthesis and Characterization

Poly(acrylic acid) homopolymers (PAA, Fig. 1, 1a). and random acrylic acid-styrene copolymers (P(S-*r*-AA), Fig. 1, 1b) were synthesized by free radical (co)polymerization of acrylic acid (AA), and styrene (S) with AA, respectively, with AIBN as initiator; benzylmercaptane (BnSH) was used as chain transfer agent [12] to control the molecular weight of the homo and copolymers [10]. The access to graft copolyelectrolytes with an ionic PAA backbone and hydrophobic polystyrene (PS) grafts (PAA-*g*-PS, Fig. 1, 1d) is given by the macromonomer technique [13]: The free radical copolymerization of methacryloyl PS macromonomers and *tert*-butyl acrylate (TBA) with AIBN as initiator and BnSH followed by hydrolyzation of the resulting poly(TBA-*g*-PS) intermediate copolymer with hydrochloride acid in dioxane [14] gives the amphipolar AA-S graft copolymers (PAA-g-PS). AA-S 2-block copolymers (PS-*b*-PAA, Fig. 1, 1c) were synthesized by sequential anionic polymerization of S and TBA with *sec*-BuLi in THF as described in literature [15] followed by hydrolyzation with hydrochloride acid in dioxane. All block and graft copolymers and the macromonomers as well as the intermediate reaction products were characterized by gel permeation chromatography (GPC) and ^{1}H-NMR. More detailed reports of the synthesis and characterization of these block [16] and graft [17] copolymers are given elsewhere.

Methods

ESA (electrokinetic sonic amplitude) measurements were carried out with an AcoustoSizer^{TM} AZR2, Colloidal Dynamics, Sydney, Australia on 2 % w/w FW 285 carbon black dispersion (pH=8) to obtain the dynamic mobility [18]. The rheology and stability measurements were performed on a 10 % w/w (pH=8) dispersion with a UDS 200 Universal Dynamic Spectrometer, Paar Physica, USA and a Lumifuge 114, L.U.M. GmbH, Germany (3000 rpm). All dispersions were prepared at 23°C according to DIN EN-ISO 8780-2 [19] in water for 2 hours with Zirconox Micro Milling Beads (2.8<∅<3.3 mm), Jyotti Ceramics, India. More detailed information about ESA and the preparation of dispersions can be found in the literature [10].

Results and Discussion

Copolyelectrolyte Synthesis and Structure

The synthesis route leads to well defined ionic hydrophobic acrylic acid (AA)-styrene (S) copolyelectrolytes with different molecular architectures and nearly the same charge density per macromolecule of about 60 to 70 mole-% AA in the copolyelectrolyte. The characteristic data of the random, block and graft copolymers employed in the investigation of the stabilization behavior are summarized in Table 1.

Table 1. Copolymer constitution as determined by ^{1}H-NMR and SEC experiments: comonomer composition and molecular weight of the PS block and graft ($M_{n,PS}$) and the copolymer

Copolymer	Type [a]	f_{AA} [b]	$M_{n,PS}$ (PD) [c]	M_n (PD) [d]	Comp. (GD) [e]
PAA	linear	100	-	2400 (1.62)	2000
P(S-r-AA)	linear	63	-	1300 (1.94)	-
PS-b-PAA	2-block	68	4880 (1.24)	8200 (1.56)	4800-b-6000
PAA-g-PS-Y	Y-graft	70	2190 (1.05)	6500 (1.56)	2700-g-1600 (0.8)
PAA-g-PS-H	H-graft	65	2190 (1.05)	8300 (1.61)	4300-g-3200 (1.6)

a) Molecular architecture
b) Acrylic acid comonomer mole-fraction f_{AA} in copolymer as obtained from ^{1}H-NMR analysis
c) Number average molecular weight $M_{n,PS}$ (polydispersity PD) of PS block or graft as determined by SEC analysis based on PS
d) Number average molecular weight Mn (polydispersity PD) of copolymer as determined by SEC analysis based on PS calibration
e) Molecular composition (number average Mn-b/g-Mn) and PS graft density (GD) as calculated from $M_{n,PS}$ and f_{AA}; GD represents the average number of grafts per individual macromolecule

As far as the polymer adsorption to carbon black is concerned which in the case of the AA-S copolymers occurs preferably by hydrophobic-hydrophobic interactions between the hydrophobic styrene units and the hydrophobic carbon black surface, PAA can be regarded as a linear ionic-hydrophilic homopolymer with only one hydrophobic benzylmercaptane (BnS-) anchor group at the end [20]. P(S-*r*-AA) exhibits a random distribution of the hydrophobic styrene anchor groups within the stabilizer macromolecule. The block copolymers consist of one polystyrene (PS) anchor block and one ionic poly(acrylic acid) (PAA) stabilizing block (2-block). The structure of the given graft copolymer depends on the graft density (GD): E. g. PAA-*g*-PS-*Y* with GD=0.8 can be considered to consist of 3-star macromolecules (*Y*-shape) since in average 4 out of 5 macromolecules contain one hydrophobic anchor graft and two electrosterically stabilizing blocks; in contrast to this, PAA-*g*-PS-*H* with GD=1.6 can be described as a graft copolymer mixture of *Y*-shape (corresponding GD=1) and gemini structure (*H*-shape corresponding to GD=2; 2 anchor and 2 stabilizing blocks) macromolecules .

Carbon Black-Copolyelectrolyte Interactions

The electrokinetic sonic amplitude (ESA) technique [21] can be used for determining the dynamic mobility μ (electrophoretic mobility in an alternating electric field) and the zeta potential as well as the size of uncoated particles [22]. Furthermore ESA is a suitable method for the investigation of pigment-polymer interaction of non-ionic polymers [23] and (co)polyelectrolytes [24, 25, 26].

As it has been demonstrated in earlier studies, the ESA method is a fast and useful tool for the in-situ investigation of adsorption/desorption phenomena [27]. The variation of the dynamic mobility μ with polymer concentration at low frequencies is only a consequence of a change of the surface charge of the pigment particles caused by the deposition of polymer on the particle surface; thus such curves can be regarded as quasi adsorption isotherms [10]. For normalization reasons and better elucidation of the pigment-polymer interactions it is advantageous to introduce the reduced dynamic mobility μ/μ_0 which relates μ of the polymer containing system to μ_0 of the pure pigment dispersion [18].

The change of the reduced dynamic mobility μ/μ_0 with time is shown in Fig. 2 for the different (co)polyelectrolytes. The particle surface charge increases due to polymer adsorption until a plateau is reached after about 15 min. The obtained reduced mobility curves are indicative of a fast physisorption of the copolyelectrolytes on the pigment surface driven by van der Waals forces or electrostatic interactions [28].

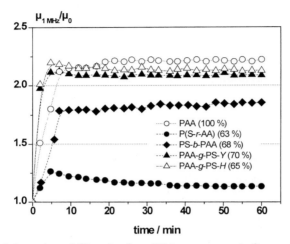

Figure 2. Reduced dynamic mobility μ/μ_0 from ESA measurements for aqueous carbon black dispersions (2 % w/w, pH=8) as a function of time after addition of (co)polyelectrolytes (10 % w/w relative to carbon black mass) with different molecular architecture; the figures in brackets give the mole-% AA in the macromolecule

For an inorganic pigment like TiO_2 it has been shown that the surface charge of the fully coated particles depends primarily on the carboxylic group content of the polymeric stabilizer [24]. This has been explained by successive adsorption steps of the copolyelectrolyte on the pigment beginning with the anchoring of the copolymer via the ionic carboxylic groups followed by a bilayer formation. However, in the case of carbon black, the distinct difference in the μ/μ_0 plateau value between the random copolymer P(S-r-AA), the 2-block copolymer PS-b-PAA and the graft copolymers PAA-g-PS-Y/H, all having nearly the same AA comonomer content, reveals effects of the polymer architecture on the electrostatic behavior of the coated carbon black (Fig. 2). The main reason for the effects of the copolymer structure on the dynamic mobility μ of the coated carbon black as compared to the coated TiO_2 (cf. [8, 25]) may be the different mechanism of anchoring of the amphipolar copolymer on the pigment surface: Whereas for the TiO_2-pigment the ionic carboxylic comonomer sequences interact with the pigment surface, in the case of carbon black it is the hydrophobic polystyrene sequence that acts as the anchor. The interaction of the organic pigment with the hydrophobic styrene units of these stabilizers results in a mono-layer adsorption and not a "solloid" and bilayer formation like on TiO_2 [8].

Assuming the PS units as anchors, the difference in the plateau value of the reduced dynamic mobility μ/μ_0 observed for the various copolyelectrolytes could be explained by the chain conformation of the adsorbed macromolecule: A more flat conformation is suggested for the P(S-r-AA) with trains and loops (see Fig. 3, conformation A) as already found for hydrophobically modified polyelectrolytes on TiO_2 [26]; this is contrasted by a more stretched brush conformation of the fully dissociated PAA sequences of the homopolymer and the block and graft copolymers resulting in a highly charged particle surface (see Fig. 3, conformation B). A comparative schematic illustration of the proposed adsorption behavior of the polymeric stabilizers is given in Fig. 3.

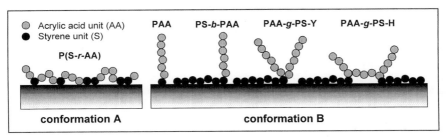

Figure 3. Adsorption behavior and conformation of ionic-hydrophobic stabilizers with different molecular architectures on the carbon black surface

Stabilization Effects of the Amphipolar Copolyelectrolytes on Carbon Black Dispersions

The viscosity of aqueous carbon black dispersions containing the polyelectrolytes was investigated to get further information about the dispersion stabilizing properties. Fig. 4 shows the viscosity behavior of carbon black in the presence of the different stabilizers, indicating the important role of the stabilizer architecture on the rheological properties of the dispersion. The viscosity of the linear homopolymer (PAA) and the random copolymer (P(S-*r*-AA)) is significantly higher as compared to the block and graft copolymers especially at lower shear rates. The lower viscosity of the carbon black dispersion with the block and graft copolyelectrolytes is indicative for well dispersed particles in the media what is a necessary condition for most applications.

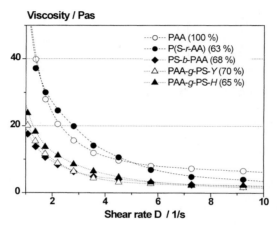

Figure 4. Viscosity of aqueous carbon black dispersions (10 % w/w, pH=8) in the presence of ionic-hydrophobic stabilizers with different molecular architectures (10 wt-% relative to carbon black mass); the figures in brackets give the mole-% AA in the macromolecule

The stabilization effect of the carbon black dispersions by the pigment surface modification with the amphipolar copolyelectrolytes and the correlation to the viscosity was investigated by the sediment volume method. From lumifuge measurements a relative sediment volume is obtained, giving information about the dispersity of the dispersion and its stabilization [29]. The results of the stabilization measurements and the viscosity at a shear rate of 1 (~3000 rpm in the lumifuge [30]) of the different dispersions are shown in Fig. 5.

From the comparison of the viscosity and the sedimentation data in Fig. 5 it is obvious that there is a good agreement between the viscosity and the sediment volume of the dispersion. For all polymeric stabilizers tested the block and graft structures provide a better performance, and the data infer that the graft structure is somewhat favorable over the block structure.

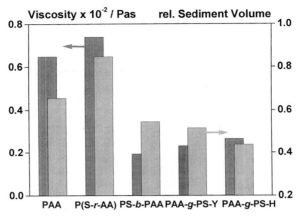

Figure 5. Viscosity (shear rate 1) and relative sediment volume of aqueous carbon black dispersions (10 % w/w, pH=8) in the presence of ionic-hydrophobic stabilizers with different molecular architectures (10 wt-% relative to carbon black mass)

This is another indication of the high efficiency of the electrosteric stabilization by a charged PAA brush and the better performance of styrene anchor blocks as compared to the mono-functionalized PAA or the random distribution within the stabilizer. Therefore it is possible to perform a tailored synthesis of copolyelectrolytes of the graft type with optimal stabilizing properties for organic as well as inorganic pigments dispersions.

Conclusion

The comparative investigation of poly(acrylic acid) and various acrylic acid-styrene based copolyelectrolytes of different polymer architectures showed distinct differences in their interactions with the carbon black dispersion. The pigment-polymer interactions as established by ESA measurements and the stabilizing properties of these stabilizers as quantified by viscosity and sedimentation measurements are strongly influenced by the charge distribution along the macromolecule and the polymer architecture but not the charge density; this is in contrast to the results obtained with TiO_2 pigments where the overall charge density (acrylic comonomer content) played an important role (cf. [8, 25]). Another difference between carbon black and TiO_2 is that "solloid" and bilayer formation occurs on TiO_2 whereas a monolayer is formed on carbon black with styrene-units as anchors. The rheology and sedimentation measurements revealed a better long time performance for block and graft copolymers as compared to poly(acrylic acid) and the random copolymer with somewhat better results for the graft structure.

In summary, these results are a further indication of the high efficiency of the electrosteric stabilization of pigment dispersions by block and graft copolyelectrolytes. By taking advantages of tailored copolymer structures and pigment-specific interactions with particular portions of the copolyelectrolytes, efficient stabilization for organic as well as inorganic pigment dispersions are accessible.

Acknowledgement

The authors like to thank S. Küchler and D. Lerche, L.U.M. GmbH for the Lumifuge measurements and discussion.

References

[1]   M. Antonietti, M. C. Weissenberger, Macromol. Rapid. Com. 18 (1997) 295
[2]   J. Clayton, Pigm. Resin Technol. 27 (1998) 231
[3]   Th. Tadros, Solid/Liquid Dispersions, Academic Press, 1987 London
[4]   D. H. Napper, Polymeric Stabilization of Colloidal Dispersions, Academic Press, 1983 New York
[5]   P. Kaczmarski, M. Tarng, J. E. Glass, R. J. Buchacek, Prog. Org. Coat. 30 (1997) 15
[6]   N.G. Hoogeveen, M. A. C. Stuart, G. J. Fleer, J. Col. Int. Sci. 182 (1996) 133
[7]   J. Schmitz, R. Höfer, Farbe&Lack 104 (1998) 22
[8]   Ch. Schaller, Th. Schauer, K. Dirnberger, C. D. Eisenbach, Farbe&Lack 11 (2001) 58
[9]   P-L. Kuo, S-C. Ni, C-C. Lai, J. Appl. Polym. Sci. 45 (1992) 611
[10] Ch. Schaller, Dissertation Uni Stuttgart, Shaker Verlag, Aachen 2002
[11] Degussa, Pigment Blacks-Technical data
[12] L. O`Brien, F. Gornick, J. Am. Chem. Soc. 77(1955) 4757
[13] R. Milkovich, M.T. Chiang (1974) U.S. Patent 3.842.050 and subsequent patents
[14] C. Ramireddy, Z. Tuzar, K. Prochazka, S. E. Webber, P. Munk, Macromol. 25 (1992) 2541
[15] X. F. Zhong, A. Eisenberg, Macromol. 25 (1992) 7160
[16] Ch. Schaller, Th. Schauer, K. Dirnberger, C. D. Eisenbach, Prog. Org. Coat. 35 (1999) 63
[17] Ch. Schaller, Th. Schauer, K. Dirnberger, C. D. Eisenbach, Adv Pol. Sci., *submitted*
[18] M. L. Carasso, W. N. Rowlands and R. W. O`Brien, J. Col. Int. Sci. 193 (1997) 200
[19] Deutsche Norm EN ISO 8780-2: Dispergierverfahren zur Beurteilung des Dispergierverhaltens1995
[20] A. V. Tobolsky, B. Baysal, J. Am. Chem. Soc. 75 (1952) 1757
[21] R. W. O`Brien, J. Fluid Mech. 190 (1988) 71

[22] R. W. O`Brien, D. W. Cannon, W. N. Rowlands, J. Col. Int. Sci. 173 (1995) 406
[23] M. Entenmann, Th. Schauer, C. D. Eisenbach, Farbe+Lack 106 (2000) 48
[24] D. Collins, W. H. Meyer, G. Wegner, H. Arndt, Th. Schauer, C. D. Eisenbach, Farbe&Lack 108 (2002) 89
[25] Ch. Schaller, Th. Schauer, K. Dirnberger, C. D. Eisenbach, Europ. Phys. J. 6 (2001)365
[26] Ch. Schaller, A. Schoger, Th. Schauer, K. Dirnberger, C. D. Eisenbach, Macromol. Symp. 179 (2002) 173
[27] Ch. Schaller, T. Schauer, K. Dirnberger, C. D. Eisenbach, Adv. Pol. Sci., *in preparation*
[28] D. H. Lee, R. A. Condrate, J. S. Reed, J. Mat. Sci. 31 (1996) 471
[29] D. Lerche, Advances in Physiological Fluid Dynamics, Narosa Publishing House, 1996 New Dehli
[30] D. Lerche, L.U.M. GmbH, Germany, *personal communication*

Macromol. Symp. 187, 707–718 (2002)

The Effect of TiO$_2$ Pigment on the Performance of Paratoluene Sulphonic Acid Catalysed Paint Systems

Elizabeth Reck, Steve Seymour*

Huntsman Tioxide, Haverton Hill Road, Billingham, TS23 1PS, UK

Summary: The work has highlighted the importance of choosing the correct TiO$_2$ pigment for paratoluene sulphonic acid (PTSA) catalysed paints. Generally, surface treated TiO$_2$ grades which are basic in nature resulted in the best optical performance in comparison to acidic surface treated grades. The relative performance of the acidic surface treated grade can be improved by increasing the PTSA level in the paint or by using a resin with a higher acidity level. It is postulated that for these TiO$_2$ grades, the higher acid levels were found to give better steric repulsion. However, overall, the basic surface treated grade retained the best optical performance in PTSA catalysed paint systems.

Keywords: acid catalysed paints, cure time, resin acidity, TiO$_2$, optical performance

Introduction

In industrial paints, the choice of TiO$_2$ generally has little effect on cure rate and degree of cure. However, in paints where cure is catalysed by an acid such as PTSA, the TiO$_2$ affects the curing mechanism. The effect the TiO$_2$ has on the cure depends on the pigment coating.

Virtually all commercial TiO$_2$ pigments are now coated. Originally, the purpose of coating was to improve durability and to lessen yellowing which occurred in certain types of paints. However, it was subsequently found, that surface treatments could be used to improve the dispersibility of pigments in different media.

As the name implies, coating involves the deposition of other matter onto the surface of the pigment particles. The coating agent must be a white, hydrated oxide; silica, zirconia, titania and alumina are commonly used. In the case of silica, it can be used to form several types of coating. It is often used in combination with alumina to form either a loose porous coating or a relatively compact coating. An even denser coating, known as a 'dense silica' coating can be formed to give a highly impermeable layer which has excellent durability. The more silica there is in the coating, generally the more acidic the pigment surface will be, whilst a pigment

  CCC 1022-1360/00/$ 17.50+.50/0

where the presence of alumina and zirconia dominate in the coating will be more basic in nature. Previous papers[1,2,3] have shown how basic TiO₂ pigments can neutralise the PTSA and hence reduce its catalytic effect. If the catalytic effect is reduced, increased paint cure time would be expected.

This paper focuses on the performance of TiO₂ pigments in PTSA catalysed paint systems and how an increased paint cure time can affect the optical paint performance.

The influence of resin acidity on cure and optical performance is also studied. In earlier papers[4,5,6] published by Huntsman Tioxide the effect of the resin acidity on TiO₂ pigments and, ultimately, the optical performance of coating was discussed.

The main finding of these earlier papers was the importance of matching the acid-basic nature of the TiO₂ pigment with the acid-basic nature of the resin. For example, TiO₂ pigment with a predominantly acidic surface offers a lot of potential reaction sites for resins with a basic nature which results in the resin molecules lying flat on the pigment surface. This gives a compact adsorbed layer with little scope for solvent entrapment between the pigment and resin. Such compact resin layers have little steric repulsion, as shown diagrammatically in Figure 1(a), and lead to poor stabilising properties and poor optical paint performance.

This paper examines the use of acid in such systems to enhance the steric repulsion between pigment particles and, hence, improve the stabilising properties. If acid (e.g. in the form of an acid catalyst) is added to a system, the acid should compete with the acidic TiO₂ pigment for the basic sites on the resin molecules resulting in fewer reactions taking place between the resin and the TiO₂ pigment. This should result in a less tightly packed resin layer around the TiO₂ pigment, giving more scope for solvent entrapment between the resin and pigment, leading to better paint stability. See Figure 1(b).

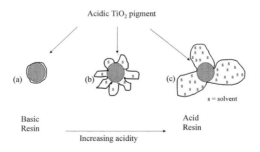

Figure 1. Resin acidity effect on an acidic TiO₂ pigment

Figure 1(c) shows the theoretical situation where so much acid has been added to the paint that the acid significantly reduces the number of basic sites on the resin. Hence the resin is only attached to the TiO$_2$ pigment to a very small degree leading to significant solvent entrapment. The resin molecules are significantly extended from the pigment surface resulting in excellent steric repulsion and good paint stability.

In a similar manner this paper also examines the use of acid when TiO$_2$ pigments which are predominantly basic are used.

If a basic resin is used with a basic TiO$_2$ pigment, there are few opportunities for reaction between the two leaving opportunities for solvent entrapment and extended resin molecules around the TiO$_2$ pigment. This should again result in excellent steric repulsion and good paint stability as shown in Figure 2(a).

However, if acid is added in increasing amounts to such a system, it will, in theory, react increasingly with the basic resin making it more acidic. This should result in more interaction with the basic TiO$_2$ pigment leading, ultimately, to poorer steric repulsion. This is illustrated in Figures 2(b) and 2(c).

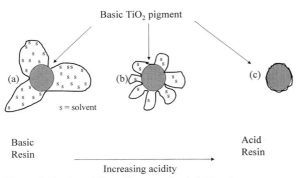

Figure 2. Resin acidity effect on a basic TiO$_2$ pigment

Part 1 - Effect of TiO$_2$ on Cure Time and Gloss in a PTSA Catalysed Paint System

To illustrate the effect of TiO$_2$ grade on cure in a PTSA catalysed system, paints were produced using 12 TiO$_2$ pigments with different acidic and basic surface treatments as shown in Table 1.

Table 1 – Surface treatments of the twelve evaluated TiO₂ pigments

Pigment name	Pigment coating	Acid-basic nature of surface coating
TR81	Alumina-zirconia	Predominantly basic
TR85	Alumina-zirconia-silica	Some acidity
R-HD2	Alumina	Predominantly basic
R-TC90	Alumina	Predominantly basic
G22/53	Alumina-zirconia-silica	Some acidity
G15/49	Alumina-zirconia	Predominantly basic
TR92	Alumina-zirconia	Predominantly basic
G63/2	Alumina-zirconia-silica	Some acidity
G9/32	Dense silica	Predominantly acidic
G22/62	Dense silica	Predominantly acidic
G9/72	Dense silica	Predominantly acidic
G63/1	Dense silica	Predominantly acidic

The cure time of the paints was determined by monitoring the viscous and elastic behaviour changes on a controlled stress rheometer during paint curing. The results are illustrated in Figure 3 where a wide range of paint cure times can be seen for the twelve pigments ranging from just below 700 seconds up to 870 seconds.

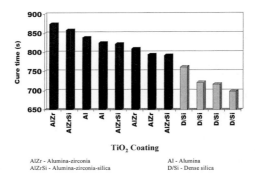

Figure 3. Effect of TiO₂ coating on the cure time of a PTSA catalysed paint system

The four pigments which have the shortest cure times in Figure 3 are all dense silica coated pigments. These also have the most acidic surfaces and the least ability to neutralise the PTSA. Hence, the cure time for these paints would be expected to be the shortest. The remaining pigments are more basic in nature and have more ability to neutralise the PTSA and hence produce longer curing times.

Following on from the results in Figure 3, the effect of cure time on gloss was studied in order to determine whether longer cure times would result in higher gloss due to better flow. Gloss values were measured and compared to the cure times. The results are given in Figure 4 where it can be seen that there is a good correlation.

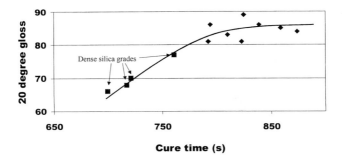

Figure 4. Effect of cure time on gloss in a PTSA catalysed paint system

Up to a point, an improvement in gloss can be achieved by increasing the cure time. However, once the cure time is more than ca. 800 seconds little improvement in gloss is evident. Obviously, for this particular formulation 800 seconds is sufficient time for optimal flow.

The square points in Figure 4 show the gloss values and cure times for the dense silica coated grades. Their shorter cure time has not helped the gloss performance. It should also be noted that the dense silica grades tended to have wider particle size distributions, which will also have a detrimental effect on the overall gloss performance.

Part 2 - Effect of PTSA Level on Cure and Optical Performance

To investigate the effect of PTSA level on cure and optical performance when using acidic and basic pigments, two further series of paints were prepared using the same formulation as above, but this time with varying PTSA levels. The first series was based on the predominantly basic pigment TIOXIDE TR81, which had resulted in the longest drying time and an excellent gloss performance in Part 1. The second series was based on G63/1, an acidic dense silica grade, which had resulted in the shortest cure time and poorest gloss performance in Part 1.

In Figure 5 the effect of PTSA level on the cure time for the two paint series is shown.

As would be expected from earlier results, TIOXIDE TR81 gives a substantially longer cure time than the dense silica grade G63/1. As the PTSA level is increased there is obviously more acid available to catalyse the cure resulting in shorter cure times for both pigments regardless of their surface treatment. The extent of reduction in cure time is similar for the two pigments.

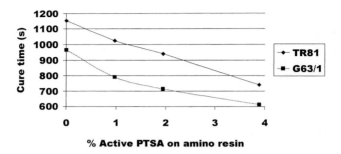

Figure 5. Effect of PTSA level on cure time

Solvent resistance and reverse impact testing of the paints related well to the cure results. The paints with 0% PTSA had a very poor solvent resistance and mechanical performance. With increasing PTSA levels, the solvent resistance and mechanical performance improved, showing G63/1, generally, to be slightly superior to TIOXIDE TR81.

In Figure 6 the effect of PTSA on the gloss performance of the two paint series is shown.

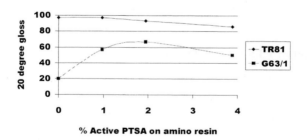

Figure 6. Effect of PTSA level on gloss performance

TIOXIDE TR81 and G63/1 reacted very differently to the increase in PTSA. The change in PTSA level affected TIOXIDE TR81 less than G63/1 and the gloss performance of TIOXIDE TR81 is significantly better than G63/1.

For TIOXIDE TR81, a slight decrease in gloss performance is noted. Referring to Figure 2, it is evident that by adding more acid to the system, poorer steric repulsion was expected for TIOXIDE TR81, which explains the poorer gloss at the higher PTSA levels. The faster cure time at the highest acid level could also be contributing towards poorer gloss.

Pigment G63/1 behaved quite differently. Regardless of the cure time decreasing with increasing PTSA level as shown in Figure 5, the gloss has initially increased quite substantially. The increased levels of acid could well be helping to improve the steric repulsion for this pigment (see Figure 1). However, at the PTSA level of 3.9% the gloss deteriorated, which suggests that the short cure time is now having a more dominant effect than the positive effect of having more acid present in the paint to improve steric repulsion.

To further investigate the results shown in Figure 6, pigment flocculation[7,8] was assessed using flocculation gradients. The flocculation data is given as a function of PTSA level in Figure 7.

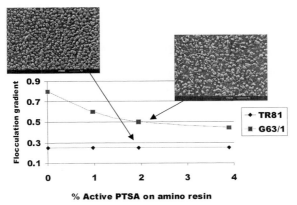

Figure 7. Effect of PTSA level on flocculation gradient

It is evident from Figure 7 that the dispersion of G63/1 is much more affected by changes in the PTSA level than TIOXIDE TR81 which has excellent flocculation stability with a consistent gradient below 0.30 suggesting that there is very little flocculation in these paints. From the results and the model in Figure 2, a slight deterioration in dispersion with increasing PTSA level would be expected for TIOXIDE TR81, however, this deterioration was not detected using the flocculation gradient technique.

For G63/1 the flocculation gradient varies from 0.80 to 0.44. The flocculation improves significantly with increasing levels of PTSA which relates well to the Figure 6 results and the model shown in Figure 1.

The illustrated micrographs for the paints with PTSA levels of 1.95% clearly support the flocculation gradient results.

In Figure 8 the opacity as a function of PTSA level is given, and, as with the gloss and flocculation gradient data, the opacity of TIOXIDE TR81 is significantly better than G63/1.

A slight decrease in the opacity level of TIOXIDE TR81 was observed as the PTSA level is increased, which agrees with the model in Figure 2. Similarly, the opacity of G63/1 generally improves with increased PTSA level, in line with the improvement in flocculation.

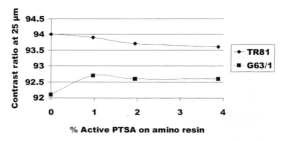

Figure 8. Effect of PTSA level on opacity performance

Part 3 - Effect of Resin Acidity on Cure Rate and Optical Performance in PTSA Catalysed Paints

Having seen how the PTSA level could significantly affect the cure rate and optical performance and how important TiO_2 choice was in this type of paint system, it was considered relevant to explore the effect of resin acidity on cure rate and optical performance.

For this purpose, a second PTSA catalysed paint system based on an alkyd resin was used. The alkyd was chosen for its high acidity (ie 20-25mgKOH/g cf 10mgKOH/g for the resin used   in Part 2). From the model illustrated in Figures 1 and 2, this high acid resin would be expected to favour G63/1 more than the medium acid resin used in Part 2. Equally, the high acid resin would be expected to favour TIOXIDE TR81 less than the medium acid resin.

Cure rate, gloss, flocculation and opacity were assessed for both paint formulations at different PTSA levels using TIOXIDE TR81 and G63/1

In Figure 9 the cure time for TIOXIDE TR81 and G63/1 are shown for both the medium acid resin and the high acid resin, where it can be seen that the paint based on the high acid resin resulted in the fastest cure times, regardless of the TiO_2 used. The high acidity level appears to help catalyse the cure.

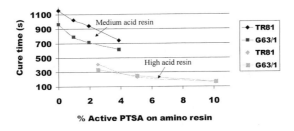

Figure 9. Effect of resin acidity and PTSA level on cure time

The other point to note, is that the high acid resin has brought the performance of G63/1 much closer to the performance of TIOXIDE TR81, compared to how the two pigments performed in the medium acid resin. The high acid resin does tend to be favouring G63/1 more than the medium acid resin. This relates well to the model in Figure 1.

As would be expected, as the PTSA level increases, more acid is available to catalyse the cure, hence the cure time decreases for both the medium and high acid resin systems.

In Figure 10 the effect of resin acidity and PTSA level on gloss can be seen for both paint formulations. The behaviour of the two pigments in the high acid resin system is similar in the tested PTSA range. G63/1 is now tending to follow the same model as TIOXIDE TR81 shown in Figure 2. This could be due to the high acid resin system having too much free acid available relative to the acidity of G63/1. In spite of G63/1 showing a similar trend and being closer in performance to TIOXIDE TR81 in the high acid resin system, the performance of TIOXIDE TR81 was still superior to that of G63/1.

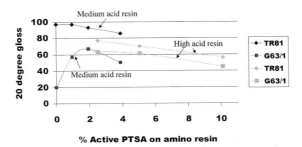

Figure 10. Effect of resin acidity and PTSA level on gloss performance

In Figure 11 the effect of resin acidity and PTSA level on pigment flocculation is shown for both paint formulations. It can be seen that the flocculation is worse in the high acid resin system than in the medium acid resin system. In both cases TIOXIDE TR81 shows the least flocculation.In contrast to the different behaviour of TIOXIDE TR81 and G63/1 in the medium acid resin system, the two pigments again showed similar trends in the high acid resin system, when the PTSA level is increased. A slight increase in flocculation was noted for both pigments.

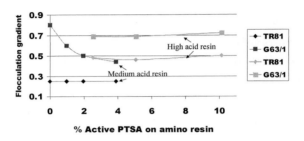

Figure 11. Effect of resin acidity and PTSA level on flocculation gradient

In Figure 12 the effect of resin acidity and PTSA level on opacity is shown for both paint formulations. It can be seen that the high acid resin reduced the performance difference between the two grades. However, TIOXIDE TR81 still performed the better.

Both pigments decreased slightly in opacity, which relates well to the slight increases in flocculation seen in Figure 11 for the high acid resin system. The fact that G63/1 is again following the same model for TIOXIDE TR81 as shown in Figure 2, again suggests that there is too much acid present in the high acid resin system relative to the acidity of G63/1.

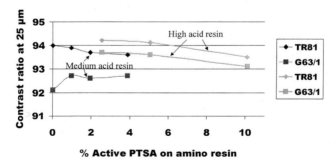

Figure 12. Effect of resin acidity and PTSA level on opacity performance

Conclusion

The paper has highlighted how the choice of TiO$_2$ affects the cure rate in a PTSA catalysed paint system. The more basic the TiO$_2$ pigment, the greater is the degree of neutralisation of the PTSA and increase in the cure time of the paint. For the paint systems tested, the basic TiO$_2$ pigments which resulted in the longest cure times also gave the best gloss performance, due to the longer cure time resulting in better paint flow.

The paper has highlighted how the choice of TiO$_2$ affects the cure rate in a PTSA catalysed paint system. The more basic the TiO$_2$ pigment, the greater is the degree of neutralisation of the PTSA and increase in the cure time of the paint. For the paint systems tested, the basic TiO$_2$ pigments which resulted in the longest cure times also gave the best gloss performance, due to the longer cure time resulting in better paint flow.

The resin acidity value was also found to be important for cure behaviour and optical performance. The high acid resin generally helped to improve the optical performance of the acidic pigment G63/1, as the steric repulsion was improved. In contrast the high acid resin generally resulted in a deterioration of the performance of the basic pigment TIOXIDE TR81. However, in spite of improving the conditions for G63/1, and deteriorating the performance of TIOXIDE TR81 by moving from a medium acid resin to a high acid resin, TIOXIDE TR81 still had the better optical performance.

Acknowledgements

The authors would like to thank the directors of Huntsman Tioxide for permission to publish this paper.

References

[1] T Entwistle and S J Gill, "*Effect of titanium dioxide pigments on the cure of thermosetting films*", Internal Huntsman Tioxide report D9200, **1985**

[2] C Bertrand, "*Titanium dioxide – Catalyst interaction: Review of results obtained with two types of high solids stoving enamel*", 15th FATIPEC Congress paper, Amsterdam, **1980**

[3] H Sander, "*The influence of titanium dioxide pigments on the rate of cure of acid catalysed one-component high solids baking paints*", Farbe und Lack, **1977**, 83, 891

[4] J E Hall, "*Pigment/resin interactions in thixotropic alkyd media*", Internal Huntsman Tioxide report D9163, **1982**

[5] K Goldsborough and J Peacock, "*The adsorption of alkyd resins by titanium dioxide*

pigments and its relation to the hiding power of alkyd paint systems", J. Oil Col. Chem. Assoc., **1971**, 54, 506-519

[6]   J B Franklin, K Goldsborough, G D Parfitt and J Peacock, *"Influence of particle charge and resin adsorption on the opacity of paint films pigmented with titanium dioxide*", Journal of Paint Technology, **1970**, 42, 551

[7]   D J Rutherford and L A Simpson, *"The use of a flocculation gradient monitor for quantifying pigment dispersion in dry and wet paint films*", Internal Huntsman Tioxide report D9186GC, **1984**

[8]   J G Balfour and M J Hird, J. Oil. Col. Chem. Assoc., **1975**, 58, 331

Macromol. Symp. 187, 719–729 (2002)

CFD Modelling of a Spray Deposition Process of Paint

Mirko Garbero, Marco Vanni and Giancarlo Baldi

Dip. Scienza dei Materiali e Ingegneria Chimica, Politecnico di Torino,

C.so Duca degli Abruzzi 24, 10129, Torino, Italy

Summary: The paint deposition process by spraying has been studied by means of Computational Fluid Dynamics in order to predict the final thickness of the coating and to determine theoretically the overspraying phenomenon. The VOF model has been used to describe the impact phenomena onto the wall and the Euler-Lagrange approach to simulate droplet trajectories on their way to the surface.
Particular attention has been devoted to the prediction of the maximum diameter reached by an impinging droplet at the end of the spreading phase. This diameter is very important in the study of the paint processes because the high viscosity and the small surface tension of paints reduce the impingement practically only to the spreading phase. Two different configurations of atomizers have been considered. The air flux provides a finer atomisation of the liquid, gives to the droplets the necessary velocity to reach the wall, but is also the main cause of overspray since the small droplets tend to follow its deviation near the wall.

Key words: CFD, simulation, drop impact , spray, spread

Introduction

Sprays are widely used in many applications as the pouring out of herbicides in agricultural field, the injection of fuel in combustion engines or the deposition of paint in coating processes. In all these processes, a detailed knowledge of the behaviour of droplets, from the outlet of the atomizer to the target surface, is very useful in order to achieve either uniform deposition, or good combustion or the desired aspect of the coating. To characterise the whole process after its atomisation, the aerodynamic transport of the droplets and the collision onto the surface are treated separately. The distinction allows us to study more deeply the phenomenon of impact, helping to better understand what happens near the surface to cover. The power growth of diesel engines achieved in recent years or the saving of paint in industrial application are two example of the usefulness of these studies.

   CCC 1022-1360/00/$ 17.50+.50/0

The outcome of the impingement of a single droplet may be its deposition on the surface or its reatomization into smaller secondary droplets with partial mass deposition. The result of the impingement is determinated by the properties of the fluid such as viscosity and surface tension as well as by the diameter of the droplet and its velocity relative to the wall. Furthermore, the surface roughness, the thickness of the liquid film and the wall temperature are important. Previous investigations [1-5] revealed that the droplets spread out on the surface and form a liquid film only if the component of moment normal to the wall is small; otherwise they form a conical sheet similar to a crown, that may disintegrate into secondary droplets.

When a droplet spreads onto a surface forms a thin liquid disk that is usually called *lamella*. First, the lamella expands very quickly and reaches a maximum radius within a short time. The kinetic and surface energy of the drop are dissipated by viscous processes in the thin sheet of liquid, and are transformed in additional surface energy. During a second stage the lamella shrinks to a small size, and in some case the recoil of the lamella may cause the drop to separate from the surface and rebound.

The thickness of the liquid and the final radius of the lamella depend above all on the viscosity of the liquid and on interfacial tensions between both liquid and gas and liquid and surface. One can calculate the dimensions of the lamella by tracking the interface between liquid and air by mean of the VOF model [6,7]. In this model a single set of momentum equations is shared by the fluids, and the volume fraction of each fluid is tracked throughout the domain. The VOF model can also include the effects of surface tension and contact angle.

To simulate a spray deposition process it is also important to know how many droplet reach the wall and where they impact. In order to achieve this aim, the droplet trajectories have been calculated by mean of the Euler-Lagrange approach. [8]

Modelling of the impact zone

If thermal effects are negligible, a dimensional analysis indicates that the impact can be described in terms of the following dimensionless parameters: [5,9]

1. Reynolds number, $\qquad$ $Re = \dfrac{\rho D V}{\mu}$

2. Weber number, $\qquad$ $We = \dfrac{\rho D V^2}{\sigma}$

3. Dimensionless surface roughness, $\qquad$ $R_s^* = 2R_s/D$

4. Dimensionless film thickness,
$$F_l^* = F_l/D$$

5. Bond Number.
$$Bo = \frac{\rho D^2 g}{\sigma}$$

Another number often used in literature is Ohnesorge number, defined as:

$$Oh = \frac{\mu}{\sqrt{\rho D \sigma}} = \frac{We^{1/2}}{Re}$$

In our case Bo and F_l can be neglected, since we have considered small droplets with low impact velocity colliding dry surfaces. The dynamic of spreading is characterized primarily by We and Re numbers (or Oh), by roughness and finally by the contact angle.

When a droplet impacts onto a surface it normally forms a thin liquid disk that is called *lamella*, which expands quickly and reaches a maximum diameter d_m: this phenomenon is called spread. Subsequently shrinks due to surface tension, and in some cases, when the elastic force is very strong in comparison to viscous dissipation, some tiny droplets are formed from the centre of the lamella. Normally after the recoiling phase, a new spread can be observed followed by a new retraction and so on till the liquid has reached its equilibrium shape [10]. The thickness of the liquid and the final radius of the lamella depend mainly on the viscosity of the liquid and interfacial tensions between liquid and gas and liquid and surface. A parameter of great interest in coating process is the maximum diameter d_m reached by the lamella after its spread. Many authors [1-3,11], have considered that the total energy owned by the drop before impact is equal to that of the lamella at its maximum diameter minus the energy dissipated by friction:

$$\underbrace{E_k + E_p + E_s}_{\text{Before impact}} = \underbrace{E_k + E_p + E_s + E_{diss}}_{\text{After impact}} \tag{1}$$

where subscripts k, p, s and *diss.* refer to kinetic, potential, surface and dissipated energy, respectively. Coupling this balance with simplified models of the spreading phase, some authors have developed simple equations for the maximum spreading ratio, that is: $B_m = d_m/D$ (D being the size of the droplet before impact. An example of this equation proposed by Moo *et al.* [12]:

$$\left[\frac{1}{4}(1 - \cos\theta) + 0.35\frac{We}{\sqrt{Re}}\right](B_m)^3 - \left(\frac{We}{12} + 1\right)(B_m) + \frac{2}{3} = 0 \tag{2}$$

Another one, that is Chandra and Avedisian [2]:

$$\frac{2We}{3Re}(B_m)^4 + (1 - \cos\theta)(B_m)^2 - \left(\frac{We}{3} + 4\right) = 0 \qquad (3)$$

In order to predict the time evolution of the size of the lamella, some expressions like $B(\tau)=h[1-\exp(-k\tau)]$, where τ is a non dimensioned time, have been proposed. One of these is: [9]

$$B(\tau) = 2.4[1 - \exp(-0.9\tau)] \qquad \text{with} \qquad \tau = t\sqrt{\frac{\sigma}{\rho R^3}}, \qquad (4)$$

Scheller and Bousfield [13] have connected a large number of experimental data according to the relationship:

$$B_m = 0.61(Re^2 Oh)^{0.166} \qquad (5)$$

The range of considered Re is from 19 to 16400, while We is between 0.002 and 0.58.

In some cases the outcome of the impact of a droplet on a surface may be its reatomization into smaller secondary droplets with partial mass deposition, and not its spreading. Such a phenomenon is called "splashing". Previous investigations [1,4-5] revealed that the droplets spread out on the surface and form a liquid film only if the component of moment normal to the wall is small; otherwise they form a conical sheet similar to a crown, that may disintegrate into secondary droplets. When a single droplet hits a surface, the non-dimensional parameter K determines whether it is completely deposited or partially splashed into secondary droplets. The parameter K can be calculated from correlations like $K=Oh\,Re_l^a$, where a is a constant. According to Mundo et al. [5], a assumes the value of 1.25, whereas for Yarin and Weiss [4], it has the value of 1.27. If K is smaller than a critical value K_{crit}, depending on the surface condition, the droplet is completely deposited and forms a liquid film on the surface. If K exceeds the critical value, a crown structure is formed around the point of impingement. The value of K_{crit} is higher for dry and smooth surfaces. The influence of surface roughness can be taken into account with the non-dimensional surface roughness number R_s^*. For low values of R_s^* K_{crit} is very high, as the out flowing fluid at the contact line between the impinging droplet and the surface cannot be redirected in a direction normal to the wall. Rising the surface roughness, K_{crit} decreases strongly at first and reach an asymptotic value for high non-dimensional roughness numbers [14]. For coating sprays, K should be small as possible in order to spread all the paint that arrives onto the surface. For this reason, droplets have small sizes and the paint is extremely viscous.

Results

Simulation of the impact of single droplets upon dry and smooth surface have been done concerning droplet different in size and velocity. The drop diameter has been varied from 25 to 200 μm, whereas the drop velocity from 5 m/s to 20 m/s. The CDF code Fluent has been used in its version 5.0 with the VOF model. Neglecting the possible instability at the rim of the lamella, axis symmetric conditions can be imposed and a 2D grid developed. The grid has been made of square cells with the edge from 1.5 μm to 2μm, depending of the droplet size. The simulations have been carried out with a constant time step equal to 10^{-9} s. This step assures a quite fast convergence from the beginning to the end of the simulation and it is proportionate to the length scale of the case. Some 3D simulation are also been performed. Regarding the characteristic of the liquid, we have considered a new paint without any organic solvent: for the viscosity we have adopted a value of 40 cP while for the surface tension a value equal to 0.03 N/m has been used. The contact angle has normally been kept constant at a value of 70°, corresponding to the advancing angle. This simplification should not lead to significant error since, in the considered case, the recoiling phase is negligible (as verified adopting a dynamic contact angle). An additional consequence is that the final size of the lamella is very close the largest diameter by the lamella after spreading. The behaviour is due to the high viscosity (40 cP) which dissipate all the inertia of the droplet and to the small surface tension which is not able to give enough elastic energy to retract the lamella. The sequence of Figure 1. confirms such a behaviour.

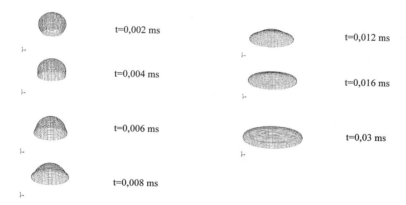

Figure 1. Impact of a 200 μm diameter paint droplet at 5 m/s colliding a smooth surface.

The figure represents the impact of a droplet of 200 μm of diameter colliding at 5 m/s. Another important aspect for the considered paint deposition process is that we have never observed any slashing. That has been also confirmed by parameter $K = \text{We}^{0,5}\,\text{Re}^{0,25}$, which is always smaller than its critic values found for smooth surfaces [5]. Figure 2. (a) reports the values of the maximum spread factor calculated by CFD as a function of Re at different Ohnesorge numbers.

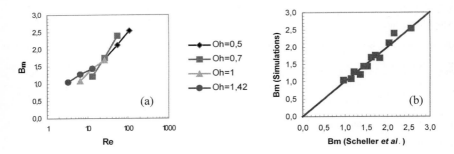

Figure 2. (a) Maximum spread predicted with CFD simulations as a function of Re at different Oh. (b) Comparison of the maximum spread obtained with CFD simulation to that predicted with the Scheller correlation.

Obviously increasing the impact velocity a larger lamella is obtained due to the greater inertia that must be dissipated. In Fig. 2. (b) are compared the results found by the simulations with the predictions of the correlation of Sheller [13]. The figure shows a good agreement between our results and the experimental study, although the range where it has been done does not cover all our tests.

Modelling of spraying phase

The prediction of the fluid flow of the continuous phase was obtained by solving the time-averaged Navier-Stokes equation in connection with a closure model for turbulence [15]. The equations of motion for a steady flow in turbulent regime, expressing the instantaneous values of the quantities by Reynolds decomposition, can be written in the following way:

$$\frac{\partial U_i}{\partial x_i} = 0 \tag{6}$$

$$\rho U_j \frac{\partial U_i}{\partial x_j} = -\frac{\partial P}{\partial x_i} + \mu \frac{\partial^2 U_i}{\partial x_j^2} - \rho \frac{\partial}{\partial x_j}(\overline{u_i' u_j'}) + S_{p,\phi} \tag{7}$$

where U_i is the average velocity component along axis x_i and u_i is the fluctuating velocity along the same axis. The solution of this equation requires the knowledge of a relationship (closure equation) between the mean velocity field and the Reynolds tensor (the term composed by the fluctuating velocities). The Reynolds tensor can be expressed as a function of the rate of strain tensor S_{ij} as follows:

$$- \rho(\overline{u_j u_i}) = 2\mu_t S_{ij} \tag{8}$$

where μ_t is the turbulent viscosity and in the k-ε model is calculated as:

$$\mu_t = C_\mu \rho \frac{k^3}{\varepsilon} \tag{9}$$

where k is the kinetic energy of the fluctuating flow, ε the turbulent dissipation rate and C_μ is constant. Of course, two equations that describe the variation of the turbulent kinetic energy k and of the turbulent dissipation rate ε are needed. They can be expressed as:

$$U_i \frac{\partial k}{\partial x_i} = -\frac{1}{\rho} \frac{\partial}{\partial x_i}\left(-\Gamma_k \frac{\partial k}{\partial x_i}\right) + \frac{2\mu_t}{\rho} S_{ij} S_{ij} - \varepsilon \tag{10}$$

$$U_i \frac{\partial \varepsilon}{\partial x_i} = \frac{1}{\rho} \frac{\partial}{\partial x_i}\left(\frac{\mu_t}{\sigma_\varepsilon} \frac{\partial \varepsilon}{\partial x_i}\right) + \frac{2\mu_t}{\rho} C_{\varepsilon 1} S_{ij} S_{ij} - C_{\varepsilon 2}\frac{\varepsilon}{k} \tag{11}$$

The empirical constants of the turbulence model are: $C_\mu = 0.09$; $C_{\varepsilon 1} = 1.44$; $C_{\varepsilon 2} = 1.92$. The additional source terms $S_{p,\phi}$ are introduced to account for the influence of droplets on the fluid flow. Detailed equations, which describe these source terms, are available in the article by Ruger et al [16]. The dispersed phase is treated by the Lagrangian approach, where a large number of droplet parcels, representing a number of real droplets with the same properties, are traced through the flow field. The representation of droplets by parcels makes it possible to consider size distribution and to simulate the measured liquid mass flow rate at the injection locations by a reasonable number of computational droplets. The trajectory of each droplet parcel is calculated solving the equation of motion for a single droplet. The basic equation of motion is the result of the force balance on the particle written in a Lagrangian reference frame. A basic equation of motion can be written neglecting the forces due to pressure gradients, virtual mass and history terms. This equation of motion has the following form, for the x direction in Cartesian co-ordinates:

$$\frac{dU_{p,i}}{dt} = F_D\left(U_i - U_{p,i}\right) + g_i \frac{(\rho_p - \rho)}{\rho_p} + F_i \tag{12}$$

Where $F_D(U_i - U_{p,i})$ is the drag force per unit particle mass and

$$F_D = \frac{18\mu}{\rho_p D_p^2} \frac{C_D \mathrm{Re}}{24} \tag{13}$$

Here, U is the fluid phase velocity, U_p is the particle velocity, μ is the molecular viscosity of the fluid, ρ is the fluid density, ρ_p is the density of the particle and D_p is the particle diameter. Where Re is the relative Reynolds number, which is defined as

$$\mathrm{Re} = \frac{\rho D_p \left| \overrightarrow{U_p} - \overrightarrow{U} \right|}{\mu}. \tag{14}$$

The drag coefficient C_D has been calculated from the following equation [17]:

$$C_D = \frac{24}{\mathrm{Re}} \left(1 + 0.15 \mathrm{Re}_p^{0.687} \right) \qquad \mathrm{Re}_p < 500 \tag{15}$$

In order to model the droplet dispersion in turbulent flow and to obtain a representation of the local velocity, the so-called eddy lifetime concept can be applied [18]. This model assumes that the droplet interacts with a sequence of turbulent eddied with randomly sampled fluctuations.

Results

Hereafter the results of the simulation of two different types of sprays normally used in coating process are reported. The droplet trajectories have been determined with the Euler-Lagrange approach implemented in Fluent. Inlet boundary conditions have been specified inside the gun for the air phase in order to achieve a better representation of turbulence at the nozzle, whereas for the paint, its mass flux, the distribution of the droplets and their inlet velocity has been prescribed immediately after the exit of the gun.

The air flow rate is one of most interesting parameters since it influences the formation of droplets, increases the impact velocity and plays an important role on the overspray phenomena. In Figures 3 and 4, the simulated air velocity vectors and the droplet trajectories for a low volume middle pressure gun are reported.

This type of gun normally operates with an air cup pressure of 0.20 MPa, consumes 300 Nl/min of air, 300 cc/min of paint with an efficiency of about 70 %. The diameter of the nozzle is 2 mm and the thickness of the circular section for the air 1.5 mm. The distance from the wall is of 20 cm and a Rosin-Rammler droplet size distribution has been used with a mean diameter of 36 µm and a spread factor of 2.11.

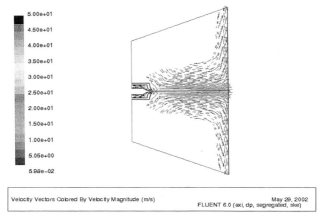

Figure 3. Velocity vectors contoured by velocity magnitude of a paint spray referring to a low volume middle pressure gun.

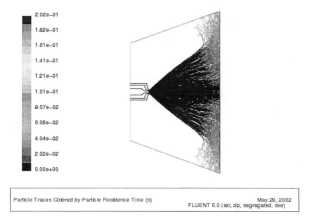

Figure 4. Particle tracks contoured by particle residence time referring to a low volume middle pressure gun.

The CFD simulation predicts a value of efficiency of about 65%, close to that given by the manufacturer. Looking at the particle tracks it appears that many droplets are deviated by the air stream and exit from the computational domain, confirming that the loss of paint is due essentially to the deviation of droplets by the air and not to splashing phenomena.

The other type of gun considered is a standard one. It differs from the previous configuration in air consumption (500 Nl/min) and cup pressure (0.35 MPa). The simulation shows the formation of a narrower cone and, even if the distance between the nozzle and the wall is now of 25 cm, it is still evident the droplet deviation. For this gun a smaller efficiency, has been found, of about 45 %. In agreement with the data of the manufactured, this could be explained with the larger air amount that forms an air pillow onto the wall, but it could also be due to the different inlet pressure of the liquid paint.

Conclusions

In this paper, a possible approach for the simulation of a coating process by spraying, using a commercial CFD code has been described. In order to characterize what happens at the wall, the impacts of a single droplet has been studied with the VOF model. The simulations have shown that all the liquid of the droplet remains on the wall during the wall impact and that a possible splash or rebound is avoided due to the high viscosity and the small surface tension of the paint.

Furthermore the friction caused by the high viscosity of the liquid dissipates all the inertia of the droplet and stops practically the impact when the lamella of liquid has reached its maximum diameter. The maximum spread and the thickness of the liquid film has been valuated.

In order to know the paint flux on the target surface, the droplet trajectories have been calculated with the Euler-Lagrange approach. Two types of spraying guns normally used in the coating industry have been considered and the effect of the air flux onto the droplet has taken into account. Since air is the main cause of the waste of paint, the overspray factor has been determined.

References

[1]   M. Rein, "Phenomena of liquid drop impact on solid and liquid surfaces", *Fluid Dyn. Res.*, **1993**, *12*, 61.

[2]   S. Chandra, C. T. Avedisian, "On the collision of a droplet with a solid surface", *Proc. R. Soc. London*, **1991**, *432*, 13.

[3]   C. D. Stow, M. G. Hadfield, "An experimental investigation of fluid flow resulting from the impact of a water drop with an unyielding dry surface", *Proc. R. Soc. London*, **1981**, *373*, 419.

[4]   A. L. Yarin and D. Weiss, "Impact of drops on solid surfaces: Self-similar capillary Waves, and splashing as a new type of kinematic discontinuity", *J. Fluid Mech.*, **1995**, *283*, 141.

[5]   C. Mundo, M. Sommerfeld, C. Tropea, "On the modeling of liquid sprays impinging on surfaces", *Atomization and Sprays*, **1998**, *8*, 625.

[6]   D. B. Kothe, R. C. Mjolsness and M. D. Torrey, "A computer program for incompressible flows with free surfaces", *Technical Report*, **1991**, *LA-12007-MS*, *LANL*.

[7]   C. W. Hirt, B. D. Nichols, "Volume of fluid (VOF) method for the dynamics of free boundaries", *J. Comput. Phys.*, **1981**, *39*, 201.

[8]   M. Ruger, S. Hohmann, M. Sommerfeld and G. Kohnen, "Euler/Lagrange calculations of turbulent sprays: the effect of droplet collisions and coalescence", *Atomisation and Sprays*, **2000**, *10*, 47.

[9]   S. Schiaffino, A. A. Sonin, "Molten drop deposition and solidification at low Weber numbers", *Phys. Fluids*, **1997**, *9*, 3172.

[10]   R. E. Ford and C. G. L. Furmidge, "Impact and spreading of spray drops on foliar surfaces Wetting", *Soc. Chem. Ind.*, **1967**, *25*, 417.

[11]   G. Trapaga and j. Szekely, "Mathematical modeling of the isothermal impingement o liquid droplets in spraying processes", *Metall. Trans.*, **1991**, *B22B*, 901.

[12]   T. Mao, D. C. S. Kuhn, H. Tran, "Spread and rebound of liquid droplets upon impact on flat surfaces", *AIChE J.,* **1997**, *43*, 2169.

[13]   B. L. Scheller, D. W. Bousfield, "Newtonian drop impact with a solid surface", *AIChE J.,* **1995**, *41*, 1357.

[14]   M. Bussmann, J. Mostaghimi, S. Chandra, "Modeling the splash of a droplet impacting a solid surface", *Phys. Fluids,* **2000**, *12*, 3121.

[15]   B. E. Launder and D.B. Spalding, "The numerical computation of turbulent flows", *J. Comput. Meth. Appl. Mech. Eng.*, **1974**, *3*, 269.

[16]   M. Ruger, S. Hohmann, M. Sommerfeld and G. Kohnen, "Euler/Lagrange calculations of turbulent sprays: the effect of droplet collisions and coalescence", *Atomization and Sprays*, **2000**, *10*, 47.

[17]   B. J. Daly and F. H. Harlow, "Transport equations in turbulence", *Phys. Fluids*, **1970**, *13*, 2634.

[18]   R. Clift and W.H. Gauvin, "The motion of particle in turbulent gas stream", *Chemeca*, **1970**, *70*, 14.

Macromol. Symp. **187,** *731–737 (2002)*

100 % Electrostatic Application of Metallic Basecoats

Dr. Bernd Biallas, Franz Stieber

BASF Coatings AG, Münster, Germany

Summary : Paint Application by electrostatic high rotation atomizers is the most efficient way of applying spray paint to car bodies in automotive OEM paint shops. Only the 2nd coat of metallic base coats usually is applied by pneumatic atomizers. The reason for this is due to color/design and process stability questions, both resulting from differences in atomisation and droplet deposition mechanisms of the two paint atomizing processes. The results in an applied metallic film are explained. A path towards a process is outlined, considering not only the spray process itself , but the whole process chain incl. :
- design
- the repair processes
- the manufacturers of plastic parts for the body
- the paint manufacturers in terms of paint reformulation and test equipment

The success of this process depends on the careful evaluation of the paint shop targets and the consideration of the total process chain.

Keywords : metallic base coats, atomization, efficiency, paint process

Introduction

In today´s application of spray paints to car bodies, 2 processes of atomization are used .

First, pneumatic atomization, usually for interior parts of the body and for the second coat of base coat on the exterior surface.

Second, atomization by electrostatic high rotation , usually for application of primers, clear coats, solid top coats and the first coat of metallic basecoats on the exterior surface.

The driving force towards electrostatic atomization is due to transfer efficiency reasons. Pneumatic atomization at the average will achieve about 40 % of transfer efficiency, while electrostatic atomization reaches about 80 %. This means advantages in paint consumption, amount of waste and emission.

Differences in color shade from both ways of applying paint result from the atomization processes taking place : pigmentation is touched severely. So in order to find solutions for reaching high transfer efficiencies, the principles of atomization have to be considered first.

          CCC 1022-1360/00/$ 17.50+.50/0

732

Atomization

Both by pneumatic atomization and electrostatic atomization droplets in range of about 2 µm to 70 µm are created. The fundamental differences between both methods, however, are the droplet deposition mechanisms on the target. In pneumatic atomization, there is a straight forward movement of droplets, with speeds up to 12 m/sec to 15 m/sec. A high flow of atomization air also is directed towards the target. At the target, this air flow is turned around, carrying many droplets with it. Due to their inertia, mainly big droplets are deposited, the small ones are lost and form the overspray. These small droplets contain less effect pigments like aluminium flakes and are enriched in resin. A separation of the paint takes place in the spray fan.

In electrostatic atomization, the droplets leave the bell radially. Shaping air and electrostatic forces turn around the droplets towards the target, missing, however part of the bigger ones, that are lost as overspray. Unfortunately, they carry many effect pigments. Velocities are much lower, about 4 m/ sec, and in the middle of the spray fan even flow back was observed.

As a consequence, the pigmentations of the films created by the two different application techniques differ very much. Films from pneumatic atomization contain much more effect pigments than those from electrostatic atomization, see Figure 1.

The change in pigmentation also has been studied by Inkpen [1].

Change of Pigmentation

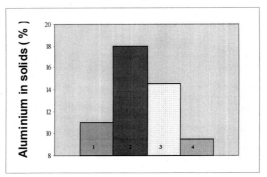

Figure 1. Aluminium content in 1 = wet sample, 2 = pneumatic applied film (270 ml/ min), 3 = pneumatic applied film (540 ml / min), 4 = electrostatic applied film (30.000 1/ min) Microscopic investigations make transparent the different coverage with aluminium.

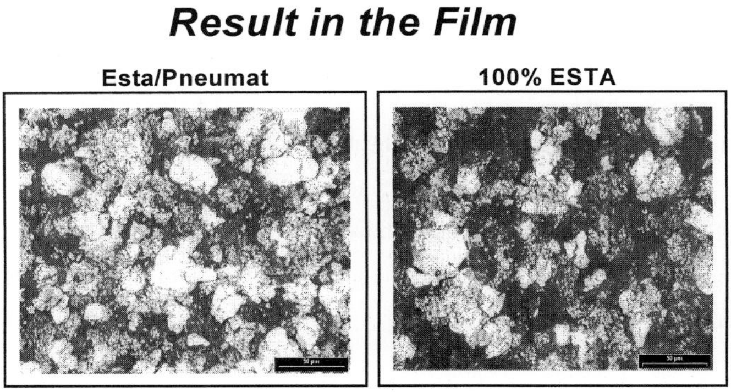

Result in the Film

Esta/Pneumat   **100% ESTA**

Figure 2. Microscopic view of metallic films : 1= applied by electrostatic plus pneumatic atomization (2 coats) , 2 = 2 coats electrostatic atomization.

But it is not only the amount of aluminium that varies. Also the particle size distribution of the aluminium flakes is changed [1] , making a compensation by reformulation more difficult.

Furthermore there is a change in color pigments : many colored metallic base coats contain pigments of different density. Due to the centrifuge effects described above, it is obvious that a change will occur here as well.

Also kinetic energies must be considered. Due to different droplet sizes and velocities, kinetic energy of droplets hitting the target from pneumatic atomization is about 50 times higher than that from electrostatic atomization. As a consequence, the parallel orientation of the metallic flakes can be influenced.

Consequences for metallic colors

It has been found, that turning towards 100 % electrostatic application, different metallic base coats react in different ways. Changing typical parameter of atomization, they react in different ways according to their pigmentation, see figure 3.

For some colors, for example dark metallics with low aluminium content, fine tuning of the parameter of an electrostatic bell will be sufficient to achieve the same color as from pneumatic application. Modern atomizers offer a progress in terms of color matching.

For other base coat colors, just mere optimization of application parameter will not be sufficient - new pigmentation is required.

When turning towards 100 % electrostatic application, these will probably be the most cases.

A coordinated optimization of application parameter and reformulation will lead to satisfactory results. For a third group of base coat colors, however, it remains very difficult to achieve color match in all angles. In this cases, the acceptable tolerances for color match will decide about the feasibility of this process.

ESTA 100: effect on colours

Influence of:	Silver	Black (Alu / Mica)	Green (Mica /Alu)
RPM 50.000 → 75.000 1/min	effect improves ••• ↑	lightness uneffected ○ →	lightness decreases (effect inverted) •• ↘
SA 200 → 350 Nl/min	effect improves •• ↗	lightness increases •• ↗	lightness indifferent ○ →
Flowrate 80 → 160 ml/min	effect decreases ••• ↓	lightness decreases (visibly darker) ••• ↓	lightness decreases (visibly darker) •• ↘
Stages (flash time) 1 stage → 2stages	effect improves •• ↗	lightness unaffected, chroma improves; • ↗	Indifferent ○ →

Strength of influence: ○ ● ●● ●●

Figure 3. Response of different colors on changes in atomization.

Investigations of the films produced by pneumatic atomization and electrostatic atomization have shown, that it is not just color itself that changes. Taking into consideration the processes described above, it is understandable that also hiding power of a film applied by electrostatic atomization is lower – due to lack of effect pigments, that to a great extend are responsible for hiding properties. This means that defects resulting from the mechanisms of electrostatic atomization must be compensated by higher pigmentation in the paint leading to higher cost for pigmentation.

Other process qualities that are effected are metallic effect and sand mark coverage.

Figure 4 gives an overview on what was observed when moving towards electrostatic atomization.

From this experience it can be stated that when turning towards 100 % electrostatic application, each individual color has to be analyzed separately. And for each individual color a fine tuning of the application parameter and the pigmentation and formulation has to be adapted in order to compensate these defects.

Observations in the wet film
- Less aluminium pigments
- less pigments with high density
- change in particle size distribution of al-flakes
- poorer orientation of plakes

Consequences in dry film
- P/B - ratio lower
- stronger color
- less hiding power
- less sand mark coverage
- Haze
- poor metallic effect

Figure 4. Observations in wet and dry films applied by electrostatic atomization compared to pneumatic application.

Paths towards ESTA 100

An option towards 100 % electrostatic atomization of metallic base coats is that this process is considered as the master. All other processes like design and color development, the application for plastic parts, the repair processes in the factory and after sales and the paint production / tinting operations have to accept that and take care of that. This process has to start with the introduction of new colors, where right from the beginning of color development those colors are created by application via electrostatic atomization. Here mainly design is involved, since ESTA-100 – colors look different.

This means the target can only be achieved by coordinated action of all partners involved.

A second option exists, when existing colors are to be transferred towards electrostatic atomization. Additionally to option 1 color match towards other lines / plants has to be considered. So each individual color of the color program has to pass an „ ESTA 100 – assessment".

Process considerations

A typical metallic base coat line will use a 2 – coat application of the base coat. It has been found out, that going towards 100 % electrostatic application, a 2-coat – process is favorable as well with respect to color match.. This means that the pneumatic zone can be used for electrostatic application.

Process comparison
(e.g.: base coat line 4.5 m/min)

Parameter criteria	standard bell+air	bell	bell + bell (on recipro, robot)	bell + bell (two full stages)
no. of machines, atomisers	9 bells + 6 guns	9 bells	9 bells + 2/3 bells	9 bells + 9 bells
booth length	8 + 5 m 13 m	8 m	8 + 5 m 13 m	8 + 8 m 16 m
investment and costs	+ / -	+	+ / -	-
maintenance, housekeeping, cleaning	+ / -	+	+ / -	-
TE % (example)	75% + 40%	65 - 75%	75% + 65%	75% + 70%
base coat consumption (example)	100 % + / -	80 % +	80 % +	80 % +
paint dump (rinse, purge, colour change	+ / -	+	+ / -	-

Figure 5. Process advantages and constrains.

The higher the number of atomizers, the higher the losses for purging. The feasible and right solutions very strongly depends on line speed that is to be realized.

Besides the topcoat line, repair processes have to be considered in detail. Using special tinted paint in the topcoat line, pneumatic repair of body parts is hardly possible. Spot repair with special repair material was found to be feasable.

Plastic parts often are painted at a different place. Also these paint processes have to be considered in order to achieve color match between hang-on parts and the car body.

Last not least the paint suppliers have to prepare themselves for this way of base coat application in development and paint production. To exploit the advantages of a more efficient basecoat paint application in the car assembly process the color development effords will increase substantially.

Conclusions

The tasks that have to be done when turning towards electrostatic atomization for metallic base coats can be deducted very clearly from the basic physical mechanisms taking place during atomization, droplet transportation and droplet deposition.

The benefit of high transfer efficiencies can only be gained by joined effort and contribution of all parties involved in the process chain.

References

[1]   Stuart Inkpen , „ Electrical and Mechanical Mechanisms for Color Variation in the Spaying of Metallic Paints", Massachusetts Institute of Technology, 1986

*Macromol. Symp. **187**, 739–751 (2002)*

Effects on the Pigment Distribution in Paint Formulations

F. Tiarks[1]*,   T. Frechen[1],   S. Kirsch[2],   J. Leuninger[1],   M. Melan[2],   A. Pfau[1], F. Richter[3], B. Schuler[1], C.-L. Zhao[4]

[1] BASF Aktiengesellschaft, Emulsion Polymers, 67056 Ludwigshafen, Germany
[2] BASF AG, Product Development Adhesives and Construction, 67056 Ludwigshafen, Germany
[3] BASF AG, New Business Development, 67065 Ludwigshafen, Germany
[4] BASF Corporation, Charlotte Technical Center, 11501 Steele Creek Rd. Charlotte, NC 28273, USA

Summary: Modern water-borne paints are applied in different areas ranging from high-gloss lacquers to flat, scrub-resistant interior paints. The pigment volume concentration (PVC) is one key-parameter adjusting the application properties. In this work formulations differing in the type of binder and dispersing agent were investigated by various techniques concerning the distribution of pigments in the liquid paints and films. The structure of the paints was analyzed by Remission Light Spectroscopy (RLS), Disc Centrifugation, Cryo-Replica Transmission Electron Microscopy (Cryo-TEM) and Cryo-Scanning Electron Microscopy (Cryo-SEM). The pigment distribution in the films was examined by means of Atomic Force Microscopy (AFM), TEM and RLS.
The tendency of the pigments to form aggregates was found to depend on both: the type of binder and the dispersing agent. Only by adjusting the properties of the binder in combination with common dispersants it is possible to get well distributed TiO_2 particles within the paint. Correlation of application properties e.g. gloss and blocking to the microscopic structure is presented.

Keywords: coatings, pigments, gloss, films, emulsion polymerization, dispersion

Introduction

The composition of modern water-borne coatings strongly depends on the desired application properties and therefore on the PVC. In general paints consist of water, a binder, pigments and filler particles. Additives like coalescents, thickeners, dispersants and defoamers are added to guarantee a sufficient stability and good application properties. The PVC is one key-parameter for describing paints: the higher the PVC the lower is the content of binder within the paint and the higher the portion of pigments and fillers. The PVC strongly determines application properties such as gloss, scrub resistance and tensile strength. Besides the PVC the pigment distribution in a paint is critical and influences the properties: gloss and hiding power at low PVC as well as scrub resistance at high PVC. Whereas in flat paints the high PVC is reached by a combination of cheap filler material and small amounts of TiO_2 (to achieve sufficient hiding power), one is limited to the solely use of TiO_2 in low PVC paints: a high portion of filler would not contribute to the opacity but increases the surface roughness and hence reduces the gloss.

To obtain maximum hiding power with a minimum amount of TiO_2 a homogeneous distribution of the pigments is essential and one of the main requirements for the formulation. The pigment distribution is determined by the colloidal interaction of the TiO_2 particles, which tend to aggregate in the paint. These clusters result in a higher surface roughness as well as a decreased hiding power due to reduced light scattering efficiency. The influence of the surface roughness on the gloss is obvious and has been described in [1]. As PVCs below 20 % are typical for gloss paints the scattering efficiency of the pigment is crucial to achieve a fully opaque coating.

TiO_2 is the favored pigment for producing opacity due to its high refractive index ($n_D \approx 2.6$). Light scattering is maximized when the particle size ranges about 250 nm. However, if the particles form clusters the scattering efficiency decreases because each cluster scatters approx. like a single, larger particle. Accordingly the scattering efficiency strongly depends on the dispergation state of the TiO_2 particles. In the process of paint formulation the break-down of agglomerates into fine particles is important as well as the stabilization of this state.[2] However, it has been shown that most clumps are dispersed in a short time by the milling process.[3] Due to this, pigment clustering in emulsion paints must involve other factors.[4]

In this work we elucidate the influence of the binder and the dispersing agent onto the pigment distribution and identify appropriate analytical tools for characterization of the dispergation state. Another crucial question to be answered is, whether the pigment distribution in the final film is predetermined in the liquid paint or whether additional agglomeration occurs during film formation and loss of water. The following samples were formulated and examined:

	Binder 1	Binder 2	Binder 3
Paste 1		P 2	
Paste 2			P 3
Paste 3			P 4
Paste 4			P 5

Table 1. Survey of the different formulations (P 1 – P 5) to examine the influence of either binder or dispersing agent onto the pigment distribution.

The difference between paste 1 and pastes 2 - 4 is the higher PVC (40 % vs. 18 %) and the dispersant. Paste 2, 3 and 4 differ in the type of dispersing agent and are optimized for high-gloss formulations. Binder 1 and 2 are straight acrylics and differ only in their comonomer system. Binder 3 is a styrene-acrylic developed for high-gloss paints. Concerning the dispersants, paste 1 contains pigment dispersant A[®], a low molecular polyacrylic acid ammonium salt. Paste 2 comprises pigment dispersant MD 20[®], a copolymer of diisobutene and maleic anhydride and paste 3 pigment dispersant Collacral LR 8954[®], a solution polymer consisting of an acrylate and acrylic acid ammonium salt. The fourth dispersant Orotan 681[®] consists of an acrylate-methacrylic acid copolymer and is added to paste 4. Overall the dispersants differ in their hydrophilicity and functional groups.

Results and Discussion

The effect of (i) binder and (ii) dispersing agent on the distribution of TiO_2-particles was investigated. The latter was studied on the liquid paint formulation and the dried film. Via Cryo-Replica TEM and Cryo-SEM it was possible to characterize the pigment distribution in the formulation. The samples of the films were investigated by means of AFM and RLS. By comparing the results gained from analysis of the wet and dried state it is possible to get an insight into the effect of film formation on the clustering of TiO_2.

Distribution of titanium dioxide pigments in the liquid state

The Cryo-Replica-technique enables to measure directly in the undiluted paint. As a reference the structure of a pigment paste with a PVC of 40 % without addition of a polymeric binder was investigated. Already in the pigment paste large aggregates of TiO_2 are formed. The dilution of the sample by different amounts of water affects the pigment distribution: with increasing dilution the size of TiO_2 agglomerates decreased. However, it is clear from this study that even at high dilution clusters are formed. In order to further separate the clusters additional interactions have to be introduced. By formulating a paint from a paste both, the dilution effect and specific interactions are introduced by the binder. Thus a good dispersion should not only break up agglomerates by quasi-diluting but additionally stabilize the small clusters by electrostatic forces against re-agglomeration. It became apparent that fast and easy analytical tools like dynamic light scattering relying on dilution of a sample are restricted in their information, as the dilution affects the cluster size. Therefore it was a major task to identify analytical tools, which enable us to detect pigment clusters in the undiluted samples.

Influence of the binder onto the pigment distribution in the liquid paint

As we concentrate on the pigment distribution in gloss lacquers PVC was not much varied. Pigment distribution at high PVC (> 40 %) is described elsewhere.[5] The focus of this work is the effect of different types of binders and the influence of additives used in the formulation.

For the first, two straight-acrylic binders were compared: binder 1 and 2 differ only concerning their comonomer compositions. Figure 1 displays results from Cryo-TEM (PVC 40 %). TiO_2 is due to its electron density of high contrast (black). The binder is removed by the sample preparation and the Pt/C that was deposited on the sample under a certain angle induces the topology contrast. These areas appear as gray to white regions. The replica of individual latex particles is detectable and indicates that no film formation occurred. As is apparent from this preliminary testing the nature of the binder has an influence on the pigment distribution. Whereas P 1 shows only small and well distributed TiO_2-clusters, P 2 shows bigger aggregates in the paint. Apparently the differences in the comonomers of the binder are essential. Similar results were obtained by Zhao et al. analyzing the effect of functional groups onto the dispergation state.[6]

a) P 1   b) P 2

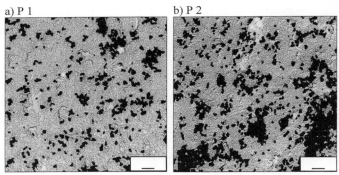

Figure 1. Use of different dispersions as binders in paint formulations and their influence on the pigment distribution analyzed by Cryo-Replica TEM. a) binder 1 leading to a fairly good TiO_2 dispergation, b) binder 2 resulting in highly agglomerated pigment particles.

Another conclusion drawn from this tests is, that the dispersing agent alone is not able to provide complete separation of all TiO_2 particles. As an effect of water-soluble oligomers on the pigment distribution is unlikely, as being shown by [5], it is the latex itself strongly interacting with the surface of the pigment resulting in differences in the pigment distribution.

After confirming that the binder influences the dispergation state, the effect of different dispersants was examined. A similar latex but optimized for high-gloss application was employed in the following testing (binder 3).

Effect of dispersing agent on the pigment distribution

The binder used in this testing was kept constant and the type of dispersant was varied. Figure 2 displays the results by Cryo-SEM (upper row) and Cryo-Replica TEM (bottom line). The difference in contrast (TiO_2 as white or black dots) is due to the different imaging techniques. In the SEM the electron-rich TiO_2 appears as white dot due to the analysis via imaging the back-scattered electrons, in the TEM micrographs TiO_2 is of high contrast (black) due to its electron density. For these samples with low PVC sophisticated Cryo-techniques have to be used. Whereas in flat paints the large $CaCO_3$ particles behave like spacers and hinder structural changes caused by water crystallization these lowly pigmented paints are more sensitive towards ice-crystals. Therefore crystallization has to be avoided which is realized by shock-freezing. A highly reliable method to avoid ice-crystal formation even by shock-freezing is the high pressure technique.[7]

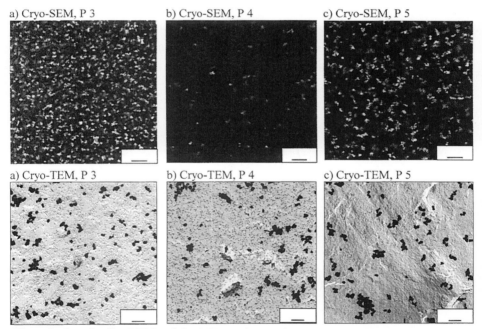

a) Cryo-SEM, P 3 b) Cryo-SEM, P 4 c) Cryo-SEM, P 5

a) Cryo-TEM, P 3 b) Cryo-TEM, P 4 c) Cryo-TEM, P 5

Figure 2. Cryo-Replica TEM and Cryo-SEM micrographs of formulations with the same type of binder but different dispersing agents. a) dispersing agent MD 20® (P 3), b) Collacral LR 8954® (P 4), c) Orotan 681® (P 5). The use of different dispersants influences the dispergation state of titanium dioxide.

As is maintained from the Cryo-Replica TEM and Cryo-SEM measurements the samples exhibit differences in the dispergation state of the TiO_2. Whereas the dispergation by use of pigment dispersant MD 20® (P 3) is relatively homogeneous and only with small clusters, the use of Orotan 681® (P 5) leads to pronounced clustering. Assuming a better interaction of the maleic acid groups in MD 20® as compared to acrylic acid in Orotan 681®, the effect of a better pigment distribution is understandable. This positive interaction enables the de-agglomeration of the clusters. The Cryo-micrographs of P 4, especially the SEM, display small agglomerates but compared to the other samples the number of detected clusters is too small, as all samples possess the same PVC. Large pigment clusters should be observed somewhere else in the sample, but were not visualized in this image.

Imaging techniques via EM are a nice tool to get an impression of the pigment distribution in the paint. However, these techniques are rather complicated, cost and time consuming and give only images of small volumes. These are not necessarily representative for the whole sample and might be misleading. An integral testing method, which allows the measurement of larger volumes is therefore important.

Consequently the paints are compared by disc centrifugation and RLS. Employing disc centrifugation the hard sphere diameter of the pigment pastes as well as the paints were determined. The formulations and pastes were diluted to a pigment concentration of 0.025 % and conditioned by ultrasound for 30 sec (Figure 3).

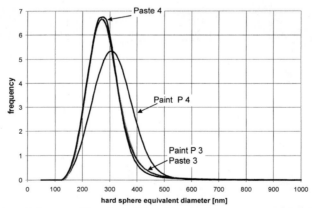

Figure 3. Results of disc centrifugation of the diluted pigment pastes and paint samples P 3 and P 4 possessing different dispersants (MD 20® and Collacral LR 8954®). The shift of the curve of P 4 to larger diameters indicates the formation of clusters.

Pastes 3 and 4 and paint P 3 exhibit an almost identical particle size distribution (approx.

270 nm) close to the primary grain size of the pigment particles (ca. 0.26 μm). Yet paint P 4 exhibits a reproducible and thereby significant shift to larger diameters (ca. 310 nm). This shift can be explained by partly agglomerated TiO$_2$ particles. The agglomeration did not occur in the paste, which already contains the dispersant, but happens after adding the binder. This demonstrates, that dispersant *and* binder influence the dispergation state of TiO$_2$ and supports the assumption of better specific interactions of functional groups in P 3 with the pigments.

This result corroborates the hypothesis that P 4 contains larger TiO$_2$ clusters that were not detected by imaging techniques due to the small picture area. However, disc centrifugation is based on diluted samples and therefore we have to backup this assertion by a second method like RLS. Advantageously RLS works in concentrated opaque samples. The qualitative extent of different particle sizes onto the remission spectrum can be drawn from the scattering factor according to Mie-Theory as the wavelength of monochromatic light and the particle size leading to optimal back-scattering depend on each other.[8]

Thus, for a given particle size distribution a large extent of small particles leads to a high re-mission in the short-wave region and a high extent of clusters yields to pronounced remission in the long-wave region. The remission spectrum is a distorted image of the original grain size distribution of the pigments. Concerning the spectra one has to keep in mind that in a pigment suspension the back-scattering decreases with an increasing pigment concentration. Paints P 3 and P 4 exhibit the same PVC and consequently their remission behavior can be compared.

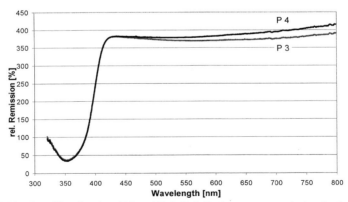

Figure 4. RLS of undiluted paints P 3 and P 4. P 4 displays more remission in the long-wave region of the spectrum, which can be seen as a hint for weakly agglomerated particles.

P 4 displays more remission in the long-wave region of the spectrum than in the short-wave compared to P 3. This effect is small but reproducible and hints to a higher content of agglomerates in the undiluted paint, analogous to the results of disc centrifugation within the diluted state. As both methods give only qualitative information, no information about the size and the volume fraction of the agglomerates can yet be provided. As can be concluded, it is the PVC of a formulation and the binder that influences the dispergation state of the pigment. Furthermore the type of dispersing agent is critical for the aggregation behavior and therefore has to be chosen carefully.

Using sophisticated imaging techniques as Cryo-EM and relatively easy methods as RLS a broad insight into the undiluted liquid state of a paint is gained. The question arises if the observed clustering is also found in the dried film and if so, how does the dispergation state influence the macroscopic paint properties.

Pigment distribution in the dried paint film

Coming from the fluid paint formulation to the dried film, suitable characterization methods determining the pigment distribution have to be evaluated. Again, one of them is RLS. Films of P 3 and P 4 were measured with this technique. The films were obtained by doctor bladeing the paints (thickness of 50 µm wet on acetate foil) and drying at ambient temperature. Regarding Figure 5 one recognizes, that the shape of the remission spectra of the paint films is much different than for the liquid paint.

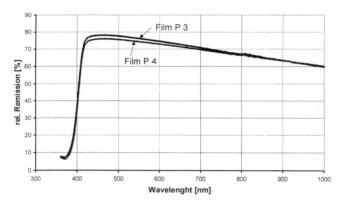

Figure 5. Analysis of the dried paint films of P 3 and P 4 by RLS. The stronger remission of P 3 in the short-wave region compared to P 4 assumes a larger amount of smaller clusters.

Whereas in the wet paint an increase in remission in the long-wave region is observed, this behavior is not detected in the spectra of the films. One possible explanation is that in the film the measurement only reflects the pigment distribution of the film surface, whereas in the paint the bulk is detected. During film formation larger aggregates tend to settle down and hence only smaller clusters or isolated pigments stay at the surface. This would explain the overall lower remission in the long-wave region of both paint films versus their liquid state. In the short-wave region the remission of P 3 is higher than that of P 4, indicating a higher amount of small clusters or single TiO_2 particles at the surface.

However this finding has to be proven by an imaging technique. One well established method for surface analysis is AFM. This technique has been employed to study the surface properties of P 3 and P 4 (Figure 6 a and b). As one clearly sees none of the paint films possesses a completely smooth surface but strong differences in the pigment clustering. Whereas P 3 displays at least pigment particles that are relatively homogeneously distributed on the surface, the analysis of P 5 reflects the inhomogeneity as revealed by Cryo-techniques. One also finds clustered particles leading to a more disturbed surface of the film.

The surface analysis by AFM of a film of P 4 displays only a small amount of pigment particles. One possible explanation could be the already mentioned tendency for aggregation leading to a pronounced segregation process meanwhile the film drying. The pigment particles gather at the bottom side of the paint film due to their higher density whereas the less dense binder particles form a polymer film on top.

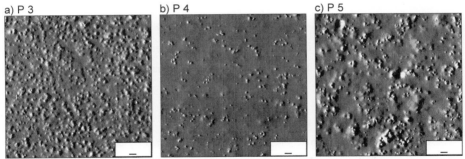

a) P 3   b) P 4   c) P 5

Figure 6. AFM of the paint films in the amplitude mode (20 x 20 μm). a) P 3, b) P 4 and c) P 5. The differences of the surface smoothness affected by TiO_2-clustering are obvious.

It can be concluded that the surface analysis by AFM fits very well the data obtained by RLS and also mirrors the image obtained by Cryo-techniques. Sample P 5 contains strongly agglomerated TiO_2 clusters due to insufficient stabilization. The large clusters of P 4 observed by RLS and disc centrifugation have not necessarily to be detected by imaging techniques. Whereas AFM analysis gives an impression of the surface of the paint samples, cross-cuts of dried paint films and subsequent analysis by TEM enables a view into the bulk structure.

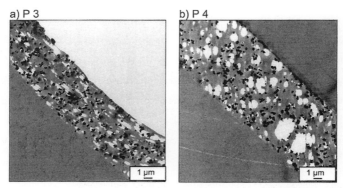

Figure 7. TEM of cross-cuts of the dried films of P 3 (a) and P 4 (b). The size of holes created by the cutting reflects the original size of pigment clusters in the polymeric matrix.

The cross sections of the film reveal the diversity of the particle distribution: P 3 contains relatively well distributed particles, whereas P 4 displays larger clusters or holes that resulted from the cutting process by clusters that fell out of the matrix. However, from the image it is quite obvious that these large clusters have not necessarily to be located at the interface to air. Due to this, surface analysis techniques alone are not sufficient to get all information about the dispergation state.

After proving that both binder and dispersing agent strongly affect the clustering of TiO_2 in paint formulations and after demonstrating that the agglomeration in the liquid state is also seen in the dried paint film, it remains to ascertain if these microscopic findings correlate to macroscopic paint properties.

Correlation to macroscopic paint properties

The correspondence to macroscopic paint properties is gained by correlation of the dispergation state and the gloss measured under an observation angle of 20° and 60° (see Table 2).

By comparing the paints the difference within the dispergation state is reflected by a lower gloss of P 4. The surface roughness of P 3 and P 4 were determined by AFM, showing that P 3 has a much higher roughness than P 4. As P 3 has a higher gloss, the surface roughness is not the only parameter being responsible for this application property. It is in addition the pigment distribution as a result of the high refractive index of TiO_2 compared to the polymer.

The haze of a film, which is often evoked by microstructures due to a bad dispergation of pigments, of P 4 is much higher than of a P 3-film and might only partly be due to a larger surface roughness. Similar is valid for P 5, as its agglomeration was established by Cryo-techniques and is seen again in the gloss and haze values.

Table 2. Application properties of the presented high-gloss formulations: gloss values (observation angle 20° and 60°) and haze, blocking after 1 day drying at room temperature and stressing with 10 kg for 24 h. Defects means film defects noticed after testing the blocking resistance. Marks for blocking and defects ranging from: 0 = excellent to 5 = poor.

sample	Gloss 20°	Gloss 60°	Haze	Blocking	Defects
P 3	62	84	81	1	0
P 4	43	76	191	4	4
P 5	47	80	124	2	0

Another important paint property is blocking. Comparing the results one has to note, that P 4 fails the blocking test and shows defects in the paint film afterwards. The blocking of P 3 and 5 was without failure and can be attributed to the more homogeneous pigment particle distribution at the film surface. The dispergation state strongly affects the macroscopic paint properties. Therefore binder and dispersant of a high-gloss formulation should always be chosen carefully.

Controlling the dispergation state by tailored dispersants and binders

The question arose, whether it would be possible to tailor the polymeric binder in order to overcome the aggregation phenomenon of pigments completely. Figure 8 displays Cryo-SEM images of paints with tailored binders by using surfactants differing in their functional groups.

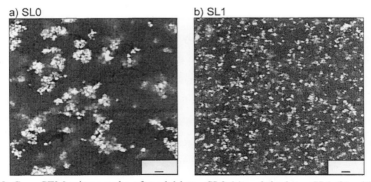

a) SL0   b) SL1

Figure 8. Cryo-SEM micrographs of model latex SL0 (a) and the latex SL1 (b) with a special surfactant. The difference in TiO_2 distribution depending on the surfactant system is obvious.

The distribution of TiO_2 in Figure 8 a) is relatively homogeneous but the pigment particles agglomerated and clusters are detected by Cryo-SEM. This means that the binder was not able to guarantee specific interactions needed for the break-up of agglomerates. The system failed to stabilize smaller clusters or even single pigment particles.

In contrast to this the addition of a special surfactant leads to the absence of pigment clusters. The TiO_2 particles are finely divided within the paint and no agglomerates are detected. This diversity in pigment distribution is again reflected by macroscopic paint properties. The gloss of SL0 is lower than that of SL1 supporting our findings in the previous chapter.

sample	Gloss 20°	Gloss 60°
SL0	10	55
SL1	27	74

Table 3. Gloss values of the formulations under an observation angle of 20° and 60°.

Apart from tailoring the binders by the use of special surfactants the copolymerization with functionalized comonomers is another suitable tool to break up pigment clusters. The obvious improvement caused by optimizing the binder was reflected in the Cryo-SEM micrographs. The dispergation state of TiO_2 as perfectly homogeneously distributed pigments throughout the whole sample can be achieved by choosing an appropriate comonomer. The dependence of the TiO_2 distribution on the binder functionalization is clearly seen in this chapter. In the best case the tailored binder completely eliminates TiO_2 clusters, which is even possible without any change of formulation additives.

Summary and conclusion

To conclude, one can state that the aggregation behavior of pigments in wet and dried paints displays parallels. Pigment agglomeration caused either by binder properties or the functionality or the dispersing agent is already visible in the liquid state and subsequently found in the dried film. The influence of the dispersing agent is obvious and an effect of clustering on corresponding macroscopic parameters as e.g. gloss was established.

The structure of paint films and liquid paints has been investigated by different techniques. The distribution of TiO_2 particles in the paint was analyzed for example by Cryo-SEM, Cryo-Replica TEM, RLS and disc centrifugation respectively. By AFM and RLS the pigment distribution on the surface of the dried paint films was examined. Cross-cuts allowed the visualization of pigment distribution in the bulk via TEM.

With this knowledge we are able to design our binders specifically, which allows the formulation of high-performance paints.

Acknowledgement

The support of L. E. Scriven (University of Minnesota, Dept. Chemical Engineering and Materials Science) regarding the Cryo-SEM measurements is gratefully acknowledged. The authors would also like to thank V. Koch, T. Frey and A. Rennig (BASF AG, Performance Chemicals Physics) for their help concerning RLS and disc centrifugation.

References

[1]   J. H. Braun, D. P. Fileds, *J. Coat. Technol.*, **1994**, *66*, 93.
[2]   J. Scröder, *Farbe & Lack*, **1995**, *91*, 11.
[3]   J. Balfour, D. Huchette, *Paint and Ink Int.*, **1995**, *8*, 52.
[4]   S. Fitzwater, J. W. Hook, *J. Coat. Technol.*, **1985**, *57*, 39.
[5]   S. Kirsch, A. Pfau, T. Frechen, W. Schrof, P. Pföhler, D. Francke, *Prog. Org. Coat.*, **2001**, *43*, **99**.
[6]   C.-L. Zhao, U. Dittrich, W. Heckmann., Y. Ma, V. Thiagarjan, E. Sutanto, L. E. Scriven, *222nd ACS National Meeting, Division of polymeric Materials: Science and Egineering*, Chicago, **2001**.
[7]   E. Sutano, U. Dittrich, C. L. Zhao, H. T. Davis, L. E. Scriven, *Proceedings of the 10th international coating science and technology symposium*, **2000**, 63.
[8]   G. Mie, *Ann. Phys.* **1908**, *25*, 377.

Macromol. Symp. **187,** *753–758 (2002)*

Air Quality Legislation in the European Union

Bruno Giordan

bruno.giordan@pipsara.com

Summary: Starting from the 1979 UN Convention on Long Range Transboundary Air Pollution, the European Union has developed both sectoral and national clean air laws. These control individual coating and printing sectors, and set national ceilings for air pollutants from all sources. Some national targets will be very difficult to achieve, but are scheduled for review in 2004, with the Commission and Parliament looking for still lower ceilings.The coatings industry cannot challenge the political drive for cleaner air, or the academic work linking air pollution to human illness and ecological damage. What we can do, in addition to developing cleaner products and technologies, is to review critically the emissions data upon which the decisions on legislation are based. This data can be very poor. Industry needs to support CEPE's work on identifying and quantifying coatings emissions, their composition and the related costs.

Introduction

Over the last four years, Europe has introduced legislation covering almost all paints and printing inks sectors. The Integrated Pollution Prevention and Control (IPPC) Directive (EC 1996), the Solvent Emissions Directive (EC 1999) and the National Emissions Ceilings (NEC) Directive (EC 2001), together with the recent proposal for a product based Directive covering decorative and vehicle refinish paints, seem to represent the community's ambitions on the control of coatings emissions. The reality is more sombre. What we have seen is only a start.

The Convention on Long-range Transboundary Air Pollution

To understand the likely direction of future developments, we must start by looking at the history of the EU's air quality programme. In 1979, the UN ECE finalised a Convention on Long-range Transboundary Air Pollution (CLRTAP (UN 1979)). This led to a number of protocols, setting targets for the control of various pollutants. The first of these protocols to be directly relevant to our industry was the VOC protocol of 1991, which set a target of reducing VOC emissions by 30 percent by 1999, based on actual emissions in 1984/90. The problem was that there was little co-ordinated effort to set up regulations to meet this goal, and by the mid-1990s it was acknowledged by the EU that the target would not be achieved.

    CCC 1022-1360/00/$ 17.50+.50/0

In the meantime, the science basis for the formation of tropospheric ozone had progressed significantly, and it was believed that larger reductions both in VOC and NOx emissions would be needed to achieve the desired standards of air quality. This led to the Gothenburg protocol of November 1999, which set targets for all pollutants involved in acidification, eutrophication and tropospheric ozone. For the EU, the reduction in VOCs was no longer 30%, but 57%: from 15353 kilotons in 1990 to 6600 kilotons in 2010.

EU Directives

In parallel with these developments of the CLRTAP agenda, the EU had developed the IPPC Directive for larger emitters, and the Solvent Emissions Directive for most smaller contained processes. It is worth pausing at this point to look at what they aim to achieve.

a. The Integrated Pollution Prevention and Control Directive

The IPPC Directive controls coatings installations with the capacity to use 200 or more tonnes of solvent per annum. For these, it sets the requirement that the process must conform with best available technique (BAT). BAT has been carefully defined in the Directive. The definition of "available" in article 2 item 11 states:

> 'available` techniques shall mean those developed on a scale which allows implementation in the relevant industrial sector, under economically and technically viable conditions, taking into consideration the costs and advantages ... ...

On the face of it, this takes reasonable account of the costs of emissions control. I will return to this question below.

b. The Solvent Emissions Directive

Turning briefly to the Solvent Emissions Directive, it is noticeable that there is no mention here of BAT. There is little doubt that the Directive does in fact reflect BAT for some sectors -- automotive OEM and vehicle refinishing are examples. It must be questioned whether either of the alternatives offered to medium-sized (5 - 15 tonnes solvent use p.a.) metal coating business -- incineration or the use of coatings with an average solids:solvent ratio of 60:40 -- are economically viable. There is clearly a risk that some painting processes will be forced to move outside the EU, to countries with a more relaxed environmental regime. This will not benefit the world ecosystem. And because air pollution is a transboundary effect, it will no even benefit the quality of air in Europe. It is significant that the UK, which has had local laws controlling VOC emissions for most of the last decade, is recording fewer severe ozone peaks, but a rising trend of background ozone concentration (NEGTAP, 2001).

c. The NEC Directive

This Directive sets an overall VOC ceiling for the EU15 of 6500 kilotons -- slightly less than the Gothenburg ceiling. This total is apportioned between member states by a process that models the impact of the emissions, looking at the effects in 50 km squares that cover the whole territory of the Union. This means that lower ceilings are set for Member States if, for whatever reason, they have worse local pollution problems. The principle of BAT as a uniform standard must therefore be abandoned. Measures that are sufficient in Finland maybe entirely unacceptable in Germany or Belgium. In fact, the ceilings are set in such a way that Germany, Belgium, probably the Netherlands and Denmark, must implement controls that go much further than BAT.

d. The next step

If the NEC Directive places a heavy burden on industry, we must also recognise that the limit of 6500 kilotons in 2010 was only accepted by the Commission with the greatest reluctance. The modelling studies showed that the target to achieve the desired air quality standards should be 5500 kilotons. There will be a review of the ceiling values in 2004, and the review process will start in the next months.

It is important to recognise that there is no separate ceiling for the coatings industry. The further reduction of 14% on average could well be achieved by tightening the regulatory screws on whichever sector -- coatings, traffic, household products -- offers the least resistance. Certainly, our industry will be competing with others for every tonne of emissions. The correct approach, of course, is to require reductions first of all in the sector faced with the lowest costs -- that is, the lowest cost of abatement per tonne of solvent. By doing this, the costs to society will be minimised, and all sectors with a positive payback will be regulated first. It is, however, unrealistic to expect this to happen automatically. The summary of recent history above shows that there has been little systematic effort either by the authorities or by our own industry to generate realistic cost data. By contrast, it is instructive to look at progress on transport emissions (the "Auto-Oil" programme), where industry does seem to have been more active and successful.

How should industry respond?

a. Emissions Inventory Data

Air quality legislation is based on emission inventories. Anthropogenic VOC emissions come from traffic, coatings, other industry solvent uses, and from household products. Data is collected by national authorities, who also forecast the future trend of emissions.

Our first problem is the quality of this data. The NEC Directive was based on official data that took no account of industry sales statistics. Industry statistics are excellent measure of local use, and hence of potential emissions. When the UK carried out a revision of both historical data and forecasts taking account of this source of data, it was seen that 2010 predictions in particular had been seriously overestimated (table 1).

Table 1. UK VOC emissions from coatings, industry (BCF) data compared with that used for the Gothenburg protocol (RAINS[1])

Coatings using sector	Net VOC emissions (tonnes)				
	BCF	RAINS	BCF	BCF	RAINS
	1990	1990	1998	2010	2010
Architectural use of paints	26500	23719	22500	15000	20332
Domestic use of paints	38854	26294	26970	12000	23003
Vehicle refinishing	23000	21662	13000	8600	10400
Manufacture of automobiles	10400	25522	7700	1600	10295
Other industrial use of paints	101285	101807	66000	44900	60825
Printing, offset	10560	3740	4236	1580	4150
Rotogravure in publication	3175	3561	1257	700	2522
Screen printing	2610	6586	2600	500	9667
Flexography and rotogravure in packaging	11520	26273	11867	2200	27139
Totals	227904	239164	156129	87080	168331

To date, however, only Germany and the UK have carried out a thorough review of the data, taking account of industry statistics. We only have partial data for specific coatings

---

[1] 'Regional Air Pollution INformation and Simulation' – further details on
http://www.iiasa.ac.at/~rains/index.html

subsectors for other Member States, but these show that the errors in the UK inventory data are likely to be quite typical for the EU15 as a whole.

It is also worth mentioning that the speciation data – the breakdown of the emissions into their individual components – is particularly unsatisfactory.

This means that the modelling studies that led to the Gothenburg protocol, and hence to the NEC Directive, may themselves be deeply flawed, as they are based on significantly inaccurate data. For the 2004 review of the Directive, we must do all that we can to ensure that the quality of the data is improved. Time is short, as the review is about to start, and we still do not have data for most EU15 Member States. Unless we treat this with a greater urgency, we will miss our chance, and risk serious structural damage to our industry.

b. Abatement Costs

Ideally, the reduction of emissions should be achieved by tackling first those for which the abatement costs are lowest. By proceeding in this manner, society ensures that the benefits of cleaner air are obtained at the lowest overall cost. We should question the validity of measures whose cost exceeds their benefit.

In this context, we need to look at costs and benefits from the point of view of society as a whole. Of course, an enterprise faced with installing BAT sees only the costs, but the community sees the benefits. These benefits come in a variety of forms. Purer air means less illness, and less premature death. It means that agricultural crops and trees are healthier, and there is less photochemical damage to, for example, painted structures. All of these benefits can be costed. The EU's ExternE research programme has developed comprehensive values for the benefits to society from reduced pollution (see http://externe.jrc.es/). Using these, Dr Holland and his team at AEAT have estimated that the abatement of one tonne of VOC emissions is worth about €2000 annually (AEAT 2002). From a knowledge of investment and running costs of abatement measures, it is then easy to calculate the payback for an individual installation.

Or lack of one. The problem lies not with the IPPC Directive -- who could oppose the adoption of BAT, taking due account of economic considerations? No, the problem lies with the NEC Directive. As explained above, this requires reductions unrelated to BAT.

There is no value to industry in challenging the political agenda – we all want clean air for ourselves and our children to breathe. What industry must do is to ensure that the basic emissions data is as accurate as possible, and that the real costs of abatement are properly documented.

Conclusions

Clean air is an ongoing project for the EU, and we can expect more regulatory initiatives. Industry must prepare itself with data on emissions and on the costs of control. Further regulation is inevitable, but we can help ensure that this is only introduced when it is properly justified.

References

- AEAT 2002: AEAT. Forthcoming. Database of Externalities of Air Pollutants. Report prepared for the European Commission, DG Environment.
- EC 1996: Council Directive 96/61/EC of 24 September 1996 concerning integrated pollution prevention and control (OJ L L 257, 10/10/1996)
- EC 1999: Council Directive 99/13/EC on the limitation of emissions of volatile organic compounds due to the use of organic solvents in certain activities and installations (OJ L 85, 29.3.1999)
- EC 2001: Directive 2001/81/EC of the European Parliament and of the Council of 23 October 2001 on national emission ceilings for certain atmospheric pollutants (OJ L 309, 27.11.2001)
- NEGTAP 2001: Transboundary Air Pollution: Acidification, Eutrophication and Ground-level Ozone in the UK, National Expert Group on Transboundary Air Pollution 2001
- UN 1979: 1979 Geneva Convention on Long Range Transboundary Air Pollution

*Macromol. Symp. **187**, 759–770 (2002)*

High Performance – Low VOC-Content: Innovative and trend-setting coating systems for industrial applications

Dr. Ulrich Christ

Emil Frei GmbH & Co., Am Bahnhof 6, D-78199 Braeunlingen, Germany

Summary: This paper describes low VOC coating systems successfully implemented and operating in two industrial branches.
For Light Alloy Wheels, the innovative coating process comprises ecofriendly chromium-free surface pre-treatment, epoxy polyester powder primer, waterborne base coat and acrylate powder clear coat.
The coating process as an example for agricultural machinery includes iron phosphatizing surface pre-treatment for steel substrates; the coating system itself consists of a water-borne stove-type dip priming operation (for all objects). Finish coat type (Plant I):
2-component water-borne PUR top coat; (Plant II): TGIC-free polyester powder top coat.
These new coating systems have proved successful in existing series production. All technical requirements and the provisions of the VOC Regulations are fully complied with.

1. Introduction

Up to this day, the wide use of ecofriendly coatings has been limited to large-series applications such as car body coating using EC, water-based fillers and water-borne base coats, or to industrial painting applications e. g. for radiator coating with one or only a few colour shades, lighting fixtures or high-bay storage and retrieval systems with EC and / or powder coatings.

As yet, more than 2/3 of the industrial paints are being used with a substantially high portion of organic solvents.

In 1999, the European Community passed the guideline 1999/13/EC that regulates the limitation of emissions of volatile organic compounds (VOC) which usually arise in the painting process. In Germany, this guideline came into force on August, 2001 as Ordinance named 31.BimSchV, i.e. Federal Emission Protection Law; now this regulation operates in all member countries of the European Community.

In order to comply with this regulation, it is necessary to either use expensive exhaust gas treatment processes (thermal post-combustion) or to generally implement low emission coating systems. From November 1, 2004, the emission limits will be applicable to new, and from November 1, 2007, to old coating facilities.

In the varied and heterogeneous field of industrial coatings with a great number of different substrates, a wide variety of parts and colour shades and gradations, individual and custom-

   CCC 1022-1360/00/$ 17.50+.50/0

ised requirements to the coating, there are presently only a few applications using environmentally acceptable paints. Although there exist in many fields of application suitable low-VOC-product technologies.[1]

This hesitant approach to the implementation of low emission coating systems can be explained by various reasons: Any product change-over entails modifications to the current equipment or machinery, hence it is a matter of investment costs; additional costs are incurred in the surface pre-treatment, since low emission water-borne powder and

EC-systems are more sensitive to the given surface finish, and quite often there are higher material costs involved.

There is, however, already a considerable number of manufacturers of industrial goods, whose business philosophy already considers and includes environmentally friendly manufacturing and coating processes. Such new technology has been used successfully and without detriment to performance. And, what is worth mentioning, these companies anticipated the new legal regulation.

This paper deals with series coating presenting to this effect two examples of such coating used in major branches of industry, where the painting and surface pre-treatment operations have been changed-over to ecofriendly, low emission products and methods. It has to be stressed that the coatings have met all requirements and the provisions of the VOC Regulations have equally been complied with.

2. Coating of Light Alloy Wheels by Components Suppliers to Automotive Industry

Today, light allow wheels are not at all regarded as the typical outfit of sports cars. In a great number of new cars they are already fitted as standard equipment. The reasons for their increasing use are reduced weight, excellent running properties and the sportive elegant appearance of the cars equipped with such wheels. Light alloy wheels are aluminium die-cast items. Features such as their pretentious, attractive exterior (gloss, metal effects, styling) and the long-term protection against corrosion, aggressive agents, loose Chipping's, adverse weather conditions, etc.) are produced by a multi-layer effect paint. World-wide, about a 100 millions light alloy wheels are produced per year.

2.1 Description of the Overall Process / Objects

The standard coating process for light alloy wheels starts with the following stages of treatment: alkaline degreasing, etching and followed by a - usually yellow - chromatising process. Together with the subsequent powder priming, this treatment provides very good adhesion and corrosion resistance. On principle, for standard wheels the top coat is applied in two steps: Metallic base coat and clear coat.

The usual 40 to 70 °C hot wheels are sprayed with the high solid metallic base coat which, in turn, is then, after a short flash-off time, overpainted wet-in-wet with a high solid clear coat. The two coat layers are then baked, cooled down and the wheel is then packed for shipment. Following is the description of an innovative and particularly ecofriendly process already used in series production:

2.2 Surface Pre-treatment

In this application case, we have as surface treatment an innovative metal-free conversion process.[2]

The process includes the following phases:

- Alkaline degreasing – Etching – Rinsing with deionised water
- Chemical treatment; a transparent layer is formed (free from chrome)
- Rinsing with deionised water
- Retained water drying

The conversion layer obtained by this process provides very good corrosion protection and paint adhesion results

2.3 Application of Priming Coat

The powder priming coat is applied by electrostatic charging with corona process. This results in a uniform coating layer and high coating efficiency along with good adhesion to the substrate and excellent corrosion protection. The viscoelastic consistency of this coating provides protection against mechanical damages.

Epoxy-polyester-powder coat (hybrid)-type
primer; FREOPOX powdercoat EKK

- Stoving conditions: 220°C/ 30 min.
- Film thickness: 80-100 μm

This type of powdercoat features excellent levelling properties, which is the prerequisite for the formation of the metallic effect expected from the subsequent base coat. Furthermore, this powder primer has a very good degassing effect as it is pervious to gas escaping from die-cast components.

2.4 Base Coat

Due to the application of the base coat, the powder-coated wheels are given an aluminium-metallic appearance. Today, there is an increased demand by consumers for light metallic colour hues. Such effects are produced by the specific making up of the coating system, on the one side, and by special aluminium pigments with variation of the particle size. The results range from very fine, light to coarse-grained sparkling effects.

The latest technological development are chromium-like, reflecting metallic effects.

Conventional, solvent-borne, acrylic-melamine-based systems show solids contents ranging:

- from 17 to 25 (% by wt.) medium solids VOC-value 800 – 700 g/l

- from 25 to 35 (% by wt.) high solids   VOC-value 700 – 625 g/l

The application is usually by electrostatic high rotation bell or disc, and/or by high volume low pressure (HVLP) spraying. Both application methods provide a minimum of over spray. In conventional wheel coating, the base coat is applied using the wet-on-wet technique and then, after some minutes flash-off time, re-coated with clear coat and stoved jointly with the latter. The application described in our example uses a base coat conforming to the latest technology, that is the water-borne technology:

-acrylate-melamine low-bake system FREIOTHERM Hydro Base coat

-solids content: 25 % by weight.

-organic solvents: 12 % by wt. VOC-value: 375 g/l

Prior to applying the base coat, the wheels are heated up to 90°C using infrared radiators in the preheating zone.

Application:

-high rotation bells, 40.000 rpm; 80 kV and automatic HVLP sprayers

-Environmental conditions: Temperature: 23-25° C; rel. humidity of the air: 55-65 %

-Stoving conditions: 130° C / 15-20 min

-Dry film thickness: 15μm

2.5 Clear Coat

The clear coat protects the metallic-base coat from colour changes and corrosion. It features excellent levelling and provides the metallic layer with gloss, thus protecting it from the influences of weather, thawing salt, fuels and other factors. Due to these stringent requirements only high-performance acrylate-based stoving systems are used. Conventional wheel coating processes use high solid clearcoats whose solids content is about 50 % by weight. As is the case with base coats, the application of clearcoats is by electrostatic high rotation bells and / or the HVLP method using the wet-on-wet-technique and stoving them together with the base coat.

The clear coat used for the present application reflects the latest state of the art:

Acrylate powder clear coat. The clear coat system is characterised by outstanding properties such as high gloss, high brilliance, levelling, good weather resistance.

The powder clear coat is applied by electrostatic charging by a corona process.

-Stoving Conditions: 190°C/25 –35 min (object temperature)

-Film thickness: 60µm

-Total film thickness of the multilayer coating system: 150-170µm

2.6 Requirements to the Coatings and Test Results / Performance

Requirement	Test Method	Requirement / Spec.	Results
Adhesion	Cross-cut test DIN EN ISO 2409	Gt 0 - 1	Gt 0
Corrosion test	CASS test DIN 50021 / 53167	240 h wb <= 1 mm	wb < 1 mm
Corrosion test	Salt spray test DIN 50021/53167	1000 h wb <= 1 mm	wb < 1 mm
Humidity test	Condensation water DIN 50019/53209	1000 h m0/g0	m0/g0
Chemical resistance	VDA 621-412	Resistance against various media	Tests passed
Weathering resistance	Accelerated weathering test VDA 621-429/430 QUV (313nm)	1000 h Gloss retention 20° angle: 90%	Tests passed (Fig. 1, 2)

Accelerated weathering - QUV-Test (B313)
Retention of gloss

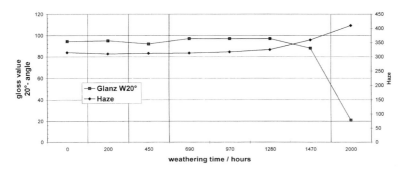

Figure 1. Light alloy wheel coating system(conventional): QUV-test results: Epoxy polyester powder primer/ High solid base coat/high solid clear coat

Accelerated weathering - QUV-Test (B313)
Retention of gloss

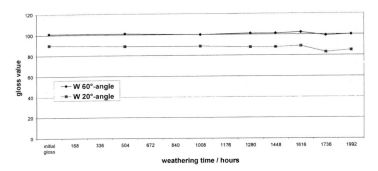

Figure 2. Light alloy wheel coating system (new): QUV-test results: Epoxy polyester powder/water-borne base coat/acrylate powder clear coat

These results show that the new coating system fulfils all requirements.

2.7 Balance of Solvents used in Previous / New Process

Below is a comparison of the quantity of solvents saved with water-based base coats and powder clear coats against conventional coatings:

Based on a yearly production rate of 2 million wheels, the balance of solvents is as follows:

	Conventional Base coat / clear coat	Water-borne Base coat / powder clear coat
Solvents content in base coat:	38.000kg	6.700kg
Solvents content in clear coat:	40.000kg	0kg
Total amount of organic solvents	**78.000**	**6.700kg**

The new coating structure entails a drastic reduction of organic solvents – more than 90%.

3. Coating Systems for Agricultural Machines

In agricultural machines industry steel is the most common material. Unlike the coating of light allow wheels, there is no standard coating process in this field of application. Depending on where the machines are intended to be used, there are coating systems suited to such purpose. The majority of these coating systems are, however, still solvent-borne. Therefore, also in this market segment the VOC Regulations give rise to reconsider the coatings systems with the intention to minimise emission values without impairment of performance.

The large number of coating systems operating in this field can be roughly classified by their corrosion protective capability and the required weather resistance

a) Low to medium requirements to corrosion protection and weather resistance

Surface pre-treatment	:	solvent degrease or blasting
Primer	:	air-drying alkyd or acrylate systems
Top coat	:	air-drying alkyd or acrylate systems

b) Medium to high requirements

Surface pre-treatment	:	blasting and/or iron phosphatizing
Primer	:	2-comp.- epoxy system
Top coat	:	2-comp.-acrylate or polyester – isocyanate (PUR) system

c) Very high requirements

Surface pre-treatment	:	zinc phosphatizing
Primer	:	KTL
Top coat	:	2-comp.-acrylate-isocyanate(PUR) system

3.1 Description of the Overall Process / Objects

The example representative for this field of use describes the painting process for soil tilling machines (plows) at **Plant I** and harvesting machines (grass / hay) at **Plant II** (new) using a water-borne 2 –layer coating system, in the first case, and, in the second case, a mixed-structured water-based primer and powder top coat-system characterized by the high perform-ance requirements. Both plants use the same sort of priming coat. For equipment-specific rea-sons, the top coat technologies used at the two plants are different.

3.2 Surface Pre-treatment

Both plants use a standard iron phosphatizing spray process.

3.3 Application of Priming Coat

The application described in this example uses a water-borne dip primer (FREIOPLAST Hydro-primer) based on a polyacrylate dispersion with a very low VOC-value of 25 g/l

(1.5 % by weight organic solvents). This priming system provides good adhesion and corrosion protection on steel and allows overspraying with 2-comp. PUR systems as well as with powder top coat systems. It features very good intercoat adhesion to liquid and powder top coats.

This primer is applied by dipping, a process that is very efficient and shows optimum results also with geometrically complicated parts.

After a flash-off time of 15 min. the primer is stoved for 30 min. at 75 – 80°C object temperature. Dry film thickness: 20-30 μm.

3.4 Top Coat Variants

At plant I, this primer system is overpainted with a 2-comp. PUR top coat and, at plant II, with a powder top coat. The components painted at different plants and with different coating technologies are all assembled at one place. Consequently, it is necessary that the top coats of the visible parts of such component harmonise with each other in their optical properties (colour hue, gloss, metamerism).

3.4.1 2-Component PUR Top Coat

Plant I uses a water-borne 2-comp. acrylate-isocyanate (PUR) top coat:(EFDEDUR 2K-Hydro-top coat).

Application process: The parent paint is homogeneously mixed with the hardener using a two-component mixer unit. The coating system is applied by air-mix spray guns.

After a flash off time at room temperature for 30 min., the coating is forcedly dried at 70°C object temperature for 1 hour.

After this, the coat is sufficiently hard for assembly. This coat system is characterised by its application safety[3] (blistering limit at more than 80 μm dry film thickness), which is important for complicated geometries where overpaintings are quite possible (dft >>60 μm) and its high weather resistance (see the following test results).

Technical data of the water-borne 2-comp. PUR top coat (RAL 3020)
(EFDEDUR –2K-Hydro top coat); all data based on spraying consistency.

Solids content	50% by wt.
Potlife	4 h
Gloss 60° angle	86
Organic solvents content	11,3% by wt.
VOC-value	228 g/l
Blistering limit	> 80μm

3.4.2 Powder Top Coat

Plant II uses a TGIC-free polyester powder top coat (FREIOTHERM powder coating) with good mechanical and weathering resistance. Gloss 60° value: 85. The application process is triboelectric charge by automatic guns.

-Stoving conditions: 180 °C/10 min. (object temperature).

-DFT of powder coat: 60-80µm; total film thickness: 100-120µm

3.5 Requirements to the Coatings and Test Results (Performance)

The following test results show that the new coating systems fulfils all requirements (see following table and fig. 3,4,5).

Require-ment	Test method	Requirement Specification	Results with PUR-top coat	Results with Powder topcoat
Adhesion	Cross-cut test DIN EN ISO 2409	Gt 0 – 1	Gt 0	Gt 0
Indentation	Erichsen indentation DIN 53156	> 3 mm	6 mm	> 3 mm
Corrosion test	Salt spray test DIN 50021/53167	240 h wb <= 2 mm	wb < 1mm	wb 1 mm
Humidity test	Condensation water DIN 50019/DIN 53209	240 h m0/g0	m0/g0	m0/g0
Weathering Resistance	QUV 313 nm	gloss retention GSB Powder top coat:300h : 50% PUR top coat: 1.000h : 50%	90 %	70%

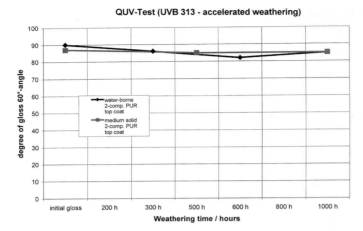

Figure 3. Comparison of weather resistance medium solid-2-comp.-PUR top coat with water-borne-2-comp.-PUR-top coat (plant I) by QUV-test

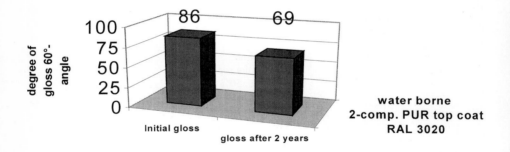

Figure 4. Outdoor weathering test of coating system plant I with water-borne top coat.

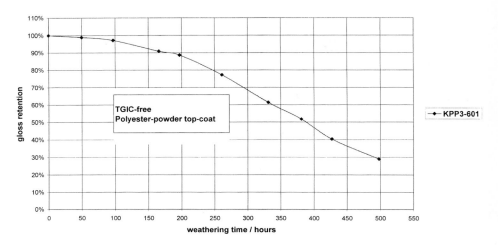

Figure 5. Weather resistance of the coating system plant II with powder top coat. QUV-test.

3.6. Balance of solvents used in previous and new coating process

Based on the same yearly production rate of machines, the balance of organic solvents as follows:

	Alkyde-primer/PUR medium solid top coat	Water-based primer/PUR Powder top coat
Primer coat	18.000kg	450 kg
Top coats	24.000kg	3.750 kg
Total amount of organic solvents:	**42.000kg**	**4.180 kg**

The new coating processes entail a drastic reduction of organic solvents – about 90 %.

4. Summary

This paper describes low VOC coating systems successfully implemented and operating in series production at two important branches of industry. It is shown that these new coatings do not only meet all requirements placed so far on conventional systems but even surpass them to a certain extent.

In the area of Light Alloy Wheels, the innovative coating process comprises ecofriendly chromium-free surface pre-treatment, epoxy polyester powder primer, water-borne base coat and acrylate powder clear coat.

The coating process for agricultural machinery includes iron phosphatizing surface pre-treatment for steel substrates; the coating system itself consists of a water-borne stove-type dip priming operation (for all objects). Finish coat type (Plant I): 2-component water-borne PUR top coat; Plant II: TGIC-free polyester powder top coat. Both coating systems provide good corrosion and weathering resistance.

These new coating systems have proved successful in existing series production. All technical requirements and the provisions of the VOC Regulations are fully complied with. Compared to previous systems, the coating technology described here entails drastic reduction of solvents and, by minimising emissions, it helps protect the environment.

References

1.)   Dören, Freitag, Stoye: Wasserlacke: Umweltschonende Alternativen für
        Beschichtungen, Techn. Akademie Wuppertal, Verlag TÜV Rheinland, 1992
2.)   L. Sebralla, JOT, 42(2002),5, p.48
3.)   W. Hovestadt, E. Jürgens, Farbe & Lack, 105 (1999), 8, p. 30

Macromol. Symp. 187, 771–780 (2002)

The Challenge of Achieving Traditional Exterior Durability Performance in Low VOC Architectural Coatings

Matthew S Gebhard*

Rohm and Haas Spring House Technical Center, Spring House PA 19477 USA

Summary: Over the years the optimization of water-borne emulsion polymers has resulted in water-borne coatings with excellent exterior durability and significantly reduced solvent emissions versus solvent-borne coatings. In recent years, pressure to further reduce emissions has necessitated the use of softer polymer compositions which can negatively impact gloss and tint retention, and dirt pick up resistance. The use of heterogeneous polymer morphologies are addressing these short comings; however care must still be taken not to affect paint durability. The development of low solvent emission thickeners and mildewcides has allowed further reduction in paint emissions.

Keywords: Volatile Organic Compounds, Architectural Coatings, Exterior Durability

Introduction

Historically the advent of water-borne emulsion polymer technology in the mid to late 1900's led to the development of water-borne paints which offered substantial reductions in the amount of volatile organic compounds (VOCs) released when the paint was applied to a substrate. This advance offered some clear benefits to the end user such as low flammability, low solvent exposure, ease of clean up, and less impact on the surrounding environment. Throughout the later part of the 20th century advances in emulsion polymer design and performance led to the development of water-borne architectural paints with excellent exterior durability characteristics. In many cases the exterior durability performance of these water-based paints and coatings exceeded the performance seen from solvent-borne alkyd paints[1], especially in the area of gloss and tint retention, and crack resistance over wooden substrates. The currently accepted performance of these water-borne architectural paints has established a high level of expected exterior durability which must be considered when developing any new low VOC binder technology.

          CCC 1022-1360/00/$ 17.50+.50/0

In recent years, legislation and public pressure to further reduce VOC emissions, has forced new constraints on polymer design which can negatively affect the exterior durability characteristics of emulsion polymer coatings. In particular, the need to form a high quality film with very limited or no volatile coalescing agents necessitates the use of softer polymer compositions. This is usually achieved by increasing the ratio of soft to hard monomers in the polymer backbone. Traditionally all acrylic polymers for architectural paints have T_g's[*] ranging from 11°C to 2°C depending on the desired balance of flexibility and hardness. These polymers typically require 5-10% coalescing solvent based on polymer solids. Eliminating the need for coalescing solvents requires lowering the T_g well below 0°C. This affects exterior durability characteristics, in particular gloss and tint retention, as well as dirt pick up resistance of the paints based on these binders. In some cases the impact can be quite significant. This ultimately reduces the service life of the coating, negating one of the benefits that many water-borne paints offer.

The use of more sophisticated polymer morphologies such as heterogeneous latex polymers[2] which contain at least two phases of differing T_g's is a classic way in which coalescent requirements of a binder are reduced without comprising surface hardness and block resistance. In fact, in many cases it is possible to develop latex binders which offer much reduced surface tack relative to an analogous homogeneous polymer. This technology has been used extensively in the development of water-borne polymers for interior clear wood finishes[3]. This technology has seen much more limited use in pigmented architectural binders, for two primary reasons:

1) The impact of a hard domain in a soft polymer can significantly reduce the binding capacity of the polymer, which is especially pronounced at high PVCs. This typically shows up as dramatically reduced abrasive scrub resistance

2) The exterior durability of paints made from these heterogeneous latex polymers is typically not very good, especially when the polymer has been designed to maximize hardness and block properties.

Low VOC Formulation Issues

In North America the push is to achieve architectural paints and coatings with as low as 50g/L

---

[*] The Tg's given are based Fox calculated Tg values not measured values

VOC[†]. In contrast, typical water-borne architectural paints sold in North America have been between 150g/L and 350g/L. One approach for addressing this gap is to reformulate with lower volatility or reduced VOC materials. In addition as paint formulations are prepared at low VOC's it is important to consider not only the contributions of coalescing solvents and glycols, but also the contributions from traditional paint additives such as film mildewcides, rheology modifiers, and surfactants. In particular mildewcides and rheology modifiers can bring in several grams per liter of VOC. In conventional VOC coatings these contributions were not significant, but at low VOC they can be the largest VOC source in the paint. It is worth while mentioning that the current VOC determination method as specified in EPA Method 24[4] does not specify a boiling point criteria for assessing whether a component is volatile, instead the method relies on a thermal desorption protocol[‡]. That being the case it has been our experience that many materials that might be assumed to be nonvolatile because of their high boiling points (>300°C) are actually surprisingly volatile in a coating.

In a typical paint formulation the contributions from the mildewcide and thickener can easily be 25g/l. When trying to hit 50g/l or lower as a VOC target these levels become problematic. Traditionally, mildewcides were provided in an organic solvent as a means of carrying these water-insoluble materials into the paint in an effective form. Simply adding 100% solid dry powders of the active ingredients to a paint causes many problems, and is often ineffective at providing mildew protection. For a typical mildewcide at 50% in an organic solvent the contribution of the organic carrier solvent is roughly 15g/L, or 33% of the allowable solvent. Recent developments in the design of water dispersed mildewcides (as represented by Rocima 342, Rocima 350 or the VOC free Rocima 371N) has reduced the VOC contribution from the mildewcide by 60-80%.

Many associative rheology modifiers also have significant levels of volatile organics such diethyleneglycol butyl ether, or propylene glycol. The solvents, which are present at 10-20% on total supplied weight, are required to reduce the as supplied viscosity to manageable levels. Another approach is to use surfactants to reduce the as supplied viscosity; however, this leads to many paint performance problems such as water-sensitivity, and surfactant staining. The

---

[†] The VOC of a coating is determined by a collection of tests known as EPA method 24, and is determined by subtracting the volume of water from the coating
[‡] 110°C for 30minutes in an ove.

development of alternative viscosity suppression technology in the early 1990's, opened the door for development of associative thickeners with extremely low VOC contributions. For example in a typical paint formulation there might be as much as 15g/l of VOC which comes from a traditional thickener which uses 20% organic solvent. As new low VOC binders are developed to address freeze-thaw and open-time issues, the trend is to be more hydrophilic and stabilized by alternative means requiring the use of more efficient associative thickeners having much higher as supplied viscosities. Thus the use of viscosity suppression technology will become increasingly more important.

As will be shown below the need for low VOC formulating capability means reducing the T_g of the base polymer. This leads to an overall reduction in tint retention of the paint. It also compromises block resistance. One intriguing way to gain back some tint retention and improve block is to use micro-voided polymers (Ropaque™ Ultra). These materials are essentially hard polymer additives which compensate for the reduction in the hardness of the base polymer. An intriguing aspect of the micro-voided polymers is they can deform under certain conditions and it has been observed that upon exterior exposure the voids can collapse causing a loss of some of their intrinsic hiding ability. This leads to an apparent increase in color strength from the colorant, thus counteracting the tint loss process. This approach has been shown to work quite well in pastel paints where there is 18-24Kg/100Liters of TiO_2. While it is true that the paints are getting darker as they age outside, the effect is quite even creating the illusion that the paint is not fading as fast. In paints with TiO_2 levels below 18Kg/100liters care must be taken to ensure that this general darkening is not large enough to cause an undesirable shift in color.

The contributions of TiO_2 choice to exterior durability have long been recognized as being very significant. Over the last several years TiO_2 suppliers have made great strides in improving the surface coatings on TiO_2 resulting in substantial improvements in gloss and tint retention. In a world of softer low VOC coatings this will become increasingly more important.

Despite many of the above formulation approaches which can either reduce paint VOC or improve exterior durability performance, the choice and design of the film forming binder is crucial to determining exterior durability performance. It is also clear that developing a low VOC binder technology based on lab properties can lead to products which have poor exterior performance. Unfortunately, there remains no substitute for evaluating exterior durability in real exterior exposures at many sites and in many formulations. This is a long process which requires several years and good planning.

Experimental Details

To address the effect of binder T_g and heterogeneous polymer morphology on exterior performance, butyl acrylate and methyl methacrylate homogeneous copolymers were prepared at varying T_g's using a conventional semi-continuous batch process, and are compared to two heterogeneous polymers which vary by T_g and the ratios of hard to soft phases. It is worth while mentioning that the process by which the polymer is made can significantly affect the film forming ability of the polymer, and thus plays a significant role in determining what polymer T_g is needed to get adequate film formation[§]. Paints at 22% PVC and 36 volume solids were prepared from these binders based on the formulation given in Table 1.

Table 1. Experimental Paint Formulation

Grind	Kgs
Tamol 731A	1.68
Tego Foamex 810	0.14
Surfynol CT-111	0.27
Ti-Pure R-706	31.68
Water	7.46
LetDown	
Water	2.40
Binder @ 50% solids	63.58
Texanol (Variable) (0-6% Based on Binder solids)	X
Surfynol CT-111	0.12
Mildewcide	0.96
Acrysol RM-2020 NPR	3.95
Acrysol RM-8W	0.56
Water	12.94

For the Blue tints 2.4Kgs Phthalo Blue per 100Liters of paint were used

The paints were then drawn down over aluminum panels and submitted to Q-Panel testing in south Florida[**]. The three properties followed were 60° gloss, L* which is a measure of dirt

[§] A discussion of these effects is beyond the scope of this paper
[**] Other sites were also chosen and the trends observed at S Fla were consistent with these other sites

pickup, and B* in a phthalo blue paint.

Experimental Results

Homogeneous Polymer Controls

Figure 1 shows the percent 60° gloss retention for a series of homogeneous BA/MMA polymers of varying T_g. As can clearly be seen the gloss retention decreases with decreasing polymer T_g. Typically the T_g of the polymer must be reduced to below 0°C to achieve adequate film formation in low VOC formulations, which results in substantially decreased gloss retention. Figure 2 is a similar plot looking at the change in b* for these paints. In this case, an increase in b* indicates that the blue pigment is losing its color strength, thus paints with larger changes in b* have poorer tint retention. As with the gloss retention there is a decrease in tint retention as the T_g of the polymer is lowered. Figure 3 also shows a similar story. As the polymer T_g is dropped, there is an increasing trend for the paint film to become darker with time (decrease in L*). This is indicative of the paints becoming dirty with exterior exposure. This data shows that simply lowering the T_g to get adequate coalescent free film formation is not an acceptable approach.

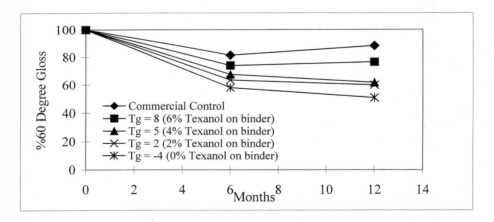

Figure 1. % 60 Degree Gloss for homogeneous latex polymers at various T_g's

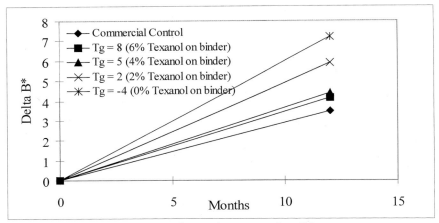

Figure 2. Delta b* for homogeneous latex polymers at various T_g's

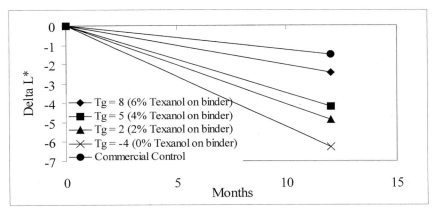

Figure 3. Delta L* for homogeneous latex polymers at various T_g's

Heterogeneous Polymers

Figure 4 shows the 60% gloss retention data for two homogenous polymer compositions (8°C and −6°C) and two heterogeneous polymer compositions. When making a heterogeneous polymer composition it often becomes necessary to further lower the T_g of the softer polymer phase in order accommodate the effect that the harder phase has on film formation. Hetero #1 represents an early attempt at using heterogeneous polymer morphology to eliminate the need for a coalescing solvent. Hetero #1 clearly has poor gloss retention. On the other hand hetero #2 shows very respectable gloss retention. These results drive home the need for assessing exterior durability for heterogonous polymer compositions. Figures 5 and 6 show the tint retention and L* (dirt pick up) data for the same polymer set in figure 4. Interestingly, in both cases the hetero polymers show very reasonable tint retention and dirt pick up resistance.

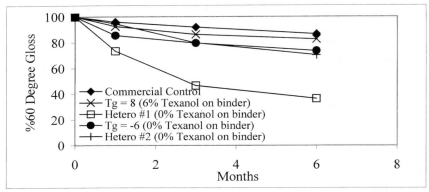

Figure 4. % 60 Degree Gloss for heterogeneous latex polymers

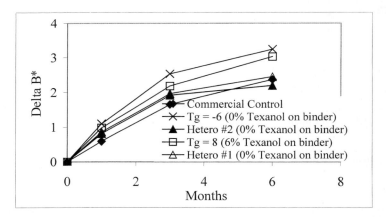

Figure 5. Delta b* for heterogeneous latex polymers

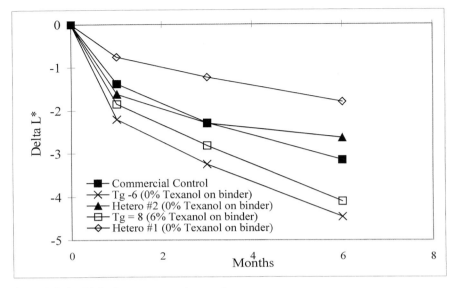

Figure 6. Delta L* for heterogeneous latex polymers

Discussion

The results presented here show that heterogeneous polymer morphologies can give substantial improvements in certain exterior durability properties; however care must be taken when designing these polymers. With appropriate design, heterogeneous polymer compositions can offer improvements in dirt pick up resistance and tint retention. Based on the dirt pick up resistance and tint retention data alone it would seem that it would be straight forward to design a multi-phase binder with exceptional exterior durability. On the other hand, if inappropriately designed a heterogeneous polymer can have very poor gloss retention, and in many cases can show very bad cracking over wooden substrates. We have found that the good results shown for hetero #2 tend to be the exception and not the rule. In fact it is quite challenging to design a heterogeneous polymer with acceptable gloss retention, and crack resistance. Aspects such as the morphology of the heterogeneous polymer particle play a significant role in the film forming capabilities of the binder and ultimately the need for coalescing solvents to achieve adequate film formation. Simply changing the T_g's of the phases is often inadequate to bring about the desired balance of exterior durability properties and general film properties. Moreover, if one is simply evaluating properties such as surface hardness, block resistance, and film forming capabilities it is highly likely that the binder will have terrible exterior durability. It is important to stress the need for assessing exterior durability when developing new binders based on heterogeneous polymer morphologies. However, with careful manipulation of the binder morphology and careful attention to optimization of film formation it is possible to design heterogeneous polymers with very respectable exterior durability.

References

1. "Organic Coatings Science and Technology" Z. Wicks, Eds, J. Wiley & Sons, New York, 1994.
2. a) J. E. Jonsson, H. Hassander, B. Tornell, *Macromolecules*, **1994**, *27*, 1932.
   b) J. E. Jonsson, H. Hassander, L.H. Jansson, B. Tornell, *Macromolecules*, **1991**, *24*, 126.
   c) C. L. Winzor, D.C. Sundberg, *Polymer*, **1992**, *33 no 18*, 3797
3. a) A.J.P Buckmann, G.C. Overbeek, T. Nabuurs, *Paint and Coatings Industry*, **August 2001**, 40.
4. "Manual on Determination of Volatile Organic Compound (VOC) Content in Paints, Inks, and Related Coating Products: Second Edition", Eds , J.J. Brezinski; ASTM, Philadelphia 1993.

Macromol. Symp. 187, 781–787 (2002)

Powder Coatings for Corrosion Protection

Andreas Schütz, Wolf-Dieter Kaiser

Institut für Korrosionsschutz Dresden GmbH, Gostritzer Straße 61-63, D-01217 Dresden, Germany

Summary: Powder coatings found a wider use in corrosion protection of steel structure. In Europe very often double-layer systems are used, based on an adhesion promoting epoxy (EP) primer and a weathering stable top coat, mostly polyester (SP) sometimes EP/SP-hybrid powders. An interesting development is the use of zinc filled EP powders as primer to offer a cathodic protection to the steel surface. Powder systems with and without zinc were compared to proved coating systems based on liquid paint materials, where powder coating systems showed results comparable to these systems. Besides many advantages of powder coatings for corrosion protection there are still some problems. The workshops carring out the powder coating have to be in control of the surface pretreatment like chromating, but espescially phosphating and the work with the chromate-free pretreatment methods for galvanized steel. As always in the field of corrosion protection it is the surface pretreatment and preparation which determines the quality of the whole coating system decisively. This problem can be solved by appropriate working. In some years the problem with the general maintenance of powder coatings after weathering and ageing will be actual. This problem should be solved because of the homogeneous coatings on larger areas. Of importance will be the adhesion on the old coating and the appearance of the maintenance coating. The touch up of smaller parts as transport damages will be much more difficult in order to the appearance.

Keywords: corrosion protection; powder coatings; surface preparation; surface pretreatment; zinc (rich) primer; adhesion; maintenance

1 Introduction

Some years ago powder coatings were typically known for all kinds of industrial finishing. But in the recent years more and more steel constructions, or parts of them, for the field of heavy corrosion protection are powder coated. The main causes for this trend to powder coatings in corrosion protection are the following:

- The demand of solvent minimization can be reached in an optimum way by using powder coatings.

- For reasons of quality assurance and cost reduction the trend to shop coating can be found generally for all kinds of steel structures.

       CCC 1022-1360/00/$ 17.50+.50/0

- The equipment of several workshops have reached dimensions that enables them for also surface preparation and powder application of large structural elements.
- Special methods for surface preparation were designed for powder coating.
- Powder systems were developed especially for corrosion protection.

Due to this trend there have been carried out several investigations on corrosion protection systems based on powder coatings, i.e. influence of surface preparation and pretreatment as well as different types of powder coatings. The results are compared to proved coating systems based on liquid paint materials.

2 EN ISO 12944

The basic European standard EN ISO 12944 „Corrosion protection of steel structures by protective paint systems" describes in part 1 the paint systems to be used according to this standard. It is also mentioned that powder coatings are excluded from this standard. This means in fact that there is in the moment no standard for the use of powder coatings in the field of corrosion protection. Since now there are only guidelines (Normally nationally but not internationally used.) which are originated from industrial finishing like GSB and Qualicoat. In Germany powder coatings are laid down in a common guideline of several federations working in the field of corrosion protection. But this guideline only deals with metal-plus-paint systems ("Verbänderichtlinie Duplexsysteme"). But as powder coatings found a wider use in the field of corrosion protection of steel parts this situation has to be thought over.

3 Surface preparation

Surface preparation plays an important role for the effectivness of corrosion protection. Besides typical methods from the field of corrosion protection like blasting and sweeping there is work on methods known from industrial finishing like chromating and phosphating. Also chrome-(IV)-free preparation methods are of high interest. For this reason the methods of industrial finishing and corrosion protection have to be optimized now. This offers chances, but may contain also a lot of problems. Investigations on the influence of the type of surface preparation on corrosion protection with powder coatings were carried out.

The momentary situation with different surface preparations and pretreatments for powder coatings can be described by the following table.

Table 1. Surface preparation and pretreatment for powder coatings in corrosion protection

substrate	surface preparation	surface pretreatment	coating	remarks
steel	blasting Sa 2½		epoxy (zinc) polyester	Sa 2½ assures in all cases good results
		phosphating	epoxy polyester	phosphating + zinc filled epoxy primer doesn´t work
galvanized steel	sweep blasting		polyester (epoxy)	only roughening, not removing
		chromating	polyester (epoxy)	excellent effectivness, but: Cr(VI) is cancerogen
		phosphating	polyester (epoxy)	several methods with different results, may causes problems
		chromate-free	polyester (epoxy)	basis: (Ti,Zr)F, promising developments

4 Testing of powder coatings

In Europe very often double-layer systems are used, based on an adhesion promoting epoxy (EP) primer and a weathering stable top coat, mostly polyester (SP) sometimes EP/SP-hybrid powders. An interesting development is the use of zinc filled EP powders as primer to offer a cathodic protection to the steel surface. But due to application the amount of zinc in the prime powder is significantly less than in typical liquid based primers, so that an active (cathodic) action is not really assured. Powder systems with and without zinc were compared to proved coating systems based on liquid paint materials, where powder coating systems showed results comparable to these systems.

In the following figures two types of powder coatings with different amounts of zinc dust in the primer are shown in transverse sections. As comparison a transverse section of a typical zinc rich primer based on liquid paint materials is also added.

784

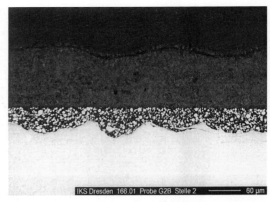

Figure. 1. EP-powder prime coat (zinc dust containing) EP/SP-powder top coat

Figure 2. EP-powder prime coat (zinc dust containing) EP/SP-powder top coat

Figure 3. EP-zinc rich primer PUR-top coat

As example for the corrosion protection properties of powder coatings some test samples after salt spray test according to ISO 7253 for the duration of 1440h (i.e. category C5-I long according to EN ISO 12944-6) are shown in the following figure.

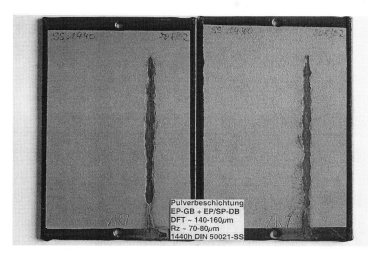

Figure 4. Powder Coating after 1440h salt spray test (ISO 7253)

Some results for loading and testing of powder coatings for corrosion protection are summerized in the following two tables.

Table 2. Powder coatings on blasted steel (R_z: approx. 45µm), weathering for 1 year with an additional loading in a salt spray chamber acc. to ISO 7253

System	No. of coats	DFT / µm	1 year weathering					
			blistering	rust	W_b / mm	adhesion / MPa	fracture face	hatch cut
1	2	120	0	Ri 0	0,5	7,2 ± 0,7	100C/Y	0/1
2		135	0	Ri 0	0	19,6 ± 4,6	100C/Y	0/0
3	1	90	0	Ri 0	0,5	5,7 ± 0,6	100B/Y	0/0
4		90	0	Ri 0	0	13,7 ± 1,5	100B/Y	0/0
			additional 720 h salt spray test					
			blistering	rust		W_b / mm	U_D / mm	hatch cut
1	2	120	0	Ri 0		1,3	1,3	0/0
2		135	0	Ri 0		1,4	2,2	0/0
3	1	90	0	Ri 0		1,3	4,4	0/0
4		90	0	Ri 0		1,4	3,8	0/0

Table 3. Comparison of corrosion protection systems based on liquid and powder paint materials (steel blasted to Sa 2½ acc. to EN ISO 12944-4)

Coating system		liquid based		powder based	
		EP zinc EP mica PUR mica	EP zincphosphate EP mica PUR mica	EP zinc SP	EP SP
	DFT / μm	240	240	200	200
ISO 7253 1440h (C5-I long)	W_b / mm	0,9	1,4	1,9	1,8
	adhesion / MPa	8,0 ± 0,1	6,1 ± 2,3	> 3,4 ± 1,2	> 2,5 ± 0,1
	fracture face	70D, 30D/Y	50D, 50D/Y	20C, 80C/Y	20C, 80C/Y
	hatch cut	0	1	0	0
ISO 6270 720h (C5-I long)	adhesion / MPa	6,3 ± 2,5	4,4 ± 0,4	11,4 ± 4,3	11,0 ± 3,1
	fracture face	80D, 20D/Y	90D, 10D/Y	30C, 70C/Y	50C, 50C/Y
	hatch cut	0-1	0-1	0	0

In table 4 some results on different surface preparation methods for galvanized steel before powder coating are shown. In all the four cases the same powder coating (SP top coat, 80μm) was applied.

Table 4. Comparison of different surface preparation methods

preparation / pretreatment		sweeping	chromating	phosphating I	phosphating II
ISO 6270	blistering	0	0	1/1	0
	hatch cut	0-1	0	3	0-1
cyclic VDA-test	blistering	2/1	2/1	2/1	2/1
	U_D / mm	2,7	2,4	3,8	2,7
	hatch cut	0-1	1	0-1	0
ΔT-test	blistering	0	0	3/3	0
	hatch cut	0-1	0	3	0

The tables verify the good performance of powder coatings in corrosion protection. The values of proved coating systems based on liquid paint materials are reached. But table 4 also shows that different phosphating methods for galvanized steel lead to different, and not always good, results. It can be concluded that intensive controlling (bath temperature, concentrations etc) is necessary to obtain good results with phosphating as surface pretreatment for powder coatings in corrosion protection.

5 Outlook and problems

Besides all the shown advantages of powder coatings for corrosion protection there will be also some problems. The workshops carring out the powder coating have to be in control of the surface pretreatment like chromating, but espescially phosphating and the work with the chromate-free pretreatment methods for galvanized steel. As always in the field of corrosion protection it is the surface pretreatment and preparation which determines the quality of the whole coating system decisively. This problem can be solved by appropriate working.

Somehow more complicated is the development of maintenance strategies for powder coatings in corrosion protection. Very actual is the touch up of transport and erection damages - surely there can´t be stoved on site. But there is an intensive search for solutions, sometimes dilettante, sometimes in very interesting and unusual ways.

In a similar way there will be a problem with the general maintenance of powder coatings after weathering and ageing. It should be also worked on solutions for that. These will be realised in a easier way, because there will be homogeneous coatings an larger areas. The important thing will be the adhesion on the old coating and the appearance of the maintenance coating. The touch up of smaller parts as transport damages will be much more difficult in order to the appearance.

Nevertheless, powder coatings are an interesting and high-grade alternativ for initial coating in corrosion protection.

Macromol. Symp. 187, 789–799 (2002)

Lead Replacement in the Molybdate Orange Colour Space

Peter Gee,* Marketing Manager, Dr. Markus A. Meier, Marketing Manager

Ciba Specialty Chemicals Inc., R-1045.2.09. CH4002 Basel, Switzerland.
e-mail: peter-1.gee@cibasc.com

Summary: Due to the toxicological concerns over lead and lead compounds, there has been an increased trend to replace lead based materials in paint systems. Change in this instance has been driven primarily by Legislation and regulations, however in many cases global companies wish to show a unified approach and some have taken both a positive environmental marketing approach over their competitors with Lead free paints and a positive move on labour relations.

The pigment manufactures have had the challenge to meet these requirements either from existing product ranges or to develop new pigments that are both commercially and technically viable. Due to the diverse application, systems and cost structures within General Industrial Paints no single product offers a universal solution.

The higher performing pigments generally meet the higher specifications technically but because of their chemical complexity in processing and structure fail to achieve the lower comparable costs against Molybdate Oranges. With less complex pigment structures limitations are identified within the technical area.

Additionally within the range of Molybdate Oranges differing grades are available treated to enhance temperature stability and chemical resistance or untreated to attain lower costs and increased saturation.

For this reason paint producers have the choice of a single product within the mid performance area that would act as a compromise or alternatively select three products to fulfil all requirements with the associated logistic problems on supply, stock inventory and quality testing.

Within the single product approach Pigment Orange 81 potentially offers the most flexible solution to the paint manufacturer in order to meet the volume market for **Mid performance** offering opacity, gloss, rheology, bleed resistance, good temperature stability and of great importance the ability to match the colour space occupied by Molybdate Orange.

For the multiple pigment approach:

High technical performance	Pigment Orange 73 offering higher saturation than Molybdate Orange but with associated costs.
Mid performance	Pigment Orange 81
Cost effective	Pigment Orange 34 and Pigment Orange 16 offer high saturation but limitations in durability.

CCC 1022-1360/00/$ 17.50+.50/0

1. Introduction

Ciba Specialty Chemicals are producers of both inorganic and organic based pigments and are uniquely positioned to provide a balanced view of potential alternatives for lead replacement in line with market demands.

Within the paint market there are a number of frequently asked questions:

> ➤ Do we need to replace Lead based on current and future legislation?
> ➤ Can we replace Lead from a technical prospective?
> ➤ How much will it cost if we replace Lead?

It is our intention to provide our customers with data in order that these questions can be addressed for their specific application.

2. Chemistry and Physical Aspects of Molybdate Oranges

Composition: $25\ PbCrO_4 \bullet 4\ PbMoO_4 \bullet 1\ PbSO_4$.

Pigment Red 104, CAS Number 77605

Pigment Red 104 is primarily a bright orange, with an inorganic composition based on a solid solution of Lead chromate, Lead molybdate and Lead sulphate. Crystal form is tetragonal.

Physical data : Density $5.41 - 6.34\ g/cm^3$

Oil absorption $5.8 - 40\ g/100g$ pigment,

depending on treatments and processing.

Alternative names: Scarlet Chrome, Molybdate Red, Moly Orange, Chrome Vermillion.

3. Why Replace Lead?

Molybdate Orange is effectively a mid to high performance product with a lower cost structure than organic pigments.

The primary disadvantage is related to toxicological concerns, which ultimately translate into higher costs for the complete chain from producer to end-user.

Aspects that should be considered in the selection of Lead for future applicational areas are:

a.) Impact on the environment for both aquatic and human life.

b.) Implementation of legislation for registration costs and testing may ultimately increase the raw material **cost** for paint producers.

c.) Implementation of legislation for the work place in the handling of materials will increase production **cost** for the pigment supplier.

d.) The negative marketing effect of toxic materials will lead to reduction in sales of Molybdate Orange and increase period **cost** based on the economy of scale.

e.) If legislation continues unabated then the long term **cost** of disposal of packaging materials could exceed the cost of the paint.

f.) If legislation continues and results in the ban in selective application areas then the **cost** of major claims could dramatically affect the profitability of the paint producer and their image for responsible care.

4. Status of Legislation – European Union

Within the European Union, Directive 98/98/EC "Adapting to technical progress for the 25th Council Directive 67/548/EEC on the approximation of the laws, regulations and administrative provisions relating to the classification, packaging and labelling of dangerous substances" indicates the classification and labelling for lead pigments. Lead-based pigments are considered toxic and dangerous for the environment and have to be labelled with the symbols T (*Skull and cross-bones*) and N *(Dead fish/dead tree)* together with the warning phrases:

 (Skull and cross-bones)   **N (Dead fish/dead tree)**

- R61:   May cause **harm** to the unborn child
- R33:   Danger of **cumulative effects**
- R40:   Possible risks or **irreversible** effect
- R50/53: Very toxic to **aquatic organism**, may cause **long-term** effects in the aquatic environment
- R62:   Possible risk of impaired **fertility**
- S53:   Avoid **exposure**; obtain special instruction before use
- S45:   In case of accident or if you feel unwell, seek **medical advise** immediately (show the label where possible)
- S60:   This material and/or its container must be disposed of as **hazardous waste**
- S61:   **Avoid release to the environment**. Refer to special instruction/safety data sheet.

Further details on the white paper will be presented later.

Within the European Union selective countries have chosen to implement more stringent levels like for example Germany

5. The Evaluation of Products for the Market

For a pigment to succeed as a replacement it should certainly achieve basic requirements of opacity for thin films and bring addition benefits in the areas of colour strength coupled with durability in paler shades and chemical resistance, therefore the following pigments were examined as potential offers:

Colour Index	Chemical type
Pigment Red 104	Lead Sulfo-Chromate-Molybdate *
Pigment Orange 81	Diketo-pyrrolo-pyrrole *
Pigment Orange 61	Isoindoline *
Pigment Orange 64	Benzimidazolone *
Pigment Orange 34	Diarylide - Pyrazolone *
Pigment Orange 16	Dianisidine *
Pigment Orange 73	Diketo - pyrrolo - pyrrole *
Pigment Orange 75	Cerium Sulfide
Pigment Orange 74	Azo
Pigment Orange 69	Isoindoline
Pigment Orange 43	Perinone
Pigment Orange 67	Qunazolone
Pigment Orange 36	Benzimidazolone

* produced by Ciba Specialty Chemicals

6. Requirements for Industrial Paints

In order to meet the requirements of the Paint maker and end user it is important to first identify the requirements for both **Paint producer and end-user**:

High saturation - Most of the shades used in Industrial paints are highly saturated for impact, Molybdate Orange covers this area well and any new pigment must cover all key shades attained by Molybdate Orange.

In addition a highly saturated product allows the paint maker great flexibility in matching and potential lower costs.

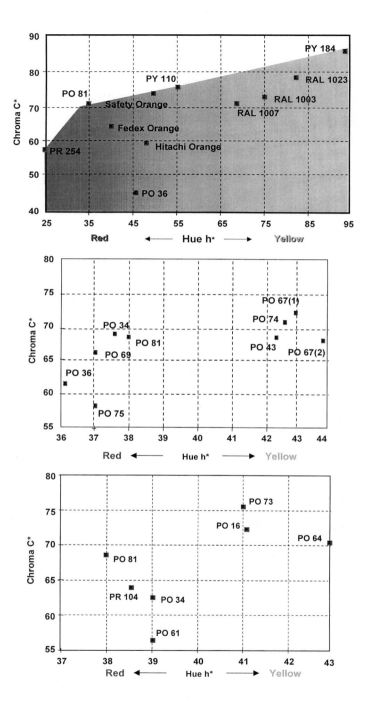

Flow - with reduction in VOC levels there is a requirement that new pigments exhibit good rheology both for lower solvent demand and production optimisation process (higher pigment loading = lower process cost). Currently Molybdate orange clearly meets that target.

High opacity - is one of the primary benefits of Molybdate Orange allowing relatively low film thickness coupled with substrate obliteration. This has traditionally been one of the weak points in replacements with organic pigments as they generally have higher tinting strength but lower opacity.

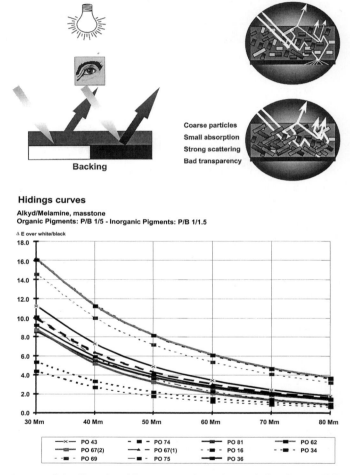

Coarse particles
Small absorption
Strong scattering
Bad transparency

Backing

Hidings curves

Alkyd/Melamine, masstone
Organic Pigments: P/B 1/5 - Inorganic Pigments: P/B 1/1.5

Δ E over white/black

	PO 43		PO 74		PO 81		PO 62
	PO 67(2)		PO 67(1)		PO 16		PO 34
	PO 69		PO 75		PO 36		

High gloss – is one of the first visual impacts on the end user, it is often to difficult achieve with organic pigments due partially to their particle size and shape, this aspect is not often noted with Molybdate Orange which generally exhibits high gloss at high pigment loadings.

Pigment	Overpaint Fastness		Gloss
	GS	Δ E	60° Angle
PO 81	**4-5**	**1.4**	**92%**
PO 73	5	0.6	87%
PO 61	5	0.5	93%
PO 64	5	0.7	93%
PO 34	4-5	1.1	90%
PO 16	4-5	1.1	73%
PR 104	5	0.7	93%
PO 75	5	0.5	87%
PO 74	2-3	8.8	90%
PO 69	5	0.7	93%
PO 43	5	0.2	74%
PO 67 (1)	2-3	9.4	93%
PO 67 (2)	2-3	9.5	92%
PO 36	5	0.5	93%

Chemical resistance - is relatively poor for traditional Molybdate Orange and the significant colour change is often observed in industrial atmospheres and in the ACE market where battery acids and hydraulic fluids are used.

Pigment	Chemical resistance			
	Acid		Alkali	
	Masstone	1/3 ISD	Masstone	1/3 ISD
PO 81	**5**	**5**	**5**	**5**
PO 73	5	5	5	5
PO 61	4-5	5	5	5
PO 64	4-5	5	1	5
PO 34	5	5	5	5
PO 16	5	5	5	5
PR 104	4	3	5	5

Intrinsic strength – becomes a more important feature in paler shades on a cost basis, within the deep shade area pigments with high intrinsic strength rely on the correct selection of an opacifying agent, a typical example being **PY184.**

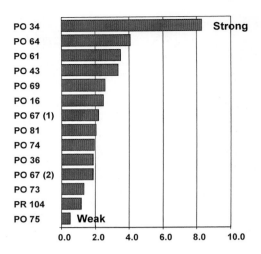

Evaluated at 1/3 International Standard Depth

Durability - accepted by the industrial but generally poor in pale shades and darkening in deep shade, although certain treated grades can offer better durability. There are benefits in the use of organic / inorganic blends of pigments in this area and in line with the expected demand for longer life coatings.

Evaluation carried out on the Atlas Weather-O-Meter (CAM 7) and assessed against the ISO grey scale after a period of **1000 and 2000 hours**

Pigment	1000 hrs WOM, alkyd melamine			
PO 81	4-5	1.5	4-5	0.8
PO 73	5	0.5	5	0.5
PO 61	4	2.7	4	0.8
PO 64	3-4	2.0	1	20+
PO 34	3d	3.5	1	20+
PO 16	3d	3.4	1-2	20+
PR 104	3-4d	3.7	4	2.0
PO 75	3	4.7	1	15.0
PO 74	3-4	1.5	4-5	1.4
PO 69	4-5	0.7	4-5	1.0
PO 67 (1)	4d	2.3	3-4	3.3
PO 67 (2)	4-5d	1.4	3-4	2.8
PO 36	5	0.7	5	0.6

2000 hrs WOM, alkyd melamine

Pigment	Masstone		1/3 ISD	
	GS	ΔE	GS	ΔE
PO 81	4-5	-	3-4	-
PO 75	2-3	5.3	1	17.5
PO 34	3d	2.9	1	27.2
PO 74	4	2.4	3	3.7
PO 69	4	0.7	4-5	1.6
PO 43	3d	3.5	3-4	1.8
PO 67 (1)	4d	1.8	2-3	4.7
PO 67 (2)	4-5d	1.3	2	8.4
PO 36	4-5	0.6	4-5	0.7

Temperature stability - Molybdate Oranges have excellent stability within the majority of temperature-cured systems (alkyd melamine, polyesters etc.) and this becomes a prerequisite for organic and inorganic alternatives.

Cost - is currently perceived as relatively low for Molybdate Orange against organic pigments and inorganic alternatives on kilo prices.

A more cost effective approach can be taken with Value-in-use.

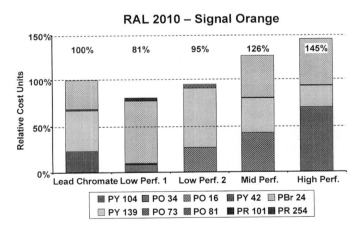

System suitability – Industrial paints are diverse with widely differing resins, solvents and cure temperature. Molybdate Orange currently offers an excellent compromise product in this area in order to maintain low inventories.

7. Systems for Industrial Paints

The market is extremely diverse in its requirements for chemical, durability and temperature stability with the following table (Table) indicating only a few of the many industrial systems on the market.

Table: Various paint systems used in the Industrial Paint market (Ac: Acetate)

Resin	Cure temp.	Chemical agent	Solvent	Application
Epoxy	RT	Amine catalysed	Ketone/ Ester	Outstanding alkali, anticorrosive & chemical resistant paints
Alkyd Melamine	130°C/ 30'		Xylene, Alcohol	Small machinery
Polyurethane PUR -1-pack	RT		White spirit, Xylene	Dries by oxidation or moisture cure; floor finish, trains, planes
Polyurethane PUR-2-pack (2-pack Acrylic)	RT - 80°C	Isocyanate cured	Butyl Ac/ Xylene, Butyl Ac/MEK	Trains, buses, planes (ACE)
Polyester PES - unsaturated styrene, 2-pack	RT	Peroxide catalyst	Monomer	Gel coats
Nitrocellulose N/C	RT		Alcohol, Ketone	Dries by solvent emission, repair paints, aerosols machinery
Chlorinated rubber	RT		Aromatic hydrocarbon	Dries by solvent emission, resistant to acid, alkali and water for anticorrosive, marine, line paints
Alkyd - short/medium	RT - 60°C		Xylene	Dries by oxidation & solvent evaporation, heavy machinery
Polyester PES - Melamine	130°C/ 30'		Butylglycol Ac, Solvesso	Exterior finishes
Polyester PES - Melamine	240°C/ 1'	Acid catalyst	Butylglycol Ac, Solvesso	Coil coated objects

8. Conclusion

Primary recommendations

	Pale	Deep
Classical approach	**Pigment Orange 34** **Pigment Orange 16**	**Pigment Orange 34** **Pigment Orange 16**
Technical limitations	Poor durability in pale shades	PO 16 requires additive for rheology.
Mid performance	**Pigment Orange 81**	**Pigment Orange 81**
Technical limitations	Can be system dependant and requires specific evaluation	Technically the best offer
High performance	**Pigment Orange 73**	**Pigment Orange 81**
Technical limitations	Only commercial at this point	Technically the best offer

Secondary recommendations

	Pale	Deep
Classical approach	-	-
Technical limitations	-	-
Mid performance	**Pigment Orange 61**	**Pigment Orange 64**
Technical limitations	Poor alkali fastness, low durability vs cost	Rheology requires additive, alkali fastness poor, positive – high saturation
High performance	**Pigment Orange 61**	**Pigment Orange 61**
Technical limitations	Dull shade vs cost, positive – high heat fastness and durability	Dull shade vs cost, positive – high heat fastness and durability

Macromol. Symp. **187,** 801–809 (2002)

Mass and Energy Flow Management in the Sector of Surface Treatment

Jutta Geldermann, Nurten Avci, Hannes Schollenberger, Frank Blümel, Otto Rentz

French-German Institute for Environmental Research (DFIU / IFARE), University of Karlsruhe (TH) Hertzstr. 16, D-76187 Karlsruhe, Germany

E-Mail: {jutta.geldermann; nurten.avci}@wiwi.uni-karlsruhe.de

Summary: Vehicle refinishing body shops are concerned by the environmental policy against photochemical air pollution caused by VOC emissions. For supporting these small enterprises the mass and energy flow model IMPROVE has been developed. Based on an LCA-approach, the process steps of vehicle refinishing are modelled. The model IMPROVE helps to disseminate the consequences of product substitution in comparison to so far used products and techniques in the body shop. For the dissemination of experiences practical guidelines have been developed. Moreover, a comprehensible tool for the Solvent Balance is being offered. Thus, various means are available for consulting and also for strategic production planning for the SME in the sector.

Keywords: Mass and energy flow management, VOC (Volatile Organic Compounds), SME (Small and Medium sized Enterprises), LCA (Life Cycle Assessment), Decision Support

Introduction

The Solvent Directive of the European Union (99/13/EC) lays down emission limits for twenty different categories of installations using solvents. Companies have to fulfil the requirements set by the Solvent directive from 2001 (new installations) respective 2005 (existing installations). Therefore, the companies are obliged to take measures in the near future. In comparison to larger companies, small and medium-sized enterprises (SME) are generally not very well informed about the measures which could help them to reduce their costs and emissions, because of their limited personal size and cash flow. Consequently, the diffusion of environmentally sound products and technologies is often insufficient. With the help of a suitable mass and energy model, geared to the special needs of small and medium-sized enterprises in the vehicle refinishing sector, the different process steps can be analysed, and the influence of product or process modifications on other process steps, can be evaluated, in order to reduce the costs and emissions caused by coating activities in body shops.

The Individual computer aided mass and energy flow model for the vehicle refinishing sector IMPROVE

Since the vehicle refinishing sector is characterised by a large number of small companies, their activities have not yet been subject to detailed research. Therefore, a practical tool for the mass and energy flow management should investigate the overall changes of emission and of costs for individual body shops, based on an integrated analysis of all processes from surface cleaning to final finishing and polishing, comprising all materials with VOC content (as gunwash, precleaner, stopper and body filler, wash primer, precoat, primer, plastics, topcoat, basecoat, clearcoat and several special coatings).

Consequently, a representative model of a typical vehicle refinishing shop has been defined in close co-operation with branch experts from manufacturers of paints and equipment and from professional associations. This model is representative for the majority of vehicle refinishing firms in Germany as regards the working structure, spraying processes and corresponding technological equipment [2; 6].

A large part of the basic data (e.g. consumption figures related to area, components and vehicles, structure of orders, working times and capacity data of the technologies, such as thermal and electrical capacity) was collected in two reference body shops. Based on these intense analyses, the mass and energy flow model IMPROVE 1.0 (individual computer aided mass and energy flow model for the vehicle refinishing sector) has been designed and implemented. The model includes 15 different spraying processes each with a maximum of 55 stages from receiving the car to handing it over to the customer and a large variety of products and technologies for each stage. The mass and energy flow model IMPROVE is implemented with the commercial software UMBERTO. The LCA software UMBERTO is based on the so-called material flow networks [4; 5]. The formal basis rests on Petri-net theory, a special type of network from theoretical informatics which with its strict systematic (see Figure 1) not only allows the setup of complex systems, but also a combined material and inventory calculation.

Figure 2 gives a screenshot of the Petri-net representing all 430 process steps in a modelled vehicle refinishing body shop. The assessment of the mass and energy flows can then be printed out, be further worked on by means of a chart editor, or be exported to following modules, e.g. spreadsheet programmes or databases, for further processing.

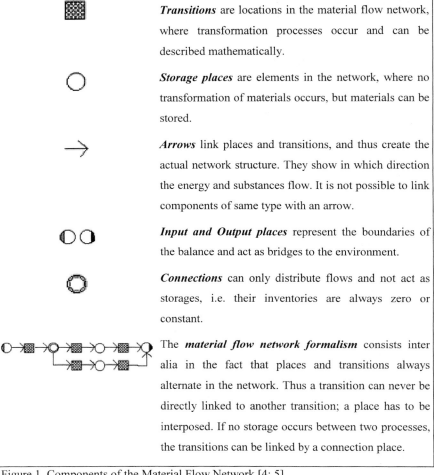

	Transitions are locations in the material flow network, where transformation processes occur and can be described mathematically.
	Storage places are elements in the network, where no transformation of materials occurs, but materials can be stored.
	Arrows link places and transitions, and thus create the actual network structure. They show in which direction the energy and substances flow. It is not possible to link components of same type with an arrow.
	Input and Output places represent the boundaries of the balance and act as bridges to the environment.
	Connections can only distribute flows and not act as storages, i.e. their inventories are always zero or constant.
	The ***material flow network formalism*** consists inter alia in the fact that places and transitions always alternate in the network. Thus a transition can never be directly linked to another transition; a place has to be interposed. If no storage occurs between two processes, the transitions can be linked by a connection place.

Figure 1. Components of the Material Flow Network [4; 5]

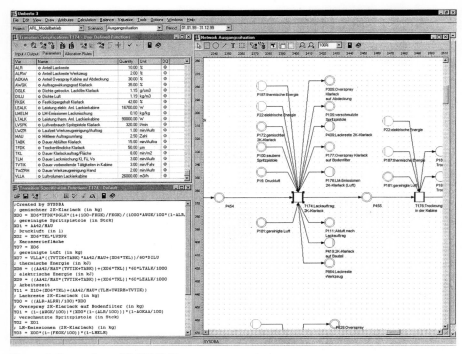

Figure 2. Screenshot of IMPROVE

For the economic and ecological analysis of the mass and energy flows, an evaluation programme has been especially designed within IMPROVE based on MS- EXCEL. Here on the one hand the individual mass and energy consumption quantities are listed according to special product groups (e.g. quantity of waste which has to be particularly monitored), and on the other they are multiplied with specific material prices, which are based on price lists from manufacturers, suppliers' invoices and billing rates. With the evaluation programme it is possible to determine the share of costs of certain product groups and the economic advantage of particular measures for reducing emissions compared to the initial situation. Figure 3 gives an example for an evaluation of the implications of material substitution and change of techniques. The results can be taken as a basis for investment decisions.

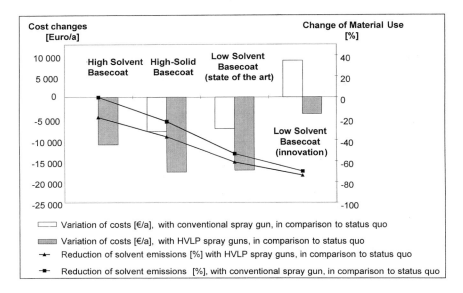

Figure 3. Implications of the use of low overspray application techniques in combination with different coating systems for the model body shop with 1 875 vehicle parts per year

The ratio of incoming and outgoing materials at each individual stage is described by a multitude of mathematical equations and parameters. The spraying processes are centrally controlled by approx. 200 parameters, such as material consumption, working time, performance characteristics, machine times and paint mixture ratios. Thus every vehicle refinishing enterprise can be depicted relatively simple and quick with regard to the individual structure of orders, working methods and the products and technologies applied, and the economic and ecological effects of measures to reduce emissions, integrated into production, can be determined in absolute figures. With this model also loops of materials, such as the reuse of distilled cleaning agents, can be taken into consideration.

Thus, IMPROVE simulates the procedure of vehicle refinishing and enables the flows of materials and energy to be determined and to be more efficiently organised in small and medium-sized vehicle refinishing firms. As the experiences from the practical use of IMPROVE in several body shops show, the concept is well accepted by practitioners, because it makes the process steps transparent with regard to cost savings and emission reduction.

While IMPROVE is conceived for the in-depths-investigation of single firms, some insights on the consequences of the introduction of emission reduction measures can be generalised.

Therefore, based on the findings of the use of IMPROVE, easy-to-use diagrams have been developed for an approximate judgements of the benefits or consequences of certain options for a specific firm. Figure 4 gives an example for the easily understood investigation about the economic consequences of the investment for a body shop, where the amortisation (payback period) for automatic cleaning devices for spray guns is derived depending on the number of painted parts per year in the considered body shop.

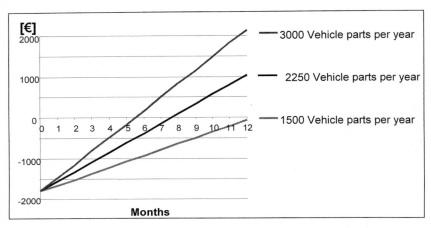

Figure 4. Diagram for simple evaluations: Determination of the amortisation for automatic cleaning devices for spray guns

Development of practical guidelines

Basically, the use of the mass and energy flow model IMPROVE is designed for the consultation of individual body shops. For immediate support for the SME affected by the new environmental legislation, a practical guideline for an improved use of water-based coating systems and other solvent reduced / free coating products (also auxiliary products) has been developed [7].

In close co-operation with nine body shops and national professional associations for vehicle refinishing, a practical changeover concept has been conceived, based on the results of IMPROVE. In addition, interviews and tests after the introduction of water-based coating systems permitted the evaluation of advantages, disadvantages and difficulties of the new introduced system. Based on these broad experiences, the elaborated guideline points out critical points, offers solution possibilities and comprehensible check-lists and names contact persons in case of difficulties.

As another further development of the IMPROVE-Approach, a tool for deriving the Solvent Management Plan according to the EU-Solvent Directive (1999/13/EC) has been conceived [8]. Based on the data contained in the mass and energy flow model IMPROVE, product information of paint producers has been analysed and revised. The developed calculation program for drawing up the solvent balance is based on the standard software MS-EXCEL for spreadsheet calculation. Also a paper based version is being offered to the firms for manual use. In addition, if a body shop uses IMPROVE, the calculated mass and energy flows can be taken as data input for the establishment of the solvent management plan.

In this comprehensive manner, specific decision support is provided for the concerned SME, as the overall concept in Figure 5 illustrates.

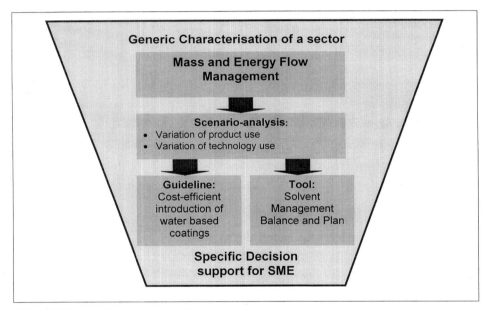

Figure 5. Comprehensive mass and energy flow management in the sector of surface treatment in SME

Conclusions

By describing the way of operating and the technological environment as well as by a complete view of shop internal material and energy flows, the mass and energy flow model IMPROVE helps to disseminate the consequences of product substitution (e.g. costs incurred, emissions caused) in comparison to so far used products and techniques in the specific body shop. Thereby possible environmental and economic reduction respectively saving potentials can be identified.

For the broader dissemination of the achieved results, practical guidelines with checklists and market analyses foster the introduction of waterborne coatings in the sector. Moreover, a calculation model for drawing up the Solvent Balance - as requested from the SME in the sector of vehicle refinishing - uses in simplified terms the data stored in IMPROVE.

The conducted research projects demonstrate, that the use of mass and energy flow models can not only support larger companies in their environmental management, but also SME. While IMPROVE was developed specifically for the processes of vehicle refinishing, further fields of application may be industrial and general commercial spraying (cf. [9]), car body construction work (e.g. welding), or lorry refinishing. In addition to the ecological and economic aspects, organisational issues can be investigated with IMPROVE, like the investigation of work time, ergonomics and capacity planning [10].

Not only the body shops will profit from the guidance notes, but also the paint manufacturers can use the mass and energy flow model IMPROVE and the guidelines for their Customer Relationship Management (CRM). Some paint producers use IMPROVE for their customer service, in order to simulate the necessary changes in the enterprises of their customers when changing the processes from conventional to waterborne products. Moreover, the results gained with IMPROVE are a valuable data source for decision support on strategic planning for future paint production and improvements in surface treatment [1].

References

[1] Avci, N.; Peters, N.; Geldermann, J.; Rentz, O.: Dynamic Mass Flow Optimisation as Decision Support for Environmental Policy. SETAC Europe 11th Annual Meeting, Madrid, May 2001.

[2] Blümel, F.: Stoffstrommanagement in der handwerklichen Autoreparaturlackierung - Zur Planung und Steuerung betrieblicher Stoff- und Energieströme in kleinen und mittleren Unternehmen der Autoreparaturlackierung. VDI-Verlag, Düsseldorf, 2001

[3] CEPE (Conseil Européen de l'Industrie des Peintures, des Encres d'Imprimerie et des Couleurs d'Art) : Technology Guidelines for Vehicle Refinishes. Brussels, 2000 http://www.cepe.org/Publications_Contenu_Technology.html

[4] Institut für Umweltinformatik Hamburg GmbH: Umberto – Das Konzept der Stoffstromnetze, in: Homepage http://www.ifu.com/software/umberto/programm/stoffstrom.htm

[5] Möller, A.; Rolf, A.: Methodische Ansätze zur Erstellung von Stoffstromanalyse, in: Schmidt, M.; Schorb, A.. (Hrsg.): Stoffstromanalysen in Ökobilanzen und Öko-Audits,

Springer-Verlag, Berlin, 1995

[6] Rentz, O., Blümel, F., Geldermann, J.; Lonjaret, J.-P., Große-Ophoff, M.: Stoffstrommanagement für kleine und mittlere Unternehmen aus dem Bereich der Autoreparaturlackierung, (*Mass and Energy Flow Management for small and medium sized companies in the sector of vehicle refinishing*) On Behalf of the Deutsche Bundesstiftung Umwelt (DBU, Osnabrück); Erich Schmid Verlag, Berlin, 2000

[7] Rentz, O.; Avci, N.; Geldermann, J.: Entwicklung eines praxisorientierten Leitfadens zum verbesserten Einsatz von Wasserlacken in kleinen und mittleren Unternehmen der KFZ-Reparaturlackierung (*Development of a practical guideline for the introduction and use of water based coatings in small and medium-sized car refinishing body shops*), On behalf of the Ministry for Environment and Traffic Baden-Württemberg (Stuttgart), 2001,
http://www.uvm.baden-wuerttemberg.de/uvm/abt4/wasserlacke/wasserlacke_reparatur.pdf
http://www.uvm.baden-wuerttemberg.de/uvm/abt4/wasserlacke/leitfaden_wasserlacken_reparaturlackierung.pdf

[8] Rentz, O.; Avci, N.; Geldermann, J.: Lösemittelbilanz und Reduzierungsplan für kleine und mittlere Unternehmen der KFZ-Reparaturlackierung (*Solvent Management Plan according to the EU—Solvent Directive (1999/13/EC) in small and medium-sized car refinishing body shops*)
On behalf of the Ministry for Environment and Traffic Baden-Württemberg (Stuttgart), 2001,

[9] Rentz, O.; Nunge, S.; Geldermann, J.: Einführung eines innerbetrieblichen Energie- und Stoffstrommanagements bei einem mittelständischen Industrielackierbetrieb für Kleinteile in Baden-Württemberg, (*Development of an energy and mass flow management model at a medium-sized enterprise for paint application in Baden-Wuerttemberg*) On behalf of the Landesanstalt für Umweltschutz, Baden-Württemberg, 2002

[10] Rentz, O.; Schollenberger, H.; Geldermann, J.: Arbeitszeitstudie für die Autoreparaturlackierung in Vertragswerkstätten (Time management in vehicle refinishing), in cooperation with Workline GmbH, on behalf of the German Association of the Automotive Industry (VDA), 2001 - 2002

*Macromol. Symp. **187**, 811–821 (2002)*

Progress in the Development of Cobalt-free Drier Systems

Johan H. Bieleman

Sasol Servo BV, the Netherlands

Summary: Driers are used in air-drying coating systems to catalyze the polymerization process. Cobalt-carboxylates are the most widely used driers. However, cobalt compounds may indirectly be implicated as carcinogen suspects as a result of studies in the U.S.A. in the national toxicology program using cobalt sulphate heptahydrate. Hence, Germany is no longer granting the Blue Angel award to cobalt-containing paints. Other transition metal carboxylates such as based on manganese or iron show much lower catalytic effects and cannot equalize cobalt as a catalyst in autoxidation polymerization reactions. The effect of various organic chelating ligands on the catalytic properties of manganese in autoxidation processes was investigated experimentally in air-drying paints. The activity of manganese is strongly effected by organic ligands. New manganese based coordination compounds enable the formulation of Co-free air-drying paints, which show good drying performances and improved color retention.

1 Introduction

Air drying alkyd paints contain besides the main constituents, alkyd resins (binders), pigment and solvents small amounts of driers (e.g. cobalt-ethylhexanoate). The driers speed up the oxidative cross-linking process of a paint film, based on the autoxidation of the unsaturated fatty acids, which are present as constituents of alkyd resins.

Driers, also referred to as siccatives when in solution, are organo-metallic compounds soluble in organic solvents and binders [ref. 1]. Chemically, driers belong to the class of metal soaps and they are added to air-drying coating systems to accelerate or promote after application the transformation from the liquid film into the solid stage within an appropriate time. The transformation occurs by oxidative cross-linking of the binder system, a process, which is catalysed by the metallic cation of the drier. The anionic part of the drier molecule serves as the carrier, to solubilize the drier in the binder system.

For decades cobalt has been the main active drier used in air-drying paints. However as the result of recent studies with cobalt sulphate heptahydrate, cobalt compounds are subject to re-classification procedures. Following the recently published Commission Directive 2001/59/ECof the European Community. Cobalt oxide is classified using following risk phrases: R 22/ R 43/ R50/ R 53 [ref. 2]. The classification for Cobalt driers have not been changed yet, as it is still unclear if the toxicity data as found for water soluble compounds like Cobalt sulphate heptahydrate, are relevant for cobalt driers.

     CCC 1022-1360/00/$ 17.50+.50/0

Drier manufacturers have initiated several test-procedures in order to collect data for the bio-availability of cobalt, used as a drier in paints. Nevertheless, the pressure to replace cobalt driers with cobalt-free alternatives is growing. In some cases cobalt-containing compositions will not be granted for environmental awards, like for the Blue Angel in Germany.

2 The drying process and the effect of driers

The drying process of alkyd paint is the result of the slow evaporation of the volatile components (physical drying) and in a second step chemical drying takes place. The oxidative cross-linking process of alkyd resins occurs via a radical reaction in which H-atoms are abstracted from the double methylene group of linoleic acid. The resulting radicals take up atmospheric oxygen and form hydroperoxides. The hydroperoxides degrade with cobalt catalyst into alkoxy and peroxy radicals. These radicals form cross-links by recombination (initially dimerization). This process is known as "the autoxidation cross-link process" (figure 1).

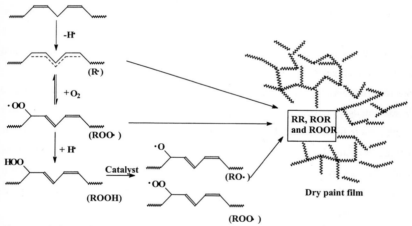

Figure 1. Schematic presentation of the authoxidation cross-linking process of alkyd resins

Apparently the term "autoxidation" has been defined as an non-catalyzed oxidation reaction of a substrate exposed to the oxygen of the air.

$$RH_2 + O_2 \longrightarrow RH\cdot + O_2^{\cdot} + H^+$$

$$RH_2 + O_2 \longrightarrow R + H_2O_2$$

Figure 2. General reactions for " autoxidation " reactions

However, autoxidation reactions occur at significant rates only in presence of a catalyst, such as a transition metal [ref. 1, 3. 4]. The autoxidation process referring to the drying of paints is accelerated by addition of driers. Without these drying catalysts the paint layer may dry only after some months; with driers this is accomplished within a few hours.

More in detail, chemical drying by oxidation can be through as a combination of four steps:

- Step 1: Induction period
- Step 2: Peroxide formation
- Step 3: Peroxide decomposition into free radicals
- Step 4: Polymerization

The induction step is measured from the time the coating is applied until the film begins to absorb oxygen from the air. The absorbed oxygen forms peroxides across the conjugated double bonds in the binder (step 2).

When the peroxides start to decompose, active cross-linking sites are formed. As crosslinking proceeds during the polymerisation, the viscosity increases rapidly.

Step 2 and 4 proceed most effectively with resins containing conjugated double bonds; however, non-conjugated but poly-unsaturated resins show also some reactivity. In such a case the multiple double bonds may cause the activation of the various methylene groups, to rearrange the position of the non-conjugated double bonds, depending on the original position of the double bonds.

The steps 1 and 2 (induction and peroxide formation) are accelerated dramatically by the presence of driers.

The mutivalent metals in the drier system act as oxygen carriers because of their susceptibility to redox reactions. Driers also activate the formation of peroxides; assumed is the multivalent metal is associated to the double bonds, increasing the oxidation susceptibility. The addition of cobalt drier reduces the energy, which is necessary for the activation of the oxygen absorption by an unsaturated resin, with a factor 10 [ref. 1, 5].

The penetration of activated oxygen into the film favors the peroxide formation. As soon as peroxides are formed their decomposition in a metal-catalysed reaction to alkoxy (RO•) and peroxy radicals (ROO•) takes place (figure 3).

Initiation:

$$RH \qquad \rightarrow \qquad R\cdot + H\cdot$$

Propagation:

$$R\cdot \ + O_2 \qquad \rightarrow \qquad ROO\cdot$$

$$ROO\cdot + RH \qquad \rightarrow \qquad ROOH + R\cdot$$

$$ROOH + M^{n+1} \qquad \rightarrow \qquad ROO\cdot + M^{n+} + H^+$$

$$ROOH + M^{n+} \qquad \rightarrow \qquad RO\cdot + M^{n+1} + OH^-$$

Termination+

$$2\ ROO\cdot \qquad \rightarrow \qquad ROOR + {}^1O_2$$

$$2\ ROO\cdot \qquad \rightarrow \qquad ROH + R{=}O + {}^1O_2$$

Figure 3. Reactions during the autoxidation cross-linking process

Drier systems for ambient cured decorative paints based on long oil alkyd resins are usually siccativated using a cobalt / zirconium / calcium combination drier. Cobalt is the active drier. However, in order to improve trough-drying, hardness and stability, auxiliary driers, like zirconium and calcium, are being used in conjunction with cobalt.

3 General characteristics of potential alternatives to cobalt driers

Essentially, drier metals can be divided into two groups: active driers and auxiliary driers. This difference should be considered arbitrary as a considerable amount of overlap exists between them.

Active driers (e.g. cobalt) promote at ambient temperatures oxygen uptake, peroxide formation and peroxide decomposition.

Auxiliary driers do not show catalytic activity themselves at ambient temperatures, but enhance the activity of the active drier metals.

Next to cobalt, various transition metals have been reported as having primary drier properties. However, only manganese has found substantial practical use.

None of the alternative driers -as metal carboxylates- resemble in performance to cobalt. The limitations of the alternatives in relation to the application in paints are summarized as follows:

Manganese	low		effectiveness,		dark	color
Iron	low		effectiveness,		dark	color
Cerium	not	effective	under	ambient	temperature	conditions
Vanadium	low			effectiveness,		color
Copper	low effectiveness, deep-green color, toxicity					

However, the reactivity as well as properties like effect on color of the same transition metal are largely related to the chemical composition of the metal compound [ref. 6]. This is understandable as "free metallic cations" do not exist in solution. The metal ions are always surrounded by anions, by solvent molecules or by other ligand groups.

Most commercial driers are pure carboxylates. However, for some applications manganese based compositions with strong chelating ligands such as bipyridine have found commercial use. The ligand plays an important role in the catalytic oxidative drying of these compounds (ref. 7). Strongly chelating ligands effect the catalytic reactivity by altering the electron density at the metallic center of a complex ion and so its redox potential.

Next section will be addressed to the performance as a drier various manganese based compositions, using organic ligands.

3 Experimental

3.1 Procedure and used materials

The effect of various driers on the drying of paints has been evaluated, using different air-drying paint systems. The composition of the long-oil (LO) paint, designated as standard paint, is shown in table 1. The drier is added to the paint during the let-down stage.

As paint systems both commercially available paints, supplied by the manufacturer for this test without drier, as well as prepared "fresh paints and varnishes" have been used.

The drying performance was determined according to ASTM or similar procedures (as indicated), using a drying recorder.

Following procedures have been used:

1: the drying recorder. Drying condition: 23 °C / 50% relative humidity. The used instrument is a straight-line recorder. Following drying stages are being considered:

        stage a – the paint flows together, the *wet-edge* time.

        stage b – a line is visible, the paint begins to polymerize: *dust-free*.

stage c – ripped film: *tack-free* or surface-dry.

stage d – surface path: through dry or *total dry*.

2: the drying was further established by the "thumb-test" and according to ASTM D1640 (wet film thickness: 60 μm).

The König hardness of the films was assessed by using the pendulum damping test according to DIN53157. A glass panel was coated with a 60 μm wet film, kept under conditions of 23 °C and 50 % RH and the hardness development in time was monitored with a König pendulum. The oscillation time measured to reduce the deflection from initial 6 ° to 3 ° is given in seconds (s/K:).

A cobalt-zirconium-calcium drier combination has been used as reference, using following metal ratio, unless otherwise indicated.

0.06 Co

0.3 Zr

0.1 Ca.

This combination is based on a mixture of commercially available grades of Co 10, Zr 18 and Ca 10 (NUODEX, Sasol Servo BV).

Both commercial as well as experimental metal complexes have been used. Details will be presented in the tables and figures.

3.2 Results and discussion

3.2.1 Co vs Mn drier

A direct comparison demonstrating the catalytic effect of Co versus Mn drier in a white house paint, formulated according to the standard paint according to table1 is shown in following table 1.

Table 1. The effect of Co vs Mn in a white LO alkyd paint.

Active* drier	Auxilliary drier*	Dust-free, Hrs	Tack-free, Hrs	Total dry, Hrs	Whiteness index	Hardness, s/K
0.08 Co	0.4 Zr 0.2 Ca	1.30	2.00	3.30	78	52
0.08 Mn	0.4 Zr 0.2 Ca	7.15	9.45	15.15	73	29

*: calculated as % metal on total resin solids

Replacing Co drier with Mn drier at same metal dosage has a detrimental effect on the drying performance, whiteness and hardness. Obviously Mn is unsuitable to replace Co drier.

For practical application a dust-free time of over 7 hours will result in adhesion of dust particles in the dried coating. Moreover, the long tack-free and total dry time increase the risk of damages to the paint film and the appearance.

3.2.2 Manganese-bipyridine complexes

Manganese complexes have been reported as effective driers [ref. 6]. Compounds of Mn and chelating compounds, like 1,10-phenantroline and bipyridine (bipy), have found widespread commercial application, for instance in urethane alkyd varnishes as well as in waterborne alkyd paints. The application of these manganese compounds instead of cobalt driers enables the formulation of light colored urethane alkyd varnishes.

Recently the active species of the Mn-bipy complex has been determined as being $[Mn_4O_2(2\text{-ethylhexanoate})_6(bipy)_2]$ [ref. 7].

A projection of the structure of the Mn-bipy complex is presented in figure 4.

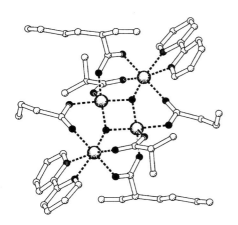

Composition:
2 Mn^{2+}
2 Mn^{3+}
6 2-ethylhexanoate
2 bipyridine
2 O^{2-}

*NUODEX FS 530
Figure 4. Structure Mn-bipy complex*.

The use of Co driers in conjunction with clear urethane alkyd resins results in dark colored varnishes and are for optical reasons less attractive. Using Mn-bipy instead of Co results in considerable improvement in color. This improvement is just visible in the liquid varnish; no differences in the hardened coating have been found. Compared to the Mn carboxylate remarkable improvement in drying speed can be noticed (table 2). The Mn-bipy complex is also applicable in waterborne coatings, based on short-oil alkyd emulsions.

Both resin systems, urethane alkyd and short-oil alkyd emulsions, have in common that the physical drying being very important and the contribution of cross-linking to the film formation and hardness is rather low, compared to systems based on LO alkyd resins.

Table 2. The effect of Mn-bipy in a furniture varnish based on urethane alkyd.

Active* drier	Auxilliary drier*	Total dry, Hrs	Color liquid varnish, Gardner
0.06 Co	0.3 Zr 0.1 Ca	2.00	15
0.06 Mn	0.3 Zr 0.1 Ca	5.30	6
0.03 Mn-bipy	0.3 Zr 0.1 Ca	1.30	6

*: calculated as % metal on total resin solids

Similar positive effects on drying rate using Mn-bipy have been recorded in the pigmented paint, based on a urethane alkyd resin (table 3). However, the Mn-bipy complex has a negative effect on the whiteness.

Table 3. The effect of Mn-bipy in a white paint, based on urethane-alkyd resin.

Active* drier	Auxilliary drier*	Total dry, Hrs	Whiteness index
0.06 Co	0.3 Zr 0.1 Ca	4.45	79
0.06 Mn	0.3 Zr 0.1 Ca	9.30	77
0.03 Mn-bipy	0.3 Zr 0.1 Ca	4.15	73

*: calculated as % metal on total resin solids

Furthermore, comparative data have been recorded in a standard white paint, based on LO alkyd resin (table 4). Although some improvement on the drying time could be noticed, the overall performance is still insufficient and needs further improvement. Like in the urethane alkyd paint, a negative effect of the bipy complex on the whiteness has been determined, also in the LO alkyd paint. Moreover, the paint film remains too soft, using the Mn-bipy drier.

Table 4. The effect of Mn-bipy in a white paint, based on LO alkyd resin.

Active* drier	Auxilliary drier*	Dust-free, Hrs	Tack-free, Hrs	Total dry, Hrs	Whitenes s index	Hardness , s/K
0.06 Co	0.3 Zr 0.1 Ca	2.00	2.30	4.45	80	36
0.06 Mn	0.3 Zr 0.1 Ca	10.00	13.00	15.15	78	29
0.03 Mn-bipy	0.3 Zr 0.1 Ca	5.00	7.00	10.00	72	21

*: calculated as % metal on total resin solids

Obviously the catalytic effectiveness of Mn is improved using bipy as ligand. However, further improvements are necessary in order to be able and use Mn-based driers instead of Co.

3.2.3 Manganese-poly-ligand complexes

A large variation in ligands and compositions of ligands have been evaluated. Bipy is a typical strong field ligand and forms stable complexes with Mn. Further improvement in drying properties could be reached in using compositions of two or more ligands: typically a strong field ligand and a weak field ligand. Next to the improved drying characteristics, another advantage of using mixed ligands is the improved storage stability and low viscosity, enabling to reach high concentrated drier compositions; the metal concentration in the complex is 6% Mn [ref. 8].

Following test results demonstrate the effect of "poly-ligand-based Mn complexes" (indicated as "Mn-plb"). The speed of drying, hardness as well as color are positively effected using the poly-ligand based Mn instead of just bipy (table 5).

Table 5. The effect of Mn-plb in a white paint, based on LO alkyd resin.

Active* drier	Auxilliary drier*	Dust-free Hrs	Tack-free Hrs	Total dry H	Whitenes s index	Hardness s/K
0.06 Co	0.3 Zr 0.1 Ca	2.00	2.30	4.45	80	36
0.06 Mn	0.3 Zr 0.1 Ca	10.00	13.00	15.15	78	29
0.03 Mn-bipy	0.3 Zr 0.1 Ca	5.00	7.00	10.00	76	21
0.03 Mn-plb	0.3 Zr 0.1 Ca	4.20	5.45	8.00	77	24

The drying speed in LO alkyd paints is still slightly inferior using the Mn-plg drier instead of Co. Further improvements can be accomplished modifying the resin composition: using medium oil alkyds or blends of LO alkyd and MO alkyd.

The drying characteristics in the MO alkyd based paints are comparable for the formulations with Mn-plg and Co as driers (table 6 and 7). The effect of using auxilliary driers in conjunction with Mn-plb is limited and does not warrant using these in the evaluated MO alkyd paints. As a consequence, the total amount of drier catalyst could be reduced from 2.4% to just 0.3% (Table 6).

Table 6. white paint based on a medium oil alkyd.

	% active drier metal in the drier	% active drier metal on solid binder	Index raw material cost drier	Dosage on paint S.b.=29.7%	Total dry, initial	Total dry, after 3 months storage at 23°C	Hardness s/König	W-I
Co-Zr-Ca, reference	1.2% Co	0.075	100	1.86	4.30	7	62	80
Mn-plg	6% Mn	0.04	60	0.40	5	6	57	81.5

Table 7. semi gloss black paint, based on a chain stopped medium oil alkyd.

	% active drier metal in the drier	% active drier metal on solid binder	Index raw material cost drier	Dosage on paint s.b. = 37%	Total dry, initial	Total dry, after 3 months storage at 23°C	Hardness s/K
Co-Zr-Ca, reference	1.2% Co	0.09	100	2.78	4	4	106
Mn-plb	6% Mn	0.09	106	1.11	9	6	113

4 Conclusion

Organic complexes of manganese offer promising features as replacement for cobalt driers.

In the formulations tested, the new Mn-poly-ligand drier provided similar total dry characteristics in the medium oil alkyd paints, without causing adverse effects in the liquid or cured coating. The new drier outperforms the traditional manganese based driers regarding drying properties.

In LO alkyd resin based paints slightly longer drying times and lowering of film hardness have been registered; it is suggested to consider re-formulation of the binder composition in case further improvements in LO alkyd systems are being required.

References

1: Bieleman, JH; *"Additives for Coatings"*, Wiley-VCH, New York/ Weinheim, published in 2000, p. 258
2: Official Journal of the European Communities, 21-08-2001
3: Middlemiss, R.G., *J. of Water-borne Coatings*, (Nov. 1985), p. 3
4: Sutton, D.A., Farmer, E.H., *J. Chem. Soc.* 122 (1943)
5: Muizebelt, W.J. et al; *J. Coat. Technol.* **1998**, *70 (876),* 83-93.; *J. Mass Spectrom.* **1996**, *31*, 545; Prog. *Org. Coat.* **2000**, *40*, 121; **1994**, *24*, 263.
6: Bieleman, J.H., *Fatipec Brussel 1996,* Congress proceedings Vol.A, S 112
7: Warzeska, S, Bouwman, E et al; *Progress in Organic Coatings,* 44 (2002) 243-248
8: Product leaflet NUODEX FS 578 DP; Sasol Servo BV

Macromol. Symp. 187, 823–834 (2002)   823

Effect of Weathering on Scratch/Mar and Mechanical Behaviour of Clearcoats

Michael Osterhold[1]*, Birte Bannert[1], Walter Schubert[1], Thomas Brock[2]

[1]DuPont Performance Coatings, Postfach, D-42271 Wuppertal, Germany

[2]FH Niederrhein, Adlerstraße 32, D-47798 Krefeld, Germany

Summary: The study examines the effect of weathering on mechanical and chemical properties of two 2K (twopack, with and without light stabilisation additives) and one 1K (onepack) clearcoats, concentrating on scratch and mar resistance and acid etch resistance. The clearcoats were investigated before and in the course of accelerated weathering.
For determining the mar resistance two different instruments were used:
    - AMTEC laboratory car wash equipment (rotating wet brush)
    - Crockmeter
Explanations for the changes in clearcoat properties caused by weathering are given by Dynamic-Mechanical Analysis, FT-IR-Spectroscopy, Universal-hardness and surface tension measurements.
The clearcoat systems investigated show a clear deterioration of their properties after short periods of accelerated weathering. This effect is more expressed for the clearcoat samples without light stabilizers.

Keywords: Scratch, Mar, Weathering, Degradation, Mechanical properties

Introduction

Over the last few years the appearance of automotive coatings has become determinant in the consideration of coating quality. Excellent appearance is expected to last during service life, but due to UV-light, acid rain and other environmental influences, the (clear-)coat properties deteriorate. Therefore, and because common evaluations for the effect of weathering on automotive topcoats (such as gloss measurements, yellowing, water spotting, blistering etc.) are not useful to obtain information according to the chemical and mechanical changes occuring in a clearcoat during weathering, car manufacturers increasingly demand, that certain tests have also to be checked after weathering.

          CCC 1022-1360/00/$ 17.50+.50/0

Materials and Methods

Following clearcoat systems were investigated:

Table 1. Investigated clearcoats

CC1	2K-system; acrylic/isocyanate; silane modified
CC1a	dto. without light stabilizers
CC1b	dto. without light stabilizers, reproduction
CC1c	dto. lowered ratio OH/NCO from 1:1,5 to 1:1
CC2	2K-system; acrylic/isocyanate
CC2a	dto. without light stabilizers
CC3	1K-system; epoxy/acid

Substrate:

Electrocoat panels (Bonder); 7,5 cm x 7,5 cm / 10 cm x 20 cm

Polypropylene for DMA

Coating Materials:

Table 2. Cure conditions

PU-Primer, black	Stoved 17 min. @ 165°C MT; filmbuild 30-40 μm
WBBC, black	Pre-dried 10 min. @ 80°C oven time; filmbuild 12-15 μm
2K-clearcoats	Stoved 20 min. @ 140°C MT; filmbuild on panels 35-45 μm on polypropylene 50-60 μm
1K-clearcoat	Stoved 35 min. @ 140°C oven time, filmbuild s.a.

Test methods:

⇨ SAE J1960-weathering, investigation of the samples after: 0 h, 250 h, 500 h, 750 h and 1000 h weathering;

⇨ Scratch/mar resistance:
   - Amtec Kistler laboratory car wash (DIN 55668, draft)
   - "AATCC-Crockmeter CM-5", dry sheet

⇨ Acid etch resistance:

   Panels were tested on a hotplate at 65 °C, one drop of 36%-sulphuric acid per minute. Minutes until first swelling / etching occurs are noted for evaluation.

⇨ FT-IR-spectroscopy (4000 cm^{-1} – 400 cm^{-1}):

- attenuated total reflection (ATR, Diamond)

- photoacoustic spectroscopy (PAS) in two different sampling depths, achieved by varying the light modulation frequency: 2,5 kHz => 7-10 µm

                                                        20 kHz => 1 µm

⇨ Dynamic-Mechanical Analysis:

   Perkin Elmer DMA 7 (tensile test on free films); 1 Hz frequency, -40 °C - 180 °C

⇨ Universal hardness:

   According to DIN 55676 (draft); hardness measured at maximum force = 30 mN

⇨ Surface tension :

   Contact angle measuring system G2 (Krüss)

⇨ Gloss measurements:

   Micro-TRI-gloss (BYK-Gardner; 20D)

Results

Scratch and mar resistance

1) Amtec Kistler laboratory car wash

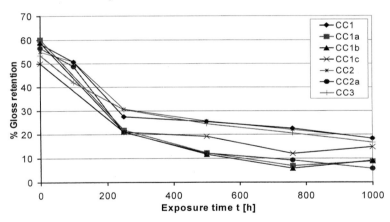

Figure 1. % Gloss retention (Amtec) throughout SAE J1960-weathering

Fig. 1 shows the residual gloss of the clearcoats after testing with Amtec Kistler laboratory car wash equipment in the course of weathering. All samples that were investigated show a significant drop of residual gloss after a short exposure time. A clear differentiation between the clearcoats with and without light-screener is noticeable. Afterwards the residual gloss decreases slightly. The difference between the samples containing light stabilizers and the clearcoats without UV-package shows, that photodegradation of non-stabilized binder systems starts after short exposure to UV-light (see FT-IR and DMA results).

2) Crockmeter

The Crockmeter scratch resistance in dependence on exposure time is shown in Fig. 2. Tested with this method, the UV-exposure results in a different effect on the clearcoat systems. Not necessarily the scratch resistance of the samples without light stabilizers drops in the course of exposure. Regarding the Crockmeter test a differentiation between the different clearcoat chemistries appears.

The silane modified system seems to be superior to the other clearcoats, because even the non-stabilised samples o this system (CC1a and CC1b) maintain a good scratch resistance.

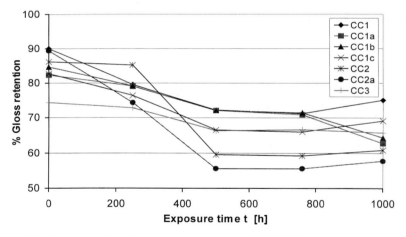

Figure 2. % Gloss retention (Crockmeter) throughout SAE J1960-weathering

Acid etch resistance

Fig. 3 shows the first discernible swelling mark after the acid test in dependence on exposure time. Swelling marks are later visible after 250 h weathering for all systems, afterwards the swelling occurs earlier. After the first period of exposure, the significant increase of T_g improves the acid resistance because acid penetration is restricted, this effect is then overlapped by the progressing photodegradation with further exposure time.

In addition, the etch resistance of the samples without light stabilizers decreases clearly after 250 hours exposure, up from 500 h the etch resistance is tendentially increasing (first etching marks), but is finally under their start values. A possible explanation could be, that because of the clear photodegradation of these clearcoats the etch resistance lessens, and with further exposure the enhanced T_g partially increases the etch resistance again. The clearcoats containing light stabilizers do not show an influence of weathering on etch resistance.

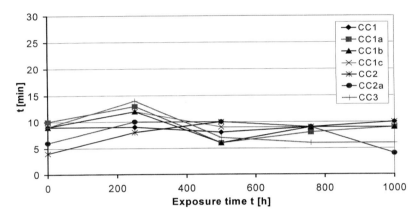

Figure 3. Occurrence of first swelling during acid test in the course of weathering

FT-IR-spectroscopy

1.) *PAS*

FT-IR spectroscopy is commonly used for the evaluation of the photodegradation rate of clearcoats during/after weathering [1]. The concentration of [(OH+NH)/CH] is increasing with exposure time because of photodegradation reactions. For the clearcoats containing no UV-screener, the photodegradation rate is clearly faster growing than for the materials containing light stabilizer as can be seen in Fig. 4 (7- 10 μm sampling depth).

In the surface region (20 kHz frequency/sampling depth approx. 1 μm) the sequence of the clearcoats is nearly the same as at 2,5 kHz frequency (sampling depth approx. 7-10 μm), but the ratio [(OH+NH)/CH] is lower in this region.

The first expectation would be vice versa because at the surface more oxygen is available, but due to low molecular photodegradation products being washed out or migrating into the film the results are explainable.

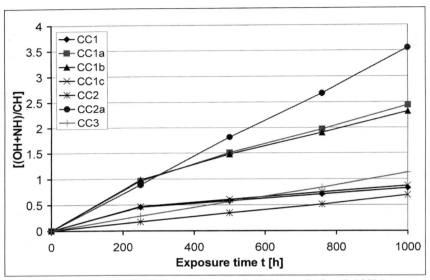

Figure 4. [(OH+NH)/CH] throughout SAE J1960-weathering (2,5 kHz)

The results of FT-IR-PAS spectroscopy explain tendencially the findings of the Amtec laboratory car wash test, but the magnitude of the drop of Amtec scratch resistance after short exposure time is extremely high.

The Crockmeter results do not correlate very well with the FT-IR-PAS results because of the good performance of the 2K-Silane material without light stabilization.

2.) *ATR*

For the evaluation of changes of the peaks in the area from 2000 cm^{-1} through 400 cm^{-1}, the ATR-technique is very helpful, because the PAS spectra show too much noise in that region.

Fig. 5 shows the spectrum of CC1a unweathered and after 500 h, 1000 h weathering. All 2K-clearcoats show a similar behaviour during exposure so one exemplary spectrum is sufficient to show the occuring changes.

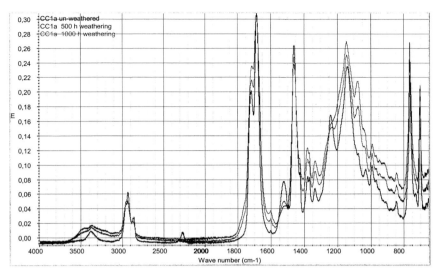

Figure 5. ATR-spectrum of CC1a unweathered, after 500 h and 1000 h weathering
Note: The area from 2000 cm^{-1} through 600 cm^{-1} is stretched.

As mentioned above, the OH,NH-peak (3700 cm^{-1} – 3100 cm^{-1}) grows with exposure time while the CH-peak decreases. The peak at 2270 cm^{-1} shows free isocyanates [2]. After 500 h weathering the peak disappears, the isocyanates react with OH-containing groups or water. The silane-modification is not discernible in the spectrum. For the further evaluation of 2K spectra the peak at 1530 cm^{-1}, known as the CNH-bending mode + CN-stretching mode of urethans, and the 1250 cm^{-1} peak, known as the CN-stretching mode of urethans, is important. The decreasing area of these peaks indicate that the amount of urethane bonds is reduced by photodegradation [3, 4]. This effect is stronger for the non-stabilized materials, but also the urethan bonds of the stabilized clearcoat diminish.

There are two main reactions described, that lead to photodegradation of urethan bonds. First, the photochemical decomposition of urethan bonds, second, the H-abstraction from the α-C and further reaction [7].

The changes in CC3´s spectrum are not as significant as those occuring in the 2K materials spectra. Except from the area from 3700 cm^{-1} through 2200 cm^{-1}, just the C=O-peak at 1730 cm^{-1} and the C-O-peak at 1160 cm^{-1} are useful for interpretation. A slight decrease in the intensity of both peaks is visible, this changes are possibly due to the photochemical splitting of ester bonds.

Dynamic-Mechanical Analysis

T_g, taken as the temperature at loss modulus′ maximum ($T_{E''max.}$) of all clearcoats that were investigated increases clearly especially in the first period of exposure (see Fig. 6). The extend of T_g-enhancement is similar for all materials, just the T_g of sample CC1c with a ratio of 1OH:1NCO increases nearly linearly.

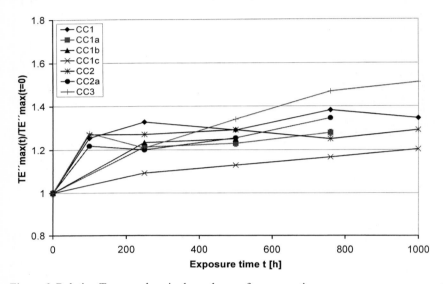

Figure 6. Relative $T_{E''max.}$ values in dependence of exposure time

This indicates that the reaction of free isocyanate groups with further OH-groups or the formation of diurea with water plays a key role in the T_g change for the 2K-systems. Surprisingly not the clearcoats without UV-screener have the clearest increase of T_g, especially CC3 shows a strong rise of T_g. This effect is due to the lower initial T_g of CC3, because during the light phase in the weathering cabinet the temperature of a black panel can reach up to 70 °C and molecule segments are then more flexible so that the chance of further T_g-increasing reactions is more likely than for the 2K-clearcoats.

All samples were additionally investigated after storing them for seven weeks at room temperature and in darkness to see, if the glass transition temperature is also increasing (increase of 6 to 11 °C, start values of T_g were between 50 to 70 °C).

The T_g-values indicate, that a certain part of the initial increase of T_g is not only caused by weathering/photodegradation, but also by after-cure reactions and the loss of residual solvents.

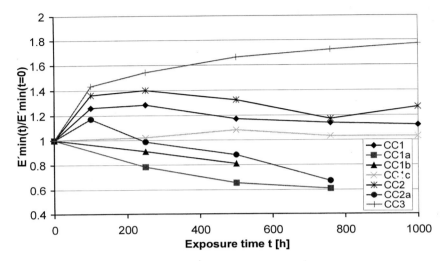

Figure 7. Relative E´min.-values vs. exposure time

The minimum data of the storage moduli $E'_{min.}$, as a measure for the crosslinking density, are given in Fig. 7 (rel. values, start values of $E'_{min.}$ were between 8 to 15 MPa). The changes of crosslinking density (XLD) of the non-stabilized materials caused by weathering differ clearly from that of the light-stabilized samples. In the first period of exposure, the XLD of all clearcoats rises, with further exposure time $E'_{min.}$ of the non-stabilized materials decreases clearly, XLD of the other (stabilized) samples, except from CC1c and CC3, decreases slightly, too. These results correlate very well with the findings of PAS-FTIR measurements. The light stabilizers, especially HALS being radical scavengers, obviously prevent the polymers from chain scission. After a longer exposure time, chain scission dominates over crosslinking reactions also in the HALS-containing samples. Only in the 1K-clearcoat CC3 crosslinking seems to be dominating, but does not prevent the coating from loosing mechanical film properties.

Universal Hardness

$HU_{Korr.}$ of all clearcoats that were investigated increases clearly with exposure time. This is due to T_g increase (further reactions, evaporation of residual solvents) and is a sign for a certain embrittlement of the clearcoats. The hardness of the non-stabilized clearcoats and CC3 show the strongest increase [7].

Surface Tension

The surface tension of all clearcoats increases clearly with exposure time, especially within the first 500 h. Surface tension of the samples containing no UV-screener is a little bit higher than of the stabilized samples at any time of exposure. An explanation for the surface tension increase might be the growing polarity resulting from photooxidation reactions (see FT-IR-results) [7].

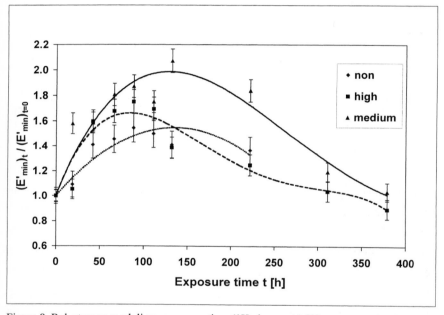

Figure 8. Rel. storage moduli vs. exposure time (1K clearcoats) [8]

In a former study, the physical behaviour of model 1K and 2K clearcoat systems containing a different amount of aromatics (non, high, medium) were investigated. As an example Fig. 9 shows the rel. storage modulus of the 1 K systems in the course of accelerated weathering [8]. It can be clearly pointed out that the largest variation (increase/decrease) in modulus can be observed in the first some hundreds hours of weathering, which was also found in the investigations described above.

Gloss measurements

The 20D gloss of the investigated materials did not change during 1000 h SAE J1960-weathering (see Fig. 9). Gloss measurements are not useful to obtain information about chemical and mechanical changes occuring during weathering. The gloss of a coating does not decrease until degradation is extremely high, but mechanical properties turn worse long before.

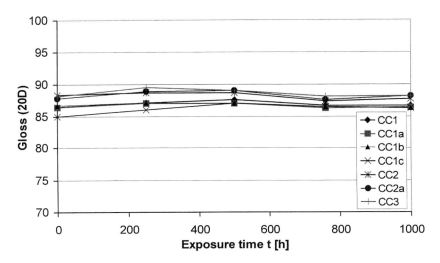

Figure 9. 20D-gloss vs. exposure time

Summary and Conclusions

The clearcoat systems investigated show a clear deterioration of their properties after short weathering periods. This effect is more expressed for the clearcoat samples without light stabilizers.

- Scratch resistance of all samples tested with Amtec laboratory car wash drops off after 250 h SAE J1960-weathering, a distribution into light stabilized / non-stabilized clearcoats is apparent

- Scratch resistance tested with Crockmeter decreases less than Amtec scratch resistance, the silane modified 2K-clearcoat performed best (also without UV-stabilization)

- Acid resistance improves at the beginning of weathering, but decreases in the further course of weathering

- Results of FT-IR-spectroscopy show, that the amount of urethan-bonds diminishes after short periods of weathering, also with the clearcoats containing light-stabilizers.

  Additionally, the ratio [(NH+OH)/CH] (referring to the area in the IR-spectra from 3700 cm^{-1} to 2200 cm^{-1}), commonly used for the evaluation of photooxidation products, increases nearly linearly.

- Glass transition temperature (T_g), taken as the temperature of $E''_{max.}$, and crosslinking density, taken as the value of $E'_{min.}$ in the rubbery-elastic zone, were determined by DMA.

  - T_g increases during weathering, especially at the beginning
  - crosslinking density of the non-stabilized clearcoat samples decreases significantly in the course of weathering; clearcoats containing HALS/UV-absorber show no (1K-CC) or just a slight (2K-CC´s) decrease of crosslinking density

- Universal-hardness and surface tension increase with weathering time
- Gloss of all investigated clearcoats does not change throughout 1000 h SAE J1960-weathering

⇨ The test methods used in this investigation give information about the chemical and mechanical changes occuring during weathering

References

[1] D. R. Bauer, Prog. Org. Coat., 23 (1993), pp. 105
[2] H. Günzler, H. Böck, IR-Spektroskopie, 2. Überarb. Aufl., Verlag Chemie, Weinheim (1983)
[3] D. R. Bauer, M. C. Paputa Peck, R. O. Carter III, J. Coat. Technol., Vol. 59, No. 755 (1987), pp. 103
[4] L. W. Hill, H. M. Korziniowsky, M. Ojunga-Andrew, R. C. Wilson, Prog. Org. Coat., 24 (1994), pp. 147
[5] S. Storp, M. Bock, Farbe+Lack, 91, No. 10 (1985), pp. 914
[6] L. Bottenbruch, R. Binsack (Hrg.), Kunststoff Handbuch, Bd. 4 Polyamide, Carl Hanser Verlag, München Wien (1998)
[7] B. Bannert, M. Osterhold, W. Schubert, Th. Brock, Europ. Coat. J., No. 11 (2001), pp. 30
[8] M. Osterhold, P. Glöckner, Prog. Org. Coat., 41 (2001), pp. 177

*Macromol. Symp. **187**, 835–844 (2002)*

Stabilizers in Automotive Coatings under Acid Attack

Ulrich Schulz*, Kirsten Jansen*, Adalbert Braig**

*Federal Institute for Materials Research and Testing, D 12200 Berlin, Germany

**CIBA SC, CH 4002 Basel, Switzerland

Summary: To examine the action of acid atmospheric precipitation on the light stabilisers in automotive coatings two common clearcoat types (TSA and 2K-PU) blended with commercially available HALS and UVA were subjected to artificial weathering (Acid Dew and Fog test and Acid-free weathering test) and natural weathering outdoors in Jacksonville, FL. Tracing gloss and haze during weathering the test results showed that the influence of acid stress on the long-term performance of the clearcoats depended on the kind of the stabilisation system.

Keywords: coatings; ageing; stabilisation; HALS; gloss; acid resistance

Introduction

Today's automotive coatings are expected to have got long-term weathering performance even when the cars are transported, stored or used in harsh environment with frequent occurrence of acid atmospheric precipitation. There are observations made with clearcoat / basecoat paint systems that surface areas where droplets of acid depositions were present during weathering suffer a faster degradation by UV radiation [1-3]. The reasons for the higher susceptibility of acid attacked surface parts to degradation by UV can be different. Firstly, higher UV sensitive species can be formed as consequence of a chemical reaction of acid with the resin. Secondly, acid deposition can introduce impurities in the coating material as so-called chromophores being able to absorb UV well and transmit the photon energy to the macromolecules easily. Thirdly, light stabilizers like hindered amine light stabilizers (HALS) and ultraviolet light absorbers (UVA), intended to protect the coating system against photo oxidation, can be attacked by acid leading to the formation of non-functioning or water-soluble derivatives. As

CCC 1022-1360/00/$ 17.50+.50/0

consequence, the acid attacked surface parts may have lost its shelter against UV radiation. Our work was devoted to the last effect. There is little information about how long the function of light stabilizers is sustained in acid-stressed environment. Usual acid ingredients of atmospheric precipitation are suggested to disable the so-called "parent HALS" to be oxidized to nitroxyl radical necessary to start the radical scavenging mechanism.[4-5] But there is not any reliable information about the progress of this reaction in the field in dependence of the type of HALS and the exposure conditions. Today's most favoured test for assessing the coating behaviour in acid-stressed environment is a 14 weeks lasting outdoor weathering test in Jacksonville, FL. This location is of particular interest for the car manufacturers because it is one of the main ports of the U.S.A. for imported cars and known for frequent occurrence of acid rain and dew generating a typical type of surface damage. Visible acid etch defects can be found immediately after the acid attacks on clearcoats using resins with relatively bad acid resistance. But it is questionable whether acid attacks to the light stabilizers manifest in a noticeable contribution to the visible coating damage caused outdoors in Jacksonville during 14 weeks. It was found in previous examinations that clearcoats using highly acid resistant resins do not experience any visible damage after 14 weeks outdoors in Jacksonville.[6] But the clearcoats may experience an invisible acid-caused damage to the light stabilizers which can lead to catastrophic failure caused by photo oxidation in further years of service. Assessing the influence of acid damaged light stabilizers on the clearcoat lifetime reliably requires to prolong the outdoor exposure test to several years. Such long exposure periods would slow down the introduction of new stabilizer formulations unacceptably. A successful use of appropriate artificial weathering tests to assess the behaviour of light stabilizers in acid-stressed clearcoats is not known till now. Therefore, aim of our examinations was finding answers to following two questions:

1. Is the influence of acid stress on the long-term performance of clearcoats dependent on the kind of the light stabilisers significantly and can such a relationship be assessed by means of an artificial weathering test, like the "Acid Dew and Fog test" (ADF test)?

2. Is a natural short test, like the 14 weeks outdoor exposure in Jacksonville, FL, able to differ the clearcoat samples in relation to the behaviour of the light stabilizers against acid attacks?

Experimental

Two different types of clearcoat were used in this study: a thermosetting acrylic resin (TSA) and a two-part polyurethane (2K-PU) as described in detail in former publication [7]. The coating systems were applied on aluminium panels and cured under production line conditions.

Table 1. Identification of the samples (HPT: Hydroxyphenyltriazine, BTZ: Benzotriazole)

TSA		2K-PU	
Sample	Stabilization system	Sample	Stabilization system
#0	unstabilized	#0	unstabilized
#1	1% HALS 1 (pK-b: 4,4 to 5,5)	#1	1% HALS 1 (pK-b: 4,4 to 5,5)
#2	2% HALS 1	#2	2% HALS 1
#3	1% HALS 2 (pK-b: ca. 10)	#3	1% HALS 3 (pK-b: ca. 12)
#4	2% HALS 2	#4	2% HALS 3
#5	2,5% HPT 1 (UVA 1)	#5	2,5% HPT1 (UVA 1)
#6	2,5% HPT 2 (UVA 2)	#6	1,5% HPT 4 (UVA 4)
#7	2,5% BTZ (UVA 3)	#7	2,5% BTZ (UVA 3)
#8	2,5% HPT 1 + 1% HALS 1	#8	2,5% HPT 1 + 1% HALS 1
#9	2,5% BTZ + 1% HALS 2	#9	1,5% HPT 4 + 1% HALS 1
		#10	2,5% BTZ + 1% HALS 1
		#11	2,5% BTZ + 1% HALS 3

As characterised in table 1 both clearcoat types were blended with commercially available HALS and UVA which were used apart in different concentration (samples 1 to 7) as well as combined with each other (samples 8 to 11). Samples without light stabilizers (samples 1) were examined for comparison. The samples were exposed to artificial weathering (Acid Dew and Fog-Test, ADF-Test in short, and Acid-free weathering test) using a weathering chamber "GLOBAL UV Test" and natural weathering outdoors in Jacksonville, FL. The changes in gloss and haze that are ones oft the most suitable optical properties to detect weathering-caused deterioration on the clearcoat surface were traced during weathering.

The cycle used for the ADF test (see table 2) was characterised by continues UV radiation and changing climatic conditions. The essential feature what the ADF cycle distinguishes from the cycles of the currently used artificial weathering tests is the spraying of an acidic deposition on the sample surface (which is a diluted mixture of sulphuric, nitric and hydrochloric acids) as the first step of the cycle. This spraying simulated the action of acid atmospheric precipitation occurring in industrial urban locations like Jacksonville.

Table 2. The 24 h weathering cycle of the ADF test

Step	Duration		Climatic conditions
1	short		spraying of the acidic deposition
2	14 hours	dry	9 h at 35°C, 30% RH + 6 h at 60°C, 5% RH with UV
3	4 hours	rain	rinsing with demin. Water, at 35°C with UV
4	6 hours	dry	at 60°C, 5% RH with UV

The cycle of the Acid-free weathering test differed from the ADF cycle only in that the acid wetting at the start was dropped.

Results of artificial weathering

In order to make differences between the stabilisation systems in the behaviour against acid attack by artificial weathering visible it was necessary, first at all, to look for the right balance of the stresses by acid and UV radiation during the test. For this reason two variants of the ADF test differing in the acidity of the solutions (pH 1,5 and pH 2,5 respectively) for the acid wetting of the samples and the exposure period (7 and 70 days respectively) were performed.

The specimens which were exposed to the test variant with pH 2,5 over 70 days experienced the tenfold higher UV radiant exposure than the other ones which were subjected to the test variant with pH 1,5 over 7 days. But the specimens experienced the same acid stress expressed by the product of concentration and action time of the acid in both test variants. The haze increase of the TSA samples after the tests was presented in dependence on the kind of the stabilization system in Figure 1. The samples did not show any differences in degradation after the short test variant using the higher acidic solution. The UV radiant exposure proved to be too small for revealing a possible acid-caused damage of the light stabilizers in the weathering results. This test variant was only able to assess the acid resistance of the resin which was the same in all of the 10 samples.

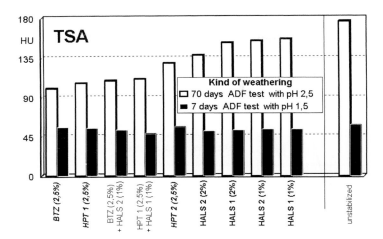

Figure 1. Haze increase (in Haze units HU) after differently long ADF test variants with equal acid stress in dependence of the kind of stabilization

The long test variant using the less acidic solution proved to be a closer approximation to the practice in terms of the ratio of the stresses by acid and UV. It caused higher ageing effects as well as a clear differentiation of the clearcoat samples suggesting that the examined stabilization systems experienced differently strong acid-caused damage which revealed after applying an ample UV radiant exposure only. Therefore, this test variant was used for our further examinations predominantly. The gloss loss of the PU-clearcoat samples was chosen for presentation of the results of the ADF test (variant with pH 2,5) in comparison to that of the Acid-free weathering test. As can be seen in figure 2 the curves for the test without acid stress in the bottom chart, characterising commonly used weathering conditions, revealed a big difference in the long-term performance between the samples containing UV absorbers or not. Finding differences in the efficiency between the UVA containing stabilisation systems each other required a very long exposure period. The curves for the ADF test in the top chart showed that the presence of acid on the surface during weathering accelerated the degradation of the coating significantly.

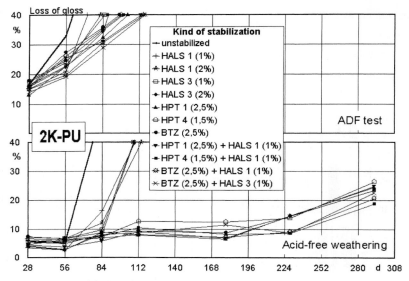

Figure 2. Gloss loss of the PU clearcoat samples upon exposure period

The acid resistance of the resin seemed to be to weak to reveal differences in the stabilizer efficiency under acid stress in the same extent as under acid-free weathering conditions. The state of the samples after 84 days of exposure was chosen to work out differences in the efficiency between the light stabilizers under acid attack. The results of the Acid-free weathering and the ADF-test as well as the differences between the two tests were presented in figure 3. Since the samples differed in the kind of stabilisation only, variations in the extent of the differences between the results of the two tests could only be caused by differences in the acid resistance of the light stabilizers. The samples were arranged in increasing order of the black balks, this means in increasing order of the susceptibility of the stabilisation systems to degradation by acid attack. In the case of the unstabilized sample (most right) the black balk characterised the acid resistance of the resin. Concerning the stabilized samples this would mean that the stabilisation system "BTZ (2,5%) + HALS 3 (1%)" (most left), showing the black balk in about the same extent, did not seem to be influenced by acid. In case of the second sample from the right, blended with 1% HALS 1 only, the gloss loss could be derived from both the acid resistance of the resin and that of the stabiliser in about the same extent. According to general experiences the HALS, known as reactive chemicals, showed the worst acid resistance. The samples using different types and different concentrations of HALS behaved as expected.

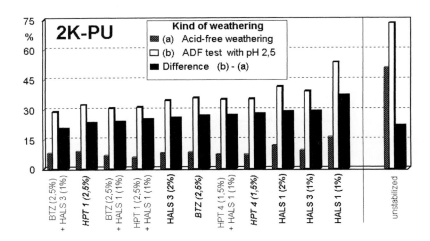

Figure 3. Gloss loss after 84 days of weathering with and without acid stress

This means, the higher the basicity and the lower the concentration the worse the acid resistance. The UVA, known as physically acting light protectors, proved to be more stabile against chemical influences. The samples containing combinations of HALS and UVA behaved better as well as worse than that containing the UA absorber singly. The influence of acid stress also revealed in different rankings in terms of the long-term performance of the samples subjected to the ADF test and the Acid-free weathering respectively.

As can be seen in figure 4 the stabilization system "HPT4 (1,5%) + HALS 1 (1%)" proved to be the worst of the light stabilizer combinations under acid stress but the best one under acid-free conditions. The ADF test results in dependence of the kind of resin were compared in figure 5. As expected the 2K-PU showed the better acid resistance. The light stabilizers behaved equivalently in both resins.

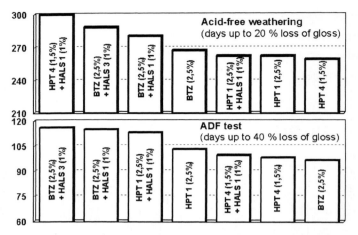

Figure 4. Rankings in long-term performance of the UVA containing PU clearcoat samples subjected to weathering with and without acid stress (ranked from the best to the worst)

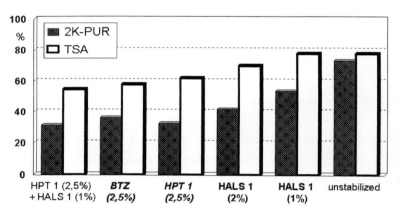

Figure 5. Gloss loss after 84 days of ADF test (pH 2,5) in dependence of the resin

Results of the natural weathering

Gloss loss and haze increase of the PU clearcoat samples after the 14 weeks lasting outdoor weathering in Jacksonville, FL were presented in the figure 6. Only the unstabilized sample differed from the other ones. Significant differences in the behaviour of the stabilised samples could be found neither for the PU nor for the TSA clears. This finding is according to the results of the Acid Dew and Fog test as can be seen in Figure 7.

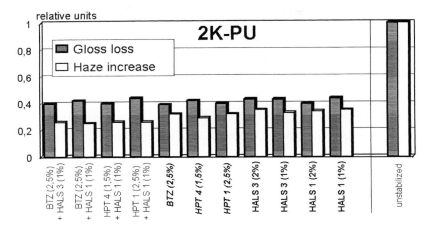

Figure 6. Changes in Gloss and Haze after 14 summer weeks in Jacksonville, FL

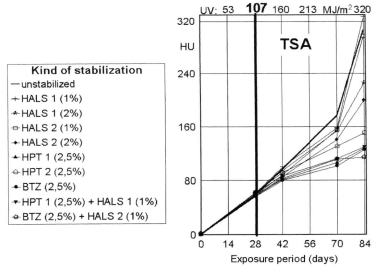

Figure 7. Haze of the TSA samples upon exposure period

The UV radiant exposure of about 107 MJ/m² applied naturally during the 14 summer weeks in Florida is marked by the vertically drown thick line in the presentation of the Haze of the TSA samples in the course of the ADF test. The ADF test was also not able to differentiate the samples after applying the equivalent UV radiant exposure (after 28 weathering days).

Conclusions

1. The influence of acid stress on the long-term performance of automotive clearcoats depends on the kind of the stabilisation system. But the lifetime of the commonly used clearcoats in acid-stressed environment, limited by the acid resistance of the resins predominantly, is mostly too short to reveal equivalent differences in the stabilizer efficiency as under acid-free environmental conditions.2. The Acid Dew and Fog test (ADF test) using a solution of pH 2,5 for simulating of the action of acid atmospheric precipitation and performed over 70 days at least proved to be able to determine the influence of acid attack on the efficiency of light stabilisers in automotive clearcoats. The pH 2,5 seems to be the optimal acidity. Higher acidity results in worse selectivity. Lower acidity requires unacceptable prolongation of the test duration.

3. The commonly used 14 weeks outdoor exposure in Jacksonville , FL proved to be too short to make differences in the resistance of light stabilizers against acid attacks visible. This means surviving 14 weeks outdoors in Jacksonville undamaged is no evidence for long-term weathering performance.

References

[1]   U. Schulz, P. Trubiroha, U. Schernau, H. Baumgart, *Progress in Organic Coatings* **2000**, *40*, 151

[2]   U.Schulz, P. Trubiroha, T. Engbert, T. Klimmasch, *Eropean Coating Journal,* **1999**, *1-2*, 18

[3]   U.Schulz, V. Wachtedorf, T. Klimmasch, P. Alers, *Progress in Organic Coatings* **2001**, *42*, 38

[4]   .Gugumus, in: *Handbook of Polymer Degradation*, 2nd ed., H. Hamid, Ed., M.Decker, Inc., New York 2001, p.68

[5]   A.V. Kucherov, J. L. Gerlock, R. R. Matheson, *Polymer Degradation and Stability*, **2000**, *69*, 1

[6]   P. Trubiroha, U. Schulz, V. Wachtendorf, K.Gaszner, *DFO-Technologie-Tage*, **2002**, *50*, 55

[7]   V. Dudler, T. Bolle, G. Rytz, *Polymer Degradation and Stability*, **1998**, *60*, 351

Durability Prediction of p-Urethane Clearcoats Using Infrared P(hoto) A(coustic) S(pectroscopy)

Leo G.J. van der Ven and Roel Leuverink

Akzo Nobel Coatings Research Arnhem, Velperweg 76, 6800 SB, the Netherlands

leo.vanderven@akzonobel.com

Summary: The durability of topcoats is dependent on a large number of factors as polymer composition, stabilization package and the conditions during the weathering process.

For obvious reasons prediction of the long-term (5-10 year) durability of coatings is very important. The rate-determining factor for the degradation of PUR coatings is photo-oxidation. The photo-oxidation rate is controlled by the polymer structure but also stabilizers as HALS has a large influence.

The prediction of the durability of clearcoats is based on tracing of the photo-oxidation rate and of the HALS longevity during exposure.

The photo-oxidation rate is measured using FTIR-PAS. The results show that degradation can be detected much earlier compared with classical methods as gloss loss. Moreover detection of differences between systems after short exposure times as well as prediction of the long-term durability are possible.

Keywords: clearcoat, durability, polyurethane, photo acoustic spectroscopy, infrared spectroscopy

Introduction

Durability prediction is in the coating industry a main issue. A lot of effort is put in research focussed on correlation between short-term artificial weathering methods and long term outdoor weathering. The idea behind these programs is the assumption that the process of degradation during weathering is only accelerated and that the mechanism does not change during the artificial exposure.

Polyurethane clearcoats are applied mainly for cars. But also new areas like the aerospace industry show interests to use these clearcoat systems. Furthermore, the warranty demands (aircraft's 8 years) are becoming more stringent. Also new environmental stresses, as acid pollution, become of more influence on the durability behavior of coatings.

To make the prediction of the long-term durability of clearcoats more reliable the classical approach of comparison artificial and outdoor weathering is not anymore sufficient. The main problem is the non-linearity of e.g. gloss loss in time (Figure 1). Cracking or gloss loss starts after already a reasonable chemical attack has been taken place.

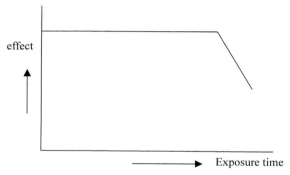

Figure 1. Non linearity effect of e.g. gloss loss/cracking in time

More knowledge about the background of the degradation is needed to overcome this dilemma.

This knowledge could be of great help to develop a service life prediction model [1].

The durability of clearcoats can be described by the following factors:

a. Photo-oxidation rate of the surface

b. Longevity of hindered amine stabilizers (HALS)

c. Longevity of UV-absorbers

These factors are related to gloss/cracking (a, b) and delamination (c). Photo-oxidation causes a more hydrophilic film, which will be more sensitive for hydrothermal stress [2].

The polymer design as well as the presence of HALS is of strong influence on this process.

This paper focuses on the measurement of the photo-oxidation rate by photo-acoustic spectroscopy. The prediction capability of this method has been investigated using several types of polyurethane clearcoats.

Experimental

Photo Acoustic FTIR Spectroscopy (PAS) was performed with an accessory of METC model 300, at 2.5 kHz using disks of 10-mm diameter. The spectra were evaluated in two ways. The relative changes of the bands in the whole spectrum were determined and the photo-oxidation product (POP) was measured. The definition of the POP value is given in Figure 2.

$$\left(\frac{(NH + OH + CH) - (CH)}{CH} \right)_{t=t} - \left(\frac{(NH + OH + CH) - (CH)}{CH} \right)_{t=0}$$

Figure 2. Definition of the Photo-Oxidation Product (POP).

The systems were exposed in the WOM (ISO 11341 Q/B as well as B/B filters), UVCON (ISO 11507, UVB-313) and in the BAT test. The BAT is a cyclic test consisting of 5 cycles of condensation/SO_2 according to ISO 3231, 16 h ventilation at ambient conditions and 100 hrs WOM according to ISO 11341 B/Q.

Two PUR clearcoats and one PUR solid color were investigated:

PUR 1   solid color (white)
PUR 2   MS clearcoat on light blue basecoat
PUR 3   HS clearcoat on HS light blue basecoat; both p-ester based

The stabilization with UV-absorber and HALS were varied.

Results and discussion

Influence of filters in WOM

The spectral distribution of using Boro/Boro or Quartz/Boro is very equal. The only difference is the short wavelength part. The Quartz/Boro filter transmits more of the wavelength between 290-310 nm. This part of the spectrum can have a big influence on the degradation rate as well as on the mechanism [3] since many polymers and crosslinker or formed crosslinks start to absorb light in this region. Figure 3 shows some results of a comparison of the two filter sets for PUR 3.

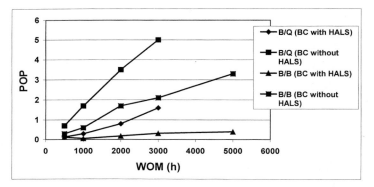

Figure 3. Comparison Quartz/Boro and Boro/Boro filters in WOM for stabilized and unstabilized PUR 3.

The photo-oxidation rate (POR) of the films exposed in the Q/B is faster as in the B/B exposure. More experiments show that the order of stability of the several systems does not change. So, the mechanism of degradation is not influenced by the filter system for these paint films.

Stabilization with HALS

Figure 4 shows the results of PUR 2 and PUR 3 with and without HALS. The protection of HALS against photo-oxidation of the clearcoat surface is obvious. Without HALS the photo-oxidation starts immediately. The stability of the p-urethane clearcoats with HALS is very good. After more than 8000 hrs of WOM exposure the POP start to increase somewhat. This can be explained by the fact that the HALS is becoming inactive during exposure which results in less protection to photo-oxidation [4]. This point is the start of a more severe degradation of the clearcoat and an indication that gloss loss/cracking will occur in the next period.

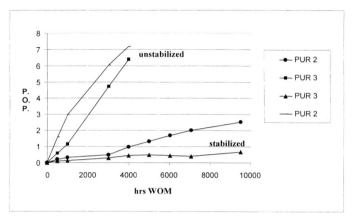

Figure 4. Influence of HALS on photo-oxidation rate of PUR 2 and PUR 3.

Comparison exposure methods

The rate of degradation has been compared for the three exposure methods. The variation is on one hand acid environment versus neutral conditions (BAT versus WOM/UVCON) and on the other hand a comparison between hard UV light (UVCON with UV-B bulbs) and soft UV-light (WOM B/Q). Figure 5 shows the POR of PUR 3.

The BAT test appears to be very aggressive. The photo-oxidation rate is very close to an unstabilized system (Fig. 4). The results suggest that either the acidic attack of the polymer is very severe or more likely, considering the same POR as the unstabilized system, the HALS activity is strongly diminished by the strong acid. ESR measurements [4] show indeed that no nitroxyl radicals are present anymore in the films exposed in the BAT test. A second conclusion from figure 5 is that for this system the difference between the WOM and the UVCON is only small.

Comparison PUR systems

Figure 6 shows the POR's of the three PUR systems during the WOM exposure. As expected the two clearcoats differ only to a less extent since the durability is dominated by the presence of the HALS. The solid color system reaches within short time a rather high level of POP and keep than constant. Several reasons can be mentioned for this behavior. Firstly the pigment particles keep the light in the upper few microns of the film. Secondly after already short time there is an equilibrium reached between the degradation of the polymer and the removal (e.g. rinsing with water) of degraded material. Thirdly the depth information of the PAS measurements could be smaller than in the case of clearcoats, since the pigments will influence the thermal diffusion of the film.

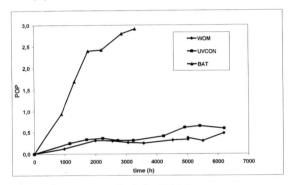

Figure 5. Comparison of exposure methods for PUR 3.

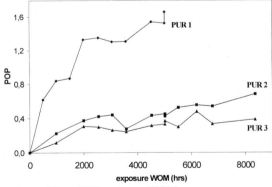

Figure 6. Comparison of three PUR systems; WOM exposure.

Relation POP and conventional parameters (gloss/cracking)

In figure 7 the POP values and the measurements of the conventional parameters gloss and cracking are plotted for the unstabilized PUR 3.

The predicting value of the POP method is clearly demonstrated. The gloss and cracking show the typical non-linear behavior whereas the POP increases from the start of the exposure indicating that the chemical degradation starts immediately. A comparison of this initial POR between systems will give a faster insight in the weathering behavior.

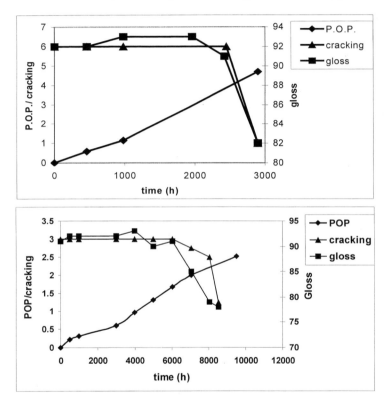

Figure 7. Comparison of POP with gloss and cracking of PUR 3; top unstabilized; bottom stabilized.

852

Conclusions

PAS measurements during weathering of p-urethane clearcoats are very useful to obtain in an early stage of the exposure information about the photo-oxidation sensitivity of clearcoats.

Differences between weathering methods, clearcoats, stabilization etc. can be quantified.

Moreover, determination of the photo-oxidation rate seems a tool to overcome the problem of non-linearity behavior of conventional parameters as gloss and cracking.

In the case of solid color p-urethane paint films the PAS method has no advantages since the photo-oxidation products reaches after short exposure time already its maximum value.

References

[1]   D.R. Bauer and J.W. Martin, *Service Life Prediction of Organic Coatings, ACS symp. Series 722*, Oxford Univ. Press, **1999**, p.1
[2]   M. Oosterbroek, R.J. Lammers, L.G.J. van der Ven and D.Y. Perera, J. of Coat. Techn., **1991**, *63*, 55
[3]   D.R. Bauer, M.C. Paputa Peck and R.O. Carter III, J. Coat. Technol., **1987**, *59*, 103
[4]   L.G.J. van der Ven, R. Leuverink, R. van Overbeek, Proceedings Athens Conference on Coatings, 1-5 July **2002**

Macromol. Symp. **187**, 853–860 (2002)

Evaluation of the Scratch Resistance with Nano- and Multiple Scratching Methods

Rolf Nothhelfer-Richter*, Eugen Klinke, Claus D. Eisenbach

Forschungsinstitut für Pigmente und Lacke e.V., Allmandring 37, 70569 Stuttgart, Germany

Summary: The scratch resistance of automotive clear coats was investigated by a single and multi scratch test procedure. New data characteristic for the reflow capability, the scratch hardness and the lateral scratch resistance were generated and evaluated. A correlation between single and multi scratch data was established.

Keywords: scratch resistance, nano-scratch test, multi-scratch test, polymer coatings, automotive clear coats, AFM

Introduction

Automotive coatings are expected to fulfil high quality and performance requirements in both the protection of the substrate and the optical appearance. The optical quality of the clear coat is impaired by damages extending only a few micrometers below the surface. Therefore single (nano) and multi scratch tests have been developed and employed in order to get a better understanding of the so-called marring process. It has been shown that the data obtained with a nano scratch test instrument, i.e., that the mechanical load at which lateral cracks occur alongside the scratch trace can be taken to rate the mar resistance [1 - 5]. However, as it will be shown in this paper, there are more characteristics to be extracted from the nano scratch test, which allow a better evaluation of the scratch resistance. In this context, the problem of how to correlate the nano scratch test data with multi scratch experiments like the Amtec-Kistler car wash simulation, the Rota-Hub scratch tester or others will be addressed as well.

Experimental

Samples

Five different automotive clear coats have been investigated. All samples were cured at

    CCC 1022-1360/00/$ 17.50+.50/0

designated curing temperature (index b) and also at a lower (index a) and a higher (index c) temperature. Curing temperatures and curing times are given in Table 1.

Table 1. Compositional characteristics and curing conditions of the samples.

Sample No.	Sample	Curing Agent	Curing Time [min]	Curing Temp. [° C]
1a	1 K			120
1b	onepack	melamine	17	140
1c	standard acrylic			160
2a	1 K			115
2b	onepack	melamine/IPDI	30	135
2c	standard acrylic			155
3a	2 K			70
3b	twopack	HDI	30	80
3c	acrylic			100
4a	2 K			115
4b	twopack	HDI	20	135
4c	polyester			155
5a	2 K			110
5b	twopack	HDI / IPDI	30	130
5c	polyester/polyacrylate			150

Scratch test devices

The single scratch test has been carried out with a CSM Nano Scratch Tester (NST); the multiple scratching was generated by using a Bayer Rota-Hub (RH) instrument. For details about methods it is referred to the literature [2, 6].

Results and discussion

The Nano-Scratch Test (NST) provides detailed information about the generation of a single scratch. An indenter of Rockwell geometry (conical, die angle 90°, tip radius 2μm) is pushed with increasing load into the sample, which is moved with a constant velocity. During the scratch process data of the tangential F_t and the normal force F_n as well as the depth are recorded. The penetration depth d_p of the indenter and the remaining depth of the scratch (residual depth d_r) are calculated using the data of a surface scan with minimal load before (prescan) and after (postscan) the scratch (Figure 1). The postscan is performed about one minute after the scratch. From this experimental procedure not only the critical values of the normal force at the first crack and the residual depth at the (arbitrary) value of 5 mN of the normal force can be taken, but additional information can be extracted from the measured curves.

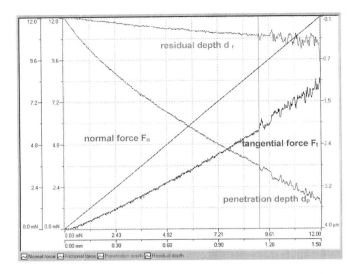

Figure 1. Original data curves as generated in the nano scratch measurement.

The scratch resistance will increase with increasing reflow capability of the sample. Therefore we define a reflow capability value $R = (d_p - d_r)/d_p$. This value will become one if the residual depth becomes zero, i.e. there is no remaining scratch observed. If there is no reflow at all, R will become zero. In Figure 2 the R-curve of sample 3b is shown as a representative example; a smoothed R-curve (named R_m), which is calculated by an error weighted averaging, is depicted besides the experimental curve. The R-curve reaches a plateau above a normal force F_n of only a few millinewtons. This reflow capability remains constant despite the occurrence of lateral cracks after a critical load of about 18 mN has been exceeded; the crack formation is reflected by the rapid fluctuations of the curve.

856

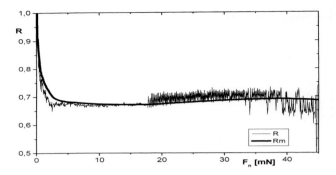

Figure 2. Reflow capability R and averaged R_m of the coating 3b (Table 1) as a function of the applied normal force F_n in the nano scratch test.

From these findings it can be concluded that the occurrence of cracks is not affecting the short time reflow capability of the sample. On the other hand, the long time reflow capability and the annealing of the sample at elevated temperatures will be affected by crack formation.

Some authors state a correlation between scratch resistance and micro hardness [7, 8]. From the nano scratch data a micro hardness $H = F_n / \frac{1}{2} A_c$ can be calculated by dividing the applied normal force F_n by the area of the indenter contacting the coating during the scratch formation (assumed to be half of the total surface area A_c of the conical shaped indenter). In a similar manner the resistance against the lateral translation Z can be defined by $Z = F_t / A_s$, where A_s is the cross-section area of the cone. Both areas can be calculated from the penetration depth using the geometric data of the indenter.

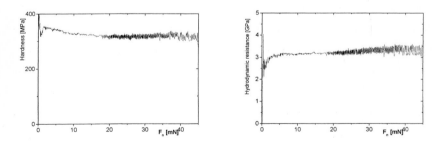

Figure 3. Micro hardness H and hydrodynamic resistance Z as obtained from nano scratch test of coating 3b.

The H and Z values show a plateau as well (Figure 3), which is also very little affected by the occurrence of cracks. Therefore the value taken at the first crack can be taken as characteristic value.

For comparison, the universal hardness was measured with a Fischerscope H 100 micro-indenter fitted with a Vickers indenter at a penetration depth of 3.5 µm (to avoid influence of the substrate). The samples show a good correlation (Figure 4); the offset is presumed to be due to the lateral movement. Further experiments shall clarify this finding. If only a ranking of samples shall be achieved, the nano scratch hardness will be sufficient.

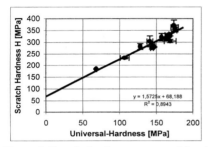 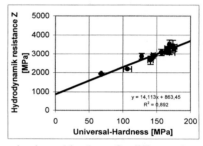

Figure 4. Correlation between scratch hardness and universal hardness for different clear coats.

The data extracted from the NST single scratch experiments are related to the Rota-Hub (RH) multi scratch test data. The scratch resistance of the coating as tested in the RH multi scratch test procedure is evaluated by the gloss loss measured before and after a multi scratch experiment. From the two gloss values the residual gloss is calculated in percent values by dividing the gloss after scratching by the gloss before scratching.

The scratch resistance is expected to increase with increasing hardness and with increasing reflow capability. A plot of R_m versus H is shown in Figure 5, where the residual gloss is indicated by different gray scaling of the points. One expects the points with high residual gloss (high scratch resistance) to lie in the upper right corner and the ones with low residual gloss to lie in the lower left corner. The results of our samples do not meet this expectations, although within the same chemical system a trend can be determined.

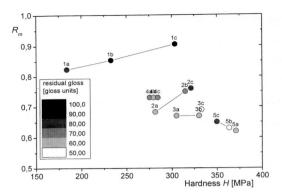

Figure 5. Reflow capability R_m versus scratch hardness H of different coatings (labels s. Table 1); the residual gloss after the multi-scratch experiment is illustrated in the filled data symbols by the gray-scaling.

About ten minutes after the scratch experiment an AFM-image was taken at the 5 mN normal force position. From this imaging the depth and the width of the remaining scratch can be obtained. As schematically illustrated in Figure 6, the scratch widens during the elapsed time (about 10 min. in this experiments) due to reflow processes. The widening can be expressed by the angle α_r of the scratch as compared to the shaft angle α_p of the indenter. During the scratch process the angle is given by the indenter tip, which has a cone angle of 90 degrees.

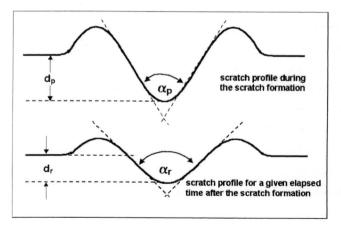

Figure 6. Schematic of the initial scratch profile during the scratch formation with a cone-shaped indentor of the shaft angle α_p and of the residual scratch profile with α_r of the scratch after a given elapsed time.

With increasing angle α_r of the remaining scratch trace, a higher residual gloss was found (Figure 7). This observation is independent from the chemical composition of the coating system. From the optical point of view the reflection of light into other directions than the reflection direction of the unscratched surface will decrease with increasing scratch angle. An angle of $\alpha_r = 180$ degree indicates a totally flat surface, i.e., the scratch has completely disappeared due to the reflow processes. The best samples are only 20 degrees below 180 degree, i.e., $\alpha_r = 160°$.

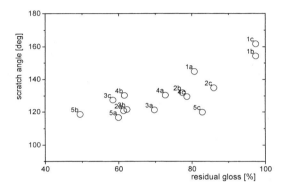

Figure 7. Scratch angle versus residual gloss as obtained for the different coatings (labels s. Table 1).

Conclusions

From the nano scratch experiments new data characteristic for the scratch resistance of the coatings can be extracted. The reflow capability was quantified by introducing a reflow value R which ranges from zero (no reflow) to one (complete reflow (healing)). According to the universal hardness a scratch hardness H and a hydrodynamic resistance Z were introduced. After leveling above a relatively low threshold value of the normal force all three values R, H and Z remain almost constant during the scratch test, even in the presence of cracks. The correlation between the NST data and the gloss measurements of the multi scratched coatings is not yet satisfying and is subject to further investigations. However, a good correlation was found between the scratch angle and the residual gloss.

Acknowledgement

The authors wish to express their thanks to the Arbeitsgemeinschaft industrieller Forschungsvereinigungen (AiF) „Otto von Guericke" e.V., (AiF). (Grant Nr. 12567) for the financial support.

References

[1] B. V. Gregorovich, K. Adamsons, L. Lin, *XXVIth International Conference in Organic Coatings* **2000**; Vouliagmeni (Athens)/Greece, 81.

[2] E. Klinke, M. Kordisch, G. Kunz, C. D. Eisenbach, *Farbe & Lack* **2002**, *108/4*, 54.

[3] U. Schulz, V. Wachtendorf, T. Klimmasch, *Congress Papers*, *6th Nürnberg Congress* **2001**, Vincentz Verlag, 259.

[4] U. Schulz, V. Wachtendorf, T. Klimmasch, P. Alers, *Progress in Organic Coatings* **2001**, *42*, 38.

[5] L. Lin, G. S. Blackman, R. R. Matheson, *Progress in Organic Coatings* **2000**, 40, 85.

[6] E. Klinke, M. Kordisch, G. Kunz, C.D. Eisenbach, Th. Klimmasch, *Farbe & Lack* **2002**, *108/2*, 52.

[7] J. L. Courter, E. A. Kamenetzky, *Congress Papers 5th Nürnberg Congress* **1999**; Vincentz Verlag, 351.

[8] E. Klinke, C. D. Eisenbach, *Congress Papers 6th Nürnberg Congress* **2001**, Vincentz Verlag, 249.

Macromol. Symp. 187, 861–871 (2002)

Quantitative Use of Ultraviolet Spectroscopy to Calculate the Effective Irradiation Dosage During Weathering

Stuart Croll*, Allen Skaja

Department of Polymers and Coatings, North Dakota State University, Fargo, North Dakota 58105, USA

Summary: The ultraviolet absorbance of a urethane coating showed typical yellowing that increased with exposure period. An effective dosage was calculated from the solar spectrum, the quantum yield for the degradation process and the ultraviolet absorption. Assuming a constant quantum yield, there is a clear acceleration of the absorption of damaging radiation because the ultraviolet absorption increases with exposure. This non-linear relationship offers possibilities on how to estimate a service lifetime. In addition, the yellowing can be analyzed as an "Urbach" tail which is usually attributed to structural disorder that introduces energy levels between the ground and excited electronic states.

Keywords: degradation, quantum yield, lifetime, Urbach, UV-vis spectroscopy

Introduction

There has been substantial work done on the chemical changes that occur during degradation of urethane coatings[1,2]. Much of this prior work determines the relative rate of appearance or disappearance of infrared bands in various polymers by comparison to another band, usually a carbon-hydrogen absorbance assuming that it is a measure of the amount of polymer material that remains. Polymers or additives may then be selected by choosing the structure that changes least[3].

The effect of ultraviolet radiation depends on the spectral distribution of the radiation, its intensity, how much is absorbed by the material, and whether any damage is done by a particular photon, i.e. the quantum yield of the degradation process. This research was done to quantify effective dosage and to determine its usefulness in estimating coating lifetimes. Effective dosage can be calculated[4]:

    CCC 1022-1360/00/$ 17.50+.50/0

$$d_{eff}(t) = \int_{\lambda_{min}}^{\lambda_{max}} E_0(\lambda,t)\left(1 - 10^{-A(\lambda,t)}\right)\phi(\lambda,t)\,d\lambda \qquad \dots 1$$

$d_{eff}(t)$ = effective dosage at a given time [W/m^2] (will vary and accumulate with exposure)
λ_{max}, λ_{min} = maximum and minimum photolytically effective wavelengths
$E_0(\lambda,t)$ = spectral irradiance, from sun or other source etc. [W/m^2/nm]
$A(\lambda,t)$ = coating absorption spectrum $\qquad\qquad \phi(\lambda)$ = quantum yield of interest
t = exposure time $\qquad\qquad\qquad\qquad\qquad \lambda$ = wavelength

One can calculate the effective damaging dosage as it varies with exposure time and which parts of the coating absorption spectrum are most important. It is a quantity that incorporates the input (irradiation), the material properties (absorbance) and an expression of the effect (quantum yield). An analogous expression could be generated for another form of damage, the corresponding material property and another controlling variable in the weather, e.g. humidity. The equation shows that it is necessary to know how the UV absorption and quantum efficiency of the crucial processes change.

Experimental Details

Polyester Synthesis

Table 1. Composition of polyester used in the model urethane polymer network.

Components	Molar ratio	Properties	
Neopentyl Glycol	3	Acid number	8.95
Trimethylol Propane	1	Hydroxyl number	166.5
Isophthalic Acid	1.714	Mol. Wt (No. avg.)	2899
Adipic Acid	1.286		

Trimethylol propane (TMP) (Aldrich), neopentyl glycol (NPG) (Aldrich), adipic acid (AA) (Aldrich), isophthalic acid (IPA) (Aldrich), and hexane diisocyanurate trimer (HDI trimer) Desmodur N3300(Bayer) were used directly as supplied. The polyester was made under nitrogen, synthesized with no catalyst: 1 hour at 160°C, 2 hour at 180°C, 2 hours at 190°C, 2 hours at 200°C, then held at 210°C until the acid value diminished below 10. The polyester had an average of 5 hydroxyls per chain, and the ratio of polyester:isocyanurate was 1:0.4.

By limiting the amount of cyanurate, this model system was crosslinked mainly through the primary hydroxyls on the polyester in order that it degrade within a practical period.

Sample Preparation

Transmission infrared absorption spectra were obtained using a Nicholet Magna 850 spectrometer. Substrates were polished silicon transmission windows[5]. These substrates are inert in the exposure environment used in this investigation as well as possessing suitable optical properties. Each of the substrates was marked so that it could be put into the spectrometer at the same position with respect to the infrared beam after each exposure period. Since the coatings were spin coated they were insensitive to their exact location in the beam and being very flat and parallel-faced they produced fringes in the infrared spectrum that could be used to monitor their thickness.

Spin coating was done at 33% solids, in a model WS-200-4NPP/RV from Laurell Technologies Corporation. For ultraviolet spectroscopy, this material was applied thinly on quartz slides (Chemglass Inc). All the samples were cured for 2 hours at 120 °C in a convection oven so that there should be very little contribution from moisture curing to form polyurea; there was no spectroscopic evidence for unreacted isocyanate after cure. These coated specimens were then exposed in intervals of a week.

Degradation Conditions

The exposure here used a QUV/SE chamber made by the Q-Panel Company. The exposure of the panels was controlled in cycles according to ASTM D4587-91; the maximum coating temperature was 60°C for 4 hours under radiation from the UVA fluorescent lamps following a period of 4 hours with condensing humidity at 55°C. The substrates used had a small diameter so all the duplicates could be accommodated in a single holder fabricated to occupy the space normally occupied by one standard 75 x 150 mm panel.

UV Spectroscopy

A Cary 500 spectrometer was used to obtain the spectra. The very small constant baseline at visible wavelengths was subtracted from each spectrum as a correction for scattering. No other scattering correction was attempted since there did not seem to be a constant slope component to the spectra and no features that would provoke Tyndall scattering were apparent in a separate study by atomic force microscopy.

Results and Discussion

A detailed discussion of the infrared analysis will be presented elsewhere. Change in fringe spacing was used as the measure of film thickness, so that the spectral changes could be corrected and quantitative deductions could be made. The carbonyl region, 1680 - 1730 cm[-1] increased as degradation proceeded, both in peak height and broadening of the peak at 1730 cm[-1]; as expected, the films were being oxidized during the accelerated exposure. In fact, over the 10 week exposure period the carbonyl peak area increased by approximately 25%. During that period, the CH stretch region around ~2890 cm-1 diminished by ~25% while the overall thickness of the films reduced to ~50% of the starting value.

UV Absorption

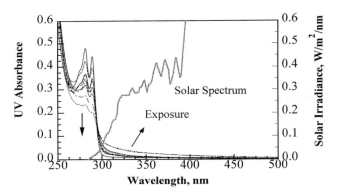

Figure 1. Ultraviolet absorption changes for a single coating in accelerated weathering, corrected for thickness changes. Part of the solar spectrum is reproduced for comparison.

Thickness changes measured from the infrared fringing was used to adjust the ultraviolet absorption so that it was function only of how the material was changing, see figure 1. Even without thickness correction, the absorption tail into the solar UV and visible region is present in the pristine material and increases with exposure. The peaks around 270 - 290 nm all diminish as the degradation increases. Here, the tail increased as degradation proceeded but without developing any appreciable structure. There are well documented cases of new chromophores generating obvious features[6, 7]; it is also commonly asserted that tails are the result of absorbing moieties being generated during degradation, but not always with any definite peaks being apparent[8, 9, 10]. Usually "yellowing," even with a new absorption peak includes a larger, underlying tail, so it is useful to seek mechanisms and descriptions for this behaviour that do not necessitate chemical changes producing specifically yellow chromophores. Predicting the lifetime of a coating polymer requires knowledge of not only chemical changes but any other processes that change the ultraviolet absorbance. The carbonyl region in infrared spectra taken during degradation also showed a broadening which is consistent with the tailing in the ultraviolet region, since the peaks around 260 - 290 nm are usually attributed to carbonyl n-π* transitions and are often seen in polyester-urethanes or acrylic polymers[11].

These samples were thin films, suitable for near ultraviolet transmission, and were visibly eroding but not changing colour perceptibly. Often, if a polymer yellows obviously, then the change of colour is used as a criterion of failure[9, 12]. It may be that the model polyester-urethane used here produced specific yellow moieties, but they were not concentrated enough to be perceptible.

Ultraviolet Absorption: Tailing

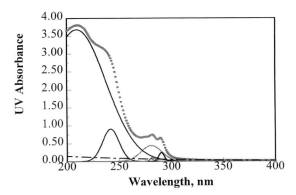

Figure 2. An example (eight weeks exposure) of the Gaussian peaks and exponential tail in the ultraviolet absorption of the polyester-urethane.

One can look elsewhere for an origin of the tailing phenomenon. Figure 2 shows an example spectrum, from a film that had been degrading for eight weeks. The figure shows the experimental data and Gaussian peaks, together with an exponential tail, that form an excellent fit to the data (the square of the regression coefficient > 0.9999). In order to match the data well, the tail had to be included when using Gaussian peak shapes. No reasonable choice of peak shapes could include the tail, although it might be part of a much larger peak that is centred at much shorter wavelengths.

The largest peak recorded was centred at ~210 nm. This peak has a much higher extinction coefficient and occurs at a wavelength where π-π* transitions, including those of carbonyl groups, are common[13]. Only the isophthalic acid component of the polyester had these peaks which also indicates that they are probably carbonyl.

The π-π* peak is so characteristically large that its own tail extends well into the visible, further than the smaller n-π* peaks. If one speculates that the tail originates from a larger, shorter-wavelength peak, then the speculation might include that only a minor broadening or shifting of that peak accounts for the increase in the tail as degradation proceeds. For energies below that corresponding to the central peak, there is an absorption tail in many materials that varies exponentially with energy, i.e. an "Urbach" tail [14]:

$$\alpha(\lambda) = \alpha_0' \cdot \exp\left[\frac{hc}{\lambda E_0} - \frac{E'_{bg}}{E_0} \right] \qquad \ldots 2$$

E_0 = construed as an energy that represents the width of the tail of localized states in a band gap

E'_{bg} = electronic band gap energy $\qquad\qquad h$ = Planck's constant

λ = wavelength $\qquad\qquad\qquad\qquad\qquad c$ = velocity of light

This equation was applied to the ultraviolet absorption spectra obtained from the polyester-urethane polymer studied here, figure 1. The exponential slopes appear not to vary with exposure period; linear regression gave a characteristic energy of ~ 0.66 eV, which is consistent with literature values for organic materials[15].

Usually, this exponential tail is attributed to thermal or structural disorder that provides electronic states that tail into the band gap[16]. Urbach found this exponential tail behavior in inorganic halides but it has since been found for many materials including glasses and amorphous semiconductors. The extended tail is a measure of a more complicated electronic density of states in which structural or compositional disorder perturb the electronic configurations. In polymers one would consider the transition between the highest occupied molecular orbital and the lowest unoccupied molecular orbital instead of a valence-conduction band gap. In the studies where this relationship, equation 2, has been used, E_0 is construed as an energy that represents the width of the tail of localized states in the band gap and must be a fraction of that band gap. This value of 0.66 eV is much less than the typical band gap in an insulator.

Ultraviolet absorbance tails examined here apparently had the same value of E_0, so the increased UV-visible yellowing cannot be explained by changes in E_0. Thus the changes must instead come from elsewhere, for example the band gap term, E'_{bg}. One could chose a peak wavelength, 210 nm, and calculate the corresponding transition energy, 5.9 eV, or the deconvoluted peak with a centre at 242 nm (see figure 2) that corresponds to an energy of 5.1 eV. However, inspection of these data did not reveal a trend in peak position with degradation interval. This may be because the spectra were obtained with polymer films thick enough to obtain accurate values near the visible region, and they had very high absorbance centred around 210 nm, so the detector's sensitivity may be limited.

An alternative approach is to employ "Tauc" gap analysis[17], an expression of the band gap used for amorphous semi-conductor materials. To use the Tauc approach, the square root of the absorbance divided by the wavelength is plotted against the reciprocal of the wavelength and the linear part is extrapolated to the abcissa to calculate the band gap [figure 3]. Here the value was found to be 4.8 eV for the unexposed polymer, diminishing to 4.3 eV as the urethane degraded. "Tauc gap" energy values that appeared to diminish a little with exposure would make the tailing absorbance increase in equation 3 as required by the experimental results. The magnitude of E_{bg} is typical of other polymers[18, 19] and somewhat lower, naturally, than the energies deduced from the peak centres. The major linear part in the Tauc plots occurred at wavelengths around 250 nm so it is the gap derived from the resultant of the peak ~242 nm and the stronger absorbance that peaks around 210 nm. The arguments presented above suggest that the tailing cannot be associated with the small n-π* peaks, so the extrapolation was not done to deduce a "Tauc gap" corresponding to those peaks.

An advantage here is that the "Tauc" gap is derived from the long wavelength side of the π-π* peak and so, again, may be connected to the processes governing the yellowing. Since the Tauc gap apparently moves slightly to lower energies as the degradation proceeds, it means that the peaks may be moving or spreading to lower energies and may thus contribute to the increased tailing. It is characteristic of a π-π* transition to shift to longer wavelengths as its "solvent" environment becomes more polar which is consistent with a defining electronic transition, E'_{bg}, to shift to lower energies as a polymer oxidises.

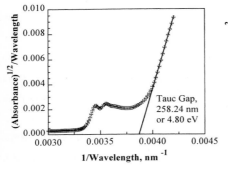

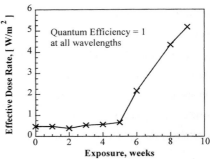

Figure 3. Example of "Tauc" plot of the ultraviolet absorption for unexposed polyester urethane.

Figure 4. Effective Dose rate calculated for exposure to the solar spectrum with a constant quantum efficiency.

Explanations of yellowing that require development of new chromophores are frequently allied to infrared analysis of degradation chemistry [20]. Such analyses find infrared bands that attributable to species developed as the degradation develops. These new species then produce smooth featureless absorption at the blue end of the visible absorption spectrum so that the material yellows[21]. The ideas presented here would account for such a smooth tailing, or yellowing, without the formation of specific chromophores.

UV Absorbance: Effective Dosage

In order to calculate the effective dosage, one needs the quantum yield for the degradation process. The calculation was carried out using the UV absorption (figure 1) and the terrestrial solar spectrum but by setting the quantum efficiency to unity at all wavelengths and exposure periods since it is, at present, unknown, but the subject of research. This calculation showed a greatly increasing dosage rate with exposure (figure 4).

Clearly, the nature of the quantum yield has a crucial rôle in determining the damaging UV dosage. There is a clear acceleration of damaging dosage. One could choose a lifetime corresponding to the start of the acceleration ~ 4 weeks in this case, or one might choose a value that represents the end, as the damage rate increases greatly, ~ 10 weeks here. This is somewhat analogous to choosing a "yield" point *versus* tensile strength in mechanical properties.

A crucial part of ultraviolet absorbance lies within the solar spectral range and is increased by degradation processes leading to further energy absorption and an accelerating rate of degradation. In such cases, the response of the system is to change its response so that it responds to a greater degree (positive feedback). A simple algebraic model of such feedback will be presented elsewhere.

The effective dosage calculation can be applied in determining the importance of the features in the UV spectrum separately. An example of the relative effect of each feature, from figure 2, is given in table 2 when it is used in equation 1 with the solar spectrum and suitable spectra for accelerated weathering chambers.

Table 2. Relative contribution to Effective Dosage of each feature in the UV absorbance of polyester-urethane film. Example chosen was exposed for eight weeks in the QUV-340 chamber.

	Solar	Xenon Arc with Borosilicate filter	Fluorescent UV-340
Whole Spectrum	1	1	1
Peak at 210 nm only	5×10^{-4}	6×10^{-4}	0.0012
Peak at 242 nm only	7×10^{-11}	3×10^{-10}	1.2×10^{-12}
Peak at 281 nm only	1.1×10^{-3}	1.6×10^{-3}	0.0012
Peak at 282 nm only	1.5×10^{-5}	5×10^{-5}	6×10^{-8}
Peak at 290 nm only	3.3×10^{-4}	5×10^{-4}	7×10^{-5}
Exponential tail only	~1	0.995	~1

Only the exponential tail contributes significantly to the calculation of the effective dosage, using a constant quantum yield. However, complete knowledge of the quantum yield is clearly very important in knowing the importance of a feature for the effective dosage of a material. One cannot discount the impact of a peak because its centre lies outside the irradiation spectrum cut-off. The table shows that the large peak at ~210 nm contributes as much as the smaller peaks closer to the solar region. This calculation also demonstrates that the borosilicate filtered Xenon arc spectrum is balanced closer to the solar spectrum than is the fluorescent UV-340, spectrum.

Conclusions

Evidently, if the lifetime of a coating depends on the amount of UV radiation that it absorbs, then the short wavelength peaks cannot be ignored merely because their centre occurs outside the solar spectrum range. Equally, "yellowing" does not necessarily require chromophores centred in the yellow part of the visible spectrum, it could easily be tailing from a peak that is further into the UV.

Yellowing is clearly a major contributor to the effective radiation dosage calculation here but detailed knowledge of quantum yield is very important in determining how much the individual absorption features contribute and will be investigated further. A contribution to "yellowing" seems to come from a broadening and diversification in the electronic transitions associated with the polymer. Further chemical changes may be necessary to reach the point of creating visible colour.

An exponential form fits the UV tailing well and corresponds to 0.66 eV, for the "Urbach" energy, that characterizes the smearing of energy levels within the electronic transition corresponding to the main ultraviolet absorption. Apparently, this energy parameter does not change with exposure in this material, but the increase in tailing may be better accounted for by the transition gap being reduced slightly as degradation continues. The approach of Tauc *et al.* may be useful as a quantification of the spreading of the larger UV peaks centred around 242 nm and 210 nm.

Effective dosage absorbed by the polyester-urethane coating accelerated as the coating degraded, similar to a "positive feedback" response. The form of the acceleration lends itself to some choices that could be used to define the lifetime of the coating. Either one could choose a point at which the adsorbed dosage accelerates noticeably, or one could extrapolate to a point when the dosage has increased vastly. If the causes for the non-linearity in dosage response of a polymer as it changes due to degradation can be understood, then a prediction of service lifetime is brought closer to realisation.

Depending on the choice of quantum yield function, effective dosage calculations demonstrate that different candidates for the source of ultraviolet radiation in accelerated exposure change the relative importance of the features in the absorption spectrum of the polymer. A Xenon arc lamp, filtered through borosilicate glass, more closely matched terrestrial sunlight than a fluorescent source.

Acknowledgements

The authors would like to thank Dr. Rasmussen of the Chemistry Department at NDSU for access to the UV spectrometer. This work was funded by the Air Force Office of Scientific Research under contract number F49620-99-1-0283.

References

[1]   C. Wilhelm, J-L. Gardette, *Polymer* **1998**, *39*, 5973
[2]   F. X. Perrin, M. Irigoyen, E. Aragon, J. L. Vernet, *Polym. Degrad. Stab*. **2000**, *70*, 469
[3]   J. L. Gerlock, C. A. Smith, V. A. Cooper, T. G. Dusbiber, W. H. Weber, *Polym. Degrad. Stab*. **1998**, *62*, 225
[4]   J. W. Martin, *Prog. Org. Coatings* **1993**, *23,* 49
[5]   L. G. J. Van Der Ven, P. J. A. Guerink, *J. Oil and Colour Chemists Assoc*. **1991**, *74,* 401
[6]   S. B. Maerov, *J. Polym. Sci: A* **1965**, *3*, 487
[7]   S. M. Cohen, R. H. Young, A. H. Markhart, *J. Polym. Sci: A-1* **1971**, *9*, 3263
[8]   A. Rivaton, F. Serre, J. L. Gardette, *Polym. Degrad. Stab*. **1998**, *62*, 127
[9]   N. S. Allen, M. Edge, M. Rodriguez, C. M. Liauw, E. Fontan, *Polym. Degrad. Stab*. **2001**, *71*, 1
[10] P. K. Das, P. J. DesLauriers, D. R. Fahey, F. K. Wood, F. J. Cornforth, *Polym. Degrad. Stab*. **1995**, *48*, 11
[11] D. R. Bauer, M. C. Paputa Peck, R. O. Carter III, *J. Coatings Technol*. **1987**, *59,* 103
[12] R. S. Davidson, *J. Photochemistry and Photobiology B* **1996**, *33*, 3
[13] H. H. Jaffé, Milton Orchin, *"Theory and Applications of Ultraviolet Spectroscopy"*, John Wiley and Sons Inc, New York, 1965
[14] S. John, M. Y. Chou, M. H. Cohen, C. M. Soukoulis, *Phys. Rev. B* **1988**, *37*, 6963
[15] O. C. Mullins, Y. Zhu, *App. Spectroscopy* **1992**, *46*, 354.
[16] S. K. O'Leary, S. R. Johnson, P. K. Lim, *J. Appl. Phys*. **1997**, *82*, 3334.
[17] J. Tauc, R. Grigorovich, V. Vancu,. *Phys. Stat. Sol*. **1966**, *15*, 627
[18] R. Mishra, S. P. Tripathy, D. Sinha, K. K. Dwivedi, S. Ghosh, D. T. Khathing, M. Müller, D. Fink, W. H. Chung, *Nucl. Inst. Meth. Phys. Res. B*. **2000**, *168*, 59
[19] S. Gupta, D. Choudhary, A.Sarma,. *J. Polym. Sci: B*. **2000**, *38*, 1589
[20] N. S. Allen, P. J. Robinson, N. J. White, D. W. Swales, *Polym. Degrad. Stab*. **1987**, *19*, 147
[21] A. Rivaton, F. Serre, J. L. Gardette, *Polym. Degrad. Stab*. **1998**, *62*, 127

Macromol. Symp. 187, 873–882 (2002)

Thermo-Analytical Characterization of Pigmented Waterborne Basecoats

Maren Buhk*, Willi Schlesing, Werner Bosch

DuPont Performance Coatings, Christbusch 25, 42285 Wuppertal, Germany

Summary: Modern waterborne basecoats contain a high degree of pigments to fulfill the wanted high brilliance and optical effects. Different kinds of pigment systems (color, pearl-effect, aluminum flakes) in a model binder matrix were investigated by Dynamic Mechanical Analysis (DMA) measurements at different pigment volume concentrations (PVC). To get more detailed information concerning the pigment binder interaction partially the binder system and the surface modification of pigments were varied. The results were correlated to technological tests (stone chip, adhesion). It can be shown, that special effects in waterborne basecoats like influences of resins or pigment-binder interaction can be investigated by means of DMA in a quite effective way. Thus it is possible to optimize one layer in a coating system in terms of its mechanical properties by reformulating the binder matrix and by a suitable surface treatment of the pigments.

Keywords: Adhesion, Dynamic Mechanical Analysis (DMA), Coatings, Pigment-Binder-Interaction, Stone chip resistance

Introduction

Dynamic Mechanical Analysis (DMA) is an important tool in characterizing polymeric materials. Visco-elastic properties, e.g. storage and loss modulus, the crosslinking density and the glass transition temperature are important values to develop coatings. For example, DMA investigations of waterborne basecoats yield information, which are necessary to optimize the binder resins with respect to the mechanical properties of coatings in a most effective way [1]. For a long time, the Dynamic Mechanical Analysis (DMA) has been used for the characterization of coating systems [2,3]. High technological demands are often made on coatings. For example the stone chip resistance and adhesion are recurring valuation criteria for the quality of a coating. Therefore different test methods are nowadays established, and the results of them are depending all the influences of the used coating layers. In pigmented binder systems, the technological characteristics are determined by the different resins, the level of pigmentation and the pigment-resin interactions. These influences can be investigated exclusively within one coating layer by means of DMA [4].

CCC 1022-1360/00/$ 17.50+.50/0

Materials and testing methods

The used materials, the test methods and test conditions were listed in Table 1.

Table 1. Used materials and test methods for investigations on pigment – binder interactions

Coatings	Waterborne basecoats		Binder types and binder ratios constant also with different pigmentations	
Coating system	For technological tests: Electro coated bonder panels with filler, basecoat and 2K clearcoat		- Original composition EC, 22-24 µm, 25' 175°C Filler, 35-40 µm, 28' 165°C Basecoat, 12-14 µm, 5' 80°C Clearcoat, 40-45 µm, 30' 145°C - Repair composition Secondary coating onto the original composition: basecoat and clearcoat	
	For dynamic mechanical tests: free basecoat films on PP panels		Basecoat, 55-60 µm, 5' 80°C + 30' 145°C	
Pigments	Type	Named	Particle size [µm]	Density [g/cm³]
	Red organic color pigment (Pigment Red 254, Diketo-pyrrolo-pyrrol)	**Color 1**	< 1	1.6
	Chromium oxid treated mica pigment	**Pearl 1**	11	3.6
	Cerous hydroxid treated mica pigment	**Pearl 2**	11	3.6
	Phosphate treated aluminum pigment	**Alu 1**	17	1.5
	Hydrosilicon treated aluminum pigment	**Alu 2**	17	1.5
Tests	Cross cut adhesion		DIN ISO EN 2409	
	Stone chip resistance		VDA (confederation of German automobile industry), test sheet 621-427, method B	
	DMA		Perkin Elmer DMA 7, 1 Hz, - 50 °C – 180 °C, 10 K/min, 10 mm x 4 mm x 0,06 mm, free films	

Thermal and technogical characterization

All investigated basecoats were formulated using the same binder composition. The technical properties of the coatings (Table 2) show that the different pigment classes, as a function of their pigment volume concentrations (PVC), lead to different property profiles. As can be seen, the repair composition (r.c.) is more sensitive to changes of the stone chip resistance with increasing PVC than the original composition (o.c.) containing only one basecoat layer. Coatings pigmented with effect pigment flakes (pearl 1, alu 1) show already at low PVC worse properties than coatings with the substantially fine distributed organic color 1 pigment.

Table 2. Technological properties of different pigmented basecoats

Pigment type	PVC	Cross cut adhesion		Stone chip resistance	
		o.c.	r.c.	o.c.	r.c.
	[%]	[c.v.]	[c.v.]	[c.v.]	[c.v.]
Color 1	0	0-1	0-1	2	2
	8	0-1	0-1	2	2
	16	0-1	0-1	2	3
	22	0-1	0-1	2	4
Pearl 1	0	0-1	0-1	2	3
	2	0-1	0-1	2	3
	4	0-1	0-1	2	3
	8	0-1	0-1	2	3
	12	0-1	0-1	2	3
	15	0-1	0-1	2	3
Alu 1	0	0-1	0-1	2	3
	3	0-1	0-1	2	3
	5	0-1	0-1	2	4
	7	0-1	0-1	2	4
	9	1	2	2-3	5
	13	3	3	3	5

o.c.: original composition, r.c.: repair composition, c.v.: characteristic values according to DIN, VDA, respectively

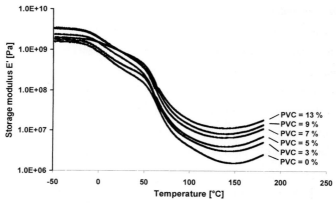

Figure 1. DMA curves of a basecoat pigmented with alu 1 at different PVC

In Figure 1 are shown the temperature courses of the storage moduli E' of basecoats with different high contents of alu 1. The shapes of the curves do not differ generally from each other. Similar curves were also observed with color or effect pigmented coatings (Figure 2). That means, that the course of differently pigmented coatings are essentially affected by the composition of the binder. As a direct consequence, that means, that the binder composition of basecoats can be investigated by means of DMA independent of their pigmentation.

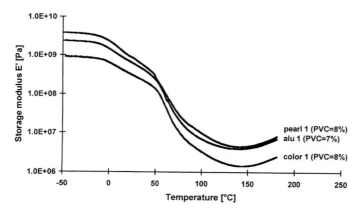

Figure 2. DMA curves of different pigmented paints at similar PVC

From these DMA investigations it could be concluded, that with the exception of the color pigmented coatings, the level of the minimum of E' (E'min) above the glass transition temperatures (Tg) rises significantly with increasing pigment content. This increasing stiffness can be explained theoretically by a pigment-resin-interaction or a pigment-pigment-interaction. The pigment-pigment-interaction seems improbable as a cause, since already with low increase of the PVC a noticeable rise of E'min occurs.

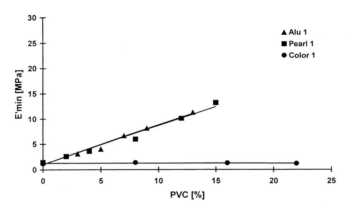

Figure 3. E'min vs PVC (1st measurement)

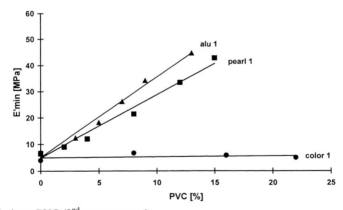

Figure 4. E'min vs PVC (2nd measurement)

In order to investigate possible pigment-resin-interactions in more detail, after the first DMA measurement the films were kept five minutes at 180 °C and measured again. Figure 3 and 4 show the dependence of E'min on the PVC of color 1, pearl 1 and alu 1 observed within the 1st and 2nd measurement. In both figures one recognizes, that with the effect pigmented systems E'min rises with increasing PVC. In addition, the courses of E'min in the 2nd measurement of the effect pigmented coatings differ noticeably from each other.

The glass transition temperature (Tg) can be taken as the temperature of the maximum of the storage modulus E" from DMA experiments (not shown here). The Tg of samples with various pigmentation heights differ neither within the 1st nor the 2nd measurements from each other. Between the 1st and 2nd measurements the Tg increases by about 8 °C. This increase of Tg occurs also with the non-pigmented samples, i.e. it is caused by a post curing of the resin system. This means that no change of the resin matrix is achieved by the differently heights of PVC.

The increase of E'min can be explained only by means of a chemical reaction of the pigment surface with components of the resin or by an adsorption of the resins onto the pigment surface. Both lead to a decreased polymer segment mobility within a thin boundary layer. This can cause an increase of the storage modulus above Tg, already observed in other systems [5, 6].

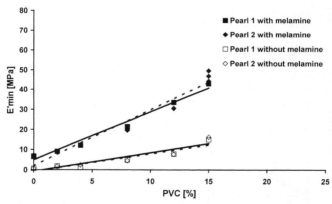

Figure 5. E'min vs PVC of pearl 1 and pearl 2 each with and without melamine

In order to check whether interactions are present with the melamine resin as a coating component, effect pigmented coatings (pearl 1 and pearl 2) without melamine resin are compared with the coatings containing melamine (Figure 5): The increase of the E'min in the coatings without melamine is noticeably lower. Thus, it seems probably that the melamine resin represents the main interaction partner for the pigment surface.

It is pointed out in Figure 5 that the slopes of E'min are independent of the pigment surface treatment in each case with and without melamine resin. Particle size measurements and scanning electron microscopic investigations have shown that these differently treated pigments have the same particle size distributions and similar surface roughnesses. As discussed above in these systems, no pigment dependent resin reactions are realized (due to constant Tg independent of the PVC), so that the slope of E'min could be explained by physical pigment-resin-interactions[7] and/or by chemical reactions in a boundary layer between pigment and resins. The latter seems less probable, since the chemically different surface treatments (Table 1) obviously does not lead to different reactivities / different interactions in the resin boundary layer and thus not to different contributions to the slope of E'min. The comparable interactions can be supported by the technical properties of these coatings (Table 3). The property profiles of the coatings pigmented with pearl 1 and pearl 2 only differ within the reproducibility of the testing methods, i.e. not very noticeably from each other.

Table 3. Technological properties of coatings containing pearl 1 and pearl 2 with and without melamine (original composition)

	Pearl 1				Pearl 2			
	With melamine		Without melamine		With melamine		Without melamine	
PVC	Cross cut adhesion	Stone chip resistance	Cross cut adhesion	Stone chip resistance	Cross cut adhesion	Stone chip resistance	Cross cut adhesion	Stone chip resistance
[%]	[c.v.]	[c.v.]	[c.v.]	[c.v.]	[c.v.]	[c.v.]	[c.v.]	[c.v.]
0	0-1	2	0-1	2-3	0-1	1	0-1	2
2	0-1	2	0-1	2-3	0-1	1	0-1	2
4	0-1	2	0-1	2-3	0-1	1	0-1	2
8	0-1	2	0-1	2-3	0-1	1	0-1	2
12	0-1	2	0-1	2-3	0-1	1	0-1	2
15	0-1	2	0-1	2-3	0-1	1	0-1	2

c.v.: characteristic values according to DIN, VDA, respectively

Comparing the E'min-slopes of paints exclusively pigmented with pearl 1, pearl 2, alu 1 and alu 2 (Figure 6) the DMA data point to a stronger pigment-resin-interaction in alu 2, while the other seemed to be comparable. This stronger interaction is supported by the technological test results (Table 4). With the original compositions containing alu 1 a noticeable adhesion loss occurs at higher pigment contents, while with alu 2 the values remain constantly good. This adhesion improving effect is even more noticeable in the repair composition.

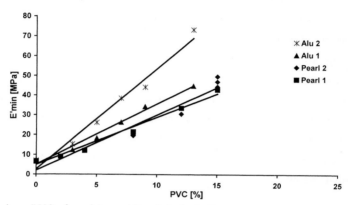

Figure 6. E'min vs PVC of pearl 1, pearl 2 and alu 1, alu 2

Table 4: Technological properties of coatings containg alu 1 and alu 2

	Alu 1				Alu 2			
PVC	Cross cut adhesion		Stone chip resistance		Cross cut adhesion		Stone chip resistance	
	o.c.	r.c.	o.c.	r.c.	o.c.	r.c.	o.c.	r.c.
[%]	[c.v.]	[c.v.]	[c.v.]	[c.v.]	[c.v.]	[c.v.]	[c.v.]	[c.v.]
0	0-1	0-1	2	3	0-1	0-1	2	2
3	0-1	0-1	2	3	0-1	0-1	2	2
5	0-1	0-1	2	4	0-1	0-1	2	2
7	0-1	0-1	2	4	0-1	0-1	2	2
9	1	2	2-3	5	0-1	0-1	2	2
13	3	3	3	5	0-1	0-1	2	2

o.c.: original composition, r.c.: repair composition, c.v.: characteristic values according to DIN, VDA respectively

As shown in Figure 4, the slopes of the E'min differ between pearl 1 and alu 1, i. e. depending on the particle sizes (pearl 1: 11 µm; alu 1: 17 µm) and the pigment classes (pearl 1: mica; alu 1: aluminum). A surface modification of the aluminum pigments (Table 1) leads in this case to a noticeable change of the slope of E'min (Figure 6). This correlates with the different technological characteristics of the coating systems (Table 4). That a specific modification of pigment surfaces effects the technical properties of the coatings, was shown earlier for color[8,9] pigments and for aluminum[10] pigments. Contrary to the pearl pigmented systems investigated here, the aluminum pigments differ noticeably by their analytical values: the surface of alu 1 is acidic with an acid number of 1.2, while the surface of alu 2 is alkaline (amine value: 5.3). As discussed above and in [1], the melamine resin is the most reactive component in the binder system. So, also chemical reactions at the pigment surface with the binder matrix can take place apart from the preferential physical melamine-pigment-interactions. Using the values found, it is probable that the alkaline alu 2 pigments can lead to stronger boundary surface reactions compared with alu 1. Due to the basic pigment surface of alu 2, the typical acid catalysis for a melamine resin crosslinking is not given. Thus a chemical interaction of the melamine itself is not to be discussed. With the system alu 2, a stronger linking of the binder matrix to the pigment surface explains the stronger slope of E'min as well as the better technical properties of the coatings in sense of better cohesion of the basecoat.

Conclusion

It could be shown that the binder matrix of coatings can be characterized by means of DMA independent of their pigmentation. Both the kind and concentration of the pigmentation have an effect on the technical properties of the coating. Beyond these effects, the properties are affected by different pigment treatments. The pigment dependent differences in the technical properties are reflected in the DMA curves, which are obtained by means of a special test and evaluation method. A variation of the binder matrix as well as the pigment surfaces showed that specific pigment-resin-interactions dominate. In contrast to the tested mica treatments it became evident that for the aluminum pigments the technical properties and DMA results of the coatings are noticeable depending on the type of surface treatment. The stronger pigment-resin-interaction of a special aluminum pigment, which was expected from the DMA results, are confirmed by the coating properties in terms of better cohesion of the basecoat systems. As already pointed out[1], the melamine resin improves the cohesion in the binder matrix. Additionally it was shown here, that the melamine resin is a preferential physical interaction partner to pigments. It also represents a chemical reaction partner in the pigment-binder boundary layer for specifically modified pigment surfaces. Thus, the technical properties of basecoats can be optimized both by a specific reformulation of the binder matrix and by a suitable surface treatment of the pigment. The DMA method we developed, permits additional statements about specific pigment-resin-interactions and facilitates thereby a more effective paint development.

Acknowledgment

The authors would like to thank Ewa Zienska, Ralf Reiferth, Timo Wiesner and Bärbel Zouari for their active assistances with these investigations.

References

[1]   W. Bosch, W. Schlesing, M. Buhk, *European Coatings Journal*, **12/1999**, 60-69
[2]   W. Schlesing, *Farbe & Lack* **99** (1993), 918-923
[3]   W. Schlesing, M.Osterhold, H. Hustert, C Flosbach, *Farbe & Lack* **101** (1995), 277-280
[4]   W. Bosch, W. Schlesing, M. Buhk, *European Coatings Journal*, **10/2001**, 60-67
[5]   A. Zosel, *"Lack- und Polymerfilme"*, Vincentz, Hanover 1996
[6]   A. Zosel, *13. FATIPEC congress 1976*, congress book, 653 pp.
[7]   C. van der Poel, *Rheol. Acta* **1**, (1958), 198
[8]   P. Thomatzek, *Farbe & Lack* **104** (1998), 26-37
[9]   Th. Batzilla, A. Tulke, *Farbe & Lack* **106** (2000), 51-61
[10] E. Jaehne, A. Henke, H.-J. P. Adler, *Coatings,* **6/2000**, 218-222

Analysis of the Environmental Parameters for Aircraft Coatings

Olga Guseva*, Samuel Brunner, Peter Richner

Swiss Federal Laboratory for Materials Testing and Research, Überlandstrasse 129, CH-8600 Dübendorf, Switzerland

Summary: Temperature, UV radiation and pollutants play a major role in the degradation of aircraft coatings. The UV radiation at flying altitude is more intense than at ground level. An analysis of the radiation spectrum was used to choose the appropriate lamp type for artificial weathering in the laboratory. For the pollutants we examined the fact that the service life of aircraft coatings was considerably reduced in the years following the eruption of the volcano Pinatubo. It is shown that sulphuric-acid-aerosol droplets dispersed in the stratosphere were the main cause for that damage.

Keywords: aircraft coatings, degradation, environmental parameters, sulphuric acid aerosol, service life prediction

Introduction

Aircraft coatings have both protective and aesthetic functions. The paint has to shield the aircraft against corrosion and environmental effects thus making them safer and more durable. The aesthetical appearance is also very important because the aircrafts carry the company logos and colours to the airports throughout the world.

Aircraft coatings are subjected to rather severe environmental conditions. The temperature, UV radiation and pollutants are known to play a major role in degradation of the airspace coatings. The air temperature differences exceed 100 °C, ranging from - 60 °C (-76 °F) at a flying altitude of 10 km [1] to over 50 °C (104 °F) in airports located in tropical desert regions.[2] Moreover, it is known that the coating surface temperature is much higher than the air temperature, especially for dark colours. [3] Apart from the temperature and UV, wetness is also a known factor influencing the life of coatings. This paper, the first one to our knowledge, gives an understanding to the phenomena of the damages of the aircraft coatings in the years from 1992 to 1995. During this

CCC 1022-1360/00/$ 17.50+.50/0

period of time the majority of aircraft coatings had only half of the expected service life: instead of usual 4 to 5 years in service the coatings became aesthetically unacceptable and lost some of their protective function much faster. It was found that the cause of these damages was the eruption of volcano Pinatubo in 1991,[4] which had injected more than 20'000'000 tons of SO_2 into the stratosphere,[5] the atmospheric layer above the clouds, in which lowest level the commercial air traffic takes place.

The purpose of the paper is the evaluation of the environmental parameters and their ranges for estimation of the service life for aircraft coating which will be the subject of the following papers. In this paper we will focus our attention mostly on the UV radiation and pollutants.

Influence of UV radiation on aircraft coatings

Sunlight is a very important factor for the ageing of aircraft coatings. It is known that the ultraviolet part is the main cause of the photo-oxidative degradation of coatings. The UV exposure data are available from several sources. One of them which gives a worldwide overview is available from satellite recording.[6] Another source, UV-intensity broadband data measured on the ground are available from the database "Baseline Surface Radiation Network" at the sites placed all over the world,[7] as shown in Figure 1. The following UV data are available from this database: UV-A global, UV-B direct, UV-B global, UV-B diffuse and UV-B upward reflected. For analysis of the ageing of aircraft coatings the UV-A data are of most interest.

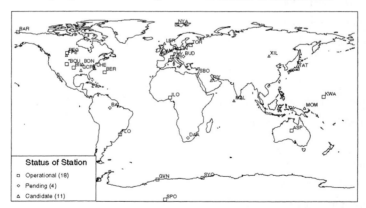

Figure 1. Stations of the Baseline Surface Radiation Network (denoted as letters)

The UV data for different wavelengths can be downloaded from Ref. 8 for some sites in the USA, Canada and New Zealand (Figure 2). These data are especially informative and useful for investigation of ageing processes in coatings when the wavelength sensitivity of a material is known.

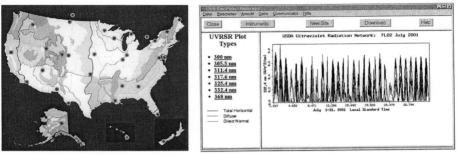

Figure 2. Test sites for which the UV intensity data are available (left). UV-intensity data for some wavelengths for one of the sites (Homestead in Florida) (right)

Aircraft coatings at a flying altitude of 10 km are subjected to more intensive radiation than the coating used on the ground: for example for a wavelength of 340 nm it is approximately 4 times more (see Figure 3).

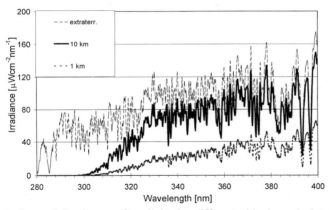

Figure 3. Spectral irradiance of sunlight for different altitudes calculated with the program MODTRAN 3.7 by the World Radiation Centre (WRC) in Davos in Switzerland[9] assuming the zenith angle of 45°, based on the measured data obtained for the site Payerne (altitude 500 m) in Switzerland in May 1998

For reliable accelerated tests for aircraft coatings one should use a suitable light source that could reproduce the sun radiation at the 10-km altitude (Figure 3, black line). Basic accelerated testing procedures have been reviewed by Wypych.[10] One of them uses high intensity sunlight focused on the sample using Fresnel mirrors with periodic exposure to moisture. The second type of accelerated test chambers employs UV fluorescent bulbs together with a dark condensing humidity cycle. The most common bulbs are the 340-UVA and 313-UVB. The third weathering chamber for coating evaluation is the one based on a xenon light source. All these procedures were discussed by Bauer[3] concerning the accelerated ageing of automotive coatings. He concluded that tests based on fluorescent bulbs at 313-UVB or on Xenon with the quartz-boro or boro-boro filters are inappropriate for predicting in-service performance. All of them contain UV light that is shorter than that observed outdoors: up to 275 nm, 280 nm and 290 nm for UVB, Xenon quartz-boro and Xenon boro-boro, respectively. Fluorescent 340 UVA bulbs are a better match to sunlight between 300 and 360 nm, however the lack of long wavelength UV and visible light has implications for some specific failure modes.[3]

For accelerated tests for aircraft coatings the choice of 340-UVA lamps seems to be the best as one can see from Figure 4: in the range 300-370 nm the intensity distribution of the 340-UVA bulb matches perfectly that at the 10 km altitude.

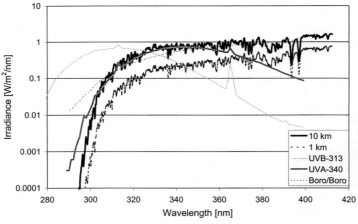

Figure 4. Spectral irradiance of sunlight at altitudes of 1 km and 10 km taken from Figure 3 compared to spectral irradiance of the 340-UVA and 313-UVB bulbs

Influence of sulphuric aerosol on aircraft coatings

For pollutants, we examine the fact that the service life of aircraft coatings was reduced by approximately 50% in the years from 1992 to 1995 following the eruption of the volcano Pinatubo in Philippines in June 1991. The eruption column reached 35 km in height and placed a giant cloud in the middle of the lower stratosphere extending to 1'100 km in diameter. During the eruption 15 to 20 megatons of SO_2 were injected into the stratosphere (Figure 5) and with the time dispersed all over the world reaching the maximal level of approximately 15 ppbv at around 26 km altitude on 21 September 1991.[11] Within the next four months the concentration of SO_2 in the stratosphere measured in the Microwave Limb Sounder (MLS) experiment on the Upper Atmosphere Research Satellite (UARS) had gradually decreased toward its usual value 0.01 - 0.05 ppbv[1] as shown in Figure 6.[12]

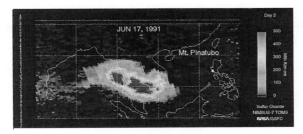

Figure 5. Distribution of SO_2 two days after eruption obtained using TOMS (Total Ozone Mapping Spectrometer)

UARS MLS Measurements of SO2 from the Pinatubo Volcano

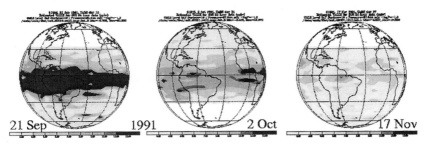

Figure 6. Change in distribution of SO_2 at 26 km from 21 September to 17 November 1991

It is known that SO_2 injected into the stratosphere is oxidised to sulphuric acid, which then forms aerosols through heteromolecular homogeneous nucleation of H_2SO_4 and H_2O vapours.[12] The detailed review on the widespread impact of the Mount Pinatubo eruption by considering the stratospheric injection and mass of the aerosol-generating sulphur gases (primarily SO_2), the transport of the eruption cloud and conversion of SO_2 to stratospheric sulphate aerosols, and the effects of this aerosol level on radiation, weather, and climate was done in Ref. 5. It was pointed out there that about 20 to 30 megaton of sulphate aerosol droplets produced by Pinatubo eruption caused the largest perturbation to the stratospheric aerosol layer since the eruption of Krakatau in 1883. The aerosol cloud spread rapidly around the Earth in about 3 weeks and attained global coverage by about 1 year after the eruption. Peak local midvisible optical depths[13] of up to 0.4 were measured in late 1992, and globally averaged values were about 0.1 to 0.15 for 2 years.

One set of satellite data that has the capability of providing information about stratospheric aerosols is obtained by the solar occultation satellite instruments, the Stratospheric Aerosol and Gas Experiment (SAGE) I and the SAGE II (see Figure 7).

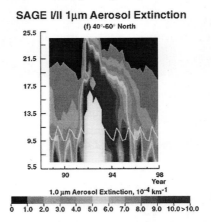

SAGE I/II 1μm Aerosol Extinction
(f) 40°-60° North

Figure 7. Aerosol concentration in the stratosphere derived from SAGE I and SAGE II.[14] The data products consist of seasonally averaged values for the 1020 nm aerosol extinction [15]

As follows from Figure 7, the aerosol content in the stratosphere at a flying altitude of about 10 km was still high in the years following the eruption up to 1997.

The question arises about the frequency of volcano eruptions that have influence on aircraft coatings. It is known that there are seldom periods without volcanic activity on the Earth.[16]

However, there is a big difference in the power of volcano eruptions. The biggest volcanoes that have had influence on the stratosphere since 1850 are shown in Figure 8.

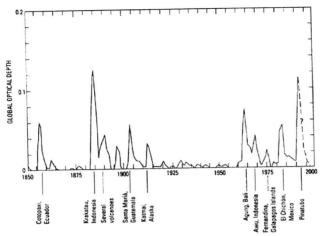

Figure 8. Estimated global optical depth at $\lambda = 0.55$ μm for the period 1850 to 1993, after Sato et al.[17] Most peaks are the result of instantaneous volcanic injections of SO_2 into the stratosphere and subsequent rapid formation and monotonic decline of sulphate aerosols.

It follows from Figure 8 that the last big volcano eruption comparable to Pinatubo was Krakatau eruption in August 1883, which also produced an aerosol veil of global extent. In fact, the maximum 20 to 30-Mt Pinatubo stratospheric aerosol loading may be not that much smaller than Krakatau's, variously estimated at between 30 and 50 Mt.[5]

Although large volcano eruptions which inject considerable amount of aerosol into the stratosphere are rather seldom events, they have dramatically influence on the aircraft coating during the following years. This aspect should be taken into account in the development of new formulations for aircraft coatings.

Quantitative description of environmental parameters for service life prediction for aircraft coatings

All the environmental parameters discussed above vary steadily depending on the day-night cycle, season, year etc. In order to use them for Service Life Prediction (SLP) for organic coatings, an averaging has to be done.

For the SLP for automotive paints Bauer used the following approach.[3,18-19] For the photo-oxidative degradation, he introduced the time unit $Photo-ox\_Extent$ [Exposure Time Equal to One Years in Florida], which is

$$Photo-ox\_Extent = C\int_0^1 I(t)e^{\frac{-E_a}{RT(t)}}dt,$$ (1)

Here the UV-light intensity $I(t)$ is multiplied with the Arrhenius term $e^{\frac{E_a}{RT(t)}}$ with the temperature T and integrated over time t. E_a is between 5 and 8 kcal/mol, the gas constant $R = 8.314\ JK^{-1}\ mol^{-1}$ and C is a constant with the unit ["Exposure Time Equal to One Years in Florida" / "Intensity" * "time"]. The influence of wetness was treated in Bauer's work,[18] with a weighting factor for each exposure site.

A similar approach was used by Jorgensen et al.[20] for the service life prediction for clear coat/coloured basecoat paint systems. The following generalised cumulative dosage model for the loss in performance, ΔP, was used:

$$\Delta P(t) = A\int_0^t [I_{UV}(\tau)]^n e^{-E/kT(\tau)}d\tau$$ (2)

where I_{UV} is the cumulative UV light dosage integrated over a bandwidth of 290-385 nm; A, n, E are fitting parameters, where E denotes the activation energy; k is the Boltzmann constant. Depending on the system, the values 3.8 - 8.4 kcal/mole-K for activation energy and 0.67 - 0.71 for n were obtained.

Based on the work above, the Arrhenius relationship should be used for the averaging of the temperature and a power law seems to be a good approximation for the UV radiation. Therefore, for the first estimation we are going to use the following averaging, where the whole time interval is represented by a sum of m subintervals for the temperature and the UV intensity, respectively,

$$\bar{T} = \sum_i^m (T(t_i)\cdot p(t_i))/\sum_i^m p(t_i) \qquad \bar{I} = \sum_i^m (I(t_i)\cdot p(t_i))/\sum_i^m p(t_i),$$ (3)

where the weighting function has the form

$$p(t) = I_{UV(at340nm)}^n \cdot e^{-[E_a/RT(t)]}\cdot \Delta t.$$ (4)

Calculations were made using the above approach with the averaging over one year. As input, monthly averaged temperature for some destination sites in the USA, Europe, Far East and Africa were taken. An assumption of flying time of 62 % was made. The value for the activation energy E_a in Eq. (4) was 153 kcal/mol, which was within the interval 129 - 180 kcal/mol (30.8 -43.1 kJ/mol) obtained by Allen et al.[21] but differed from the values of 5 - 8 kcal/mol used by Bauer[3] and Jorgensen et al.[20] The parameter n in Eq. (4) was 0.6, which was close to that used by Jorgensen et al.[20] but differed from 1, assumed by Bauer[3]. Both values were obtained as fitting coefficients for a stress-life relationship in accelerated weathering test for a reference polyurethane coating. In the following papers we will describe the procedure in details.

Using the above approach the estimated averaged value for the temperature is 19.3 °C, whereas that for the UV-parameter is 0.23 W/(m^2 nm). Both values will be used as service conditions or so-called "use level" values in the further calculations of the service life for organic coatings.

Conclusions

In this work the analysis of the environmental parameters relevant to aircraft coatings was done. The following parameters were considered: temperature, UV radiation and sulphuric aerosol. The first parameter, the air temperature, ranges from - 60 °C at flying altitude to over 50 °C in airports located in tropical desert regions.

The second parameter, UV radiation at flying altitude, is more intense than at ground level: for example, for a wavelength of 340 nm it is approximately 4 times more. From the analysis of the radiation spectrum at flying level follows that in the range 300-370 nm the intensity distribution of the 340-UVA bulb matches perfectly to that at the 10-km altitude. Therefore for the artificial weathering of the aircraft coatings in the laboratory these bulbs have to be used.

For the third parameter, pollutants, we examined the fact that the service life of aircraft coatings was reduced by approximately 50% in the years following the eruption of the volcano Pinatubo when more than 20 Megatons of SO_2 were dispersed into the stratosphere. In this work it is shown that the resulting $25-30*10^9$ kg H_2SO_4-aerosol droplets were a main cause for damage to aircraft coatings during the four years following the eruption.

Humidity as a stress parameter was excluded from considerations because the variation of this parameter across different aircrafts flying long-range routes is not so distinguishable.

All the analysis above was done to find out the quantitative description of the environmental parameters for use in the service life prediction for organic coatings and for construction of the weathering device that can simulate the ageing of the aircraft coatings in a proper way. The estimated averaged value for service-conditions or so-called "use level" value for the temperature is 19.3 °C, whereas that for the UV-parameter is 0.23 W/(m^2 nm). Both values will be in the further calculations of the service life for organic coatings.

Acknowledgements

This research was supported by the Swiss Commission for Technology and Innovation, Akzo Nobel, SR Technics and KLM. We are grateful to Dr. D. Brunner of the Institute for Atmospheric and Climate Science of the ETH Zürich and Dr. D. Schmucki of the World Radiation Center in Davos for helpful discussions.

References

[1] G. P. Brasseur, R. A. Cox, D. Hauglustaine, I. Isaksen, J. Lelieveld, D.H. Lister, R. Sausen, U. Schumann, A.Wahner, P. Wiesen, *Atmospheric environment* **1998**, *32*, 2329.
[2] The hightest temperature of 57.7 °C (136 °F) was measured in Al-Aziziyah, Libya, Africa at 112 m hight (367 ft), see Encyclopedia Britannica, chapter climate, http://preview.eb.com/eb/article?eu=109107&tocid=53252&query=climate%20extreme%20temperature.
[3] D. R Bauer, *Polymer Degradation and Stability* **2000**, *69*, 307.
[4] Communication with H.P.Roth, SR-Technics, Division Manager Dept. Materials Technology and Environment of the Zurich Airport.
[5] S. Self, J.-X. Zhao, R. E. Holasek, R. C. Torres, A. J.King, **1999**, http://pubs.usgs.gov/pinatubo/self/index.html.
[6] NASA site http://earthobservatory.nasa.gov/Observatory/datasets.html ->UV Exposure.
[7] http://bsrn.ethz.ch.
[8] http://uvb.nrel.colostate.edu.
[9] www.pmodwrc.ch.
[10] Wypych, G., *Handbook of Material Weathering*, Toronto, Canada, ChemTec Publishing 1995.
[11] http://mls.jpl.nasa.gov/joe/um_sci.html#overview.
[12] W. G. Read, L. Froidevaux, and J. W. Waters, *Geophys. Res. Lett.* **1993** *20*, 1299, http://mls.jpl.nasa.gov/joe/so2_pinatubo_21sep_2oct_16oct_17nov91.htmll.
[13] Optical depth is the mass of a given absorbing or emitting material lying in a vertical column of unit cross-sectional area and extending between two specified levels, McGraw-

Hill *Multimedia Encyclopaedia of Science and Technology* 1994.

[14] http://www-sage2.larc.nasa.gov/data/aerosol/tropospheric/

[15] Extinction is the reduction in the apparent brightness of a celestial object due to absorption and scattering of its light by the atmosphere and by interstellar dust; it is greater at low altitudes, McGraw-Hill *Multimedia Encyclopaedia of Science and Technology* 1994.

[16] T. E. Graedel, P. J. Crutzen, *Atmospheric Change. An Earth System Perspective*, New York, W.H. Freeman and Company 1993.

[17] M. Sato, J. E. Hansen, M. P. McCormick, J. B. Pollack, *Journal of Geophysical Research* **1993**, *98*, 22987.

[18] D. R.Bauer, *Journal of Coating Technology* **1997**, *69*, 85.

[19] D. R. Bauer, *Polymer Degradation and Stability* **2000**, *69*, 297.

[20] G. Jorgensen, C. Bingham, D. King, A. Lewandowski, J. Netter, K. Terwilliger, *NREL/* CP-520-28579, **2000**, http://www.doe.gov/bridge.

[21] N. S. Allen, H. Katami in *"Polymer durability, Advances in Chemistry Series 249"*, editors: R. L. Clough, N. C. Billingham , K. T: Gillen, American Chemical Society 1996, p. 537.

Influence of Raw Materials on the Formulation of Interior Emulsion Paints from the Point of View of EN 13300

Wolfgang Könner

KRONOS INTERNATIONAL, INC., Technical Service Department, Peschstrasse 5, D-51373 Leverkusen, Germany

Summary: The prior standard for interior emulsion paints, DIN 53778, was replaced by EN 13300 in April 2001. The new standard includes test standards ISO 11998 (Wet scrub resistance) and ISO 6504-3 (Contrast ratio). The new procedures differ greatly from the old test standards. This paper describes the influence of binders, film-forming agents, dispersants, extenders and pigments on the classification of interior emulsion paints in relation to EN 13300 compared to the prior standard.

The polymer dispersion tests showed that both the old and new test standards indicate similar influences. However, this is not the case with additives. Extenders also have varying effects that are linked to their hardness or other characteristics. The opacity to ISO 6504-3 is affected in various ways by the extenders and the selection of a suitable titanium dioxide pigment.

Keywords: new standards; indoor emulsion paints; wet scrub resistance; wash and scrub resistance; contrast ratio

Introduction

Just under 2.4 million metric tons of emulsion paint were manufactured in Europe in 2001, with Germany contributing roughly 830,000 tons to this total, including 580,000 tons of emulsion paint for indoor use.

Minimum requirements for interior emulsion paints were formerly specified in DIN 53778 Part 1. These essentially provided for data on the assessment of the wash and scrub resistance, and of the opacity as expressed by the contrast ratio. The test specifications were described in standards DIN 53778 Parts 2 and 3. This German Industrial Standard (Deutsche Industrie Norm) was replaced by European Standard EN 13300 in April 2001. The methods in the new test specifications, including ISO 11998 (Wet scrub resistance) and ISO 6504-3 (Contrast ratio), differ greatly from DIN 53778 Parts 2 and 3. This paper describes the influence of raw materials on the classification of interior emulsion paints in relation to EN 13300 compared to the prior standard.

1. Comparison of the test methods

1.1. Determination of wash and scrub resistance vs. wet scrub resistance (DIN 53778 Part 2 vs. ISO 11998)

DIN 53778 Part 2 specifies the application of a uniform coating with a dry film thickness of 100 ($\pm$ 5) μm. After 28 days of storage in a standard climate, the coating is scrubbed down to the substrate using a conditioned brush with a known abrasion factor. According to DIN 53778 Part 1, an interior emulsion paint was classified as wash-resistant if it withstood 1,000 scrub cycles, and as scrub-resistant if it withstood 5,000 scrub cycles, without being worn down to the substrate. One weak point was considered to be the visual estimation of when the

CCC 1022-1360/00/$ 17.50+.50/0

coating was worn through, which differed from person to person. Consequently, transfer in an ISO standard was rejected.

To determine the wet abrasion according to ISO 11998, the coating material is preferably applied with a 400 μm doctor blade. Again after 28 days in storage, the wetted film is scrubbed with an abrasive pad using a maximum of 200 scrubs. After drying, the difference in weight is determined and the wet abrasion calculated in μm based on the dry film density. The advantages of this method are the shorter operating time of the scrubbing device (roughly 5 minutes compared to over 2 hours for testing scrub-resistant coatings according to the prior standard) and the more precise determination that results from using a prescribed number of scrubs and calculating the weight difference.

EN 13300 divides paints into the following five classes based on their wet abrasion:

Classification	Wet abrasion	
Class 1	< 5 μm	at 200 scrubs
Class 2	5μm and < 20 μm	at 200 scrubs
Class 3	20 μm and < 70 μm	at 200 scrubs
Class 4	< 70 μm	at 40 scrubs
Class 5	70 μm	at 40 scrubs

1.2. Comparison of contrast ratio determination according to the old and new standards (DIN 53778 Part 3 and ISO 6504-3)

The weakness of DIN 53778-3 was that the determination of the contrast ratio with a spreading weight of 300 g/m² resulted in virtually no differentiation between the emulsion paints. In other words, virtually every emulsion paint achieved a contrast ratio of at least 98% at this high spreading weight and could thus be classified as opaque. For this reason, emulsion paints were applied to Morest cards with a doctor blade gap of 100 or 150 μm in order to allow differentiation. However, different solids contents, rheological properties and film application speeds resulted in different dry film thicknesses at the same blade gap, meaning that a comparison based on contrast ratio or opacity was subject to error. Therefore, a more precisely defined method was required to determine opacity.

The two methods described in ISO 6504-3 are based on the observation that the contrast ratio is a roughly linear function of the reciprocal spreading rate over a limited film thickness range. If the contrast ratio is determined for a given spreading rate, which EN 13300 requires manufacturers to state as an average value in m²/l, reproducible results can be obtained for multiple film thicknesses by interpolating the measured values [1].

Because the wet film thickness usually cannot be determined precisely, the mass per unit area of the respective coating is determined in the methods described and the corresponding wet film thickness calculated accordingly. The density and solids content of the coating must be known for this calculation. Although determining these values means additional laboratory work, being able to calculate the wet film thicknesses is a great advantage compared according to DIN 53778-3.

EN 13300 divides paints into the following four classes based on their contrast ratio:

Classification	Contrast ratio
Class 1	99.5%
Class 2	98.0% and < 99.5%
Class 3	95.0% and < 98.0%
Class 4	< 95.0%

2. Influence of binders

The binder plays a key role when formulating emulsion paints. It ensures that the pigments and extenders are bound together and that the emulsion paint adheres to the substrate. Thirty-seven polymer dispersions were tested in a series of trials according to DIN 53778 Part 2, ISO 11998 and other standards.

The formulation shown in Table 1 was used as the base formulation. The pigment volume concentration (PVC) was 78%. Paints containing polymer dispersions with a minimum film-forming temperature (MFT) higher than 5 °C had 1% Texanol as the film-forming agent (FFA). The solids content of the paints was held constant when switching polymer dispersions.

Table 1. Base formulation

Formulation	Parts by weight
Water	179.5
In-can preservative	2.0
Dispersant	0.5
Dispersant (45%)	3.0
Defoamer	2.0
Film-forming agent *	10.0
Cellulose paste (3%)	90.0
Titanium dioxide	220.0
CCP 0.3 μm	20.0
CCN 2.5 μm	118.0
CCN 5.0 μm	155.0
Talcum	70.0
Diatomaceous earth	20.0
Polymer dispersion (54.4%)	110.0
Total	1000.0
PVC (%)	78.0

*The film-forming agent was replaced with water in formulations whose binder displayed an MFT < 5 °C

CCP- calcium carbonate precipitated
CCN- calcium carbonate natural

Every binder was tested with a special emulsion-paint pigment and a general-purpose pigment. Both TiO_2 pigments were produced by the sulphate process and had different surface treatments and oil absorption values. The oil absorption of the special emulsion-paint pigment is 34 g/100 g pigment, while that of the general-purpose pigment is 18 g (Table 9, Pigments 1 and 4).

Figure 1 compares the wet abrasion test according to ISO 11998 with the wash and scrub resistance test according to DIN 53778 Part 2. Thirty-two paints were classified as scrub-resistant, 30 (94%) of which displayed a wet abrasion of < 20 μm. Two paints, each containing special emulsion-paint pigment 1, were assigned to EN 13300 Class 3 based on their wet abrasion values of 21.3 μm and 23.4 μm.

On the other hand, of the 32 wash-resistant formulations, two paints also containing special emulsion-paint pigment 1 were assigned to Class 2 with wet abrasion values of 19.6 μm and 18.6 μm. The remaining 30 wash-resistant formulations range between 20 μm and 70 μm wet abrasion and are assigned to Class 3 according to EN 13300. However, there are seven paints that do not fulfil the minimum requirement of 1,000 scrubs, but are still assigned to Class 3 according to EN 13300 because their abrasion is less than 70 μm. Three paints yielded abrasion values > 70 μm.

Triple determinations were carried out. The range of variation in determining the wet abrasion was ± 2.5 μm for the paints assigned to Class 2, and deteriorated to ± 5 μm for the paints in Class 3.

ISO 11998 and DIN 53778 Part 2 yielded matching results for 85% of the paints tested (Fig. 1), in that wash-resistant formulations can be assigned to Class 3 and scrub-resistant ones to Class 2 according to the new standard.

Analysis of the chemical binder composition showed that emulsion paints containing styrene acrylics have the lowest abrasion at an average of 17.6 μm, while those with pure acrylics have a mean wet abrasion of 24.1 μm and those with terpolymers based on vinyl acetate have a mean wet abrasion of 35.3 μm (Figure 2). Consequently, the good pigment-extender binding capacity of styrene acrylics is characterised by comparatively low wet abrasion values. Among the terpolymers based on vinyl acetate, a polymer dispersion gave outstanding results with being on the same level as the styrene acrylics and pure acrylics: One difference compared to the other vinyl acetales is its rather broad particle size distribution. This seems to result in a coating film of unusual compactness.

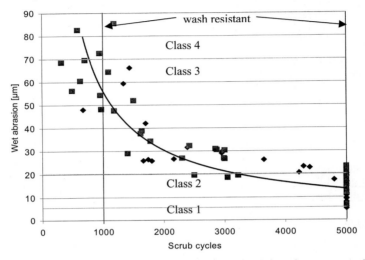

Figure 1. Correlation between of ISO 11998 and DIN 53778 based on a general-purpose pigment ♦ and a special emulsion-paint pigment　in different polymer dispersions

Twenty-one binders with an MFT below 5 °C were tested. They yielded a mean wet abrasion of 44.9 μm with the special emulsion-paint pigment and 27.5 μm with the general-purpose pigment. The lower wet abrasion values with the general-purpose pigment can be attributed to the lower oil absorption and the associated lower binder demand. The wet abrasion of the 16 binders with an MFT above 5 °C averaged 24.4 μm with the special emulsion-paint pigment and 13.7 μm with the general-purpose pigment.

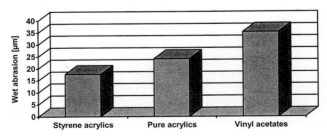

Figure 2. Mean wet abrasion with various polymer dispersions

Polymer dispersions with an MFT above 5 °C generally have a higher glass transition temperature and better mechanical properties. This results in harder films with lower wet abrasion (Figure 3).

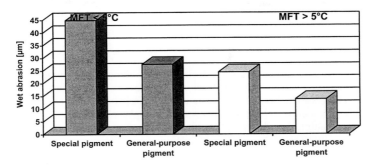

Figure 3. Mean wet abrasion of polymer dispersions with MFT < 5 °C and MFT > 5 °C

Because there is a growing trend towards VOC-free coatings in interior emulsion paints, the mechanical superiority of binders with an MFT above 5 °C can only be exploited if some VOC content is accepted.

3. Influence of film-forming agents

The lowest temperature at which a binder can form a film is decisive for the formulation of coatings. If the MFT of the polymer dispersion is higher than 5 °C, substances must be added to lower the MFT. This task is handled by film-forming agents (FFA). The higher the MFT, the greater the quantity of FFA that must be added.
A well spread styrene acrylic with a MFT of 20 °C was used as the polymer dispersion in the base formulation. Texanol, Lusolvan FBH and Kristallöl 30 were added at 1% each as FFAs.

3.1. Comparison of film-forming agents

The scrub resistance increases (DIN 53778) in the order Kristallöl 30, Lusolvan FBH and Texanol (Table 2). The selected products also displayed significantly different abrasion values in previous in-house tests conducted by KRONOS. Upon an exchange of FBH the performance of a coating can be improved by some 500 cycles.

Table 2. Test results for film-forming agents

	DIN 53778	ISO 11998	ISO 6504-3
Film-forming agent	Scrub cycles	Wet abrasion [μm]	Spreading rate [m²/l] for Class 1
Texanol	2,590	12.5	< 10.00
Kristallöl 30	2,060	14.0	< 10.00
Lusolvan FBH	2,300	14.4	< 10.00

The wet abrasion (ISO 11998) of the test formulation containing 1% FFA by weight varies only by 2 µm. That the difference is not significant unlike the results after the old standard may be due to the much shorter test time (5 minutes instead of more than 2 hours). The wash liquid thus leads to considerably less swelling of the coating film. As a result, the water sensitivity of the coating film is of less importance in the wet abrasion test according to ISO 11998 than in the scrub resistance test according to DIN 53778 Part 2. There is no evidence of an influence on the opacity according to ISO 6504-3.

4. Influence of dispersants

The mere objectives of dispersants are to facilitate the wetting of the pigments and extenders and to stabilize the state of dispersion achieved. How excessive dispersant can have a negative effect on wash and scrub resistance is shown in [2]. Other in-house tests conducted by KRONOS have indicated that especially sodium polyphosphate lowers this resistance.

The base formulation containing a styrene acrylic was used to test the influence of dispersants. The first paint contained a combination of sodium polyphosphate and the sodium salt of a polymeric carboxylic acid. One of the dispersants was used alone in each of the other two paints. The total concentration of the solids content of the dispersant was 0.3% by weight referred to the proportion of pigment and extender, or 0.185% w/w for the overall formulation.

Comparison of dispersants

The best abrasion values are obtained with the paint containing the sodium salt of a polymeric carboxylic acid. The abrasion values deteriorate as the quantity of sodium polyphosphate increases. As in the FFA test, however, the wet abrasion varies 2 µm only and no differentiation is possible. This can probably also be attributed to the shorter action time of the wash liquid. There is no evidence of an influence on the opacity according to ISO 6504-3.

Table 3. Test results for dispersants

	DIN 53778 Part 2	ISO 11998	ISO 6504-3
Dispersant	Scrub cycles	Wet abrasion [µm]	Spreading rate [m²/L] for Class 1
Carboxylic acid	2,500	13.0	< 10.00
Polyphosphate/carboxylic acid	2,290	13.8	< 10.00
Polyphosphate	2,100	15.1	< 10.00

5. Influence of extenders

In terms of quantity, extenders are the main constituents of interior emulsion paints. Thus, they also influence the optical properties of the paint directly - via their brightness and tone - and indirectly - via their effect on the distribution of the titanium dioxide pigment. They also affect the other properties, such as the ease of application, storage stability, mud cracking, sheen, wet abrasion and scrub resistance.

The styrene acrylic selected resulted in scrub-resistant paints, i.e. differentiation was not possible using DIN 53778 Part 2. For this reason, the PVC of the base formulation was increased from 78% to 80% by reducing the binder quantity, in order to allow differentiation.

5.1. Comparison of extenders
The extenders were compared to one another in the following groups, depending on their particle shape and size: fine extenders (with a separate investigation comparing different fine calcium carbonates), coarse extenders, coarse lamellar extenders and matting agents.

5.1.1. Influence of very fine extenders
The quantity of very fine extenders added was only 2%, as higher concentrations can lead to an increase in internal film tension and thus also to an increase in mud cracking. Furthermore, the high binder demand can result in a decrease in the number of scrubs. Precipitated calcium carbonate, precipitated aluminium silicate, natural kaolin and a calcined kaolin were tested.
Despite the low concentration used, the above tested extenders have a marked effect on the test results. The number of scrub cycles varies between 1,580 and 2,590 (Table 4). The good performance of precipitated calcium carbonate compared to the other extenders could be explained by its lower oil absorption and the associated lower binder demand. The emulsion paints with natural kaolin and those with calcined kaolin withstood fewer scrub cycles than the paints with precipitated aluminium silicate. This is a remarkable result because the oil absorption of the aluminium silicate is by approx. 3-fold of the kaolines.
Comparing the kaolins to one another shows that the formulation with the calcined kaolin achieves more scrub cycles than that containing natural kaolin. Since the wet abrasion varies by a maximum of only 2.5 µm, the results do not reveal a clear trend.

Table 4. Test results of very fine extenders

	Precipitated aluminium silicate	Precipitated calcium carbonate	Natural kaolin	Calcined kaolin
Particle shape	Roundish	Cigar-shaped	Fine-lamellar	Amorphous
Mean particle size [µm]	0.035	0.3	0.5	1.4
Oil absorption [g/100 g]	150	26	48	55
Whiteness	95	97	88	91
Brightness L*	97.5	97.4	97.4	97.3
Tone (white) b*	2.25	2.31	2.39	2.38
Tinting strength L *	57.1	56.2	57.2	56.3
Tone (grey) b*	-2.32	-2.59	-2.50	-2.54
Porosity Δ L*	5.7	5.1	4.4	5.6
Sheen (85° gloss)	1.6	1.4	1.5	1.3
Scrub cycles, DIN 53778	2,370	2,590	1,580	2,100
Wet abrasion [µm], ISO 11998	13.2	12.5	14.3	15.0
Spreading rate [m²/l] for Class 1, ISO 6504-3	< 10.80	< 10.00	< 10.00	< 9.30

According to ISO 6504-3, the reproducibility of the opacity test is 4%. Consequently, a formulation with precipitated aluminium silicate has an advantage over paint containing precipitated calcium carbonate or natural kaolin in terms of the spreading rate, which is 0.8 m²/l higher for Class 1. In contrast, a slight disadvantage results with calcined kaolin. This could be attributed to the calcination process, in which the natural kaolin loses its lamellar structure and becomes amorphous. Experience has shown that the distribution of titanium dioxide pigment in the coating film is somewhat less uniform as a result.

5.1.2. Influence of different fine calcium carbonates

The following calcium carbonates were tested separately in the base formulation: 11.8% limestone powder with a mean particle size of 0.7 μm, marble powder with 0.9 μm and 2.5 μm, and chalk with 3.0 μm. The emulsion paint containing chalk withstands the most scrub cycles, followed by that containing the coarser marble powder and that with limestone powder. The wet abrasion was not clearly different. It is conceivable that the chalk generates lubricious abraded material and the scrub brush therefore removes less of the coating film. This results in comparatively good abrasion values. In contrast, the coarse abrasive pad with the 135 g mount is much more abrasive than the scrub brush. The other calcium carbonates display higher wet abrasion in keeping with the low abrasion values.

Table 5. Test results of fine calcium carbonates (CCN)

	Limestone	Marble	Marble	Chalk
Particle shape	Trigonal rhombohedral	Trigonal rhombohedral	Trigonal rhombohedral	Amorphous
Mean particle size [μm]	0.7	0.9	2.5	3.0
Oil absorption [g/100 g]	20	20	19	17
Whiteness	93.5	95.0	94.5	84.5
Brightness L*	97.5	97.3	97.4	96.2
Tone (white) b*	2.48	2.17	2.31	3.07
Tinting strength L *	58.2	58.3	56.2	55.8
Tone (grey) b*	-2.53	-2.30	-2.59	-2.61
Porosity Δ L*	3.6	4.4	5.1	4.4
Sheen (85° gloss)	1.8	1.7	1.5	1.5
Scrub cycles, DIN 53778	1,950	1,690	2,590	3,120
Wet abrasion [μm], ISO 11998	13.4	14.4	12.5	13.2
Spreading rate [m²/l] for Class 1, ISO 6504-3	< 8.40	< 10.00	< 10.00	< 10.20

As regards the opacity according to ISO 6504-3, only the formulation containing limestone powder shows a somewhat poorer spreading rate, this probably being attributable to the relatively low porosity of the coating surface and the lower dry-hiding effect associated with it.

5.1.3. Influence of relatively coarse extenders

The following were tested separately as relatively coarse extenders: 15.5% marble powder with a mean particle diameter of 7 μm and 5 μm, and cristobalite with 5 μm. The paint with the coarser marble powder displayed the highest scrub values and the lowest wet abrasion due to its low oil absorption. It was followed by the 5 μm marble powder and, after a fairly large interval, by the cristobalite. Although the mean particle size of this marble powder and the cristobalite are the same, the upper particle size range of the cristobalite is lower. This is manifested in a higher oil absorption and thus explains the lower scrub value. However, the advantage of the 5 μm marble powder is no longer apparent when it comes to wet abrasion. It can be assumed that this is due to the greater hardness of the cristobalite (Mohs hardness 6-7; marble: Mohs hardness 3).

As regards the opacity according to ISO 6504-3, both 5 μm extenders are on the same level. However, the formulation with the 7 μm calcium carbonate has a much lower spreading rate. In addition to the poorer titanium dioxide dispersion, this can also be attributed to an effect best described by the term "optical window". The 7 μm extender has an upper particle size range of 30 μm. If two or more of these coarse extender grains are on top of one another they extend from the coating surface all the way to the substrate. Because there are virtually no differences in the refractive indices of the extenders and the polymer dispersion, light rays are not scattered at their boundary surfaces, but rather pass through them, as through a window. The result is lower opacity. On the other hand, the upper particle size range produces a somewhat rougher surface, so that the sheen can be reduced.

Table 6. Test results of relatively coarse extenders

	Cristobalite	CCN, marble	CCN, marble
Particle shape	Trigonal rhombohedral	Trigonal rhombohedral	Trigonal rhombohedral
Mean particle size [μm]	5	5	7
Oil absorption [g/100 g]	27	16	15
Whiteness	94	94	93.5
Brightness L*	97.4	97.4	97.3
Tone (white) b*	2.25	2.31	2.34
Tinting strength L *	56.9	56.2	55.6
Tone (grey) b*	-2.24	-2.59	-2.72
Porosity Δ L*	5.9	5.1	5.0
Sheen (85° gloss)	1.3	1.5	1.1
Scrub cycles, DIN 53778	990	2,590	2,750
Wet abrasion [μm], ISO 11998	12.7	12.5	9.6
Spreading rate [m²/l] for Class 1, ISO 6504-3	< 10.00	< 10.00	< 7.70

5.1.4. Coarse lamellar extenders

Talcum, mica and an intergrowth of mica, quartz and chlorite were tested as lamellar extenders at a proportion of 7% (w/w) referred to the base formulation. The number of scrub cycles increases and the wet abrasion decreases in the order mica, talcum and mineral intergrowth. While talcum and mica are relatively soft, with Mohs hardness values of 1 and 3, respectively, the mineral intergrowth has a higher overall hardness due to the quartz content (Mohs hardness: 7).

As a result of their brightness, the extenders influence the optical properties of the paints. The dark mica produces a brightness value in the emulsion paint that is a good unit lower than that of the talcum. This higher light absorption contributes to the elevated opacity of the paint containing mica, as determined according to ISO 6504-3. In contrast, the formulations containing talcum and the mineral intergrowth require lower spreading rates to achieve the highest opacity class.

Table 7. Test results of coarse lamellar extenders

	Talcum	Intergrowth	Mica
Particle shape	Coarse lamellar	Coarse lamellar	Coarse lamellar
Mean particle size [μm]	13	25	27
Oil absorption [g/100 g]	32	22	37
Whiteness	91.0	77.5	Not available
Brightness L*	97.4	96.9	96.3
Tone (white) b*	2.31	2.41	2.70
Tinting strength L *	56.2	55.5	56.2
Tone (grey) b*	-2.59	-2.69	-2.29
Porosity Δ L*	5.1	5.0	6.2
Sheen (85° gloss)	1.4	1.5	1.5
Scrub cycles, DIN 53778	2,590	4,630	750
Wet abrasion [μm], ISO 11998	12.5	7.3	22.9
Spreading rate [m²/l] for Class 1, ISO 6504-3	< 10.00	< 7.80	< 12.00

5.1.5. Influence of diatomaceous earth and cellulose fibre
On account of the irregular particle shape of diatomaceous earth and the fibrous structure of cellulose fibres, both have a matting effect. Diatomaceous earth is more effective in reducing sheen.

Table 8. Test results of diatomaceous earth and cellulose fibre

	Diatomaceous earth	Cellulose fibre
Particle shape	Chalk-like	Fibrous
Mean particle size [μm]	6	40 x 20
Oil absorption [g/100 g]	110	120
Whiteness	88	84
Brightness L*	97.4	97.3
Tone (white) b*	2.31	2.38
Tinting strength L *	56.2	56.0
Tone (grey) b*	-2.59	-2.68
Porosity Δ L*	5.1	4.0
Sheen (85° gloss)	1.5	1.8
Scrub cycles, DIN 53778	2,590	3,200
Wet abrasion [μm], ISO 11998	12.5	14.0
Spreading rate [m²/l] for Class 1, ISO 6504-3	< 10.00	< 6.40

The cellulose fibre leads to the higher scrub values. However, lower wet abrasion is achieved with diatomaceous earth. It cannot be ruled out that the soft cellulose fibres are virtually plucked out of the coating surface by the abrasive pad, whereas the scrub brush does not work the coating as hard. The decrease in opacity when using cellulose fibre is pronounced. This may be due to the fibre size, which could cause the "optical window" effect as experienced with the coarse calcium carbonates.

6. Influence of titanium dioxide pigments

No other raw material lends coatings such high opacity and tinting strength as titanium dioxide. This is due to the high refractive index of the white pigment and the great difference in refractive index compared to the polymer dispersions.

The five titanium dioxide pigments characterised in Table 9 were tested with four selected polymer dispersions in the base formulation with a PVC of 80%. Pigments 1 and 2 are highly surface-treated special emulsion-paint pigments, while TiO_2 grades 4 and 5 can be characterised as general-purpose pigments with less surface treatment. Pigment 3 is an intermediate grade, as indicated by the inorganic surface treatment and oil absorption.

Each of the four binders forms a film at temperatures below 5 °C and is thus suitable for VOC-free coatings.

Table 9. Characterisation of the titanium dioxide pigments

	Al_2O_3 [%]	SiO_2 [%]	ZrO_2 [%]	Oil absorp. [g/100 g]	Density [g/cm^3]	C [%]
Pigment 1	4.0	10.0	-	34	3.7	-
Pigment 2	4.4	9.4	-	41	3.6	-
Pigment 3	4.7	3.4	-	27	3.8	0.10
Pigment 4	3.1	-	0.4	18	4.1	0.19
Pigment 5	3.9	-	-	16	4.1	0.19

6.1. Analysis according to DIN 53778-2 and ISO 11998

The scrub resistance test according to DIN 53778 Part 2 and the wet abrasion test according to ISO 11998 yield matching results. As a rule, the paints with more scrub cycles also have lower wet abrasion (Figures 4 and 5). The best values are achieved with general-purpose pigments (paint-grade pigments) 4 and 5. Special emulsion-paint pigments 1 and 2 result in the lowest scrub values and the highest wet abrasion. Pigment 3 occupies the middle field among the styrene acrylics.

The polymer dispersions used have a very strong influence on the differentiability of the titanium dioxide pigments tested. The pigments tested with the vinyl acetate-ethylene polymer dispersions having a fairly broad particle size distribution (0.1 to 0.55 µm) display only slight differences of approximately 300 scrub cycles and 3 µm for wet abrasion. A much stronger pigment influence is observed with the styrene acrylics. Differences of up to 3,000 scrub cycles and up to 17 µm for wet abrasion are found between the special emulsion-paint pigments and the paint-grade pigments in this case.

The vinyl acetate-ethylene benefits pigments 1 and 2, while pigments 4 and 5 do not display any advantages at all in this formulation. In contrast, pigment 3 behaves indifferently. This binder achieves higher scrub values and lower wet abrasion with pigments 1 and 2 than the three styrene acrylics. The wash-resistant coatings (approx. 1,700 scrub cycles) display a wet abrasion of only approx. 19 µm and are thus assigned to Class 2 to EN 13300. In contrast, the styrene acrylics offer advantages in combination with general-purpose titanium dioxide pigments 4 and 5. These formulations achieve more scrub cycles than paints formulated with vinyl acetate-ethylene. Styrene acrylic 1 yields particularly positive results in terms of wet abrasion and the number of scrub cycles.

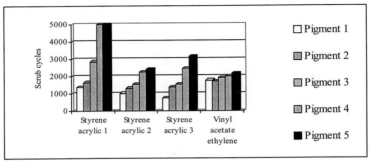

Figure 4. Wash and scrub resistance according to DIN 53778 Part 2

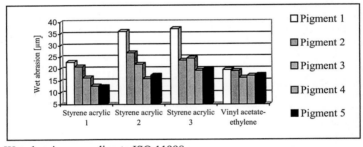

Figure 5. Wet abrasion according to ISO 11998

6.2. Analysis of opacity according to ISO 6504-3

The larger the surface area that can be coated opaquely with one litre of coating material, the more economic the coating is. Titanium dioxide pigments with a fairly high oil absorption value can be expected to achieve greater opacity due to the higher resulting porosity and the associated stronger dry-hiding effect. This is confirmed in some cases by the considerably higher spreading rates of the paints pigmented with pigments 1, 2 and 3 (Figure 6). Comparing general-purpose titanium dioxide pigments 4 and 5 shows that pigment 4, which was produced by the sulphate process, leads to a higher spreading rate in every paint tested. This is an often observed phenomenon which is commonly explained by the lower brightness and less pronounced bluish undertone of this kind of pigments. The following argument may explain the good results of the combination of the vinyl acetate-ethylene polymer dispersion with pigments 1, 2 and 3: the pigments are very highly surface-treated with Al_2O_3 and SiO_2 (less in pigment 3). In emulsion paints with a high extender content, this surface treatment acts like a "spacer" between the pigment particles and thus prevents excessively dense packing of the titanium dioxide particles, which would otherwise lead to a decline in the optical performance of the pigment. In conjunction with the broader particle size distribution (0.1 to 0.55 µm) of the vinyl acetate-ethylene polymer dispersion, it would appear that this results in a better packing density for the interior emulsion paint produced.

Pigment 3 particularly recommends itself as a good compromise between low wet abrasion values and high opacity.

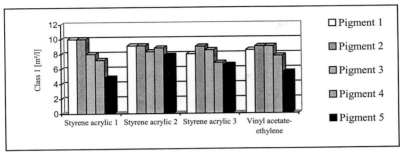

Figure 6. Spreading rate [m²/l] with different TiO₂ pigments

Summary

Thirty-seven polymer dispersions were tested in accordance with DIN 53778 Part 2, and ISO 11998. The results indicated that the old and new standards are comparable. Styrene acrylics displayed the lowest wet abrasion, followed by pure acrylics and terpolymers based on vinyl acetate. One dispersion in the latter group with a fairly broad particle size distribution had a particularly low wet abrasion value. Polymer dispersions with a minimum film-forming temperature (MFT) above 5 °C display lower wet abrasion than those with an MFT below 5 °C.

The results of tests with film-forming agents (FFA) and dispersants reveal distinct differences between the standards. The influence of raw materials on the scrub values determined by the old standard is much stronger than in the wet abrasion test according to the new standard. While there are only signs of a trend in the new standard, the water swellability and sensitivity of the raw materials in the coating hardly play a role any more due to the shorter action time of the wash liquid.

Extenders affect the results of tests according to EN 13300 in a variety of ways. The particle shape and particle size of extenders influence the dispersion of titanium dioxide pigments and thus play a part in the opacity according to ISO 6504-3, which is also affected by their brightness and tone, as well as dry-hiding. The higher the oil absorption or binder demand of the extenders, the greater the dry-hiding effect and thus the opacity caused by them, while the scrub and wet abrasion resistance deteriorate at the same time. Greater opacity can also be caused by stronger light absorption combined with lower extender brightness, as is the case with mica.

Relatively hard extenders apparently result in lower wet abrasion than the number of scrub cycles would suggest. This may be attributable to the greater abrasiveness of the abrasive pad, which is much harder than the scrub brush on fairly soft extenders in the coating film, such as chalk and cellulose fibres. The lowest wet abrasion value was achieved with a mineral intergrowth.

Different interior emulsion paint formulations are possible as regards the titanium dioxide pigment and the binder. In combination with styrene acrylics, general-purpose titanium dioxide pigments are the recommended choice because the low binder demand results in low wet abrasion values. Of the two general-purpose pigments, titanium dioxide 4 produced by the sulphate process displays definite advantages in terms of opacity according to ISO 6504-3. Another option is to combine a vinyl acetate-ethylene polymer dispersion having a fairly broad particle size distribution with a highly surface-treated pigment. The wet abrasion of this

908

combination is far lower than with the styrene acrylics tested. As the high binder demand of these pigments brings about a marked dry-hiding effect, very high opacity according to ISO 6504-3 is achieved in this way. When using this binder, however, the best results for wet abrasion and opacity are obtained with pigment 3, which is less surface-treated than pigments 1 and 2. Good compromises can also be achieved with this pigment in combination with the styrene acrylics.

References

[1] pr DIN ISO 6504-3, November 2001
[2] KRONOS Titandioxid in Dispersionsfarben, 1989

Time-Temperature Effects in Polymer Coatings for Corrosion Protection as Analyzed by EIS

Gordon Bierwagen*, Lingyun He and Dennis Tallman[1]

Department of Polymers & Coatings and Department of Chemistry[1] , North Dakota State University, Fargo, ND 58105   USA

Summary: More information is contained in Electrochemical Impedance Spectroscopy (EIS) results at steady-state vs. Temperature data sets than just the low frequency limit/polymer coating film resistance data most often cited. We have analyzed various EIS-Temperature data sets for several coating films in immersion and present the results considering the coating film dielectric constant vs. frequency and temperature. Water up-take can be analyzed by these methods and can be used to estimate the role of this process in the failure of corrosion protective coatings. For coating films that have been subjected to cyclic exposure, analysis of the dielectric constant vs. frequency data resulting from EIS data on these films indicate that cyclic exposure contributes significantly to 'physical aging' of the coating polymer.

Introduction

Barrier properties are a key factor in organic coating to protect the metal substrate from the corrosion. In general it means the ability of the intact organic coating to prevent corrosive species such as water, ion and other reactants to penetrate down to the coating layer. The quantitative measurement of these barrier properties is of interest of the coating scientist to evaluate the protection performance of the organic coating under corrosive environment. For corrosion of a metal substrate to occur, corrosive species such as water, oxygen and electrolyte must be present at metal/coating interface. Organic coatings provide much of their protection against corrosion by blocking the other species necessary for the corrosion process from the metal surface.

Electrochemical impedance spectroscopy (EIS) has been widely used in the characterization of the organic coating in exposure to a corrosion causing environment. The low frequency impedance modulus ($|Z|_{low\ f}$) has been considered as almost equivalent to the coating resistance in the stages of exposure of a coating/metal system prior to any significant metal

CCC 1022-1360/00/$ 17.50+.50/0

damage, and therefore one measure of the barrier properties of the film.[1] By the method, the changes in coatings during prolonged exposures can be monitored to evaluate the life time of the coating.[2] Moreover, through some assumption on the physical processes involved in the film degradation and corrosion, equivalent circuit fitting can be performed on EIS data so that one can calculate intrinsic properties of the coating such as the such as coating resistance, coating capacitance, charger transfer resistance and double layer capacitance.

Recently, our laboratory has begun to explore the interpretation of the EIS frequency data besides using solely the low frequency regime characterize to the barrier properties of coatings by examining EIS data with respect to coating polymer dielectric properties over the entire measurement frequency range. First, we will describe how dielectric constant measurement characterizes the barrier properties and the changes in the polymer during thermal cycling testing, second, we try to introduce how dielectric constant information can be extracted from EIS data.

The dielectric constant of a material can be represented by a complex quantity is given in the following equation:

1. $\qquad \varepsilon^* = \varepsilon' + j\,\varepsilon''$

where ε' is referred to the real or storage part of the dielectric permittivity, which measures the amount of the alignment of the dipoles and ε'' is referred to loss factor which represents the energy required to align the dipoles and move the ions. ε' is proportional to the capacitance and ε'' is proportional to the conductance. Tan δ is the ratio of ε'' to ε', and the ionic conductivity can be derived from the ε''.[3]

In the previous work, we have observed that the barrier properties have an abrupt decrease when the temperature was over T_g of the coating through thermal cycling EIS testing.[4] The relaxation process of the polymer reflects the mobility of the polymer chain, which is also highly dependent on the temperature, through the dielectric constants measurement under various the temperature, the relationship between the barrier properties and the relaxation process of the coating and the influence of the temperature can be well related.

Impedance data can be related to the dielectric constant[5] through the following equations. The complex impedance is given by

2. $\qquad Z = Z' + jZ''$

Using the relation between admitance and impedance as functions of frequency, and defining ω as the angular frequency and ε_0 as the dielectric constant of free space, one has the following equations for calculating the real and complex parts of the dielectric constant from the complex impedance as:

3. $$\varepsilon'(\omega) = \frac{-Z''(\omega)}{Z'(\omega)^2 + Z''(\omega)^2} \cdot \frac{1}{\omega\varepsilon_0}$$

4. $$\varepsilon''(\omega) = \frac{Z'(\omega)}{Z'(\omega)^2 + Z''(\omega)^2} \cdot \frac{1}{\omega\varepsilon_0}$$

These calculations of the real and imaginary part of the dielectric constant are most accurate in the high frequency regime because at the high frequency range, the impedance behavior is dominated by the capacitance values, and the influence from the ionic or electric conductance could be minimized. We will use them for the entire range of ω assuming the high resistance values

Experimental

Sample Preparation

Aluminum alloys 2024-T3 panels from Q-Panel Lab Products of size 7.62 x 7.62 x .864cm were used as the substrate. There were three coatings applied to the substrate, the first one examined was a BASF e-coat with plasma polymer pretreatment (e-coat) and the second coating studied was a $SrCrO_4$ pigmented epoxy-polyamide primer(PEPP), the detail of how to make coating samples could be referred in our previous paper[6]. The thirdtwo packages clear epoxy coating comprising Epon 1001 CX-75 (Shell Chemicals) and Ancamide 2353(Air Product) were cast on the panel using a drop-down bar. Before the application of the coatings, the panels were pretreated by alkaline cleaner Turco 4215-S. After air dry for one week, the film final thickness was 26.6±1.2 , 28.0±0.9 and 300±30 μm respectively measured by an electromagnetic digital coating thickness gauge Elcometer 345 (Elcometer Instrument Ltd.).

Experimental Procedure

The thermal cycling testing protocol and the experimental set-up can be found in our previous paper[6]. The protocol has been modified a little to fit the requirement of the experimental, which is shown as in following schematic. There were total three runs applied to the samples. For thermal EIS testing, a Gamry CMS300 System with PC4 Potentiostat was used for the impedance measurements. A three electrodes system is use by with a platinum counter electrode and a silver/silver chloride reference electrode in the Corrocell™, and the test samples as working electrode. The scanning frequency range is from 5,000 Hz to 0.01 Hz with 10mV (rms) applied AC voltage in open circuit mode. The testing solution is dilute Harrision's solution (0.05% NaCl and 0.35% $(NH_4)_2SO_4$) and the exposure area for testing is 8.30cm^2. The major differences of two protocols used in this study lie on (1) the testing temperature of the first and the second coating samples were rt, 35, 55, 75 and 85°C, the more testing temperatures were set for the third coating including rt, 35, 45, 55, 65, 75 and 85°C. (2) The testing frequency for the first and second coating was from 0.1Hz to 5000Hz, and the testing frequency was lower down to 0.01Hz to 5000Hz for the third coating system. DSC measurements by a Perkin-Elmer 7 Series Thermal Analyzer were also performed on the samples before and after thermal cycling testing. In these measurements, the temperature scan was conducted under N_2 from -50°C to 150 °C with a heating rate of 10 °C /min.

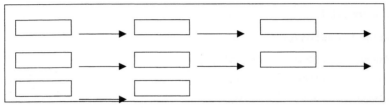

Figure 1. Schematic of Thermal Cycle Test Protocol

Results and Discussions
Wet State T_g and Dry State T_g

The initial purpose of the thermal cycling testing to develop a fast and reasonable corrosion accelerated evaluation protocol for organic coatings, and the results can be found in the previous paper.[6] As we had observed before in our laboratory,[4] the glass transition temperature, T_g, for wet coating films could be determined from the EIS data. To do this, we examined plots of $|Z_{low\ freq}|$ vs. $1/T_{absolute}$ (°K), and determined the T_g from the break point in the slopes of the data.

The low frequency impedance modulus $|Z|_{0.01Hz}$ was considered as the barrier properties index, usually, the temperature dependent properties showed an Arrhenius relationship. Thee plot of the log ($|Z|_{0.01Hz}$) vs. $1/T*1000$ for three runs of thermal EIS results shows a discontinuities at a very specific temperature, as shown in figure 2, and these points are identified as the T_g transition.

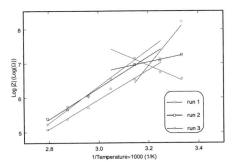

Figure 2. T_g determination from EIS results: low frequency $|Z|$ vs 1/Temperature for 2K epoxy coating

The T_g obtained from the EIS testing was about $43.72 \pm 1.76°C$, which was about $20°C$ lower than the T_g of dry film measured by DSC, which was $66.°C$. This was attributed to the water absorption and plasticization during the thermal cycling testing. The T_g depression was also considered as the index to evaluate the barrier properties of the coating because the T_g depression was proportional to water intake until the sample was saturated with water.[6] By this method, the fast evaluation of the protective coating could be made upon the T_g depression during the thermal cycling testing.

Roll-off Frequency and Polymer Relaxation

As mentioned before, the barrier properties are often characterized by the low frequency impedance modulus $|Z|_{0.01Hz}$. If one assuming a simple Randles circuit model, the $|Z|_{0.01Hz}$ is roughly equal the resistance of the coating. It decreases during the thermal cycling testing because the mobility of the polymer chain increases sharply with heating over T_g. At same time, the water diffuse into the coating system and the ingress of water causes strong plasticization of the polymer, which improve the polymer mobility even further, making the water diffusion easier. For samples exposed under thermal cycling testing, the polymer chain mobility and water absorption/plasticization both influence the barrier properties of the coating. To properly consider these mutual effects, a parameter which include both the effect of coating resistance and capacitance should to be considered.

In thermal cycling testing, the impedance modulus $|Z|$ is changes from being primarily resistive behavior to capacitive behavior as the testing frequency increases. It also decreases as the immersion temperature increases. To identify the transition from resistive to capacitive behavior, the break point frequency ($f_{45}°$ - the frequency that the phase angle between the real and imaginary components of the impedance reaches 45°), sometimes also defined as the roll-off frequency, is used to identify where the transition behavior changes in the Bode plots.

In Figure 3 an example of a Bode plot of the samples under thermal cycling testing is shown, in which the inflection points/break point frequencies are marked.

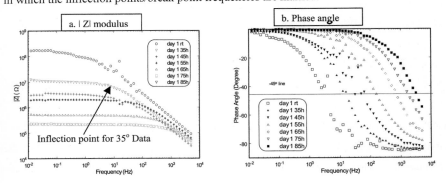

Figure 3. The Bode plots of a.)impedance modulus and b.) phase angle to show the Inflection point

By this method, we can obtain a series of break point frequency values under various exposure temperatures during thermal cycling testing. The plot of the break point frequency vs. 1/T of three coating systems also identifies the discontinuity at the T_g transition, at which the frequency was called the roll off frequency, three thermal runs are summarized as shown in figure 4. We can calculate directly the frequencies at which the phase angle is 45° for each individual temperature run.

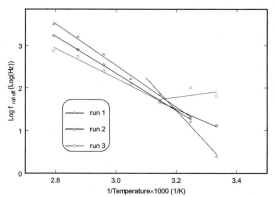

Figure 4. Log (break point)(Hz) vs 1/ $T_{Absolute}$ (1/K) for 2K epoxy coating systems

Analysis of Results from e-coat and PEPP coat

There were only four roll off frequency data points obtained for the e-coat, at 35,55,75,and 85°C, and no roll off frequency could be found on the EIS data measured at room temperature. It is very difficult to find the transition point by this data. Taking the temperature where roll-off frequency is 1Hz was found tobe a good estimate of the T_g of a good coating system tested in our lab.[7] When the frequency reached 1 Hz, the temperature obtained was close to the T_g result estimated from the EIS $|Z|_{low\ f}$ vs. 1/T data. Similarly, the transition point which corresponded the wet state T_g of PEPP coat could be observed. They are pretty close to the EIS results. The relaxation frequency for three runs were found to be 2000, 560 and 500 Hz respectively. The large value of relaxation frequency demonstrated the poor barrier properties of PEPP coating although the decrease of relaxation frequency implied the improvement of barrier properties of PEPP coat during thermal cycling testing. We have measured the T_g of the samples prior to immersion (dry state) and after immersion (wet state), the thermal cycling by DSC and we have used $|Z|_{0.01Hz}$ to calculate the T_g of the wet stage of samples. T_g results of three coating systems from these three different methods are shown in Table 1 below.

Table 1. Comparing T_g results for e-coat, PEPP coat and 2K epoxy coat from different methods

e-Coat				
T_g from the DSC (°C) (Average of three runs)	Prior testing (dry): 61.9 No wet state data			
T_g from the plot $|Z|_{0.01Hz}$ vs. 1/T(°C)	1st run 45.5	2nd run 46.5	3rd run 45.5	Average 45.8
T_g from the plot $f_{roll\ off}$ vs. 1/T(°C)	1st run 48.5	2nd run 48.5	3rd run 50.6	Average 50.0
PEPP Coat				
T_g from the DSC (°C) (Average of three runs)	Prior testing (dry): 85.7 No wet state data			
T_g from the plot $|Z|_{0.01Hz}$ vs. 1/T(°C)	1st run 66.0	2nd run 56.0	3rd run 49.6	Average 57.2
T_g from the plot $f_{roll\ off}$ vs. 1/T(°C)	1st run 64.8	2nd run 52.7	3rd run 51.7	Average 56.4
2K plain epoxy coating				
T_g from the DSC (°C) (Average of three runs)	Prior testing (dry): 66		After testing (wet): 44	
T_g from the plot $|Z|_{0.01Hz}$ vs. 1/T(°C)	1st run 42.06	2nd run 44.36	3rd run 44.76	Average 43.72
T_g from the plot $f_{roll\ off}$ vs. 1/T(°C)	1st run 43.15	2nd run 40.28	3rd run 45.37	Average 42.93

The T_g results from both EIS calculations were close to the T_g of wet film after thermal cycling measured by DSC. Thermal cycling EIS has been proved a powerful technology to detect the T_g of wet state sample by identifying the abrupt change in barrier properties as measured by |Z|. There are some differences between the T_g results obtained from $|Z|_{0.01Hz}$ and $f_{roll\ off}$ or $f_{45}{}^O$, the possible reason for which is the contribution of coating capacitance incorporated into $f_{roll\ off}$ calculation. Besides the wet state T_g obtained from the thermal EIS results, the roll-off frequency at wet stage, which was 60Hz for run 1 testing of clear epoxy coating, 1Hz for E-coat and 2000Hz for run 1 testing of PEPP coat respectively provides the information about polymer chain relaxation. As in the previous studies[8,9] , a 1Hz roll off frequency was found for epoxy coatings tested by thermal cycling, its reciprocal was proportional to the polymer relaxation time and it was a characteristic of the barrier property of coating system.

Barrier properties and physical ageing

In this section, the dielectric constant results from thermal EIS testing of three different coatings used for corrosion protection were discussed.

The dielectric constant data calculated from EIS results vs. frequency at various temperatures are presented in Figure 5 (real portion and imaginary portion of dielectric constant) for e-coat data, the rest of the data is discussed in summary form below. Among these coating systems, the e-coat shows superior corrosion protection because of its excellent barrier properties, the PEPP coating had poor performance because its high pigment volume concentration. The plain epoxy coating did not provide much corrosion protection as it only a polymer that was not designed for corrosion protection use in un-pigmented form. For dielectric constants data, the dot curves stand the first thermal run data, the dot-line curves stand for the second run data and the line curves stand the third run data. The data measured at same temperature was drawn in same color in the plots.

There is no significant change in the dielectric constant data between runs for e-coat sample, which implied barrier properties of the coating kept constant during three thermal runs. The ε' and ε'' increased significantly for PEPP coating after first run when temperature was below 55°C.. The increase of dielectric constants could be attributed to the plasticizing effect on molecular chain by H_2O and the increase caused by water uptake[10]. When the temperature reached 75°C, no increase was found because the system was saturated with water, but an obvious plateau was observed that is often related to the appearance of the delaminated area. An increase of ε' and ε'' is also observed for plain epoxy coating when temperature was below 45°C. No apparent increase was found when temperature was over 45°C. No shoulder or plateau is found for this sample. We are still seeking an explanation for these observations.

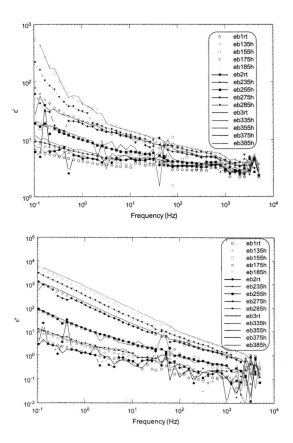

Figure 5. Real Portion (ε′) and Imaginary Portion (ε″) of Dielectric Constant of e-coat vs. Frequency for Various Thermal Conditions.

The barrier properties of samples could be clearly measured by ε′ and ε″ derived from EIS , and the changes on barrier properties due to temperature cycling in immersion were mainly affected by the plasticising effect of water on molecular chain and the water uptake. Further, the thermal cyclic exposure contributed significant of physical ageing of the coating sample.

Conclusions

The barrier properties of corrosion protective organic coatings during the thermal cycling testing were measures by EIS. The $|Z|_{0.01Hz}$ and roll-off frequency from EIS testing can be used as indexes to evaluate the barrier properties of the coatings. The plot of them against the reciprocal of the temperature showed discontinuity, which represented the T_g of the polymer plasticized by the diffused water The roll-off frequency which includes the effects came from both the resistance and capacitance seems to be a characteristic index of coating barrier properties. Analyzing the EIS data via the dielectric constant data clealy illustrates the effect of water plasticization and water uptake on barrier properties of coatings.

Acknowledgements

This project was supported by US Air Force Office of Scientific Research under Grant No. F49620-99-1-0283, Lt. Colonel P. Truelove - Program Manager.

References

[1] Delucchi, S. Turri, A. Barbucci, M.Bassi, S.Novelli and G.Cerisola, *J. Polymer Science: Part B: Polymer Physics* **2002**, 40, 52-64

[2] Gordon Bierwagen, Junping Li, Lingyun He, and Dennis Tallman "Fundamentals of the Measurement of Corrosion Protection and the Prediction of Its Lifetime in Coatings," Chapter 14 in *Proceedings of the 2nd International Symposium on Service Life Prediction Methodology and Metrologies*, Monterey, CA, Nov. 14-17, 1999, ACS Symposium Series # 805, J. Martin and D Bauer, ed., ACS Books, Washing ton, DC (2001) 316-350

[3] D.Kranbuehl, D. Hood, LmcCollough, H. Aandahl, N. Haralampus, W. Newby & M. Eriksen, "Frequency Dependent Electromagnetic Sensing for Life Monitoring of Polymers in Structural Composites During Use," *Progress in Durability Analysis of Comosite Systems,* A.H. Cordon, K.Reifsnider, Eds., A.A.Balkema Publ., Rotterdam (1996) pp.53-59

[4] J.Li, C.S.Jeffcoate, G.P.,Biewagen, D.J.Mills & D.E.Tallman, *Corrosion,* **1998**, *54*, 763-771

[5] J.R. Macdonald, ed., *"Impedance Spectroscopy"*, Wiley, NY (1989)

[6] Gordon P. Bierwagen*, L. He, J. Li, L. Ellingson & D.E.Tallman, *Prog. Organic Coatings,* **2000**,*39* 67-78

[7] Private communication, S. Duval

[8] Gordon Bierwagen, "EIS Studies of Coated Metal in Accelerated Exposure," presented at EIS 2001, Marilleva-Trento, Italy, June 17-22, 2001; to be published in *Prog. Organic Coatings*

[9] Junping Li, Ph.D. Dissertation North Dakota State University, Fargo, ND USA (2001)

[10] S.Duval, Y. Camberlin, M.Glotin, M.Keddam, F.Ropital and H. Takenouti , Abstract n331, 198th Meeting of The Electrochemical Society, Phoenix, October 2000

Thickening Mechanism of Associative Polymers

Alicia Maestro*, Carmen González, José María Gutiérrez

Chemical Engineering and Material Science Dept., University of Barcelona, C/ Martí i Franquès 1, 08028 Barcelona, Spain

Summary: Steady state viscosity and viscoelasticity of HMHEC solutions were studied. Viscosity increases with concentration due to a reinforcement of the micellar network. High shear rate viscosities are independent of temperature. Two relaxation processes were observed, the long one related to the lifetime of the hydrophobic junction and the short related to rapid Rouse-like relaxations of the free chains. When SDS is added, mixed micelles form that reinforce the network up to an optimum [SDS]/[HMHEC] ratio. Above this ratio, the micelles in excess isolate the polymer chains, the long relaxation process disappears and Rouse-like relaxations occur, corresponding to rapid movements of free chains.

Keywords: rheology; viscoelastic properties; HMHEC; associative polymers; thickener

Introduction

When a latex paint is formulated, a careful adjustment of the final rheology is needed. The high-shear-rate viscosity must be high enough to give sufficient film build and hiding but low enough to avoid extreme brush or roller drag. Viscosity at low shear rates must be high enough to prevent settling during storage and sagging after application, but has to allow the leveling of the film. It implies that a shear-thinning behavior is generally required. Latex coatings require the addition of thickeners to achieve the proper rheology. Water-soluble polymers modified through the addition of hydrophobic groups are used as viscosity modifiers in a variety of water-borne technologies including paints, inks, and cosmetics[1-5]. These copolymers typically consist of a hydrophilic backbone with a small number of hydrophobic groups either randomly distributed along the polymer chain or present as terminal groups. These polymers are termed associating polymers (AP) because they thicken the aqueous medium due to the aggregation of hydrophobes into surfactant-like micelles. The hydrophobes of the same chain can be joined to the same micelle, forming a loop[6], but they can also be joined to different micelles. If micelles are formed by hydrophobes of different molecules and molecules have hydrophobes in different micelles, hydrophilic backbones act as bridges between micelles. Then a three-dimensional network is extended and viscoelastic properties of the solution are enhanced. A full formulated paint has a big quantity of components that can interact with the thickener and modify the rheology. So, for a good understanding of the thickening mechanism it is necessary to study simpler systems, as water solutions of APs. The effect of the addition of other components, as surfactants, can be then studied.

APs commonly used are hydrophobically modified ethoxylated urethanes (HEUR), hydrophobically modified cellulose derivatives, for example the hydrophobically modified hydroxyethyl cellulose (HMHEC), and hydrophobically modified alkali swellable polymers (HASE). Their water solutions have been described to be shear-thinning and viscoelastic[7-11]. Viscoelasticity of HEURs end-capped with alkyl groups has been widely studied. A single Maxwell model can describe linear viscoelasticity of their solutions. The relaxation time has been identified as the disengagement time of a hydrophobe from the micelle[8]. Unlike what happens with end-capped HEURs, a single Maxwell model cannot describe HMHEC linear viscoelastic behavior, and a generalized Maxwell model with a relaxation time spectra has to be used to fit viscoelasticity[12], through the equations

$$G' = \int_{-\infty}^{+\infty} H(\ln \lambda) \frac{(\omega\lambda)^2}{1 + (\omega\lambda)^2} d \ln \lambda \tag{1}$$

$$G'' = \int_{-\infty}^{+\infty} H(\ln \lambda) \frac{\omega\lambda}{1 + (\omega\lambda)^2} d \ln \lambda \tag{2}$$

where G' and G" are storage and loss moduli, ω is the frequency and $H(\ln\lambda)$ is the spectra of relaxation times λ.

Rheology of AP solutions can be drastically modified through the addition of surfactants, because they interact with the hydrophobic groups of the polymer. This effect is strongly dependent upon the ratio surfactant/AP[13]. In some cases there is a range of surfactant concentrations where a phase separation occurs into a non-viscous supernatant and a viscoelastic hydrogel precipitate[4,14-15]. In other cases, an increase on viscoelasticity is observed up to an optimum surfactant concentration, followed by a hard decrease at higher surfactant concentrations[4-5,15-16]. This behavior has been attributed to the formation of surfactant-polymer mixed micelles. As a result, a higher number density of micelles exists and they are closer together. This permits the conversion of looping chains to bridging chains in such a way that the network reinforces and viscoelasticity increases, up to an optimum surfactant concentration. Over this concentration, the increased number of surfactant micelles progressively lowers the probability of more than one hydrophobe from different polymer chains in a given micelle, in such a way that bridges break, micelles become isolated and the network dissipates. Once the network is disintegrated the AP responds to surfactant in a manner essentially identical to that of a water-soluble polymer.

In this communication, viscoelasticity and shear-thinning behavior of HMHEC water solutions are presented and related with the thickening mechanism. Then, the influence of the addition of an anionic surfactant, the sodium dodecyl sulfate (SDS), is studied.

Materials and methods

HMHEC (molecular weight $M_W = 560000$; hydrophobes = hexadecyl) and SDS were supplied by Aldrich. Solutions were prepared with Millipore MilliQ water. Oscillatory measurements were carried out with a Haake RS100 Rheometer in the control shear stress mode. Previous stress sweep were developed to find the linear range. Shear-thinning experiments were performed with a RS150 Rheometer in the controlled shear rate mode. Various double cone and cone and plate sensors were used depending on the viscosity range of the samples.

Results and discussion

HMHEC water solutions

Figures 1 a) and b) show the steady state viscosity of HMHEC solutions at several concentrations and temperatures. A shear-thinning behavior is observed for all cases, except at the lower concentration, were a slight shear-thickening is observed, that is attributed to loop-to-bridge shear-induced transitions as molecules elongate under shear[9]. At higher concentrations shear-thickening is not observed probably because when micelles are close enough the number of bridges is not limited by the distance between them and elongation of chains does not favors the formation of new bridges.

Figure 1 a) shows that concentration increases viscosity in all the range of shear rates, as expected, since a higher number of micelle-like aggregates favors the formation of bridges and enhances the network. It can be observed in Figure 2 b) that viscosity at low shear rates decreases with temperature T, and tends to be independent of T at high shear rates. While at low shear -in the Newtonian range- the association structure and, as a consequence, the viscosity, depends on Brownian motion and, consequently, on temperature, at higher shear rates Brownian motion can be neglected in front of shear. Therefore, viscosity tends to be independent of temperature.

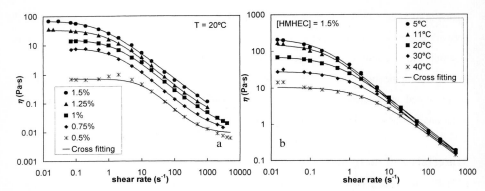

Figure 1. Steady state curves of HMHEC aqueous solutions a) at $T = 20°C$ and several concentrations; b) at [HMHEC] = 1.5% and several temperatures.

Experimental data shown in Figure 1 have been fitted through the Cross model (Equation 3), where η_0 and η_∞ are the low and high shear viscosities, subindex c indicates the shear rate at which equilibrium viscosity is in the middle of η_0 and η_∞ and the exponent m is around 0.9. As it can be seen, the fitting is quite good, except at 0.5%, because the shear-thickening behavior cannot be described by the Cross model.

$$\eta(\dot{\gamma}) = \eta_\infty + \frac{\eta_0 - \eta_\infty}{1 + \left(\dfrac{\dot{\gamma}}{\dot{\gamma}_c}\right)^m} \tag{3}$$

Viscoelasticity of HMHEC solutions cannot be fitted through a single Maxwell model. Figure 4 shows that two differentiated relaxation processes occur, one at intermediate (long relaxation process) and the other at high frequencies (short relaxation process). It can be seen in this Figure that oscillatory results can be fitted through the generalized Maxwell model if $H(ln\lambda)$ is supposed to have the shape of a logarithmic distribution around a mean relaxation time[12], which is identified with the inverse of the exit rate of an hydrophobe from its micelle[8]. This corresponds to the long relaxation process. An addend which is proportional to the frequency has to be used in Equation 2 to fit loss modulus at the higher frequencies. It has been related to rapid Rouse-like relaxations of the free chains[12].

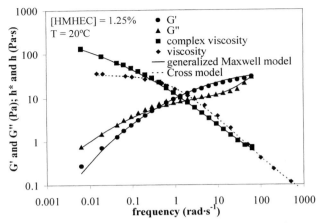

Figure 2. Frequency sweep and steady state curves of 1.5% HMHEC aqueous solution at T = 20°C.

The steady-shear viscosity has been added to the figure for comparison with the oscillatory complex viscosity, η^*. It can be seen that the Cox-Mertz rule, that describes the case in which the magnitude of the complex viscosity and the steady shear viscosity superimpose when shown in the same graph, is not satisfied, a common occurrence for associating polymers[17]. Moreover, zero shear and zero frequency viscosities do not coincide. Higher differences are obtained at lower concentrations.

HMHEC-SDS water solutions

When SDS, an anionic surfactant, is added to the solution, phase separation occurs in a range of [SDS]/[HMHEC] ratio at HMHEC concentrations below 1.2%. Above this concentration, a high enhancement of viscoelasticity functions is observed for the same ratio of surfactant/AP. Addition of more SDS above this optimum ratio produces the disappearance of the two phases existing for HMHEC concentrations below 1.2% and a progressive decrease of viscoelasticity functions for all the cases. Figure 3 shows the increase of the complex viscosity curve η^* at a concentration of SDS that corresponds to the optimum ratio, and the subsequent decrease when more surfactant is added, for a solution of 1.25% HMHEC.

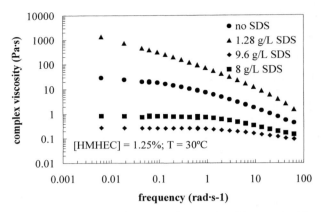

Figure 3. complex viscosity curves η^* for HMHEC = 1.25% without and with several quantities of SDS.

Zero frequency viscosity has been calculated through extrapolation of empirical fittings of the oscillatory viscosity, for solutions where no phase separation occurs. Results are shown in Figure 4 as the reduced viscosity $\eta_0/\eta_{0(no\ SDS)}$ *vs.* [SDS]/[HMHEC] for several concentrations of HMHEC. It can be seen that curves superimpose independently of the concentration of the polymer. The maximum is reached around [SDS]/[HMHEC] = 0.08.

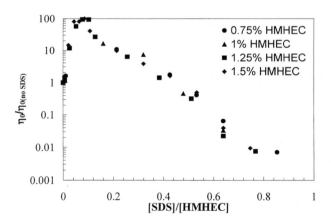

Figure 4. Low frequency reduced viscosity $\eta_0/\eta_0\ _{(no\ SDS)}$ *vs.* [SDS] / [HMHEC] ratio for all concentrations of HMHEC tested.

It is well established[8-9,18] that associating polymers thicken the medium because of hydrophobic associations into micelle-like aggregates that are bridged through hydrophilic backbones forming a three-dimensional network. Hydrophobic groups of the same molecule can associate to the same micelle, forming a loop that does not participate in the network, or can be joined to different neighboring micelles, bridging them and enhancing the network. At relatively low concentration of micelles, distances between them are long and for this reason few bridges can exist, and many hydrophobic groups are forming loops. When SDS is added, it associates with the hydrophobes of the polymer and forms mixed micelles. The formation of mixed micelles releases hydrophobes to form additional micelles and, as a result, the average separation between them decreases and looping chains can change into bridging ones. This contributes to the enhancement of the network and the increase of viscoelasticity. The phase separation observed for HMHEC solutions below 1.2% is attributed to the strong enhancement of interpolymer association. The big forces between micelles tend to bring them close, and then the more diluted solutions expel the excess of water into a supernatant fluid. When HMHEC concentration is increased, the volume of the watery supernatant fluid progressively decreases, up to 1.2% HMHEC, where there is no more water excess.

Above the optimum ratio, an increasing amount of surfactant molecules causes the formation of more mixed micelles in which the average number of involved polymer side chains progressively decreases, and as a result the number of bridges between them diminishes, until a situation is reached at which every side chain is covered by its own SDS micelle. Then, the solutions respond in a manner essentially identical to that of a water-soluble polymer. Figure 5 shows that at the higher [SDS]/[HMHEC] ratios tested the long relaxation process has disappeared, indicating that no more interpolymer associations occur. Storage and loss modulus, G' and G", superimpose at high frequencies with a slope of 0.5 or run parallel with a slope around 2/3. The shape of this curves indicates that solutions relax through a Rouse-like[19] or a Zimm-like[20] model, corresponding to short relaxation processes related to rapid movements of free polymer chains into the solvent.

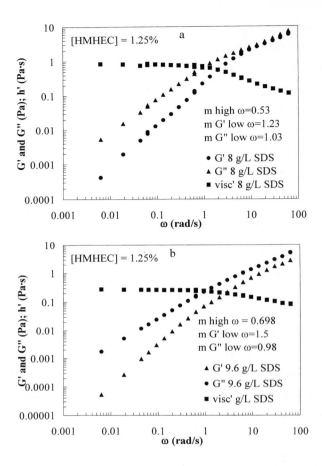

Figure 5. Frequency sweeps of HMHEC-SDS water solutions at high ratios of [SDS]/[HMHEC].

Conclusions

HMHEC water solutions present a shear-thinning behavior that can be fitted to the Cross model. The steady state viscosity increases with the concentration in all the range of shear rates, because of the reinforcement of the micellar three-dimensional network. Only the viscosity at low and intermediate shear rates is affected by temperature, decreasing with it due to an increase of Brownian motion. At higher shear rates, Brownian motion can be neglected compared with the effect of shear and temperature does not affect viscosity.

These solutions present two relaxation processes. The long one can be fitted through the generalized Maxwell model if the spectra of relaxation times is supposed to have the shape of a logarithmic distribution of relaxation times around a mean relaxation time that can be identified with the exit rate of an hydrophobe from its micelle.

When SDS is added, a phase separation or an increasing of viscoelastic parameters occur up to an optimum [SDS]/[HMHEC] ratio, due to the enhancement of interactions produced by the formation of mixed micelles. At higher ratios, the excess of micelles produces the breakdown of the network because micelles become progressively poorer in HMHEC hydrophobes, up to the total isolation of polymer chains, when each hydrophobic group is joined to a different SDS micelle. Then, HMHEC molecules can move freely around the solution, in such a way that Rouse and Zimm behaviors are observed.

References

[1] W. H. Wetzel, M. Chen, J. E. Glass, *Adv. Chem. Ser.* **1996**, *248*, 163.
[2] A. Yekta, B. Xu, J. Duhamel, H. Adiwidjaja, M. A. Winnik, *Langmuir* **1993**, *9*, 881.
[3] M. R. Tarng, M. Zeying, K. Alahapperuma, J. E. Glass, *Adv. Chem. Ser.* **1996**, *248*, 450.
[4] U. Kästner, H. Hoffmann, R. Dönges, R. Ehrler, *Colloids Surf.* **1996**, *112*, 209.
[5] A. Yekta, J. Duhamel, H. Adiwidjaja, P. Brochard, M. A. Winnik, *Macromolecules* **1995**, *28*, 956.
[7] J. E. Glass, D. N. Schulz, C. F. Zukoski, *ACS Symp. Ser.* **1991**, *462*, 2.
[8] T. Annable, R. Buscall, *J. Rheol.* **1993**, *37*, 695.
[9] K. C. Tam, R. D. Jenkins, M. A. Winnik, D. R. Basset, *Macromolecules* **1998**, *31*, 4149.
[10] L. Karlson, F. Joabson, K. Thuresson, *Carbohydr. Polym.* **2000**, *41*, 25.
[11] G. Kroon, *Prog. Org. Coat.* **1993**, *22*, 245.
[12] A. Maestro, C. González, J. M. Gutiérrez, *J. Rheol.* **2002**, *46*, 127.
[13] R. Tanaka, J. Meadows, P. A. Williams, G. O. Phillips, *Macromolecules* **1992**, *25*, 1304.
[14] V. Kumar, C. A. Steiner, *Colloids Surf. A* **1999**, *147*, 27.
[15] L. Piculell, F. Guillemet, K. Thuresson, V. Shubin, O. Ericsson, *Adv. in Coll. Interf. Sci.* **1996**, *63*, 1-21.
[16] J. P. Kaczmarski, M. R. Tarng, Z. Ma, J. E. Glass, *Colloids Surf. A* **1999**, *147*, 39.
[17] B. Xu, A. Yekta, M. A. Winnik; *Langmuir* **1997**, *13*, 6903.
[18] R. D. Jenkins, C. A. Silebi, M. S. El-Aasser, *ACS Symp. Ser.* **1991**, *462*, 222.
[19] P. E. Rouse, Jr., *J. Chem. Phys.* **1953**, *21*, 1272.
[20] B. H. Zimm, *J. Chem. Phys.* **1956**, *24*, 269.

Electrochemical Impedance Spectroscopy for Characterization of Coatings with Intrinsically Conducting Polymers

P.T. Nguyen[*], U.Rammelt, W.Plieth*

Institute of Physical Chemistry and Electrochemistry, Dresden University of Technology, Dresden, Germany

Summary: The method of electrochemical impedance spectroscopy (EIS) was applied to investigate the behaviour of a thin intrinsically conducting polymer film (ICP) deposited on a metal substrate. Especially the conductivity, the redox properties, the anion release properties, and the corrosion protection of a coating with and without ICP film on an iron or steel substrate were studied. Combined with other electrochemical methods, the reactions taking place at an injured surface area of the coated iron were studied. The corrosion protection mechanism of polythiophene could be explained.

Keywords: intrinsically conducting polymer, electrochemical impedance spectroscopy, corrosion protection, mild steel

I.   Introduction

The application of intrinsically conducting polymers (ICP) for corrosion protection of metals was suggested and very intensive studies were done in recent years. However, the protection mechanism is unclear and even the question, is there a protection, is not yet answered.

In this paper results are presented of an investigation of thin conducting polymer films (polythiophene, polymethylthiophene) with and without topcoat. The questions related to the corrosion protection mechanism of such systems are:

- Is the oxygen reduction catalyzing the reversible oxidation/reduction behaviour of the ICP [1-5]

- Can a defect be repaired by formation of a passivating oxide film [3-5]

- How can the properties of conducting polymer films be changed under the coating in contact with an electrolyte [6-8]

To answer the above questions, a serie of measurements was made using the method of electrochemical impedance spectroscopy (EIS) combined with other electrochemical methods like cyclic voltammetry and open circuit potential measurements. Based on the results, models of the corrosion protection mechanism of coatings with polythiophene were dicussed.

The equivalent circuit of an electrode covered with a film of an ICP and a topcoat

An electrode with the described coating is completely inert except if the coating has a defect. With a defect, the corrosion reaction of iron on the defect is measured additionally. An organic coating with defect can be modelled with an equivalent circuit shown in Fig.1. The parameters C_c and R_c measured at high frequencies correspond to the organic coating, whereas the parameters C_{dl} and R_{ct} measured at low frequencies are related to the electrochemical corrosion reaction at the metal substrate. The coating capacitance C_c depends on the properties of the coating (thickness, composition, structure, dielectric constant), the coating resistance R_c is determined by the defect area. The larger the defect area the smaller R_c. Due to the corrosion process taking place at the interface metal/coating, the adhesion of the coating decreases. The coating is gradually delaminated. The area of the defect is enlarged, R_c decreases. [9-10]

The conducting polymer itself is not explicitly represented in the equivalent circuit, but the elements of the equivalent circuit are influenced by the conducting polymer, especially the double layer capacitance C_{dl} and the charge transfer resistance R_{ct}. If the conducting polymer protects the iron in the defect area, R_{ct} is very high and the potential of the electrode is shifted to more positive values. In case a passive layer is formed, R_{ct} and C_{dl} correspond to the resistance and the capacitance of the oxide layer.

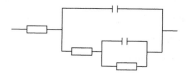

Figure 1. Equivalent circuit of a polymer coated metal according to [10]

Equivalent circuit of an electrode with an ICP film

The ICP film can be analyzed using a transmission line model introduced by Pickup et. al [11. This model can be simplified, if only the change of the conductivity is observed. A simplified equivalent circuit is given in Fig. 2

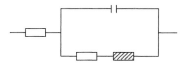

Figure 2. Simplified equivalent circuit of an electrode with a conducting polymer film

The polymer capacitance C_{CP} is related to the redox process taking place in the polymer film and the conductivity is reflected by the resistance R_{CP}, Z is a complex impedance mostly related to a diffusion process in the ICP.

II.   Results and discussion

1. Film formation with adhesion promoter

The coating process is based on the pretreatment of the substrate surface with 10^{-3}M 2(3-thienyl)ethylphosphono acid used as adhesion promoter. The synthetic details and the characterisation of the adhesion promoter were described elsewhere [12]. Then a polymethylthiophene (PMT) film is built up galvanostatically in a solution of the monomer. The adhesion promoter forms an ultrathin film with a structure like a self assembled monolayer (SAM). This SAM layer is covalently bond to the substrate, on which PMT can be deposited by electropolymerisation without any metal dissolution [13].

The redox properties of the film were investigated with cyclic voltammetry on PMT films grown on steel with SAM monolayer and on Pt. In both cases, the CV peaks show that the characteristic redox property of the PMT film on steel and on Pt are the same [6].

2. Corrosion protection of an ICP film without topcoat

The corrosion of a PMT covered steel electrode was investigated by EIS during immersion in aeraeted 0.1 M NaCl. Comparing the steel samples with and without PMT film, it is shown that the PMT film reduces the corrosion rate of mild steel but cannot completely protect the metal against corrosion. This is in agreement with potentiodynamic curves of steel with and without PMT [13]. The PMT film shifts the corrosion potential to more positive values and reduces the corrosion rate. However, the corrosion potential of PMT coated steel is only 100 mV more noble than that of bare steel and remains within the potential range where iron is dissolved. Passivation of the metal substrate caused by the positive value of the redox potential of PMT, an often discussed protection mechanism [2,14-17], is not observed.

3. Corrosion protection of an ICP film with topcoat

For the corrosion test with topcoat, two standard sample configurations were chosen and exposed in neutral aerated 0.1 M NaCl solution. The first sample was a mild steel with an alkyd coat, the second sample was a mild steel with a PMT film and an alkyd topcoat, for comparison. A small hole was made by a needle in order to simulate a defect. Measuring the open circuit potential, it was observed if the steel substrate was in the passive or in the active state. The PMT, even combined with a topcoat, caused no passivation of the steel substrate, the corrosion potential remained within the active range of iron dissolution. In combination with PMT, only a potential shift of +100 mV was observed.

The results show: the thin PMT film cannot protect mild steel against corrosion. It needs a topcoat to protect the samples. However in many cases the topcoat interacts with the PMT film which is reduced. This can be seen by overcoating a PMT film with conventional clear coating systems. The color of PMT changes from blue (oxidized, conducting state) to red (reduced, non conducting state), (see Table 1).

It means that the electroactivity of the PMT film gets lost by the interaction with the coating. Only with polyurethane (pure or containing an acidic hardener) the blue color of the PMT film shows that no reduction reaction takes place.

Table 1. PMT film overcoated with some conventional coatings

clear coating system	color of the PMT film
epoxy 2K (ZD73-0000) with hardener	red
epoxy 2K (ZD73-0000) without hardener	red
epoxy with acrylic acid as hardener	red
Alkydal F26/60% in xylol	red
alkyd	red
waterborne coating (pH =11)	red
polyurethane 2K + hardener ZH62	red
polyurethane without hardener	blue
polyurethane + acrylic acid as hardener	blue

The influence of the pH of aqueous electrolytes on the electroactivity of PMT films was studied seperately. Figure 3 shows results of EIS measurements of PMT films on Pt in 0.1 M LiClO$_4$ with different pH values.

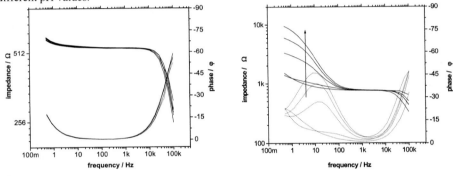

Figure 3. EIS measurements of PMT films on Pt in 0.1 M LiClO$_4$ at different pH values. The phase angle is indicated by dotted lines.

At lower pH (< 9.7) the polymer is still conductive. The resistance is low (about 10 Ω) and it does not change in 90 minutes. At pH=11 the resistance is changed, it increases to 500 Ω after 60 minutes and to 5000 Ω after 120 minutes. It can be seen that the pH value has a strong influence on the conductivity of PMT films. After the 120 minutes immersion at pH = 11, the PMT film is immersed in an acidic electrolyte (pH = 2) in order to verify, whether the PMT film becomes conductive again. The loss of conductivity is irreversible. A change of the pH from 11 to 2 gives only a small decrease of the resistance. The reduced PMT film cannot be reoxidized by oxygen, as was found for PANi [1-2]. The results of EIS measurements are summarized in Figure 4.

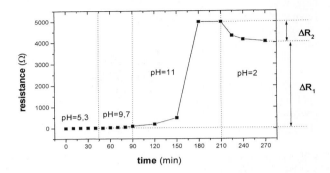

Figure 4. Resistance of PMT films on Pt at pH = 5.3, 9.7, 11.0 and 2.0 obtained from impedance measurements

A possible explanation is given by Wang [8]. According to these studies, PMT is attacked by nucleophilic anions and water. The formation of C=O bonds in aromatic rings disturbs the conjugated system of the polymer, so that the polymer is deactivated. The structure is irreversibly changed by the nucleophile attack. The higher the concentration of nucleophile in the electrolyte, the faster the polymer is deactivated. The concentration of nucleophiles depends on the pH value of the electrolyte.

For further studies of the influence of PMT electroactivity on the corrosion protection of mild steel an experiment was developed where the reaction on the steel surface and the reaction in the conducting polymer film could be observed separately. The principal electrode configuration is shown in Fig.5. The conducting polymer film was formed on a platinum electrode. A separate mild steel electrode could be connected or disconnected with the platinum electrode. This electrode was covered by an insulating epoxy film, which was scratched in order to simulate a small defect. The behaviour of both electrodes was investigated under various conditions by EIS.

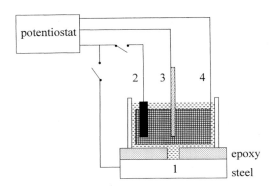

Figure 5. Scheme of the electrochemical cell:(1) mild steel/epoxy, (2) Pt/PMT, (3) SCE reference electrode, (4) Pt counter electrode, exposure in 0.1 M LiClO$_4$, aerated and deaerated,

The behaviour of the Pt/PMT electrode is shown in Figure 6. In deaerated solution the resistance of the PMT film is very low (approx. 50 Ω) and remains nearly constant after coupling with the steel/epoxy electrode for 120 minutes, i.e. PMT is in the conductive state but does not oxidize the steel. After 2 hours the solution is saturated with air and now the conductivity of PMT decreases, if the Pt/PMT electrode is connected with the steel/epoxy electrode. A second time constant is observed in the frequency range of 1 kHz to 10 Hz and the resistance increases to 5 kΩ. The reduction of PMT is caused by iron corrosion in the O$_2$ containing solution.

PMT remains conductive in aerated LiClO$_4$ solution, if the Pt/PMT electrode is not connected with the steel/epoxy electrode [18]. After connection the resistance increases very fast and rises continuously up to 22 hours. It means that in the hole O$_2$ reduction takes place as cathodic process causing iron dissolution and, simultaneously, reduction of PMT.

The behaviuor of the mild steel/epoxy electrode is shown in Figure 7. The high frequency part is related to the coating parameters (C$_c$, R$_c$). The low frequency part can be attributed to the electrochemical corrosion reaction taking place at the steel substrate (R$_{ct}$, C$_{dl}$). The value of C$_c$ is not changed during exposure in deaerated and aerated solution. The value of R$_c$ is about 400 kΩ and remains nearly constant during coupling with the Pt/PMT electrode in deaerated solution (Fig. 7a). In contrary, in aerated solution R$_c$ is decreased to 20 kΩ after 30 minutes due to the corrosion of mild steel in the hole (Fig.7b). The size of the hole is markedly increased and the bottom of the hole is covered with the corrosion products. In the high frequency range, the double layer capacitance C_{dl} increases because of the increasing defect area. The charge transfer resistance decreases significantly indicating the strong dissolution of iron in the defect.

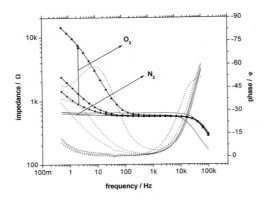

Figure 6. EIS measurements of a Pt/PMT electrode coupled to a steel/epoxy electrode after (—) 0, 30, 60, 90, 120 minutes in deaerated solution and then after (■)30, (●)60 minutes, (▲)22 hours in aerated solution. The phase angle is indicated by dotted lines.

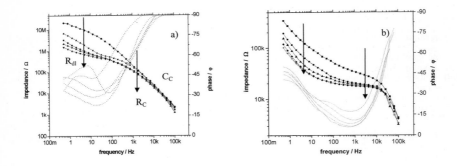

Figure 7. Bode plots of a mild steel/epoxy electrode in 0.1 M LiClO₄, a) deaerated, b) aerated after coupling with a Pt/PMT electrode. The phase angle is indicated by dotted lines.

The process could be principally described as follows:

anodic reaction $\qquad$ $Fe \rightarrow Fe^{2+} + 2e$ $\qquad\qquad$ (1)

cathodic reaction $\qquad$ $\frac{1}{2} O_2 + 2e + H_2O \rightarrow 2\ OH^-$ $\qquad\qquad$ (2)

additionally $\qquad$ $PMT_{ox} + 2e \rightarrow PMT_{red} + 2A^-$ $\qquad\qquad$ (3)

$$2OH^- + Fe^{2+} \xrightarrow{} Fe(OH)_2 \xrightarrow{} Fe(OH)_3 \longrightarrow Fe_2O_3 \qquad (4)$$

$$\qquad\qquad\qquad +O_2 \qquad\qquad\qquad -H\ O$$

The reaction (3) is irreversible, because the resistance of PMT increases continuously, i.e. the reduced PMT film provides no electrons for the oxygen reduction. After 22 hours connection between both electrodes, rust is observed in the hole of the epoxy covered steel electrode and the potential of the steel electrode changes from +0.3 V to –0.6 V(vs.SCE), immediately after disconnection. The bare steel surface within the hole cannot be passivated by the PMT film in O_2 containing solution.

4. Possible protection of ICPs

The results have shown that ICP films shift the electrode potential towards more positive potentials, but the shift does not lead to passivation. But, a noteworthy synergism exists between PMT film reduction, oxygen reduction and iron dissolution. The experiments have demonstrated that there is no catalytic effect of the conducting polymers, neither for the oxygen reduction nor for the metal passivation, contrary to statements in the literature. The potential of iron is remaining in the active dissolution region. A passive oxide layer cannot be formed under these conditions.

Combination of the adhesion promoting SAM layer and a film of conducting polymers as a primer coating, increases the corrosion protection of secondary coatings, because this combination gives good protection against delamination.

Recently a possible protection mechanism was suggested in the literature [19-20]: The layer of the conducting polymer is separating the process of metal dissolution from the process of oxygen reduction. This would prevent the local pH increase at the metal surface and the subsequent delamination.

IV.   Conclusion

It was shown that EIS is an useful method to investigate the corrosion protection of a combination of organic coatings with conducting polymer films.

An ultrathin adherent ICP film like PMT could reduce the corrosion rate of mild steel and improve the corrosion protection by a topcoat. The electroactivity of PMT gets lost if the topcoat is reducing the PMT.

An experiment was performed separating the PMT film and the mild steel/epoxy electrode. The experiment showed, that the ICP film (more specific, the polymethylthiophene) could not passivate the mild steel electrode. Only by a complex combination of oxygen reduction, iron dissolution and ICP reduction an interaction between ICP film and mild steel was observed.

References

[1]   D.W.DeBerry, *J. Electrochem.Soc.* **1985**, 132, 1022.
[2]   N. Ahmad, A.G.MacDiarmid, *Synth. Met.* **1996**, 78, 103.
[3]   B.Wessling, *Adv. Mater.* **1994**, 6, 226.
[4]   B. Wessling, J. Posdorrfer, *Electrochim. Acta.* **1999**, 44, 2139.
[5]   J.L Camelet, J.C. Lacroix, S. Aeiach,. P.C. Lacaze, *J. Electroanal. Chem.* **1998**, 445, 117.
[6]   PT. Nguyen, *PhD thesis*, in prepared.
[7]   J. Wang, *Electrochim. Acta.* **1997**,42, 2545.
[8]   U. Rammelt, G. Reinhard, *Prog. Org, Coat.* **1994**, 24, 309.
[9]   J. N. Murray, *Prog. Org. Coat.* **1997**, 31, 375.
[10]   M. Kendig, F. Mansfeld, S. *Tsai, Corr. Sci.* **1983**, 23, 317.
[11]   X.Ren, P.G.Pickup, *J. Electroanal. Chem.*, **1997**, 420, 251.
[12]   I.Mäge, E.Jähne, A.Henke, H.-J.Adler and M.Stratmann, *Progr. Org. Coat.* **1998**, 34, 1
[13]   U.Rammelt, P.T.Nguyen and W.Plieth, *Electrochim. Acta.* **2001**, 46, 4251.
[14]   W.-K. Lu, R.L. Elsenbaumer, B. Wessling, *Synth. Met.* **1995**, 71, 2163
[15]   M. Fahlman, S. Jasty, A.J. Epstein, *Synth. Met.* **1997**, 85, 1323.
[16]   W. Su, J.O. Iroh, *Electrochim. Acta.* **1999**, 44, 2173.
[17]   W. Su,. J.O. Iroh, *Electrochim. Acta* **2000**, 46,1.
[18]   P.T.Nguyen, U.Rammelt and W.Plieth, *Electrochim. Acta*, **submitted**
[19]   T. Schauer, A. Joos, L. Dulog, C.D. Eisenbach, *Farbe&Lack* **1999**, 105, 52
[20]   P.J. Kinlen, V. Menon, Y. Ding, *J. Electrochem.Soc.* **1999**, 146, 3690.

Studies on the Thermal Ageing of Organic Coatings

Ute Holzhausen[1], Dr. Sieghard Millow[1], Prof. Dr. H.-J. P. Adler[2]

[1] Institut für Lacke und Farben e.V., Magdeburg, (Institute for Varnishes and Paints, incorporated society), Germany

[2] Dresden University of Technology, Institute of Macromolecular Chemistry and Textile Chemistry, Dresden, Germany

Summary: The ageing of organic protective coatings depends on environmental impacts and is associated with a variety of ichemical and physical processes. The results of former thermal ageing studies indicate that evaporation and polymer degradation are the decisive processes for changes in coating properties.
For the verification of these assumptions the impact of
- the thermal exposure temperature T_L and
- the length of the exposure time t_L
on properties of clear lacquer coatings was determined by further studies.
The gradation of T_L was aimed at creating conditions where on the one hand only evaporation processes occur, while on the other hand evaporation and polymer degradation reactions occur in parallel. The objective was to find the time t_L and the temperature T_L at which degradation starts to prevail.

1   Introduction

The durability and service life of coatings is limited mainly by ageing. During ageing, a large number of irreversible physical and chemical processes such as the diffusion, migration and evaporation of softening plasticizing substances, as well as post-polymerization, post-linkage, autoxidation, polymer degradation, relaxation and post-crystallization can occur, which are largely dependent on environmental factors. These ageing processes manifest themselves in a number of different ways. Despite the large volume of research which has already been carried out in this field, the processes which effect changes in the properties of coatings are still disputed, and the processes underlying ageing have still not been satisfactorily explained.

Earlier studies have found that almost all characteristic values and coating properties change during thermal exposure at 100°C in accordance with a logarithmic function in the form of equation 1-1 where the length of exposure is t_L (|1| - |9|).

Characteristic value = k * ln t_L + c   **eq. 1-1**

(where k and c are empirically determined constants)

CCC 1022-1360/00/$ 17.50+.50/0

940

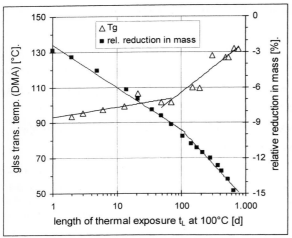

Figure 1-1: Tg-DMA and relative reductions in mass (gravimetrically determined) of a thermosetting, solventbased Acrylate-Melamine-coating in dependence on the length of exposure at 100°C

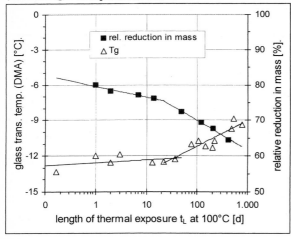

Figure 1-2: Tg-DMA and relative reductions in mass (gravimetrically determined) of a drying, waterbased Acrylate-dispersion-coating in dependence on the length of exposure at 100°C

Figures 1-1 and 1-2 summarize the values determined depending on length of exposure t_L for Tg (DMA), the glass transition temperature established by DMA, as well as those for the gravimetrically determined relative reductions in mass of two coatings with different binding agents, different curing/drying mechanisms and different solvents.

Closer examination revealed that at a certain point a change almost always occurred in the logarithmic function. The constant k acquired a different value. The changes were predominantly such that the properties or characteristic values changed more rapidly from this point on than they would have according to the old function.

The point at which the functional correlation changed depended primarily on the coating material (the binding agent), and less on the characteristic value or property being observed.

Further studies were aimed at identifying the causes underlying the changes in coating properties which were ascertained, and which were suspected in evaporation processes and in degradation processes of the polymer matrix.

2   Approach to a Solution

In order to establish the reasons behind the aforementioned changes in properties, external environmental factors were to be varied in such a way that ageing processes and phenomena could be ascribed to particular agents. One option was to select a number of graded thermal exposure temperatures ranging from ambient temperature to 140°C.

This gradation of thermal exposure temperatures T_L meant that the thermal energy recorded was either less than or greater than/equal to the activation energy for post-linkage or degradation processes. This would result either in evaporation processes alone, or in evaporation processes in parallel with post-linkage and/or polymer degradation reactions. This would then by necessity be reflected in a change in functional dependencies, for instance between Tg and the exposure temperature, and/or between internal stresses and the exposure temperature.

The studies had to be restricted to purely thermal loading, thus avoiding a parallel input of other forms of energy, as the effects of the graded energy input would otherwise have been masked.

3   Method

The characteristic values and properties of, for instance, water-dilutable clear varnish coatings were determined, as well as how they changed depending on length of exposure t_L and exposure temperature T_L. The following exposure temperatures T_L were selected: 25°C, 40°C, 60°C, 80°C, 100°C, 120°C and 140°C.

The *1st water-dilutable coating* was a 2K-PUR system (PUR-BS) which cures at room temperature. It was composed of the following principal elements:

- Binding agent: (parent component) based on an OH-functional polyacrylate resin (secondary),
- Curing agent: hydrophilic trimeric HDI,
- various auxiliary agents.

The *2nd water-dilutable coating* (AY-BS (wb)) was air-drying and was composed as follows:

- Binding agent component: styrene-acrylate copolymer emulsion,
- Film forming agent: butyl glycol,
- various auxiliary agents.

The changes in the coating properties caused by thermal exposure were established at various time intervals by:

- thermal analysis,
- radiometric tests,
- ATR-FTIR spectroscopy,
- gravimetric analyses and
- physico-technical examinations.

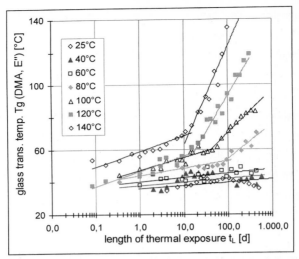

Figure 4.1-1: Glass transition temperature Tg-DMA of the coating PUR-BS in on the temperature and length of exposure

4   Results

4.1   Studies into thermal exposure of the PUR-BS coating

4.1.1   Determination of physical and mechanical properties

Since the glass transition temperature, one of the most important substantive coefficients of polymer materials, is very well suited to describing structural changes of the polymer matrix, particular importance was ascribed to it.

Figures 4.1-1 and 4.1-2 summarize the values established for the Tg (DMA) of the PUR-BS coating at different lengths of exposure t_L and exposure temperatures T_L, together with their correlations.

The glass transition temperatures of PUR-BS rise as thermal exposure increases. Both the Tg (DMA) and the Tg of the DSC analysis (not illustrated here) show a functional dependence of t_L in the form of equation 1-1. Whilst when T_L=25°C, 40°C and 60°C the functional dependencies between Tg (DMA) and t_L did not vary over the entire period of the study, when T_L=80°C, 100°C, 120°C and 140°C there was a change in the logarithmic function. In all four cases, the constant k acquired a greater value. The point at which this functional change occurred depended on exposure temperature T_L. For T_L=80°C, the functional change occurred after ca. 85 days. This change occurred considerably earlier, after ca. 43 days, at T_L=100°C. At exposure temperatures of 120°C and 140°C, these changes were observed even earlier, after ca. 16 days and ca. 13 days respectively.

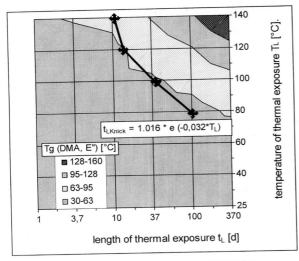

Figure 4.1-2: Correlation at the coating PUR-BS between the exposure temperature, the length of exposure and Tg (DMA) and respectively the functional change of Tg

Figure 4.1-2 is a graphic representation in the form of a 2D contour profile showing the correlations between Tg (DMA), the exposure temperature and the length of exposure. This also clearly demonstrates that Tg (DMA) is dependent on both T_L and t_L. Superimposed on the contour profiles is a line showing the dates when the functional correlation of Tg (DMA) changed for the respective exposure temperature. One can observe an exponential dependency in the general form of equation 4.1-1. between the point in time t_{LKnick}, at which the functional change occurs and a kink (*Knick*) appears in the curve, and exposure temperature T_L (the empirically determined function is also given in Figure 4.1-2).

$$t_{LKnick} = a * e^{-b*T_L}$$

eq. 4.1-1

(where a and b are empirically determined constants)

As thermal exposure at lower temperatures increases in duration, the point of sudden functional change is displaced.

The degree of coincidence between the values of the exponentially determined function and the actual data is about 95%. Given this functional dependency, one cannot exclude the possibility that the change in the logarithmic function, the kink in the Tg (DMA) curve, might also be observed at temperatures of 60°C (after $t_L \approx 615d$).

The ATR-FTIR spectroscopic studies shed additional light on the degradation processes with PUR-BS. Depending on t_L and T_L, a clear reduction in the extinction of the absorption band of the NC oscillation of the urethane group could be recorded at $1531 cm^{-1}$. Figure 4.1-3 shows the curves of the relative extinction changes for the NC oscillation at exposure temperatures $T_L=80°C$ and $T_L=120°C$.

It is significant that

- a linear functional dependency of the logarithm of exposure length was found, even for a relative reduction in extinction,

- at a point dependent on T_L, there is a change in the functional dependencies between the relative reduction in extinction and t_L, giving the curve a kink. This results in a change of function – similar to that observed with Tg (DMA) – such that the extinctions change more rapidly after the kink than they would do without this functional change.

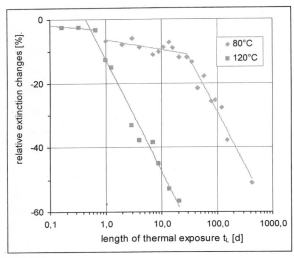

Figure 4.1-3: Relative extinction changes of the absorption band of the NC oscillation of the coating PUR-BS in dependence on the length of exposure at T_L=80°C and T_L=120°C

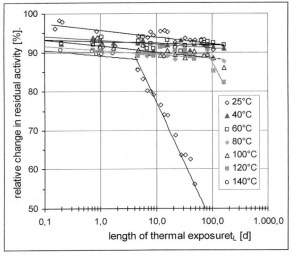

Figure 4.1-4: Relative C-14- residual activity of the coating PUR-BS in dependence on the temperature and length of exposure

At T_L=140°C the relative extinction changes were even greater, compared with those at T_L=120°C, resulting in the absorption band of the NC oscillation no longer being at all detectable after thermal exposure of 270d.

New absorption bands of functional groups or molecule skeletons formed during ageing were not detected, regardless of t_L und T_L. This suggests that the degradation process with PUR-BS coating occurs selectively with the formation of volatile products resulting from the decomposition of the urethane grouping. Low-molecular organic (e.g. amines) and inorganic (e.g. NH_3, CO_2) compounds would be conceivable. Characteristics similar to the Tg path were found when reducing the coating mass during thermal exposure.

The results of the retention of C-14 marked ethane-1,2-diol in PUR-BS show that the relative changes in residual activities, corresponding to the pulses per second calibrated on the coating masses, of PUR-BS were constant throughout the entire period of the study, taking into account measured value variations, with the exception of samples exposed to temperatures of 140°C and 120°C. It is highly probable that the functional hydroxyl group of the C-14 marked ethane-1,2-diol reacted with the isocyanate curing agent to the urethane group, and that the ethane-1,2-diol was consequently incorporated in the network through chemical bonding. This is supported by the high relative residual activities of ≈92%, which are above average for this study in our experience (and correspond to pulse rates of 11,000 pulses/s * g), the non-dependence on the temperature and length of exposure and the small reduction in relative residual activities observed in particular towards the beginning of thermal exposure in the period from 0.01d (15 min) to 0.1d (140 min) (Figure 4.1-4).

As already mentioned, a change in the functional correlation between the relative residual activity (calibrated pulse rate) and t_L was recorded for PUR-BS after ca. 4 to 5 days at $T_L=140°C$ and after ca. 85d at $T_L=120°C$. From this point on, the values for the residual activity changed more rapidly in accordance with a logarithmic function of the form of equation 1-1. This functional change is probably due to polymer degradation.

One must qualify this, however, by allowing for the fact that the chemical composition of the PUR-BS samples for the retention studies and those of the samples for the other studies are not identical, because the dihydric alcohol ethane-1,2-diol formed a constituent part of the network. This can lead to a difference in thermal stability.

Studies of C-14 marked PUR-BS samples aged at temperatures of 25°C, 40°C, 60°C, 80°C, 100°C and 120°C did not show any evidence of the growing dominance of polymer degradation processes relating to the C-14 ethyl ester grouping up to an exposure length of ca. 120 days.

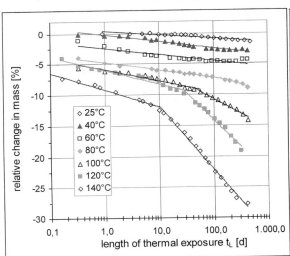

Behaviour which was similar in particular to the DMA-Tg path was found when the coating mass was reduced during thermal exposure.

The results of the gravimetric mass analyses in Figure 4.1-5 show that the reduction in mass also proceeded in accordance with the general logarithmic function set up in equation 1-1.

The functional correlation between the reduction of the coating mass and the length of thermal exposure did not

Figure 4.1-5: Relative reductions in mass (gravimetrically determined) of the coating PUR-BS in dependence on the temperature and length of exposure

change at exposure temperatures of $T_L \leq 60°C$, and changed only at temperatures of $T_L \geq 80°C$. In every case the constant k again acquired a higher value when the function changed.

The high relative reductions in mass of more than 20% at exposure temperatures of 120°C and 140°C cannot merely be explained by the release of the solvent and of the auxiliary agents. This gives added weight to the suspicion that highly volatile polymer degradation products had already formed and been released.

The rise in Tg and the formation or increase of internal stresses (self-contained stresses) are also closely related in the case of PUR-BS (see Figure 4.1-6).

Whilst at exposure temperatures $T_L = 25°C$, 40°C and 60°C the diminution in volume which accompanies the reduction in mass could be compensated for by flow processes to the extent that expansive internal stresses even built up in one phase of thermal exposure (over the entire period of the study at 25°C, and up to $t_L \approx 100d$ at 40°C and up to $t_L \approx 30d$ at 60°C), this phenomenon could no longer be observed at temperatures T_L greater than or equal to 80°C. At these temperatures contractive internal stresses built up in the PUR-BS coating as the Tg rose. The point at which contractive internal stresses begin to build up depends on the exposure temperature in the case of PUR-BS too.

The rise in Tg, the build-up or increase of contractive internal stresses and in particular the reduction in extinction of the absorption band of the NC oscillation of the urethane group lead one to anticipate increasing embrittlement or increased hardness of the PUR-BS in the course of thermal exposure and depending on the temperature. Figure 4.1-7 exhibits the changes in the mechanical coating characteristic Erichsen cupping, in relation to the length and temperature of exposure.

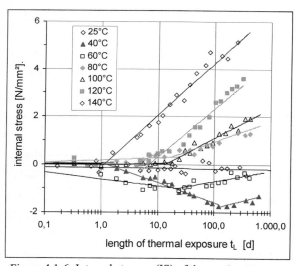

The cupping values are almost constant at temperatures of 25°C, 40°C and 60°C during the period of the thermal exposure study. At $T_L \geq 80°C$ the coating becomes less flexible and more brittle. The points in time at which this property changes are also dependent on exposure temperature and correlate well with those from the ATR-FTIR spectral analysis (t_{LKnick} at 80°C ca. 25d to 30d, t_{LKnick} at 100°C

Figure 4.1-6: Internal stresses (IS) of the coating PUR-BS in dependence on the temperature and length of exposure

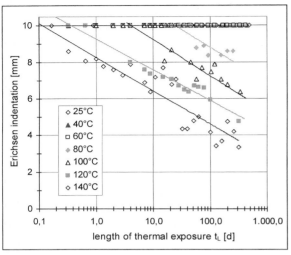

Figure 4.1-7: Erichsen indentation values of the coating PUR-BS in dependence on the temperature and length of exposure

ca. 4d to 5d, t_{LKnick} at 120°C ≈10h and t_{LKnick} at 120°C ≈3h). Like all the other characteristic values of PUR-BS, the cupping values during thermal exposure also change in accordance with a logarithmic function of the form of equation 1-1. The values of constants k at $T_L \geq 80°C$ are again almost exactly identical; they average out at 0.81.

The damage sustained by PUR-BS in the case of a rapid, discontinuous loading, such as that to which the coating is subjected during the impact test, was again dependent on the temperature and length of exposure. The resistance of the coating to cracking under impact-generated deformation was very high and almost constant over the entire period of thermal exposure at T_L=25°C, 40°C and 60°C (drop height 95 cm – 100 cm).

As with the results of Erichsen's cupping test, the coating became less elastic at exposure temperatures greater than or equal to 80°C, with cracking under impact-generated deformation.

4.2   Studies into thermal exposure of the AY-BS coating

4.2.1   Determination of physical and mechanical properties

Thermo-analytical studies into thermal exposure of the AY-BS coating, which is based on a styrene acrylate copolymer emulsion, showed that thermal exposure again led to an increase in Tg values. As found previously with the PUR-BS coating, the Tg changes with AY-BS also follow a logarithmic function of the type of equation 1-1.

It is striking that a change in the function only occurs again at a temperature of 100°C or above, and at a relatively late point in time (after ca. 105d, after ca. 35d at 120°C and after ca. 20d at 140°C) (*Figure 4.2-1*). From this point on, the changes in the Tg values are again greater than they would be according to the previous function. These functional changes in the Tg path were again recorded simultaneously with the DMA analysis and almost simultaneously with the DSC analysis. As with PUR-BS, the appearance of this functional change was dependent on both the length of exposure and the exposure temperature.

948

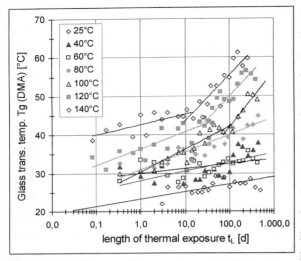

Figure 4.2-1: Glass transition temperature Tg-DMA of the coating AY-BS in dependence on the temperature and length of exposure

The results on the retention of C-14 marked ethane-1,2-diol in AY-BS show that the pulse rates of AY-BS calibrated on the coating mass, the residual activities, are logarithmically dependent on the length of exposure. At the beginning of thermal exposure, in the period from 0.01d (15 min) to 1d, a sharp drop in residual activities was recorded depending on the temperature, and was ascribed to the phase of film formation (drying). This phase is completed after ca. 1-2 days' exposure at T_L=25°C, after ca. 5 hours at T_L=40°C, after ca. 2.5 hours at T_L=60°C, after ca. 2.5 hours at T_L=80°C, after ca. 45 minutes at T_L=100°C and within the first fifteen minutes at T_L=120°C and T_L=140°C.

The end of the drying phase is marked by residual activities in the region of 250 to 350 pulses/s per gram AY-BS.

Surprisingly enough, a change in the functional correlation for the residual activities of AY-BS

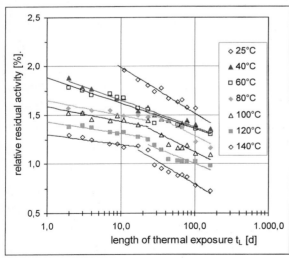

Figure 4.2-2: Relative C-14- residual activity of the coating AY-BS in dependence on the temperature and length of exposure

depending on t_L at $T_L \geq 80$°C was also recorded. In every case the constant k acquired a greater value, i.e. the diffusion and evaporation of the C-14 marked ethane-1,2-diol occurred more rapidly from this point onwards (the point at which the kink appears in the curve) than would have been the case with the previous old function (Figure 4.2-2). The kink appeared after ca. 15 days at T_L=140°C, after ca. 20 days at T_L=120°C, after ca. 22 to 25 days at T_L=100°C and after ca. 28 to 30 days at T_L=80°C. The cause of this functional change could not be determined. One suspects that chemical and/or structural changes in AY-BS favour the diffusion and evaporation of ethane-1,2-diol.

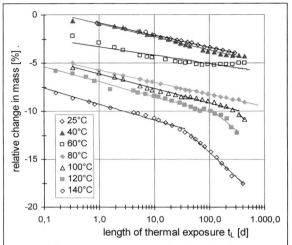

Figure 4.2-3: Relative reduction in mass (gravimetrically determined) of the coating AY-BS in dependence on the temperature and length of exposure

Figure 4.2-3 exhibits the values of the gravimetrically determined relative reduction in mass of AY-BS, related to the mass of the coating following conditioning under ambient conditions. This demonstrates that the functional dependencies between the reduction in mass and the length of exposure change in the course of thermal exposure at temperatures equal to or exceeding 100°C. The points at which these changes occur almost coincide with those for the glass transition temperature, and are equally dependent on the length and temperature of exposure.

The smaller changes in Tg values in comparison with PUR-BS, the later build-up and smaller increase in contractive internal stresses and the relatively small loss in mass of no more than 17 % lead one to expect evidence of embrittlement appearing later in AY-BS during thermal exposure. This was confirmed by Erichsen's cupping values. These showed that the coating retained a constant high level of flexibility throughout the entire period of the study regardless of the length and temperature of exposure.

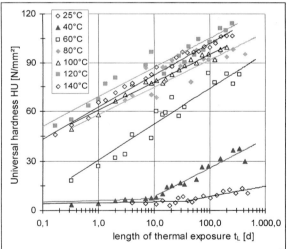

Figure 4.2-4: Universal hardness of the coating AY-BS in dependence on the temperature and length of exposure

As might have been anticipated, the characteristic values of pendulum hardness (attenuation 'hardness') and universal hardness (HU) also exhibited a functional dependence on the temperature and duration of exposure in accordance with a logarithmic function (from 60d at T_L= 25°C and from 12d at T_L=40°C). But surprisingly, no changes in functional correlations occurred in the period of the study.

The value of the constants k of all logarithmic functions determined for HU was of about the same magnitude from $t_L \approx 12d$ (*Figure 4.2-4*).

5 Summary and Evaluation of Results

The most important findings from the results presented are as follows:

All characteristic values of the coatings included in the study, and not just the glass transition temperature Tg, change, independently of the mechanism of film formation and of the binding agent and solvent base, according to the length of thermal exposure t_L in accordance with a logarithmic function of the following form:

$$Characteristic\ value = k * ln\ t_L + c \hspace{2cm} \text{eq. 5-1}$$

(where k and c are empirically determined constants)

Their rates of change thus become increasingly small.

In the logarithmic time-scale, changes in these logarithmic functions occur for many characteristic values, depending on exposure temperature T_L and length of exposure t_L; the values of k and c change without transition. This gives the curves a kink. A typical curve consists of two curve branches which constantly merge.

The new functions are predominantly such that an acceleration in the change of characteristic values occurs in relation to the previous function, and the characteristic values and properties deteriorate more rapidly.

Functional change is dependent not only upon T_L and t_L as stated above, but also upon the binding agent of the coating.

The functional change occurs for all characteristic values of a binding agent at the same exposure temperature after about the same length of exposure, leading one to suppose that there must be a common cause for this. As the values change spasmodically with the exposure temperature, we are very probably dealing here with temperature dependent reactions, suggesting that new chemical reactions are taking place.

There often exists between the length of exposure t_{LKnick}, the point at which a kink occurs in the curve, and the exposure temperature T_L a functional dependency of the following exponential function

$$t_{LKnick} = a * e^{-b*T_L}$$

eq. 5-2

(where a und b are empirically determined constants).

One can conclude from these results that physical and chemical ageing processes lead to embrittlement and the formation of internal stresses in organic surface protective coatings.

With studies depending on the length and temperature of exposure, the first appearance of a kink in the curve of a characteristic value indicates a greater dominance of chemical ageing processes at these and higher exposure temperatures.

At exposure temperatures where there is no functional change, physical ageing processes predominate. These are caused primarily by evaporation of low-molecular coating components.

It is especially significant that, contrary to the assumption made by the Arrhenius equation on activation energy, the introduction of new chemical degradative reactions is not only dependent on temperature, but also on time.

A knowledge of the functions for changing characteristic values might form the basis of an assured and effective assessment of the long-term characteristics of coatings, because it has been shown that the value of the constants in the logarithm function is dependent on the characteristic value, and in many cases is of an equal magnitude both before and after the kink for all temperatures to which a coating is exposed.

References

[1] F. Müller: farbe + lack 103 (1997) 3, p. 148
[2] iLF Information No. 1, 1994, p. 86
[3] T. Frey: farbe + lack 101 (1995) p. 1006
[4] H. G. Peters: XX. Fatipec-Kongress Nizza, 1990, p. 292
[5] H. G. Peters: Ergebnisbericht zum Forschungsvorhaben FKZ 513A/0022B, sponsored by the Ministry of Education and Cultural Affairs of the State of Saxony-Anhalt
[6] E. V. Schmidt: farbe + lack 94 (1988) p. 616
[7] E. V. Schmidt: farbe + lack 96 (1990) p. 108
[8] U. Holzhausen, S. Millow: farbe + lack 105 (1999) No. 7, p. 118
[9] S. Millow: Ergebnisbericht zum Forschungsvorhaben FV-Nr. 12286 BR, sponsored by the Federal Ministry of Economics

On Large Color Differences in Non-Euclidean Color Spaces

Hans G. Völz

Deswatines-Straße 78, D-47800 Krefeld, Germany

Summary: The color difference formulas CIE94 and CMC are only applicable to small color differences. For this reason, three papers have been written in which a basis for Euclideanization of these systems and, thus, for the calculation of large color differences was established. The original articles gave the equations for the Riemann spaces that were used to determine by calculus of variation the geodesics for acceptability. Several examples were shown. Subsequently, a direct method of transforming the Riemann space into a Euclidean space was published. With additional calculations, this method could also be applied successfully to the CMC system. This was also proven by examples. Several flaws that surfaced in both systems were listed and corrected (missing upper application limit, missing warning regarding non-Euclideanicity, lack of standardization, missing invariance for the event that reference and sample were transposed).

Introduction

Systems for the measurement of color differences, such as CIELAB and, subsequently, CIE94 and CMC were established as needed for industrial purposes (acceptability). However, they also play a role in colorimetric research (perceptibility). Since these systems are only applicable to the measurement of small color differences, a basis for calculation of larger color differences in the non-Euclidean CIE94 and CMC color spaces was published in three papers [1, 2, 3]. Following is a description of the essential trains of thought that led to these results. The mathematical formulas and programming instructions do not fit the scope of this report. They can be found in the main articles in Farbe [1, 2, 3].

Non-Euclidean Color Spaces

1. CIE94 Color Space

1.1 Calculation Based on Geodesics

While the CIELAB formula expands a Euclidean space in cylinder coordinates, the CIE94 formula [4] has the metrics of a Riemann space, characterized by the so-called line element [1]. The color space is greatly simplified because the long axes of the sectional ellipses in the CIELAB a*,b* plane are always radially oriented in the C^*_{ab}

© WILEY-VCH Verlag GmbH, 69469 Weinheim, 2002

direction. In Riemann spaces, the geodesic shown by the integral over the line element, not the straight line, is the shortest distance between two given points. This minimum distance was found with the algorithm of the calculus of variations.

In CIE94, the factors of the CIELAB formula become dependent on the chroma C^*_{ab} of the color locus. This causes the sectional ellipses in the plane to become larger as they move away from the gray axis. Solving and calculating the integrals become more difficult. However, the equations can still be explicitly described. Examples with selected color loci were calculated on the basis of set conditions (see Geodesics, Figures 1 and 2).

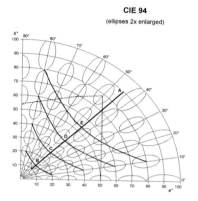

Fig. 1 Color space according to CIE94. Geodesics (1st Series) on the a*,b* plane of the CIELAB color space. The dotted line defines the limit of commercially producible colors for the lightness plane L* = 50 (according to L. GALL [7]).

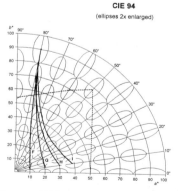

Fig. 2 Color space according to CIE94. Geodesics (2nd Series) on the a*,b* plane of the CIELAB color space. (See Fig. 1 regarding the meaning of the dotted line).

Furthermore, a CIE94($k_L : k_C : k_H$) space was shown with the use of ellipsoids, which represent perceptibility as stated by WITT [5]. In this case, the parameter factors k_L, k_C, k_H assume values other than 1 since the basic conditions have been abandoned (see Figure 3). The examples provided for this color space showed the greatest disagreement by far between the color differences calculated with the delta formula and those calculated with the geodesics.

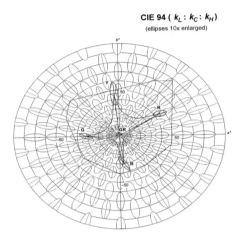

CIE 94 ($k_L : k_C : k_H$)
(ellipses 10x enlarged)

Fig. 3 Color space according to CIE94 ($k_L : k_C : k_H$). WITT ellipses and geodesics on the a*,b* plane of the CIELAB color space. (See Fig. 1 regarding the meaning of the dotted line).

1.2 Calculation by Euclideanization

Another article then described how the Riemann CIE94 space can be transformed into a Euclidean space [2]. For this purpose, three new coordinates for chroma, hue angle, and lightness were introduced. These were used to calculate the color differences in the known manner. It was found that two coordinates on the a*,b* plane suffice for the calculation.

The results were compared with those obtained on the basis of actual geodesics, and very good agreement was found. In addition, the new formulas were compared with suggestions previously offered by other authors.

2. CMC Color Space

The two publications cited initially show how geodesics in the CIE94 space can be calculated and how the CIE94 space can be transformed into a Euclidean space. This seemed to suggest that the latter method could also be applied to the CMC space [3].

First, obvious flaws in the CMC formula [6] were determined: the lack of standardization and the missing invariance guarding against transposition of the two color loci. Corrective steps were proposed. Subsequently, the line element was established which, at the same time, determines its limiting ellipsoids in the color space. Based on the new Euclidean line element, the new lightness and the new chroma always depend on the old lightness and the old chroma alone. The third coordinate, the new hue angle, was somewhat more difficult to calculate. To accomplish this, the entire color wheel had to be divided into sectors in which different integrals had to be formed.

Then, the color differences were calculated from the new coordinates by the known method. This included the calculation of specific average values for the CIELAB coordinates adapted to the CMC system. The necessary mathematical steps for the graphic presentation were provided. Finally, calculation samples based on examples given in the earlier papers were supplied so that comparisons could be made (see Figures 4 and 5 and Figure 6 after Euclideanization of Figure 5).

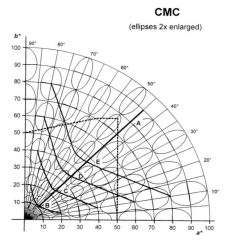

Fig. 4 Color space according to CMC. Geodesics (1st Series) on the a*,b* plane of the CIELAB color space. (See Fig. 1 regarding the meaning of the dotted line).

CMC

(ellipses 2x enlarged)

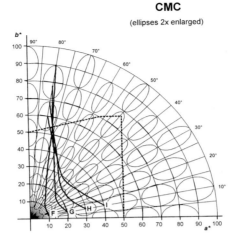

Fig. 5 Color space according to CMC. Geodesics (2nd Series) on the a*,b* plane of the CIELAB color space. (See Fig. 1 regarding the meaning of the dotted line).

CMC

(circles 2x enlarged)

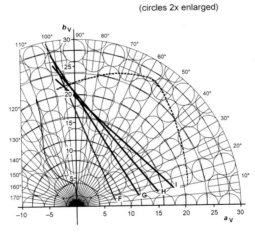

Fig. 6 Color space according to CMC. Geodesics (2nd Series) on the a_v,b_v plane of the CMC color space. (See Fig. 1 regarding the meaning of the dotted line).

Conclusion

In all three papers, flaws in the cited standards were pointed out, and corrective steps were proposed. Regarding CIE94 and CMC this concerns mainly the following:

- An upper application limit must be given.
- A warning concerning non-Euclidicity must be included.
- The formulas must be standardized based on the minimum color difference.

Due to the invariance against transposition of reference and sample, the arithmetical average values of the coordinates must be used.

References

[1] Völz H. G., Die Farbe 44 (1998), pp 1
[2] Völz H. G., Die Farbe 44 (1998), pp 97
[3] Völz H. G., Die Farbe 44 (19989/2000), pp 1
[4] Commission Internationale de l'Eclairage, Industrial Color Difference Evaluation. CIE Tecnical Report 116, CIE Zentral-Büro, Wien, 1995
[5] Witt, K., Col. >Res. Appl. 15 (1990) pp. 189
[6] Britische Norm BS 69 23 (1988)
[7] Gall L., Farbe+Lack 98 (1992), pp. 863

J